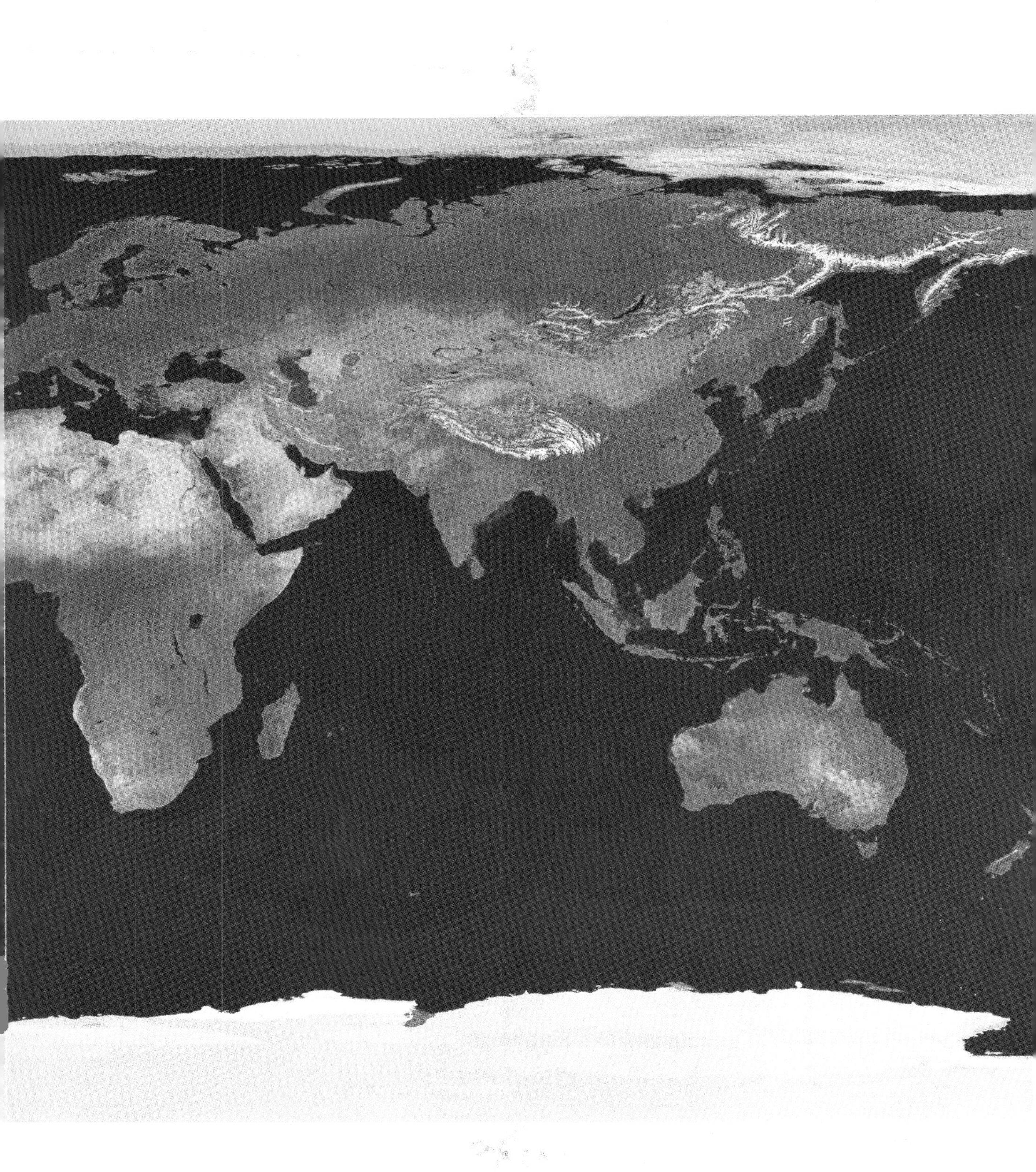

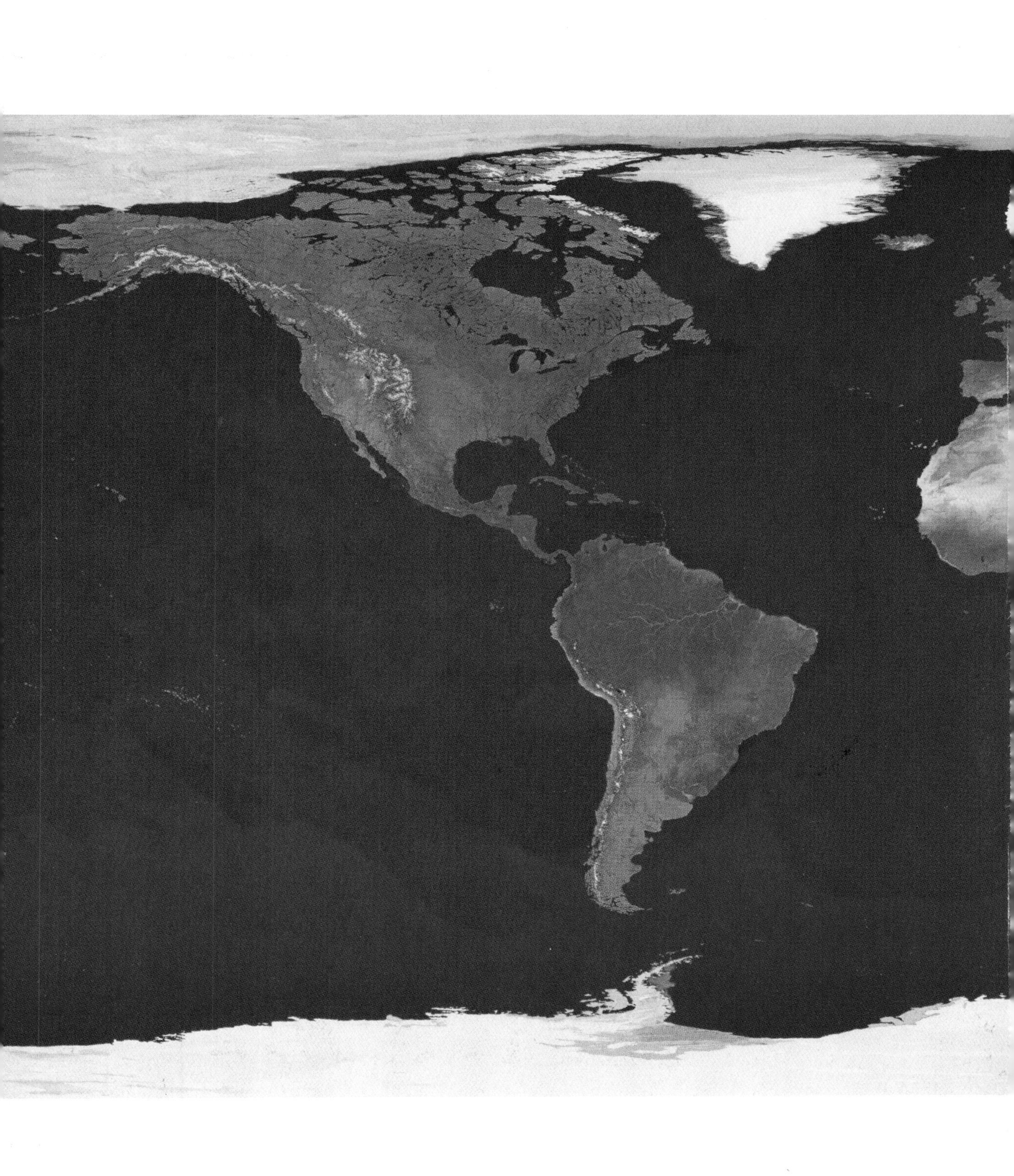

People, Places, and Environment

An Introduction to Geography

William H. Renwick
James M. Rubenstein

Miami University, Oxford, Ohio

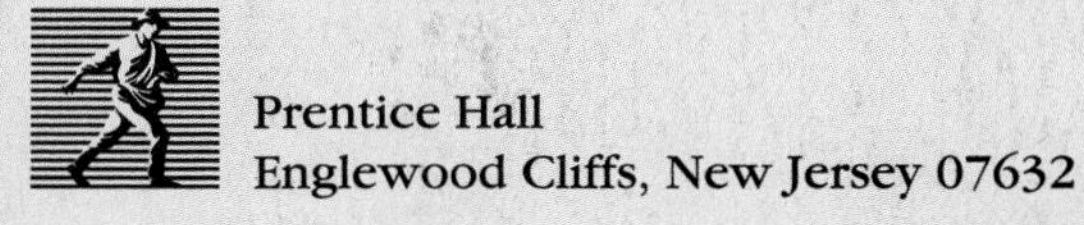

Prentice Hall
Englewood Cliffs, New Jersey 07632

Library of Congress Cataloging-in-Publication Data

Renwick, William H.
People, places, and environment : an introduction to geography /
William H. Renwick & James M. Rubenstein
p. cm.
Includes bibliographical references and index.
ISBN 0-02-399311-1
1. Geography. I. Rubenstein, James M. II. Title.
G128.R46 1995
910—dc20 94-25277
CIP

Editorial production supervision: Barbara Marttine Cappuccio
Editor-in-chief: Paul Corey
Editor-in-chief of development: Ray Mullaney
Marketing manager: Leslie Cavaliere
Development editors: Patricia Nealon/Fred Schroyer
Interior designer: Becky Bobb
Cover designer: Maria Lange
Managing editor: Kathleen Schiaparelli
Director of production and manufacturing: David W. Riccardi
Manufacturing buyer: Trudi Pisciotti
Electronic page make-up: David Tay
Photo editor: Lorinda Morris-Nantz
Photo researcher: Tobi Zausner

Chapter opening photograph credits:
All photographs copyrighted by the individuals or companies listed. All other photograph credits are listed next to photos within chapters. Any photos that do not have credits listed were taken by the author, William Renwick.

Cover photograph: Galen Rowell
Insert: Side one, Tom Van Sant/The Stock Market
Side two, World Ocean Floor by Bruce C. Heezen and Marie Tharp. Copyright 1980 by Marie Tharp. Reproduced by permission of Marie Tharp, 1 Washington Ave., South Nyack, NY 10960.
Chapter 1 Viviane Moos/The Stock Market
Chapter 2 Earth Satellite Corporation/Science Photo Library/Photo Researchers, Inc.
Chapter 3 Bruce Thomas/The Stock Market
Chapter 5 The Stock Market
Chapter 7 Reuters/Bettman
Chapter 8 Gerd Ludwig/Woodfin Camp & Associates
Chapter 9 Reuters/Bettman
Chapter 10 Paul Steel/The Stock Market
Chapter 11 William Waterfall/The Stock Market
Chapter 12 Craig Hammell/The Stock Market
Chapter 13 Viviane Moos/The Stock Market

Printed in the United States of America
10 9 8 7 6 5 4 3 2 1

ISBN 0-02-399311-1

Prentice-Hall International (UK) Limited, *London*
Prentice-Hall of Australia Pty. Limited, *Sydney*
Prentice-Hall Canada Inc., *Toronto*
Prentice-Hall Hispanoamericana, S.A., *Mexico*
Prentice-Hall of India Private Limited, *New Delhi*
Prentice-Hall of Japan, Inc., *Tokyo*
Simon & Schuster Asia Pte. Ltd., *Singapore*
Editora Prentice-Hall do Brasil, Ltda, *Rio de Janeiro*

Brief Contents

Contents

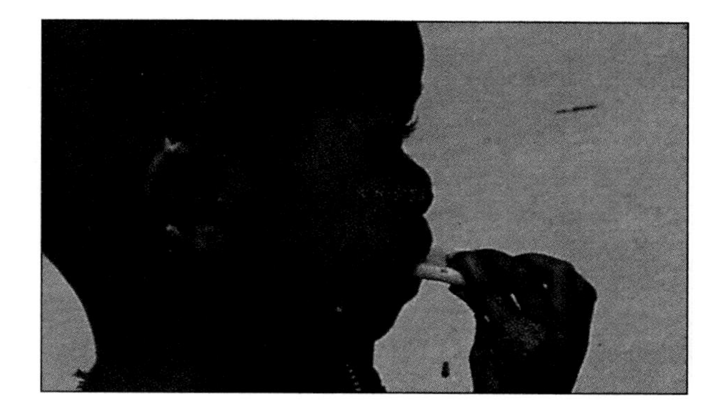

1

Basic Concepts of Geography 1

6

Natural Resources 212

7

Population 256

8

Culture 294

9

Political Geography 344

10

Development 386

11

Agriculture 426

Preface

Geography is the study of where things are located on Earth's surface and the reasons for the location. The word *geography*, coined by the ancient Greek scholar Eratosthenes, is based on two Greek words. *Geo* means "earth," and *graphy* means "to write." Geographers ask three simple questions: where, why, and so what. Where are people and environments located on Earth's surface? Why are they located in particular places? What is the significance of their location?

Geography at the Turn of the Millennium

Geography is enjoying a renewal of meaning and interest as a unifying focus of inquiry for the twenty-first century. Three themes underlie contemporary geographic thought: *interconnectedness, change,* and *historical context.*

The world of the late 1990s is *interconnected* to a greater extent than ever before. Because commodities are traded on world markets, a drought in Australia or a crop failure in Russia has immediate repercussions in grain markets around the world. Industrial development in China, deforestation in Indonesia, and air pollution control in the United States have a common output: the emission of tons of carbon dioxide into the atmosphere. Rapid transport of materials and information make possible long-distance transfer of production. Companies maintain headquarters in one location, research in a second location, production in a third location, markets in a fourth location, and financial records in a fifth location. Satellites permit instantaneous diffusion of news events and culture around the world.

In the contemporary world, *change* is the rule rather than the exception. Climate is no longer regarded as constant and unchanging: scientists and planners eagerly await each year's temperature and precipitation data, like financiers watching fluctuations in the stock market. The size of the world's population increases rapidly, as does the productivity of the world's farms, but forest cover is decreasing. A couple of dozen new countries have emerged from the breakup of the Soviet Union and its satellites. Once-divided Europe has become the world's largest economy, the European Union. The industrial restructuring that began in the 1970s continues to upend traditional economic relationships, as former agrarian nations like South Korea and Mexico become integrated in world trade.

The fast pace of change and the strong links between once remote places distinguish the contemporary world from the world of the mid-twentieth century, but they do not separate it. In fact, *historical context* has become even more important in understanding the present. The collapse of the world order established in the late 1940s, after World War II, has caused a rejuvenation of suppressed cultural identities and conflicts. Rapid cultural change brought by telecommunication and global trade has strengthened nationalist forces. The speed of change in the global environment is unprecedented in modern time, and requires intensified examination of prehistoric environmental changes as examples and analogs for the present.

Divisions within Geography

Because geography is a broad subject, some specialization is inevitable. At the same time, one of geography's strengths is its diversity of approaches. Rather than being forced to adhere rigorously to established disciplinary laws, geographers can combine a variety of methods and approaches. This tradition stimulates innovative thinking, although students who are looking for a series of ironclad laws may be disappointed.

Human versus physical geography. Geography is both a natural and a social science. When geography concentrates on physical features, such as climate, soil, and vegetation, it is a natural science. When it studies cultural features, such as language, social customs, and industries, geography is a social science. This division is reflected in some colleges, where physical geography courses may carry natural science credit and human geography courses social science credit.

This book is concerned with geography from both natural science and social science perspectives. The distinction between physical and human geography reflects differences in emphasis, not an absolute separation. Human and physical geographers, while continuing their specialized analyses, recognize the importance of an integrated, global view. This book reflects an emerging holistic approach in geography.

Topical versus regional approach. Geographers face a choice between a topical and a regional approach. The topical approach, which is used in this book, starts by identifying a set of important issues to be studied, such as climate change, formation of landforms, population growth, and economic restructuring. Geographers using the topical approach examine the location of different aspects of the topics, the reasons for the distribution, and the significance of the observed relationships.

The alternative approach is regional. Regional geographers start by selecting a portion of Earth and studying the environment, people, and activities within the area. The regional geography approach is used in courses on Europe, Africa, Asia, or other areas of Earth. Although this book is organized by topics, geography students should be aware of the locations of places in the world. One indispensable aid in the study of regions is an atlas, which can also be used to find unfamiliar places that may pop up in the news.

Descriptive versus systematic method. Whether using a topical or a regional approach, geographers can select either a descriptive or a systematic method. Again, the distinction is one of emphasis, not an absolute separation. The descriptive method emphasizes the collection of a variety of details about a particular location. This method has been used primarily by regional geographers to illustrate the uniqueness of a particular location on Earth's surface. The systematic method emphasizes the identification of several basic theories or techniques developed by geographers to explain the distribution of activities.

This book emphasizes systematic methods because a descriptive book would contain a large collection of individual examples not organized into a unified structure. However, some description is important: a completely systematic approach suffers because some of the theories and techniques are so abstract that they lack meaning for the student.

Features

In writing this book, our intent has been to provide students with an understanding of the mechanisms that connect distant parts of Earth, the causes of global change in the 1990s, and the importance of the historical context of that change. The text includes several key features that facilitate such study:

1. Chapters are organized by major geographic topics, but each chapter contains references to other chapters dealing with similar or related topics. For example, global warming is discussed or referred to in chapters on weather, climate, the biosphere, landforms, and industry. Impacts of international trade appear in chapters on natural resources, culture, development, agriculture, industry, and urbanization.
2. Each chapter is subdivided into four or five topics. The key concepts in each topic are identified in a *chapter outline* at the beginning of the chapter and summarized at the end. A list of *key terms* appears at the end of each chapter, with page references to a *glossary*.
3. Each chapter ends with a section on *Critical Issues for the Future*. These sections identify important unanswered questions, continuing debates, and topics where geography can make significant contributions toward solving major problems.
4. Two sets of questions are given at the end of each chapter. *Questions for Study and Discussion* highlight the key topics of the chapter, and help students check that they

have understood the material. *Thinking Geographically* is a set of questions that invite students to look more deeply into topics covered in the chapter and apply geographic concepts to the world around them.

5. A special appendix on scale and map projections enhances the text discussion of the subject. We are grateful to Phillip Muehrcke, Professor of Geography at the University of Wisconsin-Madison, and former president of the American Cartographic Association, for his clear explanation of the subject.
6. Ancillary material has been prepared to help instructors use this text. The Instructor's Manual includes a review of each chapter's objectives, test questions related to the text, projects and exercises for students to do at home or in a laboratory environment, and blank base maps. A computerized test bank and slides and transparencies of selected maps and illustrations are also available.

Acknowledgments

The successful completion of a book like this requires the contributions of many people. We would like to gratefully acknowledge the help we received.

A number of people reviewed portions of the manuscript at various stages in the revision process and offered excellent suggestions. These reviewers included:

- Carolyn L. Cartier, University of Oregon
- Miroslaw Grochowski, Rutgers University
- Doc Horsley, Southern Illinois University at Carbondale
- Tim Kubiak, Eastern Kentucky University
- Robert B. Mancell, Eastern Michigan University
- Vincent P. Miller, Indiana University of Pennsylvania
- Adrian A. Seaborne, University of Regina (Canada)
- Robert W. Wales, The University of Southern Mississippi
- Eugene M. Wilson, University of South Alabama

This book was produced at a time when Macmillan Publishing Company and Prentice Hall Publishing Company were completing a merger that has created the country's dominant publisher of geography books under the banner of Paramount Publications. We are proud to stand among the first geography books from the new publisher. If the work done by the combined Macmillan and Prentice Hall team to produce this book is any indication, American geographers can expect and (demand) that bigger is better.

We especially wish to thank Paul Corey, whose vision and understanding of geography while at Macmillan has sustained this and other projects of ours over the years. We have been pleased to work for the first time with Barbara Marttine, Production Editor, and Ray Mullaney, Editor-in-chief of Development. Ray has a special ability to maintain a close eye on a project so that daily progress is made. This project had not one but two development editors, Fred Schroyer and Patricia Nealon. Their hard work and diligence kept the project operating on an incredibly tight timetable. Our photography editor, Tobi Zausner, provided us with inspired choices; we hope you are stimulated by the book's photographs. We would also like to thank Kelly Cooney, Kenneth Guttman, Andrew Johns, and Nicole Monroe at Miami University for their help in preparing map data.

Bill especially thanks Debra, Peggy, Oliver, and Levi for putting up with the seemingly endless rattle of the keyboard clicking away at four in the morning. Jim especially thanks his wife Bernadette Unger, who won an election while this book was in production.

About the Authors

Dr. William H. Renwick received his Ph.D. from Clark University in 1979. He taught at UCLA and Rutgers University and worked as an environmental consultant before joining the Miami University faculty in 1978. At Miami he teaches physical geography, natural resources, and hydrology. His research is on geomorphology and environmental management, especially soil erosion, sediment yield, and water quality. He is co-author of *Exploitation, Conservation, Preservation: A Geographical Perspective on Natural Resource Use* (Wiley).

Dr. James M. Rubenstein received his Ph.D. from Johns Hopkins University in 1975. His dissertation on French urban planning was later developed into a book entitled *The French New Towns* (Johns Hopkins University Press). In 1976 he joined the faculty at Miami University, where he is currently professor of geography. Dr. Rubenstein's college textbook entitled *The Cultural Landscape: An Introduction to Human Geography* has been a bestseller through four editions. Besides teaching courses on Urban and Human Geography and writing textbooks, Dr. Rubenstein also conducts research in the automotive industry and has a recently published book on the subject entitled *The Changing U.S. Auto Industry: A Geographical Analysis* (Routledge). Originally from Baltimore, he is an avid Orioles fan and follows college lacrosse.

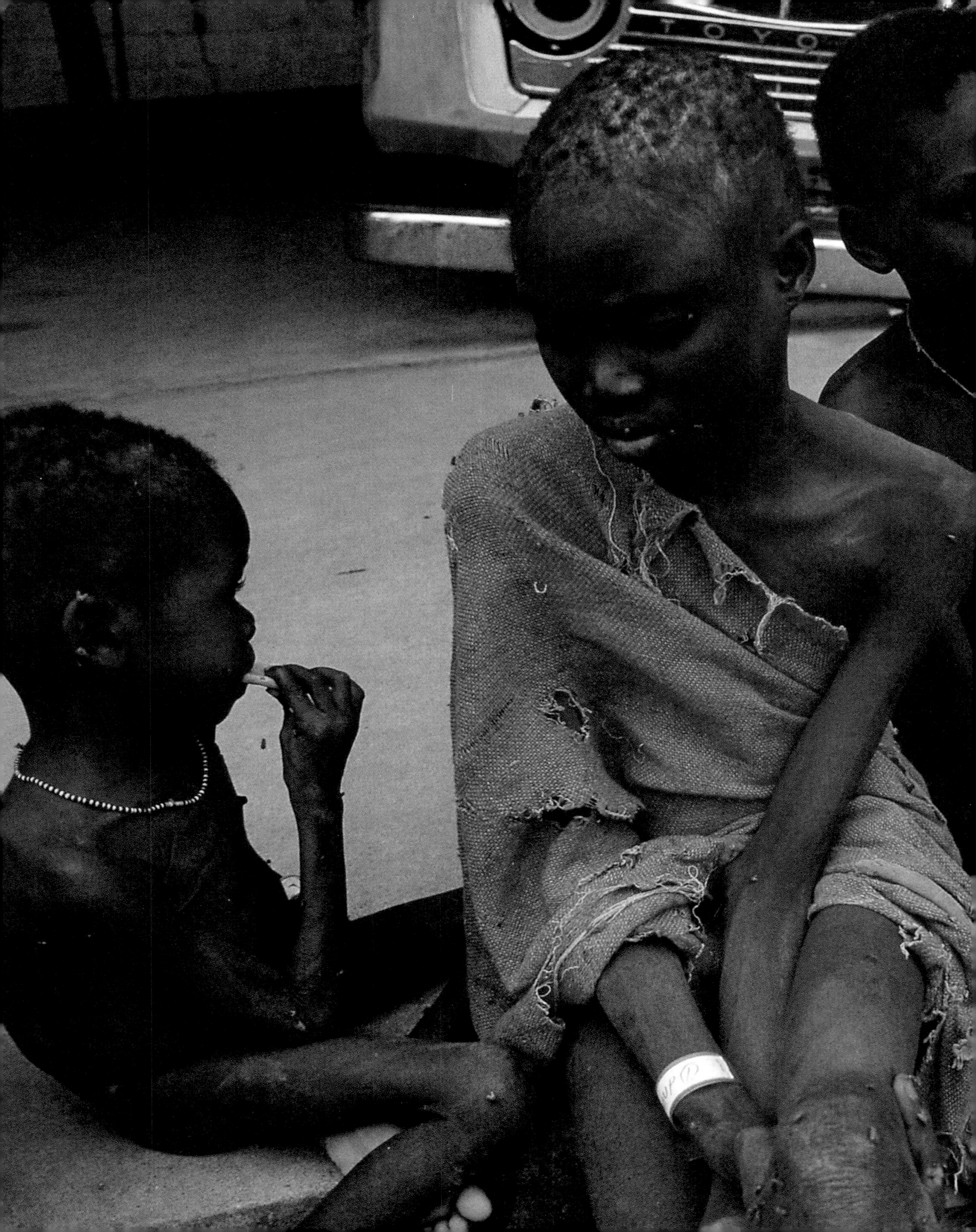
TOYO

1 Basic Concepts of Geography

- What is geography?
- Area analysis
- Spatial analysis
- Physical and human geography systems

Sudan—war and famine.

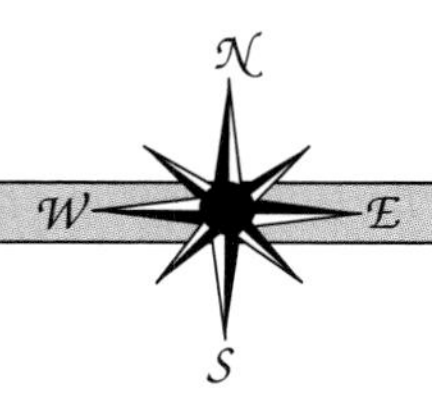

1. What is geography?
Geography is the study of where things are located on Earth's surface and the reasons for the location. Maps, as tools of the discipline, graphically depict the locations of places and help us interpret underlying patterns.

2. Area analysis
Geographers describe the locations of human and environmental phenomena and identify areas or regions that share trends and features.

3. Spatial analysis
Geographers study the distribution of objects across Earth's surface and the processes by which human and environmental phenomena move from one place to another.

4. Physical and human geography systems
The physical environment results from interaction among four systems: a biotic system (living organisms), and three abiotic systems (land, sea, and air). The biotic system includes patterns of human culture, such as beliefs, social patterns, and material traits. The physical environment influences human actions, but people have the ability to make choices.

INTRODUCTION

Satellite imagery has given us new views of Earth and has opened new opportunities for exploring our planet. A remarkable example is the Geosphere image which appears in this book's foldout (in the front of a book). This composite picture was assembled by the Geosphere Project of Santa Monica, California. Thousands of images were recorded over a ten-month period by satellites of the National Oceanographic and Atmospheric Administration (NOAA). The images then were electronically assembled, much like a jigsaw puzzle, to form this dramatic view of the entire planet.

Images such as this help geographers study the distribution of vegetation and landforms that give distinctive character to different places in the world. For example, note the wide band of tan color extending across North Africa, southwestern Asia, and Central Asia. This is a desert region, known by several names, including the Sahara, Arabian, Thar, Takla Makan, and Gobi. It appears tan because it is too dry for green vegetation to form a continuous cover.

Shown in white are the ice-covered regions of Antarctica and Greenland, the Himalayas and other high mountain areas, and the floating ice cap in the Arctic. Deep green tones in such regions as west-central Africa and northern South America indicate predominantly forest vegetation.

Geographers are not content just to observe and describe the characteristics of places. Geographers are scientists who work to understand *underlying processes* that explain observed patterns. For example, it is reasonable to assume that the dry desert lands of North Africa and Asia must result from a lack of water. But what *process* deprives this area of moisture? The answer is atmospheric circulation, which results from energy flow, water movement, and other environmental processes. This is one of many phenomena that geographers analyze.

Geographers also study how human behavior can transform the character of a place. A good example is the Sahel area in North Africa, along the southern margin of the Sahara desert. The sparse vegetation of the Sahel results in part from human efforts to feed the rapidly growing population. People here are reducing the natural vegetation by clearing land to grow crops and by allowing many domestic animals to graze. Human interaction with the environment is causing *desertification*—in other words, it is helping to convert the area to desert.

The Geosphere Project's composite image is a much more accurate view of Earth than those depicted by ancient and medieval maps, such as the one on page 7, drawn in 1607 by Flemish cartographer Pieter van den Keere. This old rendering was drawn from hearsay reports by a man who never flew. In contrast, the modern satellite image accurately displays the outlines of coasts, positions of land masses, and locations of mountain ranges and deserts.

And yet in some respects, the Geosphere Project's image is not realistic. For example, it distorts the shape and size of Greenland and Antarctica. Some distortion is unavoidable when a three-dimensional sphere-like Earth is portrayed in only two dimensions, just as an orange becomes distorted if you mash it flat. The image also portrays a cloudless Earth, which does not exist. In reality, at any moment clouds are swirling above some portions of Earth's surface, and some vegetation may be covered with snow. And don't forget that, at any moment, half of the planet is in darkness! The thousands of individual images that were combined to make the composite Geosphere Project image all were taken on clear days, unobstructed by clouds or darkness.

The Geosphere Project image also fails to show the remarkable range of human activities on Earth's surface. Humans have divided Earth's land masses into approximately 200 countries, but the political boundaries separating these states are invisible from space. People have built cities, factories, farms, dams, and roadways, but they cannot be seen in the image.

Despite these shortcomings, the Geosphere Project's image is invaluable because it allows us to view the planet as a unified whole. Geographers devote much of their attention to understanding the unique character of different places on Earth—and they always keep in mind the fact that environmental processes and human behavior have worldwide impact.

What Is Geography?

A common notion of geography is that it consists of memorizing countries and their capitals, or climates and crop types, or exports and imports. People also commonly associate geography with photographic essays of exotic places in popular magazines. These are indeed important facets of geography, but they only scratch the surface. Contemporary geography is much more.

Geography studies where things are on Earth's surface, the reasons for their location, and their significance. The very word *geography* indicates this. Coined by the ancient scholar Eratosthenes, it is based on two Greek words, *geo-* meaning "Earth," and *-graphy* meaning "to write." Geographers ask three basic questions:

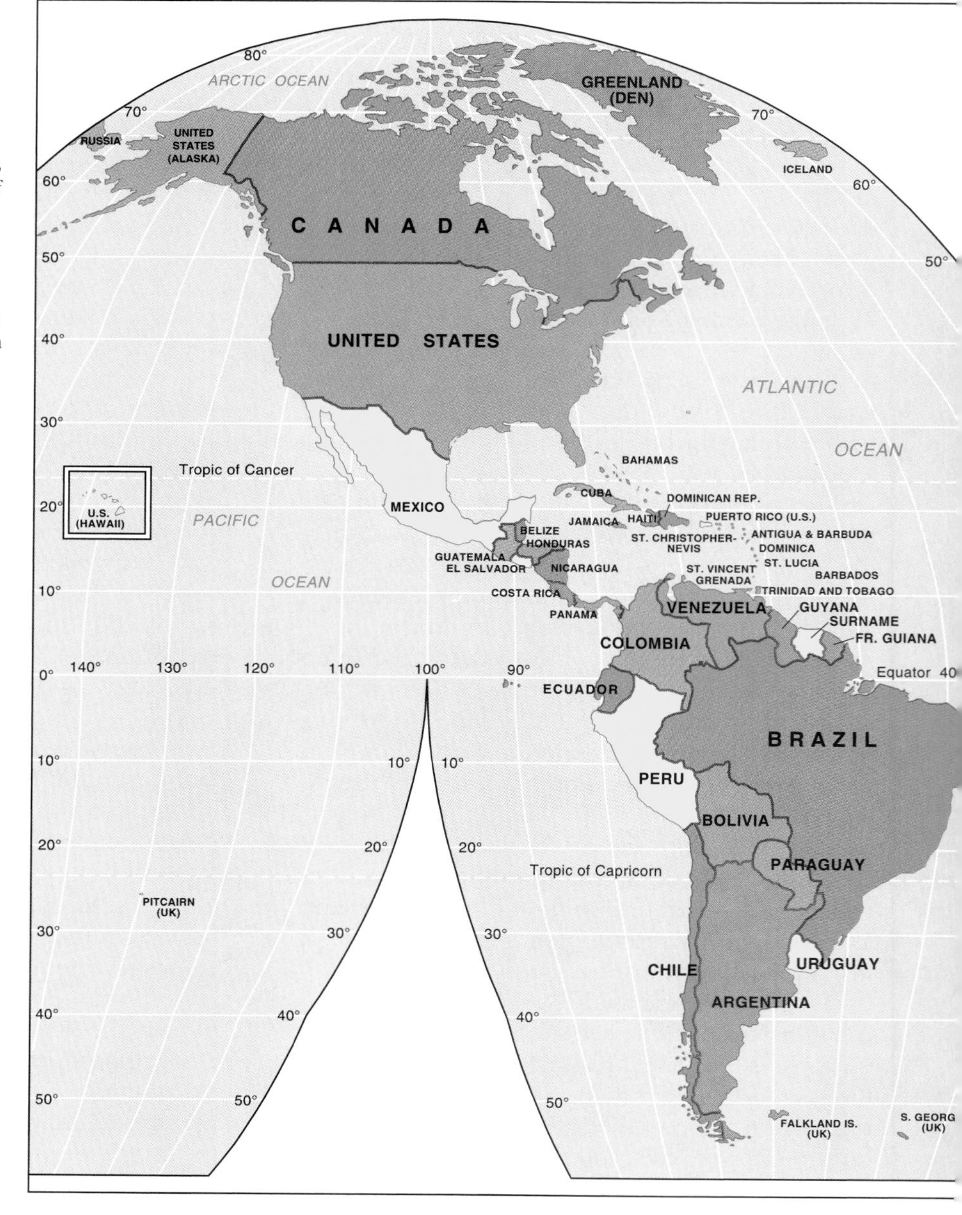

FIGURE 1–1
The world map reveals one of Earth's most significant features—the political boundaries that separate its five billion inhabitants. The names of the states evoke images of different environments, peoples, and cultures. However, the political boundaries are only one of the many patterns that geographers observe on Earth. Geographers study the distribution of a wide variety of social and environmental features and trends—population growth, agricultural practices, and climates—many of which transcend political boundaries. As scientists, geographers also try to explain why these patterns exist on Earth.

- *Where* do people and environments occur on Earth's surface?
- *Why* are they located in particular places?
- *What is the significance* of these locational patterns?

Like other scientists, geographers apply their knowledge to analyze contemporary issues, in this case those that arise from the interaction of human and environmental trends and features.

Geographers are observers of patterns. Political boundaries are a good example of one of many patterns they observe. People have divided Earth's land surface into nearly 200 countries, ranging in size from giant Russia, which occupies one-sixth of Earth's land area, to microstates such as Andorra, Monaco, and San Marino (Figure 1–1). Geographers study patterns of how people organize their societies to occupy land.

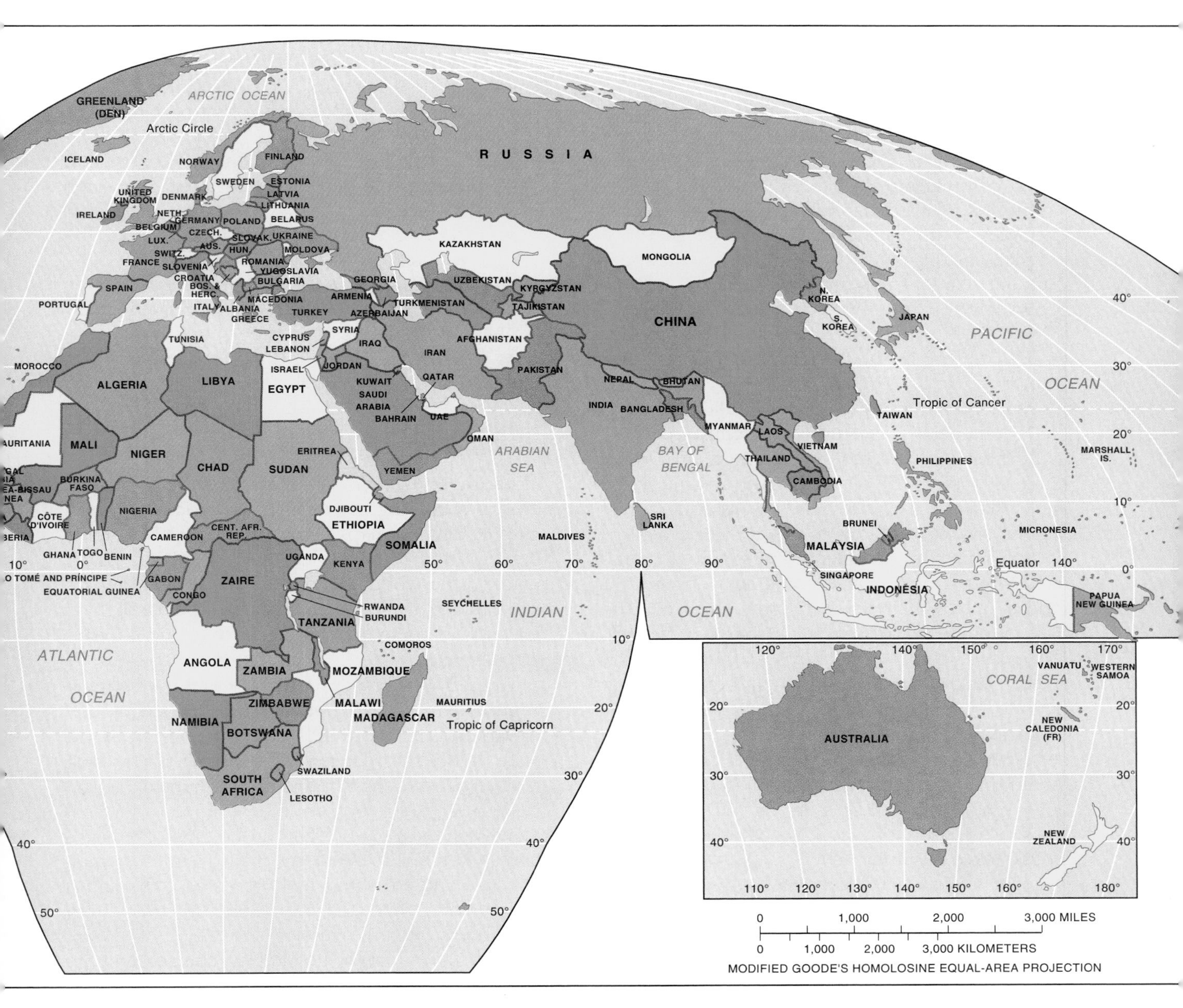

Geographers study the location and distribution of a wide variety of human activities and physical features, most of which transcend political boundaries. These include climate, landforms, agriculture, languages, religions, industry—see this book's chapter titles. As scientists, geographers try to explain the processes underlying the observed patterns and to interpret these with geographical perspectives. Geographers document and measure changes in social, economic, and environmental conditions. Knowledge of these changes, which have been especially rapid in the twentieth century, is critical to understanding and managing global problems in the twenty-first century.

Geography is a *way of thinking* about urgent problems. Geographers document human trends such as population growth in Africa, depletion of energy resources in the United States, and disputes among followers of different religions in Asia. For example, to understand hunger in Somalia and its neighboring East African countries, geographers examine relationships among climate, population growth, environmental degradation, political unrest, and other characteristics. Geographers explain unrest in the Middle East through such phenomena as the distribution of energy resources, differences in religious beliefs, and conflicting strategies for fostering economic growth.

Because geographers are trained in a broad range of topics that relate to Earth's surface, they are particularly well equipped to understand interactions among different forces affecting a place. For example, an insufficient food supply is a problem in much of Africa. Geographers observe that food supply is hampered in some areas of Africa by environmental factors, such as lack of water or fertile land, but in other areas by human factors, such as lack of storage facilities and efficient distribution systems. Farmers in Africa might plant more crops if they could count on getting the food to customers in the cities. Geographers believe that a meaningful perspective on why humans might behave as they do has a strong and compelling basis in the study of place. No other scientific discipline takes this approach.

Some specialization is inevitable in a broad subject such as geography. Three prominent divisions are physical geography, human geography, and regional geography. Physical and human geography emphasize topical approaches. When geography concentrates on topics related to physical features, such as climate, soil, and vegetation, it is a natural science. When it studies human topics, such as languages, industries, and cities, geography is a social science. Regional geographers start by selecting a portion of Earth, such as Europe, Africa, or Asia, and studying the environment and people within the area.

Contemporary Themes in Geography

All geographers collect and analyze geographic information, but they focus on different topics and employ different analytical approaches. Most contemporary geographers employ three analytical methods:

- *Area analysis* is a traditional approach that integrates the geographic features of an area or a place.
- *Spatial analysis* or locational analysis emphasizes interactions among places.
- *Geographic systems analysis* encompasses earth science and culture-environment traditions. It emphasizes understanding of environmental and human systems and interactions among them.

Let us look briefly at each of these methods.

Area analysis. Geographers have a long tradition of surveying, describing, and compiling geographic data on maps of an area. Maps are graphic descriptions of *places* on Earth's surface. Each place in the world occupies a unique *location* and features a distinctive combination of human behavior and environmental processes that give it a special character.

Neighboring places can be combined into a **region,** which is an area of Earth defined by one or more distinctive trends or features, such as climate, agriculture, language, or population growth rate. *Area analysis* is a method of organizing the study of Earth's peoples and environments through identification of regions and description of similarities and differences among them. Certain human activities and environmental processes give regions their unified character and distinguish them from other areas of Earth's surface.

Some introductory geography courses emphasize area analysis by organizing much of the syl-

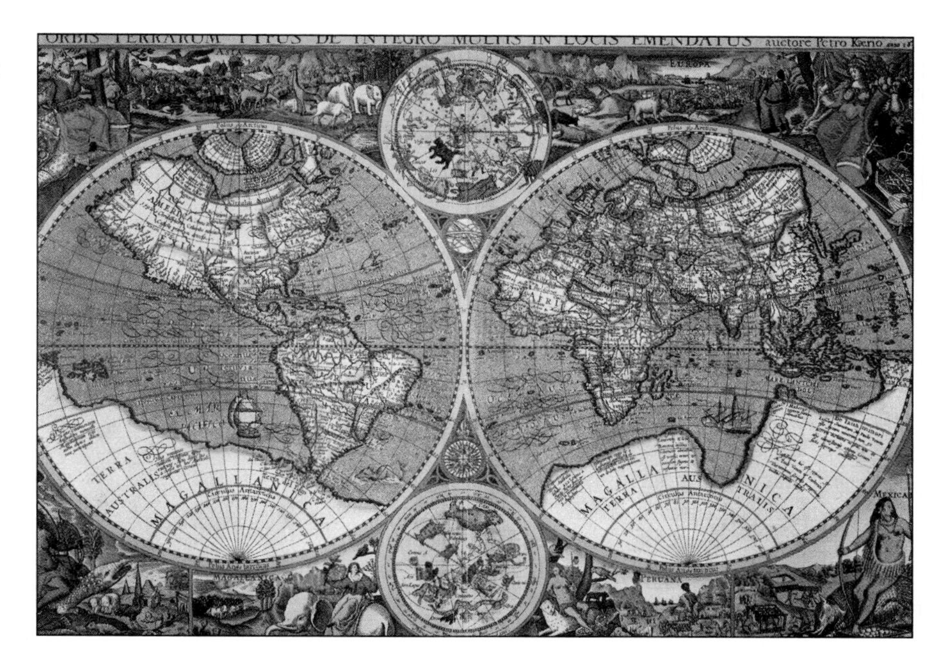

A world map drawn in 1607 by Pieter van den Keere. (The Granger Collection)

labus around regions of the world, such as Latin America, East Asia, and Sub-Saharan Africa. That approach selects a portion of Earth and studies the environment, people, and activities within the region. Although this book is organized by topics rather than regions, you should be aware of the locations and characteristics of places in the world.

Spatial analysis. While accepting that each place or region is unique, geographers recognize that specific human activities and environmental processes are rarely confined to one location. A second method of geographic inquiry, *spatial analysis*—also known as locational analysis—looks for patterns in the distribution of human actions and environmental processes.

The distribution of a human activity or an environmental condition changes over time. Geographers study the *movement* of people, goods, ideas, economic or industrial resources, and natural materials like water across Earth's surface. A place is connected to all other places through processes of human and energy movements. Examples abound: people in one place originate an idea and then "move it"—communicate it—to people in other places. Humans migrate to new lands and bring with them their behavior; they clear vegetation and extract resources; they import and export their products and services; they exchange their currencies and lend their labor forces; they travel from place to place for work, pleasure, or survival.

A place may also be connected to other places through processes of environmental movements. The movement of energy influences a region's distinctive climate; the movement of water carves landforms; the circulation of the atmosphere creates unique weather effects; pollutants discharged at one point in a river move downstream and affect water quality there.

Geographic systems analysis. A third method of geographic inquiry views Earth as a set of interrelated environmental and physical systems. A **system** is an interdependent group of items interacting in a regular way to form a unified whole. As an *earth science,* a field that also includes geology, meteorology, and oceanography, geography studies individual *environmental systems*, such as the climatic processes that produce precipitation, the hydrologic processes that determine what happens to the rain when it reaches Earth's surface, and the characteristics of the river valleys receiving the water runoff. Geographers also examine interrelationships among *physical systems*, such as the impact of increased precipitation levels on an area's hydrology, soils, landforms, and vegetation.

Humans are increasingly important agents of change in Earth's physical environments. The *culture-environment* approach examines human systems across the planet's surface and analyzes interactions among these human systems. Human population and rates of reproduction vary among places, as do humans' cultural preferences, their methods of obtaining food and earning a living, and their living arrangements.

Geography differs from other sciences in its emphasis on *human-environment interactions*. People modify the physical environment in which they live, and in turn they are influenced by environmental conditions. Geographers ask several questions concerning human-environment interactions. How have people changed their environment, and how have environmental processes changed people? Why have people tried to change the environment, and how has the environment influenced people? What are the consequences of each process?

This book views Earth as a series of distinct yet interrelated environmental and human systems. Area analysis and spatial analysis provide basic concepts and framework for systematic study of physical and human patterns and processes.

Development of Geographic Study

From the earliest times, geographers mapped the known world and described the people and environments found in different places.

Geography in the ancient world. As noted, the originator of our word *geography* was Eratosthenes (276?–?195 B.C.). For nearly a half-century, he headed the most prestigious library of the ancient world, located in Alexandria, Egypt. Eratosthenes not only accepted the concept that Earth was round, as few did in his day, but also calculated its circumference to within an amazing 0.5 percent accuracy.

In one of the first geography books, Eratosthenes described the known areas of the world and divided Earth into five climatic regions—a torrid zone across the center, two frigid zones at the extreme north and south, and two bands of temperate climate between the torrid zone and each of the frigid zones. He also prepared one of the earliest maps of the known world.

Greeks were concerned with geographic concepts for hundreds of years before Eratosthenes invented the term. In the sixth century B.C., Thales of Miletus applied principles of geometry to measuring land area. His disciple, Anaximander, argued that the world was shaped like a cylinder, and he made a map of the world based on information from sailors in Miletus, an ancient port city in present-day Turkey.

Aristotle (384–322 B.C.) was the first Greek philosopher to demonstrate that Earth was spherical. He did so by noting that all matter tends to fall together toward a common center, that during an eclipse Earth's shadow on the moon is circular, and which stars are visible changes as one travels north or south.

Hipparchus (190?–?125 B.C.) drew imaginary lines on Earth's surface to create reference points for the location of places. To this day, we depend on his concept of north-south meridians and east-west parallels.

The best-known geographer in ancient Rome was Strabo (63? B.C.–?A.D. 24). His seventeen-volume work *Geography* was an exhaustive description of the known world, including two introductory volumes and eight volumes on Europe, six on Asia, and one on Africa. Strabo regarded Earth as a sphere at the center of a spherical universe.

The career of Ptolemy (A.D. 100?–?170), of Alexandria, Egypt, represented the culmination of geographic concepts in the ancient world. In the second century A.D., the Roman Empire controlled an extensive area of the known world, including much of Europe, northern Africa, and western Asia. Taking advantage of information collected by Roman merchants and soldiers, Ptolemy wrote an eight-volume *Guide to Geography*. He also prepared a number of maps, which were not exceeded in quality for more than a thousand years.

Geography also developed in China, independent of European studies. The oldest Chinese geographical writing, from the fifth century B.C., describes the economic resources of the country's different provinces. Phei Hsiu, known as the "father of Chinese cartography," produced an elaborate map of the country in A.D. 267.

Geography in the Middle Ages. After Ptolemy, little progress in geographic thought was made in the ancient world. Following the collapse of the Roman Empire in the fifth century A.D., the word *geography* disappeared from European vocabulary. During the Middle Ages (roughly A.D. 1100-1500), geographic inquiry continued outside of Europe.

Beginning in the seventh century, Muslim armies controlled much of northern Africa and southern Europe and eventually reached as far east as present-day Indonesia in Southeast Asia. Muslim writers such as Edrisi (1099?–1154), ibn-Batuta (1304?–?1378), and ibn-Khaldun (1332–1406) gathered accurate knowledge about the locations of coastlines, rivers, and mountain ranges in the conquered areas.

Revival of geography in Europe. Geographic thought enjoyed a resurgence in Europe in the seventeenth century, inspired by exploits of European explorers to establish trading routes and gain control of resources elsewhere in the world. *Geographia Generalis*, written by the German Bernhardus Varenius (1622–1650), stood for more than a century as the standard treatise on systematic geography. Varenius also wrote a description of Japan but died before he could complete a more comprehensive work on regional geography.

The German philosopher Immanuel Kant (1724–1804) placed geography within an overall framework of scientific knowledge. Kant argued that all knowledge can be classified logically or physically. For example, a *logical classification* organizes plants and animals into a systematic framework of species, based on their characteristics, regardless of when or where they exist. A *physical classification* identifies plants and animals that occur together in particular times and places. Descriptions according to time comprise history, and descriptions according to place comprise geography. History studies phenomena that follow one another chronologically, while geography studies phenomena that are located beside one another.

In the late twentieth century, geography has matured into a vibrant and varied science, addressing every spatial aspect of our planet, using techniques ranging from the traditional to the frontiers of technology. In *Geography in America*, a book about the profession, the authors discuss the present branches of geography. We list them here to give you an idea of the breadth and depth of our profession.

- Agricultural and land use geography
- Biogeography (geography of life forms)
- Cartography (map making)
- Climatology
- Coastal and marine geography
- Cultural geography
- Energy geography (fuels)
- Environmental geography
- Gender geography
- Geographic education
- Geomorphology (geography of landforms)
- Gerontological geography (aging and elderly people)
- GIS (geographic information systems)
- Hazards geography (from AIDS to tornadoes)
- Historical geography
- Industrial geography
- Mathematical/statistical/modeling geography
- Medical geography
- Native American geography
- Physical geography
- Political geography
- Population geography
- Recreation, tourism, and sport geography
- Regional development and planning geography
- Regional geography (such as Africa, Latin America, Asia)
- Remote sensing
- Socialist geography
- Transportation geography
- Urban geography
- Water-resource geography

Each of these branches contains further specialties. For example, cultural geographers may focus on language, religion, social customs, etc. Some branches overlap, enriching one another. As we noted, *Geo-* means "Earth," and *-graphy* means "to write."

Maps

Geography is most immediately distinguished from other disciplines by its reliance on maps. For centuries, geographers identified the locations of continents, elevations of mountains, and courses of rivers, and depicted the information on maps.

A **map** is a two-dimensional—in other words flat—representation of Earth's surface, or a portion of it. Maps are basically scale models of the real world, made small enough to work with on a desk or wall, and made flat because three-dimensional models are expensive and difficult to reproduce. Maps range from here's-how-to-get-to-the-party sketches to precise, sophisticated, computer-generated, engineering-quality images. Photographs taken from airplanes are a kind of map, as are images obtained from satellites. Maps range from the functional sketch to the visually appealing and intellectually stimulating to both professional geographers and casual users.

A map is a learning device. We look at a map to learn where in the world something is found, especially in relation to a place we know, such as a town, body of water, or highway. Maps help us to find the shortest route between two places and to avoid getting lost along the way.

Geographers do not merely look at maps, they *interpret* them. Geographers extract information from maps to explain patterns of human behavior and the physical environment. A series of maps can show dynamic processes of change by depicting differences among time periods in such features as extent of snow cover or number of people with AIDS. Patterns on maps suggest interactions among different features of Earth. Geographers also use maps to communicate conclusions or explanations about human or physical processes, such as the influence of climate type on vegetation and agriculture. Placing information on a map is one of the principal ways that geographers share their evaluation of data or critical analysis of patterns.

A map is different from a photograph because it is a selective representation of Earth, an artistic creation based on scientific principles. The science of making maps is called **cartography.** To communicate geographic concepts effectively through maps, cartographers must create them properly and assure that users know how to read them. To create an accurate map several decisions must be made. The two most important are *scale* and *projection*.

Scale. The first decision a cartographer faces is how much of Earth's surface is to be presented in the map. Is it necessary to show the entire globe, or just one continent, or a country, or a city? A world map must omit many details because there simply is not enough space. Conversely, a map may depict only a small portion of Earth's surface but provide considerable detail about a particular place.

The level of detail and the amount of area covered on a map depend on its scale. The scale of a map is the same concept as the scale of a model car or boat; **scale** is the relation between the length of a feature on a map and the length of the actual feature on Earth's surface. For example, if a one-foot long roadway on a map is really 24,000 feet long on the ground, the map scale is 1:24,000.

Cartographers usually present scale in one of three ways: a fraction ($\frac{1}{24,000}$) or ratio (1:24,000), a written statement ("1 inch equals 1 mile"), or a graphic bar scale (as in Figure 1–2, lower right).

A *fractional scale* shows the numerical ratio between distances on the map and Earth's surface. A scale of 1:24,000 or $\frac{1}{24,000}$ means that one unit (inch, centimeter, foot, finger length) on the map represents 24,000 of the same unit (inch, centimeter, foot, finger length) on the ground. The unit chosen for distance can be anything, as long as the units of measure on both sides of the ratio are the same. The 1 on the left side of the ratio always refers to a unit of distance *on the map*, while the number on the right always refers to the *same unit* of distance *on Earth's surface*.

The *written scale* describes the relationship between map and Earth distances in words. For example, the written statement "1 inch equals 1 mile" on a map means that one inch on the map represents one mile on Earth's surface. Again, the first number always refers to map distance, and the second to distance on Earth's surface. Here the units can be different—such as inch and mile—for ease of use.

A *graphic scale* usually consists of a bar line marked to show distance on Earth's surface. To use a bar line, first determine with a ruler the distance on the map in inches or centimeters. Then hold the ruler against the bar line and read the number on the bar line opposite the map distance on the ruler. The number on the bar line is the equivalent distance on Earth's surface.

A map's scale can be any ratio the mapmaker desires. Here are three examples:

1. A flower bed could be mapped at 1:1 scale by drawing the flower bed at exactly the same size on a large sheet of paper.
2. A city could be mapped at 1:250,000 scale, where one inch on the map represents 250,000 inches (about 4 miles) on the ground.
3. Earth could be mapped at 1:100,000,000 scale, where one inch on the map represents 100,000,000 inches (about 1,600 miles) on the ground. This is the scale of Figure 1–1.

When comparing map scales, remember that the smaller the fractional scale the larger the area represented (example 3 above), and the larger the fractional scale the smaller the area covered (example 1 above). A world map uses a smaller scale than a city map, because it covers a larger area. A large-scale map is suitable for detailed information about a small area (Figure 1–2).

Projection. Earth is approximately the shape of a sphere and is accurately represented in the form of a globe. However, a globe is an extremely limit-

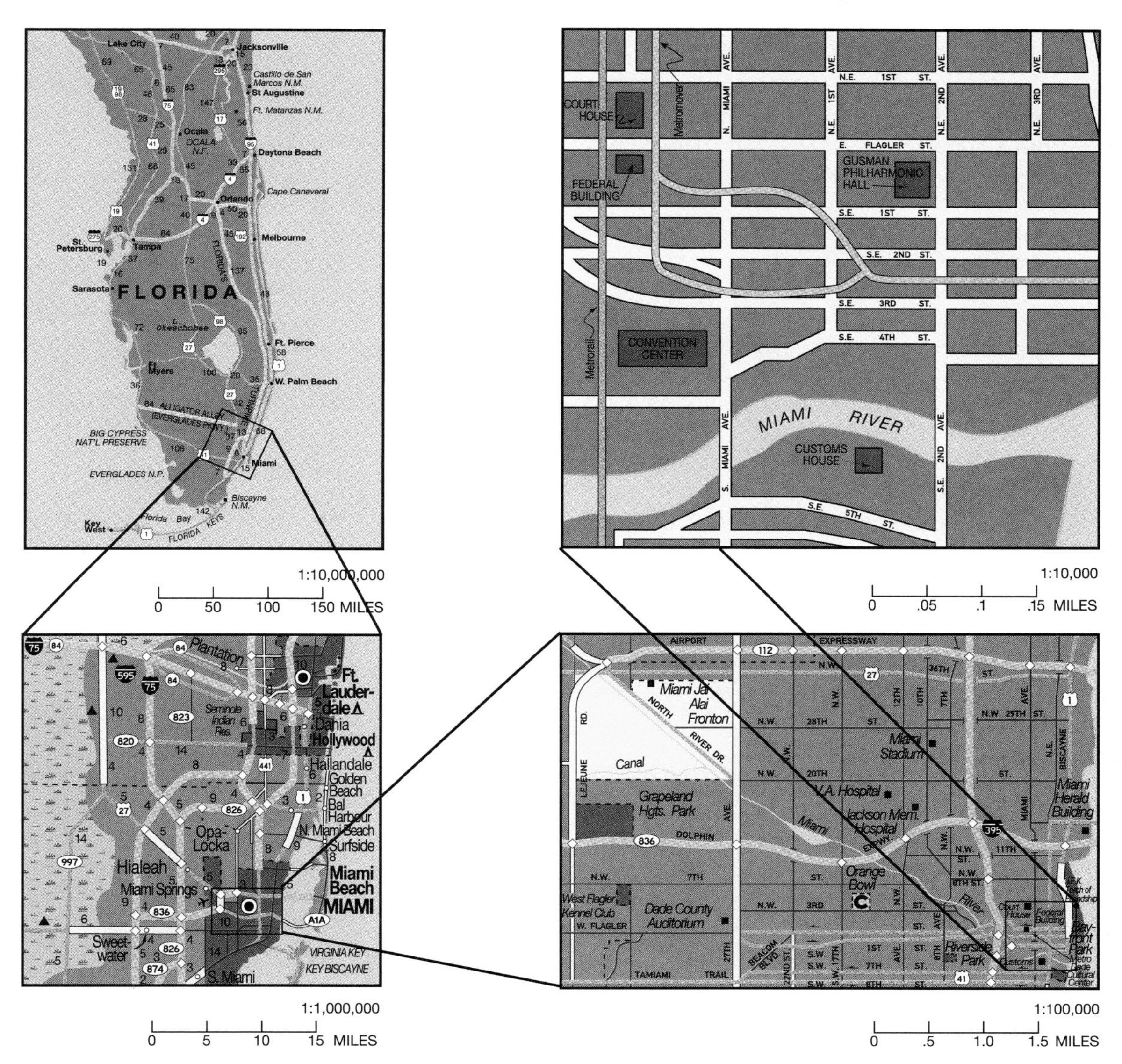

FIGURE 1–2
The map of Florida in the upper left has a fractional scale of 1:10,000,000. Expressed as a written statement, one inch on the map represents ten million inches (about 158 miles) on the ground. The bar line below the map displays the scale in a graphic form. The map of South Florida in the lower left has a scale of 1:1,000,000; one inch on the map equals about 16 miles. The map of the Miami metropolitan area in the lower right has a scale of 1:100,000; one inch on the map equals about 1.6 miles. The map of downtown Miami in the upper right has a scale of 1:10,000; one inch on the map equals about 0.16 miles or 833 feet.

ed tool to communicate information about Earth's surface. A small globe doesn't have enough space to display detailed information, while a large globe is too bulky and cumbersome to use. A globe is difficult to write on, photocopy, mail, or carry in the glove compartment of a car!

Consequently, most maps are flat. However, every cartographer faces a problem when drawing a spherical Earth on a flat piece of paper: some distortion unavoidably results. Transferring locations on Earth's surface to locations on a flat map is called **projection.** Cartographers have invented hundreds of projections, but none is free of some distortion.

The problem of distortion is especially severe for world-scale and other small-scale maps. Four types of distortions can result:

1. The *shape* of an area can be distorted, so that it appears more elongated or squat than in reality.
2. The *distance* between two points may be increased or decreased.
3. The *relative size* of different areas may be altered. One area may appear larger than another on a map but in reality be smaller.
4. The *direction* from one place to another can be distorted.

Most of the world maps in this book, such as Figure 1–1, are *equal-area projections*. The primary benefit of this type of projection is that the relative size of the land masses on the map are the same as in reality. The projection also minimizes distortion in the shape of most land masses, although areas toward the North and South poles, such as Greenland and Australia, become more distorted. These areas are sparsely inhabited, so distorting their shape usually is not important.

To preserve the approximate size and shape of land masses, however, the map is forced into other distortions:

- Note that the two hemispheres (eastern and western) are separated into two pieces, a characteristic known as *interruption*.
- Observe that the meridians (vertical lines), which in reality converge at the North and South poles, do not always converge on the map. Also, they do not form right angles with the parallels (horizontal lines).

In contrast, we use uninterrupted projections to display information in Figures 1–4 and 1–5, for example (ahead on pages 16 and 17); the Robinson projection was used for Figure 1–4, while Figure 1–5 utilizes the Mercator projection. The Robinson projection is a compromise that somewhat distorts both shape and area. Because it is uninterrupted, the Robinson projection helps display processes that can take place over the Atlantic Ocean, such as wind circulation, but it is less useful for depicting human activities, because land areas are relatively small and somewhat distorted.

The Mercator projection has several advantages: shape is distorted very little, direction is consistent, and the map is rectangular. Its greatest disadvantage is that area is grossly distorted toward the poles, making high-latitude places look much larger than they actually are. For example, compare the sizes of Greenland and South America in Figures 1–1 and 1–5. Figure 1–1 is more accurate. See the appendix for more information about map projections.

Geographic Information Technology

Having largely completed the great task of accurately mapping Earth's surface, which required several centuries, geographers have turned to new technologies to learn more about the characteristics of places. Two important technologies developed during the past quarter-century are remote sensing from satellites (to

A Landsat satellite image of Lake Garda, in northern Italy. The Alps are to the north, the Piedmont region to the south. Pink areas are towns. (Geospace/Science Photo Library/Photo Researchers, Inc.)

collect data) and geographic information systems (computer programs for manipulating geographic data). These technologies help geographers create more accurate and complex maps and measure changes over time in the characteristics of places.

Remote sensing. The acquisition of data about Earth's surface from a satellite orbiting the planet or other long-distance methods is known as **remote sensing.** Geographic applications of remote sensing include *mapping* of vegetation and other surface cover, *gathering data for large unpopulated areas*, such as measuring the extent of the winter ice cover on the oceans, and *monitoring changes*, such as weather patterns and deforestation.

Landsat satellites, launched by the United States between 1972 and 1984, were the first satellites that were intended primarily for mapping characteristics of Earth's surface. Passive sensors in the *Landsat* satellites measure the amount of radiation emanating from the Earth's surface in particular wavelengths. Sensors primarily measure colors of visible light, although some measure infrared (heat) energy. In a few cases, active sensors like radar send radiation to Earth and measure the radiation that is reflected back to the satellite.

Remote sensing satellites operate by scanning Earth's surface, much like a television camera scans an image in the thin lines you can see on a television screen. The sensor is moved across the landscape in a line, then moved slightly to scan other lines in succession. At any moment, a sensor is recording the amount of energy from only one place, an area called a picture element or *pixel.* A map created by remote sensing is essentially a grid containing many rows of pixels.

The smallest feature on Earth's surface that can be detected by a sensor is determined by the size of the pixel. This is called the *resolution* of the scanner. Early *Landsat* sensors had a pixel size of 59 meters by 59 meters (194 feet by 194 feet), compared to 30 meters by 30 meters (98 feet by 98 feet) on later *Landsat* versions and 10 meters by 10 meters (33 feet by 33 feet) and 20 meters by 20 meters (66 feet by 66 feet) on the French *Spot* satellite system. The importance of a smaller pixel size is that smaller objects can be detected.

Weather satellites take a broader view, and thus have very large pixels, covering several kilometers on a side. This way, they can rapidly map a large area such as a continent. Weather forecasters need data about a large area very quickly, because weather systems change rapidly. The sensor used to produce the Geosphere Project image, with a resolution of about 1 kilometer (0.6 mile), can map the entire Earth once every two days.

GIS. A **GIS** is a computer system that stores, organizes, analyzes, and displays geographic data (Figure 1–3). Each piece of information about a location is stored in a separate computer file that represents an "information layer." A single layer can be displayed by itself or combined with other layers to show relations among different kinds of information. Powerful desktop microcomputers have increased the use of GIS.

Geographers use GIS to analyze both environmental and social processes. An environmental example is predicting the amount of water that will flow into a river from melting snow. This can be predicted by combining a layer showing snow cover in the river's headwaters with layers depicting surface temperature and topography. A human geography example is combining a street map with a population map to determine the number of people living within walking distance of a proposed bus route.

Area Analysis

Every place on Earth has a unique location. Geographers identify where places are located, describe the special features of these places, and group the identified places into distinctive regions.

Location

Fundamental to geographic thinking is the principle of location. Just as historians study the logical sequence of human activities in time, so do geographers study the logical arrangement of human activities in space. **Location,** which is the position of anything on Earth's surface, can be identified both by name and according to mathematical principles.

Name. The simplest way to describe a particular location is by referring to its name, because inhabited places on Earth's surface have been named by somebody. Geographers call the name given to a

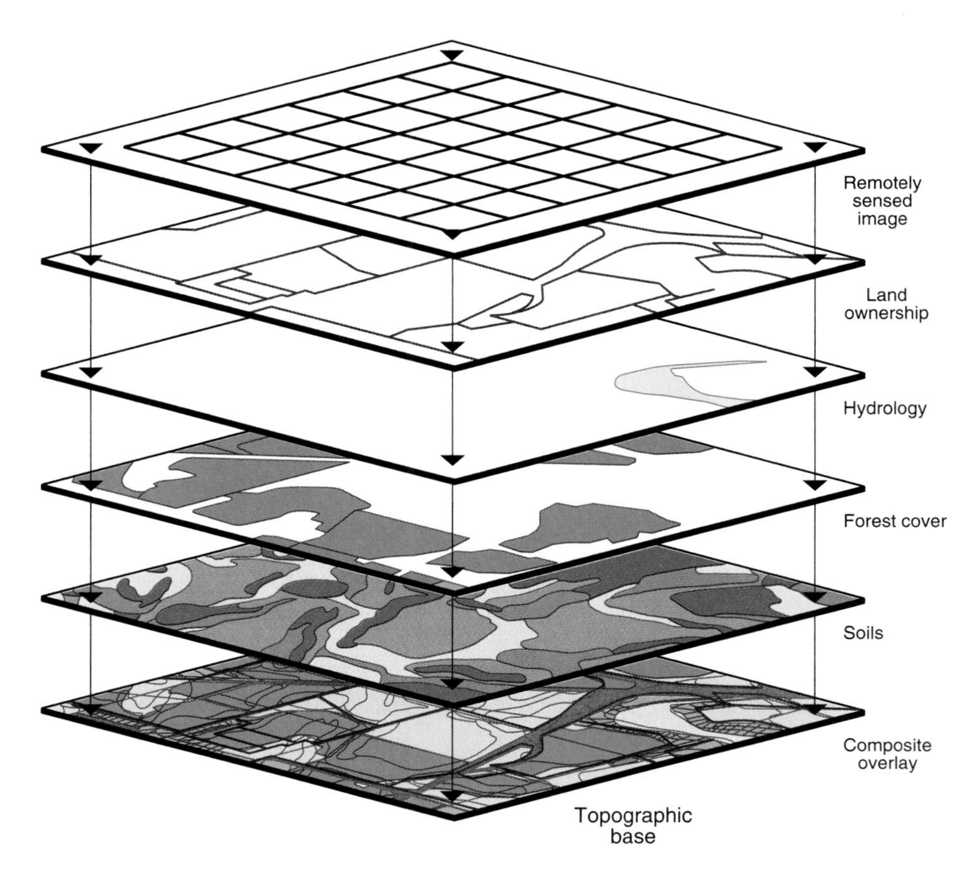

FIGURE 1–3
A geographic information system (GIS) involves storing information about a location in layers. Each layer represents a different piece of human or environmental information. The layers can be viewed individually or in combination.

portion of Earth's surface the **toponym,** or nominal location, of a place.

The Board of Geographical Names, operated by the U.S. Geological Survey, was established in the late nineteenth century to be the final arbiter of names placed on maps produced in the United States. In recent years, the board has been especially concerned with removing offensive place names, such as those considered racially or ethnically derogatory.

Some place names derive from features of the physical environment. Trees, valleys, bodies of water, and other natural features appear in the place names of most languages. For example, the capital of the Netherlands, 's Gravenhage (more commonly called The Hague), means "the prince's forest;" Aberystwyth, Wales, means "mouth of the river Ystwyth," while 22 kilometers (13 miles) upstream lies the tiny village of Cwmystwyth, which means "valley of the Ystwyth." The name of the river, Ystwyth, is the Welsh word for meandering, descriptive of a stream that bends like a snake.

The community with perhaps the longest name in the world is a town in Wales called *Llanfairpwllgwyngyllgogerychwyrndrobwllllantysiliogogogoch*. The name means *the church of St. Mary's in the grove of the white hazelnut tree near the rapid whirlpool and the Church of St. Tisilio near the red cave!* The town's name originally encompassed only the first twenty letters, but when the railway came through in the nineteenth century, the townspeople lengthened it. They decided that signs with the longer name in the railway station would attract attention and bring more business and visitors to the town.

A place name can tell us a great deal about the social customs of early inhabitants. Some settlers select place names associated with religion, such as a saint, while other names derive from ancient history such as Athens, Attica, and Rome. A place name may also indicate the origin of the settlers. Place names commonly have British origins in North America and Australia, Portuguese origins in Brazil, Spanish origins elsewhere in Latin America, and Dutch origins in South Africa.

Places can change names as a result of political upheavals. For example, following World War II, Poland gained control over territory that was formerly part of Germany and changed many of the place names from German to Polish. Among the larger cities, Danzig became Gdańsk, Breslau became Wrocław, and Stettin became Szczecin.

Names associated with communism have been changed in recent years throughout Eastern Europe, in many cases reverting to the names used before the Communists gained power. For example, in Olomouc, a city of 100,000 in the Czech Republic, Lenin Street has been changed to Liberty Street, Red Army Square to Lower Square, and Liberation Street to Masaryk Street (for the first president of democratic Czechoslovakia between 1919 and 1935). Gottwaldov, a nearby city named for a communist president of Czechoslovakia, reverted to its former name Zlín.

Sometimes a place name is so symbolic that its use can cause political difficulty. When Yugoslavia's southernmost republic declared independence in 1991, its leaders wished to call the new country Macedonia, the same name it had as a republic within Yugoslavia. But Greece felt threatened by the new country's use of the name Macedonia, because Macedonia was also the name of Greece's northernmost region. As the home of Aristotle and Alexander the Great, ancient Greek Macedonia was an important cultural hearth for Greek and Western civilization. Older Greeks recalled the Communists' promise: if their side won the civil war in Greece in 1948-1949, they would annex Greek Macedonia into a new federation comprising the neighboring countries of Yugoslavia and Bulgaria.

Greece suggested that the new country be called the Slavic Republic of Macedonia, but this was rejected because only 64 percent of the inhabitants of the new country were Slavs; using the word "Slavic" would offend Albanians and Turks, who comprised 21 percent and 5 percent of the new country's population, respectively. Lack of agreement on a name for the new country delayed diplomatic recognition by other countries and financial support to help it achieve economic development.

Mathematical location. If needed, the precise location on Earth's surface can be described mathematically according to a coordinate system known as latitude and longitude. The universally accepted numbering system of latitude and longitude derives from two series of lines—more precisely, arcs—drawn on the globe. These arcs are known as parallels and meridians (Figure 1–4).

A **meridian** is an arc drawn on a globe between the North and South poles. Every meridian has the same length and the same beginning and end points. The location of each meridian is identified on Earth's surface according to a numbering system known as **longitude.** One meridian, which passes through the Royal Observatory at Greenwich, England, has been designated by international agreement as the "starting point" for numbering the meridians. It is labeled 0 *degrees* (0°) longitude and is called the **prime meridian.**

The meridian on the opposite side of the globe from the prime meridian is 180° longitude. All other meridians have numbers between 0° and 180° and are designated "east" or "west" to show which direction they are from the prime meridian. For example, New York City is located at 74° *west* longitude, while Lahore, Pakistan, is 74° *east* longitude. San Diego is located at 117° *west* longitude, while Tianjin, China, is at 117° *east* longitude.

The second set of arcs drawn on a map are **parallels,** which are circles drawn around the globe parallel to the equator at right angles to the meridians. The numbering system used to indicate the location of parallels is called **latitude.** Earth's equator is the middle parallel, halfway between the North and South poles. The equator is 0° latitude, the North Pole is 90° north, and the South Pole is 90° south. New York City is located at 41° *north* latitude, while Wellington, New Zealand, is at 41° *south* latitude. San Diego is located at 33° *north* latitude, while Santiago, Chile, is at 33° *south* latitude.

We can determine the mathematical location of a place more precisely if necessary. Each degree is divided into 60 *minutes* (′), and each minute in turn is divided into 60 *seconds* (″). For example, the official mathematical location of Denver, Colorado, is 39°44′ north latitude and 104°59′ west longitude.

The State Capitol building in Denver is located at 39°42′52″ north latitude and 104°59′04″ west longitude. Latitude and longitude is an especially useful location system for navigation on the sea. The *nautical mile* is derived from this coordinate system—it is the distance on the ground represented by one minute of latitude, or 1,852 meters (6,076.115 feet).

Time zones. Longitude plays an important role in calculating time. Earth is divided into 360° of longitude, including from 0° to 180° west and 0° to 180° east longitude. Every 24 hours, Earth rotates through the full 360°. Therefore, Earth turns 15° each hour (360° divided by 24 hours). The time at the prime meridian or 0° longitude is designated **Greenwich Mean Time** (GMT). Because Earth rotates eastward, your clock advances one hour from GMT for each 15° traveled east from the prime meridian. For each 15° you travel west from the prime meridian you set your clock one hour earlier than GMT (Figure 1–5).

The meridian for 75° west longitude runs through the eastern United States, so time in the eastern United States is 5 hours earlier than Greenwich Mean Time. (The 75° difference in longitude divided by 15° per hour = 5 hours.) When the time is 11 A.M. GMT, the time in the eastern United States is therefore 5 hours earlier, or 6 A.M. The 48 coterminous U.S. states and the Canadian provinces share four standard time zones, known as Eastern, Central, Mountain, and Pacific. Most of Alaska comprises the Alaska Time Zone, which is 9 hours earlier than GMT. The Hawaii Time Zone is 10 hours earlier than GMT. Canada has an Atlantic Time Zone, which is 4 hours earlier than GMT, and Newfoundland Time Zone, which is 3½ hours earlier than GMT.

In Figure 1–5, you can see that the **International Date Line** for the most part follows

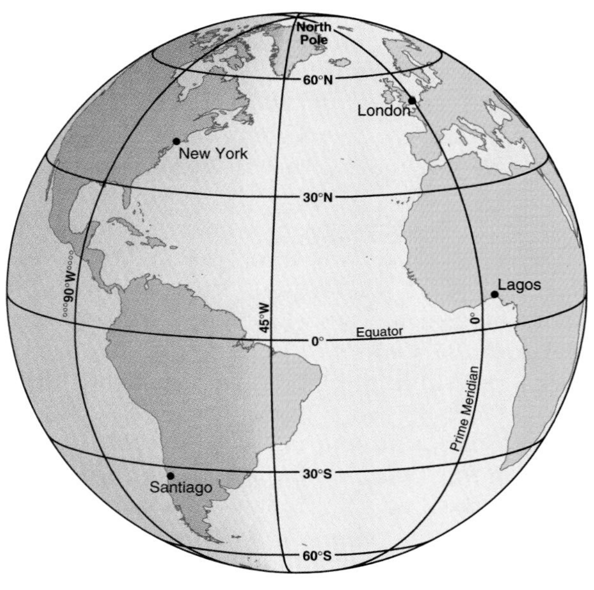

FIGURE 1–4
Meridians are arcs that connect the North and South poles. The meridian running through Greenwich, England, is the prime meridian, or 0° longitude. Parallels are circles drawn around the globe parallel to the equator. The equator is 0° latitude, while the North Pole is 90° north latitude.

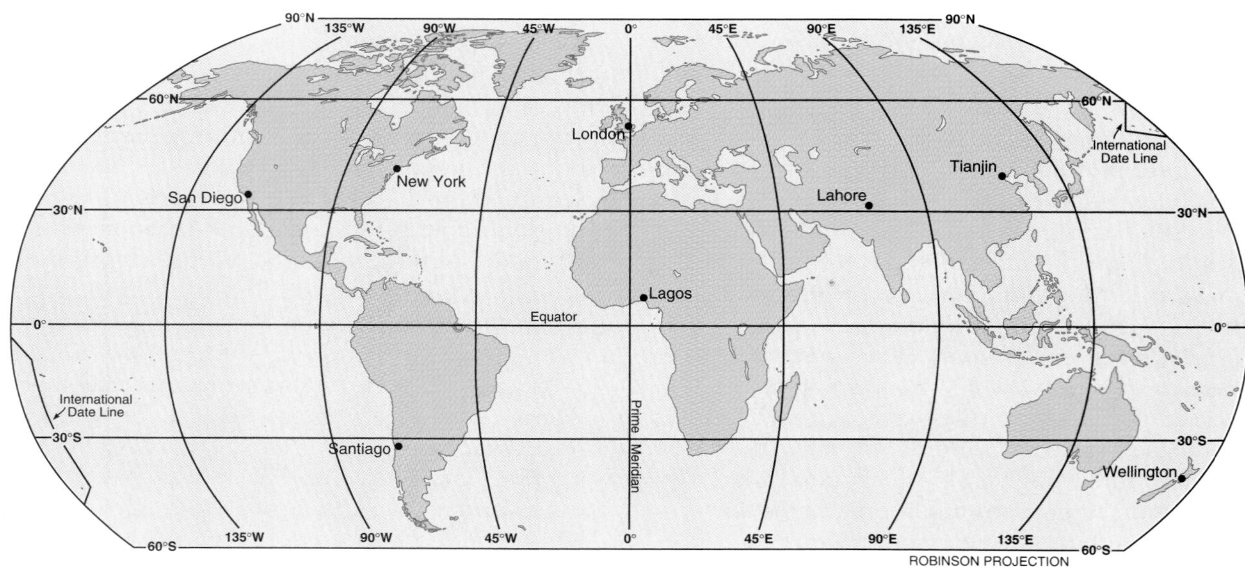

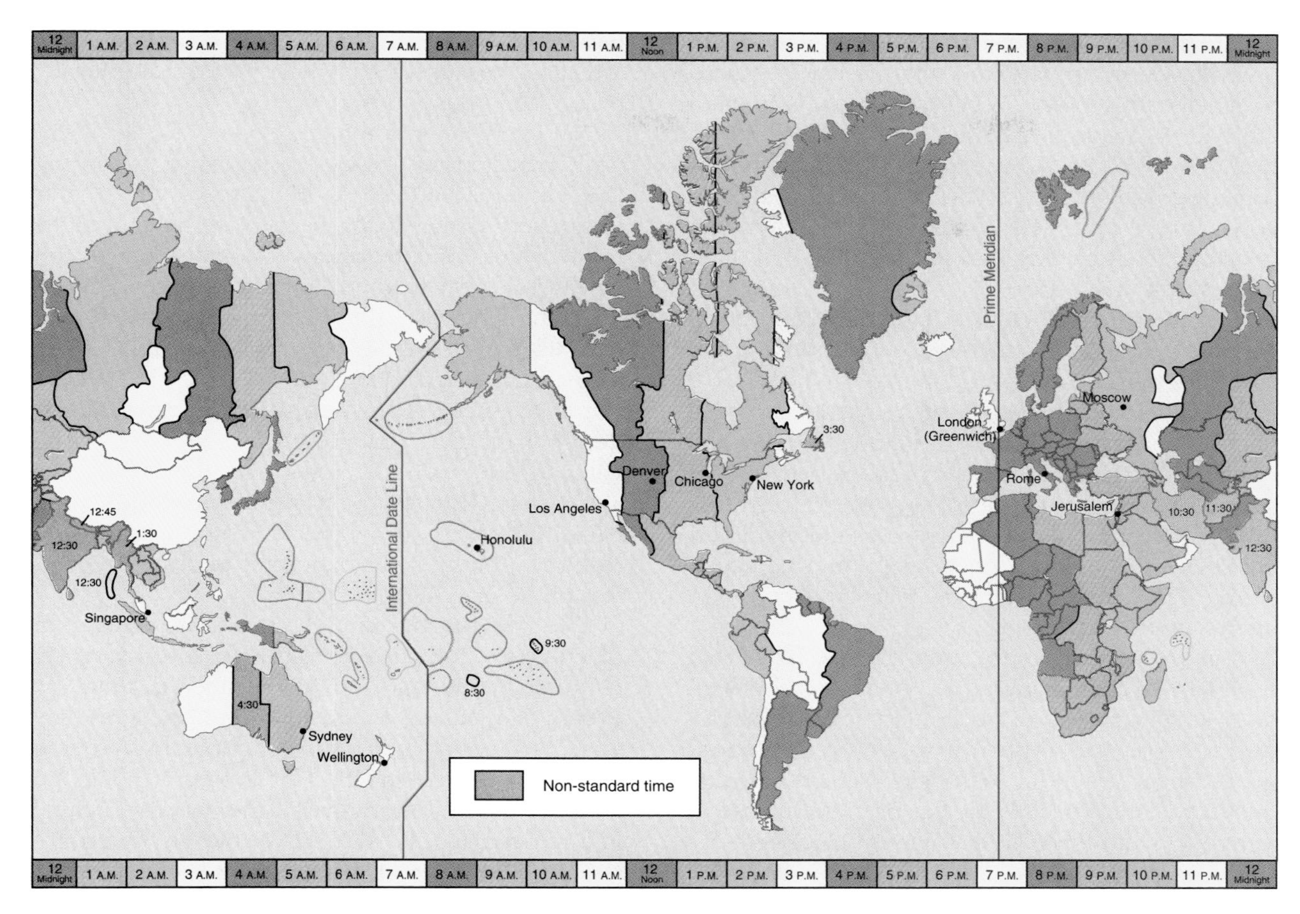

FIGURE 1–5
Greenwich Mean Time (GMT) is the time near the prime meridian, or 0° longitude. The Pacific Time Zone, which encompasses the western part of the United States and Canada, is eight hours behind GMT because it is situated near 120° west longitude.

180° longitude, although it deviates in several places to avoid dividing land areas. When you cross the International Date Line heading east (toward America) the clock moves back 24 hours, or one entire day. When you cross it going west (toward Asia) the calendar moves ahead one day. When the time in New York City is 2 P.M. Sunday, it is 7 P.M. Sunday in London, 8 P.M. Sunday in Rome, 9 P.M. Sunday in Jerusalem, 10 P.M. Sunday in Moscow, 3 A.M. Monday in Singapore; and 5 A.M. Monday in Sydney, Australia. Continuing farther east, it is 7 A.M. *Monday* in Wellington, New Zealand, but 9 A.M. *Sunday* in Honolulu, Hawaii, because the International Date Line lies between New Zealand and Hawaii.

Regions

Geographers combine neighboring places into regions. When Julius Caesar wrote 2,000 years ago that Gaul was divided into three parts, he cited a good example of area analysis. Where people and environments within an area display similarities and regularities, it constitutes a region. One region's cultural, economic, and physical characteristics will differ from those of other regions. Identifying the differences and similarities between regions is fundamental to understanding the modern world.

In the past, regional geographers identified an area of Earth's surface and described in careful detail as many of its characteristics as they could uncover. Contemporary area analysis can be traced to nineteenth-century French geographers, including Paul Vidal de la Blache (1845–1918) and Jean Brunhes (1869–1930). A regional studies method—sometimes called the *cultural landscape* method—was later adopted by several American geographers, including Carl Sauer (1889–1975) and Robert Platt (1880–1950).

According to the cultural landscape approach, geographers should start by closely observing the various environmental and human aspects of a particular place, since each place has its own distinctive landscape based on a unique combination of social relationships and physical processes. The science of geography, according to this method, involves sorting out the relationships among different social and physical phenomena in a particular study area, since all objects in a landscape are interrelated.

A more contemporary area analysis approach may start by identifying an important trend or feature, such as growth of population, level of precipitation, or consumption of energy. Then, geographers search for reasons to explain why that trend or characteristic is greater or more intense in one region and lower or less intense elsewhere.

Formal region. Geographers identify three types of regions: formal, functional, and vernacular. A **formal region**—also known as a *uniform* or *homogeneous* region—is an area in which the selected trend or feature is present throughout.

Geographers typically employ formal regions to explain broad global or national patterns. A formal region may display a distinctive cultural characteristic (such as language or religion), an economic trend (such as predominant type of agriculture practiced or average income), or an environmental characteristic (climate, level of precipitation, or soil type). Although the selected characteristic used to distinguish the formal region may be quantifiable, it is often designed to illustrate a general concept rather than a precise mathematical distribution.

Some formal regions, such as countries or local government units within countries, may be easy to identify. Montana is an example of a formal region, characterized by a government that passes laws, collects taxes, and issues license plates with equal intensity throughout the state. We do not have difficulty identifying the formal region of Montana, because it has clearly drawn and legally recognized boundaries.

Not everyone in a formal region possesses identical characteristics. Not every farmer living in the U.S. or Canadian wheat belt grows wheat, nor does every farmer living in the U.S. ranching area raise cattle. Nonetheless, we can distinguish the wheat belt as a region in which the predominant agricultural activity is wheat farming. Similarly, we can distinguish formal regions within the United States characterized by the fact that the people tend to vote for Republican candidates. Republicans may not get all 100 percent of the votes in these regions, nor in fact do Republicans always win. However, in a presidential election the candidate with the largest number of votes receives all of the electoral votes of a state regardless of the margin of the victory.

The previous examples show one danger in identifying a formal region: the need to recognize cultural, economic, and environmental diversity even while making a generalization. Problems may arise because a minority of people in a region speak a language, practice a religion, or possess a resource different from that of the majority. People in a region may play distinctive roles in the economy but hold different positions in society based on gender or ethnicity.

Functional region. A **functional region**—also known as a *nodal* region—is an area in which an activity has a focal point. The characteristic dominates at a central node and diffuses towards its outer edges with diminishing importance. Geographers often use functional regions to display economic information. The region's node may be a shop or service, and the boundaries of the region mark the limits of the trading area of the activity.

An example of a functional region is the circulation area of a newspaper. A newspaper dominates circulation figures in the city in which it is published. Farther away from the city, the percentage of people who choose that newspaper declines, while the percentage of people who read a newspaper published in another city increases. At some point, the circulation of the newspaper from the second city matches the circulation of the original newspaper. That point is the boundary between the nodal regions of the two newspapers.

A functional region is unified through a **network**, which is a system of connected lines forming an interrelated chain. One well-known example of a network in the United States is a television network, such as ABC, CBS, or NBC, which comprises a chain of stations around the country simultaneously broadcasting the same program, such as a football game. The United States is divided into several hundred functional regions based on television markets. Every market for television advertising has an *area of dominant influence (ADI)*, the region in which the preponderance of viewers are tuned to that market's stations.

An ADI is a good example of a functional region, because the characteristic—people who are viewing a particular station—is dominant at the center and diffuses towards the periphery. For example, everyone in Des Moines, Iowa, who wishes to watch a program on NBC tunes to Channel 13. In Omaha, Nebraska, 225 kilometers (140 miles) to the west, everyone watching NBC is tuned to Channel 6 (Figure 1–6). With increasing distance eastward from Omaha, Channel 6's signal gets weaker and Channel 13's gets stronger. The percentage of people watching NBC declines for Channel 6 and increases for Channel 13.

The boundary between the Omaha and Des Moines ADIs is the point where the viewership of channels 6 and 13 is equal, near the Cass/Adair county line. Other functional regions in Iowa centered around NBC affiliates include Sioux City's Channel 4 in the northwest, Davenport's Channel 6 in the east, Waterloo's Channel 7 in the northeast, and Rochester, Minnesota's, Channel 10 in the north.

Vernacular region. A **vernacular region,** or *perceptual region*, is one that people *believe* to exist, as part of their cultural identity. These regions emerge from concepts that people use informally in daily life, rather than from scientific models developed through geographic thought.

As an example of vernacular regions, Americans frequently use the terms *sunbelt* and either *frostbelt* or *rustbelt* to distinguish two regions in the country. A number of important characteristics distinguish the sunbelt from the frostbelt, including more temperate climate and higher levels of population and economic growth. Sunbelt refers to the southern and western parts of the United States, while frostbelt or rustbelt refers to the northern and eastern parts.

Analysts have difficulty fixing the precise boundary between the two regions. At a conference called *The Sunbelt: A Region and Regionalism in the*

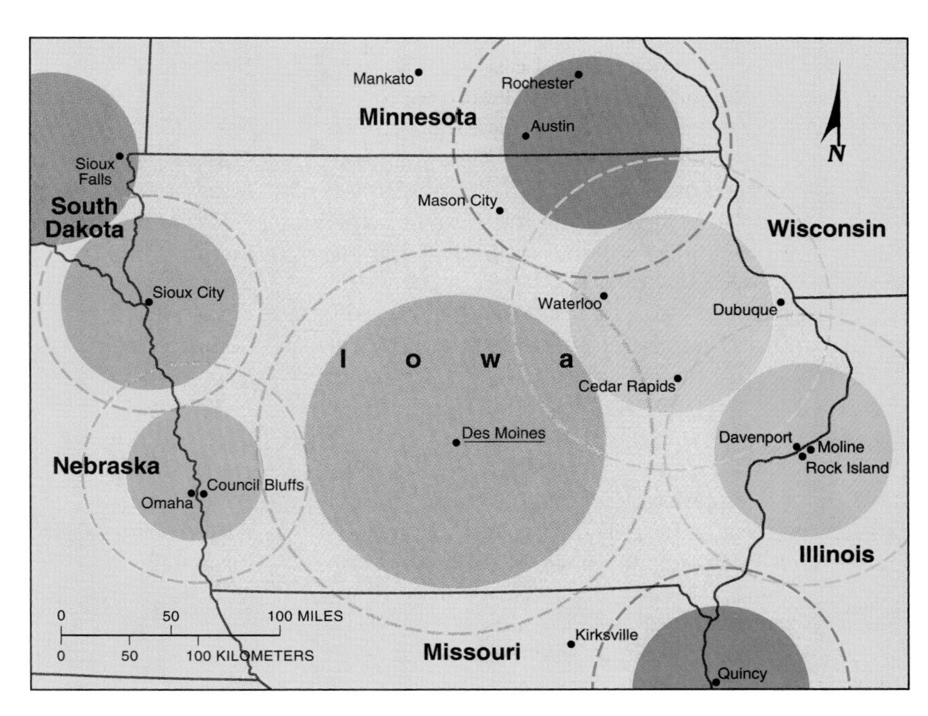

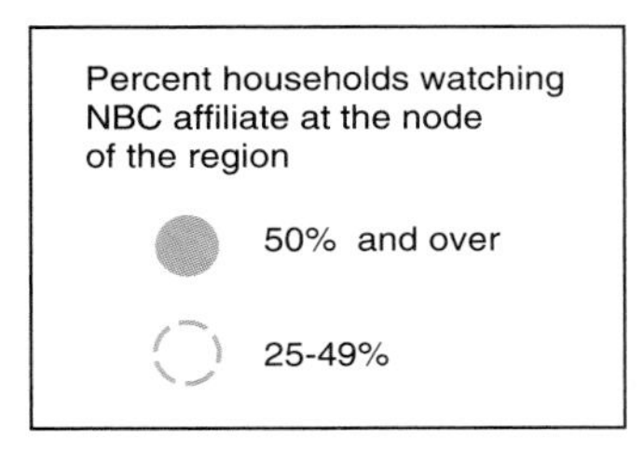

FIGURE 1–6
The map shows the regions of dominance for different television stations within Iowa. These areas are known as areas of dominant influence (ADIs). In several cases the node of the functional region—the location of the television station—is in an adjacent state. Functional regions frequently overlap state or national boundaries. When new housing, highways, and cable television lines are constructed, a county on the periphery of an ADI may be reallocated to another functional region.

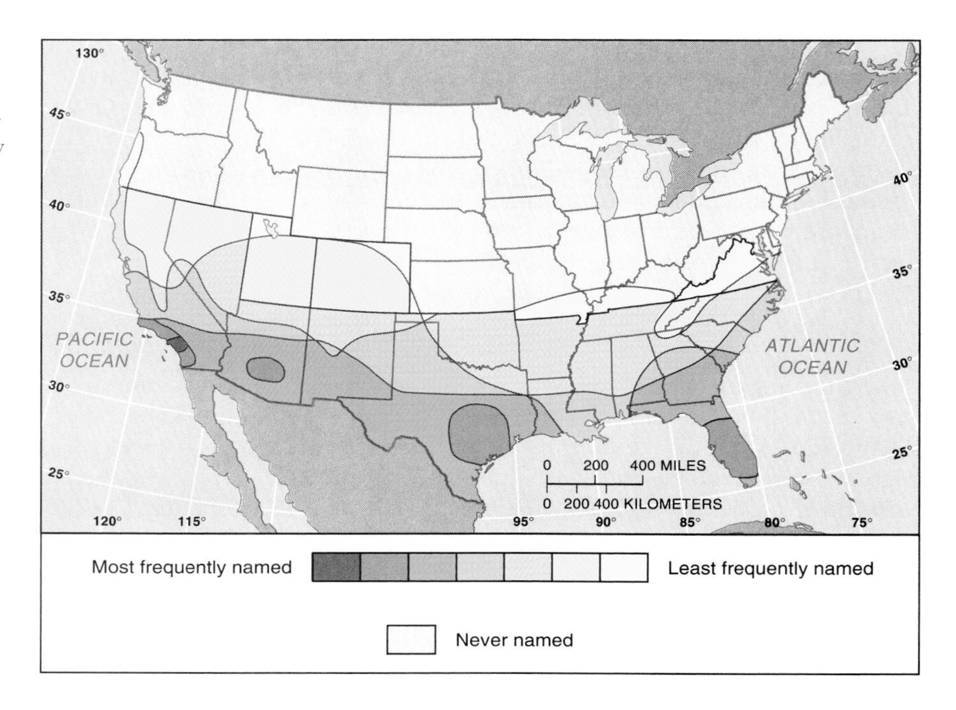

FIGURE 1–7
Geographers at a recent conference disagreed on the definition of the sunbelt, an example of a vernacular region. Respondents most frequently mentioned southwestern California between Los Angeles and San Diego, followed by south Florida and south central Arizona between Phoenix and Tucson. Some geographers cited areas as far north as Oregon and Virginia. (Adopted from Stanley D. Brunn, "Sunbelt USA," *Focus 36* (Spring 1986): 35. Used by permission of The American Geographical Society.)

Making? participants were given blank outline maps of the United States and asked to delineate the sunbelt. Respondents most frequently cited southern California, from Los Angeles to San Diego, as part of the sunbelt. Other areas of the United States that most participants considered part of the sunbelt included southern Texas, southern Florida, and central Arizona. A few considered the sunbelt to reach as far north as Oregon or Virginia (Figure 1–7).

Perceptual regions can play a critical role in organizing daily life. For example, students at one university were shown a map of the campus divided into squares and asked to indicate in which squares they felt safe walking alone at 10:30 P.M. When combined, the responses portrayed a campus divided into regions that were widely regarded as safe and regions that were widely regarded as dangerous. Such studies can also determine if perceptions of safety are uniform among groups of students or vary by age, gender, and ethnicity.

Regional integration. A region gains uniqueness from possessing a distinctive combination of human and environmental characteristics. Not content with merely identifying these characteristics, geographers search for relationships among them. Geographers recognize that characteristics are integrated with each other.

Geographers examine the integration of various features within regions. For example, geographers divide the world into uniform regions, grouped according to countries that are relatively developed economically and those that are developing (sometimes called less developed countries, or LDCs). The relatively developed regions, such as Europe, North America, and Japan, are located primarily in the northern latitudes, while the developing regions are concentrated in the southern latitudes. This north-south regional split underlies many of the world's social and economic problems.

A variety of characteristics—such as per capita income, literacy rates, televisions per capita, and hospital beds per capita distinguish relatively developed from developing regions. Geographers demonstrate that the distribution of one characteristic of development is associated with the others.

It is the geographer's job to sort out the associations among various human and environmental characteristics. For example, a geographer may conclude that political unrest in the Middle East, Eastern Europe, and other areas derives from the mismatch of political boundaries to the spatial distributions of important human and physical features, such as language, religion, and resources.

In some cases, geographers can build models to show that one characteristic is causally related to another. For example, regional differences in population growth rate are caused primarily by differences in birth rates. However, geographers often hedge their bets: they recognize that one characteristic must be associated with others, even if the relationship cannot be modeled precisely. Geographers may have difficulty in constructing exact models of cause and effect, since they integrate many human and environmental features in an effort to explain a region's distinctiveness.

Example: Integrating information about cancer. Recognizing that various human and environmental characteristics are integrated helps us to understand social issues. For example, the percentage of people who die each year from cancer differs among regions within the United States. The mid-Atlantic region has the highest levels, with Maryland ranked first among the fifty states, followed by Delaware; the rate in Washington, D.C., which is adjacent to Maryland, is higher than in any state (Figure 1–8).

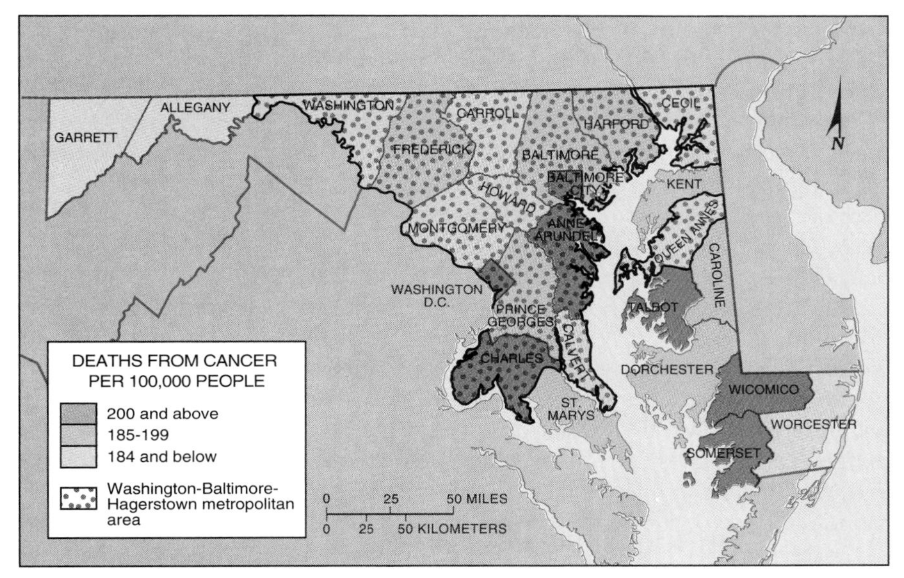

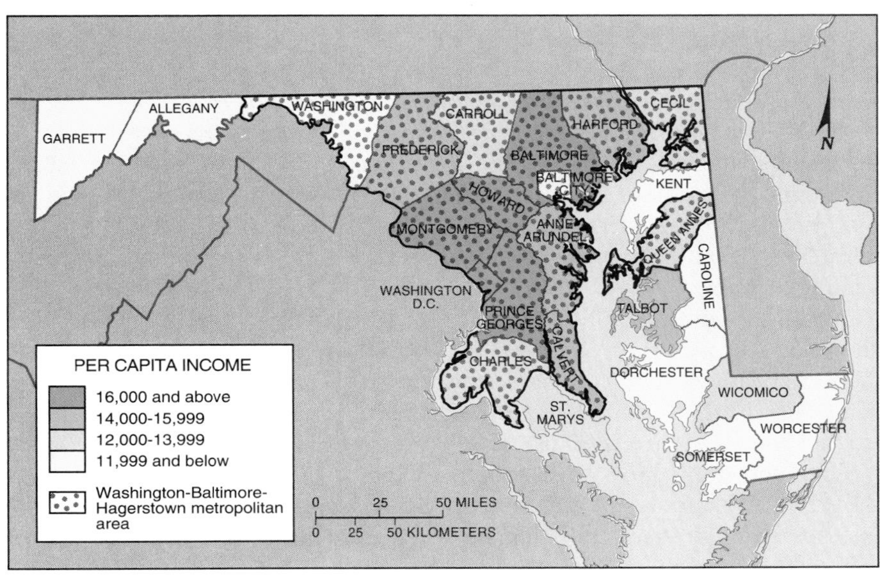

FIGURE 1–8
(Top) Maryland has the highest incidence of cancer in the United States, although the rate varies widely within the state.
(Bottom) The highest rates are in areas where incomes are relatively low, including the large cities of Baltimore and Washington and rural counties on the Eastern Shore of the Chesapeake Bay. Chesapeake Bay pollution may also play a role in the relatively high cancer rates on the Eastern Shore.

Why does Maryland have the highest cancer rate among the fifty states? Mapping the location of cancer victims among Maryland's major subdivisions (twenty-three counties plus an independent city of Baltimore), as well as the District of Columbia, shows sharp internal variations. The death rate from cancer in Baltimore City is more than 50 percent higher than in Garrett County.

The map of cancer rates by county in Maryland does not communicate useful information to someone who knows little about the location of human activities and environments within the state. By integrating other spatial information, we can begin to see factors that may be associated with regional differences in cancer. Such integration can be accomplished using a GIS.

Let's take a scientific approach. We can divide the state into counties that comprise part of the Washington-Baltimore metropolitan area and the counties that are not included in a metropolitan area. (Chapter 13 explains how counties are classified as metropolitan or nonmetropolitan.) Thus far, this division does not appear to be very helpful in explaining the location of cancer victims, because we can find high and low rates among both metropolitan and nonmetropolitan counties. However, within the metropolitan areas a pattern emerges: the highest cancer rates are in the cities of Baltimore and Washington, while the suburban counties surrounding the two cities have lower rates.

Once we recognize that the cities have higher cancer rates than the suburbs, we can combine that information with other features. People in the cities of Baltimore and Washington are more likely than suburbanites to have low incomes and low levels of education. As a result, people living in the cities may be less able to afford medical care to minimize the risk of a cancer-related fatality and may be less aware of the cancer risks associated with lifestyle choices such as smoking and consuming alcohol.

Among nonmetropolitan counties, Maryland shows a sharp division between the west, where rates are relatively low, and the east, where rates are relatively high. Income and education do not explain the difference, because levels are lower in most of the nonmetropolitan counties when compared to the levels in the metropolitan areas. Instead, we must attempt to integrate other economic and environmental factors into our explanation.

People on the Eastern Shore of the Chesapeake Bay may be especially exposed to cancer-causing chemicals, because relatively high percentages are engaged in fishing and farming compared to people living in the mountainous western counties. The nearby Chesapeake Bay is one of the nation's principal sources of shellfish, and many Eastern Shore residents work in seafood-processing industries. But the Chesapeake Bay also suffers from runoff of chemicals from Eastern Shore farms, which make heavy use of pesticides, as well as discharges of waste from factories, for the most part located in the metropolitan counties on the western side of the Bay. Prevailing winds also carry pollutants eastward from industries in the metropolitan areas. Exposure to these environmental conditions may increase the rate of cancer among Eastern Shore residents.

Spatial Analysis

Similarities and differences among places result from interaction among people and environments. Geographers search for regularities in the location of people and environments, measure them, and identify processes producing the regularities.

Distribution

The regular arrangement of a phenomenon across Earth's surface is known as **spatial distribution.** Spatial distribution has three important properties: density, concentration, and pattern.

Density. The frequency (number of times) with which something exists within a given unit of area is its **density.** The phenomenon being measured could be people, dwelling units, plants, rivers, or virtually any other object. Examples of units of area include square miles, acres, and hectares (Figure 1–9).

Geographers frequently calculate the *arithmetic density*, which is the total number of objects in an area. Arithmetic density in population studies involves two measures: the number of people and the area. For example, the United Kingdom's population is about 58 million, and the country's land area is about 243,000 square kilometers (94,000 square miles). Dividing population by area, we get about 239 persons per square kilometer (617 per-

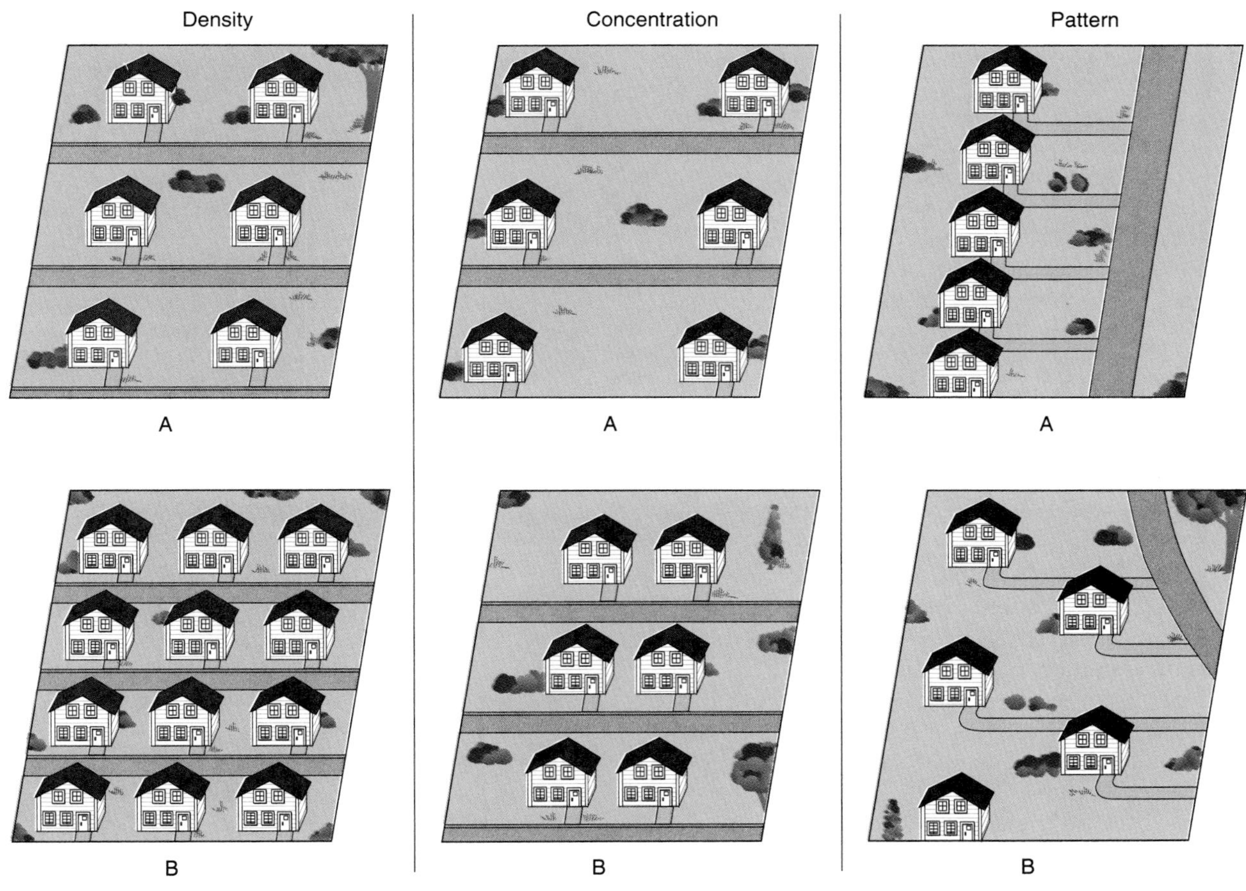

FIGURE 1–9
Spatial distribution is represented in three ways: density, concentration, and pattern. Assume that the area in all figures represents 1 acre. (Left) The density is six houses per acre in A and twelve houses per acre in B. (Center) The houses are dispersed in A and clustered in B. Note that while the concentration changes, the density is the same in both figures. (Right) Five houses are arranged in a linear pattern in A and form an irregular arrangement in B.

sons per square mile), the population density of the United Kingdom (see Chapter 7).

The most populous country in the world, China, with approximately 1.2 billion inhabitants, is by no means the densest. The arithmetic density of China is approximately 122 persons per square kilometer (315 persons per square mile), only one-half as high as in the United Kingdom. Although China has about twenty times more inhabitants than the United Kingdom, it also has nearly forty times more land.

High population density is not necessarily related to poverty. The Netherlands, one of the world's wealthiest countries, has an arithmetic density of approximately 371 persons per square kilometer (961 persons per square mile). One of the poorest African countries, Mali, has an arithmetic density of 7 persons per square kilometer (18 persons per square mile).

Geographers measure population density in other ways, depending on the subject being studied. Geographers concerned with the relationship between population growth and food supply often calculate the *physiological density* (number of persons per area suitable for agriculture) and *agricultural density* (number of farmers per area of farmland). Urban geographers frequently use *housing density*, which is the number of dwelling units per area. *Drainage density*, which is the length of a river per unit of area, measures the ability of runoff to shape the landscape.

Concentration. The spread of something over a given study area is **concentration.** If the objects in a given area are close together, they are considered *clustered.* If they are relatively far apart, they are considered *dispersed.* To compare concentration of two study areas, they should have the same number of objects and the same area.

Geographers use the concept of concentration in a number of ways. For example, one of the major changes in the distribution of the U.S. population is greater dispersion. The total number of people living in the United States is growing slowly—less than 1 percent per year—and the land area is not changing, but the concentration of population is changing from relatively clustered in the Northeast to a more even dispersion across the country.

Concentration is not the same as density. One study area with relatively high density could have a dispersed population, while another study area with the same density could have a clustered population.

We can illustrate the difference between density and concentration by the change in the distribution of major league baseball teams in North America. In 1900, the major leagues had 16 teams, a distribution that remained unchanged for more than half a century. Beginning in 1953, the following six of the 16 teams moved to other cities:

- Braves—Boston to Milwaukee in 1953, then to Atlanta in 1966
- Browns—St. Louis to Baltimore (Orioles) in 1954
- Athletics—Philadelphia to Kansas City in 1955, then to Oakland in 1968
- Dodgers—Brooklyn to Los Angeles in 1957
- Giants—New York to San Francisco in 1957
- Senators—Washington to Minneapolis (Minnesota Twins) in 1960

These moves resulted in a more dispersed distribution. Before the moves, seven teams were clustered in the three northeastern cities of Philadelphia, New York, and Boston, compared to only three teams after the moves. In 1953, no team was located south or west of St. Louis, but after the moves teams were located on the West Coast and the Southeast for the first time (Figure 1–10).

In addition to the shifts by established teams, the major leagues expanded between 1960 and 1993 from 16 to 28 teams. The new teams selected the following locations:

- Angels—Los Angeles in 1961, then to Anaheim (California) in 1966
- Senators—Washington in 1961, then to Dallas (Texas Rangers) in 1972
- Mets—New York in 1962
- Astros—Houston in 1962
- Royals—Kansas City in 1969
- Padres—San Diego in 1969
- Expos—Montreal in 1969
- Pilots—Seattle in 1969, then to Milwaukee (Brewers) in 1970
- Blue Jays—Toronto in 1977
- Mariners—Seattle in 1977

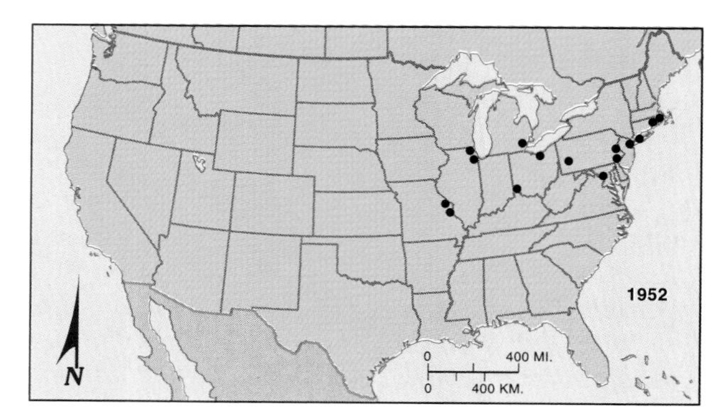

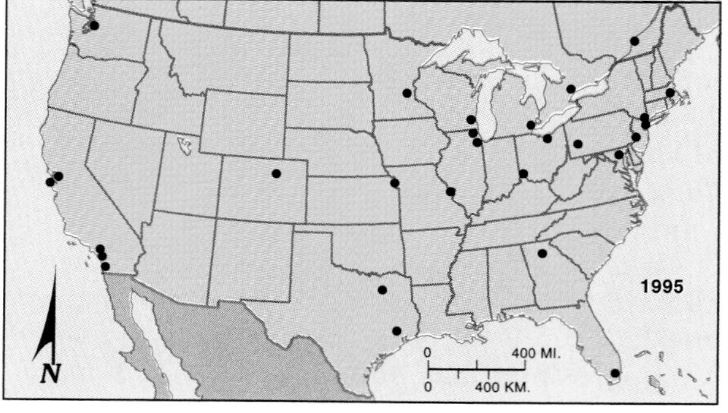

FIGURE 1–10
The changing distribution of North American baseball teams illustrates the difference between density and concentration. The figures show that density of baseball teams in North America has increased from 16 per North American land area in 1952 to 28 per North American land area in 1995. At the same time, the distribution has changed from clustered in the northeastern part of the United States to more dispersed across the United States and southern Canada.

- Marlins—Miami (Florida) in 1993
- Rockies—Denver (Colorado) in 1993

Thus, the *density* of major league teams in North America increased from 16 to 28 at the same time the *distribution* became more dispersed across the United States and Canada.

Pattern. The third property of distribution is the **pattern**. Can distribution of a trait be depicted on a map in a somewhat orderly manner, or does it appear without shape? Some phenomena are organized in a regular, or geometric, pattern, while others are distributed randomly. Geographers observe that many objects form a linear distribution, such as the arrangement of houses along a street or stations along a subway line. A regular pattern suggests that causal factors may explain the observed arrangement.

Objects are frequently arranged in a square or rectangular pattern. This is a clear phenomenon in urban geography. For example, many American cities contain a regular pattern of streets, known as a grid pattern, which intersect at right angles at uniform intervals to form square or rectangular blocks. Another example is the quilt-like pattern of farm fields seen when flying over the U.S. Midwest.

The distribution of baseball teams also follows a regular pattern. With two exceptions, the teams are located in North America's largest metropolitan areas. Milwaukee, Wisconsin, and Kansas City, Missouri, have teams but two larger metropolitan areas, Tampa, Florida, and Phoenix, Arizona, do not.

The pattern of forests and farmland can be critical to the composition of species within the community. A wooded area arranged in a long narrow strip along a stream or around the perimeter of a field may contain species that prosper in edge conditions, while species that require the deep shade of the forest interior would be less abundant. Geometric arrangements of river channels and their associated valleys indicate conditions in the underlying rock that have caused the river valleys to form distinctively.

Movement

To interpret spatial distribution patterns, geographers employ **spatial association**. In this concept, the distribution of one phenomenon across the landscape is systematically related to the distribution of another. For example, the distribution of livestock in Africa's Sahel region is related to the distribution of watering holes. The spatial association could take the form of a verifiable causal link—livestock need water—while in other cases the relationship may not be quantifiable.

A residential area of Milpitas, California. The distribution of houses in this area follows a rectangular pattern set by the arrangement of streets. (Stock Market)

Spatial associations among people and environments *which are at different locations* require *movement*. **Spatial interaction** describes the movement of physical processes, human activities, and ideas within and among regions. Interaction between people and the environment can occur at five dimensions in space; in order of increasing complexity, they are point, line, area, volume, and time.

Any two points in space are separated by **distance,** which is the measurement of amount of separation between two places. Interaction between two points in space is likely to be limited. A person located at one point in space can have interaction with a tree located at another point. If the distance between the two points is small, the human can look at the tree or draw a picture of it. If the distance is greater, the person's interaction could consist of possessing an image (photograph or drawing) of the tree. Typically, the farther away one object is from another the less likely the two are to have spatial interaction. Contact diminishes with increasing distance and eventually disappears. This trailing-off process is called **distance decay.**

Meaningful spatial interaction requires a line connecting the two points in space. The line could be a road connecting the point where the person is located to the point where the tree is rooted. Alternative types of interaction are possible: a person could pick the fruit from the tree, sit under the tree, or chop down the tree to build a house or a fire. A line could also serve as an obstacle to spatial interaction between two points in space. The line could be a fence or an international boundary.

A complex network permits spatial interaction between groups of people located in many areas and physical features located in many areas. Rivers and their tributaries form drainage networks. Water and sediment are carried downstream through these networks, linking areas upstream and downstream. Polluted runoff in one part of the drainage network could adversely affect water quality elsewhere in the network.

Airlines in the United States have adopted distinctive networks known as "hub-and-spokes." Under the "hub-and-spokes" system, airlines fly planes from a large number of places into one hub airport within a short period of time and then a short time later send the planes to another set of places. The network enables travelers originating in relatively small cities to reach a wide variety of destinations by changing planes at the hub airport (Figure 1–11).

Spatial interaction becomes even more complex in three-dimensional spaces. Communities of plants and animals both breathe the air above Earth's surface and dig into the ground below. Finally, spatial interaction changes over time. One generation may regard an area of trees as a source of food, another as a building material, and the next as a place for recreation.

Dallas-Fort Worth airport, Texas. The airport (abbreviated DFW) is one of American Airlines' major hubs. Dozens of airplanes arrive within a few minutes and take off about an hour later. By scheduling service this way, an airline can offer passengers flights to many cities with one change of plane at the hub. (Gabe Palmer/The Stock Market)

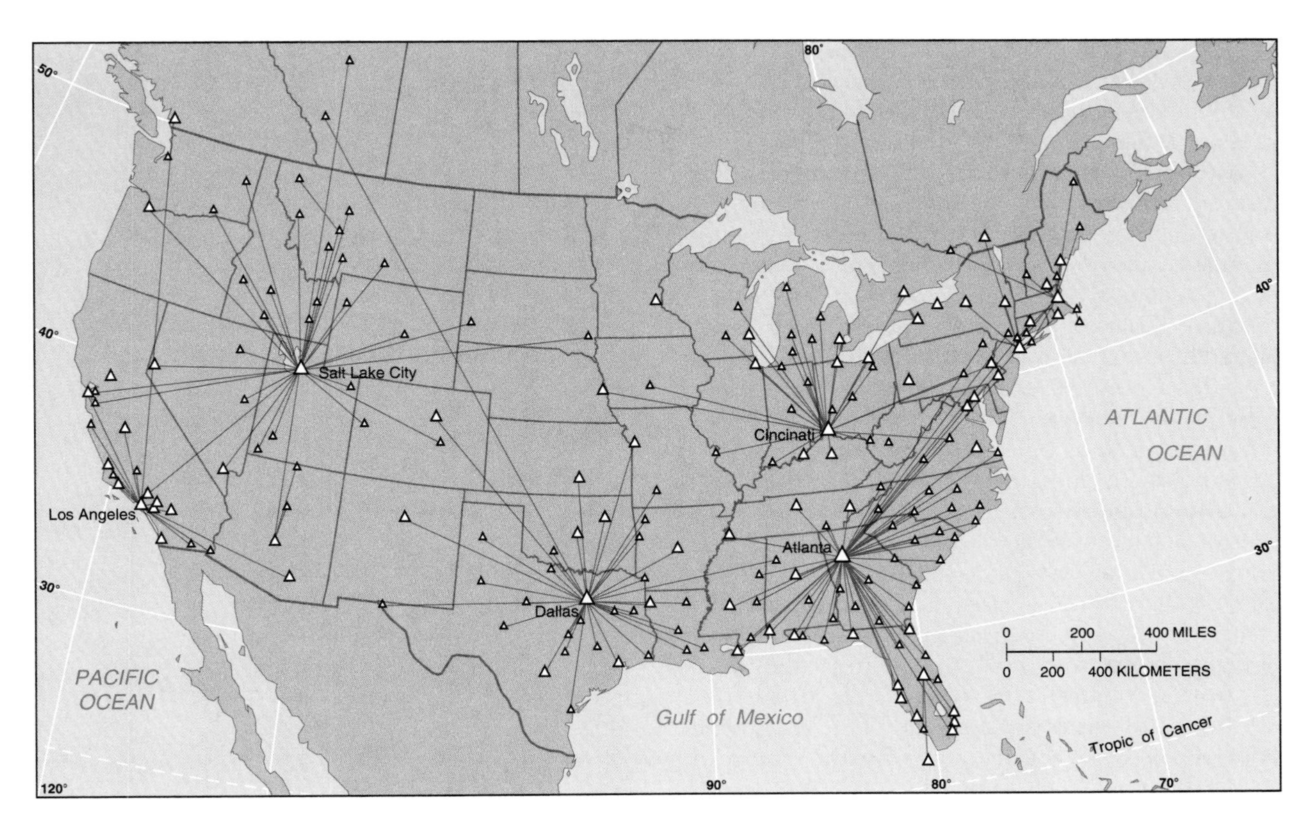

FIGURE 1–11
Delta Airlines' routes within the continental United States and Canada form a network, based on a principle known in the airline industry as "hub and spokes." Each triangle represents an airport served by Delta; the size of the triangle represents the number of cities connected from that airport by nonstop service on Delta. The lines connect each airport to the city to which it sends the most nonstop flights. Most flights originate or end at one of the company's hub cities, especially Atlanta, Cincinnati, Dallas-Fort Worth and Salt Lake City. Los Angeles, New York, Boston, and Orlando serve as smaller hubs for Delta's domestic service.

Diffusion. The process of spread of a feature or trend over time across Earth's surface is **diffusion.** Any innovation, such as use of a tool or development of a technology, originates at a node known as a **hearth.** A hearth emerges because a group of people is willing to try something new and has the ability to allocate resources to nurture the innovation. To develop a hearth, the group must also have the technical ability to achieve the desired idea and the economic structures, such as the agricultural capacity or the financial institutions, to implement the innovation.

Geographers observe two basic types of diffusion: relocation diffusion and expansion diffusion. **Relocation diffusion** is the spread of a feature or trend through bodily movement of people or physical phenomena *from one place to another*. People migrate for a variety of environmental, political, or economic reasons. When they move, they carry with them their cultural characteristics, such as language, religion, and social customs. The most commonly spoken languages in North and South America are Spanish, English, French, and Portuguese, primarily because several hundred years ago, Europeans who spoke those languages comprised the largest number of migrants.

The process of relocation diffusion was also at work when a large quantity of radioactive thorium-230 was released into New Mexico's Puerco River in 1979. A flood wave following the collapse of a uranium tailings pond near Church Rock, New Mexico, caused relocation of the contaminants along an 80-kilometer reach of the river.

Expansion diffusion is the process of spread of a feature or trend among people or physical phenomena from one area to another in a snowballing process. Three types of expansion diffusion are hierarchical diffusion, contagious diffusion, and stimulus diffusion.

Hierarchical diffusion is the spread of a feature or trend from one person or node of innovation to another through bypassing of other persons or areas. Innovation may originate at a particular place, such as a large urban center, and not diffuse to more isolated rural areas until much later. Hierarchical diffusion may result from the spread of ideas from political leaders or socially elite people to others in the community. On the other hand, an innovation such as *hip hop* (or *rap*) music originated with low-income African-Americans.

Contagious diffusion is the widespread diffusion of a feature or trend throughout the population or physical system. As the term implies, this form of diffusion is analogous to the spread of a contagious disease, such as influenza. The spread of gypsy moths from the northeastern United States is another example of contagious diffusion. **Stimulus diffusion** is the spread of an underlying principle or trend, even though on the surface a feature apparently fails to diffuse.

Whether phenomena diffuse through relocation or expansion diffusion is a critical contemporary issue for many countries and even many geographers, economists, and politicians. How should today's relatively poor countries promote development? One alternative—the *international trade approach*—assumes that a country's economy develops as a result of relocation diffusion of economic practices from relatively developed to developing countries. A second alternative—the *self-sufficiency approach*—assumes that development is achieved through a process of expansion diffusion inside a country. The health and welfare of a people are vitally affected by the ability of a country's leaders to judge which alternative best suits its region. These alternatives are discussed in more detail in Chapter 10.

The process of diffusion helps us to understand the distribution of acquired immunodeficiency syndrome (AIDS) within the United States. During the early 1980s, New York, California, and Florida were the nodes of origin for the disease within the United States. Half of the fifty states had no reported cases, while New York City, with 3 percent of the nation's population, contained more than one-fourth of the AIDS cases. In the neighboring state of New Jersey, AIDS cases dropped with increasing distance from New York City. A decade later, the disease had spread to every state, although California and the New York City area remained the focal points (Figure 1–12).

The AIDS quilt on display in Washington, DC. The quilt was assembled as a memorial to people who have died of this disease. By increasing awareness of AIDS, the creators of the quilt hope to slow diffusion of the disease. (Alon Reininger/Contact Press Images/The Stock Market)

Physical and Human Geographic Systems

Geographers know that environmental and social conditions are interrelated. *Systems theory* helps geographers to understand regularities that they observe in the physical environment and in human behavior.

Systems theory views Earth as a set of systems that work together to form a *whole* rather than as a collection of isolated, unrelated parts. Geographers studying a particular physical or human process

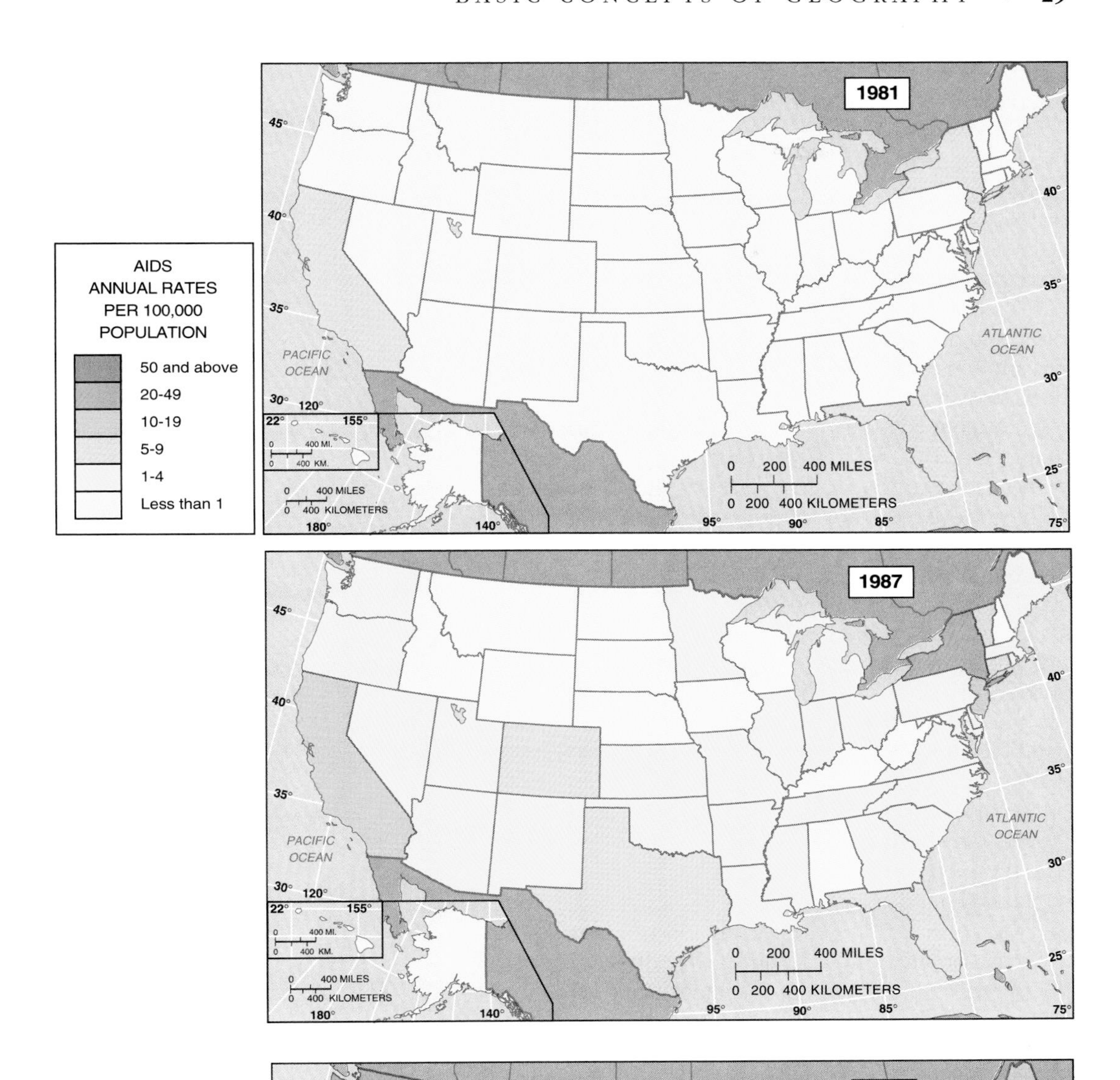

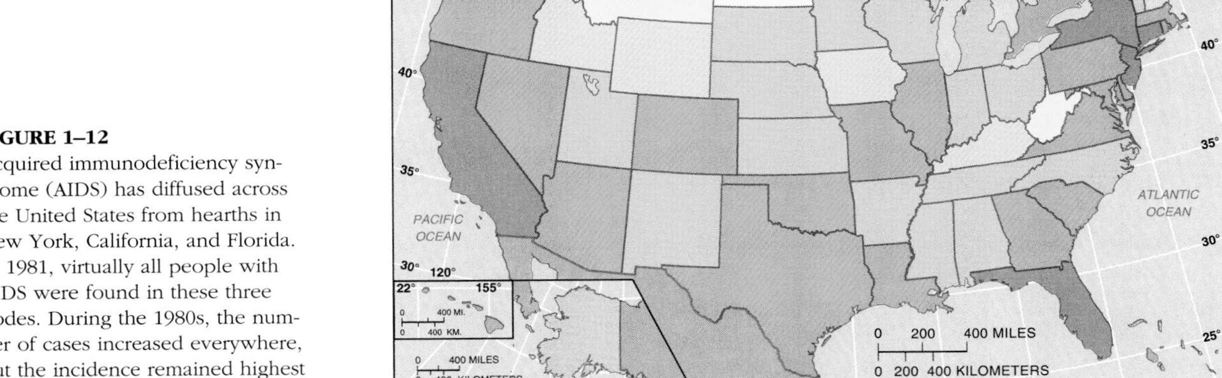

FIGURE 1–12
Acquired immunodeficiency syndrome (AIDS) has diffused across the United States from hearths in New York, California, and Florida. In 1981, virtually all people with AIDS were found in these three nodes. During the 1980s, the number of cases increased everywhere, but the incidence remained highest in the three original nodes.

keep in mind that the individual process always interacts with others to form a global system.

The study of geographic processes is complex because most of the systems studied are open. An **open system** is one that exchanges matter with other systems. A plant is an example of an open system: it takes in sunlight, carbon dioxide, water, and nutrients as inputs and produces outputs of oxygen and carbohydrates (sugars) through the process of photosynthesis. Through the process of respiration, a plant consumes carbohydrates and releases carbon dioxide, water, and heat into the environment.

A **closed system** is entirely self-contained by being shut off from the surrounding environment, with the exception of energy flows. Earth is essentially a closed system in terms of physical matter, such as air, water, and solid material. Earth receives energy from the sun, and it radiates heat energy back to space (see Chapter 2).

Since Earth formed 4.6 billion years ago, no significant matter has entered the system (just meteors and comet debris), and no significant matter has left (just some hydrogen and helium gas). Physical matter cannot be created or destroyed, although it can be transformed. For example, petroleum when burned is transformed from a liquid to a mixture of gases—but the original atoms in the petroleum are still there. Recycling is important, because Earth contains limited quantities of resources.

The parts of a system are related to each other through positive and negative feedback loops. **Feedback** is the return of a system's output as an input, producing a circular flow of information. A **negative feedback** slows or discourages response in a system, while a **positive feedback** amplifies or encourages response in the system.

Snow cover and temperature are related by a positive feedback loop (Figure 1–13). As the amount of snow cover increases, the greater is the amount of sunlight reflected back to space. Reflection of sunlight back to space keeps temperatures low and reduces snowmelt. With reduced snowmelt, the snow cover increases, completing the positive feedback loop.

Negative feedbacks dominate in natural systems, encouraging long-term equilibrium and balance. A system that maintains its general character over a given time frame is said to be in **equilibrium.** A system is in **steady-state equilibrium** if its rates of inputs and outputs are equal, and the amounts of energy and matter are stable or are fluctuating around a stable average.

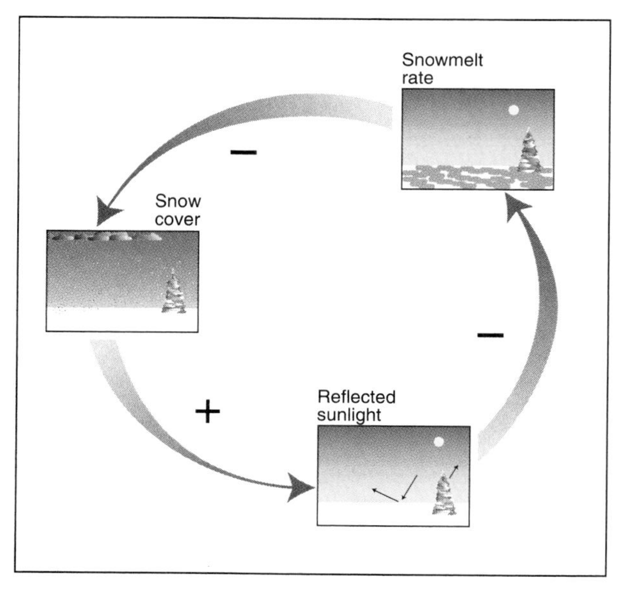

FIGURE 1–13
Feedback in an environmental system. A negative sign indicates that as the component increases the other decreases; a positive sign indicates that a change in one component produces a change in the other component in the same direction. In this example, an increase in the amount of snow cover causes reflection of more sunlight back to space (a positive relationship). Reflection of more sunlight back to space keeps temperatures lower (a negative relationship). Lower temperatures causes the snow cover to increase (a negative relationship). The two negative signs cancel each other, making the loop a positive feedback system.

An unchecked positive feedback can create a runaway condition that produces instability and ultimately destroys the system, as well as other interrelated systems. Because human behavior and physical processes are interrelated, geographers debate whether such human actions as burning petroleum in automobiles, cutting down forests in the Amazon, or giving birth to babies in East Africa are contributing to positive feedbacks in environmental systems.

Earth's Physical Systems

Geographers study natural processes in terms of four open systems. Three of the four systems are **abiotic,** composed of nonliving or inorganic matter: the *atmosphere (air)*, *hydrosphere (water)*, and *lithosphere (crust)*. The fourth is the **biotic** system (also called the *biosphere*), which encompasses all of Earth's living organisms (Figure 1–14). Because

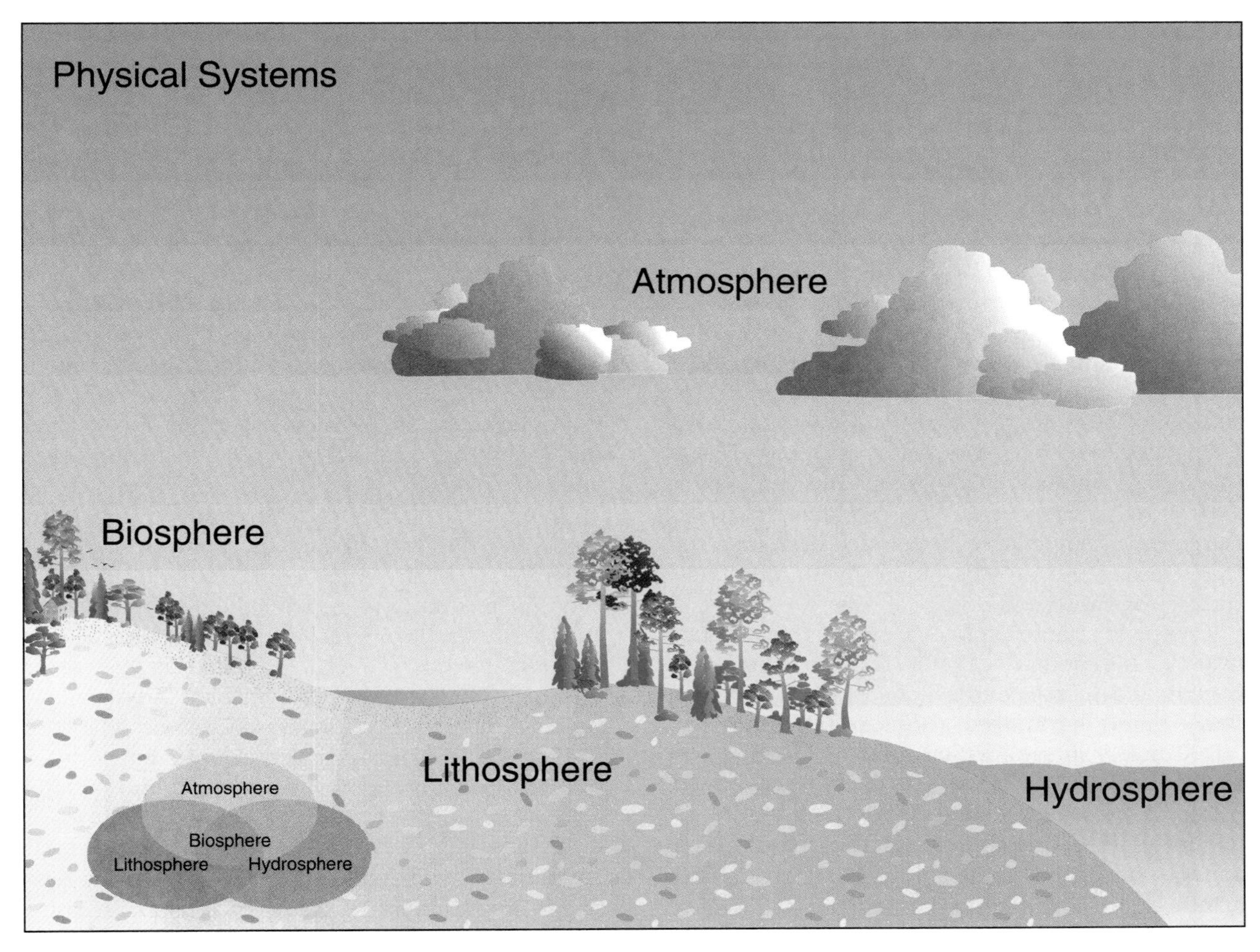

FIGURE 1–14
Geographers regard natural processes as consisting of four open systems, including three composed of nonliving matter—the atmosphere (air), hydrosphere (water), and lithosphere (Earth's crust)—plus the biosphere, which comprises all of Earth's living organisms. The four systems are interrelated and overlap with each other.

living organisms cannot exist except through interaction with the surrounding physical environment, the biotic system includes parts of the three abiotic systems. Organic matter is composed of materials from all three of the inorganic systems.

Abiotic systems. Surrounding the globe are three abiotic systems—air, water, and Earth's crust. The chemical composition of the three layers is distinctive, an inheritance from their origins in the distant geologic past. Because of differences in the density of the three types of substances, they form concentric shells.

The **atmosphere** is a thin layer of gases surrounding Earth to an altitude of less than 480 kilometers (300 miles). Pure dry air in the lower atmosphere contains about 78 percent nitrogen and 21 percent oxygen by volume. It also includes about 0.9 percent argon (an inert gas) and 0.036 percent carbon dioxide (a crucial percentage, as you will see in upcoming chapters).

Air is a mass of gas molecules, held to Earth by gravity creating pressure. Variations in air pressure from one place to another cause winds to blow, create storms, and control precipitation patterns.

The **hydrosphere** is the water realm of Earth's surface, including the oceans, surface waters on land (lakes, streams, rivers), ground water in soil and rock, water vapor in the atmosphere, and ice in glaciers. Water's chemical composition is expressed in the familiar formula H_2O, meaning that a molecule of water consists of two atoms of hydrogen bound to one atom of oxygen. The water molecule is extremely stable in the surface environment of Earth because the hydrogen and oxygen atoms are strongly bonded.

Water can exist as a vapor, liquid, or ice. Over 97 percent of the world's water is in the oceans in liquid form. The oceans sustain a large quantity and variety of marine life, both plants and animals. Sea water supplies water vapor to the atmosphere, which returns to Earth's surface as rainfall and snowfall. These are the most important sources of freshwater, which is essential for the survival of plants and animals. Water gains and loses heat very slowly, so oceans also moderate seasonal extremes of temperature over much of Earth's surface. Oceans also provide humans with food and a surface for transportation.

Earth is an almost spherical body, approximately 6,400 kilometers (4,000 miles) in radius. It is constructed in concentric spheres. The *core* is a dense, metallic sphere about 3,500 kilometers (2,200 miles) in radius. Surrounding the core is a *mantle* about 2,900 kilometers (1,800 miles) thick. A thin, brittle outer shell, the *crust*, is 8 to 40 kilometers (5 to 25 miles), thick. The **lithosphere** consists of Earth's crust and a portion of upper mantle directly below the crust extending down to about 70 kilometers (45 miles).

Powerful forces deep within Earth bend and break the crust to form mountain chains and shape the crust to form continents and ocean basins. The shape of Earth's crust influences climate: If the surface of Earth were completely smooth, then temperature, winds, and precipitation would form orderly bands at each latitude.

Biotic system. The **biosphere** consists of all living organisms on Earth and portions of the three abiotic systems. The atmosphere, lithosphere, and hydrosphere function together to create the environment of the biosphere, which extends from the depths of the oceans through the lower layers of atmosphere. On the land surface, the biosphere includes giant redwood trees, which can extend up to 110 meters (360 feet), as well as the microorganisms that live many meters down in the soil, deep caves, or rock fractures.

These four "spheres" of the natural environment interact in many ways. Plants and animals live on the surface of the lithosphere, where they obtain food and shelter. The hydrosphere provides water to drink, and physical support for aquatic life. Most life forms depend on breathing air, and birds and people rely on air for transportation. All depend on inputs of solar energy.

The three abiotic systems interact in the biosphere. The biosphere is where atmospheric conditions affect the lithosphere, and the land surface influences the properties of the adjacent atmosphere. Similarly, energy and matter continually flow between the sea surface and the lower layer of the atmosphere and between the hydrosphere and lithosphere.

Humans also interact with each of these four "spheres." We atrophy and die if we are without water. We pant if oxygen levels are reduced in the atmosphere, and cough more if the atmosphere contains pollutants. We need heat, but excessive heat or cold is dangerous. We rely on a stable lithosphere for building materials and fuel for energy. And of course, we count on the rest of the biosphere for food.

Plants and animals interact with each other through exchange of matter, energy, and stimuli. An ecological system, or **ecosystem,** is a group of organisms and the nonliving physical and chemical environment with which they interact. An ecosystem is an example of an open system, because inputs and outputs cross back and forth between it and other ecosystems. **Ecology** is the scientific study of ecosystems. Ecologists study the interrelationships among life forms and their environments in particular ecosystems, as well as among various ecosystems in the biosphere.

Earth's Cultural Systems

Geographers and other social scientists cannot predict precisely how people will obtain food, with whom they will fight, or how many babies they will bear. However, because people tend to behave consistently, systems methods can help geographers understand cultural patterns across Earth's surface.

Culture includes the distinctive *customary beliefs*, *social forms*, and *material traits* (goods) of a group of people. Geographers are interested in the systems that result from the three key aspects of this definition.

Customary beliefs. Differences among places result in part from a people's distinctive customary beliefs. The cultural landscape is our unwitting autobiography, according to geographer Peirce Lewis, because it reflects in a tangible form our tastes, values, aspirations, and fears.

Unlike other living organisms, humans can assign meanings and values to their physical environment and formulate goals and purposes to modify it. Geographers study how people develop meaning—such as positive or negative feelings—about their homes, streets, factories, offices, schools, parks, and other places in their everyday lives.

People derive different meanings from the environment and hold different objectives in making use of it. Some human impacts on the environment are trivial, and some are based on deep-seated cultural values. Why do we plant our front yards with grass, water it to make it grow, mow it to keep it from growing tall, and impose fines on those who fail to mow often enough? Why not let dandelions grow, or pour concrete instead? Why does one group of people consume the fruit from deciduous trees and chop down the conifers, while another group chops down the deciduous trees for furniture while preserving the conifers as religious symbols? Powerful cultural beliefs are at work.

Defining something as a "natural resource" is in reality a *cultural appraisal*. People with one kind of culture might concentrate their settlements on flattish uplands, whereas another people in the same area might cluster in the valleys. Some people believe that the physical environment is created by supernatural forces and therefore regard it with great respect, awe, or fear, while others consider that the environment is something to be opposed, tamed, or even plundered.

Customary beliefs are especially strong in population growth and religion. The process by which the human population increases is significant to geographers, because during the past half-century the world's population has been increasing at a more rapid rate than ever before in history, and most of that growth has been clustered in relatively poor regions. Population grows when the number of births exceeds the number of deaths. With diffusion of medical technology around the world, differences in death rates between relatively developed and developing countries have narrowed. Today, a region's population growth is determined primarily by human decisions concerning how many babies to bear.

Geographers disagree about the reasons that people in some regions with rapid population growth suffer from lack of access to an adequate food supply. Some geographers emphasize the limits on the capacity of the physical environment to provide enough food in some regions, while others emphasize people's lack of control over the production and distribution of food in the society (see Chapter 7).

Humans hold especially strong religious beliefs. Political unrest and wars have been clustered in places where the territories inhabited by various religious groups do not coincide with the boundaries of countries and governments (for example, Northern Ireland, the Middle East, South Asia, and Southeastern Europe). Geographers recognize that people living in two places may share religion and other customary beliefs because of interaction, while differences among groups arise because of isolation. When people migrate they take along their customary beliefs. In the contemporary world, communications and transportation systems enable groups in dispersed locations to interact without migration.

Social forms. Groups of people create specific systems to transform the physical environment in accordance with their cultural preferences. Humans place farms, cities, and other structures systematically on Earth's surface in accordance with their culture's prevailing social forms, such as governments and arrangement of settlements.

Humans cluster their cultural and economic activities in a familiar social form, the *settlement*. In some societies, most people live in small rural settlements surrounded by farm fields, while in others most people live in urban areas. Geographers study the processes that distribute settlements across Earth's surface and within particular societies. Geographers also analyze processes by which cultural groups and economic activities are sorted out within settlements.

Geographers adopt different perspectives to understand patterns in human settlements, such as the clustering of low-income minority groups in inner-city neighborhoods. Humanist geographers may try to understand conditions in the inner city, such as how gang members occupy space and how graffiti marks a gang's territory. Such studies attempt to expose errors and prejudices that middle-class people often hold concerning social forms in inner-city minority communities.

According to socialist geographers, human settlement patterns such as segregated inner-city neighborhoods result from the formation of social classes. Middle-class residents of suburbs possess greater access than do inner-city residents to means for producing wealth and institutions for distributing wealth, such as banks and government agencies.

Material traits. A third important cultural system is a society's production and distribution of goods, such as food, clothing, and shelter. In poorer, less economically developed societies, most people are farmers, growing food for themselves, their families, and their neighbors. In wealthier societies with relatively developed economies, only a small percentage of people are farmers because they have the technology to produce enough food for the entire population. Freed from the need to farm, most people in relatively developed societies purchase food and other goods with wages earned by working in a factory, office, or shop.

Some economic geographers apply systems approaches to identify suitable sites for farming, manufacturing, retailing, and services. Concepts from geography's tradition of spatial analysis can be applied to calculate the best location for an economic activity and the size of its market area.

Others who specialize in industrial geography emphasize that economic systems result from changes in methods of production. Two hundred years ago, during the industrial revolution, production clustered in large factories, influencing the growth of large cities. In recent years, many companies have adopted more flexible methods of production to meet the demands of an increasingly fragmented market. This restructuring has led companies to relocate some production in small towns, as well as in developing countries, where the cost of labor is lower.

At a global scale, some geographers emphasize the role of political and economic systems that produce differences in possession of material goods. The world economy is seen as a unified system, with different societies playing different roles. Relatively developed countries that compete with each other to control material wealth form a central trading core, while developing countries occupy peripheral locations in the global economy.

Human-Environmental Interaction

Geographers recognize that physical environment and human systems are intimately associated (Figure 1–15). For example, humans have a limited tolerance for extreme temperature and precipitation levels and thus tend to avoid living in places that are too hot, too cold, too wet, or too dry. As an exercise to illustrate this, compare the maps of global climate and population distribution (ahead in Figures 3–1 and 7–1). Note that relatively few people live in the dry (B) and cold (E) climate regions.

Population growth is a problem in places where the climate and other characteristics of the physical environment may limit human ability to produce food. However, people can adjust to the capacity of the physical environment by adopting new technology, consuming different foods, migrating to new locations, and allocating food differently. Therefore, technology becomes a part of the human-environment relationship.

Environmental determinism. Two important schools of thought on the human-environment relationship include *environmental determinism* and *possibilism*. Two nineteenth-century German geographers, Alexander von Humboldt (1769–1859) and Carl Ritter (1779–1859), concentrated on how the physical environment affects social development. Their view is called **environmental determinism.** They argued that the scientific study of social and environmental processes is fundamentally the same, although physical scientists have made more progress in formulating general laws than social scientists. According to Humboldt and Ritter, geographers should apply laws from the natural sciences to understanding relationships between the physical environment and human actions.

Other influential geographers adopted environmental determinism thinking in the late nineteenth and early twentieth centuries. The German geographer Friedreich Ratzel (1844–1904) and his American student, Ellen Churchill Semple (1863–1932), claimed that geography was the study of the influences of the natural environment on people. Another early American geographer, Ellsworth Huntington (1876–1947), argued that climate was a major determinant of civilization. For instance, the temperate climate of maritime northwestern Europe, according to Huntington, pro-

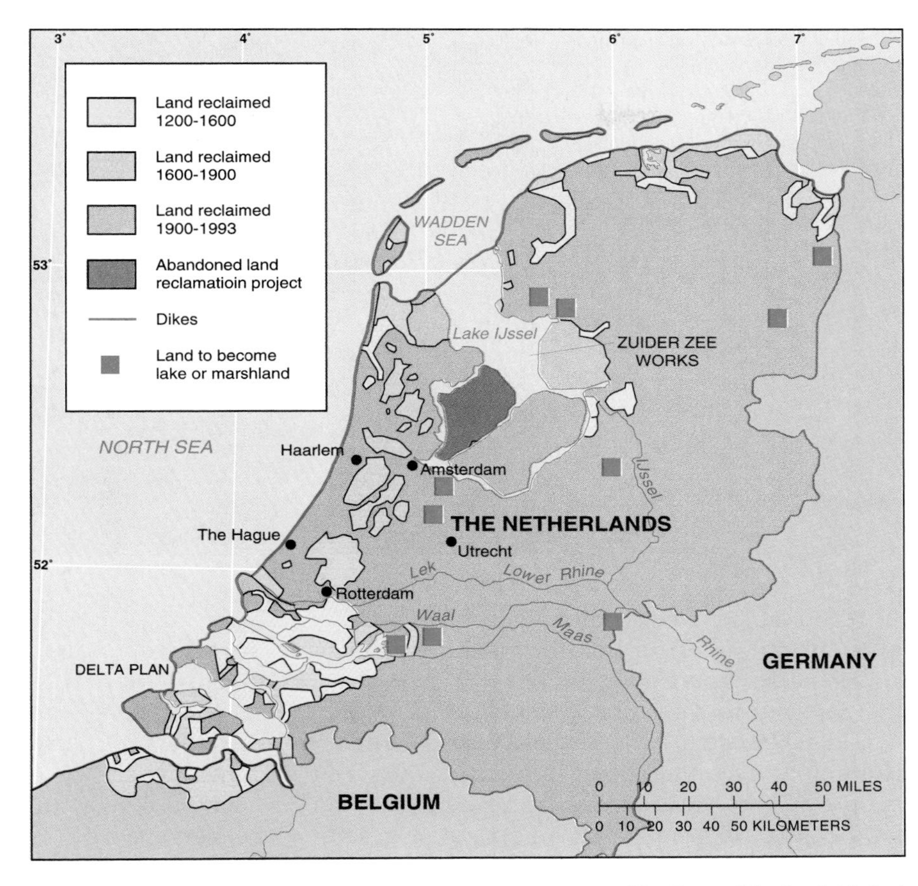

FIGURE 1–15
Because more than half of the Netherlands lies below sea level, most of the country today would be under water if it were not for massive projects to modify the environment. Since the thirteenth century the Dutch have reclaimed more than 4,500 square kilometers (1,800 square miles) of *polders*, which are pieces of land created by draining water. Most polders are reserved for agriculture, although some are used for housing, and one contains Schiphol, one of Europe's busiest airports. The second distinctive modification of the environment is the construction of massive *dikes* to prevent the North Sea from flooding much of the country. The Dutch have built dikes in two major locations, the Zuider Zee project in the north and the Delta Plan project in the southwest. With these massive projects finished, attitudes toward the environment have changed in the Netherlands. A plan adopted in 1990 calls for returning 26 square kilometers of farms on polders to wetlands or forests. Widespread use of insecticides and fertilizers on Dutch farms has contributed to contaminated drinking water and other environmental problems.

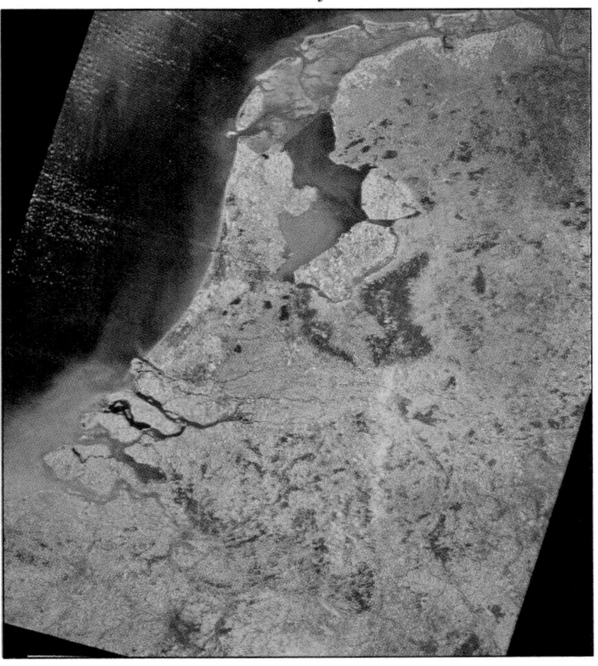

(Jim Brown/The Stock Market)

duced greater human efficiency, as measured by better health conditions, lower death rates, and higher standards of living.

Although modern geographers reject environmental determinism, they recognize that climate and other physical conditions of a location do influence human activities, especially production of food. From one generation to the next, people learn that different crops thrive in different climates. Rice, for example, requires a good deal of moisture, whereas wheat survives on a limited amount of water and grows poorly in very wet environments. On the other hand, wheat is more likely than rice to be grown successfully in colder climates.

Possibilism. According to **possibilism**, the physical environment may limit some human actions, but *people have the ability to adjust to the physical environment.* A person can choose a possible course of action from many alternatives in the physical environment.

A barge carries thatch on a canal through a polder in the Netherlands. Thatch (plant stalks) is a traditional material for roofing in this country. Historically, windmills played a key role in pumping out water to build polders. (World View/Science Photo Library/Photo Researchers, Inc.)

Technical knowledge can influence the capacity of a cultural group to transform its environment and increase its possibilities. Some societies have the capacity to transform the environment to a considerable extent while others do not. Modern technology has altered the interaction between people and the environment, because humans can modify the physical environment to a greater extent than in the past. For example, air conditioning has increased the attraction of living in warmer climates while more elaborate insulation and heating have permitted survival in colder climates.

A people's level of wealth can influence its attitude toward modifying the environment. A farmer who uses a tractor may regard a hilly piece of land as an obstacle to avoid, while a farmer with only a hoe may regard a hill as the only opportunity to produce enough food for survival through painstaking efforts to cultivate the land by hand.

Geographers are concerned that a society possessing a relatively high capability to transform the environment for the short-term benefit of people may also have the capability to adversely affect environmental quality. Human actions can deplete scarce environmental resources, destroy irreplaceable environmental resources, and utilize environmental resources inefficiently. Refrigerants in air conditioners that have increased the comfort of residents of warmer climates have also increased the amount of chlorofluorocarbons in the atmosphere, thereby damaging the ozone layer and contributing to global warming.

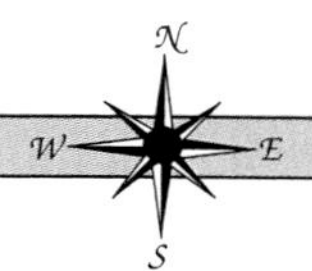

CONCLUSIONS: CRITICAL ISSUES FOR THE FUTURE

The world of the twenty-first century will present new and difficult challenges to citizens, businesses, and governments. Critical issues include:

- Global-scale environmental change, including deforestation, soil degradation, increased concentrations of atmospheric CO_2, probable global warming, and decreased atmospheric ozone levels.
- Rates of population growth in some regions that strain cultural systems, deplete scarce resources, and produce environmental changes.

- Increasing disparities in wealth and threat of conflict between rich and poor countries and within individual countries.
- Systems of food production and distribution that do not meet the nutritional needs of many people in developing countries.

Confronting problems of this magnitude will require a great deal of creativity and hard work from all segments of society, including geographers. Geography offers several unique perspectives for understanding contemporary issues.

First, geography emphasizes interdisciplinary approaches to contemporary issues. Because geographers are trained partly as natural scientists and partly as social scientists, they have a wide breadth of understanding that makes geography a unique discipline. Geographers work well in interdisciplinary teams analyzing major problems that involve combinations of human and physical factors. Because they can communicate with specialists in a variety of natural and social sciences, geographers are often asked to take a leadership role in group efforts.

Second, geographers have a global perspective. A century ago, a famine or a war in a distant part of the world had little consequence outside the immediate region, but today local events have worldwide impact. New information technology and transportation systems have created global markets for many commodities. What happens in one place has ramifications in other areas and often throughout the world. Concern over an endangered owl in the U.S. Pacific Northwest causes a rise in lumber prices not only throughout the United States, but in Canada and East Asia, as well. A 1986 accident at Chernobyl, a nuclear power plant in the Soviet Union (now Ukraine), spread radioactive fallout around the world, and diffused fear of nuclear power.

Third, geography has a strong component of applied research. Since the earliest days of the discipline, geographers have worked on problems of practical significance, such as finding new routes for commerce and evaluating the natural resources of sparsely inhabited lands. Geographers emphasize studying everyday phenomena about places. They are aware of conditions in the world around them, and are accustomed to applying their knowledge to them.

Chapter Summary

1. What is geography?
Geography is the study of people, their natural environments, and how they interact to form the patterns we observe on Earth's surface. Geographers utilize maps to depict the location of places and to interpret underlying patterns. Remote sensing from satellites and GIS are two recently developed tools that help geographers more fully analyze patterns on Earth's surface.

2. Area analysis
Location is the position on Earth's surface an object occupies. Location can be specified mathematically or by place-name. Each place on Earth has a unique character, but places can be combined into regions according to shared trends or features.

3. Spatial analysis
Geographers study the distribution of objects across Earth's surface and processes by which human and environmental phenomena move from one place to another. Three properties of spatial distribution include density, concentration, and geometric pattern. Movements of matter, people, and information result in interactions among places along networks through diffusion processes.

4. Physical and human geographic systems
Earth's physical environment results from a process of interaction among four open systems. The three abiotic systems include the atmosphere, hydrosphere, and lithosphere. The three abiotic systems interact with each other in the biosphere. Cultural systems include customary beliefs, social forms, and material traits. Some geographers regard the distribution of human activities as the result of underlying regularities in cultural systems while others emphasize the diversity and uniqueness of human behavior in explaining cultural systems. The physical environment influences human actions, although humans can adjust and choose a course of action from many alternatives.

Key Terms

Abiotic system, 31
Atmosphere, 31
Biosphere, 32
Biotic system, 31
Cartography, 10
Closed system, 30
Concentration, 24
Contagious diffusion, 28
Culture, 32
Density, 22
Diffusion, 27
Distance, 26
Distance decay, 26
Ecology, 32
Ecosystem, 32
Environmental determinism, 34
Equilibrium, 30
Expansion diffusion, 28
Feedback, 30
Formal region, 18
Functional region, 18
GIS, 13
Greenwich Mean Time, 16
Hearth, 27
Hierarchical diffusion, 28
Hydrosphere, 31
International Date Line, 16
Latitude, 15
Lithosphere, 32
Location, 13
Longitude, 15
Map, 9
Meridian, 15
Negative feedback, 30
Network, 19
Open system, 30
Parallel, 15
Pattern, 25
Positive feedback, 30
Possibilism, 35
Prime meridian, 15
Projection, 12
Region, 6
Relocation diffusion, 27
Remote sensing, 13
Scale, 10
Spatial association, 25
Spatial distribution, 22
Spatial interaction, 26
Steady-state equilibrium, 30
Stimulus diffusion, 28
System, 7
Toponym, 13
Vernacular region, 19

Questions for Study and Discussion

1. What is geography? Describe differences among physical, human, and regional geography.
2. What contributions did ancient and medieval geographers make to the development of geographic thought?
3. What are the three ways to indicate scale on a map?
4. What are the four types of distortions that can result from map projections?
5. How do remote sensing and GIS contribute to geographic study?
6. What is the difference between a meridian (or longitude) and a parallel (or latitude)?
7. What is the difference between formal regions, functional regions, and vernacular regions? Give examples of each of the three types of regions.
8. What is the difference between density and concentration?
9. What do geographers mean by the concept of spatial interaction?
10. What is the difference between relocation diffusion and expansion diffusion?
11. What is the difference between an open system and a closed system?
12. What are the four natural systems? Which of the systems are biotic, and which are abiotic?
13. What three elements are included in the concept of culture?
14. What is the difference between environmental determinism and possibilism?

Thinking Geographically

1. Cartography is not simply a technical exercise in penmanship and coloring; nor are decisions confined to scale and projection. Mapping is a politically sensitive undertaking. If you were a resident of New Zealand, what changes might you suggest to the world map projections used in this book? Look at how maps in this book distinguish the borders of India with China and Pakistan and of Israel with its neighbors. Can you identify other logical ways to draw these boundaries?
2. Imagine that a transportation device (perhaps the one in *Star Trek*) would enable all humans to travel instantaneously to any location on Earth. What impact might that invention have on spatial interaction?
3. When earthquakes, hurricanes, or other so-called "natural" disasters strike, humans tend to "blame" nature and see themselves as innocent victims of a harsh and cruel nature. To what extent to environmental hazards stem from unpredictable nature, and to what extent do they originate from human actions? Should victims blame nature, other humans, or themselves for the disaster? Why?
4. Geographic approaches, such as the area analysis, spatial analysis, and systems analysis, are supposed to help explain contemporary issues. Find a story in your newspaper which can be explained through application of geographic concepts.
5. According to chaos theory, systems contain variations that cannot be fully explained or predicted. For example, a daily weather forecast may call for a 30 percent chance of showers. Despite new technology such as satellite images and computer models, scientists are unable to predict systems behavior with precision. Do you believe that the unpredictable behavior that some systems display supports chaos theory, or do you believe that our ability to explain a system may increase as we gain more knowledge of how it works? Cite examples to support your view.

Suggestions for Further Readings

Abler, Ronald F., Melvin G. Marcus, and Judy M. Olson, eds. *Geography's Inner Worlds*. New Brunswick, NJ: Rutgers University Press, 1992.

Blaut, J. M. "Diffusionism: A Uniformitarian Critique." *Annals of the Association of American Geographers* 77 (March 1987): 48–62.

Bodman, Andrew R. "Weavers of Influence: The Structure of Contemporary Geographic Research." *Transactions of British Geographers New Series* 16 (1991): 21–37.

Brown, Lawrence A. *Innovation Diffusion: A New Perspective*. London: Methuen, 1981.

Brunn, Stanley D. "Sunbelt USA." *Focus* 36 (Spring 1986): 34–35.

Claval, Paul. "The Region as a Geographical, Economic and Cultural Concept." *International Social Science Journal* 39 (May 1987): 159–172.

Cohen, Saul B., and Nurit Kliot. "Place-Names in Israel's Ideological Struggle over the Administered Territories." *Annals of the Association of American Geographers* 82 (December 1992): 653–680.

Constandse, A. K. *Planning and Creation of an Environment*. Lelystad, The Netherlands: Rijksdienst voor de IJsselmeerpolders, 1976.

Eldridge, J. Douglas, and John Paul Jones, III. "Warped Space: A Geography of Distance Decay." *Professional Geographer* 43 (November 1991): 500–511.

Entrikin, J. Nicholas, and Stanley D. Brunn, eds. *Reflections on Richard Hartshorne's 'The Nature of Geography.'* Washington, D.C.: Association of American Geographers, 1989.

Espenshade, Edward B., Jr., ed. *Goode's World Atlas*. 18th ed. Chicago: Rand McNally and Company, 1990.

Forman, R.T.T., and M. Godron. *Landscape Ecology*. New York: John Wiley, 1986.

Gaile, Gary L., and Cort J. Willmott, eds. *Geography in America*. New York: Merrill/Macmillan, 1989.

Gardner, Lytt I., Jr., et al. "Spatial Diffusion of the Human Immunodeficiency Virus Infection Epidemic in the United States, 1985–87." *Annals of the Association of American Geographers* 79 (March 1989): 25–43.

Gleick, James. *Chaos*. New York: Penguin, 1987.

Gould, Peter R. *The Geographer at Work*. Boston: Routledge and Kegan Paul, 1985.

Graf, William L. "Fluvial Dynamics of Thorium-230 in the Church Rock Event, Puerco River, New Mexico." *Annals of the Association of American Geographers* 80 (September 1990): 327–342.

Gross, Jonathan L., and Steve Rayner. *Measuring Culture*. New York: Columbia University Press, 1985.

Hagerstrand, Torsten. *Innovation Diffusion as a Spatial Process*. Chicago: University of Chicago Press, 1967.

Haggett, Peter. "Prediction and Predictability in Geographical Systems." *Transactions of British Geographers New Series* 19 (1994): 6–20.

Hamm, Bernd, and Martin Lutsch. "Sunbelt v. Frostbelt: A Case for Convergence Theory?" *International Social Science Journal* 39 (May 1987): 199–214.

Hartshorne, Richard. *The Nature of Geography*. Lancaster, PA: Association of American Geographers, 1939.

James, Preston E. *All Possible Worlds: A History of Geographical Ideas*. New York: Bobbs-Merrill, 1972.

Janelle, Donald G., ed. *Geographical Snapshots of North America*. New York: Guilford, 1992.

Johnston, R. J. *Philosophy and Human Geography*. 2d ed. London: Edward Arnold, 1986.

__________, ed. *The Dictionary of Human Geography*. 2d ed. Oxford: Basil Blackwell, 1985.

Mabogunjie, Akin L. "Geography as a Bridge Between Natural and Social Sciences." *Nature and Resources* 20 (1984): 2–6.

Malanson, George P., D. R. Butler, and S. J. Walsh. "Chaos Theory in Physical Geography." *Physical Geography* 11 (1990): 293–304.

Meinig, D. W., ed. *The Interpretation of Ordinary Landscapes*. New York: Oxford University Press, 1979.

Mikesell, Marvin W. "Tradition and Innovation in Cultural Geography." *Annals of the Association of American Geographers* 68 (March 1978): 1–16.

Monmonier, Mark. *How to Lie with Maps*. Chicago: University of Chicago Press, 1991.

Noronha, Valerian T., and Michael F. Goodchild. "Modeling Interregional Interaction: Implications for Defining Functional Regions." *Annals of the Association of American Geography* 82 (March 1992): 86–102.

Norton, William. *Explorations in the Understanding of Landscape: A Cultural Geography*. Westport, CT: Greenwood Press, 1989.

Peet, Richard. "The Social Origins of Environmental Determinism." *Annals of the Association of American Geographers* 75 (September 1985): 309–333.

Penning-Rowsell, Edmund C., and David Lowenthal, eds. *Landscape, Meanings and Values*. London: Allen and Unwin, 1986.

Phillips, Jonathan D. "The Human Role in Earth Surface Systems: Some Theoretical Considerations." *Geographical Analysis* 23 (October 1991): 316–331.

Reed, Michael, ed. *Discovering Past Landscapes*. London: Croom Helm, 1986.

Rhoads, Bruce L., and Colin E. Thorn. "Contemporary Philosophical Perspectives on Physical Geography with Emphasis on Geomorphology." *Geographical Review* 84 (January 1994): 90–101.

Rice, Bradley R. "Searching for the Sunbelt." *American Demographics* 3 (March 1981): pp. 22–23.

Rowntree, Lester B., and Margaret W. Conkey. "Symbolism and the Cultural Landscape." *Annals of the Association of American Geographers* 70 (December 1980): 459–474.

Santos, Milton. "Geography in the Late Twentieth Century: New Roles for a Theoretical Discipline." *International Social Science Journal* 36 (November 1984): 657–672.

Shannon, Gary W., and Gerald F. Pyle. "The Origin and Diffusion of AIDS: A View from Medical Geography." *Annals of the Association of American Geographers* 79 (March 1989): 1–24.

Shannon, Gary W., and Rashid L. Bashshur. *The Geography of AIDS*. New York: Guilford Press, 1991.

Smallman-Raynor, M., A. Cliff, and P. Haggett. *London International Atlas of AIDS*. Oxford: Blackwell Publishers, 1992.

Solot, Michael. "Carl Sauer and Cultural Evolution." *Annals of the Association of American Geographers* 76 (December 1986): 508–520.

Strahler, Arthur N. "Systems Theory in Physical Geography." *Physical Geography* 1 (January 1980): 1–27

Tuan, Yi-Fu. "Cultural Pluralism and Technology." *Geographical Review* 79 (July 1989): 269–279.

Unwin, Tim. *The Place of Geography*. New York: Halstead Press, 1992.

Wagner, Philip L., and Marvin W. Mikesell, eds. *Readings in Cultural Geography*. Chicago: University of Chicago Press, 1962.

Zimmerer, Karl S. "Human Geography and the 'New Ecology': The Prospect and Promise of Integration." *Annals of the Association of American Geographers* 84 (January 1994): 108–125.

We also recommend these journals: *American Cartographer, Annals of the Association of American Geographers, Antipode, Applied Geography, Area, Canadian Geography (Géographie canadien), Geographical Analysis, Geographical Review, Geography, Journal of Geography, Photogrammetric Engineering and Remote Sensing, Professional Geographer, Progress in Human Geography, Remote Sensing of Environment, Transactions of the Institute of British Geographers.*

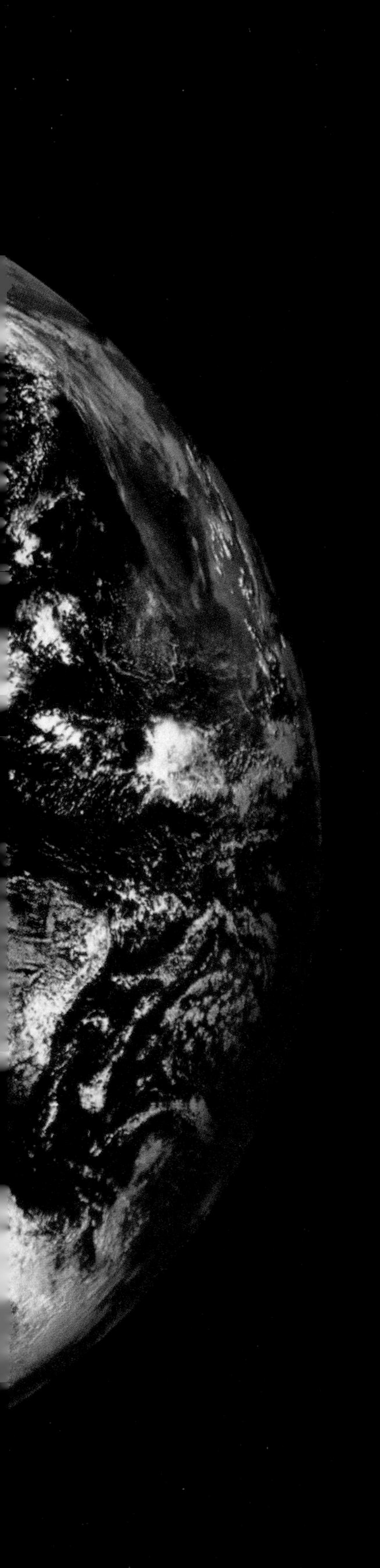

2 Solar Energy, Weather, and Climate

- Solar energy
- Heat transfer between the atmosphere and Earth
- Earth's energy budget
- Precipitation processes
- Circulation patterns

The weather on September 9, 1983

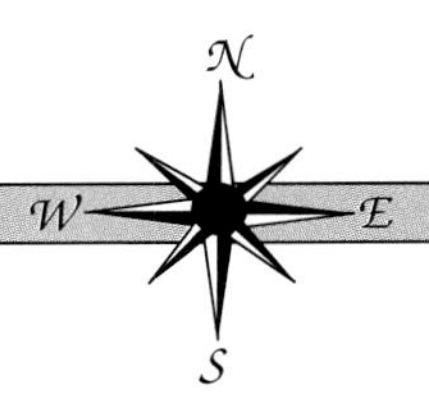

1. Solar energy

Earth's weather is powered by energy radiated from the sun. Because of Earth's constantly changing position—its tilt on its axis and its daily rotation—the amount of energy reaching any place on Earth varies according to the angle of the sun, the length of day, and the season of the year. Given a specific latitude and time of year, receipt of solar energy differs in amount from that reaching other places on the planet at the same time.

2. Heat transfer between the atmosphere and Earth

Four heat transfer processes operate in the atmosphere—radiation, latent heat exchange, convection, and conduction. Each process affects temperature, and each affects movement of heat energy from one area of the planet to another.

3. Earth's energy budget

Energy from the sun travels through space, through Earth's atmosphere, and to Earth's surface, from which it is reradiated eventually back into space. This continual energy movement is described by Earth's energy budget. It is dominated by flows of shortwave radiant energy from the sun and longwave radiation back into space.

4. Precipitation processes

Precipitation occurs when warmer air rises and is cooled, and water vapor in the air condenses at the cooler temperature. Three different processes cause air to rise. These are convectional precipitation, orographic precipitation, and frontal precipitation.

5. Circulation patterns

Atmospheric and oceanic circulation patterns underlie precipitation. Pressure variations cause winds to blow from areas of high pressure to areas of low pressure. Winds are modified by the Coriolis effect, caused by the rotation of the planet.

INTRODUCTION

Weather influences virtually all aspects of natural environments and human activity, from clothing to agriculture. This chapter explores the mechanisms that cause weather. Weather conditions—such as storms, snowfall, clear skies, and warmth or cold—are caused by radiant energy received from the sun. This solar energy circulates from tropical latitudes to the polar regions, and from lower levels to upper levels of Earth's atmosphere, and eventually back into space.

A satellite view on page 42 shows Earth's weather on September 9, 1983. Dark areas are free of clouds, while white areas show high, thick clouds, where precipitation has formed. Contrast this figure with the entirely cloud-free composite view at the beginning of Chapter 1, which was constructed by the Geosphere Project from GIS images taken over several months. One of the most prominent cloudy areas in the single-day image is a storm, known as a tropical cyclone, or hurricane, located in the low latitudes of the Atlantic Ocean. As a hurricane moves away from the equator, it develops a spiraling, rotating motion that strengthens its winds, in much the same way that ice skaters can spin rapidly by drawing in their arms and legs.

Another prominent storm visible in the image is a mid-latitude cyclone, which is a storm that develops in mid-latitudes along what is called a front—a boundary between warm air and cold air. Particularly dramatic contrasts in weather can develop along cold fronts, including thunderstorms and in extreme cases, tornadoes. Knowledge of global patterns of energy circulation enables weather forecasters to predict the approximate path of a hurricane and to forecast cyclones, tornadoes, and sunny weather, all of which affect our lives. In this chapter we look at the patterns of energy flow and the atmospheric processes that create the weather and influence life differently at various places on Earth.

The patterns of clouds and clear skies on the image represent a snapshot of weather conditions at one instant in time. When averaged over many years, these weather conditions over an area constitute climate, the subject of the next chapter. Recurring weather patterns define climate regions of dryness, of frequent rain, of strong seasonal temperature contrasts, and of warm, moderate temperatures throughout the year. Although models of climate can explain average seasonal conditions, they are not able to predict the weather on any given day. In addition, human activity may be causing changes in the climate, and the consequences can be far-reaching.

Solar Energy

Circulation of energy in the atmosphere can be likened to that of an automobile engine, which uses *energy* from gasoline to perform *work* in the form of motion. An engine releases energy inside cylinders through the burning of gasoline. This concentrated energy drives the vehicle and is dissipated through the engine's exhaust and friction.

The fuel driving the atmosphere is *solar energy,* which heats the air, creating wind and ocean currents. These movements carry heat and moisture from one part of the planet to another before the energy is dissipated into space. In an automobile, motion takes the form of pistons moving back and forth, shafts and gears turning, and wheels spinning. In the atmosphere, air is in motion from one part of the globe to another, sometimes in steady flows and sometimes in intense spinning or vertical movements that produce storms.

The Sun—Source of Earth's Surface Energy

The sun is a large thermonuclear reactor, in which hydrogen atoms combine to produce helium. Most important to us is a byproduct of this reaction: the release of vast amounts of energy. The sun radiates this energy into space in all directions, and Earth intercepts a tiny fraction of it. Yet, this small, steady energy flow is sufficient to power circulations of the atmosphere and oceans, and to support all life on Earth. The modest percentage that Earth intercepts from the sun still is 10,000 times greater than the combined energy that humans generate!

As Earth revolves along its elliptical orbit around the sun each year, it receives this energy across a void averaging 150 million kilometers (93 million miles). The fairly constant distance between Earth and the sun—varying by only about 3 percent through the year—helps keep the average temperature of Earth hospitable to living things (Figure 2–1).

Earth's Tilt and Latitude

The amount of solar energy intercepted by a particular area of Earth, which is called **insolation,** depends on two features:

- The *intensity* with which the solar radiation strikes.
- The *number of hours* during the day that the solar radiation is striking.

Intensity of solar radiation. The intensity of solar radiation at a particular place on Earth depends partly on the angle at which the sun's rays hit that place. The **angle of incidence** is the angle at which

Photovoltaic cells converting solar radiation to electricity. The cells are oriented perpendicular to the sun's rays to maximize the intensity of intercepted radiation. (Courtesy of PG&E)

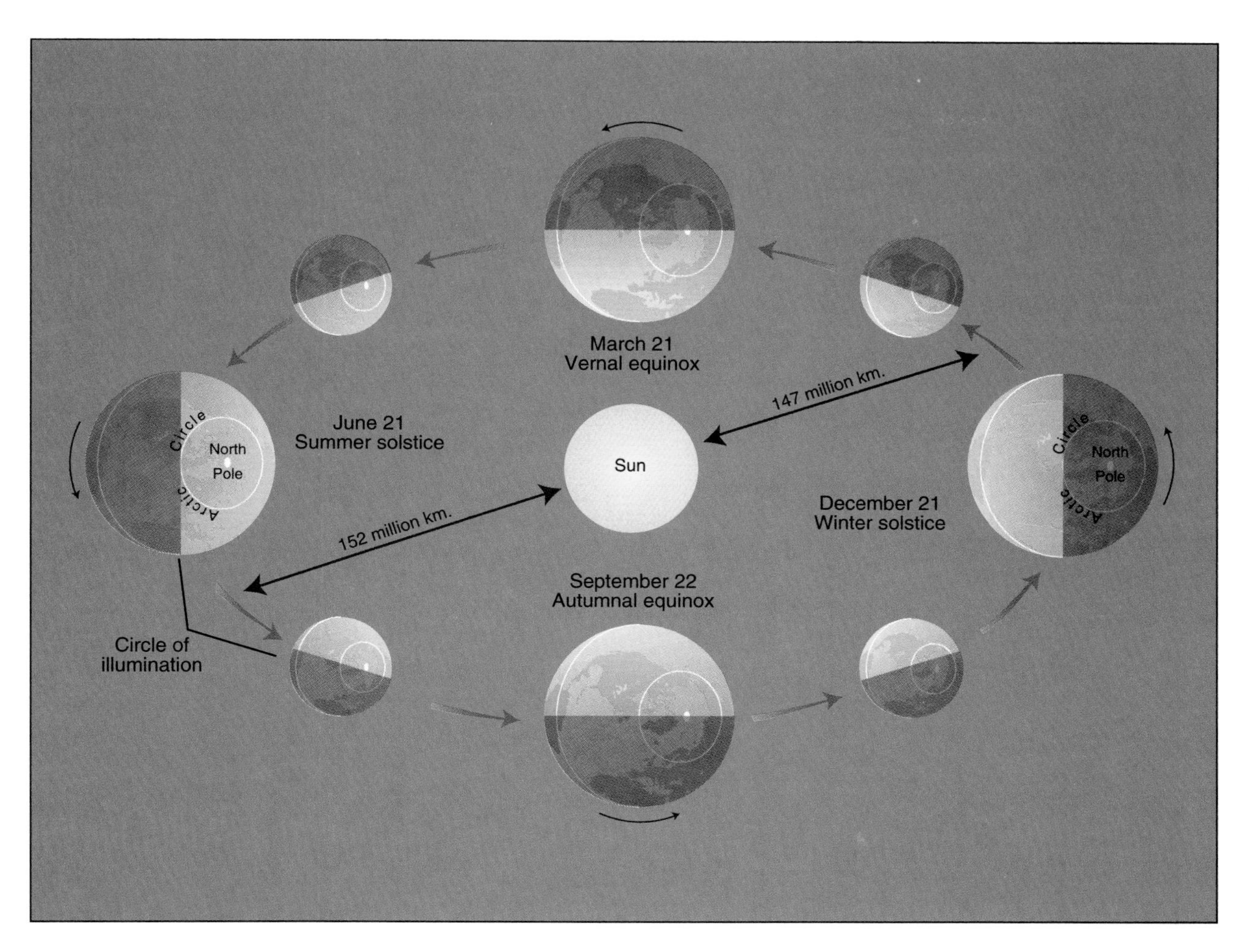

FIGURE 2-1
Earth's orbit around the sun. Earth revolves around the sun in an orbit that follows an imaginary plane known as the plane of the ecliptic. The plane of the ecliptic intersects all points on Earth's orbit around the sun and passes through the sun as well. Earth's path on this plane is an ellipse. As it revolves around the sun, Earth rotates on its axis of rotation, which is the line passing through the North and South poles perpendicular to the equator. The line passing through the North and South poles is tilted at an angle of 66.5° away from the plane of the ecliptic, or in other words 23.5° away from a line perpendicular to the plane of the ecliptic. This tilt is constant throughout the year, no matter where Earth is in its orbit. The tilt causes the Northern Hemisphere to be more exposed to the sun's radiation from March to September (Northern Hemisphere summer), while the Southern Hemisphere receives more radiation from September to March (Southern Hemisphere summer and Northern Hemisphere winter).

solar radiation strikes a particular place at a point in time. The angle of incidence at a particular place on Earth varies within a day and among seasons.

To illustrate the angle of incidence, consider what happens when a 1 square meter beam of sunlight strikes a spherical object such as Earth (Figure 2–2). If this beam strikes Earth's surface perpendicularly (that is, at an angle of 90°), the surface it touches also will have an area of 1 square meter. At an angle of less than 90° (that is, an oblique angle), then the area illuminated by that beam is greater than 1 square meter. But the *intensity of the radiation is reduced,* because the beam's energy is spread over the larger area.

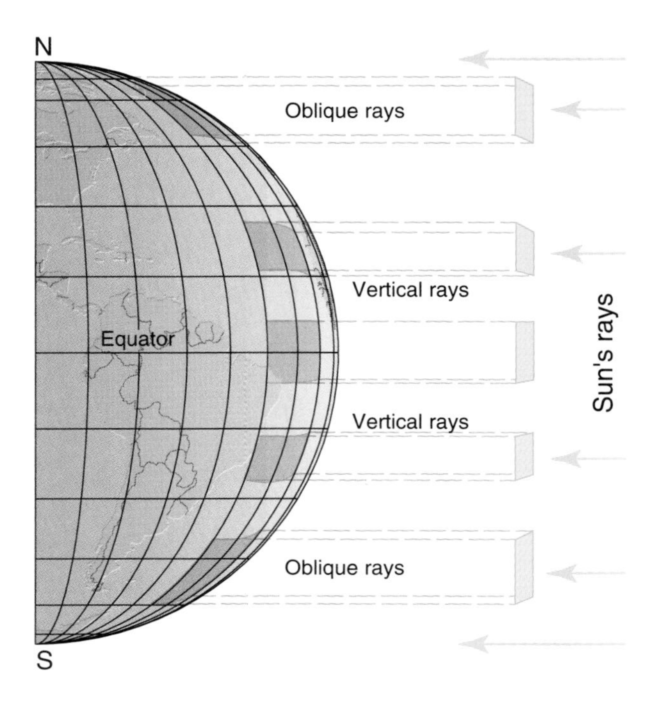

FIGURE 2–2
The angle of incidence. The angle at which the sun's radiation strikes Earth's surface, known as the angle of incidence, varies. The angle of incidence affects the concentration of solar radiation (insolation) that a place receives. Radiation is most concentrated when the sun's rays are perpendicular or vertical to Earth's surface, as occurs when the sun is directly overhead. As shown, a place receiving more perpendicular insolation has this energy concentrated over a smaller area, making the area warmer. A place receiving more oblique (that is, less perpendicular) insolation has it spread over a larger area, making the area cooler. (From R. W. Christopherson, *Geosystems,* 2d ed., Macmillan Publishing Co., 1994)

As the angle is reduced even further, energy is distributed over a still larger area and the intensity of radiation becomes less and less. Compared to the level of radiation at a 90° angle, the intensity of energy is about 86 percent at an angle of 60° and about 50 percent at an angle of 30°. Thus, the intensity of radiation per unit of surface area is affected greatly by the angle of the sun's rays, or the position of the sun in the sky. When the sun is overhead, sunlight is more intense and the level of energy received is more concentrated than when the sun is low in the sky.

Daily and seasonal variations in solar intensity. On a daily basis, the sun is most intense at noon, when the sun is highest in the sky, and least intense at sunrise and sunset. Beach-goers are warned to take this daily variation into account: they are more likely to get sunburned around noon than early in the morning or late in the afternoon. Why? Because insolation varies at different times of day.

The angle at which the sun's energy strikes Earth's surface varies with season, as well as within a day. Throughout the year, the area of Earth's surface where the sun is overhead keeps shifting, due to Earth's tilt and continual revolution around the sun, as was shown in Figure 2–1. Imagine our tilted Earth slowly revolving around the sun through the four seasons of the year, and how this shifts the overhead point of the sun. In the Northern Hemisphere, during the summer, the sun is higher in the sky with more intense radiation, whereas in the winter the sun is lower in the sky with less intense radiation. In the Southern Hemisphere, the opposite is the case, and the seasons are reversed there.

As Figure 2–1 illustrates, Earth rotates once every 24 hours around its axis of rotation, which is an imaginary line passing through the North and South poles. The axis of rotation is inclined 23.5° away from being perpendicular to the sun's incoming rays. This axial tilt of 23.5° remains constant, regardless of Earth's point in its orbit or its daily rotation. This axial tilt is a key factor that determines how much insolation any point on Earth receives.

The amount of insolation a place receives in a particular season depends on its latitude. At noon on the **vernal equinox** (March 20 or 21) and the **autumnal equinox** (September 22 or 23), the perpendicular rays of the sun strike the equator, and the sun is directly overhead at the equator (Figure 2–3). At places along the **Tropic of Cancer** (23.5° north latitude) and the **Tropic of Capricorn** (23.5° south latitude), the intensity of radiation is reduced to 92 percent of the level at the equator. At places along latitude 45° north and south, the intensity of solar radiation is 71 percent of the level at the equator, and at places along latitude 60° north and south, the intensity is only 50 percent of the level at the equator. At the North and South poles on the equinoxes, the sun is seen on the horizon, and its direct radiation is spread so broadly that little is received.

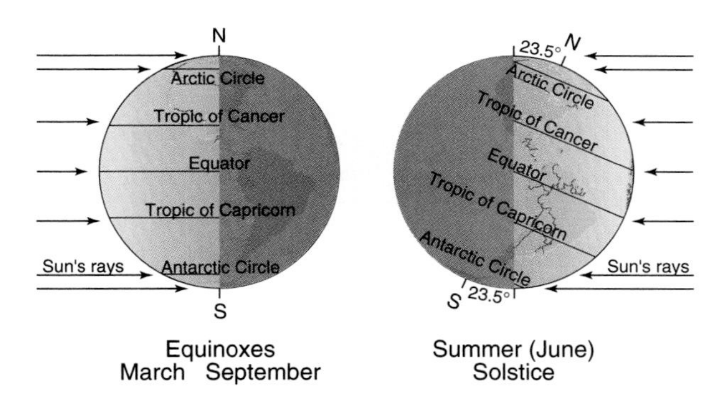

FIGURE 2–3
The angle at which solar radiation strikes places at a particular latitude varies by season. These globes show how Earth receives solar radiation (insolation) on the first day of spring and fall (the equinoxes) and on the first day of Northern Hemisphere summer.
(a) On the *vernal and autumnal equinoxes,* the noon sun is directly overhead at the equator. The higher the latitude (toward either pole), the lower the sun is in the sky. At the poles the sun is seen to lie on the horizon.
(b) On the *solstice,* the noon sun is directly overhead of places at latitude 23.5°. This figure shows solar elevation angles at noon on the Northern Hemisphere summer solstice. Note that the sun appears well above the horizon in the Northern Hemisphere, but it appears very low in the Southern Hemisphere. These patterns are reversed six months later during summer in the Southern Hemisphere (the winter solstice in the Northern Hemisphere). (From R. W. Christopherson, *Geosystems,* 2d ed., Macmillan Publishing Co., 1994)

From the vernal equinox until the autumnal equinox (late March until late September), the sun is directly overhead at places along some latitude in the Northern Hemisphere. The northernmost latitude receiving direct insolation is the Tropic of Cancer (23.5° north latitude), where the sun is directly overhead at noon on the **summer solstice** (June 20 or 21). From the autumnal equinox until the vernal equinox, places in the southern latitudes receive direct insolation, and the sun is directly overhead of places along the Tropic of Capricorn (23.5° south latitude) at noon on the **winter solstice** (December 21 or 22).

Day length. The total amount of heat that a particular place on Earth receives in a day is determined by the *amount of time* during which the sun's energy strikes the place, as well as the intensity with which it strikes. Places on the equator always receive 12 hours of sunlight and 12 hours of night. But in higher latitudes, the amount of daylight varies considerably with the seasons. For example, London, England, at 51° north latitude, receives about 16 3/4 hours of daylight at the summer solstice, but only about 7 1/4 hours of daylight at the winter solstice. In London, the sun is 62.5° above the horizon at noon on the summer solstice, compared to only 15.5° above the horizon at noon on the winter solstice.

Midnight sun. On the day of the summer solstice, poleward of 66.5°, the sun shines the entire 24-hour day, and the sun appears to travel across the horizon without ever setting. (Brian Stablyk/Allstock)

Variations in the length of day from place to place result from the 23.5° tilt of Earth's axis away from the perpendicular. Figure 2–3b illustrates the illumination of the globe at the Northern Hemisphere summer solstice in June. Because the North Pole is located at Earth's axis of rotation, it maintains the same position with respect to the sun throughout the 24-hour rotation of Earth at the solstice. At the summer solstice, if you stand at the North Pole, you can watch the sun travel in a circle around you at a constant elevation of 23.5° above the horizon. On the day of the solstice, the North Pole is in full sunlight for the entire 24 hours, a phenomenon called the *Midnight Sun*.

Distance from the Sun. During the Northern Hemisphere summer, the angles of incidence of the sun's rays are relatively high, and the days are relatively long. Yet, Earth as a whole receives somewhat less overall radiation at this time because it is farther from the sun, due to Earth's elliptical orbit. Earth is closest to the sun when the Southern Hemisphere has relatively long days and receives relatively high angles of incidence.

The Southern Hemisphere receives somewhat more radiation during its summer (between September and March) than does the Northern Hemisphere during its summer (between March and September), because the angles of incidence are highest and days are longest in the Southern Hemisphere at the same time that Earth is closest to the sun. However, seasonal variations in the amount of energy that a place receives are affected mainly by solar elevation angle and amount of daylight.

Spatial and Seasonal Variations in Radiation Inputs

The amount of solar radiation reaching places at a particular latitude varies through the year because of seasonal changes in the angle of incidence, day length, and distance from the sun. The tropical areas at low latitudes generally are warm throughout the year because they receive large amounts of insolation in every season, whereas places at high latitudes have strong seasonal contrasts in the amount of sunshine and level of temperature (Figure 2–4).

Because Earth's axis of rotation is tilted, seasonal variations in the amount of incoming solar radiation are less at places near the equator than at places in higher latitudes. Places at higher latitudes receive somewhat more solar radiation than do places at lower latitudes during the summer, but they receive much less during the winter. From one season to another, the changing distribution of solar energy increases the ability of plants and animals to survive, especially at latitudes above 50° with harsh winters.

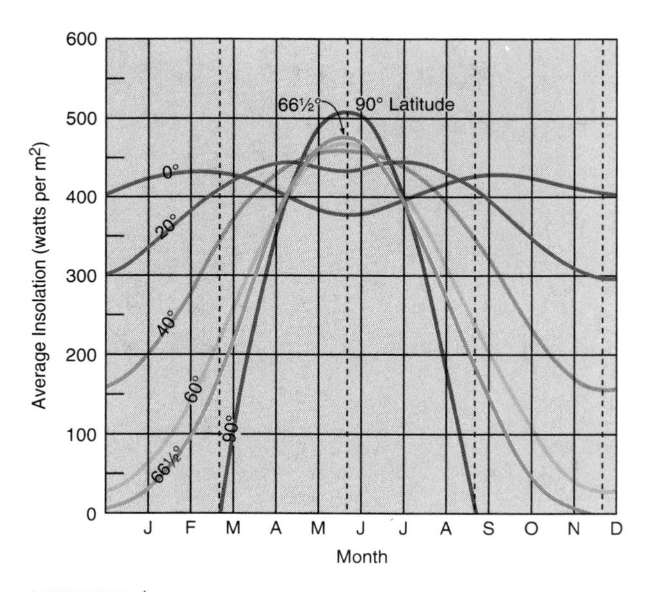

FIGURE 2–4
Solar radiation at different latitudes by season. Because Earth's axis of rotation is tilted, seasonal variations in the amount of incoming solar radiation are less at places near the equator than at places in higher latitudes. Places at higher latitudes receive somewhat more solar radiation than do places at lower latitudes during the summer, but they receive much less during the winter.

Heat Transfer Between the Atmosphere and Earth

Have you ever noticed that on a winter day you feel colder sitting inside near a window or exterior walls than near the building's interior walls, even though air temperatures at both places are about the same? Have you ever wondered why an air temperature of 20°C (68°F) feels comfortable, whereas a water temperature of 20°C feels chilly? Certainly you have experienced the cooling effects of moisture and air blowing on your skin. All of these effects on the temperature of your skin result from processes of *heat transfer*. Similar processes of heat transfer operating in the atmosphere as well as on your skin are responsible for movement of vast amounts of energy from place to place on Earth.

In this section we will look at what happens to solar energy as it passes through Earth's atmosphere to the surface and then is returned to space. On its journey, it drives daily weather worldwide, causing gentle breezes and hurricanes, rain showers and snowfalls, freezing and thawing, climates and ocean currents.

Four Heat Transfer Processes

Four heat transfer processes—conduction, convection, radiation, and latent heat exchange—affect the temperature of Earth as a whole, as well as individual areas. All four processes are illustrated in Figure 2–5.

Conduction. The transfer of heat, molecule-to-molecule, through materials is **conduction.** Generally, heat is conducted rapidly through many solids and liquids, but much more slowly through gases because their molecules are farther apart. When you burn your hand on a hot iron, conduction is the reason.

Convection. The rising of warmer, less dense portions of fluids is **convection.** Through convection, hotter water in the bottom of a pan rises, and a hot air balloon rises in cooler air. Home heating systems like baseboard electric heaters and hot water systems rely on convection to distribute heat. They heat the air immediately around them by radiation: the air expands, rises, and spreads through the room.

Radiation. Energy transmitted by electromagnetic waves, including radio, television, light, and heat, is **radiation,** or radiant energy. You feel heat *radiating* from a fire. Heat travels at the speed of light from the fire to your skin, which senses it. You do not feel heat from a fire by conduction, because molecules in air are too far apart to transfer heat energy rapidly from one to another. Radiation generally travels through materials and the vacuum of space, weakening with distance. Radiant energy allows us to see the sun's light energy, feel its heat energy, and "listen" to its radio energy.

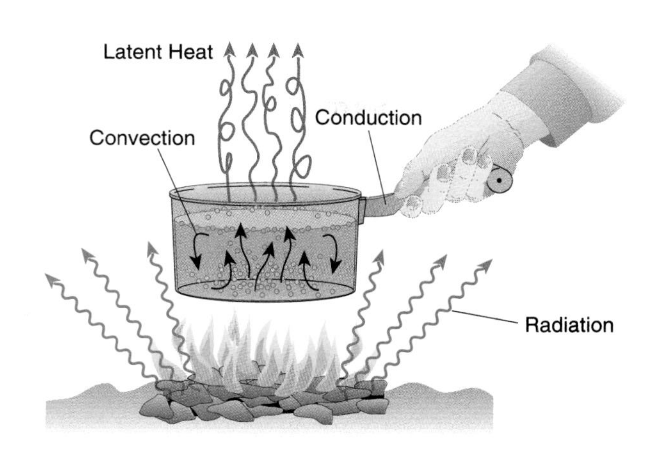

FIGURE 2–5
A pan of water over a fire illustrates all four important heat transfer processes. Heat is *radiated* from the fire to the pan, *conducted* to the water, *convected* throughout the pan, and transferred to the air as *latent heat.*

Radiant energy waves have different lengths. The **wavelength** is the distance between successive waves, like waves on a pond (Figure 2–6). Wavelength affects the behavior of the energy when it strikes matter—some waves are reflected and some are absorbed. Figure 2–6 shows the *spectrum* of wavelengths and what we call them.

Latent heat exchange. The exchange of energy necessary to change water from one of its states to another—solid, liquid, or gaseous—is **latent heat exchange.** Water stores large amounts of heat when it evaporates. When it condenses, it releases this heat into its surroundings. A familiar example is steam from a teakettle condensing on your finger, causing a painful burn.

Radiation and Reradiation

The sun's energy comes to Earth as radiant energy, for this is the only way energy can travel through the vacuum of space. Two ranges of wavelengths—called shortwave and longwave—are most important for understanding how solar energy affects the atmosphere. Most insolation is **shortwave energy,** with wavelengths between 0.2 and 5 microns (a **micron** or micrometer is one millionth of a meter). Wavelengths visible to the human eye account for a

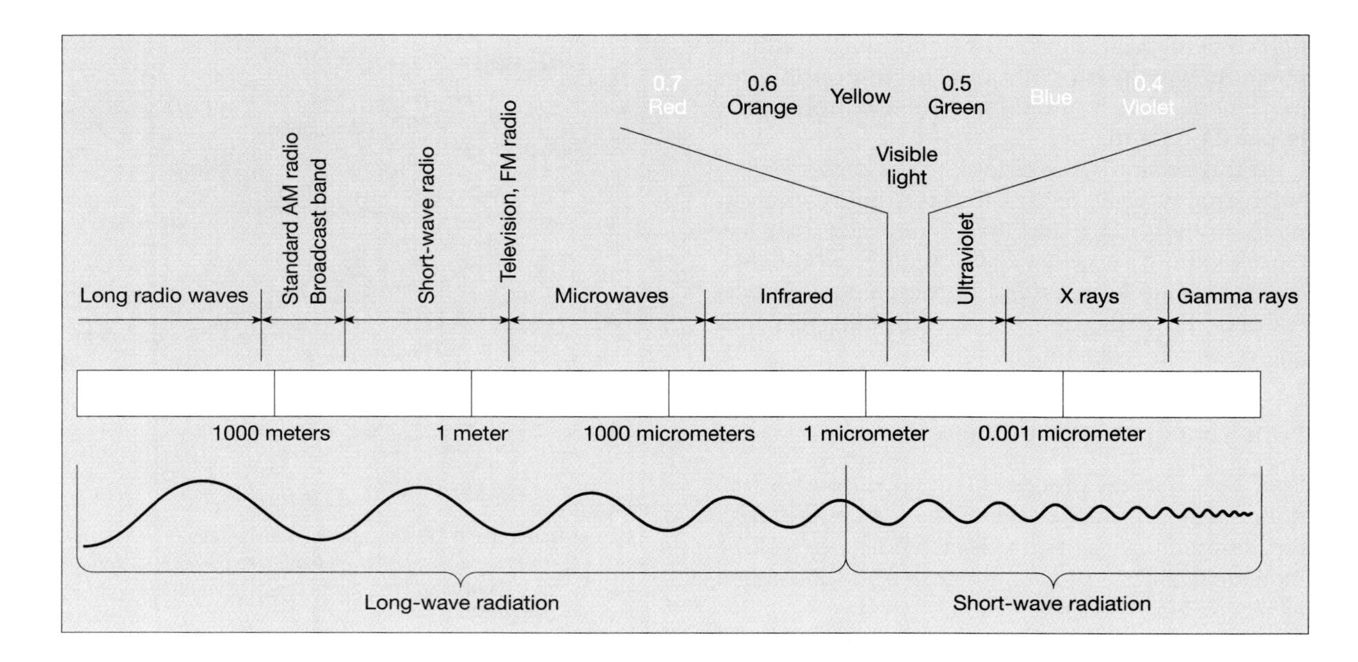

FIGURE 2–6
The spectrum of wavelengths. Natural electromagnetic radiation occurs in a wide range of wavelengths, from several kilometers to less than one millionth of a micrometer (also known as a micron, which is one millionth of a meter). We name the type of radiation according to its wavelength ("radio waves," "heat," "light," X-rays"), but otherwise these forms of radiation are essentially the same. The human eye is sensitive to energy with wavelengths of 0.4 to 0.7 microns; within this range we distinguish wavelengths of energy as different colors. (From E. J. Tarbuck and F. K. Lutgens, *Earth Science,* 7th ed., Macmillan Publishing Co., 1994)

Freezing rain occurs if rain falls when the ground temperature is below 0°C (32°F). Under this condition, the rain freezes when it touches cold objects like these trees. When the water freezes, latent heat is given up by the water, and either conducted into the tree or radiated to the atmosphere. (Sotographs/The Stock Market)

small portion of this shortwave energy, from about 0.4 to 0.7 microns (Figure 2–6). In contrast, most of the energy reradiated by Earth is **longwave energy,** in wavelengths between 5 to 30 microns.

It is very important that most energy arriving from the sun is shortwave while most energy reradiated by Earth is longwave. The shortwave energy from the sun easily passes through the atmosphere to reach Earth's surface, but when the surface reradiates longwave energy, much of it is blocked and absorbed by the atmosphere. The blockage of outgoing longwave energy causes Earth's atmosphere to heat.

All matter emits radiation, even very cold objects in space. The particular wavelength and amount of energy emitted varies according to the composition and temperature of that matter. A hot object emits more energy at a shorter wavelength than a cool object. Therefore, a very hot body like the sun radiates shortwave energy, whereas Earth—a much cooler body—reradiates longwave energy (Figure 2–7). The sun, at about 5,727°C (10,340°F), emits 200,000 times more energy per unit of surface area than Earth, which averages about 7°C (45°F).

Some objects absorb relatively large amounts of radiation and are consequently warmed. Objects such as snow that absorb relatively small amounts of radiation may reflect it and send it off in another direction like a mirror. Or, they may allow the radiation to pass through unchanged, like visible light through glass.

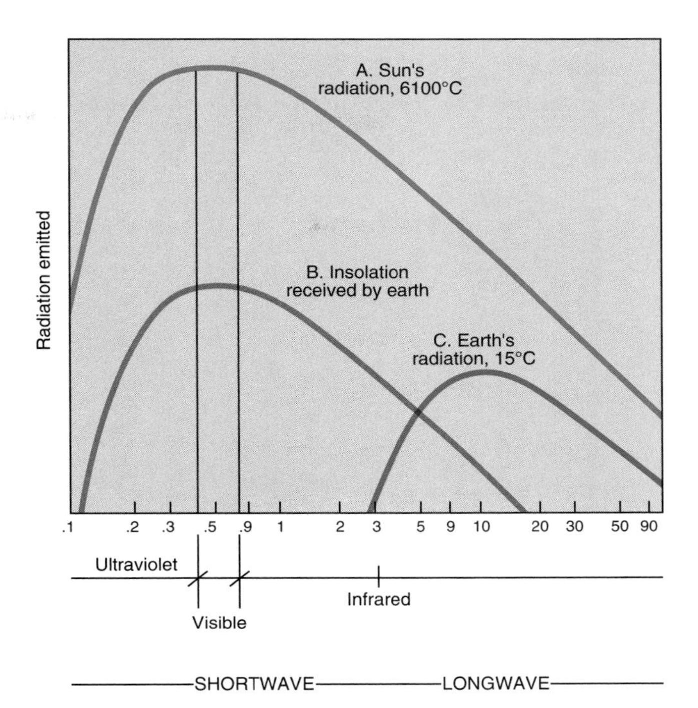

FIGURE 2–7
Intensity of radiation produced by the sun and Earth. The horizontal axis shows the wavelength spectrum. The vertical axis shows the relative power of the radiation being emitted.
Curve A: Because the sun is very hot (about 6,000°C or 10,340°F), it emits most of its radiation in relatively short wavelengths, including the wavelengths of visible and ultraviolet light.
Curve B: Only a small fraction of this insolation strikes Earth, where it is converted to heat energy, warming the surface, oceans, and atmosphere.
Curve C: Because Earth is much cooler than the sun (average surface temperature is 7°C or 45°F), it emits radiation mostly in longer infrared wavelengths. This important fact is essential to the greenhouse effect.

Absorption of radiation. As energy from the sun passes through the atmosphere, some wavelengths are absorbed, warming the atmosphere, while others pass through or are reflected, either to be absorbed elsewhere or to travel back into space. Matter in the atmosphere and at Earth's surface—such as air, water, clouds, soil, and leaves—absorbs, reflects, or transmits varying amounts of energy.

Table 2-1 lists the substances most important to energy flow in the atmosphere. Although their patterns of absorption, transmission, and reflection of radiation may be complex, these materials affect shortwave energy coming from the sun and longwave energy being reradiated from Earth and the atmosphere. Nitrogen, oxygen, and argon—which together constitute about 99.9 percent of the atmosphere—do not have a great effect on radiant energy flows because they are transparent to nearly all shortwave and longwave radiation.

In contrast to these more abundant gases, a group of substances known as **greenhouse gases** plays an important role in the atmosphere despite their low concentrations. The more important greenhouse gases are carbon dioxide, ozone, water vapor, CFCs, and methane. These are relatively transparent to incoming shortwave energy, but absorb outgoing longwave energy. The presence of these trace gases keeps the atmosphere much warmer than would be the case otherwise. If all of these gases vanished from the atmosphere, we

TABLE 2–1

Generalized behavior of some important materials with respect to shortwave (solar) and longwave (terrestrial) energy

Material	Percent of Atmosphere	Behavior with Respect to Shortwave Radiation	Behavior with Respect to Longwave Radiation
Nitrogen	78	Transparent	Transparent
Oxygen	21	Transparent	Transparent
Argon	1	Transparent	Transparent
Ozone	Trace; variable	Absorbs some ultraviolet	Major absorption bands
CO_2	0.036	Transparent	Major absorption bands
Methane	Trace; variable	Transparent	Major absorption bands
Chlorofluorocarbons (CFCs)	Trace; variable	Transparent	Major absorption bands
Water vapor	Trace; variable	Transparent	Major absorption bands
Liquid water (including clouds)	Trace; variable	Reflects some, absorbs most	Opaque (absorbs virtually all)
Soil/rock	—	Reflects some, absorbs most	Opaque (absorbs virtually all)
Vegetation	—	Reflects some (green), otherwise absorbs	Opaque (absorbs virtually all)

would exist on a very chilly UV-irradiated planet! The **greenhouse effect** is atmospheric warming that results from the passage of incoming shortwave energy and the capture of outgoing longwave energy, similar to the way a greenhouse works. The greenhouse effect causes the atmosphere to act like the glass roof of a greenhouse over Earth.

The Process of Latent Heat Exchange

We can distinguish between two types of heat—sensible and latent:

- **Sensible heat** is detectable by your sense of touch—it is heat you can feel, from sunshine or a hot pan. The atmosphere, oceans, rocks, and soil all have sensible heat, for you can feel their relative warmth or coldness.
- **Latent heat** is "in storage" in water and water vapor. You cannot feel latent heat, but when it is released it has a powerful effect on its immediate environment.

Latent—which means "hidden"—is a good word to describe the heat that controls the state of water, for it is invisible stored heat. When ice melts, it must absorb heat energy from its surroundings. This is why ice melting in your hand feels so cold—it is absorbing heat from your hand. The heat becomes stored in the meltwater as latent heat.

Latent heat also is stored in water vapor. If you ever have had a finger scalded by steam, you know the startling amount of latent heat that was stored in the vapor and conducted from the condensing water into your finger! Latent heat exchange is what happens when water changes state (Figure 2–8).

Here is an analogy that describes the concept of latent heat. Consider a glass containing a mixture of ice and water on a hot summer afternoon. Because the ice water is much cooler than the air and other objects around it, heat energy is absorbed into the water. The heat begins to melt the ice, but as long as the water contains *some* ice, its overall temperature remains at 0°C (32°F). The water is absorbing enough heat to convert part of it from one state to another, but not enough heat to change its temperature. The heat the water absorbs to melt the ice becomes *latent* heat, because it cannot be *directly sensed* as temperature.

Latent heat of vaporization. When water is converted from liquid to vapor, additional heat is required—in fact much more is needed than is required to melt ice. The **latent heat of vaporiza-**

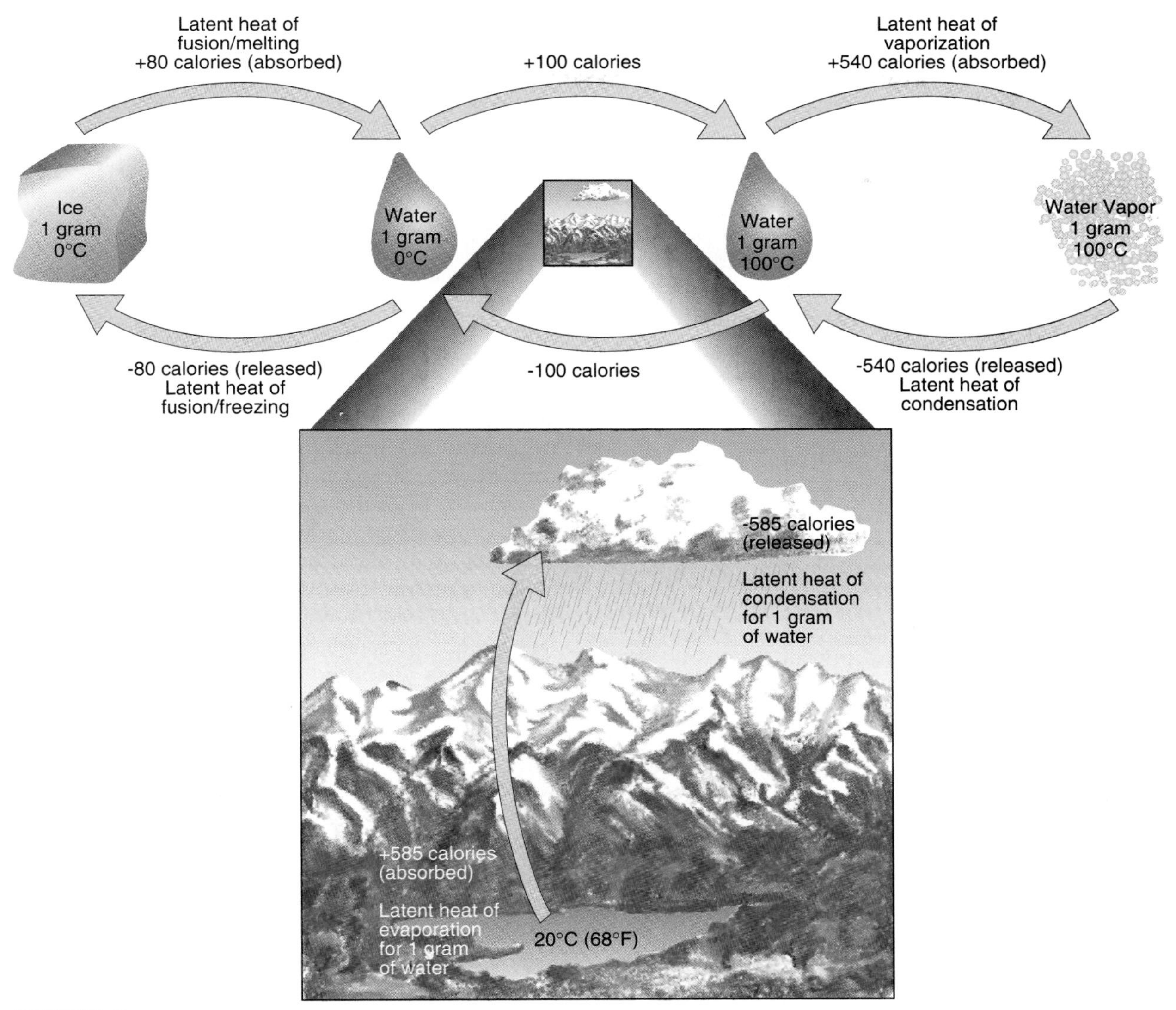

FIGURE 2–8
Latent heat is the heat involved in changing the state of water, through such processes as melting, freezing, vaporization, and condensation. When water changes form (or phase), large amounts of heat are either stored or released. For example, when a gram of 0°C (32°F) ice melts, 80 calories of latent heat must be absorbed from the ice's surroundings to do so. (This is why ice chills things that it touches.) When a gram of water vapor condenses, 540 calories of latent heat are released. (This is why steam can "scald" you.) Observe that the transition between liquid and vapor involves greater amounts of heat per gram than freezing and melting. (From R. W. Christopherson, *Geosystems,* 2d ed., Macmillan Publishing Co., 1994)

tion—the process that turns liquid to vapor—is about 540 calories (heat units) per gram at 100°C, compared to about 80 calories per gram for the latent heat of melting.

In the ice water example, a portion of the heat used to melt the ice came from sensible heat—heat you can sense, the heat expressed as the temperature of the environment around the glass. The remainder came from the latent heat released by water vapor condensing to liquid on the surface of the glass. Thus, the melting of the ice water involved three transfers of heat:

Latent heat was converted to *sensible* heat by condensation of water vapor on the glass.
Sensible heat was transferred directly from the warm air to the cool glass.
Sensible heat was converted to *latent* heat in the melting of the ice.

Now let us transfer this knowledge of latent heat to the much larger scale of the atmosphere. The amount of energy involved in latent heat transfers in the atmosphere is vast, especially for major weather systems and hurricanes. Hurricane forecasters say that a hurricane "gathers strength," as it moves over the ocean. What they really are referring to is *latent heat.*

Insolation supplies shortwave energy to warm the seawater. As it evaporates, countless molecules of water vapor—each with its own latent heat—hover above the water, waiting to join the storm as it passes over. As the hurricane accumulates this water vapor, it also is gathering an enormous amount of latent heat energy. The vapor rises in the rotating storm, becomes cooled, and condenses. Latent heat is released as sensible heat, a process roughly analogous to adding gasoline to a forest fire.

The moisture in one major hurricane at its peak was estimated to weigh 27 trillion (27,000,000,000,000) metric tons (30 trillion tons). The latent heat released by this condensing water vapor in a single day roughly matched the total amount of energy consumed in the United States for six months. This provides some insight into why hurricanes are dangerous.

One reason that Earth is such an extraordinary planet is that it is the only one we know on which water exists in the environment in solid, liquid, and gaseous forms. Water is continually being converted from one state to another because of transfers of energy. For example, energy transferred to ice makes it melt, and energy transferred to water makes it evaporate. Conversely, energy removed from water makes it freeze, and energy removed from water vapor makes it condense to a liquid

Convection and Advection

Liquid water changes to vapor primarily at Earth's surface. This water vapor is carried aloft by rising air. In the atmosphere, vapor is converted back to liquid (clouds and rain) or solid (ice clouds and snow) through the formation of *precipitation,* which we will discuss later in the chapter. This movement of water back and forth between the surface and the atmosphere acts like a great conveyor belt for energy, picking up latent heat at the surface, carrying it aloft, and releasing it in the atmosphere. The conveyor is driven by convection.

Recall that convection is movement in any fluid, caused when part of the fluid is heated. The heated portion expands and becomes less dense, and therefore rises up through the cooler portion. Convection causes the turbulence you see in boiling water and in turbulent clouds overhead. As air is warmed, it expands and becomes less dense. In becoming less dense—that is, weighing less per unit of volume—warm air rises above cooler, denser air, just as a hot-air balloon rises through the cooler air surrounding it.

Think of an island on a sunny day (Figure 2–9). Solar energy passes down through the atmosphere and is absorbed by the sandy surface. As the surface warms, it reradiates longwave energy, some of which is absorbed by the air just above the ground. The water surface becomes less warm because some of the radiation is reflected, some is used to evaporate water, and some penetrates the water to warm lower depths of the sea. The air therefore grows much warmer over the island than over the water. This warmer air over the island expands, and grows less dense and begins to rise. As the warm air rises, cooler air descends to take its place. The cooler air is then warmed by the surface, and the convection process continues.

Convection also causes *horizontal* movements of air. Large horizontal transfers of air—and the latent heat contained in water vapor in the air—are major components of Earth's energy system. Horizontal transfer of this type is called **advection.** Heat is advected from tropical areas toward the poles when warm winds blow poleward. Heat in ocean currents moving toward the polar regions is another form of advection.

Let us look at a small-scale example. During the daytime, warm air rising over the land allows cooler air from the sea to advect over the land. In other words, wind in some coastal areas blows from the sea toward and over the land. On a large scale, during summertime over Eastern Europe and Central Asia, the season's heating causes winds to

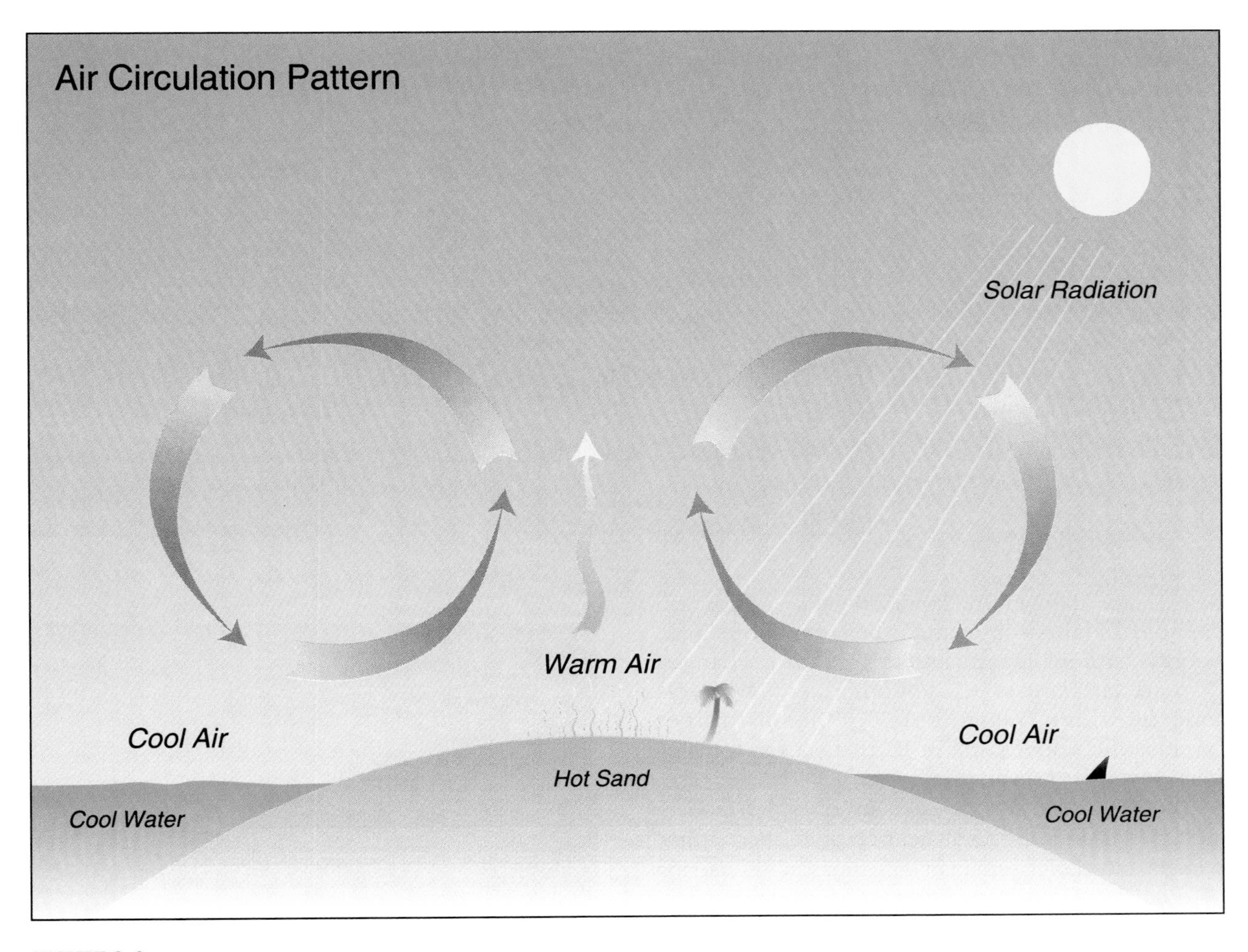

FIGURE 2–9
Convection in the atmosphere occurs when air is warmed from beneath, expands, and rises. Cooler air descends to replace it. A common example is where land (such as an island, the Florida peninsula, or Cape Cod) is surrounded by water. Convection is strongest in the daytime when the surface is warmed by the sun. Because of a stronger convection process, the wind tends to be gustier in the daytime than at night.

blow onto the land from over oceans to the south and east. During winter, because the land area cools more than the sea, cold air sinks over the land, warmer air rises over the ocean, and the winds tend to blow from Central Asia toward the warmer Pacific Ocean.

Heat storage

Heat is stored in two ways. You have learned about latent heat being stored in water. The other type of storage is simpler: heat absorbed by Earth's surface, which is temporarily stored as sensible heat (heat we can feel) in water, the soil, and rocks. Storage of sensible heat depends on the ability of particular materials to absorb and transfer the heat beneath the surface. To be stored for very long, heat must be moved below the surface, because if it were to remain at the surface it simply would be radiated back to the atmosphere.

Specific heat is the amount of energy required to raise a substance's temperature by a given amount. Different substances have different specific heats. Most important to us is that water has a very high specific heat compared to land (rock, soil, trees). This means that water stores

A convective storm over the Seychelles Islands in the Indian Ocean. (Stan Osolinski/The Stock Market)

heat better than land. Therefore, the oceans are capable of absorbing and releasing more heat energy without significant temperature changes than are land areas. In addition, a very large volume of water is available for heat storage, because the upper parts of Earth's oceans transfer heat downward through mixing.

The great heat-storage capacity of water moderates coastal climates. Areas near oceans have warmer winters and cooler summers than areas in the interior of continents, because during the winter oceans release heat the waters absorbed during the summer.

In higher latitudes, places near the oceans, such as Western Europe, have relatively mild climates compared to inland locations. This is true in part because heat absorbed in the tropics is advected to higher latitudes in ocean currents. One example is the Gulf Stream, which is part of an ocean current that flows from the tropical western Atlantic north to the coast of Western Europe. The presence of the Gulf Stream moderates climate even in southern Iceland, which is quite livable despite being near the Arctic Circle.

Rocks and soil can also conduct heat downward and store it beneath the surface. In mid-latitude areas, which have strong seasonal contrasts, soils absorb heat by warming during the spring and summer and release heat in the autumn and winter. In most cases, storage of heat is limited to the top 2 or 3 meters (6 to 9 feet) beneath the surface; the depth of heat storage in the oceans is much greater.

The Gulf Stream, a flow of warm water northward in the western Atlantic, is clearly visible in this satellite image. Large amounts of heat are carried poleward in ocean currents like this. (Photri)

Earth's Energy Budget

The important energy exchange processes that underlie Earth's weather patterns and climatic system include solar radiation (insolation), latent heat exchange, convection, advection, and heat storage in the land and water. How are these processes related?

We can relate them with an *energy budget* for Earth's atmosphere. Environmental scientists use Earth's energy budget in the same way that financial accountants use a money budget—to identify places (accounts) where things are stored and to monitor transactions in and out of the accounts. Just as an accountant can track a firm's income and expenses, an environmental scientist can track movements of heat energy. An energy budget uses a common unit of measure for all the "accounts" in the environmental system, and a balanced energy budget assures that all of Earth's processes are accounted for.

Elements of an Energy Budget

The energy budget divides Earth's environment into three accounts: the atmosphere, space (including the sun), and Earth's surface (including solid land and liquid oceans). Figure 2–10 illustrates the average annual flows among the energy budget's three accounts.

In Figure 2–10, we view the total insolation to planet Earth as 100 percent (or 100 units). About 31 percent of this energy is immediately reflected back to space and never absorbed. The reflection, called **albedo,** is the "Earthshine" you would see if you could look at Earth from space, just as we see the moon's albedo as "moonshine." The brightest objects are clouds and snow-covered surfaces. Most of the reflection is from clouds; the rest is reflection from molecules in the air and from the ground surface.

Of the 69 percent that is not reflected back to space, 24 percent is absorbed by the atmosphere, leaving an average of 45 percent absorbed by the land and oceans, as shown. So, on average, the insolation budget is 55 percent reflected or absorbed into the atmosphere, and 45 percent absorbed by land and sea. However, this distribution varies widely from one day to the next. On cloudy days, far more than 55 percent of insolation is reflected and absorbed by the atmosphere. But if the weather is clear, as much as 80 percent of the incoming solar radiation at the top of the atmosphere may be absorbed by the surface.

The energy absorbed by Earth's surface and atmosphere is eventually returned to space, but only after many exchanges between the surface and the atmosphere. Earth radiates longwave energy, which is absorbed by the atmosphere or escapes to space through the relatively transparent "window" in the atmosphere between 8 and 13 microns (see Figure 2–11). The atmosphere radi-

Longwave radiation from the ground to the atmosphere during the nighttime cools air near the ground. When the relative humidity reaches 100 percent, condensation occurs, and water droplets accumulate on solid surfaces such as this grassy area. (Charles Krebs/The Stock Market)

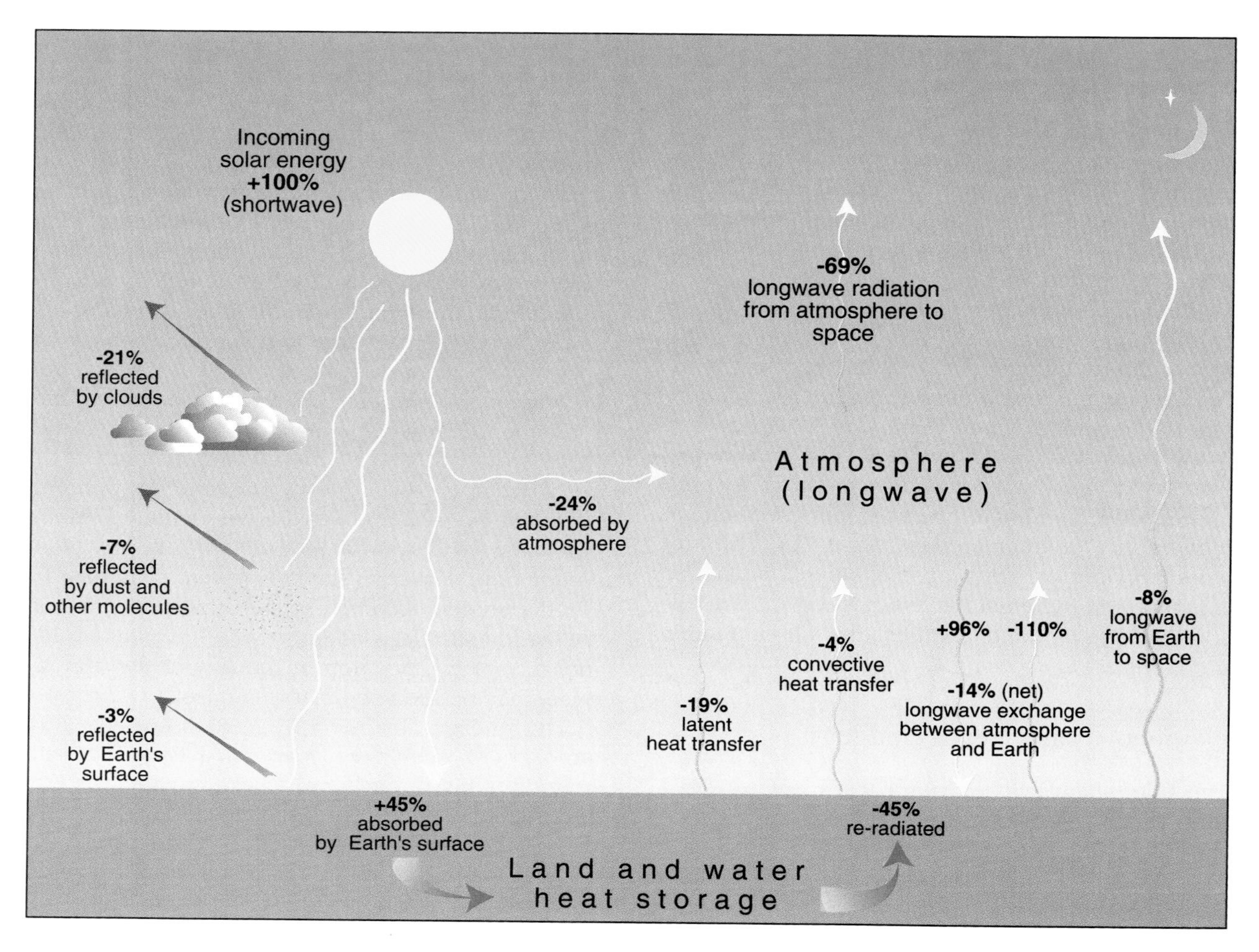

FIGURE 2–10
The average annual energy budget for Earth, simplified to show how 100 units of solar energy are reflected, absorbed, or reradiated by our planet's complex systems. If you trace the incoming shortwave energy (100 units), you will see that eventually all of it is reradiated. Note that the exchanges of energy among the atmosphere, Earth, and space are predominately longwave (heat).

ates longwave (heat) energy in all directions to space and to Earth.

Importance of earth's energy budget. When we discuss Earth's general climate patterns in the next chapter, we should take two major conclusions about Earth's energy budget with us:

1. *The energy budget is dominated by flows of longwave radiation.* Two-thirds of the energy received at the surface arrives as longwave energy from the atmosphere, compared to less than one-third in direct shortwave solar radiation. Longwave energy arrives at Earth's surface 24 hours a day, helping to reduce the gap in surface temperature between day and night.

2. *Radiant energy flows are strongly influenced by atmospheric composition.* Especially important are dust, ozone, and greenhouse gases (carbon dioxide, water vapor, methane). Greenhouse gases constitute less than one-tenth of 1 percent of the atmosphere, but they have a profound influence on atmospheric temperature. Therefore, small changes in the amounts of greenhouse gases in the atmosphere can significantly alter Earth's energy budget.

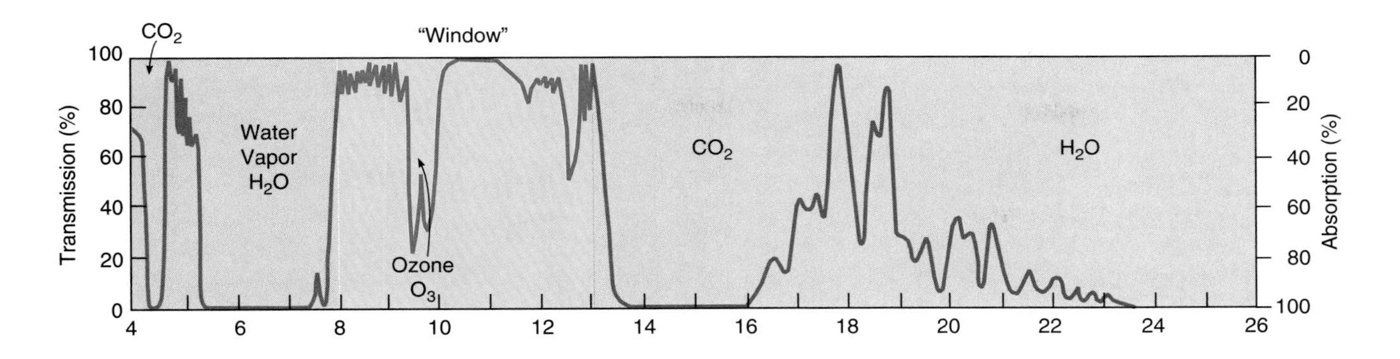

FIGURE 2–11
The atmospheric "window." Earth's atmosphere is transparent to incoming solar energy, but it absorbs much of the longwave energy that is reradiated by Earth. The graph shows the percentage of radiation that is absorbed at various wavelengths by three important gases—water vapor (H_2O), ozone (O_3), and carbon dioxide (CO_2). The higher the curve, the higher the percentage of energy being transmitted at a particular wavelength. Note that the atmosphere is most transparent at a band between about 8 and 13 microns. This band forms a "window" through which the majority of Earth's reradiated energy escapes back to space.

Local Differences in the Energy Budget

The energy budget is an important tool for understanding Earth's atmosphere because it depicts annual flows for Earth as a whole. An average annual energy budget can obscure important differences among *places* and among *seasons.*

Differences among places. Energy budgets vary by latitude. Incoming shortwave radiation is concentrated in the tropics, because the sun is closer to overhead than in higher latitudes (Figure 2–12). The tropics receive more shortwave insolation than they lose longwave energy. Higher latitudes are the opposite: they lose more longwave energy than they receive shortwave insolation. Regional imbalances occur because energy is transferred from the tropical areas to high latitudes through advection of latent heat in water vapor, and through advection of sensible heat in the atmosphere and the oceans. These transfers help to cool the tropics and warm the high latitudes.

Latent heat exchange must have water to occur. Variations in available water, due to an area's proximity to the ocean or its annual rainfall, help to explain variations in weather and climate. Deserts heat faster in bright sunshine than do oceans. Oceans remain cooler because energy is absorbed from them to evaporate water. Similarly, temperatures are higher in cities, where vegetation is limited and soil is covered by pavement and buildings, than in surrounding rural areas, where more heat is used to evaporate water.

Differences among seasons. The insolation a place receives varies through the year. During the spring and summer, Earth's surface and atmosphere are warmed by large amounts of incoming solar energy (Figure 2–13). Incoming energy that exceeds outgoing energy is stored as sensible heat in soil, rock, and water. As Earth's surface and atmosphere warm up, they radiate more heat energy, so that by late summer the outgoing radiation is relatively high. During the autumn and winter, radiation goes out in greater amounts than it comes in, but as Earth and its atmosphere cool, the amount of outgoing radiation also decreases.

Any given area of Earth does not receive its greatest amount of solar energy during its warmest month. For example, in North America, the summer solstice is in late June, but July and August are generally warmer. The lag reflects the time needed to warm the ground, oceans, and air in summer and to cool them in winter.

As we have learned, the total amount of radiation coming to Earth's surface is the energy arriving at the top of the atmosphere minus energy

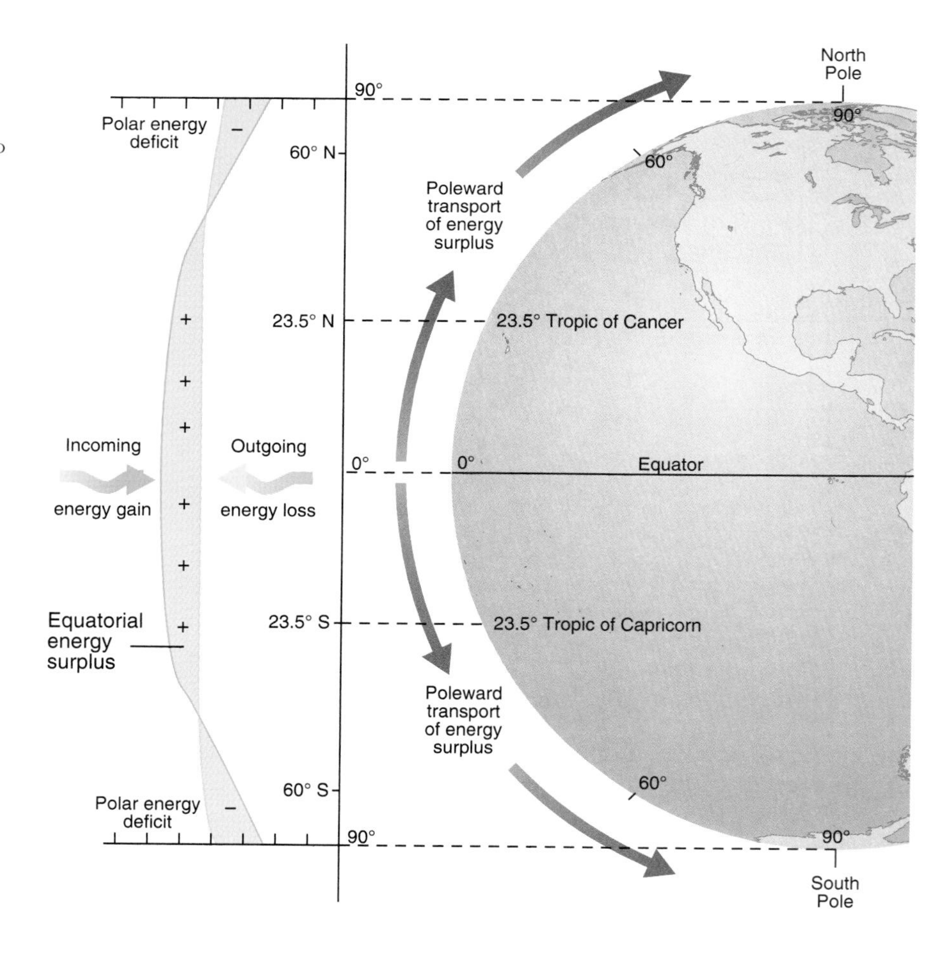

FIGURE 2–12
Energy budget by latitude. In the tropics, where insolation is most concentrated, more energy is received from the sun than is lost to space by longwave radiation. However, in high latitudes, more energy is lost (reradiated) back to space than is gained. How is this possible? Advection (horizontal movement) of air and ocean currents move excess heat from the tropics to high latitudes. (From R. W. Christopherson, *Geosystems,* 2d ed., Macmillan Publishing Co., 1994)

losses within the atmosphere due to reflection and absorption. At a global scale, tropical locations receive more than twice as much solar radiation as polar regions. However, the distribution of energy to regions of greater or lesser sunshine does not correspond simply to variations in latitude. A particular region's typical energy budget is strongly influenced by the distribution of land and water across Earth's surface (Figure 2–14).

For example, the level of sunshine in the interior of the Amazon basin near the equator is much less than in northern Africa and the southwestern United States, located about 30° north latitude. If this seems paradoxical, consider that the insolation reaching the surface is greater in deserts because of their generally clear skies, whereas along the equator, clouds reflect incoming energy, diverting much of it from reaching the surface.

Global Warming

In addition to seasonal variations, the energy budget displays longer-term warming trends. Many scientists believe that over the next few decades or centuries the atmosphere and oceans will absorb more energy than they reradiate. Global warming is likely to result from increased concentrations in the atmosphere of some greenhouse gases, especially CO_2, CFCs, and methane (Figure 2–15). CFCs are also believed to be responsible for declining concentrations of ozone in the stratosphere (see Focus Box 2-1).

CFCs, water vapor, carbon dioxide (CO_2), methane (CH_4), and ozone are only trace constituents of the atmosphere, yet they are heavy hitters in controlling atmospheric energy movements. Carbon dioxide is the most prominent absorber of longwave

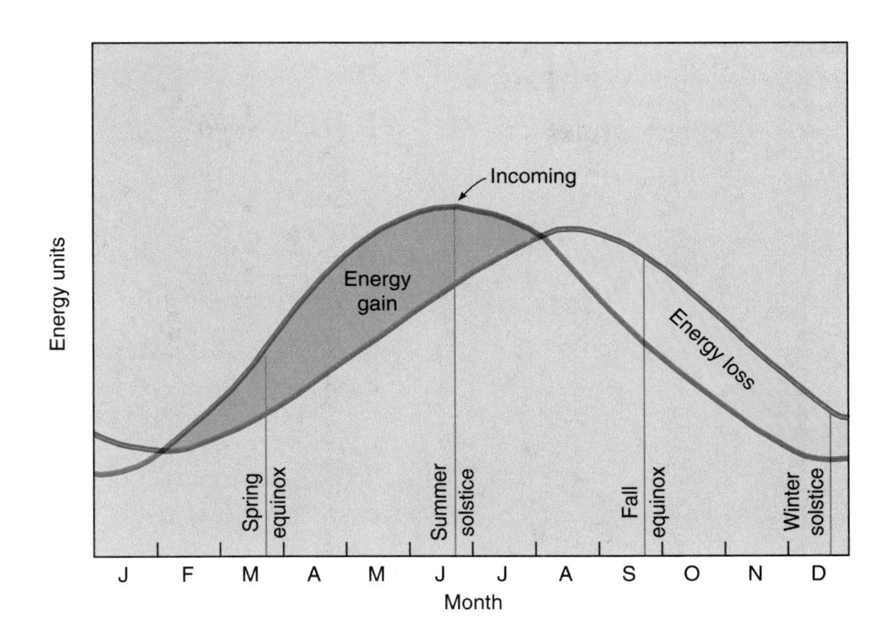

FIGURE 2–13
Energy gain and loss by season. This figure shows the annual cycle of incoming and outgoing radiation in a Northern Hemisphere mid-latitude environment, such as the United States. In spring and early summer, more radiation is received than lost, causing the atmosphere to heat. In the autumn and winter, more radiation is lost than is gained. The times of maximum and minimum temperature are usually one to two months later than the solstices due to "thermal lag." This lag is caused by time required for Earth materials (land, oceans) to absorb insolation and reradiate it.

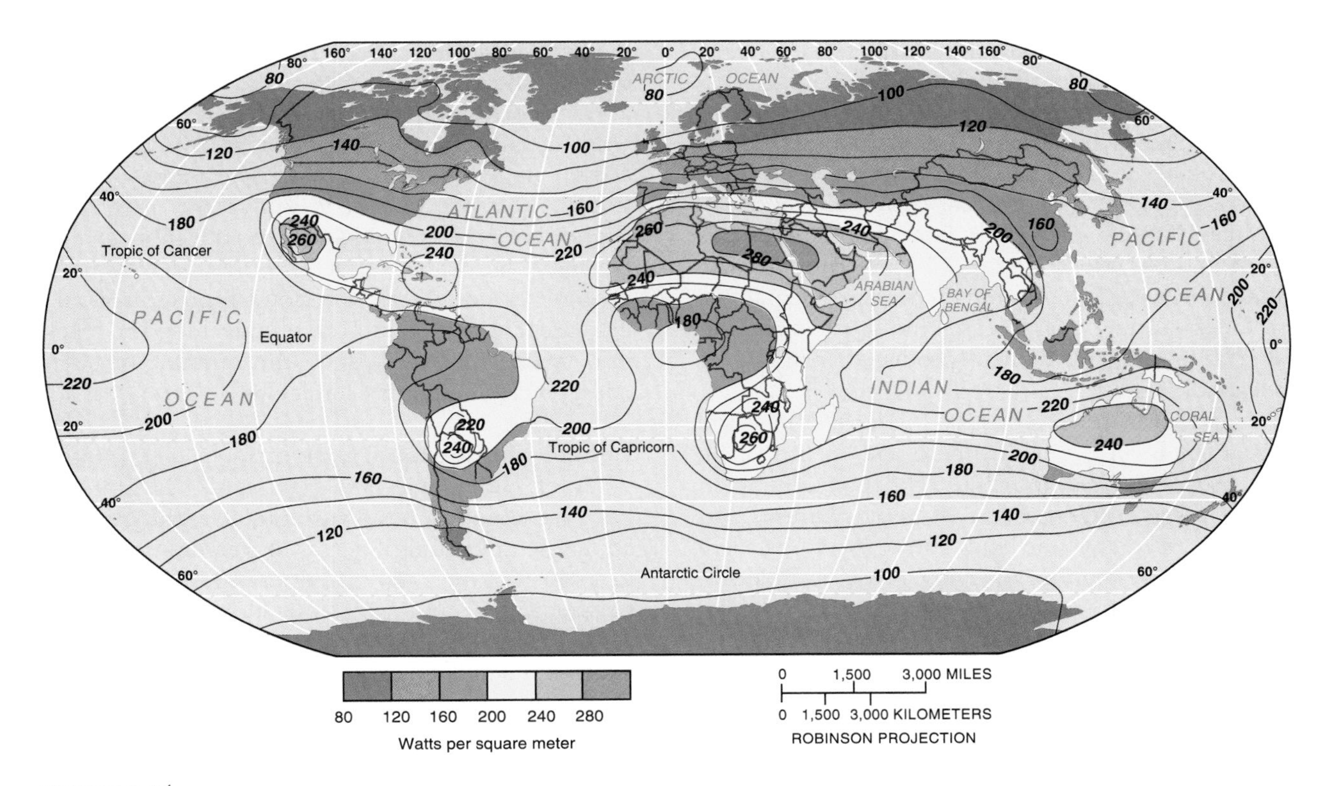

FIGURE 2–14
Mean annual shortwave radiation arriving at Earth's surface. The level varies widely with latitude and atmospheric conditions. Arid areas such as the Sahara of northern Africa receive large amounts of radiation because of the lack of clouds. In the equatorial zone and in some parts of the mid-latitudes, clouds reflect much energy, reducing amounts of energy received at the surface. (From R. W. Christopherson, *Geosystems,* 2d ed., Macmillan Publishing Co., 1994)

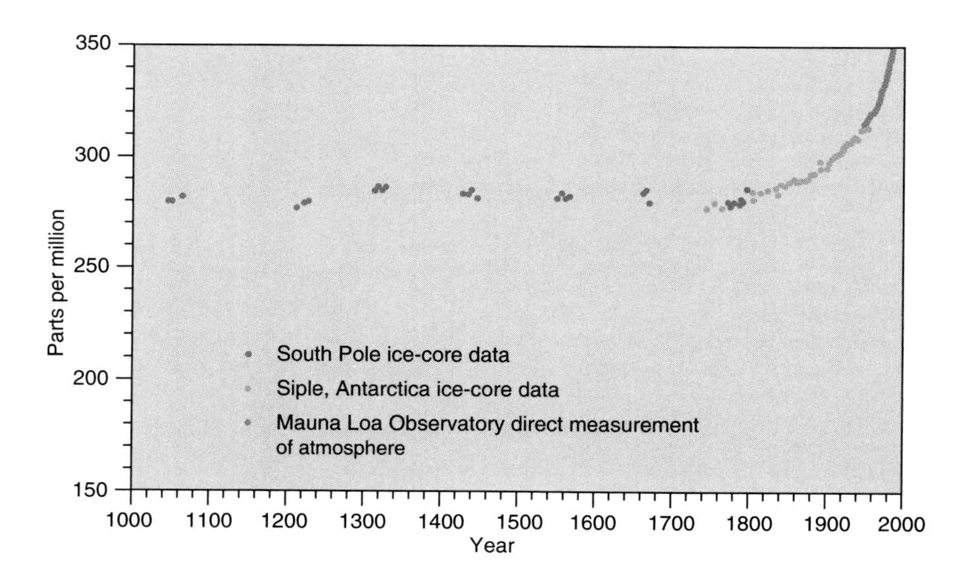

FIGURE 2–15
Concentration of carbon dioxide in the atmosphere. Carbon dioxide concentrations are now about one-third higher than two centuries ago, almost entirely as a result of burning fossil fuels in factories, cars, and homes.

energy (see Figure 2–11). Not all wavelengths in the longwave range are absorbed: the atmosphere is transparent at between about 8 and 12 microns. This gap acts as an "open window in the greenhouse," allowing large amounts of heat to escape.

Concentrations of greenhouse gases are increasing because of human activity. Carbon dioxide is of particular concern. Its major source is burning of fossil fuels, including coal, petroleum, and natural gas. Fossil fuels contain large quantities of carbon, which when burned is returned to the atmosphere as CO_2. Among the three principal fossil fuels, burning coal releases the greatest amount of carbon per unit of heat produced; natural gas releases the least.

In addition to combustion of fossil fuels, the CO_2 content in the atmosphere also is increasing because of *deforestation*. CO_2 is continually taken up by plants in the process of photosynthesis and released as plants die and decay. As forests are destroyed the number of living plants is reduced. Agricultural activity can also increase CO_2 in the atmosphere by reducing the amount of organic matter in soil. However, the effect of changing vegetation and soils on CO_2 generation is much smaller than the release of carbon through fossil-fuel burning.

Methane (CH_4) is formed by decaying organic matter, such as in sanitary landfills and the digestive systems of cattle. The atmospheric concentration of methane has been increasing in the past few decades, probably as a result of a rapid growth in the numbers of cattle, which humans are breeding and raising for food.

Scientists do not agree about the precise impacts on global climate from increased concentrations of greenhouse gases. The expectation that atmospheric temperatures will rise around the world is based on the fact that greenhouse gases absorb more heat as they become more concentrated. The prediction is uncertain, however. Higher temperatures might evaporate more water, placing more water vapor in the air and resulting in more clouds. Increased cloud formation would increase Earth's albedo, or energy reflected back into space, and so reduce the amount of solar energy reaching Earth's surface.

Debate about global warming. Determining whether global warming has started is difficult, because climate is so variable. Global average annual temperatures increased steadily from 1880 to 1940 (Figure 2–16). Temperatures fell slightly between 1940 and 1975, but have risen sharply since 1975.

Some scientists conclude that temperature varies randomly from year to year. Others argue that global warming has been steady during the past century, with the exception of a few cool intervals that possibly have been associated with major eruptions of volcanoes.

If global warming truly is in progress, you will feel some of the effects in your lifetime. Global warming would slowly increase the melt-

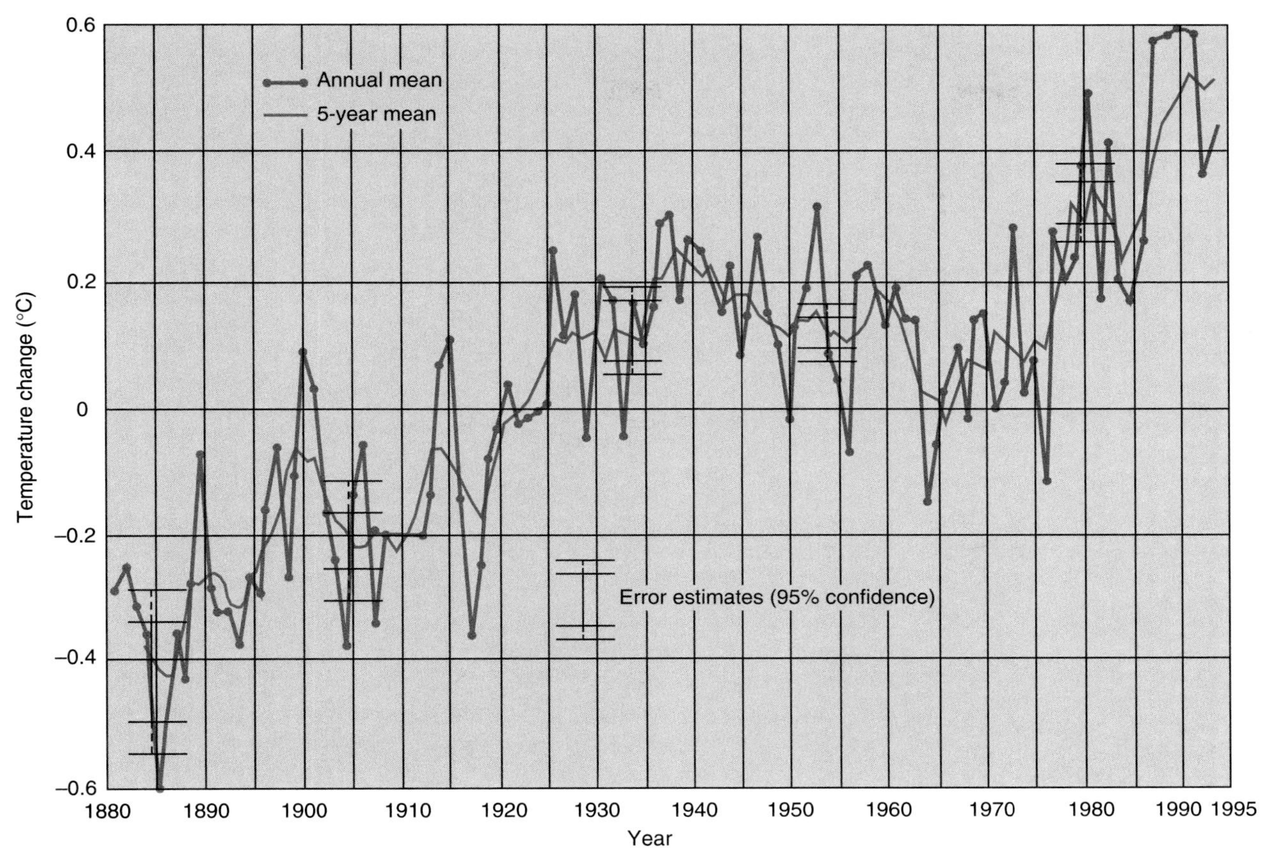

FIGURE 2–16
Annual mean temperatures for the entire Earth's surface. Global mean temperatures have risen by about 1C° (1.8F°) during the twentieth century. However, scientists disagree about the cause of the trend. Some conclude that annual temperatures have varied randomly during the twentieth century, but others argue that the graph displays evidence of global warming. (From R. W. Christopherson, *Geosystems,* 2d ed., Macmillan Publishing Co., 1994)

ing of the polar ice caps and glaciers worldwide, gradually raising sea level. Parts of many coastal cities eventually would be inundated—although residents would have plenty of time to respond.

Global warming also could change flow patterns in the oceans and atmosphere, shifting climates worldwide. Shifts in climates would affect ecosystems, agriculture, and housing needs. Some humid areas could gradually convert to deserts and vice versa. Regions that produce much of the world's food could become less productive, while other regions could become more fertile. A wealthy country could become poorer as its food production declines, while a poorer country could enjoy newfound wealth.

Precipitation

People rely on precipitation to be "normal." Normal rains and snowmelt are necessary for consistent agriculture, to feed Earth's nearly 6 billion humans. People who live on lowlands along rivers rely on "normal" precipitation to keep the river within its banks and out of their basements. Plants and animals are adapted to a "normal" amount of moisture for their environment. However, "normal" doesn't always happen.

During the late 1980s, California suffered its most severe drought in recorded history. Reservoirs were drained, domestic water use was

FOCUS BOX 2-1

Ozone—Our UV Shield in the Stratosphere

Most oxygen molecules are made of two oxygen atoms (O_2). A rarer form, called **ozone,** combines three oxygen atoms into each molecule (O_3). In the atmosphere near Earth's surface, ozone is generated by lightning and other electric sparks. At ground level, ozone is not beneficial to us, being highly corrosive, irritating mucus membranes, and damaging some materials.

Ozone also exists in the upper atmosphere, where it is essential to protecting life on Earth. It forms a rarefied "ozone layer" between 18 and 50 kilometers (10 to 31 miles) elevation. Solar energy helps form the ozone in this layer. The ozone layer is beneficial because it absorbs incoming ultraviolet (UV) radiation. Without this shield, intense solar UV radiation would bombard Earth, damaging living cells. UV exposure is one cause of increasing skin cancer in people.

The amount of ozone in the ozone layer is decreasing. Satellites monitoring the layer's thickness reveal that it has been thinning over Earth's poles, actually creating a "hole" over the South Pole. This thinning is a serious problem, and some scientists are predicting a substantial increase in skin cancers. The *UV Index* in television weather reports is a response to this concern, warning people to stay inside during the daily maximum insolation period, or to use a "UV-blocking" cream.

Why is the ozone layer thinning? Pollutants are contributing to chemical reactions that break down ozone into ordinary O_2 oxygen. The most important pollutants are **chlorofluorocarbons (CFCs),** which are large synthetic molecules of chlorine, fluorine, and carbon atoms. CFCs have remarkable thermal properties that make them valuable as coolants in air conditioners, refrigerators, and in many industrial products. Unfortunately, once discharged into the air, these molecules gradually rise into the ozone layer, where each molecule may break up 10,000 molecules of ozone.

Because of this international problem, CFCs manufacture is being banned. But the ban is not yet universal; it is being phased in to allow manufacturers time to develop replacement refrigerants. CFC molecules will continue to be pumped into the atmosphere for years, although at a lower rate. Those CFCs already in the ozone layer will remain active for decades.

restricted, and irrigation was curtailed. Similarly, much of the U.S. Midwest suffered a severe drought in 1988 that heavily damaged crops and lowered the level of the Mississippi River so much that barges became stranded. In 1990 and 1993, the same regions experienced record wet years, with floods on the Mississippi, Missouri, and other rivers. The southeastern United States had a drought in 1991 and exceedingly wet summers with widespread flooding in 1991 and 1994.

Such extremes of precipitation are rare during a typical human lifetime, but actually occur quite frequently viewed over thousands or millions of years. And, at any given time, while most places are experiencing "normal" precipitation, some place on Earth is experiencing unusual precipitation. Because variations in the amount of rainfall sooner or later affect us all, unusually wet or dry periods make headlines.

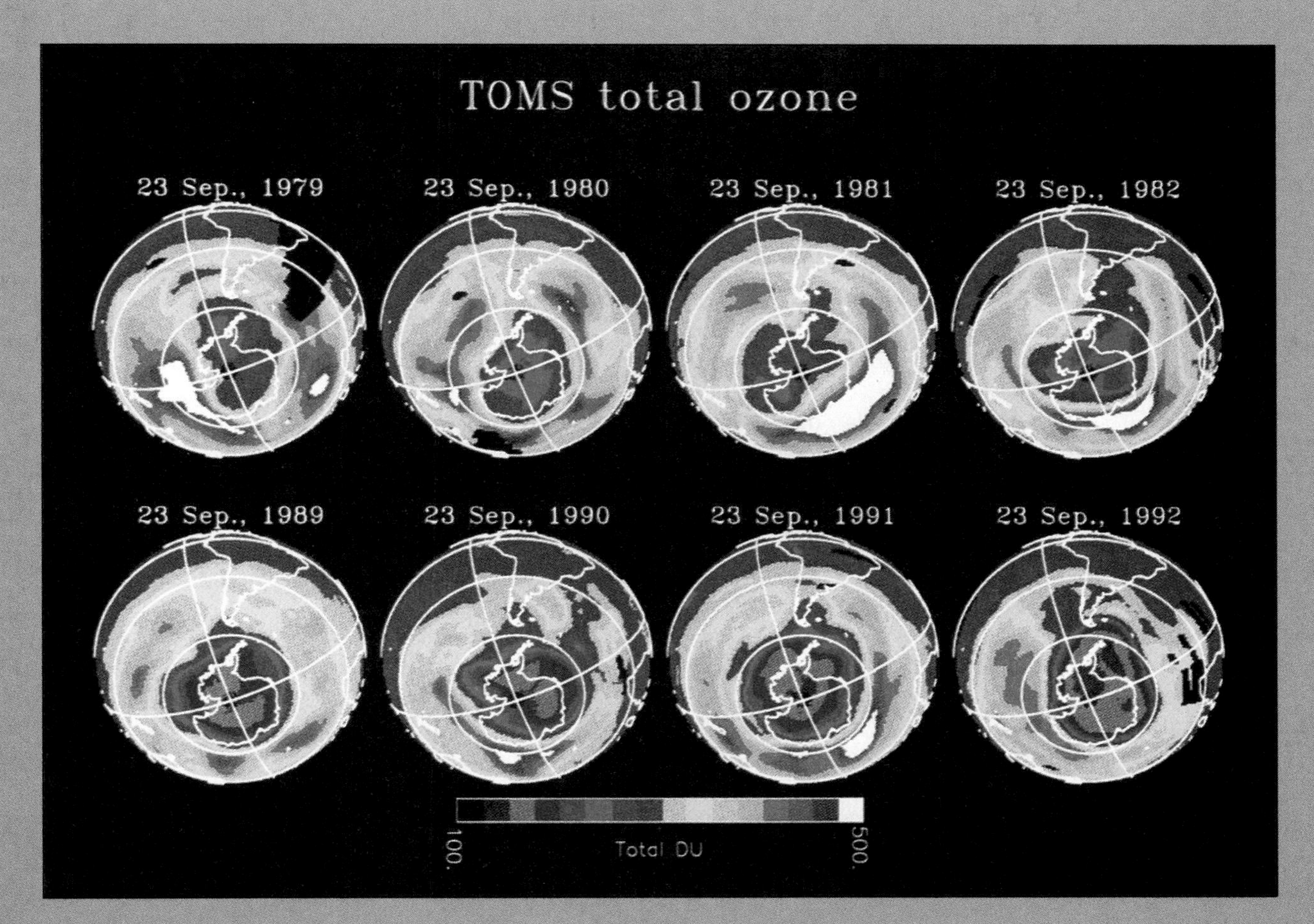

Data from the Total Ozone Mapping Spectrometer (TOMS) aboard a weather satellite reveal the deepening of the Antarctic ozone hole between 1979–82 (top row) and 1989–92 (bottom row). Ozone concentrations over Antartica vary seasonably in response to variations in isolation. The hole, indicated by blue and purple shading, is most prominent in September, the Antartic spring. Ozone depletion has also been observed in the mid-latitudes, and in the Northern Hemisphere, although not as frequently or severely as over Antartica. (Images courtesy of Goddard Space Flight Center, NASA)

Humidity and Condensation

From the standpoint of Earth's energy budget, precipitation is both a *part* of energy flow and a *consequence* of it. Recall the "conveyor belt" of rising air that carries water vapor, with its latent heat, high enough to become cooled, condense, and cause precipitation. Precipitation is part of the flow of energy, carrying the latent heat aloft and liquid water back to Earth. To understand this, let's look at humidity and condensation.

Air contains water in gaseous or vapor form. Air may hold very little water vapor, as in dry desert air, or it may be filled with water vapor, as in a steamy jungle. Although a lot of space exists between gas molecules in the air, the room available for water vapor molecules is limited. This limit varies according to air temperature. We measure the water vapor content of air by the *pressure* that the water molecules exert. The maximum amount of water vapor that air can hold is called the **saturation vapor pressure** (Figure 2–17).

Precipitation begins when moisture in the air exceeds this amount.

Relative humidity tells us how "wet" air is. Relative humidity is the *actual* water content of the air expressed as a percentage of how much water the air *could* hold at a given temperature. For example, if a 30°C (86°F) sample of air contains half the water vapor that it could hold at that temperature, its relative humidity is 50 percent. But, if cooled to 22°C (71°F), that same air sample with the same amount of water vapor would be three-fourths saturated, at 75 percent relative humidity. And if cooled to 15°C (60°F), that same air sample would be saturated, at 100 percent relative humidity.

When air cools, its relative humidity rises, and if the air is cooled enough, it can become saturated. If cooled still further, it becomes *supersaturated* and contains more water than it can hold in a vapor state. Condensation results from supersaturated air.

Condensation in the atmosphere produces clouds. Clouds form when water vapor molecules condense around tiny particles of dust, sea salt, pollen, and other impurities. They contain small droplets of liquid water or particles of ice—depending on temperature—that are too light to fall. But if these water droplets or ice particles continue to grow as more moisture from the air condenses on them, eventually they become too large to be supported by air currents, and they fall to the surface as rain or snow.

Relative humidity tends to fluctuate daily with changing air temperatures, because the amount of water vapor in the air holds steady but the daily temperature rises and falls. Thus, relative humidity typically is lower in the warm afternoon and higher at nighttime.

Precipitation Caused by Air Movement

The movement of air causes precipitation in three ways:

- *Convection*—in which air warmer than its surroundings rises, expands, and cools by this expansion.
- *Orographic uplift*—in which wind forces air up and over mountains.
- *Frontal uplift*—in which air is forced up a boundary (front) between cold and warm air masses.

Precipitation from convection. On a warm, humid summer day, the sky is clear in the morning, and the sun is bright. The sun warms the ground quickly, and the air temperature rises. Most of the warming of the air takes place close to the ground, because the humid air is a good absorber of longwave radiation, which is being reradiated from the ground.

As the air near the ground warms, it expands, becomes less dense, and rises through the sur-

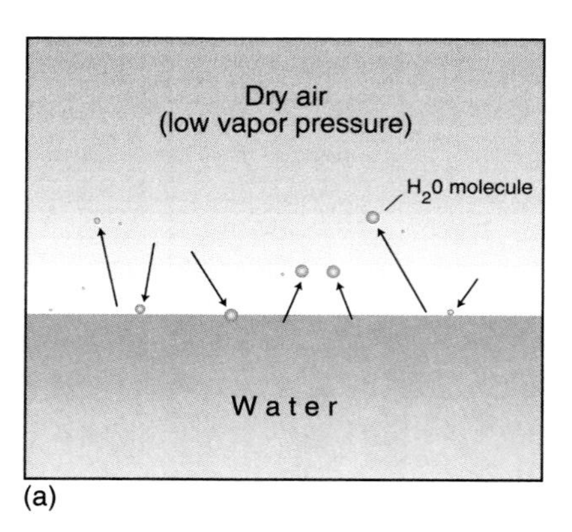

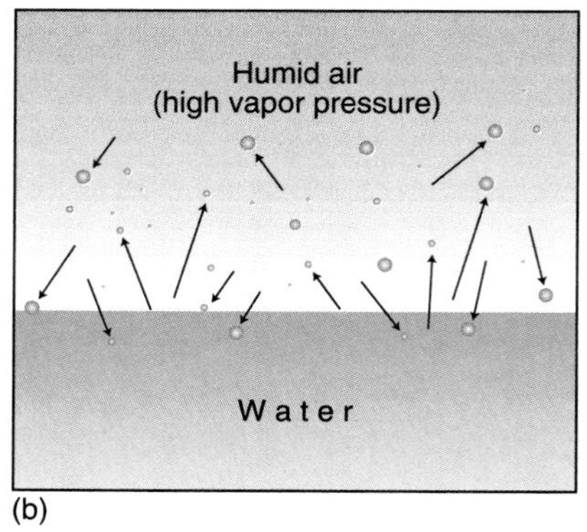

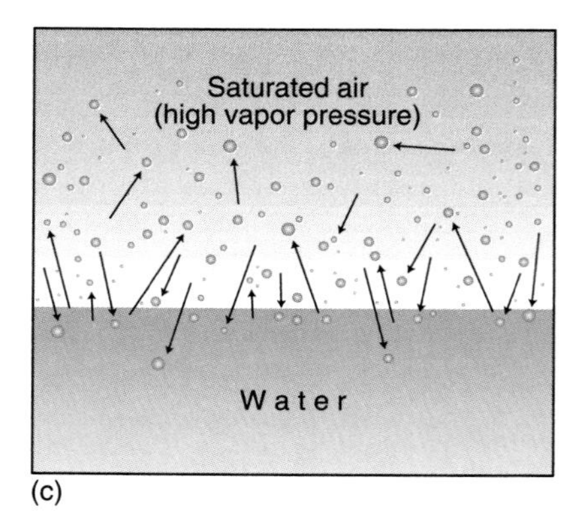

FIGURE 2–17
Vapor pressure. The arrows show movements of molecules of water.
(a) Dry air holds very little water vapor.
(b) As heat energy is added to water, more of its molecules become energetic enough to break free of the liquid water and enter the atmosphere.
(c) When air becomes saturated with water vapor (100 percent humidity), a continuing exchange of water molecules occurs between the air and water.

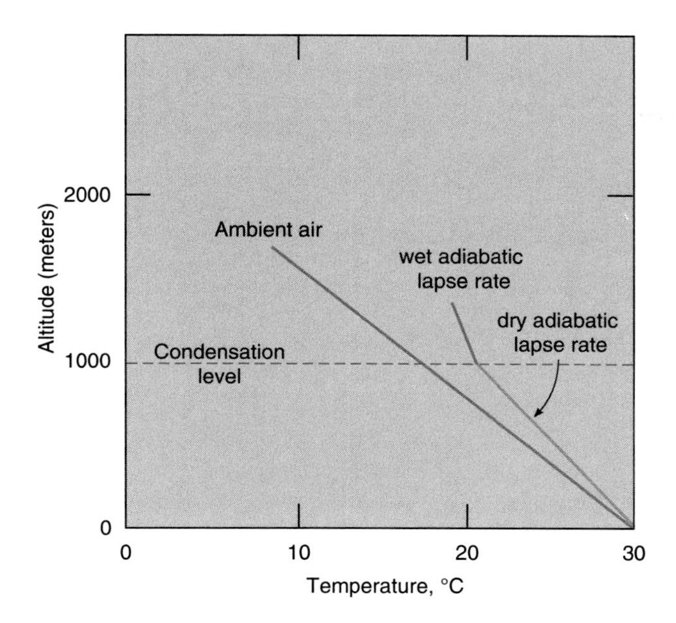

FIGURE 2–18
Adiabatic cooling. As air rises, it cools at a rate of about 1°C per 100 meters (5.4°F per 1,000 feet). The temperature of the rising parcel of air relative to the temperature of the air around that parcel determines its behavior. If the parcel is warmer than surrounding air, it tends to rise, and therefore is unstable. If it is the same temperature or cooler, it resists movement, and therefore is stable. In this example, the air parcel is warmer, and thus unstable. Condensation in the air above 1,000 meters (3,300 feet) elevation causes the air to be warmed by the release of latent (or stored) heat as sensible heat (which can be felt). This further contributes to the instability of the air. Intense thunderstorms can form under such conditions.

rounding cooler air above, like a hot air balloon. Convection is in progress, conveying humid air higher into the atmosphere.

Because air is a gas, it is compressed by the weight of overlying air. When it rises, air has less weight above it, and the lower pressure allows the air to expand. Compressing a gas causes an increase in temperature, while expanding it causes a decrease. The decrease in temperature that results from expansion of rising air is called **adiabatic cooling;** the word *adiabatic* means *"without heat being involved."*

As it rises, the air rapidly cools adiabatically (by expansion) at a rate of about 1°C for each 100 meters (5.4°F per 1,000 feet) elevation. When the air reaches several hundred meters above the ground it has cooled to the point where it is saturated, and clouds begin to form (Figure 2–18).

Rising warm air may mix with the surrounding cooler air, slowing the convection. If the convection were driven only by heating at the ground surface, it would probably not be strong enough to cause water droplets to grow big enough to fall as precipitation.

But as soon as condensation begins, latent heat—another important source of energy—is released. Latent heat further warms the rising column of air and makes it less dense than the surrounding cool air. The process is self-reinforcing, as the cloud grows rapidly, with strong vertical motion and rapid condensation. The result can be the gusty winds and intense rain of a thunderstorm.

The Sierra Nevada mountains in California, seen from the east. Because of orographic effects, the mean annual precipitation near the crest of the Sierras exceeds 100 centimeters (40 inches). But in the foreground, Owens Valley is in a rain shadow, and has mean annual precipitation of less than 30 centimeters (12 inches). (Michele Burgess/The Stock Market)

Thunderstorms are responsible for a large portion of the world's precipitation. In tropical climates, where strong insolation makes temperatures high, all that is needed for intense daily thunderstorms is a source of humidity. In mid-latitude climates, thunderstorms occur mostly in the summer, because higher temperatures allow the air to hold the moisture needed for latent heat to add to strong convection. Thunderstorms are especially common in situations where some other factor is present to favor uplift, the trigger that starts the self-reinforcing growth of the storm. Mountains and fronts, especially cold fronts, often provide the trigger.

Precipitation from orographic uplift. Precipitation occurs from **orographic uplift** when the wind forces air to rise over mountains. As the air rises it cools adiabatically (by expansion), and precipitation results. After air has moved up the *windward* side of a mountain and over the top, it

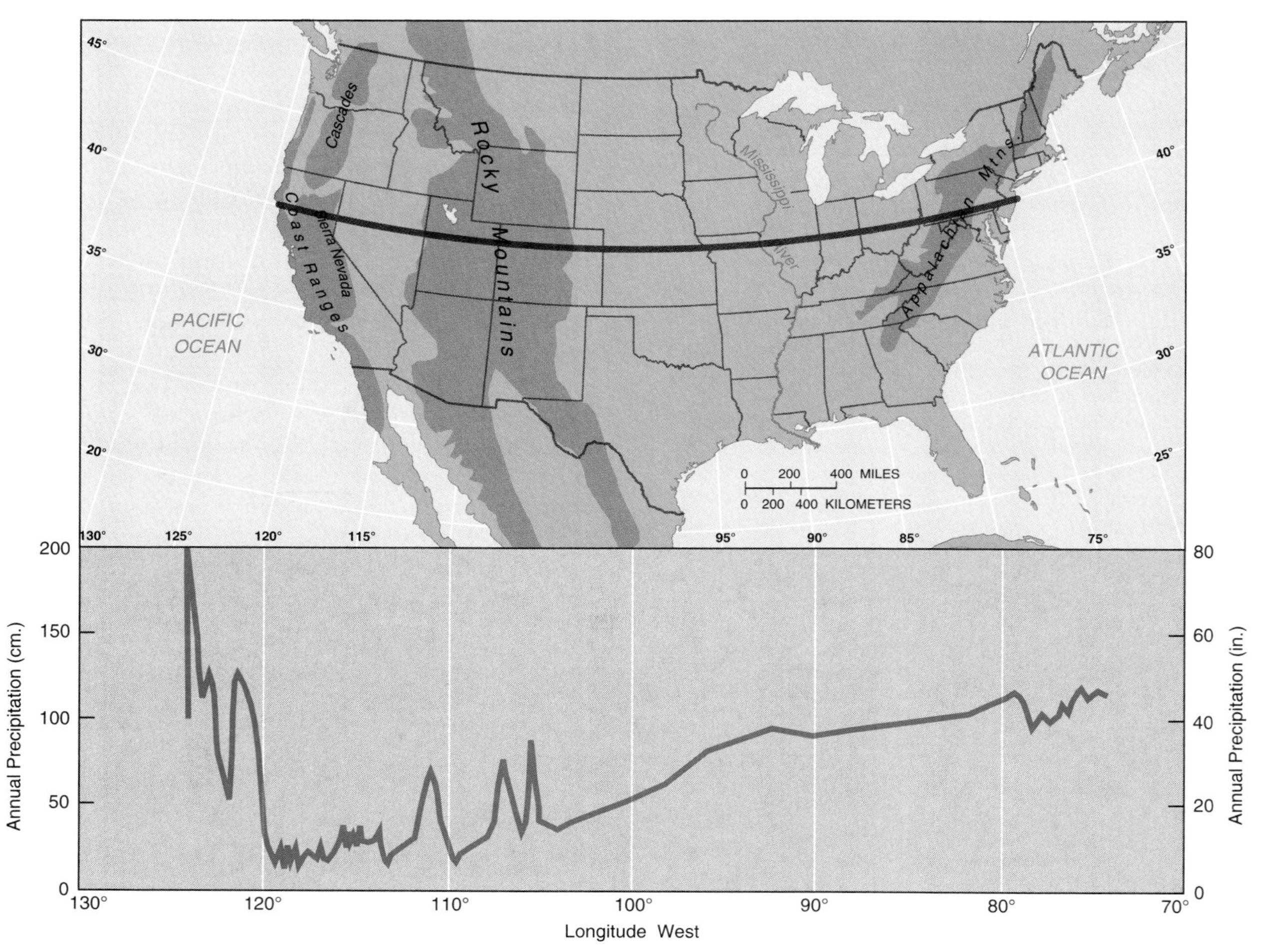

FIGURE 2–19
Cross-section of elevation and precipitation in the United States along latitude 40°N. In the western United States, high rainfall totals occur near the coast, where orographic lifting causes rain on the western side of coastal mountain ranges. Rainfall amounts are much less on the east side of the Sierra Nevada mountains, because most of the available water has been removed from the air by orographic lifting and precipitation. Another peak of rainfall is seen over the Rocky Mountains, with low rainfall to the east of that range, for the same reason. Higher rainfall totals in the eastern United States are possible because these areas receive moisture from the Atlantic Ocean and the Gulf of Mexico. A small orographic effect is evident in the Appalachian Mountains.

then descends on the *leeward* side. As it does so, its relative humidity drops significantly. The leeward side of a mountain range is often much dryer than the rainy windward side. A dry region on the leeward side of a mountain range is called a *rain shadow*. Some of the world's major deserts are arid because they are situated on the leeward side of a mountain range.

The western United States provides an excellent example of orographic effects on rainfall (Figure 2–19). Moist air from the Pacific Ocean, the region's major source of moisture, travels from west to east, producing precipitation as it rises first over coastal mountain ranges in Washington, Oregon, and California, and then the Sierra Nevada Mountains. Some of the highest average U.S. rainfall totals occur in the Sierras, but east of the mountain range lie the rain-shadow deserts of Nevada and Utah. The region is arid because it is cut off from Pacific moisture by the Sierras and from Atlantic moisture by the Rocky Mountains.

As you can see, orographic precipitation also occurs in the Rocky Mountains, but the amount is less than in the Sierras because the Rockies are more isolated from moisture sources. Immediately east of the Rocky Mountains is another rain shadow area, the Great Plains. Crossing the Plains, precipitation generally increases eastward as moist air is more likely to move in from the Gulf of Mexico.

In the eastern United States, air is forced to rise over the Appalachians, producing another orographic effect. The Gulf of Mexico and Atlantic Ocean supply moisture for ample precipitation east of the Mississippi Valley. However, the eastern side of the Appalachians does not experience a rain shadow, because the region has moisture sources on both sides: the Gulf of Mexico to the southwest and the Atlantic Ocean to the east.

A basic pattern in what you just read is that weather systems move from west to east across the United States, forming moister western slopes of mountains and dry rain-shadow areas to their east. Why do weather systems move from west to east? We will answer this important question when we discuss global circulation later in this chapter.

Precipitation from frontal uplift. **Frontal precipitation** forms along a front, which is a boundary between two air masses. An **air mass** is a large region of air—millions of square kilometers—with distinctive characteristics of temperature, pressure, and humidity. An air mass acquires these characteristics from the land over which it forms.

In North America, air masses that form over central Canada tend to be cool (because of Canada's relatively high latitude) and dry (because of the region's isolation from moisture sources). This is called a *continental polar* air mass. In contrast, air masses formed over tropical water, such as over the Gulf of Mexico, tend to be warm and moist, and are called *maritime tropical*.

When a cool air mass, such as continental polar air from Canada, meets a warm air mass, such as maritime tropical air from the Gulf of Mexico, a boundary or front may form between them. Because cool air is relatively dense, it tends to sink, while less dense warm air tends to rise (Figure 2–20). A **front** is a region characterized by ascending air, cloudiness, and precipitation.

As air masses migrate across Earth's surface, the fronts move with them. When a cold air mass advances against a warmer one, the boundary is called a **cold front.** The advancing cold, denser air wedges beneath the warm air, forcing it to rise, generating clouds and usually precipitation (Figure 2–20a). A cold front can move quite fast and generate intense thunderstorms if the warm air is sufficiently moist.

The passage of a well-developed cold front in central and eastern North America includes very distinctive weather. Before the front arrives, the air is warm and moist, with the wind typically from the south or southwest. As the front arrives, intense precipitation falls, and in the summer thunderstorms usually form. As the front passes, the wind shifts to the west or northwest and the temperature falls quickly. Then the sky clears, and cool dry air arrives, with blue skies and bright sunshine.

In the opposite situation, when a warm air mass advances against cooler air, the boundary is a **warm front.** The warm air rides up over the cool air as though it were a gentle ramp (Figure 2–20b). Precipitation along a warm front is less intense than along a cold front, and a warm front often passes without significant precipitation. In winter, a warm front sometimes causes freezing rain, as rain formed when the warm air aloft falls

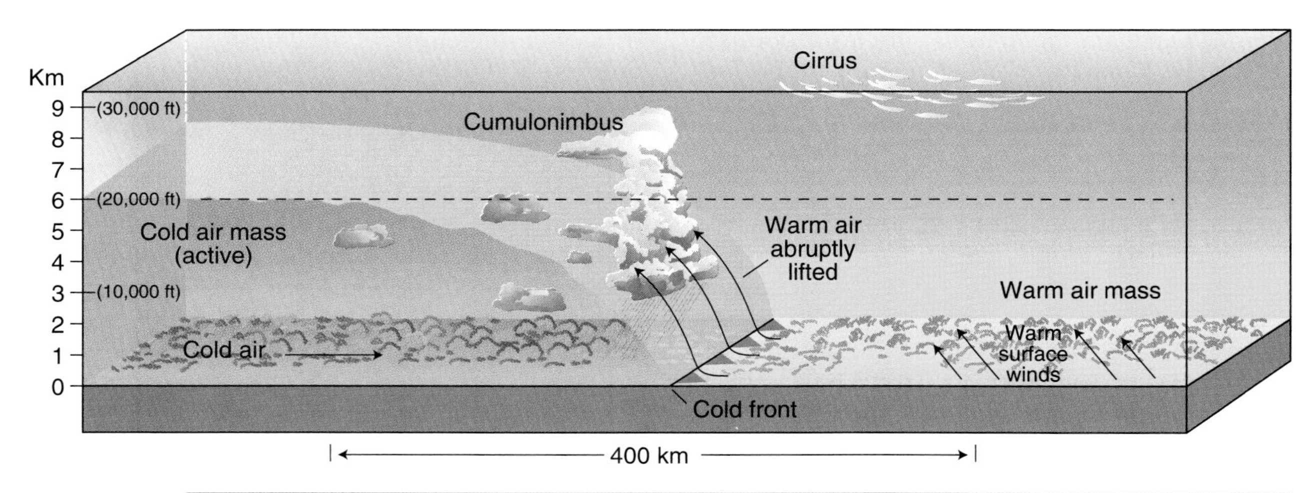

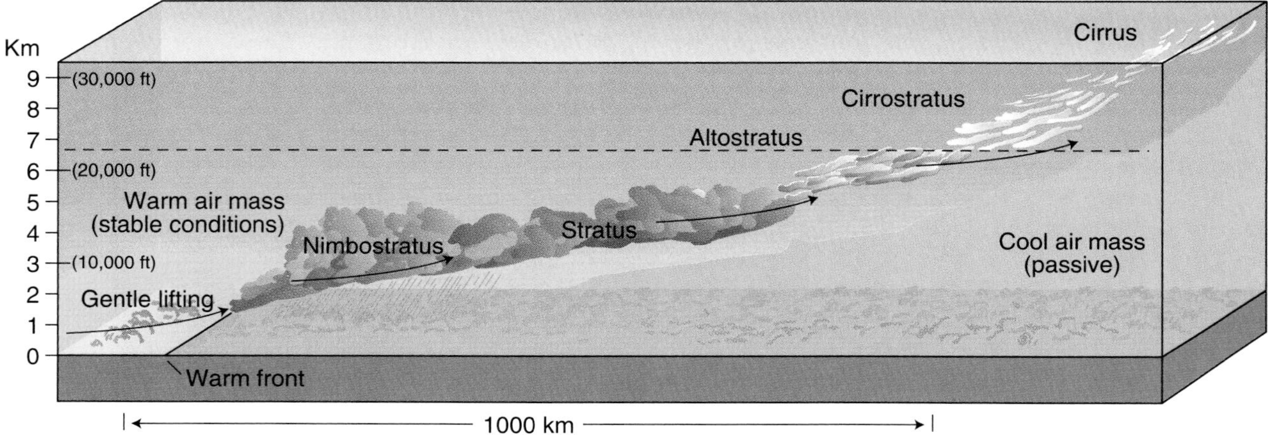

FIGURE 2–20
Fronts are boundaries separating warm and cold air.
(a) A cold front generates precipitation as the cold air drives under the warm air like a wedge, lifting the warm air.
(b) In a warm front, warm air rides up over cold air, causing precipitation.
(From R. W. Christopherson, *Geosystems,* 2d ed., Macmillan Publishing Co., 1994)

through colder air below, freezing as it reaches the ground. Air masses and the front between them may become stalled and not move, creating a **stationary front.** Note the cold, warm, and stationary fronts depicted on the weather map (Figure 2–21).

Because a cold front normally involves rapid vertical motion in warm and relatively humid air, the clouds that form along it are typically **cumulus clouds,** which are tall, puffy clouds with billowy tops. In warm fronts, the warm air rises gradually over the cold air, and the stable layering of warm air over cold produces broad, flat layers of clouds known as **stratus clouds.**

Circulation Patterns

Air is made of molecules, so air has mass. Earth's gravity attracts this mass of air to the surface, giving it weight. We think of air as a light-weight substance, but the thickness of air over your head—480 kilometers (300 miles) thickness—has tremendous weight, pressing on you and Earth's surface at an average of about 1 kilogram per square centimeter (14.7 pounds per square inch). This means that directly above each square centimeter of Earth is a column of air that weighs on average about 1 kilogram (Figure 2–22).

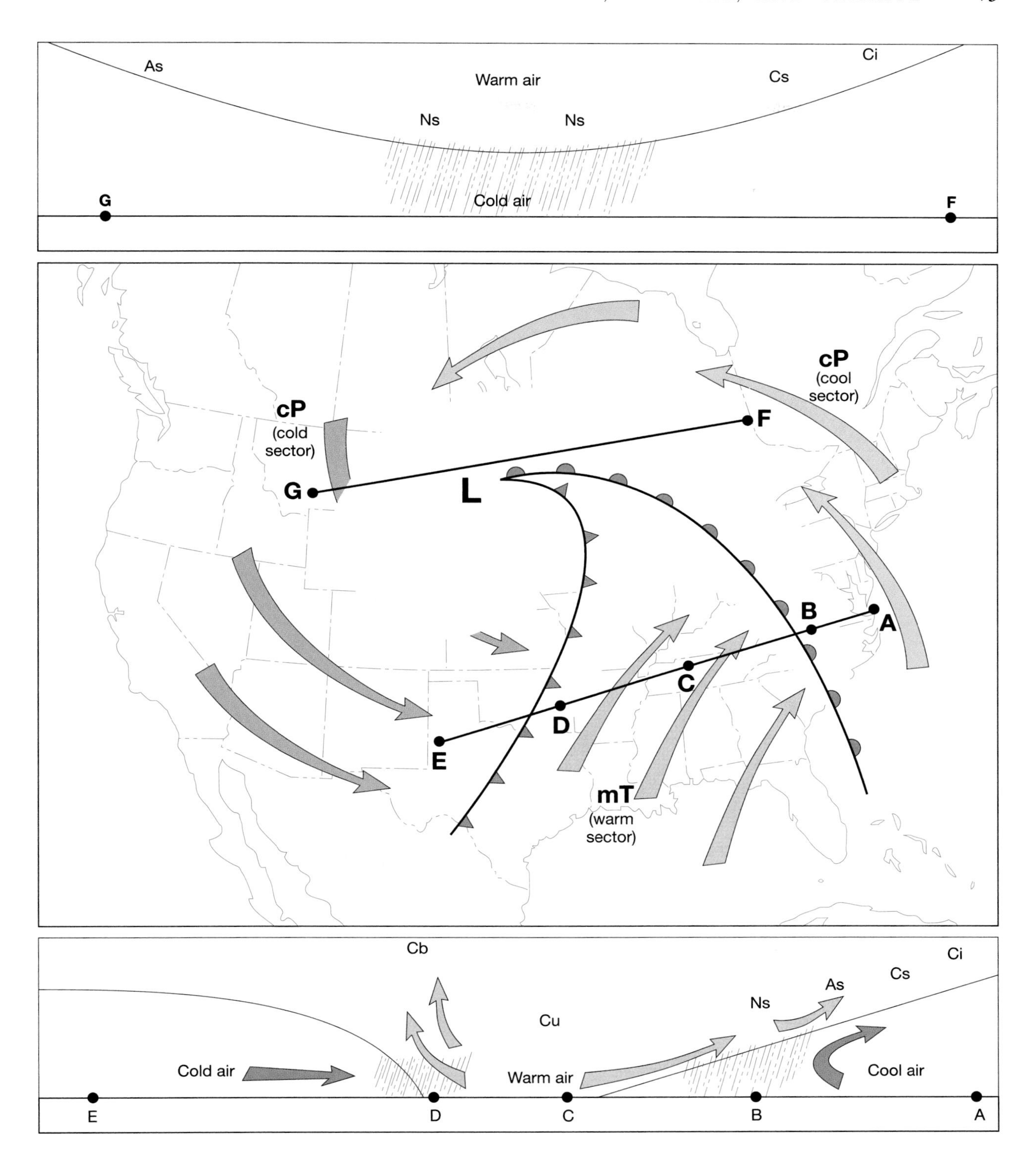

FIGURE 2–21
Patterns of winds, clouds, and precipitation around an idealized mid-latitude cyclone. (From E. J. Tarbuck and F. K. Lutgens, *Earth Science,* 7th Ed, Macmillan Publishing Co., 1994)

Cumulus clouds developing over the South Florida peninsula. The view is to the south, with the Atlantic Ocean on the left and Gulf of Mexico on the right. The land is warmer than the adjacent Atlantic Ocean and Gulf of Mexico, and this warmth stimulates convection and cloud formation. Note the absence of clouds over Lake Okeechobee, the large lake in the center of Florida, because it is cooler, like the ocean. (NASA/Science/Photo Researchers, Inc.)

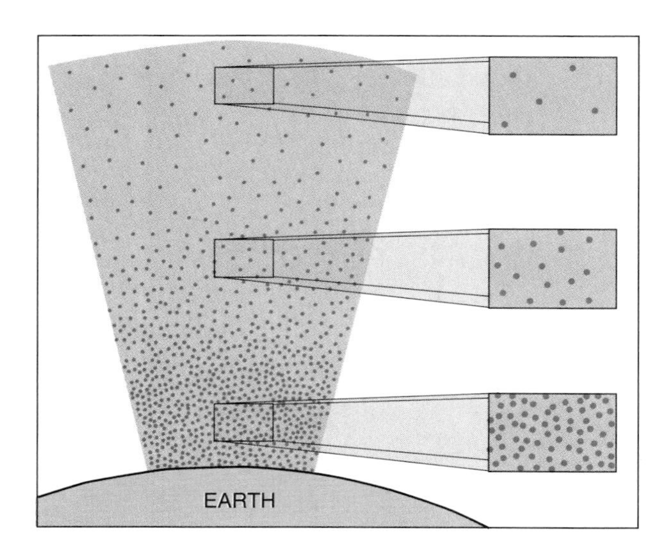

FIGURE 2–22
Visualizing the density of air (greater near Earth's surface) and atmospheric pressure. (From R. W. Christopherson, *Geosystems,* 2d ed., Macmillan Publishing Co., 1994)

Atmospheric pressure varies with altitude because the higher you go, the less air exists above you. Thus, atmospheric pressure is greater at sea level than in "mile-high" Denver or atop Mount Everest (8.8 kilometers or 5.5 miles high). Because sea level is a surface that is at virtually the same height worldwide, scientists use the average atmospheric pressure at sea level as a world standard. Atmospheric pressure is measured with a **barometer.** The average atmospheric pressure at sea level, as read on a barometer, is 1,013.2 millibars (29.92 inches of mercury).

Pressure and Winds

At any location, atmospheric pressure varies with conditions. For example, the air above the island in Figure 2–9 is warmer than the air around it, and therefore it is less dense. Because warm air is less dense, it weighs less, and the atmospheric pressure over the island is lower than the pressure over the cool sea. As the warm, lighter air over the island rises, higher pressure over the cool sea forces air

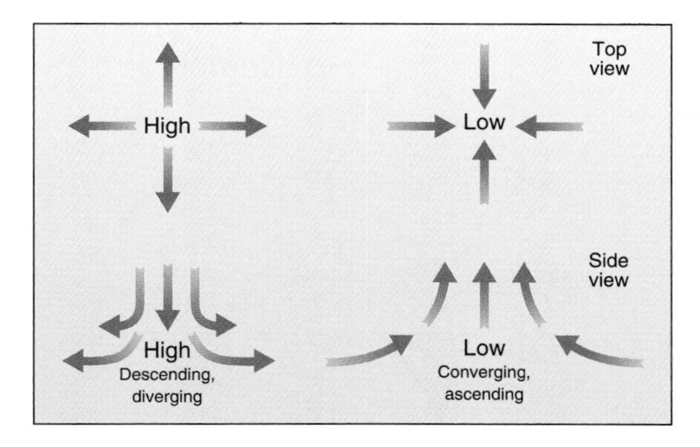

FIGURE 2–23
(a) Wind movement is partly determined by pressure. Air ascends in regions of low pressure.
(b) Air descends in regions of high pressure. Wind blows from areas of high pressure to areas of low pressure.
(From R. W. Christopherson, *Geosystems,* 2d ed., Macmillan Publishing Co., 1994)

horizontally toward the island to replace the rising warm air. The difference in pressure between the two places is the **pressure gradient.**

Differences in pressure produce wind. In Figure 2–23a, if pressure at the surface is low, air ascends vertically, and horizontal winds converge toward the area of low pressure. But if pressure at the surface is high (Figure 2–23b), air descends, and winds blow away (diverge) from the area of high pressure. Air moves from the top of a rising column of low pressure toward the top of a descending column of higher pressure air, in a complete cycle of circulation.

Coriolis effect. If Earth did not rotate, winds would simply blow in a straight line from areas of high pressure to areas of low pressure. However, on our real spinning planet, winds follow an indirect, curving path. This deflection of wind (and any other object moving above Earth's rotating surface) is called the **Coriolis effect.**

To understand Coriolis effect, imagine two people throwing a ball back and forth on the back of a flatbed truck (Figure 2–24a). Both people and the ball are moving at the same speed and *in the same direction,* so they do not change position relative to one another. The ball game is the same as if the truck were not moving.

On the merry-go-round (Figure 2–24b), when the ball is thrown in a straight line, the turning platform moves the other person away from where the ball arrives. As they throw the ball back and forth, the ball follows its expected straight-line path, but the receiver has been moved by the merry-go-round. To the players, it *appears* that the ball has followed a curving path. Similarly, as Earth rotates beneath moving air, the air *appears* to move in a curve.

Near the equator, Earth is like the flatbed truck in Figure 2–24a; the Coriolis effect is zero at the equator. But toward the poles, rotating Earth is more like a spinning disc, and the Coriolis effect is strongest in the polar regions. This is shown in Figure 2–24c, where a plane flies from the North Pole in a straight line toward the equator, but arrives far to the west of its intended destination because Earth rotated eastward during the hours that the plane was in flight.

The net result of Coriolis effect is a *deflection to the right* in the Northern Hemisphere (and a deflection to the left in the Southern Hemisphere). The effect is significant only when considerable distance is involved: a pilot must consider it in setting the flight path of a several-hour flight across the ocean, but not for a model airplane across a backyard. The Coriolis effect helps weather forecasters understand the behavior of storms.

Figure 2–25 shows the result when Coriolis effect is added to the vertical air movements in high- and low-pressure areas. It puts a spin on the winds, adding a spiral motion to create cells which rotate clockwise and counterclockwise. In the Northern Hemisphere (Figure 2–25a), high-pressure cells are rotating clockwise, and low-pressure cells are rotating counterclockwise. The opposite patterns hold in the Southern Hemisphere (Figure 2–25b). To summarize, differences in air pressure create winds, and Earth's spin curves their flow.

Global Circulation

Understanding global circulation patterns is very important. These patterns control our weather, climate, agriculture, and travel. At the global scale, atmospheric circulation operates like the convection system over the warm island we described in Figure 2–9. Beginning at the equator, we can identify four zones of circulation patterns: Intertropical convergence zone, subtropical high pressure zones, Mid-latitude low pressure zones, and polar high pressure zones.

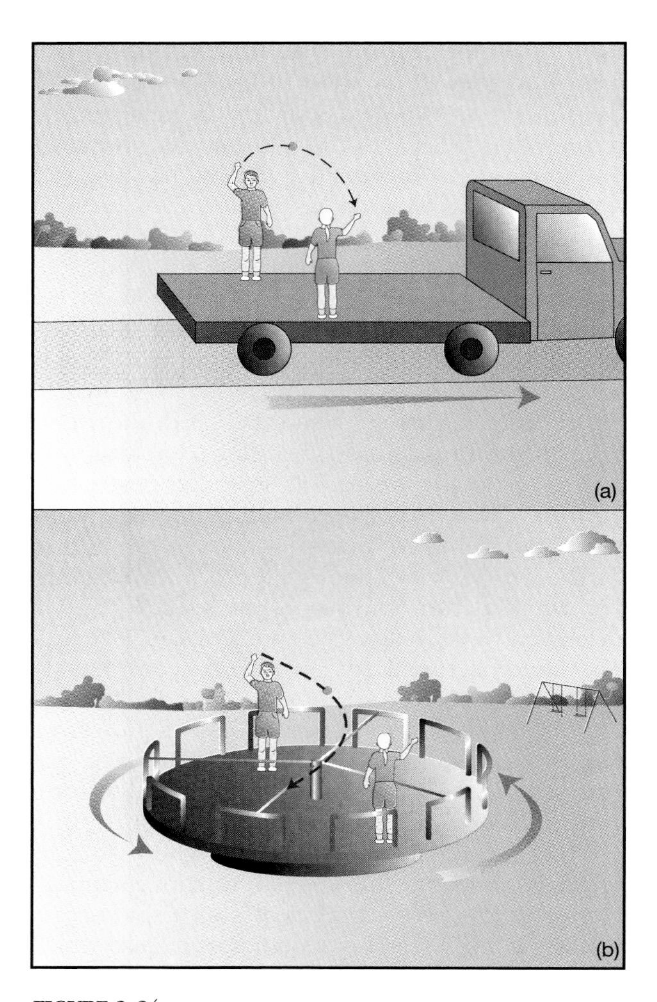

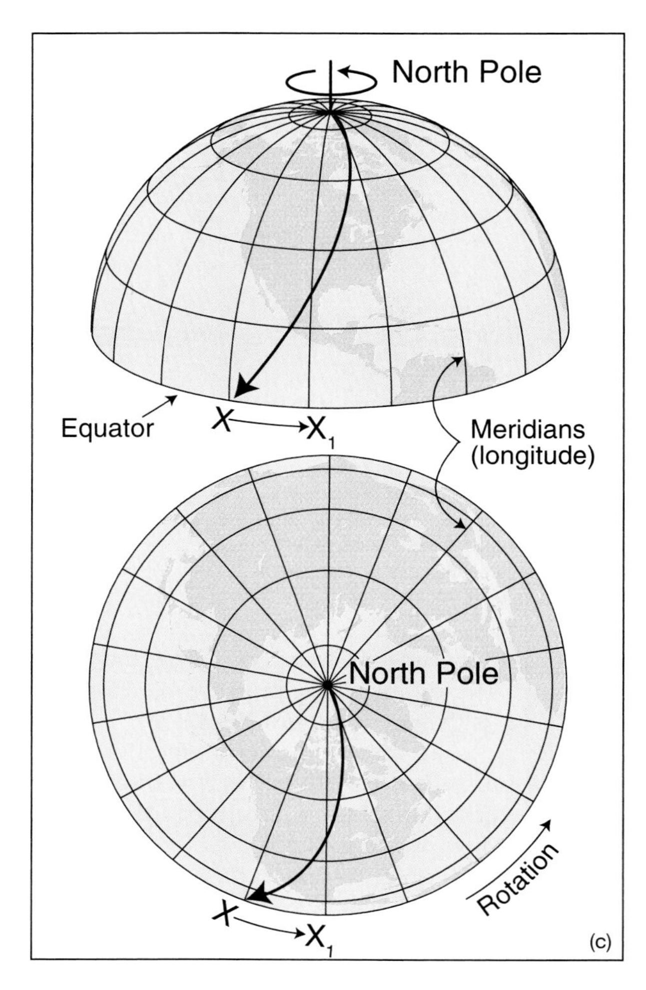

FIGURE 2–24
Coriolis effect.
(a) When two people on a moving flatbed truck throw a ball to each other, the people and the ball are moving at the same speed and in the same direction, so they do not change position relative to one another.
(b) When two people on a merry-go-round throw a ball to each other, the ball follows a straight-line path but the people have moved. The ball appears to the people to be following a curve. This apparent curving of the ball demonstrates the Coriolis effect.
(c) When an airplane flies in a straight line from the North Pole to the Equator it would actually land west of its intended destination because Earth had rotated eastward during the flight.
(From E. J. Tarbuck and F. K. Lutgens, *Earth Science,* 7th ed., Macmillan Publishing Co., 1994)

Intertropical convergence zone. In the tropics, dependable year-round inputs of solar energy heat the air, expanding it and creating low pressure. As a result, convective rising of air occurs daily above Earth's equator. This forms the **intertropical convergence zone (ITCZ),** so-called because it is a zone between the Tropics of Cancer and Capricorn where surface winds converge (Figure 2–26).

As air rises in the ITCZ, convectional precipitation occurs, usually as afternoon thunderstorms. Air converges toward the equator at the surface, replacing the rising air. The Coriolis effect deflects this moving air to the right in the Northern Hemisphere to form the Northeast Trade Winds and to the left in the Southern Hemisphere to form the Southeast Trade Winds. (Winds are named for

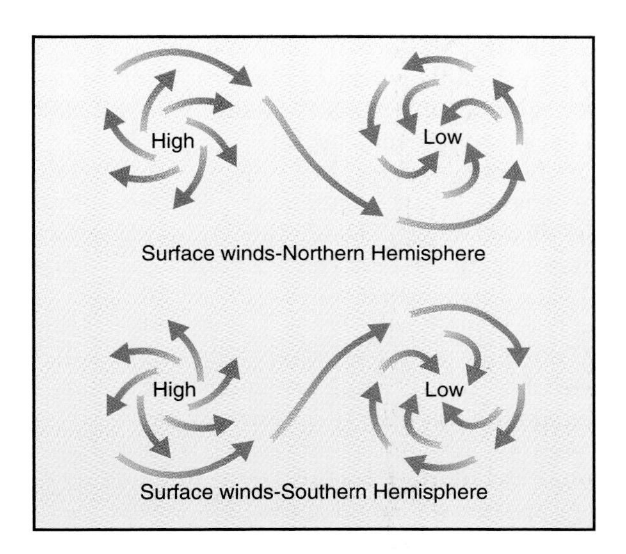

FIGURE 2–25
Wind movement with Coriolis effect added.
(a) The Coriolis effect, caused by Earth's rotation, deflects wind to the right of its expected path in the Northern Hemisphere. This causes spiraling circulation around high- and low-pressure centers, in the directions shown.
(b) The same pattern reversed for the Southern Hemisphere. (From R. W. Christopherson, *Geosystems,* 2d ed., Macmillan Publishing Co., 1994)

the direction from which they blow.) Aloft, air circulates away from the ITCZ both northward and southward, toward subtropical latitudes. This air has lost most of its moisture in the daily rainfalls, and it is now warm and dry.

Subtropical High Pressure Zones. Warm, dry air that spreads poleward from the ITCZ descends at about 25° north and south latitudes. This creates zones of high pressure that are especially strong over the oceans. These **subtropical high pressure zones** are areas of dry air, bright sunshine, and little precipitation.

This descending dry air associated with subtropical high-pressure cells creates an arid climate, so most of the world's major desert regions are on land in this zone, at about 25° north and south latitudes. Further, most are on the western edges of continents, including deserts in northern Mexico and southwestern United States, the Atacama in Peru and Chile, the Sahara in northern Africa, the Kalahari in southern Africa, and the Australian desert. The eastern sides of these continents are under the same high-pressure cells, but they tend

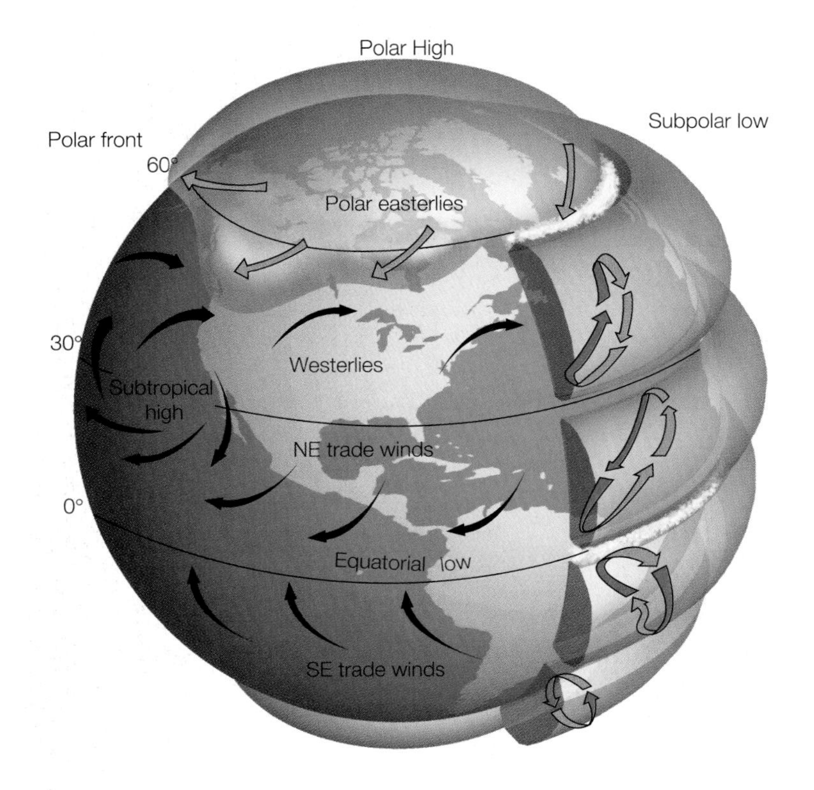

FIGURE 2–26
General circulation of the atmosphere. This simplified illustration of average atmospheric circulation patterns shows the major circulation zones by latitude: intertropical convergence zone (ITCZ), subtropical high-pressure zones, mid-latitude low-pressure zones, and polar high pressure zones. (From E. J. Tarbuck and F. K. Lutgens, *Earth Science,* 7th ed., Macmillan Publishing Co., 1994)

to be more humid, because the circulation brings in warm, humid tropical air.

On the poleward sides of the subtropical high-pressure cells, circulation is toward the poles. But, these winds are deflected by the Coriolis effect, so winds prevail from the southwest in the Northern Hemisphere and from the northwest in the Southern Hemisphere.

Mid-latitude low pressure zones. Poleward of the subtropical high-pressure zones are the **mid-latitude low pressure zones.** These lower-pressure areas experience convergence of warm air blowing from subtropical latitudes and cold air blowing from polar regions. The warm and cold air masses collide in swirling low-pressure cells that move along the boundary between the two air masses, which is known as the ***polar front.*** Winds in these regions generally blow from west to east, and are therefore called *westerlies*.

Polar high pressure zones. In the polar regions, the intense cold caused by meager insolation creates dense air and high pressure. Because the air is so cold, the **polar high pressure zones** contain very little moisture, and convection and precipitation are limited. (If you wonder where all the snow and ice around the poles comes from when the regions receive so little snowfall, it results from snow accumulating over many thousands of years.)

Seasonal Variations in Global Circulation

Global circulation patterns vary with the seasons. We can illustrate typical conditions during January and July (Figure 2–27).

Global circulation in January. During January, Earth's average atmospheric pressure and winds are arranged into broad zones according to latitude (Figure 2–27a). Conditions in the Northern and Southern Hemispheres roughly mirror each other. The ITCZ is plainly visible, generally at about 15° to 20° south latitude.

High-pressure cells predominate in the subtropical regions just north and south of the ITCZ, especially over the subtropical oceans. The Southern Hemisphere, where it is summer in January, displays the clearest tendency for high pressure over the oceans and low pressure over land. The Northern Hemisphere has a high pressure region in eastern Asia, as well as in the eastern Pacific and Atlantic oceans.

In the mid-latitudes, less continuous low-pressure regions in the Northern Hemisphere appear most clearly over oceans. The mid-latitude low pressure zone is much more consistent in the Southern Hemisphere, because of the absence of land between 40° and 70°. Such a monthly average map obscures the presence of storms, but it is useful to show the general condition of the atmosphere during Northern Hemisphere winter.

Global circulation in July. By July, January's major pressure and wind zones have moved northward. In some cases they have moved from being over land to water areas, or vice versa (Figure 2–27b). The ITCZ, which was south of the equator in January, is almost entirely in the Northern Hemisphere during July, as far north as 30°N latitude in southern Asia. The subtropical highs are still present over the oceans, although they are strengthened in the Northern Hemisphere and weakened in the Southern.

The most notable change in the subtropics and mid-latitudes is the replacement of high pressure over Asia with low pressure. In fact, this region has the world's highest average pressure during January, and the world's lowest average pressure during July. As a result, wind directions in the vicinity are reversed. This seasonal reversal of pressure and wind produces a **monsoon circulation,** in which winter winds from the Asian interi-

FIGURE 2–27
Worldwide average atmospheric pressures and winds during Northern Hemisphere winter in January and summer in July.
(a) In January, the intertropical convergence zone (ITCZ) is generally south of the equator, especially over land areas. In the Northern Hemisphere, the subtropical high-pressure cells are weak or absent, a large and intense high-pressure cell dominates interior Asia, and deep low-pressure areas are over the North Pacific and North Atlantic.
(b) In July, the ITCZ has shifted to the north, and low pressure has replaced high pressure over interior Asia. Weak low pressure also is seen over North America, and the subtropical highs are prominent. In the Southern Hemisphere, the mid-latitude low-pressure belt between
40° and 60° south latitude is very strong.
(From R. W. Christopherson, *Geosystems,* 2d ed., Macmillan Publishing Co. 1994)

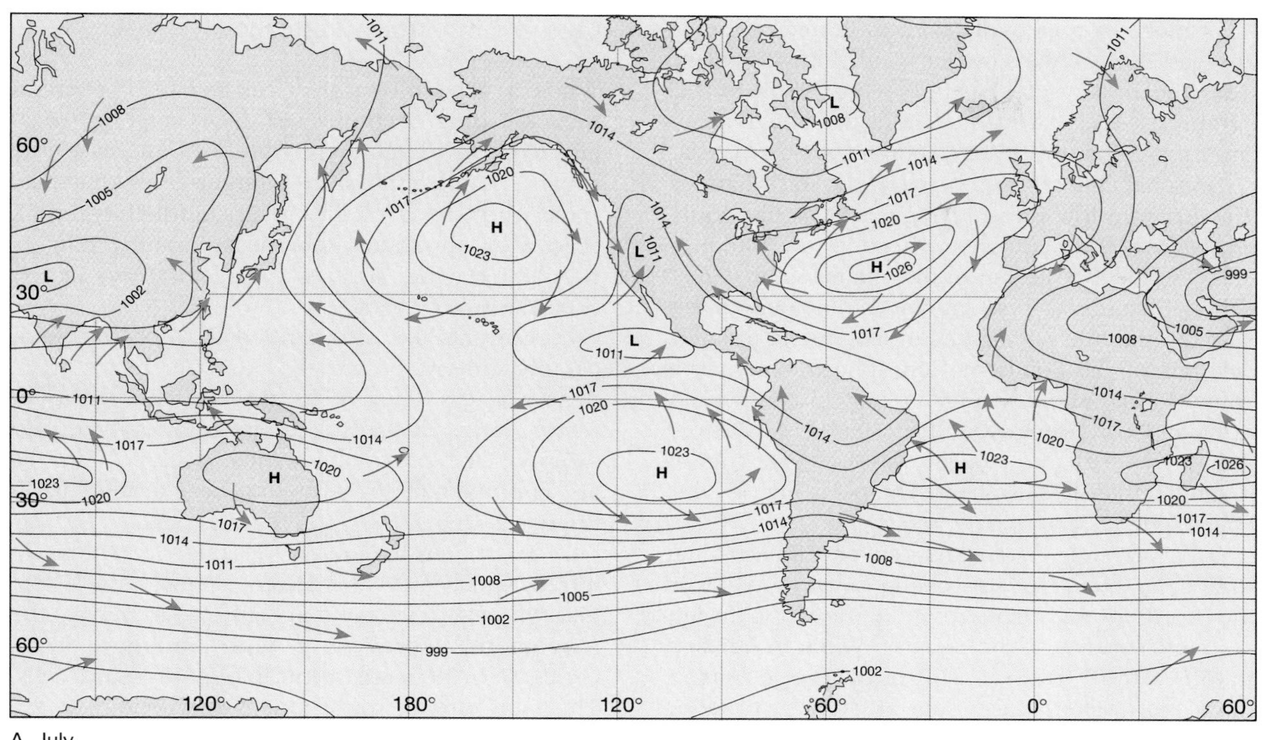

A. July

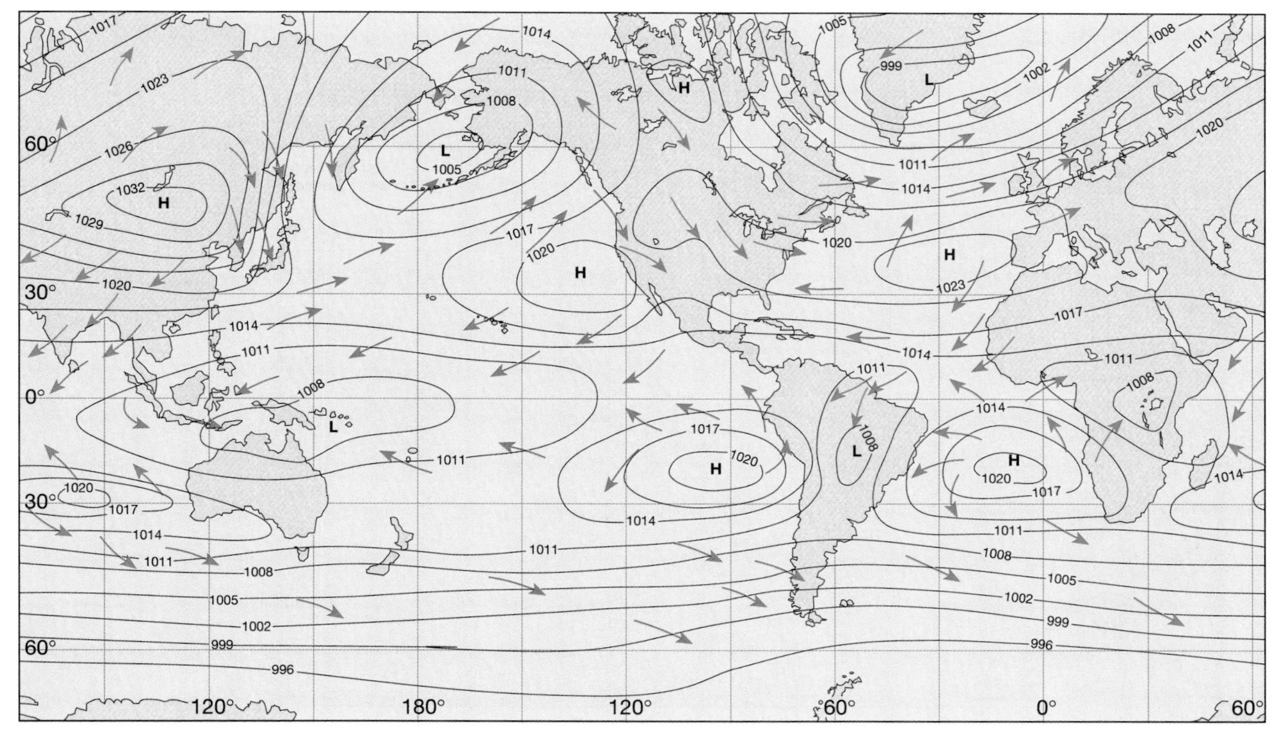

B. January

or produce extremely dry winters in most of south and east Asia, while summer winds blowing inland from the Indian and Pacific oceans result in wet summers. These wet summers are the famous "monsoon season" of heavy rains in southern Asia.

Seasonal changes also occur in the mid-latitudes. The extreme low-pressure regions over the northern oceans in January are replaced in July with generally weak low pressure and inconsistent winds, although storms with strong winds may periodically pass through the regions. As in January, in the mid-latitudes of the Southern Hemisphere, where it is winter in July, the low-pressure system is especially deep and consistent around the globe, because of the absence of large land masses to break it up as in the Northern Hemisphere. The polar high is similarly strong in July.

With the shift in seasons between January and July, solar radiation also changes. During July in the Northern Hemisphere, solar radiation inputs are highest north of the equator, while in January, insolation is greater south of the equator. Circulation patterns also shift seasonally, bringing strong seasonal contrasts in weather as wind direction and zones of frequent storms move north and south.

Ocean Circulation Patterns

When a wind blows over the ocean, it exerts a frictional drag on the sea surface, creating *waves* and *currents*. Continuing wind adds energy to the waves, building them into larger ones that may travel thousands of kilometers until they break onto a distant shore. And the continuing drag of prevailing winds causes broad *currents* in the ocean's surface layers.

Currents are also formed through the vertical movement of water. Currents move from denser to less dense areas of water; the density of water varies according to its temperature and salinity. Like prevailing winds, oceanic currents are important in redistributing heat around Earth (Figure 2–28).

Prominent features of oceanic circulation are **gyres,** which are wind-driven circular flows. Note that the pattern of gyres in Figure 2–28 mirrors the movement of prevailing winds in Figure 2–27. Gyres form beneath tropical high-pressure cells. The Gulf Stream forms the western limb of the gyre in the North Atlantic.

Ocean currents circulate warm water from low equatorial latitudes to higher latitudes; they

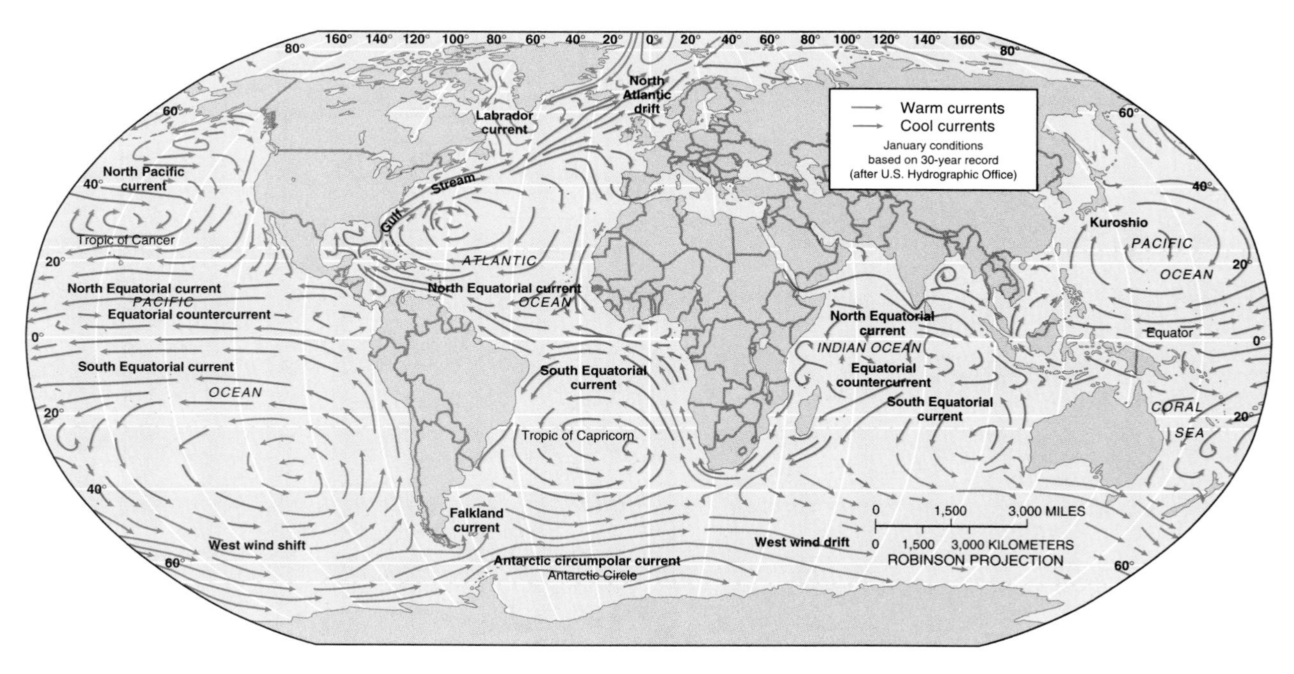

FIGURE 2–28
Oceanic circulation, showing major currents.

carry heat poleward by advection. Conversely, cool currents travel toward the equator, most notably along the west coasts of subtropical land areas. Because cold currents cool the lower portions of the atmosphere above them, rates of evaporation and convection are reduced. With less moisture available, adjacent land masses may be very dry. The Atacama desert of Peru and Chile owes its aridity to this effect.

El Niño. Although they generally behave more consistently that atmospheric winds, oceanic flows can vary in speed and direction. An important example of a varying ocean current is **El Niño,** the spanish term for *the Child.* The term refers to the Christ child, because El Niño occurs around Christmastime. El Niño is a circulation change that occurs every few years in the eastern tropical Pacific Ocean. The normal cool flow westward from South America slows and sometimes is replaced by a warm-water flow eastward from the central Pacific. The change in ocean currents is linked to a change in atmospheric circulation patterns in the Pacific Ocean known as the Southern Oscillation; the two phenomena together are referred to as El Niño-Southern Oscillation (ENSO).

Peruvian fishermen—who named El Niño—felt its effects during the 1970s. The reversed flow of water deprived fish in the eastern Pacific of their food and contributed tom the collapse of the Peruvian anchovy industry. Meteorologists also have linked ENSO events to flooding in the United States, droughts in Australia, and reduced rainfall in India.

Storms: Regional-Scale Circulation Patterns

ENSO and Asian monsoons are examples of variations in circulation that affect large areas of Earth. Storms affect smaller regions, although a large storm can affect an area of as much as several thousand kilometers.

A **cyclone** is a storm in which winds rotate around a large low-pressure area. Winds rotate counterclockwise in the Northern Hemisphere and clockwise in the Southern Hemisphere. We will look at two types of cyclones: tropical cyclones (hurricanes or typhoons) and mid-latitude cyclones.

Tropical cyclones. Tropical cyclones are intense, rotating convective systems that develop over warm ocean areas in the tropics and subtropics, primarily during the warm season. Storms with wind velocities exceeding 119 kilometers per hour (74 miles per hour) are called **hurricanes** in North

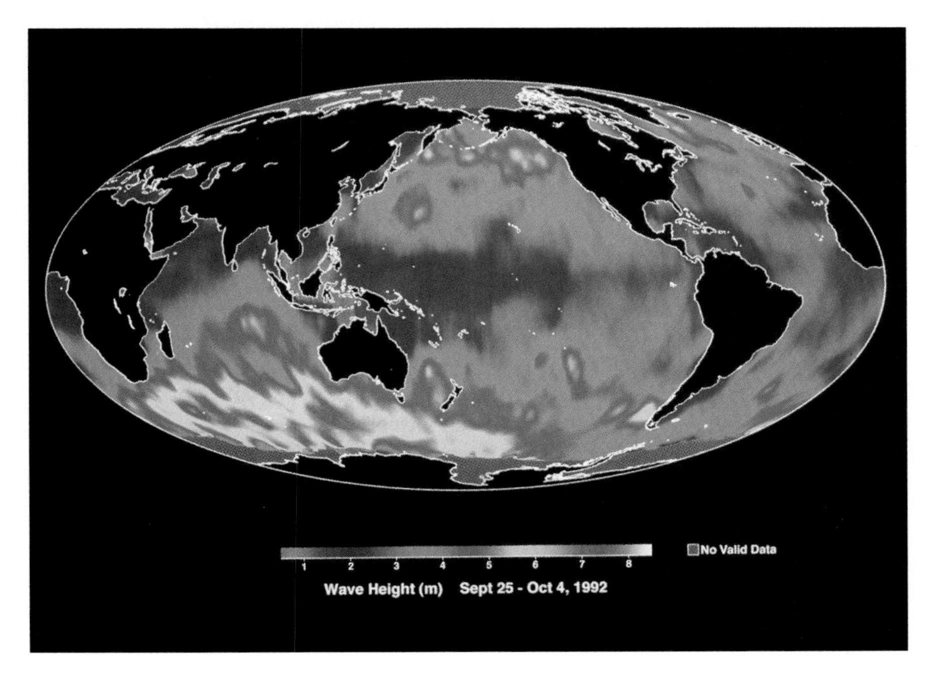

Distribution of wave heights between September 25 and October 4, 1992. Wave heights indicate wind speed. Magenta areas have wave heights of less than 1 meter (3 feet); red areas have wave heights of 7-8 meters (23–26 feet). The area of large waves in the southern Indian Ocean results from intense storms at the time these data were collected. Another area of storms is visible in the North Pacific. The calmest areas are in protected waters, such as the Gulf of Siam and the western Caribbean Sea. The image was made from radar altimetry data collected by the *TOPEX/Poseidon* satellite (Earth Satellite Corporation/Science Photo Library/Photo Researchers Inc.)

America, *typhoons* in the western Pacific, and *cyclones* in the Indian Ocean.

A cyclone typically develops in the eastern portion of an ocean, within the trade-wind belt. It begins as an area of low pressure (rising air) and converging winds, drawing in warm, moisture-laden air. The humid tropical air contains a great deal of energy, especially latent heat in the water vapor, although as sensible heat, as well. As energy is drawn into the developing storm, convection is intensified by condensation and the release of latent heat. This is an example of positive feedback, as discussed in Chapter 1. Latent heat feeds convection, which draws in more moisture, which releases more latent heat, and so on. The center of low pressure grows more intense, and as wind speed grows, the Coriolis effect causes a spiraling circulation, counterclockwise to the Northern Hemisphere and clockwise in the Southern Hemisphere.

Tropical storms move with the general circulation over the subtropical Atlantic, Pacific, and Indian oceans, from west to east. As it travels across the warm ocean surface, a storm often intensifies. By the time it reaches the western portion of the ocean, a storm can have an atmospheric pressure as much as 10 percent lower than average sea-level pressure, and winds can exceed 150 kilometers per hour (90 miles per hour).

Because a hurricane thrives on warm, moist air, it is most intense over ocean areas during the warm season. It "starves" over land, because it loses its source of energy. Further, the smooth ocean surface favors development of high winds, because the hills and trees of land areas slow the wind by friction.

Hurricanes pose the greatest threat to humans in tropical and subtropical coastal areas, on the western margins of the oceans, and in southern Asia, where monsoon circulation draws air northward from the Indian Ocean. When a hurricane strikes land, the combination of intense wind and extremely low pressure causes a **storm surge,** an area of sea in the center of the storm that may be elevated several meters. The surge carries large waves crashing onto low-lying coastal areas, with potentially devastating results. In fact, the majority of hurricane deaths and damage result from the storm surge. Adding to this hazard, tornadoes commonly are spawned as hurricanes come ashore.

The coastline most vulnerable to tropical cyclones may be in the Ganges River delta, on the Bay of Bengal, including Bangladesh and Bengal in eastern India. Land in the delta is low-lying, with elevation of generally less than 2 meters (6 feet), and densely populated Banglasdesh has very limited resources for coping with threat of a storm surge. Although warning systems alert people in advance of approaching storms, they often lack transportation to reach high ground. Several severe storms have struck the area in recent decades, including one in 1970 that killed about 300,000 people.

Mid-latitude cyclones. Mid-latitude cyclones dominate the weather in mid-latitude regions. The mid-latitude cyclone forms over colder, drier land areas that cannot provide the immense volume of heat and water vapor needed to generate hurricane, as is the case in the tropics. A **mid-latitude cyclone** is a center of low pressure that develops along a polar front. It moves from west to east along that front, following the general circulation in the mid-latitudes. Mid-latitude cyclones are much more common than hurricanes and much less intense.

In a mid-latitude cyclone, air is drawn toward the center of low pressure from both the warm and cold sides of the polar front. A warm front develops where warm air is drawn toward cold, typically on the eastern side of the storm. On the western side, the spiralling motion causes cold air to drive under the warm air, forming a cold front. As the center of low pressure moves eastward, the fronts move with it, bringing precipitation to areas over which they pass. The passage of a front at the surface is usually marked by a significant change of temperature, precipitation, and shifting winds. The repeated passage of such storms creates highly variable weather conditions, alternating between cold and warm air, as the polar front moves back and forth across the land.

The Weather On September 9, 1983

Global and regional circulation patterns just described are visible in satellite images of Earth, such as the opening image of this chapter, also shown in Figure 2–29. The image was taken in September, late in the Northern Hemisphere sum-

FIGURE 2–29
A satellite view shows Earth's weather on September 9, 1983, deliberately selected to show weather patterns as clearly as possible. This is the same image as the chapter opening.

mer. At this time of year, the ITCZ generally is north of the Equator; in this image it is at about latitude 10°N. The Amazon and much of the South Pacific have spotty low clouds, distinguished from higher clouds in this image by their darker color. September is the month when tropical cyclones are most common in the Northern Hemisphere, and two are visible in this image. A well-developed one is over the eastern Caribbean Sea. A smaller one, separated from its ocean source of energy, has moved inland over the lower Mississippi Valley of the United States.

Several mid-latitude cyclones are visible in this image. Two lie along the polar front in the Northern Hemisphere, one centered over central Canada and the other in the North Atlantic, northwest of Europe. Each has a cold front extending to the southwest and a warm front extending southeast from the center of low pressure.

Over the United States, a tropical cyclone has acquired some of the characteristics of a mid-latitude cyclone, and it is linked to a front extending eastward across the Atlantic. A band of high pressure and clear weather lies between the polar front and the tropical cyclone. Another strong cold front appears over Argentina and the southern Atlantic, likely associated with a Southern Hemisphere mid-latitude cyclone just off the eastern edge of the image.

A false-color image of Europe as seen from the *NOAA-9* weather satellite on August 9, 1987. The spiral-shaped pattern of clouds results from counterclockwise circulation around a low-pressure cell centered over Ireland. The cloud band extending southwestward over northwestern France and Portugal is a cold front. A warm front extends across the North Sea. (European Space Agency/Science Photo Library/Photo Researchers, Inc.)

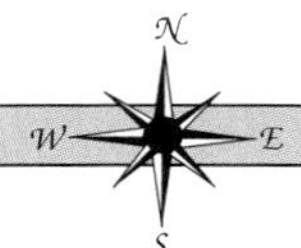

CONCLUSIONS: CRITICAL ISSUES FOR THE FUTURE

The daily weather we experience is a product of a localized pattern determined by Earth's energy budget. Important variables include the amount of solar energy arriving in a day and the current circulation pattern, such as whether cold air is streaming in from the polar regions or a whether a low pressure system is overhead. When a weather pattern over a given part of the world is consistent and predictable over a number of years, it is known as a climate.

Circulation patterns, which are driven by solar energy, exist because energy is not evenly distributed around the globe. Storms are weather phenomena of particularly intense motion, creating severe winds or precipitation on the ground; large amounts of solar energy are transported in storms, and ultimately are dissipated in the upper atmosphere to be returned to space.

Earth's energy budget is almost certainly undergoing change, in response to discharge into the atmosphere of higher levels of carbon dioxide and other human-derived substances. Changes in the energy budget almost certainly will cause changes in weather and climate, but because atmospheric circulation is so complex, the specific nature and timing of the changes are unknown. Whether the Great Plains of North

America become drier, wetter, warmer, or cooler depends more on the paths that storm tracks take across the continent and the airflow patterns across that area than on the global-scale energy budget. To better understand spatial patterns and potential changes in the world weather map, the next chapter examines world climates.

Chapter Summary

1. Solar energy

Solar radiation (insolation) received by places at a particular latitude varies with season and by day length. Daily and seasonal differences in intensity are caused by variations in the angle at which the sun's rays strike places at a particular latitude. This variation in insolation causes variation in the weather. The annual total of radiation is highest in low latitudes, while day length and angle of incidence cause greater seasonal variations at high latitudes.

2. Heat transfer between the atmosphere and Earth

Exchange of energy among Earth, the atmosphere, and space (including the sun) is achieved through shortwave and longwave radiation, latent heat exchange, and convection. Energy arrives from the sun as shortwave energy and is reradiated by Earth as longwave radiation. The atmosphere is relatively transparent to short wavelengths, but it absorbs longwave energy. This causes the atmosphere to retain heat, much like a greenhouse.

3. Earth's energy budget

Insolation absorbed by Earth's surface and atmosphere is eventually returned to space, but only after many exchanges have taken place between the ground and sky. Human activities may be leading to global warming through emission into the atmosphere of greenhouse gases, especially carbon dioxide.

4. Precipitation processes

Precipitation forms when humid air rises and is cooled, and the water vapor in it condenses. Three processes create most precipitation: convection, orographic uplift over mountain ranges, and interaction between cold and warm air masses along fronts. Storms such as hurricanes and thunderstorms are forms of convective precipitation, driven by surface heating and release of latent heat. Orographic rainfall occurs on the windward sides of mountain ranges, and drier areas are formed to the leeward sides. Frontal precipitation occurs mostly in mid-latitude low-pressure zones.

5. Global Circulation patterns

Wind blows from areas of high pressure to areas of low pressure,under the influence of the Coriolis effect, which results from Earth's rotation. Global circulation patterns include bands of low pressure in the tropics, high pressure feeding trade winds in the subtropics, low pressure in the mid-latitudes, and high pressure at the poles. Convectional precipitation dominates the tropics, while the main form of precipitation in the mid-latitudes is frontal.

Key Terms

Adiabatic cooling, 69
Advection, 56
Air mass, 71
Albedo, 59
Angle of incidence, 46
Autumnal equinox, 48
Barometer, 74
Chlorofluorocarbon (CFC), 66
Cold front, 72
Conduction, 51
Convection, 51
Coriolis effect, 76
Cumulus clouds, 73
Cyclone, 81
El Niño, 81
Front, 72
Frontal precipitation, 71
Greenhouse effect, 54
Greenhouse gases, 53
Gyre, 80
Hurricane, 81
Insolation, 46
Intertropical convergence zone (ITCZ), 77
Latent heat, 54
Latent heat exchange, 51
Latent heat of vaporization, 54
Longwave energy, 53
Micron, 51
Mid-latitude cyclone, 84
Mid-latitude low pressure zones, 78
Monsoon circulation, 78
Orographic uplift, 70
Ozone, 66
Polar high-pressure zones, 78
Polar front, 78
Pressure gradient, 75
Radiation, 51
Relative humidity, 68
Saturation vapor pressure, 67
Sensible heat, 54
Shortwave energy, 51
Specific heat, 57
Stationary front, 73
Storm surge, 83
Stratus clouds, 73
Subtropical high-pressure zones, 77
Summer solstice, 49
Tropic of Cancer, 48
Tropic of Capricorn, 48
Vernal (spring) equinox, 48
Warm front, 73
Winter solstice, 49

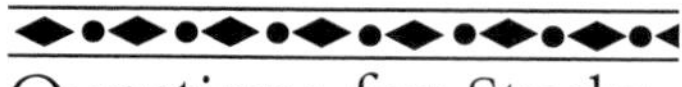

Questions for Study and Discussion

1. What processes are responsible for the major energy exchanges between Earth's surface and the atmosphere?
2. How do nitrogen, oxygen, carbon dioxide, water vapor, ozone, and dust affect the transmission of shortwave and longwave radiation in the atmosphere? How do each of these gases modify the exchanges of energy among Earth, the atmosphere, and space?
3. Why, in terms of energy availability and convective processes, do hurricanes happen: a) mostly over oceans; and b) in late summer and autumn?
4. Under what atmospheric conditions is convection most intense? What conditions inhibit convection?
5. How do climatic conditions in coastal areas differ from those in continental interiors? Why do these differences occur?
6. What are the minimum and maximum daily amounts of solar radiation arriving at the top of the atmosphere at your latitude? When do these occur? How do angle of incidence and day length affect these seasonal differences?

7. What are the major features of a mid-latitude cyclone? What sequence of weather would one expect to observe as a mid-latitude cyclone passes?
8. Describe the four heat transfer processes.
9. Outline Earth's energy budget.

Thinking Geographically

1. On a day when indoor temperatures are substantially different from outdoor temperatures (as in winter in a mid-latitude climate), measure the indoor air temperature near a window and near an interior wall. Sit near each of these places and see how warm or cold you feel. Do air temperature differences fully account for the differences in comfort? What is the role of radiant energy exchanges in how warm or cool we feel?
2. For a two-week period, keep a daily journal of the weather including such things as air temperature, wind direction, cloudiness and precipitation. For the same period, clip the weather map from a daily newspaper. Then compare the two records.
3. In what ways do topography and/or land cover affect the weather where you live?
4. At your library, find daily climatological records for a place that interests you. (For locations in the U.S. they are published monthly as "Local Climatological Summary" by the National Oceanographic and Atmospheric Administration, or NOAA). Examine a few months' records of daily maximum and minimum temperatures and daily precipitation. You may wish to enter them in a computer spreadsheet. Look for evidence of typical patterns of precipitation in relation to temperature, and try to explain these patterns in terms of the weather systems that cause them.

Suggestions for Further Readings

Arntz, Wolf E. "El Niño and Peru: Positive Aspects." *Oceanus* 27 (Summer 1984): 36–39.

Balling, Robert C. Jr., and R. S. Cerveny. "Long-Term Associations Between Wind Speeds and the Urban Heat Island of Phoenix, Arizona." *Journal of Climate and Applied Meteorology* 26 (June 1987): 712–16.

Cerveny, Randall S. "Meteorological Assessment of Homer's Odyssey." *Bulletin of the American Meteorological Society* 74 (June 1993): 1,025–34.

Chagnon, Stanley A. "Inadvertent Weather Modification in Urban Areas: Lessons for Global Climate Change." *Bulletin of the American Meteorological Society* 73 (May 1992): 619–27.

Downton, Mary W. and K. A. Miller. "The Freeze Risk to Florida Citrus. Part II: Temperature Variability and Circulation Patterns." *Journal of Climate* 6 (February 1993): 354–63.

Entekhabi, Dara, I. Rodriguez-Iturbe, and R. L. Bras. "Variability in Large-Scale Water Balance with Land Surface-Atmosphere Interaction." *Journal of Climate* 5 (August 1992): 798–813.

Gallo, K.P., A. L. McNab, and T. R. Karl. "The Use of NOAA-AVHRR Data for Assessment of the Urban Heat Island Effect." *Journal of Applied Meteorology* 32 (May 1993): 1,563–77.

Gates, D. M. *Energy and Ecology*. Sunderland, MA: Sinauer, 1985.

Givoni, Baruch. "Climatic Aspects of Urban Design in Tropical Regions." *Atmospheric Environment 26B* (September 1992): 397–406.

Givoni, Baruch. "Impact of Planted Areas on Urban Environmental Quality: A Review." *Atmospheric Environment* 25B (No. 3, 1991): 289–99.

Gray, William M. "Strong Association Between West African Rainfall and U.S. Landfall of Intense Hurricanes." *Science* 249 (14 September 1990): 1,251–56.

Hamdan, M. A., and N. Gazzawi. "The Effect of Clouds on Solar Radiation." *Energy Conversion and Management* 34 (January 1993): 29–32.

Harrington, John A., R. S. Cerveny, and R. C. Balling, Jr. "Impact of the Southern Oscillation on the North American Southwest Monsoon." *Physical Geography* 13 (October-December, 1992): 318–30.

Holland, Greg J., and M. Lander. "The Meandering Nature of Tropical Cyclone Tracks."*Journal of the Atmospheric Sciences* 50 (May 1993): 1,254–66.

Kerr, Richard A. "The Weather in the Wake of El Niño."*Science* 240 (13 May 1988): 883.

Knight, Charles A. "Precipitation Formation in a Convective Storm." *Journal of the Atmospheric Sciences* 44 (1 October 1987): 2,712–26.

Kuhn, J. R., K. G. Libbrecht, and R. H. Dicke. "The Surface Temperature of the Sun and Changes in the Solar Constant."*Science* 242 (11 November 1988): 908–11.

Landsberg, H. E. "Weather, Climate and You." *Weatherwise* 38 (October 1986): 248–53.

Leathers, Daniel J. and D. A. Robinson. "The Association Between Extremes in North American Snow Cover Extent and United States Temperatures" *Journal of Climate* 6 (July 1993): 1,345–55.

Li, Bin, and R. Avissar. "The Impact of Spatial Variability of Land-Surface Characteristics on Land-Surface Heat Fluxes." *Journal of Climate* 7 (April 1994): 527–37.

Longhetto, A., L. Giacomelli, and C. Giraud. "A Study of Correlation Among Solar Energy, Atmospheric Turbidity and Pollutants in Urban Area." *Atmospheric Environment* 26B (March 1992): 29–43.

Maddox, John. "Two Gales Do Not Make a Greenhouse."*Nature* 343 (1 February 1990): 407.

Manabe, S. and A. J. Broccoli. "Mountains and Arid Climates of Middle Latitudes." *Science* 247 (12 January 1990): 192–95.

Meisner, Bernard N. and L. F. Graves. "From Temperature to Hurricanes: apparent temperature." *Weatherwise* 38 (August 1985): 211–13.

Montgomery, Michael T, and B. F. Farrell. "Tropical Cyclone Formation." *Journal of the Atmospheric Sciences* 50 (January 1993): 285–310.

Pearce, Fred. "Fire and Flood Greet El Niño's Third Year." *New Scientist* 141 (15 January 1994): 9.

Rasmusson, Eugene M. and P. A. Arkin. "A Global View of Large-Scale Precipitation Variability." *Journal of Climate* 6 (August 1993): 1,495–1,522.

Roman, Charles T. "Hurricane Andrew's Impact on Freshwater Resources." *BioScience* 44 (April 1994): 247–55.

Schlatter, Thomas. "A Hurricane Primer." *Weatherwise* 46 (October/November 1993): 40–42.

Sharifi, M. B., K. P. Georgakakos, and I. Rodriguez-Iturbe. "Evidence of Deterministic Chaos in the Pulse of Storm Rainfall." *Journal of the Atmospheric Sciences* 47 (April 1990): 888–93.

Simons, Paul. "Why Global Warming Could Take Britain by Storm." *New Scientist* 136 (7 November 1992): 35–38.

Stoll, Matthew J. and A. J. Brazel. "Surface-Air Temperature Relationships in the Urban Environment of Phoenix, Arizona." *Physical Geography* 13 (April-June, 1992): 160–79.

Stuller, Jay. "Stormy Weather: Climate Forecasting Remains Partly Cloudy." *Sea Frontiers* 39 (July/August 1993): 25–29.

Walker, Jearl. "Searching for Patterns of Rainfall in a Storm." *Scientific American* 252 (January 1985): 112–13.

Williams, J. M., F. Doehring, and I. Duedall. "Heavy Weather in Florida: 180 Hurricanes and Tropical Storms in 22 Years." *Oceanus* 36 (Spring 1993): 19–26.

Williams, R. T., M. S. Peng, and D. A. Zanikofski. "Effects of Topography on Fronts." *Journal of the Atmospheric Sciences* 49 (15 February 1992): 287–305.

Zehnder, Joseph A. and P. R. Bannon. "Frontogenesis Over a Mountain Ridge." *Journal of the Atmospheric Sciences* 45 (15 February 1988): 628–44.

We also recommend these journals: *Atmospheric Environment; Bulletin of the American Meteorological Society; Journal of the Atmospheric Sciences; Monthly Weather Review; Physical Geography; Progress in Physical Geography; Weatherwise.*

3 Climate Patterns and Change

- Classifying climate
- Earth's climate regions
- Patterns of climate through time
- Future global warming

The Ganges Delta, Bangladesh

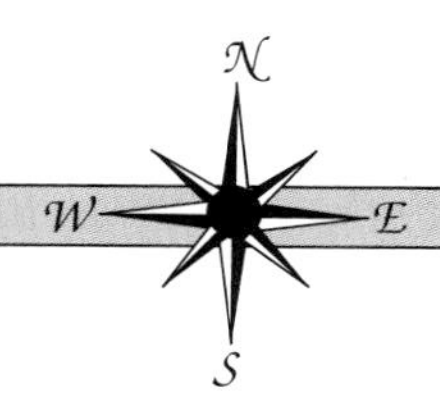

1. Classifying climate
A place's climate is formed by a pattern of typical weather conditions—temperature, precipitation, wind, storm patterns, humidity—and seasonal variations in these conditions, averaged over time. Weather patterns must be monitored for several decades to detect variations in climate. Vegetation and terrain are also used to classify climate.

2. Earth's climate regions
Earth can be divided into five main climate regions designated by the letters A, B, C, D, and E. Climates designated A are humid tropical climates; B climates are dry; mid-latitude climates are divided into warm (C) and cold (D) types; and E climates are found in polar regions.

3. Patterns of climate through time
Climate has been extremely variable during the past 3 million years. Global average temperatures have been as much as 10C° (18F°) lower than the present average climate. Causes include changes in Earth's orbit around the sun, continental drift, and changes in the concentration of greenhouse gases in the atmosphere.

4. Future global warming
Earth's climate is warming and will likely continue to warm in part as a result of human actions. Scientists are uncertain about the extent of warming, its impact, and whether it can be prevented. Continued use of fossil fuels in homes, industries, and motor vehicles is believed to contribute to global warming.

INTRODUCTION

Chapter 2 explored the atmospheric processes that produce weather—the daily and seasonal differences in temperature and precipitation within particular regions. Geographers also study temperature, precipitation, and other atmospheric phenomena that persist for much longer periods than a thunderstorm or a heat wave. **Climate** is the totality of weather conditions over several decades or more, *a place's weather pattern over time*. The vegetation, natural resources, and human activities of a particular region of Earth are heavily influenced by its climate.

At first glance, describing the climate of a place may appear to be a straightforward task of calculating annual averages of temperature and precipitation. In reality, climate is extremely difficult to describe, because air temperature and precipitation change—not only day-to-day, but year-to-year—and such factors as windiness, humidity, and occurrence of storms must also be considered.

Climate changes over time. Earth has alternated between warming periods and cooling periods. Glaciers have covered large areas during cool periods, and they have melted during warm periods. The human population has survived by adapting to the alternating periods of warming and cooling. Ironically, human activity now may be the cause of climatic changes that could fundamentally alter Earth.

People are affected by various elements of climate. For example, the number of hours of sunshine in a day and the degree of cloudiness are important to anyone who is active out-of-doors—gardeners, farmers, hunters, builders, vacationers at the beach, and residents of homes heated by solar energy. Amounts of rainfall and snowfall make great differences in how we build our homes and roads from place to place. The windiness of a place is important in designing structures and harnessing wind power. Pollutants in the air can have both short-term and long-term effects on human and animal health. Counts of pollen and mold spores are important to those having allergies.

FIGURE 3–1
This world map of climate regions is largely based on the Köppen system of classifying climates.

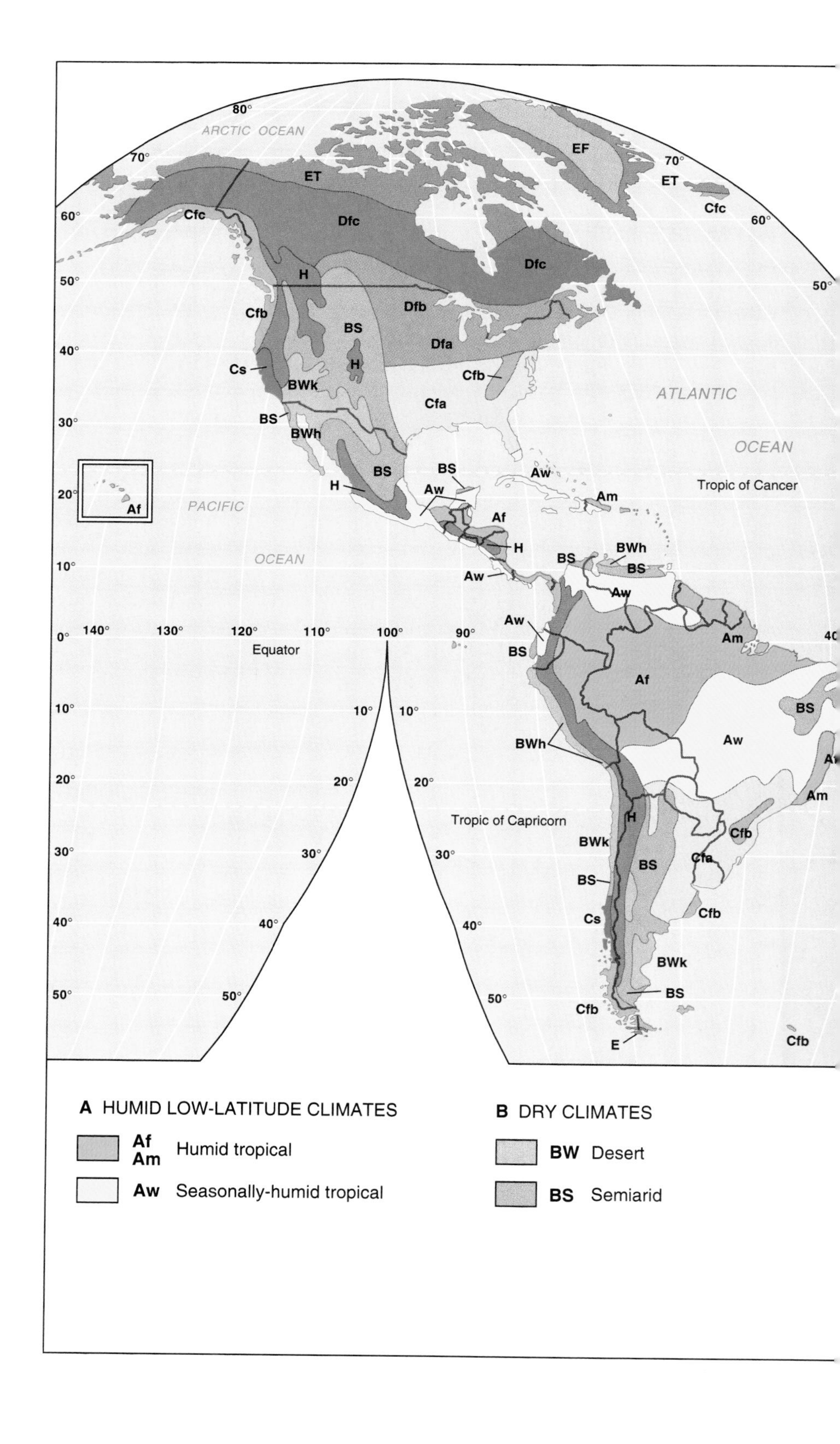

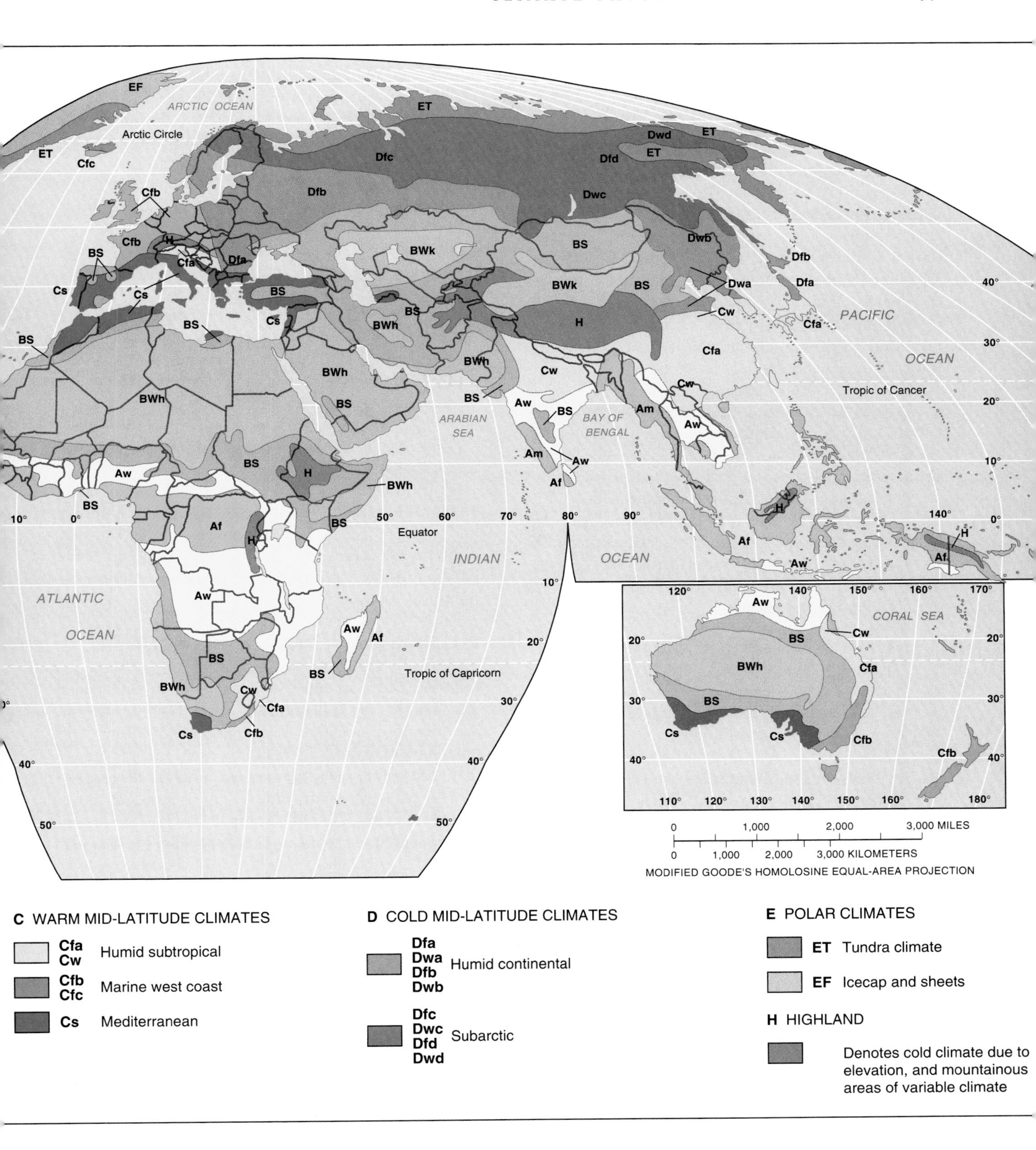
ARCTIC OCEAN
Arctic Circle
EF
ET
Cfc
Cfb
Dfc
Dfd
Dwd
Dfb
Dwc
Dwb
Dfa
Dwa
BWk
BS
Cs
Cfa
H
BWh
Cw
Am
Aw
Af
PACIFIC
OCEAN
Tropic of Cancer
ARABIAN SEA
BAY OF BENGAL
Equator
INDIAN
OCEAN
ATLANTIC
OCEAN
Tropic of Capricorn
CORAL SEA
0 1,000 2,000 3,000 MILES
0 1,000 2,000 3,000 KILOMETERS
MODIFIED GOODE'S HOMOLOSINE EQUAL-AREA PROJECTION
C WARM MID-LATITUDE CLIMATES
Cfa Cw Humid subtropical
Cfb Cfc Marine west coast
Cs Mediterranean
D COLD MID-LATITUDE CLIMATES
Dfa Dwa Dfb Dwb Humid continental
Dfc Dwc Dfd Dwd Subarctic
E POLAR CLIMATES
ET Tundra climate
EF Icecap and sheets
H HIGHLAND
Denotes cold climate due to elevation, and mountainous areas of variable climate

Classifying Climate

Knowing a place's climate allows analysis and planning by geographers, climatologists, geologists, social scientists, government, and industry. Naming and describing specific types of climate gives people a vocabulary essential to communicating information about places.

The most widely used climate classification system was devised in 1918 by German geographer Wladimir Köppen (Figure 3–1). Köppen identified five basic climate types and subdivided each to reveal important distinctions. Climatologists have modified Köppen's classification many times, but have retained its essential elements.

Köppen's scheme uses uppercase letters *A*, *B*, *C*, *D*, and *E* to define the major climate types. Letters indicate the latitudes at which the climates occur, with *A* climates nearest the equator and *E* climates toward the poles. *A*, *C*, *D*, and *E* climates are distinguished primarily by *temperature* (with *A* the warmest); *B* climates are distinguished primarily by *precipitation*. The higher elevations of mountains, classified as *H*, or highlands, display complex climates not easily depicted on a world map. As Figure 3–1 shows:

- **A** or tropical climates are warm all year.
- **B** or dry climates have limited moisture, as calculated by a formula.
- **C** or warm mid-latitude climates have cool winters and warm summers.
- **D** or cold mid-latitude climates have cold winters and mild summers.
- **E** or ice climates are cold all year.

Climate Measures

To understand a region's distinctive climate, the two most important measures are *air temperature* and *precipitation*. Vegetation is also associated with distinctive climate.

Air temperature. In everyday speech, we refer to "hot" or "cold" climates. Climatologists usually record high and low temperatures for each day at designated weather stations and calculate the average daily temperature as the average of the two extreme readings. A region may be warm throughout the year, with average daily temperatures rarely falling below 20°C (68°F), whereas a cold region rarely has average daily temperatures above 10°C (50°F). Some regions experience seasonal variations, while others are cold or warm throughout the year. As we will see in Chapter 7, humans avoid living in the hottest and coldest regions of Earth.

Air temperature varies with elevation, even a few meters above the ground surface. A very general rule of thumb is that temperature drops about 6.4C° per 1,000 meters (3.5F° per 1,000 feet), an effect you can feel simply by climbing a mountain. Ground surface temperature in a desert can reach 70°C (160°F) during the daytime, while air just 2 meters above the ground is less than 55°C (130°F); at night, the desert ground may be significantly cooler than the air above it. To minimize the effects of the ground surface on temperature records, air temperature is usually measured at a height of about 1.4 meters (55 inches) inside a ventilated shelter that shades the thermometer and other instruments from direct sunlight.

Topography and proximity to an ocean also affect air temperature (Figure 3–2). In both summer and winter, the higher elevations of the Rockies and Appalachians are cooler than adjacent lowlands, and the Pacific Ocean moderates winter temperatures along the west coast of the United States.

Precipitation. The level of precipitation is extremely variable between places and over time. A thunderstorm can unleash heavy rain in one place and a light sprinkling just a short distance away. Worldwide, precipitation can range from a few centimeters a year (1–2 inches) in desert areas to more than 200 centimeters (80 inches) in tropical areas. A few tropical mountain areas have recorded more than 10 meters (33 feet) of rainfall per year. Humans cluster in regions with moderate levels of precipitation and avoid regions that are very wet or very dry.

Precipitation usually is measured by collecting rain or snow in a cylindrical container that is marked in millimeters or hundredths of an inch. Snowfall normally is recorded as an amount of liquid water, rather than the depth of snow accumulated on the ground, because the amount of water in snow varies widely with the snow's texture. Weather stations typically report daily precipitation, although the instruments are capable of recording the amount for other periods of time, such as a minute or an hour.

Figure 3–3 shows worldwide average precipitation. Note how the levels correspond to phenome-

A farmer near South Bend, Indiana, displays corn stunted by the drought of summer 1988. The drought only lasted a few months, but because it occurred during the summer growing season, it caused major crop damage. (The Stock Market)

na you learned about in Chapter 2: the rainy ITCZ around the equator, the mid-latitude high-pressure areas with their dry air, the mountain rain-shadow areas, and the dry polar regions.

There is much more to precipitation than just measuring rainfall. Plants demand water, and thus play a key role in what happens to the precipitation. Therefore, most climate classification systems consider precipitation *in relation to what a region's vegetation needs.* Just as people demand water from the ground, so do plants. In fact, a single tree may remove 100 or more liters of water from the soil in a single day.

Plants take water up through their roots and evaporate it through their leaves, releasing it into the atmosphere as water vapor. This process is called **transpiration** (roughly comparable to perspiration in people). The amount of water that *could* evaporate from a damp soil or be transpired by plants if all of their needed moisture were available, is called **potential evapotranspiration (POTET).** POTET indicates the moisture demand that plants make on their environment. Evapotranspiration is a fundamental part of the hydrologic cycle, which will be discussed in detail in Chapter 5.

Because heat energy must be available to evaporate water, warmer climates make possible a greater POTET. In tropical climates, where temperatures typically exceed 20°C (68°F) throughout the year, annual POTET can exceed 150 centimeters (60 inches). In contrast, annual POTET is only a few centimeters in polar regions, and 50 to 90 centimeters (20 to 35 inches) in mid-latitude climates.

In distinguishing between arid and humid climates, measurement of POTET is important. Parts of tropical East Africa and central Minnesota both receive the same average annual rainfall of about 80 centimeters (30 inches). Yet, trees are scattered in East Africa and abundant in Minnesota. The reason is that East Africa is hotter and therefore has a much greater potential evapotranspiration than central Minnesota. East Africa's POTET exceeds its annual precipitation, so it cannot support as many trees.

Over large areas of the Upper Midwest of the United States, as much as 25 centimeters (10 inches) of rain fell during the month of June 1993, and wet weather continued into July. The result was the most severe flooding in this part of the Mississippi River in recorded history. (Randy O'Rourke/The Stock Market)

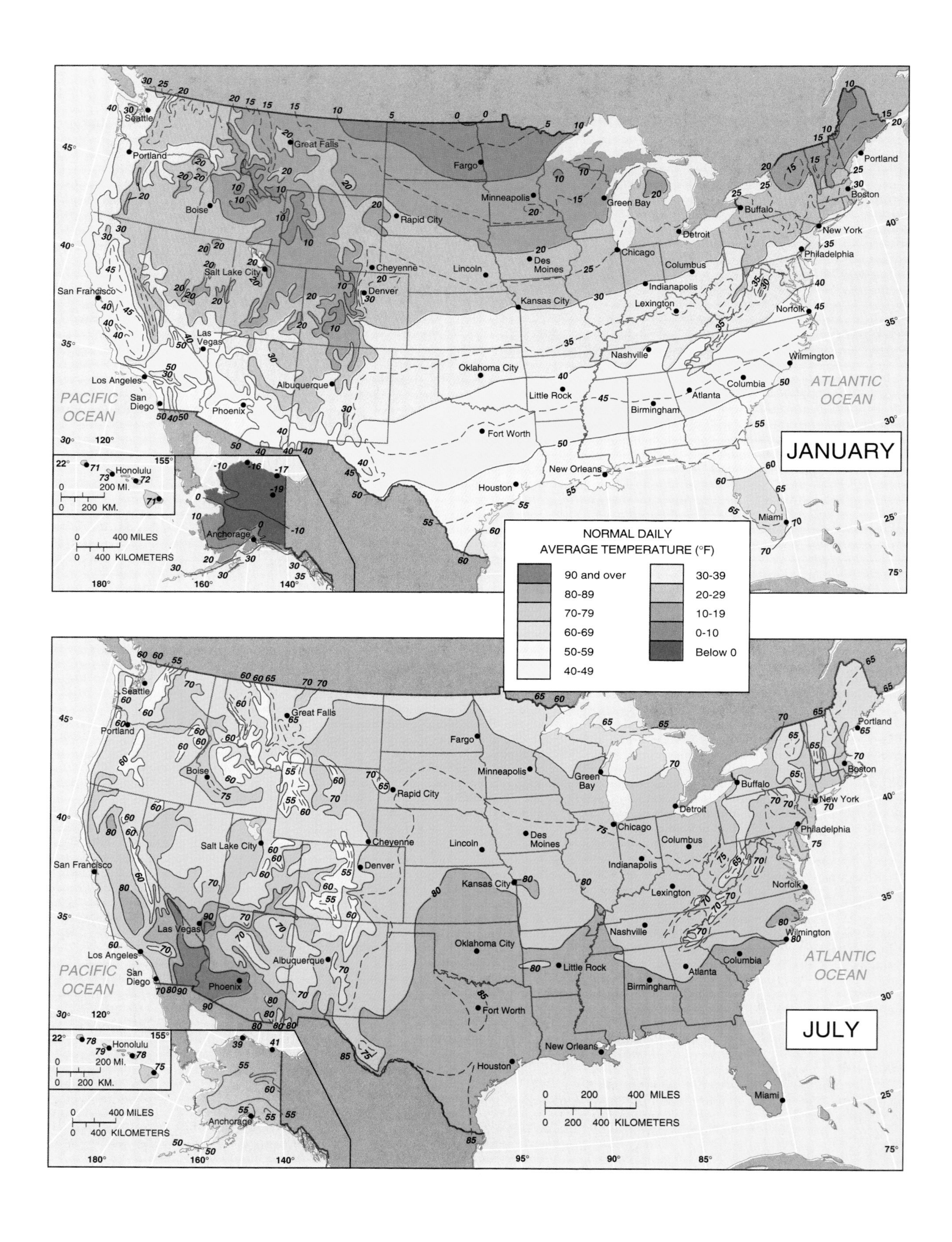
JANUARY
JULY
NORMAL DAILY AVERAGE TEMPERATURE (°F)
90 and over
80-89
70-79
60-69
50-59
40-49
30-39
20-29
10-19
0-10
Below 0
PACIFIC OCEAN
ATLANTIC OCEAN
Seattle
Portland
Boise
Great Falls
Rapid City
Fargo
Minneapolis
Green Bay
Detroit
Buffalo
Boston
New York
Philadelphia
Chicago
Des Moines
Lincoln
Cheyenne
Salt Lake City
San Francisco
Denver
Kansas City
Columbus
Indianapolis
Lexington
Norfolk
Las Vegas
Nashville
Wilmington
Oklahoma City
Los Angeles
Albuquerque
Little Rock
Atlanta
Columbia
San Diego
Phoenix
Birmingham
Fort Worth
New Orleans
Houston
Miami
Honolulu
Anchorage
0 200 MI.
0 200 KM.
0 400 MILES
0 400 KILOMETERS
0 200 400 MILES
0 200 400 KILOMETERS

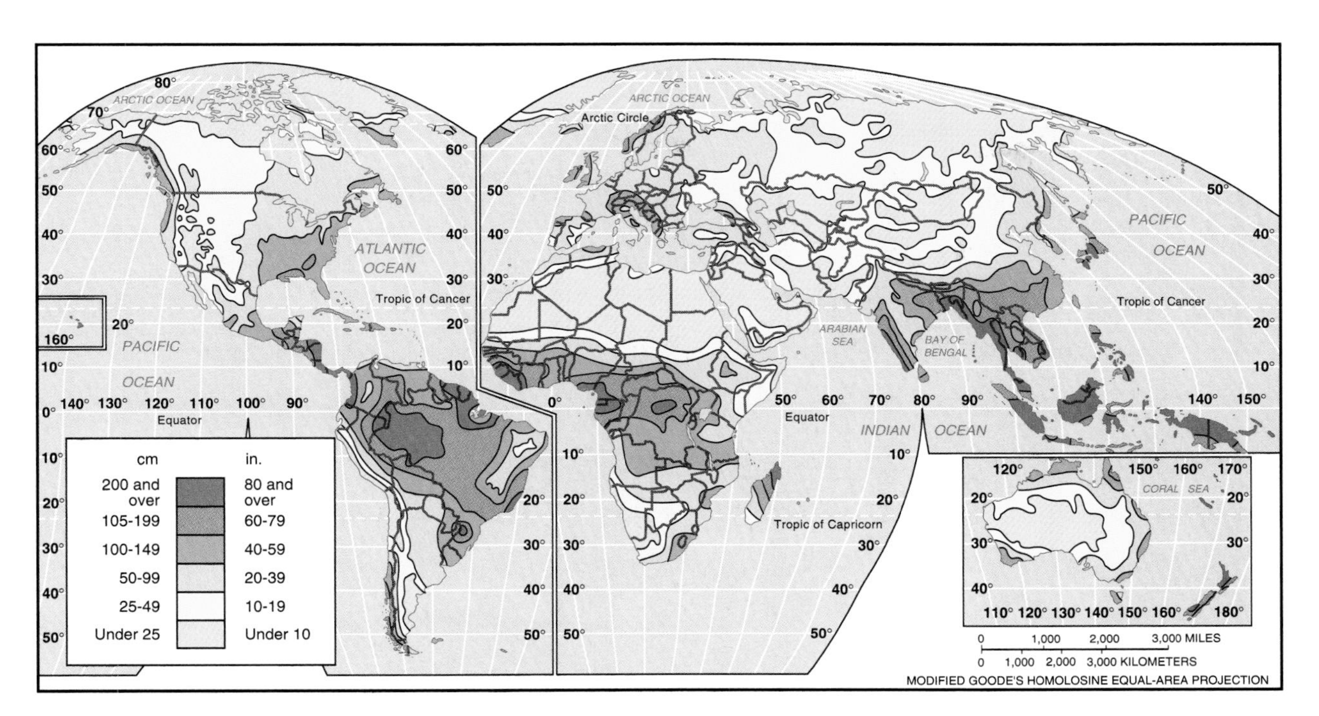

FIGURE 3–3
Average annual precipitation worldwide. Generally, climates with higher temperatures require greater rainfall to maintain vegetation.

Figure 3–4 shows average annual precipitation in North America. The highest annual totals occur in the mountains along the Pacific Coast in Washington and British Columbia. In this region, frequent low-pressure systems bring moisture-laden air from the Pacific. This air is lifted orographically over the Sierra Nevada and Cascade mountains, causing high precipitation in the mountains and a rain shadow over the Great Plains. The low precipitation in the southwestern United States and northern Mexico results from the subtropical high-pressure zone of the eastern Pacific as well as the rain shadow which is isolated from moisture sources.

In the eastern part of North America, precipitation varies from north to south—the area in the north receives less rainfall, whereas places in the south receive much more. This north-south gradient is caused primarily by distance from the region's prime moisture source, the Gulf of Mexico. In addition, the lower temperatures in the north reduce the amount of moisture the air can hold.

FIGURE 3–2
Normal daily average temperatures in the United States during January and July, based on data between 1931 and 1960.
(a) In January, the lowest temperatures occur in the interior of the continent and in areas of high elevations. Higher temperatures occur in the south and along the West Coast.
(b) In July, the highest temperatures occur in the interior of the continent, especially in the desert southwest. Temperature differences between northern and southern portions of the United States are less than in January, and storm systems traveling along the polar front are less intense as a result. Be careful when trying to interpolate between temperature lines, because large changes may occur over short distances, depending on elevation, slope, soil type, vegetation, and urban development.

Climate Is an Average of Highly Variable Weather

The climate of a region is identified by averaging weather statistics on temperature, precipitation, cloudiness, wind, atmospheric pressure, and other factors collected at weather stations over a long period. However, reliable weather records have been compiled for only a brief period of Earth's history (less than 100 years in most of the world), and at a limited number of stations around the world. Climatologists must base their classifications on these limited data.

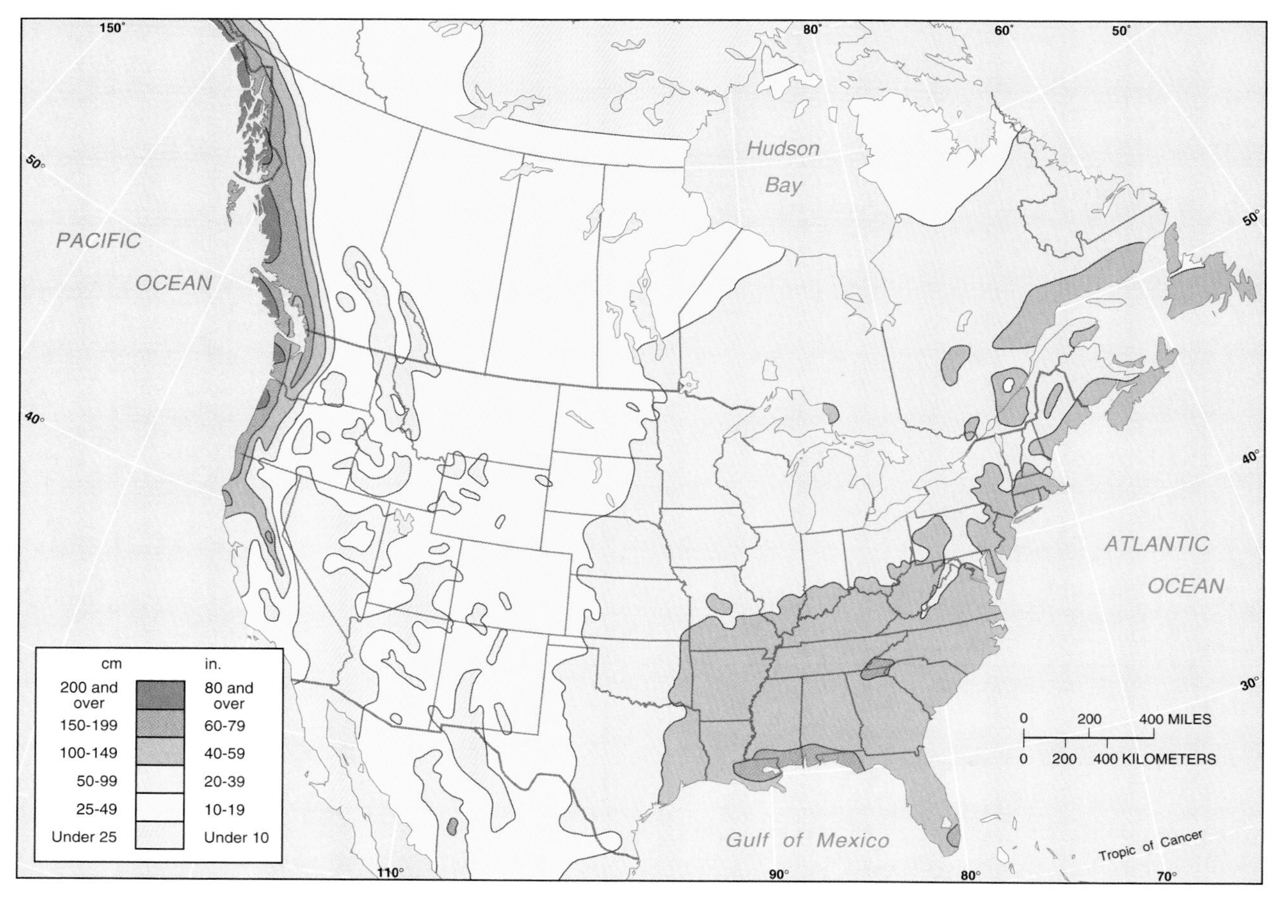

FIGURE 3–4
Average annual precipitation, United States and Canada. Precipitation levels change sharply over short distances in the west because of orographic effects. The western edge of the Rockies receives high rainfall, whereas the western plains are arid as a result of the rain-shadow effect. The highest annual precipitation totals are along the West Coast and in the southeastern United States, which are close to sources of moist air (the Pacific Ocean and the Gulf of Mexico).

Weather data collected at one place over a short period illustrate the difficulty that climatologists face. Figure 3–5 shows temperatures in your authors' home town of Oxford, Ohio, for the month of April 1993:

- Temperatures averaged +10.3°C (+51°F) that month. But when examined on an hour-by-hour basis, air temperature actually was within 5C° (9F°) of this average only 60 percent of the time. The remaining 40 percent of the hourly temperature readings were much warmer or cooler than the average.
- Daily high temperature averaged 15.5°C (60°F), but ranged from 2°C to 25°C (36°F to 77°F).
- Daily low temperature averaged 4.8°C (41°F), but ranged from −2°C to +13°C (28°F to 55°F).

Given these variations, what can you learn about Oxford's climate? Not much, other than you need a variety of clothes, because temperature varies wildly during April!

Interpreting precipitation data for April 1993 also is difficult:

- Oxford received 119 millimeters (4.7 inches) of precipitation during the month.
- Precipitation was recorded during 77 hours in April (the month of April has a total of 720 hours).

- Rain fell on thirteen of April's thirty days, but rainfall was not continuous on those days.

Again, what does the rainfall data tell you about Oxford's climate? Little, except that rainfall appeared to be relatively common, occurring about 10 percent of the time during April 1993. You can see that even detailed data like this for an entire month do not permit us to describe Oxford's climate. Climatologists need at least several decades of data to develop a meaningful average.

Daily weather is both predictable and random. The satellite image on page 42 displays some consistent patterns: low-pressure cells and fronts swirling above the mid-latitudes, clear subtropical zones, and the intertropical convergence zone (ITCZ) of cloudiness and precipitation that girdles Earth around the equator. The image also shows unpredictable elements of daily weather: cells and fronts can take distinctive paths and are of different intensities.

The longer the period over which we observe weather and calculate averages, the more consistency we see. Seasons are more consistent than the daily weather. We know that mid-latitude winters generally are cool and summers hot. Latitudes bring some consistency to weather, too: tropical areas always are warm, mid-latitude deserts always are dry. But variability is always present: as even a casual watcher of *The Weather Channel* knows, every season seems to bring a new high or low temperature record or level of precipitation.

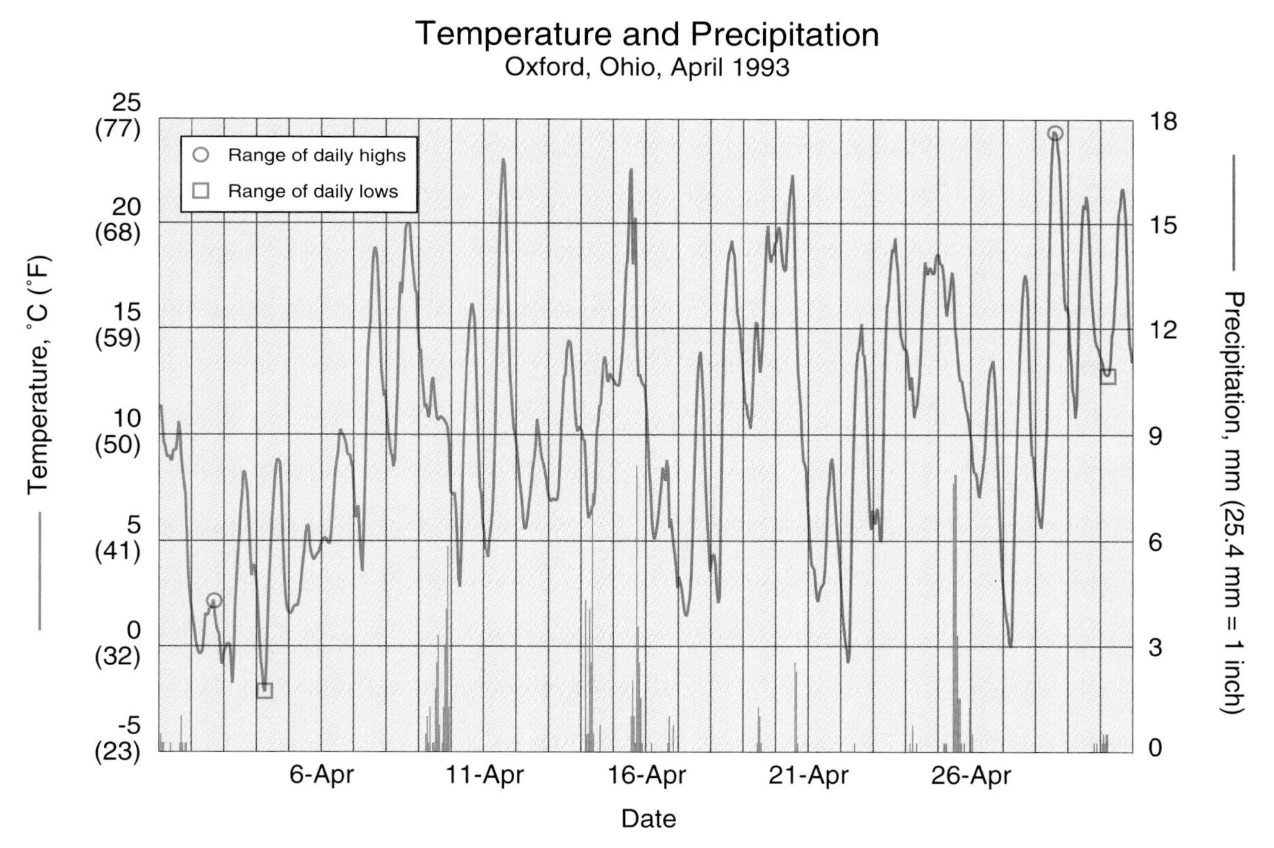

FIGURE 3–5
Hourly temperature and precipitation data for your authors' home town of Oxford, Ohio, during April 1993. Variability is the most obvious feature of any climatic record, including this one. Temperature ranged from −2°C to +24°C (29°F to 76°F). Precipitation totaled 119 millimeters (4.7 inches) for the month, mostly falling with the passage of cold fronts, shown in the large temperature drops.

Temperature and precipitation records are broken frequently because most weather records go back only a few decades, not because Earth's climate is changing dramatically. With reliable records available for only a few decades, we have difficulty in conclusively proving that more subtle changes may be occurring in Earth's climate, such as global warming.

Patterns on a map of climate—that is, *long-term average* weather—differ from those on a weather map for a single day. Averaging conditions over a long period of time—perhaps thirty years—removes some of the variability from daily weather records and thereby sharpens our focus on longer-term geographic patterns. Over the longer run, we can see the relationship between climate and such geographic features as latitude, proximity to the sea, and presence of mountains.

Climate varies less than daily weather during our lifetimes, but Earth's climate does change over hundreds and thousands of years. Evidence that large-scale climate changes have occurred in the past lends credibility to the argument that Earth is again in a period of warming.

Vegetation-Based Climate Boundaries

What temperature forms the boundary between warm and cool? What level of moisture qualifies a climate as humid? Answering these questions is like naming colors: We can generally distinguish between red and orange, but how do we put a boundary between them? Similarly, climates seldom change abruptly across a precise boundary. In reality, one grades into the next in a broad transition zone.

Climatologists look at the distribution of plants to help them draw boundaries between climate regions. This is possible because certain types of plants prefer particular climates. For example, semiarid climates favor shrubland and grassland vegetation, whereas more humid climates favor forest vegetation.

Most climate classifications place boundaries where predominant vegetation changes. This approach identifies boundaries between climate regions that coincide with other environmental features, rather than arbitrary temperature and precipitation levels.

Earth's Climate Regions

Take a moment to review Figure 3–1, because this entire section refers to it. Climate zones form rough horizontal bands around Earth, following lines of latitude. Tropical (A) and dry (B) climates predominate in the low latitudes, warm (C) and cold (D) climates in the mid-latitudes, and ice (E) climates in the higher latitudes. *These horizontal bands show the powerful influence of temperature on climate.* As explained in Chapter 2, higher latitudes receive less insolation than lower latitudes.

Figure 3–1 also shows differences in climates between eastern and western sides of continents, as well as between coastal and inland portions. *This distribution shows the importance of precipitation to climate.* Notice, for example, that western portions of China and the United States have drier climates than the eastern portions of these countries. The distribution of dry climates closely matches regions with low precipitation, as shown on Figure 3–3. A region has a distinctive climate in part because of access to sources of moisture at least part of the year. The following section summarizes the essential features of Köppen's five major climate regions.

Climate regions are divided into several subtypes. The subtypes are identified by adding one or two letters after the letter which designates the major climate region. Keep in mind that enormous variation can occur within each climatic region, and that regions are separated from each other by broad transitions, not sharp boundaries.

The important subtypes are described in a series of boxes beginning on page 105. The description of each subtype includes a climograph depicting monthly average temperature and precipitation for a typical place in the climate region, as well as a map showing the location of the place.

Humid Low-Latitude Climates

Low-latitude climates are warm throughout the year: even in the coldest month, the average temperature exceeds 18°C (64.4°F).

Because low-latitude areas are warm throughout the year, temperatures vary more within a single day than from one month to another. This is the opposite of prevailing conditions in mid-latitude locations, such as North America, where temperature differences between winter and summer are far greater than between day and night.

Low-latitude humid climates (A) lie mostly within 10° north and south of the equator, but can extend to 20° north and south. These areas, under the rainy ITCZ, include the steamy jungles of the world's tropical rain forests. The warm tropical air can hold a large amount of moisture, resulting in heavy precipitation.

High temperatures and rainfall mean that low-latitude climates can support abundant plant growth and productive agriculture. With adequate water supply and fertile soil, two or three harvests per year are possible. About one-third of the world's people live in the humid climates.

The high temperatures mean that a great deal of energy is available to evaporate water, so POTET is very high. Because precipitation is greater than evapotranspiration during most of the year, the tropics generally are humid.

Low-latitude humid climates include these main subtypes:

- **Humid tropical climates** (Af) are humid throughout the year. See page 105.
- **Seasonally humid tropical climates** (Am and Aw) have distinct wet and dry seasons. See page 106.

Dry Climates

Dry climates cover 35 percent of Earth's land area, although less than 15 percent of the world's inhabitants live in the region. Dry lands are generally located in bands immediately to the north and south of the low-latitude humid climates. The most extensive region of dry lands extends from North Africa to Central Asia; the African portion is known as the Sahara (an Arabic word which means *desert*). Sand dunes cover only a small portion of the Sahara and other dry lands; more common is a cover of stone or gravel.

Dry climates are distinguished from other climates by precipitation rather than temperature. Potential evapotranspiration exceeds precipitation in the dry lands. The precise level of precipitation that qualifies a region's climate as dry depends on a formula related to temperature. Higher rainfall totals are necessary to classify a cold place as dry than to classify a warm place as dry.

The dry climates are divided into two major subtypes:

- **Desert climates** (BW) have precipitation much less than of POTET. See page 107.
- **Semiarid climates** (BS) have more precipitation than desert climates, but still less than POTET. See page 108.

Humid Mid-Latitude Climates

Mid-latitude climates include warm (C) and cold (D) climates. In the mid-latitudes, seasonal differences in the amount of insolation produce variations in temperature from one month to the next, as you learned in Chapter 2. In summer the mid-latitudes receive more radiation per day than equatorial areas. But in winter, insolation is cut to one-half or even one-third of summer levels. Consequently, the mid-latitudes experience a distinct cool winter season. Even where frost is confined to only one or two winter months, plants and animals that thrive in the tropics cannot survive winter in the mid-latitudes. Winter restricts life in the mid-latitudes to organisms that can tolerate at least occasional freezing conditions.

Precipitation in the mid-latitudes is heavily influenced by the polar front—the boundary between warm tropical air and cold polar air along which mid-latitude cyclones form and travel. The significance of these storm systems is that daily weather tends to be more variable in the mid-latitudes than in the tropics. This climatic zone experiences rainy spells separated by dry spells, and rainfall can occur at any time of day rather than concentrated in the afternoon and evening.

The cold mid-latitude climates differ from the warm mid-latitude climates according to average temperature. D climates—as the names imply again—are colder than C climates, and they usually are found at somewhat higher latitudes than C climates.

Warm mid-latitude climates (C). In a C climate, the average temperature of the coldest month is between 0°C (32°F) and 18°C (64.4°F). In addition, the average monthly temperature exceeds 10°C (50°F) at least eight months of the year.

In many mid-latitude environments, such as the south-central United States, seasonal precipitation patterns are influenced by temperature. Because warmer air holds more moisture, greater rainfall occurs in summer than in winter. In the southeastern United States, mid-latitude cyclones are less frequent in summer than in winter. But summer rainfall is more intense because more water is present in the air, so winter and summer receive similar amounts of rainfall.

The warm mid-latitude climates include several subtypes:

- **Humid subtropical climates** (Cfa and Cw) have hot summers. See page 109.
- **Marine west coast climates** (Cfb and Cfc) feature mild winters and cool summers. See page 110.
- **Mediterranean climates** (Cs) have dry summers. See page 111.

Cold mid-latitude climates (D). Cold mid-latitude climates (D) have long, cold winter seasons, with some warmth in the summer. The coldest month averages below 0°C (32°F). Four of the months have an average temperature above 10°C (50°F). Although colder than C climates, D climates generally receive less precipitation than C climates.

About 21 percent of the world's land area is influenced by D climates, although only about 13 percent of the world's inhabitants live in the cold climates. D climates do not develop in the Southern Hemisphere, because large land masses do not exist at the latitudes associated with D climates.

The cold mid-latitude climates (D) include several subtypes:

- **Humid Continental climates** (Dfa, Dfb, Dwa, and Dwb) have cold winters and mild summers. See page 112.
- **Subarctic climates** (Dfc, Dfd, Dwc, Dwd) have colder winters than humid continental climates. See page 113.

High-Latitude Climates

High-latitude polar climates (E) are characterized by two important features: low average temperatures and extreme seasonal variability. The average temperature of every month is less than 10°C (50°F).

Annual insolation is relatively low in these climates, but the amount varies during the year. During winter, no insolation reaches the North and South poles, whereas for a few days at midsummer, the poles can receive the highest daily total on Earth.

- **Tundra climates** (ET) have at least one month with temperatures between 0°C (32°F) and 10°C (50°F). See page 114.
- **Ice-cap climates** (EF) do not have any months with temperatures averaging above freezing. See page 115.

Highland Climates (H)

Climate changes with elevation on a mountain. The lower elevations of a mountain may have a tropical or warm climate, while conditions are as cold as the polar climates at high elevations. The complexity of climate patterns makes it virtually impossible to depict climates accurately on a world map, such as Figure 3–1. The H designation indicates high local variability, not simply cold conditions.

The boxes on the following pages illustrate and describe some of the major climate regions.

Low-Latitude Humid Climates

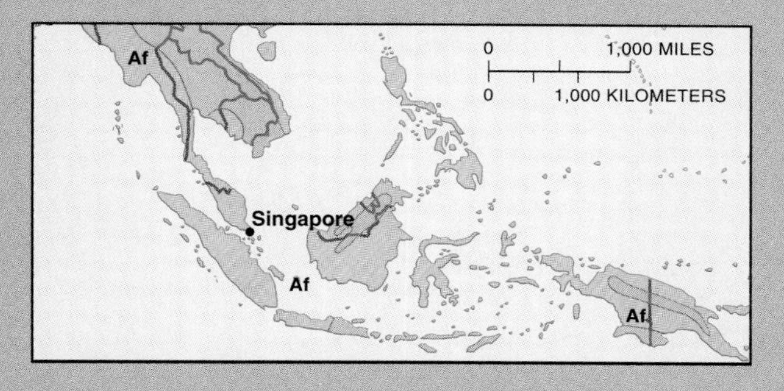

HUMID TROPICAL CLIMATES (Af)

Humid tropical climates (designated Af) generally lie within 10° of the equator. The most extensive areas are in the Amazon River basin of South America, equatorial Africa, and the islands of Southeast Asia.

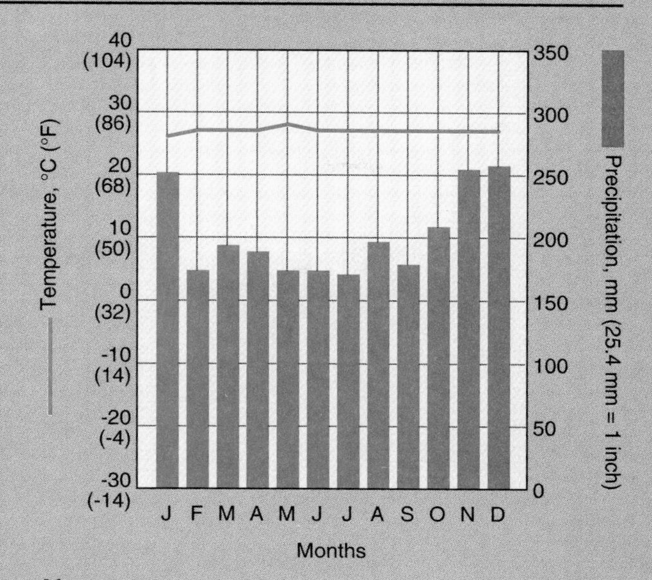

Af
Precipitation: 241.3 cm
Temperature Range: 2°C
Lat/long: 1°10'N, 103°51'E

A thunderstorm in Majuro, Micronesia, a tropical humid climate. Frequent intense rainfall is characteristic of this climate. (Salmoiraghi Micronesia-Majuro/The Stock Market)

Humid tropical climates are influenced by the ITCZ most of the year, so rainfall in these areas is primarily convectional. Atmospheric instability results from intense daily insolation, and the high moisture content of the atmosphere favors severe convectional storms. After clear mornings, rain may fall in the afternoon or evening nearly every day. Under these conditions, the annual totals of rainfall exceed 200 centimeters (80 inches) in many areas, and POTET is generally 120 to 170 centimeters (48 to 68 inches) per year. In oceanic areas of this climate, tropical cyclones are generated, forming hurricanes or typhoons. Southeast Asia, the Pacific islands of Oceania, and the Caribbean are especially prone to these tropical storms.

The small country of Singapore typifies a humid tropical climate. Temperatures are warm throughout the year, averaging between 26°C and 28°C (79°–82°F). Annual precipitation totals about 240 centimeters (94 inches), and every month brings at least 150 millimeters (6 inches). Annual precipitation substantially exceeds evapotranspiration, so the area is very humid, and the vegetation is lush.

Low-Latitude Humid Climates

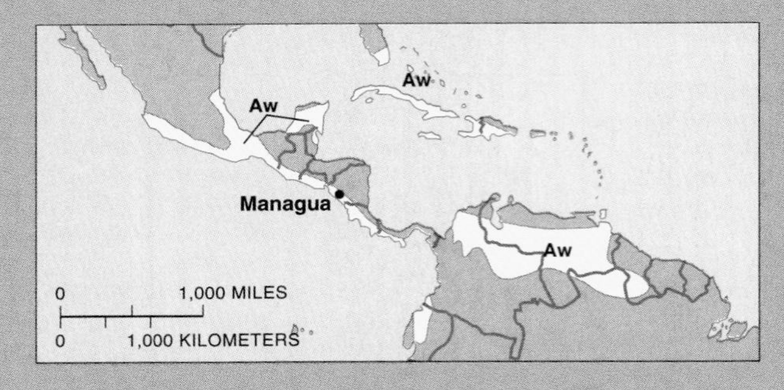

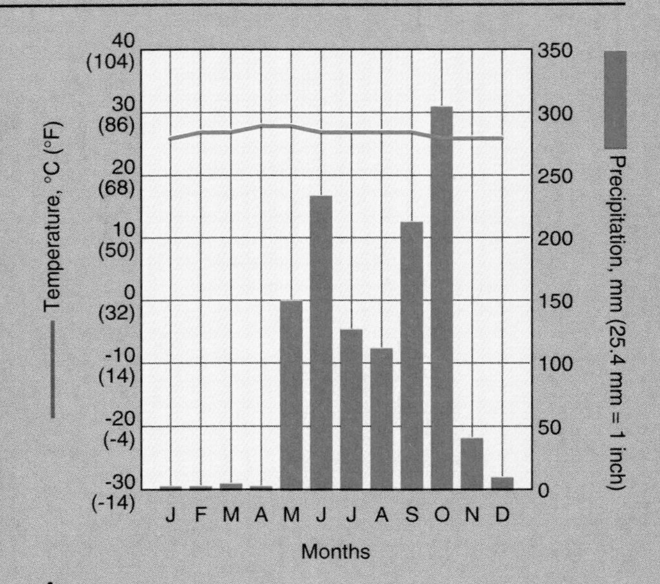

Aw
Precipitation: 120.4 cm
Temperature Range: 2°C
Lat/long: 1°0'N, 86°20'W

SEASONALLY HUMID TROPICAL CLIMATES (Am, Aw)

Many areas of the humid tropics have a distinct dry season, with rainfall concentrated in part of the year. In Central and South America and Africa, the dry season is caused by seasonal shifts in the location of the ITCZ. The ITCZ moves north during the Northern Hemisphere summer (May–October), and south during the winter (November–April), and rainfall shifts with it.

In Managua, Nicaragua, rainfall is heavily concentrated in the six-month period from May through October, with little rain falling in the other 6 months. Annual rainfall totals about 120 centimeters (48 inches), roughly half that of Singapore. Average monthly temperatures vary between 25° and 29°C (77–84°F).

Precipitation is also seasonal in South and Southeast Asia because of the monsoon circulation pattern. Monsoon rains are critical to agriculture in Asia. The timing of this circulation change varies from year to year, and if the rain arrives too early or too late, crop yields may be severely affected.

Some of the highest average rainfall totals in the world (more than 5–10 meters or 16–33 feet) occur on the southern slopes of the Himalayas, where mountains cause orographic lifting. In winter, however, intense high pressure develops over central Asia, and the winds in southern and eastern Asia blow from the north and northwest. The air warms as it descends from the Tibetan Plateau, resulting in extremely dry conditions.

Savanna vegetation is characteristic of seasonally humid tropical climates. In the background, the top of Mt. Kilimanjaro, at 5,895 meters (19,340 feet) elevation, has an icecap climate. (Michele Burgess/The Stock Market)

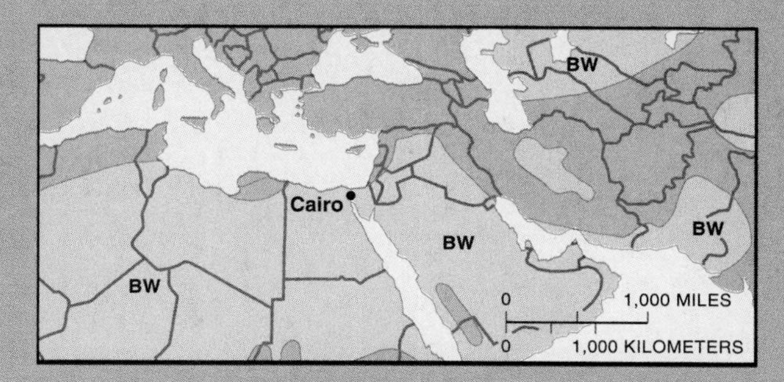

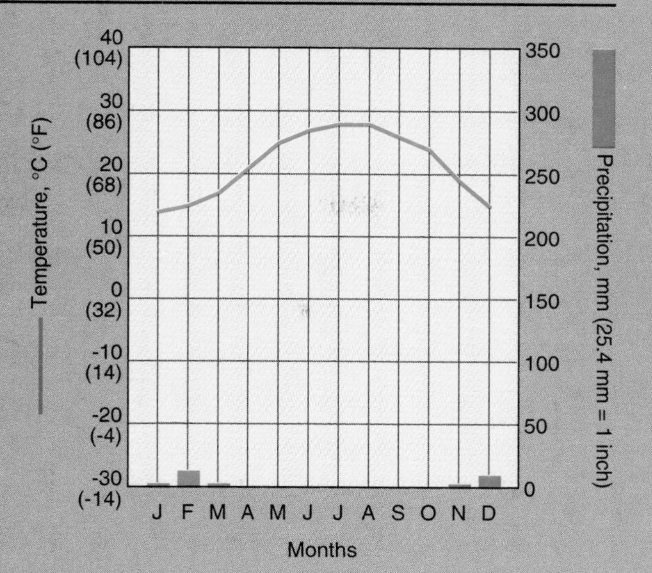

BW
Precipitation: 3.0 cm
Temperature Range: 14°C
Lat/long: 30°1'N, 31°14'E

DESERT CLIMATES (BW)

The most extensive regions of desert climates are found in the low latitudes on the western sides of continents. The deserts of northwestern Mexico and the Atacama Desert of coastal Peru border the Pacific; the Sahara and Kalihari deserts of Africa border the Atlantic; and the Australian Desert borders the Indian Ocean. These BW climate regions result from air descending in the subtropical high-pressure zones over the eastern parts of the oceans (in other words, off the west coasts of the continents).

Cold ocean currents also increase the aridity of these areas. The cold ocean surface cools the lower layers of the atmosphere, inhibiting evaporation and convection, and therefore reducing precipitation.

Warm BW deserts such as Cairo, Egypt, commonly experience high temperatures exceeding 40°C, which is 104°F. Incoming sunlight is not reflected by clouds, and water is not available to absorb the excess heat through evaporation (latent heat). These high temperatures cause great demand for water, so the rainfall supports little vegetation. BW climates have the greatest differences between annual precipitation and POTET. Annual POTET exceeds 150 centimeters (60 inches) in many BW regions, but as precipitation totals less than 20 centimeters (8 inches), the demand greatly exceeds supply.

A desert environment in Tunisia. Potential evapotranspiration in this warm desert is much greater than precipitation, and only vegetation adapted to severe moisture stress can survive. (Richard Steedman/ The Stock Market)

Dry Climates

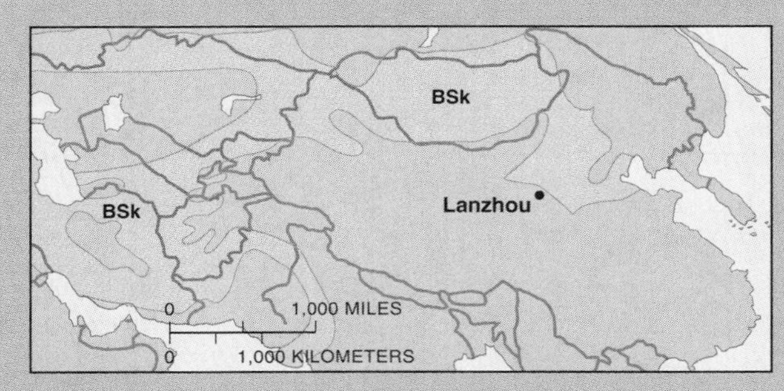

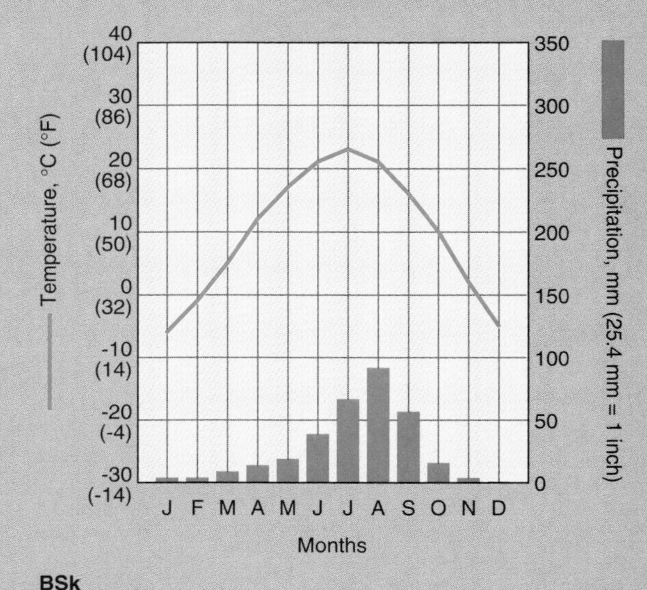

BSk
Precipitation: 31.2 cm
Temperature Range: 29°C
Lat/long: 36°4'N, 103°44'E

SEMIARID CLIMATES (BS)

Not all the world's dry lands are barren of vegetation. In many areas, sufficient rain falls to support plants at least during some seasons. Semiarid climates (BS in the Köppen scheme) are climates with enough moisture to support some vegetation. Annual precipitation is only 20–50 centimeters (8–20 inches). Semiarid climates typically lie in transitional zones between deserts and humid regions.

Some semiarid climates are found in the mid-latitudes, in interior regions of continents cut off by rain-shadow effects from sources of moisture. In Asia, for example, a mountain belt stretching east-west from the Caucasus to China blocks Indian Ocean moisture from the interior. The Rockies in North America and Andes in South America similarly block Pacific moisture from interior regions.

Seasonal temperature contrasts are high in mid-latitude BS climates. Long summer days and bright sunshine warm the ground and, because of lack of water, this heat is not dissipated through evaporation. But unlike tropical deserts, mid-latitude semiarid climates are cool part of the year, so that plants require less moisture to survive. In extremely dry semiarid areas, such as interior basins of the western United States or western China, agriculture is possible with water imported from wetter mountain areas.

Grasses often prosper in semiarid conditions. Grains such as wheat thrive in the grasslands (or *steppes*) of North America and central Asia. These climates also support extensive grazing activities.

Semiarid climates have enough moisture for significant plant growth at least part of the year, but moisture is frequently short. Particularly well-adapted to semiarid climates are grasses, such as this area in the Caucasus mountains of Georgia, in southwestern Asia. (D. Thomas/ Photo Researchers, Inc.)

Warm Mid-Latitude Climates

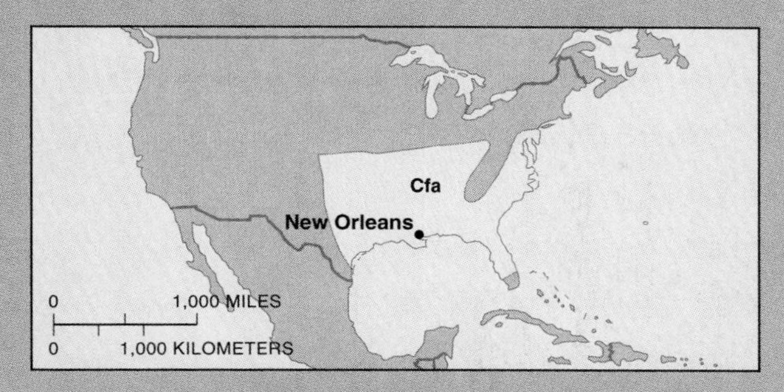

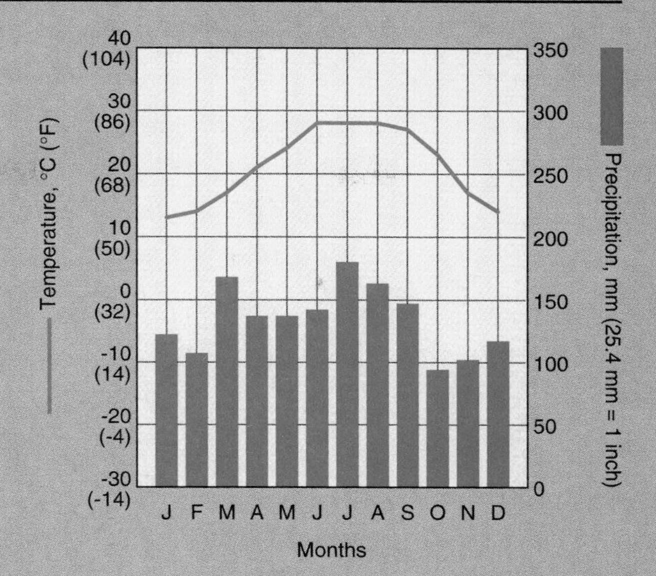

Cfa
Precipitation: 161.5 cm
Temperature Range: 15°C
Lat/long: 30°0'N, 90°5'W

HUMID SUBTROPICAL CLIMATES (Cfa, Cw)

Humid subtropical climates cluster on the eastern sides of continents between about 25° and 40°, and on the western sides between about 35° and 50°. Cfa climates are relatively warm most of the year. Cool winter temperatures bring occasional frost in low latitudes, and snowfall occurs poleward, but the ground does not remain frozen or snow-covered for long.

The warm, humid conditions of Cfa climates support a long growing season and highly productive agriculture, sometimes permitting two harvests in a year. Plants adapted to tropical humid climates cannot survive the winter frost. Instead, most Cfa climates have deciduous species that lose their leaves in autumn and become dormant in winter. Eastern China, Western Europe, southeastern United States, and parts of Brazil and Argentina are the largest areas of Cfa climates.

Precipitation can be concentrated in summer or distributed evenly through the year. Increased summer rainfall results from greater moisture in the warmer air, as in New Orleans.

The subtropical winter-dry climates (Cw), found in parts of South and Southeast Asia, have a distinct winter dry spell because of the monsoon circulation. Agriculture and water supplies do not suffer during the winter dry season, though, because temperatures are lower then, and therefore POTET is low.

Humid subtropical climates support lush vegetation, such as this wetland forest in Louisiana. Many of the world's most productive agricultural regions are in this climate. (Randy O' Rourke/The Stock Market)

Warm Mid-Latitude Climates

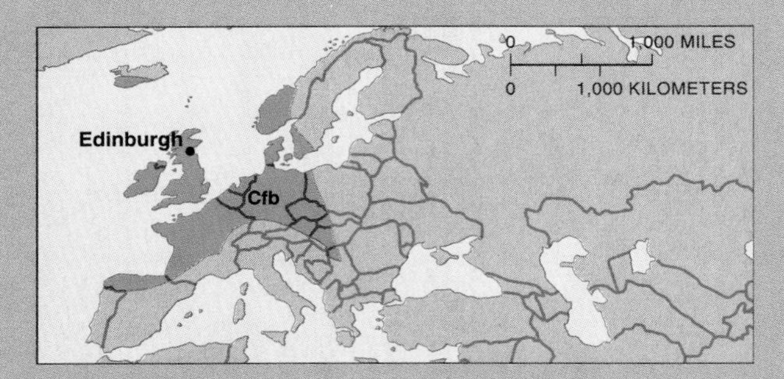

Cfb
Precipitation: 65.8 cm
Temperature Range: 12°C
Lat/long: 55°57'N, 3°12'W

MARINE WEST-COAST CLIMATES (Cfb, Cfc)

The west coasts of continents between about 35° and 65° north and south latitudes have mild climates with little seasonal changes in temperature and plentiful moisture year-round. Major regions include western North America between northern California and coastal Alaska, and southern Chile, as well as places in northwestern Europe such as Edinburgh, Scotland.

Marine climates are moderated by ocean temperatures. Typical summer temperatures are 15° to 25°C (60° to 77°F), about 10C° (18F°) cooler than locations further inland at these latitudes. Similarly, average winter temperatures of –5°C to +15°C (23°F to 59°F) are milder than inland winters at the same latitudes.

Because of mild temperatures, some plants stay green all winter, even when snow occasionally falls. Maritime influence on temperature is so great that winter temperatures normally associated with subtropical latitudes are found as far poleward as 55°. Kodiak, Alaska, in a maritime location at 57° north latitude, has the same January average temperature as Richmond, Virginia, which is at 37° north latitude, in a humid subtropical region.

The North Atlantic and the northeastern and southern Pacific are known for storms which bring plenty of moisture to the adjacent land areas. Mid-latitude cyclones are frequent, especially in winter. The low temperatures mean that air contains less moisture than at lower latitudes, so drizzle is common. Annual rainfall is about 70 to 120 centimeters (28 to 47 inches). Where mountains are near the coast, as in western North America, Chile, and Norway, orographic effects contribute to high rainfall, 2 to 3 meters (80 to 120 inches) per year.

Normandy, France has mild winters despite its relatively high latitude (49°N). (David Frazier/The Stock Market)

Warm Mid-Latitude Climates

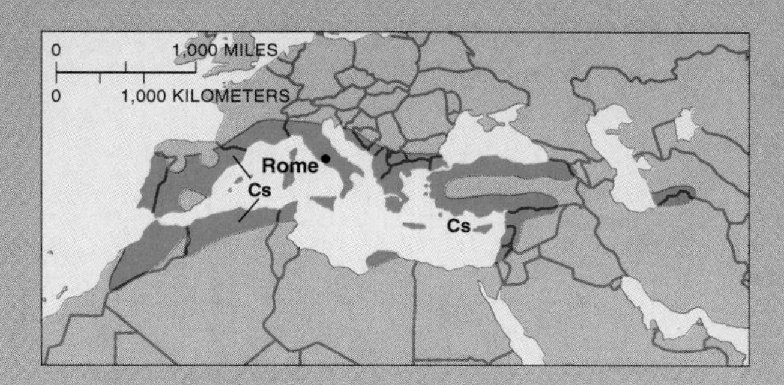

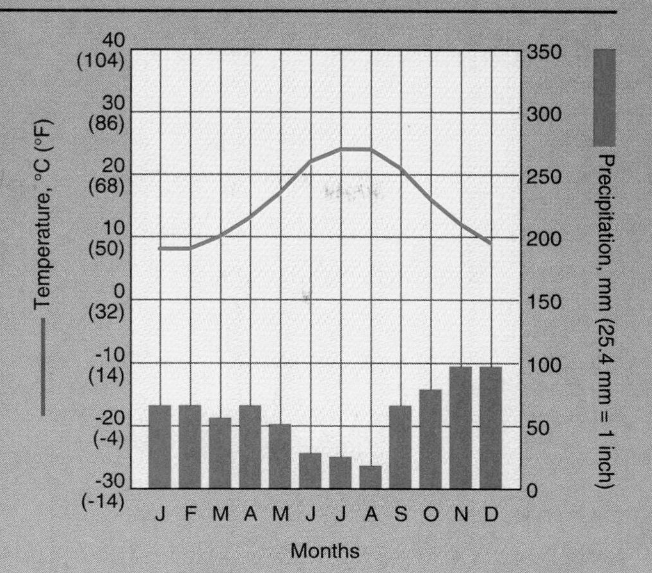

Csa
Precipitation: 71.4 cm
Temperature Range: 16°C
Lat/long: 41°54'N, 12°30'E

MEDITERRANEAN CLIMATES (Cs)

Mediterranean climates occur on the western margins of continents. As the name indicates, this climate is typical of the lands bordering the Mediterranean Sea in southern Europe and northern Africa. Similar climates are found in extreme southern Africa, southwestern Australia, Southern California, northern Mexico, and central coastal Chile.

Precipitation is seasonal, because of movement of the subtropical high-pressure zones. In summer, these zones move poleward and bring aridity. In winter, they move toward the equator and are replaced by more frequent storms of the mid-latitude low-pressure zone.

In Mediterranean climates, cool, rainy winters bring occasional frost or snow, with frequent mid-latitude cyclones. Warm, dry summers have 20°C to 35°C (68° to 95°F) temperatures.

Mediterranean climates have significant water-availability problems. Precipitation occurs in winter, when demand for water among plants and animals is relatively low. Very little rainfall occurs in the summer, when POTET is highest. Plants with roots that can reach deeper soil moisture or ground water stay green during the summer, but shallow-rooted grasses dry out. Because water is scarce in the summer, many plants grow mostly in the winter, even though sunlight is less then. Agriculture in Mediterranean climate regions requires storage of winter precipitation in reservoirs for irrigation in summer.

Seasonal variations in water availability change the appearance of the landscape in Mediterranean climate regions, such as coastal California, (left) winter (right) summer. (Connie Coleman/Allstock)

Cold Mid-Latitude Climates

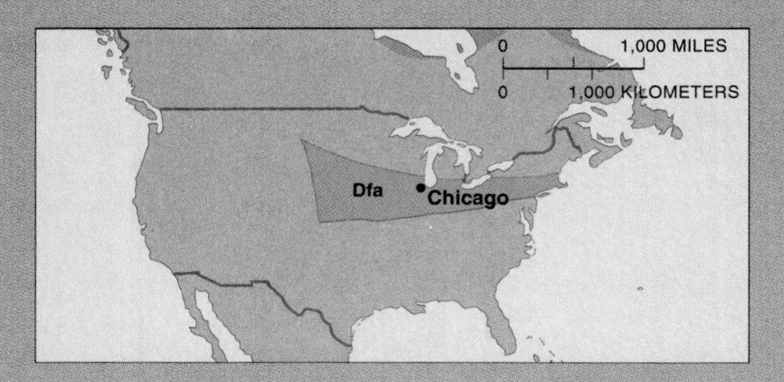

HUMID CONTINENTAL CLIMATES (Dfa, Dfb, Dwa, Dwb)

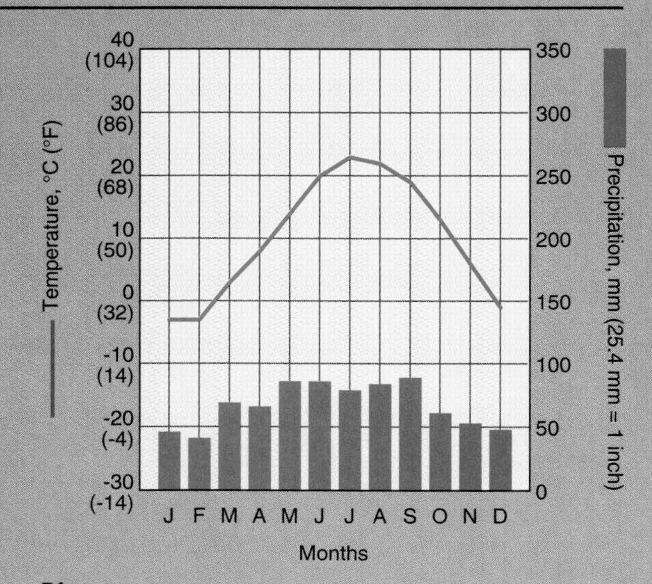

Dfa
Precipitation: 80.8 cm
Temperature Range: 26°C
Lat/long: 41°45'N, 87°40'W

Humid continental climates are so-named because they are located in the interior of continents, remote from the oceans. These climates occur between about 35° and 60° latitude, in the interior and eastern portions of Northern Hemisphere continents. Many eastern coastal areas are also classified as humid continental climates, because prevailing westerly winds extend the climate's influence from the interior to them.

Humid continental climates display strong seasonal contrasts in temperature. Summer temperatures are warm: because the days are long, and the sun angles are high, the region receives more solar radiation during summer than tropical climates. But in winter, when days are short and the sun is low in the sky, temperatures often fall below freezing. In more southerly humid continental areas, such as New York City, January temperatures may average close to freezing, but frost and snow still occur. Farther north, in places like Moscow, Russia, winter may bring many weeks of subfreezing temperatures.

In most humid continental climates, precipitation occurs throughout the year from frequent mid-latitude cyclones. Typical annual rainfalls are 60 to 150 centimeters (24 to 60 inches). With an annual POTET of 50 to 90 centimeters (20 to 35 inches), precipitation is usually ample for plants. In winter, plants need little water, but in summer they demand every drop that falls. During summer dry spells, farm fields may require supplemental irrigation.

Humid continental climates have strong seasonal contrasts, as seen in these two views of a birch woodland in summer and winter. (Gregory K. Scott/Photo Researchers, Inc.)

Cold Mid-Latitude Climates

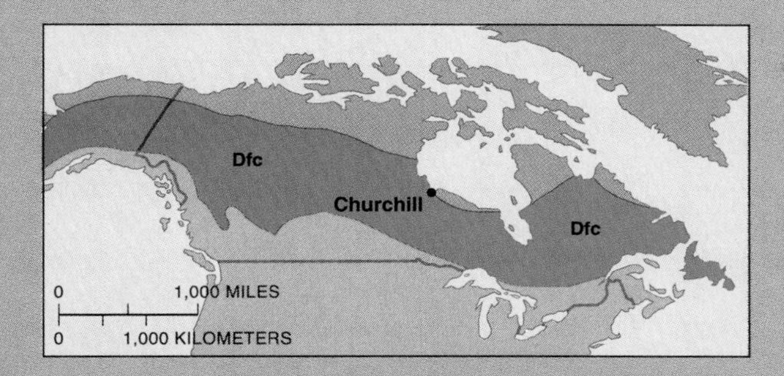

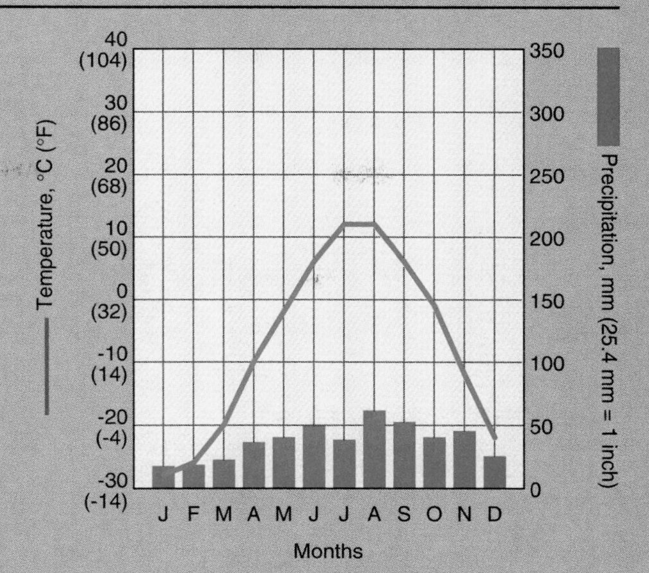

Dfc
Precipitation: 44.3 cm
Temperature Range: 40°C
Lat/long: 58°45'N, 94°5'W

SUBARCTIC CLIMATES (Dfc, Dfd, Dwc, Dwd)

Like the humid continental climates, subarctic climates are essentially limited to the Northern Hemisphere, because of little land mass at these latitudes in the Southern Hemisphere.

Winters are extremely cold in subarctic climates, −20°C to −10°C (−4°F to +14 °F), and temperatures remain well below freezing for many months. Summers bring mild temperatures approaching those of much warmer climates; 10°C to 20°C (50°F to 68°F), with warm spells above 30°C (86°F). Temperatures rise rapidly during summer, because these areas receive 15 to 20 hours of daylight.

Annual rainfall is low, generally 20 to 50 centimeters (8 to 20 inches), an ample level for plants because virtually no evapotranspiration occurs for six months of the year. Winter precipitation usually is locked up as snow and ice until the spring melt releases this water to plants. Summer water shortages are rare, even though POTET may exceed precipitation during part of the summer.

The growing season is short, generally June to September. The extensive regions of spruce and pine covering much of northern Canada and Siberia are called boreal forests, and the subarctic climates are sometimes called **boreal forest climates**.

Winter near Hudson Bay, Canada, located in a subarctic climate region. (Stephen J. Krasemann/Photo Researchers, Inc.)

High-Latitude Climates

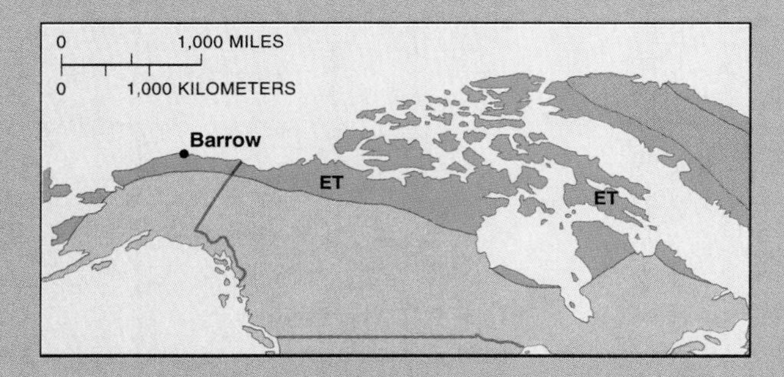

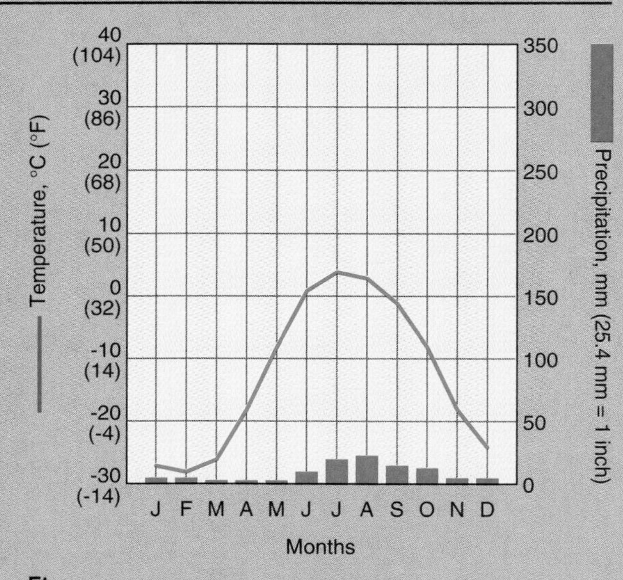

Et
Precipitation: 11.0 cm
Temperature Range: 32°C
Lat/long: 71°28'N, 156°0'W

TUNDRA CLIMATES (ET)

At its northern limit, the boreal forest gives way to a treeless landscape in which vegetation survives the long winter by staying close to the ground, where it is protected from winter's extremes. This low-lying, hardy vegetation, called tundra, gives the climate its name.

Tundra temperatures are low all year, rising above freezing only for a brief, cool summer. The annual average temperature is below freezing. Winter temperatures below −20°C (-4°F) are common, typically staying below freezing for 3 to 5 months. Mid-summer daytime highs rarely exceed 10°C to 15°C (50°F to 59°F), even though the sun can shine 20 to 24 hours a day. The ground may be permanently frozen, a condition is known as **permafrost**.

Annual precipitation rarely exceeds 30 centimeters (12 inches). Most falls as snow, covering the ground for months except where bared by the wind. Although annual rainfall is slight, water is plentiful because evapotranspiration rates are low. Most of the year water is present as ice rather than liquid. In the brief summer when the ground thaws, the water may not be able to drain from the soil because of permafrost below the surface; saturated soils are therefore common.

Flowers bloom during the brief summer growing season in Ellesmere Island, northern Canada, located in a tundra climate region. (Stephen J. Krasemann/Photo Researchers, Inc.)

High-Latitude Climates

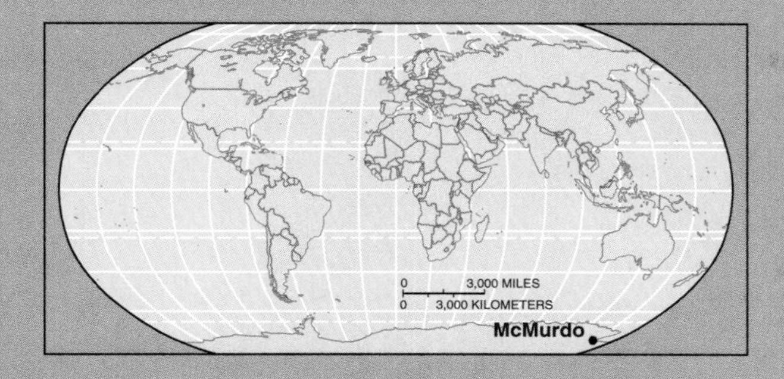

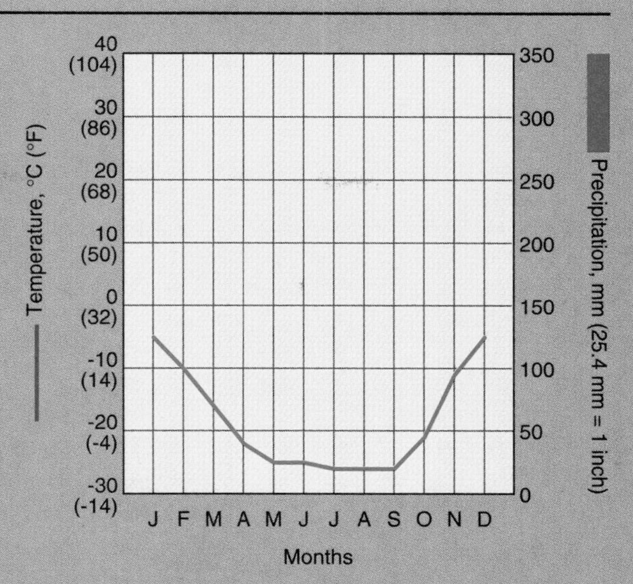

Ef
Precipitation: 0 cm
Temperature Range: 21°C
Lat/long: 77°0'S, 170°0'E

The edge of the Antarctic ice cap. In ice-cap climates, temperatures are only rarely above freezing, and glaciers form from the accumulation of snow. Virtually all of Antarctica has an ice-cap climate, with little glacial melting. The glaciers flow to the sea, forming icebergs. Global warming could accelerate the rate of glacial melting, raising the level of the seas. (Robert W. Hernandez/Photo Researchers, Inc.)

ICE-CAP CLIMATES (EF)

Near the poles are climates where even the warmest month averages below freezing. Virtually all of Antarctica and most of the Arctic Ocean have an ice-cap climate. Permafrost is extensive and thick beneath a year-round cover of snow or ice. The extremely low temperatures prevent the air from holding much water, so annual precipitation usually is less than 10 centimeters (4 inches). Precipitation falls as snow and compacts into a thick mass of glacial ice. Cores of ice extracted in Antarctica indicate average annual precipitation of at most a few centimeters. Because of low precipitation, as well as high winds, Antarctica contains desertlike valleys that are free from snow or ice cover.

Past Climate Change

So far, you have seen that climates vary from place to place. Climates also vary over time. The world climate map (Figure 3–1) is based on averages over periods of a few decades in the mid-twentieth century. However, the world climate map has not always looked like this. Considerable change has taken place through the years—and will continue in the future.

Humans have long been able to modify environments, sometimes radically, by building settlements, clearing forests for agriculture, domesticating animals, manufacturing goods, damming rivers, extracting resources, and filling the air, seas, and land with waste. Earth's land surface is profoundly different today from what it was only 200 years ago. The forests of the eastern United States were almost completely removed by 1900, and the tall-grass prairie of the Great Plains was replaced by agricultural crops. Only recently have people realized that humans, in addition to modifying Earth's surface, probably have set in motion global-scale climatic changes. Debate continues about the nature of these changes.

To interpret the significance of *future* climate change, we must understand patterns of *past* climate changes. Studying past climate changes can give us valuable clues in determining whether recent changes are unique because of human actions.

Climatic Change Over Geologic Time

Viewed over Earth's geologic time (4.6 billion years), the climate of the last 3 million years has been quite exceptional. This period, which includes our present time, is known to geologists as the **Quaternary Period**. Earth has experienced more climatic variability during the past 3 million years than in most of the previous 200 million years. Periods of similar variability probably occurred prior to 200 million years ago, but the record of Earth's earlier history is not very clear.

Climate during the Quaternary Period has included intervals in which global average temperature was as much as 10C° (18F°) cooler than the present, and intervals in which the climate was warmer than today (Figure 3–6). Earth's climate has shifted between warm and cool periods between 10 and 30 times during the Quaternary Period, although the exact number of glacial periods is not known.

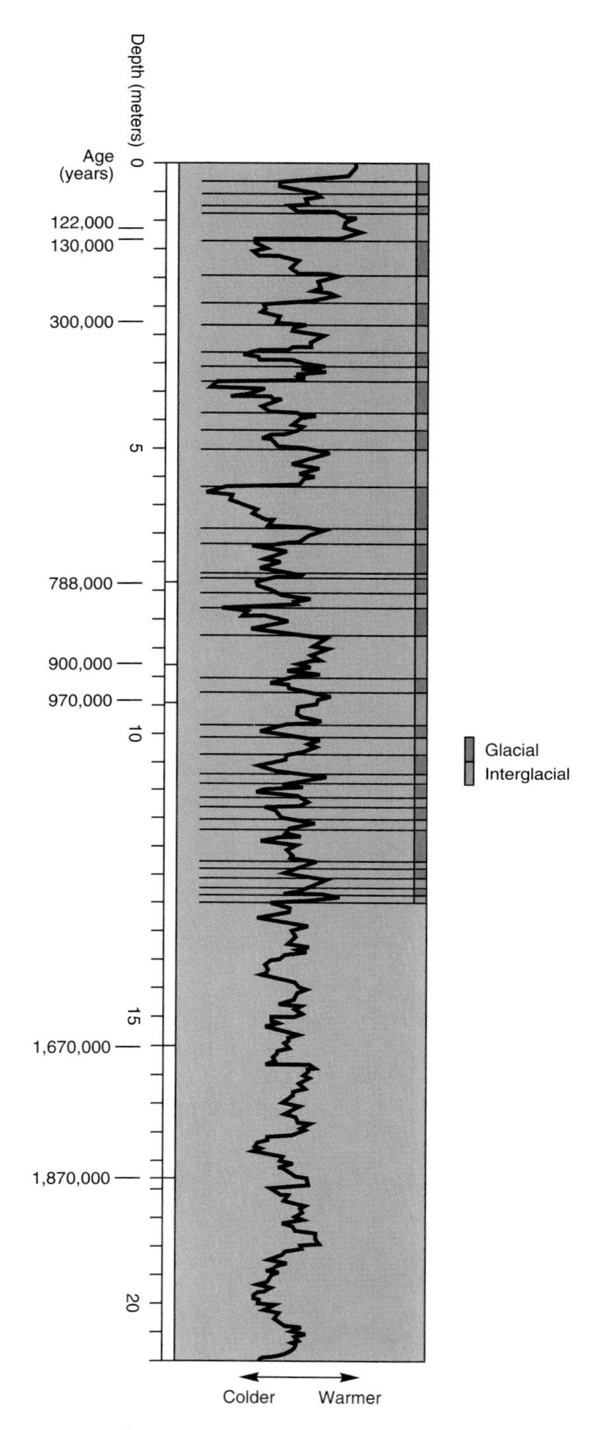

FIGURE 3–6
Earth's temperature for the past 1.9 million years. Evidence comes from samples taken from drilling beneath the ocean floor. In geologic terms, this is a short time period—only the last 0.04 percent of Earth's 4.6 billion-year history. Several episodes of continental glaciation have been separated by warmer periods, including the present day.

During the cooler periods, great continental ice sheets—like those covering much of Antarctica and Greenland today—extended over much of North America, northern Europe, and northern Asia. These ice sheets were giant glaciers, formed from accumulating snow that could not melt in the chilly climate. For an idea of their size, the ice sheets covering Greenland and Antarctica today are 3 kilometers (1.8 miles) thick! In everyday language we describe periods of heavy glaciation as the Ice Ages.

Even unglaciated areas—those not covered by glaciers—were much different than today. Areas that today are covered with deciduous forest were frozen tundra, and lakes existed in places that now are deserts. The cool periods were separated by warmer times, some considerably warmer than today. Subtropical climates extended well into the mid-latitudes.

The last Ice Age reached its maximum only about 18,000 years ago—within the period of human archaeological records. At that time global average temperatures were about 5C° (9F°) cooler than the present. The melting of the ice back to its present extent was completed only about 9,000 years ago. So, Earth now is in a period of relatively warm climate.

Recent evidence from drilling into Greenland's ice sheet suggests that Earth's climate in the past 10,000 years has been remarkably stable in comparison to that of the preceding 150,000 years. Perhaps this stability helped humans to flourish. Prior to 8,000 B.C., the global average temperature may have changed by as much as 10C° (18F°) in only a few decades!

Earth's climate and people during the past 18,000 years. To better understand the great variability in climates in recent geologic time, let's consider the climates of the world 18,000 years ago as the last great Ice Age was ending. At that time, much of the Northern Hemisphere north of about 40°N was ice-covered; the coverage was particularly extensive over North America and northern Europe. (See Figure 3–7, which is helpful in this discussion.) Extensive grasslands spanned Asia and North America, where forests now grow. Sea ice extended over much of the ocean surrounding Antarctica, and most of Argentina was covered with ice and snow. Some tropical and subtropical areas were much drier than today.

Sea level was about 85 meters (280 feet) lower than at present. Many areas that today are under water were exposed. For example, most of Indonesia was connected to the Asian mainland, and New Guinea was joined to Australia. Alaska and Russia were connected, as were Great Britain and France. These land bridges were important migration routes for animals and humans. North America's first human inhabitants may have walked across from Asia by way of Alaska.

Beginning about 16,000 years ago, Earth's climate started to warm dramatically. Glaciers melted, and the water they contained drained into the oceans, raising sea level. The melting was particularly rapid about 15,000 to 12,000 years ago. Extreme cold renewed glaciation briefly between about 12,000 and 11,500 years ago, but by 9,000 years ago the glaciers had receded nearly to their present positions. By 6,000 years ago, world temperatures were about as warm as the present, perhaps even warmer in some areas.

Within the past 1,000 years climate has varied, though less dramatically than in the past. Recent climate changes have been important in European history. For example, between A.D. 800 and 1000, the Vikings—seafaring pirates and adventurers from present-day Scandinavia—extensively explored the North Atlantic Ocean and established settlements in Greenland and possibly North America. What has this to do with climate? Lack of sea ice during those centuries as a result of especially warm temperatures aided the Vikings' exploration.

Cooling occurred in the Middle Ages, beginning about A.D. 1200. The period from about 1500 to 1750, when temperatures were especially cool, is known as the "Little Ice Age." Glaciers advanced in Europe, North America, and Asia. Since the early 1700s, and especially after 1900, climates have warmed relatively steadily. Most of Earth's glaciers have been shrinking since the early 1800s. Short cooling intervals occurred in the 1800s, including one in 1884–1887 caused by the eruption of the volcanic island of Krakatau in Indonesia. Krakatau spewed so much ash and sulfur dioxide into the atmosphere that it prevented some shortwave insolation from reaching Earth's surface. The 1930s and 1940s were relatively warm, cooling occurred from about 1945 to 1970, and the 1980s were warm again.

Important to us is that these climatic fluctuations have occurred well within the period of human settlement of Earth. As noted, glacial melting from the most recent Ice Age was particularly rapid

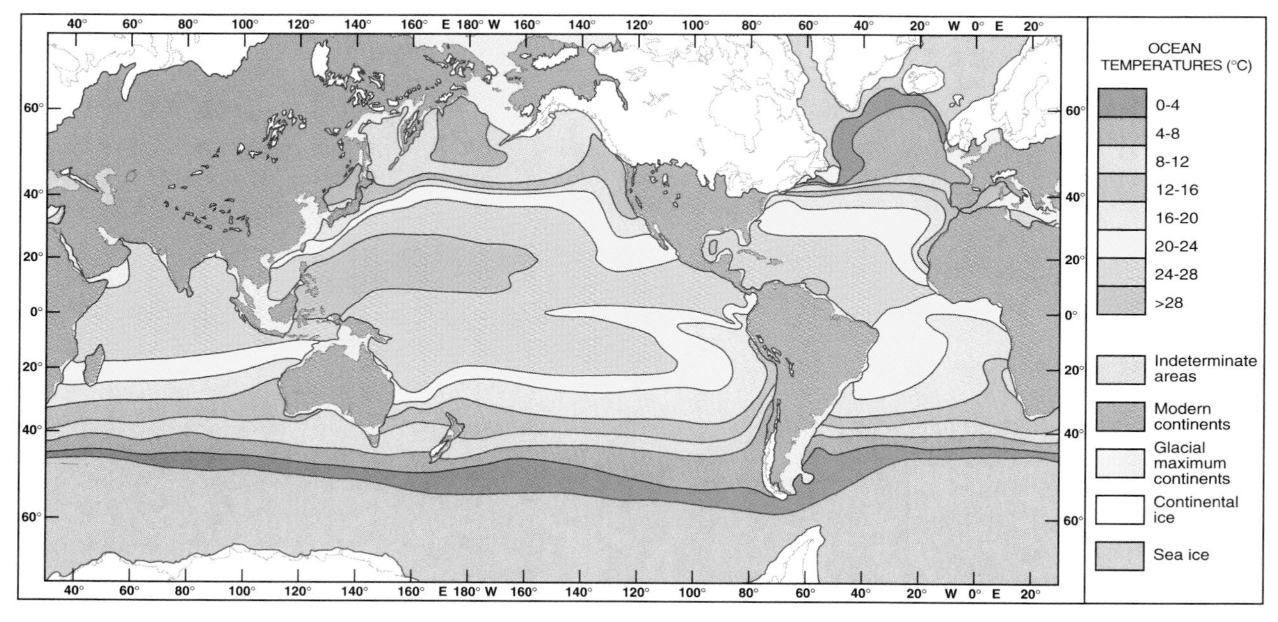

FIGURE 3–7
Ocean temperatures in August and the locations of major glaciers (white areas) about 18,000 years ago. Glaciers blanketed Canada, the northern United States, northern Europe, all of Iceland, and southern South America. Earth's land area was much greater—and the area covered by oceans smaller—because storage of water in glaciers lowered sea level. This increased land area compared with the present is shown in cream color. Important land areas in the past include southeastern Asia, South America, and the land bridge between Russia and Alaska. It was across this land bridge that early Native Americans may have migrated from Asia. A similar bridge connecting Great Britain also encouraged migration. The recently opened English Channel Tunnel (known as the Chunnel) permits humans to travel freely between Great Britain and France by surface transportation for the first time in several thousand years. On this projection, the apparent area of glaciation is exaggerated because the projection does not show areas equally. High-latitude areas appear larger in relation to low-latitude areas. (From R. W. Christopherson, *Geosystems,* 2d ed., Macmillan Publishing Co., 1994.)

about 15,000 to 10,000 years ago. Archaeologists have found that agriculture and cities probably developed about 10,000 to 8,000 years ago. Most scholars of early human culture recognize that climate change was fundamental to development of civilizations, influencing the availability of water for crops, and perhaps driving major migrations.

The period of glacial melting and warming between 15,000 and 5,000 years ago coincided with expansion of human settlement in northern Europe and North America. Settlement extended gradually northward on the European mainland, and was well-established throughout much of the British Isles by 2,000 B.C.. Meltwater from the glaciers raised sea level, inundating low-lying areas and separating land areas. The English Channel filled, isolating the British Isles from Europe, and the land bridge connecting North America and Asia was submerged, halting migration from Asia into North America.

Modern human inventions such as heating, housing, transportation, and industry have made us less dependent on day-to-day conditions. But we remain intimately controlled by the dynamics of the atmosphere, both short term (weather) and long term (climate). Many believe that as a result of human actions climate will change much more rapidly in the future.

Possible Causes of Climatic Variation

Why does climate vary? Many potential causes exist, but it is difficult to say which are important and which we can ignore. Understanding the causes of climatic change is important, for if we learn that climate is affected by human activities, such as burning fossil fuels, then we can take action to limit these effects. On the other hand, if climatic change is entirely governed by natural processes, then we need not be as concerned about increasing CO_2

concentrations in the atmosphere. We will summarize three possible causes of climatic change: astronomical factors, geologic processes, and human modification of Earth's surface and atmosphere.

Astronomical hypotheses. In Chapter 2 you saw that inputs of solar energy drive atmospheric circulation, and thus climate. Changes in insolation could alter climate. Humans know little about variations in the amount of solar radiation reaching Earth. But we certainly know that astronomical factors could cause both long-term and short-term variations in insolation.

In the long run—over tens of thousands of years—we know that the geometry of Earth's revolution around the sun fluctuates (Figure 3–8). For one thing, Earth's orbit varies in shape on a 100,000-year cycle, taking Earth closer to and farther from the sun. Also, Earth spins like a top, but a slightly unbalanced one, and because of this imbalance it wobbles slightly and very slowly as it spins. To picture this, imagine a fast-spinning top that is winding down and losing momentum. The top begins to wobble slightly, its axis no longer stays vertical, and it starts to wander. Earth is doing the same thing, although each wobble takes about 26,000 years. This means that the tilt of Earth's axis is not always 23.5°; it can vary between 22° and 24°. This doesn't sound like much, but recall that seasons where you live vary annually because the angle of insolation shifts. Other variations in the geometry of Earth's orbit around the sun also occur, over periods of tens of thousands of years.

Variations in solar energy input that result from these wobbles are called **Milankovich cycles,** after an astronomer Milan Milankovich, who hypothesized about their link to climate. The timing of major cold and warm periods over the past 2 million years roughly corresponds to these variations in Earth's-sun geometry. The Milankovich hypothesis does not explain shorter-term climatic change, however.

One short-term astronomical factor that may cause variation in the sun's radiation output is sunspots. **Sunspots** are relatively cool regions on the surface of the sun that vary in number and appear and disappear over a cycle lasting 11 years. (The number of sunspots may also vary over longer periods.) Sunspots affect the output of solar energy. Some climate variations appear to have periods close to the 11-year sunspot cycle. The sunspot cycle also affects ozone concentrations in the upper atmosphere.

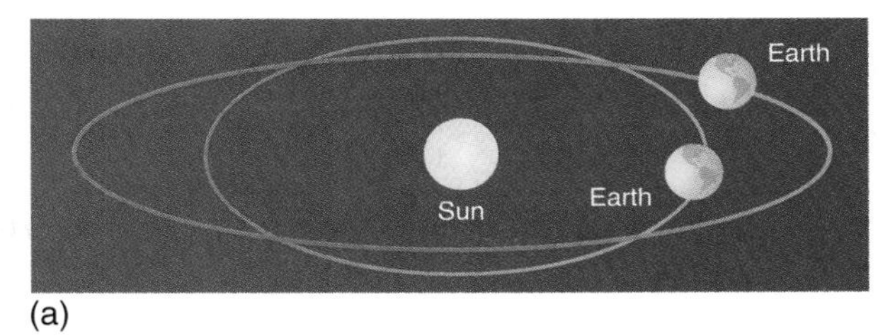

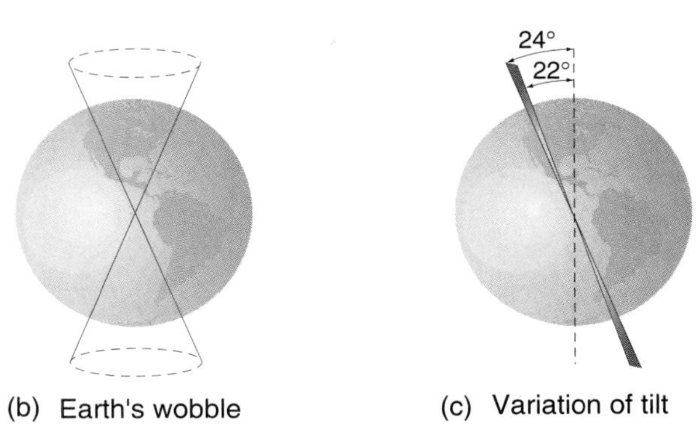

FIGURE 3–8
Climate may vary with long-term cycles that are part of Earth's revolution around the sun.

Geologic hypotheses. Geologic factors also may cause long-term and short-term climatic change. One long-term mechanism is *continental drift*, or *plate tectonics*. This remarkable phenomenon will be discussed further in Chapter 5, but briefly, **continental drift** is the movement of plates of Earth's crust at rates of up to a few centimeters per year. This action makes the continents migrate, so after millions of years a continent might move great distances, opening up oceans or draining them. As we learned in Chapter 2, the locations of continents and oceans affect atmospheric circulation, including the ITCZ, the monsoon circulation of Asia, mid-latitude cyclones, and the stormy zone around Antarctica.

Plate tectonic movements have also caused the formation of major mountain ranges, such as the Himalayas, Andes, and Rockies. Continental movements are ponderously slow, so they cannot explain variations within the Quaternary Period. But they might help explain differences between the Quaternary Period and the previous 200 million years.

Volcanic eruptions can influence climate for a few years by injecting large amounts of dust and gases—especially sulfur dioxide—into the upper atmosphere. These gases reduce the amount of solar radiation filtering through the atmosphere to Earth. The eruption

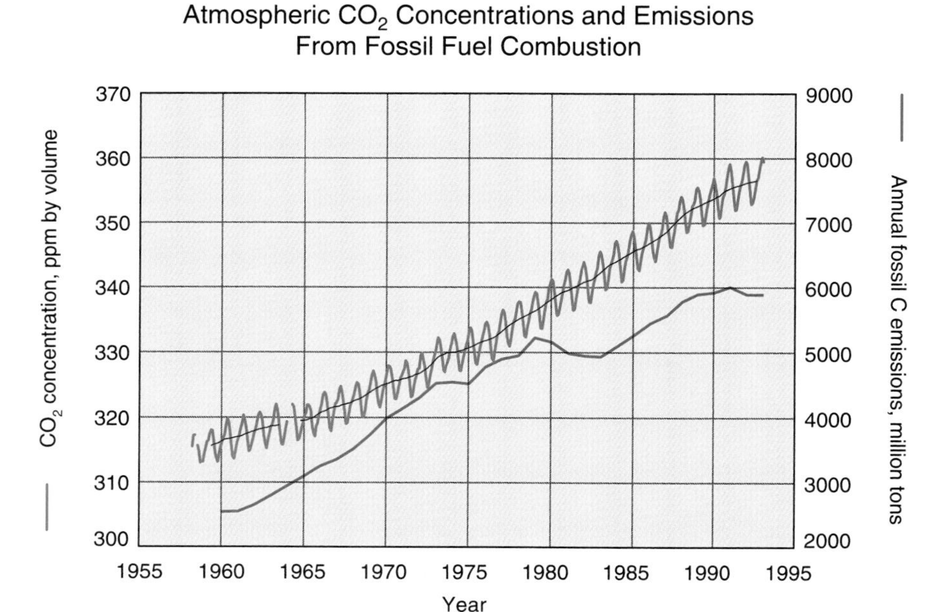

FIGURE 3–9
Atmospheric carbon dioxide concentrations and emissions from fossil fuel combustion.

of Krakatau volcano in Indonesia in 1883 caused a marked cooling between 1884 and 1887, and the 1991 eruption of Mt. Pinatubo in the Philippines may have been a factor in the unusually cool weather of 1992. Past periods of more frequent volcanic eruptions may have lowered temperatures at other times.

Human causes: Disturbing the land surface and atmospheric composition. Processes like sunspot cycles and continental drift are beyond human control. However, humans are active participants in changing the climate. Two important ways that humans influence climate are altering the atmosphere and removing vegetation.

The carbon dioxide content of the atmosphere has increased dramatically since the start of the industrial revolution in the late eighteenth century (Figure 3–9). Most atmospheric scientists believe that elevated CO_2 levels cause warming, because CO_2 is a greenhouse gas.

Analyses of air trapped in Antarctic ice reveals that CO_2 was higher during past warm periods, and lower during glacial periods (Figure 3–10). Dead plant matter lying in swamps and in the soil decays during periods of warm climate, releasing more carbon to the atmosphere. During cool periods carbon accumulates on Earth's surface, reducing atmos ɛric CO_2. Also, the oceans may store more

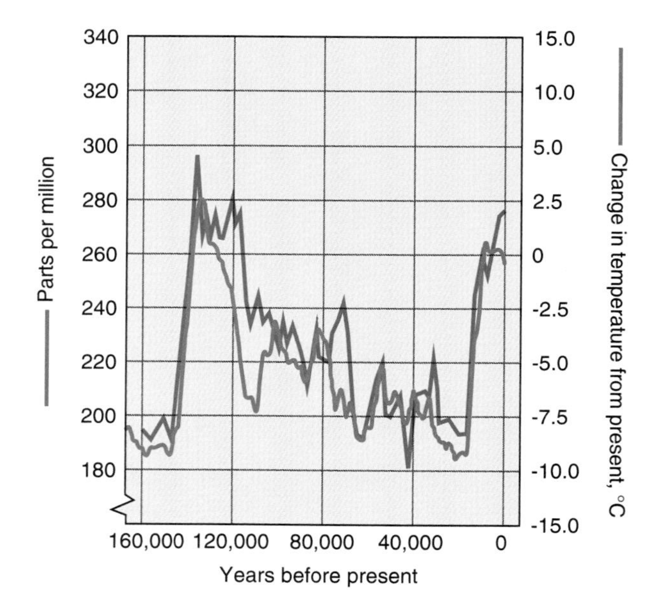

FIGURE 3–10
Antarctica has a dry climate, so the small annual snows accumulate slowly over the millennia. Glacial ice accumulates slowly, and buried layers include samples of the atmosphere at the time the snow accumulated. Vertical sample cores from the ice provide a record of past climate and atmospheric chemistry. Atmospheric CO_2 and temperature have varied together, with low CO_2 at times of low temperatures, and high CO_2 during warm periods. The links between temperature and CO_2 are complex, and not well understood.

carbon when they are cold and less when they are warm. Past changes in CO_2 concentrations thus may have been a result, not a cause, of changes in temperature. But in any event, we can see a very close link between climate and the biosphere.

In addition to altering atmospheric composition, humans influence climate by removing vegetation. Reduction in vegetation cover by deforestation or overgrazing of grasslands may alter climate. Deforestation could reduce the water returned to air as vapor, affecting humidity and precipitation. Replacing vegetated surfaces with bare ones may increase the amount of solar radiation reflected back to space (albedo), resulting in local cooling or reduced convective rainfall.

A city is usually much warmer than surrounding rural areas, forming what is known as an **urban heat island.** The pavement and buildings in cities store heat during the day and release it at night. The absence of plants in cities reduces evapotranspiration.

Future Global Warming

Forecasting weather beyond the next five days is very difficult, so prediction of future climate is an immense challenge. At the present state of the art, we cannot plan for the next ten or twenty years with confidence. Please keep this in mind, and understand the dilemma that scientists and government officials face when planning for global warming.

Evidence of global warming. During the twentieth century, Earth's temperature has increased about 0.5C° (0.9F°). Is this rise the result of humans discharging more CO_2 into the atmosphere? Or does the warming represent the natural end of the "Little Ice Age"? If the twentieth-century warming trend continues at the same rate for just 1,000 years—a very short period in geologic time—Earth will have warmed as much in the next 1,000 years as it has in the past 18,000 years.

Many climate predictions are based on the assumption that atmospheric CO_2 will increase from about 275 to about 550 parts per million in the atmosphere. (In other words, a random sample of one cubic meter of air would contain 550 cubic centimeters of CO_2 molecules, and 999,450 cubic centimeters of nitrogen, oxygen, argon, water vapor, and other trace gases; the CO_2 concentration would equal 0.055 percent of air.)

The doubling of CO_2 is based on extrapolations of current growth in fossil fuel use. Doubling atmospheric CO_2 concentrations could increase global average temperature by 1° to 5C° (1.8° to 9F°). For comparison, global average temperatures during warm periods between ice ages were 5° to 10C° (9° to 18F°) higher than during ice ages. Warming would not be the same everywhere. Much greater warming is expected at high latitudes than low latitudes, so Canada, northern Europe, and northern Asia could expect warming of 2° to 4C° (4° to 7F°).

We do not know how fast Earth will warm, because we don't know how rapidly atmospheric CO_2 content will increase, nor how quickly climate responds to increased CO_2. If fossil fuel consumption grows at the rate of 1960–1990, doubling of atmospheric carbon dioxide will occur by about the year 2050. But if fossil fuel consumption slows as it did in the early 1990s, the doubling might come much later.

The rate of global warming may depend on the kind of fossil fuel we burn. When burned, coal releases much more CO_2 than oil, for each unit of energy obtained. Earlier in the twentieth century, many industries and power plants switched from coal to oil to reduce air pollution. But world coal consumption has increased much more rapidly than petroleum consumption during the past two decades, as petroleum supplies dwindle.

Consequences of global warming. We can only guess about the spatial impacts of global warming. Will storm tracks shift? How will a place's current levels of rainfall, snowfall, and temperature change? Will the seasons change? How will ecosystems be affected? What effect will this have on sea level, worldwide energy use, and food supply?

One serious effect would be a worldwide rise in sea level of perhaps 1 to 5 meters (3 to 16 feet). People living near coasts would face danger from rising seas. The danger would not be from constant inundation, because sea level would rise very gradually over years, allowing people time to relocate, raise structures, or build dikes. The danger would be from occasional severe storms that would cause sudden flooding farther inland. The Dutch have shown that well-built dikes can hold back the sea (see Chapter 1), but poorer countries cannot afford such protection.

Global warming could reduce water supply in some regions. Consider a mid-latitude environment such as eastern Nebraska that averages 60 centimeters (24 inches) precipitation and 55 centimeters (22 inches) evapotranspiration. The 5 centimeters (2 inches) precipitation not transpired by plants flows into streams and rivers. A small increase in evapotranspiration, due to a warmer climate, could sharply reduce water flow to the region's streams and rivers.

Semiarid regions and populated subhumid areas, such as southeastern Europe, depend on river flow for irrigation, drinking water, and waste removal. These areas might suffer severe water shortages if warming increases evapotranspiration and decreases river flow. On the other hand, if this warming brings greater precipitation, agriculture may be helped rather than harmed. Agricultural production might be especially helped in areas currently receiving little precipitation.

Should we try to halt global warming? Although global warming is widely believed to be in progress, and many scientists believe humans are significant contributors, little agreement exists on whether we should try to stop it. Those who argue for immediate action emphasize the potentially severe consequences of warming in some areas. But it is hard to convince people to spend money to prevent an event that has not yet occurred.

Reducing CO_2 concentrations will be difficult because we depend on fossil fuels in our daily lives, and energy producers employ many people and earn billions of dollars a year. Significantly reducing fossil fuel use is possible only if we consume less energy or shift to alternative energy sources. Would people prefer to cut back on their use of coal-generated electricity, or spend money on new ways to produce electricity? Either alternative is expensive and inconvenient (see Chapter 6).

Another reason for not acting to curb global warming is the belief that it is easier to adapt to climatic change than to prevent it. People already adjust to changing weather, commodity prices, and technology from year to year, so why shouldn't they be able to adjust to climate change too?

We must remember, though, that humans are not the only life on the planet. We are just part of many interacting ecosystems. Some animals and plants cannot adapt to a climate change, and humans could be responsible for extinction of these species. Even though humans may successfully adapt to climate change, should we not halt global warming for the sake of other species?

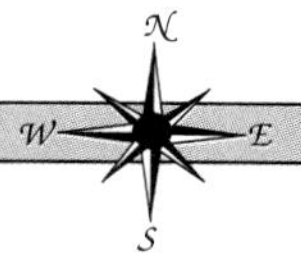

CONCLUSIONS: CRITICAL ISSUES FOR FUTURE GLOBAL WARMING

In this chapter you have seen that climate is weather statistically averaged over time. Weather we experience continually changes, by the hour and day. Over the centuries, Earth's climate has alternated between warming and cooling periods. Earth may now be in a period of warming, but unlike the past, human actions may be an important cause of the warming.

We have also seen that many different climate types occur on Earth's surface. They occur in patterns that are closely related to the circulation patterns described in Chapter 2. Keep in mind this distribution of climates as we examine Earth's vegetation in Chapter 5. A place's distinctive pattern of temperature and moisture controls its vegetation. Climate influences where humans have chosen to settle (Chapter 7), and the crops they have chosen to grow for food (Chapter 11). Climate is critical to both the amount of food that people can produce in a place and the type.

Because climate exerts strong influences on both the natural environment and human activity on Earth's surface, the possibility of climate change in the near future is of great significance. Until recently, people thought of climate as stable and benign. We now see evidence of significant climate change and expect it to continue, and we see humans as a major factor in the function of Earth's systems.

Chapter Summary

1. Classifying climate
Climate is weather and its seasonal variations averaged over time. The most important variables in defining climate are temperature and precipitation. Climates are classified according to precipitation in relation to potential evapotranspiration (POTET), or water demand by vegetation.

2. Earth's climate regions
The Köppen scheme classifies climate into five main categories: A, B, C, D, and E.

- Humid tropical (A) climates occur in the tropical low-pressure zone and are dominated by the intertropical convergence.
- Arid climates (B) predominate in the subtropics, generally on the western sides of continents, and in continental areas isolated from moisture sources.
- Warm mid-latitude climates (C) occur in subtropical areas on the eastern sides of continents and on west coasts at higher latitudes.
- Cool mid-latitude climates (D) occur in continental areas, mostly in the Northern Hemisphere.
- Polar climates (E) occur at high latitudes.

3. Patterns of climate through time
Earth's climate has varied greatly within the past 3 million years. Average global temperatures have been both higher and much lower than today. Glaciers covered much of Earth's surface during cold periods, the last of which ended about 18,000 years ago. Several causes for these variations have been proposed, including changes in Earth's orbit around the sun, geologic factors such as plate tectonics and volcanic eruptions, and changes in the composition of the atmosphere, some human-caused.

4. Future global warming
Evidence suggests that humans have induced global warming by burning fossil fuels which release CO_2 into the atmosphere. Global warming could increase evapotranspiration and flooding of low-lying areas as sea levels rise. To some scientists, past trends in climate suggest that warmer temperatures are part of a natural cycle, and intervening to prevent global warming would be unwise, because of the disruption to the economy and the belief that humans can adapt to changing climate.

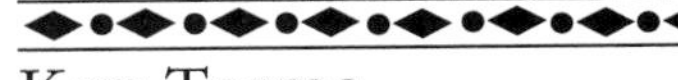

Key Terms

Boreal forest climate, 113
Climate, 93
Continental drift, 119
Desert climate, 103
Humid continental climate, 104
Humid subtropical climate, 104
Humid tropical climate, 103
Ice-cap climate, 104
Marine west-coast climate, 104
Mediterranean climate, 104

Questions for Study and Discussion

1. What are the key differences between weather and climate? How long a record of weather is necessary to adequately describe the climate of a place?
2. In what way does temperature affect the definition of a climate as humid or arid? Why?
3. Why are vegetation regions used to define many major climate types?
4. For each of the eleven climate types described in this chapter, describe how the characteristics of the climate relate to the general circulation model described in Chapter 2.
5. Describe major variations of global average temperature that have occurred during: (a) the past 1,000 years; (b) the past 10,000 years; (c) the past 100,000 years.
6. What are the major astronomical, geologic, and human-related hypotheses regarding the causes of climatic variation? At what time scale does each hypothesis operate?
7. What are the major causes of global warming? What are some of its likely effects on human activities?

Thinking Geographically

1. This is a good group-research activity: Obtain a list of daily record high and low temperatures for your location, since record-keeping began, if possible. Enter the data in a spreadsheet or program that will perform sorting, and sort the records by year. What percentage of record high temperatures occurs in the second half of the period of record? What percentage of record low temperatures occurs in the second half of the period of record? Can you spot any trends?
2. Compare the map of climates (Figure 3–1) with the world population map (Figure 7–1). In what climate regions are the greatest concentrations of people found? Why?
3. How might an increase in average annual temperature of 5C° (9F°) affect your day-to-day life?

Suggestions for Further Reading

Barnola, J. M., D. Raynaud, Y. S. Korotkevich, and C. Lorius. "Vostok Ice Core Provides 160,000-Year Record of Atmospheric CO_2." *Nature* 329 (1 October 1987): 408–14.

Brazel, Sandra W. and R. Balling, Jr. "Temporal Analysis of Long-Term Atmospheric Moisture Levels in Phoenix, Arizona." *Journal of Climate and Applied Meteorology* 25 (February 1986): 112–17.

Bunyard, Peter. "The Significance of the Amazon Basin for Global Climatic Equilibrium." *The Ecologist* 17 (No. 4/5, 1987): 139–41.

Changnon, Stanley A. and K. E. Kunkel. "Assessing Impacts of a Climatologically Unique Year (1990) in the Midwest." *Physical Geography* 13 (April–June, 1992): 180–90.

Climap Project Members. "The Surface of the Ice-Age Earth." *Science* 191 (19 March 1976): 1,131–37.

Fovell, Robert G. and M-Y. C. Fovell. "Climate Zones of the Coterminous United States Defined Using Cluster Analysis." *Journal of Climate* 6 (November 1993): 2,103–35.

Gleick, P. "Climate Change, Hydrology and Water Resources." *Reviews of Geophysics* 27 (1989): 329–44.

Graumlich, Lisa J. "Precipitation Variation in the Pacific Northwest (1675–1975) as Reconstructed from Tree Rings." *Annals of the Association of American Geographers* 77 (March 1987): 19–29.

Groisman, Pavel Ya. and D. R. Easterling. "Variability and Trends of Total Precipitation and Snowfall Over the United States and Canada." *Journal of Climate* 7 (January 1994): 184–205.

Hansen, James E. and A. A. Lacis. "Sun and Dust Versus Greenhouse Gases: An Assessment of Their Relative Roles in Global Climate Change." *Nature* 346 (1990): 713–19.

Imbrie, John and K. P. Imbrie. *Ice Ages: Solving the Mystery*. Cambridge: Harvard, 1986.

Isaac, G.A. and R.A. Stuart. "Temperature-Precipitation Relationships for Canadian Stations." *Journal of Climate* 5 (August 1992): 822–30.

Kerr, Richard A. "Did the Roof of the World Start an Ice Age?" *Science* 244 (23 June 1989): 1,441–42.

Kheshgi, Haroon S. and B. S. White. "Does Recent Global Warming Suggest an Enhanced Greenhouse Effect?" *Climate Change* 23 (February 1993): 21–39.

Manning, J.C. *Applied Principles of Hydrology*. Columbus: Merrill, 1987.

Michaels, Patrick J. and D. E. Stooksbury. "Global Warming: A Reduced Threat?" *Bulletin of the American Meteorological Society* 73 (October 1992): 1,563–77.

Muller, Robert A. "A Perspective on the Climate of Regions." *Physical Geography* 12 (July–September, 1991): 252–59.

Oliver, John E. "The History, Status and Future of Climate Classification." *Physical Geography* 12 (July–September, 1991): 231–51.

Raynaud, D., J. Jouzel, M. Barnola, J. Chappellaz, R. J. Delmas, and C. Lorius. "The Ice Record of Greenhouse Gases." *Science* 259 (12 February 1993): 926–33.

Robinson, Peter J. "Trends in the Relationship Between Monthly and Daily Temperatures." *Physical Geography* 13 (July–September, 1992): 191–205.

Schneider, S.H. "The Greenhouse Effect: Science and Policy." *Science* 243 (10 February 1989): 771–81.

Schneider, S.H. "Will Sea Levels Rise or Fall?" *Science* 356 (5 March 1992): 11–12.

Schule, Wilhelm. "Anthropogenic Trigger Effects on Pleistocene Climate"? *Global Ecology and Biogeography Letters 2* (March 1992): 33–36.

Shukla, J., C. Nobre, and P. Sellers. "Amazon Deforestation and Climate Change." *Science* 247 (16 March 1990): 1,322–25.

Stolarski, Richard, R. Bojkov, L. Bishop, C. Zerefos, J. Staehelin, and J. Zawodny. "Measured Trends in Stratospheric Ozone." *Science* 256 (17 April 1992): 342–49.

Sundquist, Eric T. "The Global Carbon Dioxide Budget." *Science* 259 (12 February 1993): 934–41.

Williams, Kay R.S. "Correlations Between Palmer Drought Indices and Various Measures of Air Temperature in the Climatic Zones of the United States." *Physical Geography* 13 (October–December, 1992): 349–67.

Xue, Yongkang and J. Shukla. "The Influence of Land Surface Properties on Sahel Climate. Part 1: Desertification." *Journal of Climate* 6 (December 1993): 2,232–45.

Young, Kenneth C. "Reconstructing Streamflow Time Series in Central Arizona Using Monthly Precipitation and Tree Ring Records." *Journal of Climate* 7 (March 1994): 361–74.

We also recommend these journals: *Climatic Change; Geografiska Annaler* Series B; *Journal of Hydrology; Nature; Oceanus; Physical Geography; Progress in Physical Geography; Science.*

4 Landforms

- Plate tectonics
- Slopes and streams
- Ice, wind, and waves
- The dynamic planet

The Karawanken Mountains, Slovenia.

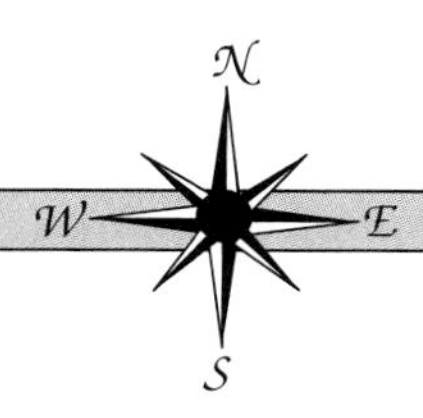

1. Plate tectonics

Major landforms of the world are shaped through a combination of endogenic and exogenic processes. *Endogenic* (internal) processes move portions of Earth's surface horizontally and vertically. Earth's crust consists of several moving pieces, known as tectonic plates. The movement of these pieces causes earthquakes, volcanic eruptions, and formation of mountain ranges at plate boundaries.

2. Slopes and streams

Exogenic processes shape Earth's surface externally. Rocks are broken into smaller pieces through exposure to air and water. The pieces then are moved downhill by gravity or are carried by streams, wind, or ice to other locations. Streams play a major role in shaping landforms. They erode material from some places, transport it downhill, and deposit it elsewhere.

3. Ice, wind, and waves

Glaciers—rivers of ice—carve landforms in regions under continuous snow cover, such as mountaintops and the poles. In deserts, where vegetation is scarce, the wind moves material to shape landforms. Along coastlines, waves caused by wind blowing across the ocean surface cause intensive erosion. This rapidly changes landforms in these areas.

4. The dynamic planet

Earth's surface is continually changing. Rates are faster in areas of strong erosive processes and weak materials, and slower where processes are weak but rocks are strong. Change on Earth's surface is rapid enough to affect human settlements and natural resources. In particularly dynamic environments, natural hazards include floods, earthquakes, volcanoes, and landslides.

INTRODUCTION

Travel from your home in any direction, and you likely will see a variety of landforms—hills, valleys, flatlands, depressions, gulleys, and perhaps a mountain or beach. This chapter looks at Earth's rocks, soil, and surface landforms, which together comprise one of the four "spheres," the lithosphere. **Geomorphology** is the study of landforms and the processes that create them.

Earth's lithosphere appears to be fixed and unchanging, because we see it in contrast to the other three spheres, which change rapidly. The atmosphere changes daily—hot to cold, wet to dry, windy to still. We can see seasonal changes in the biosphere—plants and animals appear, grow, reproduce, and die. The hydrosphere is visibly dynamic—rivers flood, streams dry up. But the lithosphere seems unchanging. Landforms like plains, hills, and valleys do not change noticeably in a lifetime; soil is soil; rock outcrops seem as permanent as monuments.

But landforms, in their remarkable variety, did not just appear as we see them at some distant time. Landforms constantly change, although often imperceptibly slowly. Like the atmosphere, the lithosphere is driven by continual transfers of energy and matter, and it interacts with the other three spheres. Some of the energy that the lithosphere receives comes from deep within Earth: the heat in Earth's core drives movement on the planet's surface.

Occasionally, landforms change quickly, a reminder to us that the lithosphere is quite dynamic—earthquakes shake the land, volcanoes spew forth hot gases and molten rock, and ocean waves pound the shore. Humans insist on getting in the way of these dramatic changes in landforms by building settlements in places where earthquakes, volcanoes, and hurricanes are likely to occur. Because of the rarity of these events, people discount their importance and thus become vulnerable when they do strike.

As with the other spheres, our actions have profoundly changed the lithosphere. Agricultural practices are depleting soil fertility. In semiarid areas, overgrazing by animals has produced desertlike conditions. We have dammed rivers, diverted them for irrigation, dredged them for navigation, and polluted them with eroded soil. And we have spewed our pollutants indiscriminately into all four spheres.

Plate Tectonics

Earth's landforms include rugged mountain ranges arranged in linear patterns that extend for thousands of kilometers. Three especially prominent mountain ranges on Earth's land surface include:

- The Rockies of North America, connecting through Central America to the Andes of South America.
- The Himalayas, extending across Asia.
- The north-south system of mountains in Eastern Africa.

Large as these highly visible mountain ranges are, none rank as the world's longest. That title belongs to a mountain system beneath the oceans, the interconnecting mid-ocean ridges that total more than 64,000 kilometers (40,000 miles) in length. Take a moment to study the chapter-opening illustration, a remarkable view of Earth minus its 71 percent of ocean cover, and you can see these mid-ocean ridges and the other mountains described.

Earth's mountains—as well as its valleys, hills, and depressions—are built through a combination of endogenic and exogenic processes. **Endogenic processes** (or internal processes) are forces that cause movements beneath Earth's surface, such as mountain-building and earthquakes. These internal mechanisms move portions of Earth's surface horizontally and vertically, raising some parts and lowering others. The internal actions that build Earth's features are simultaneously attacked by **exogenic processes** (or external processes), which are forces of erosion, such as running water, wind, and chemical action.

Endogenic and exogenic forces work together to continually move and shape Earth's crust (Figure 4–1). Endogenic processes form rocks and move them to produce mountain ranges, ocean basins, and other topographic features. As these rocks become exposed, exogenic activities go to work. They erode materials, move them down hillslopes, and deposit them in lakes, oceans, and other low-lying areas. We will examine all of these processes.

For millennia, people believed that Earth's continents and oceans were fixed in place for all time,

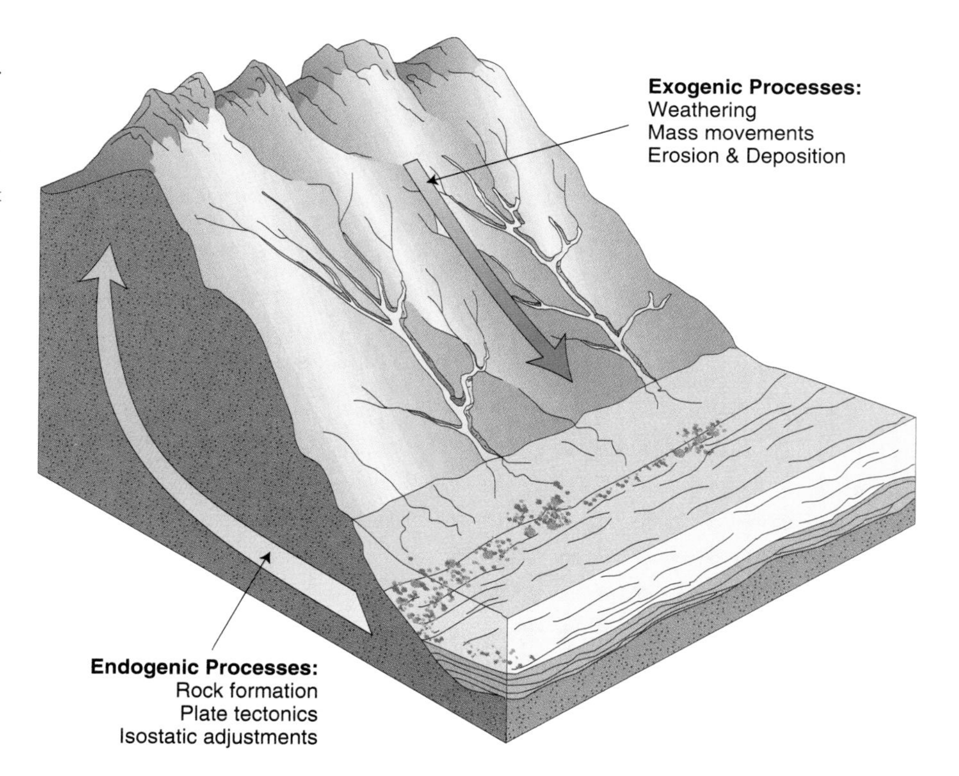

FIGURE 4–1
Endogenic and exogenic processes. The landforms around you are the product of interaction between endogenic and exogenic processes. Endogenic processes involve movement of Earth's crust and rock formation through tectonic action. Exogenic processes wear down rocks at the surface.

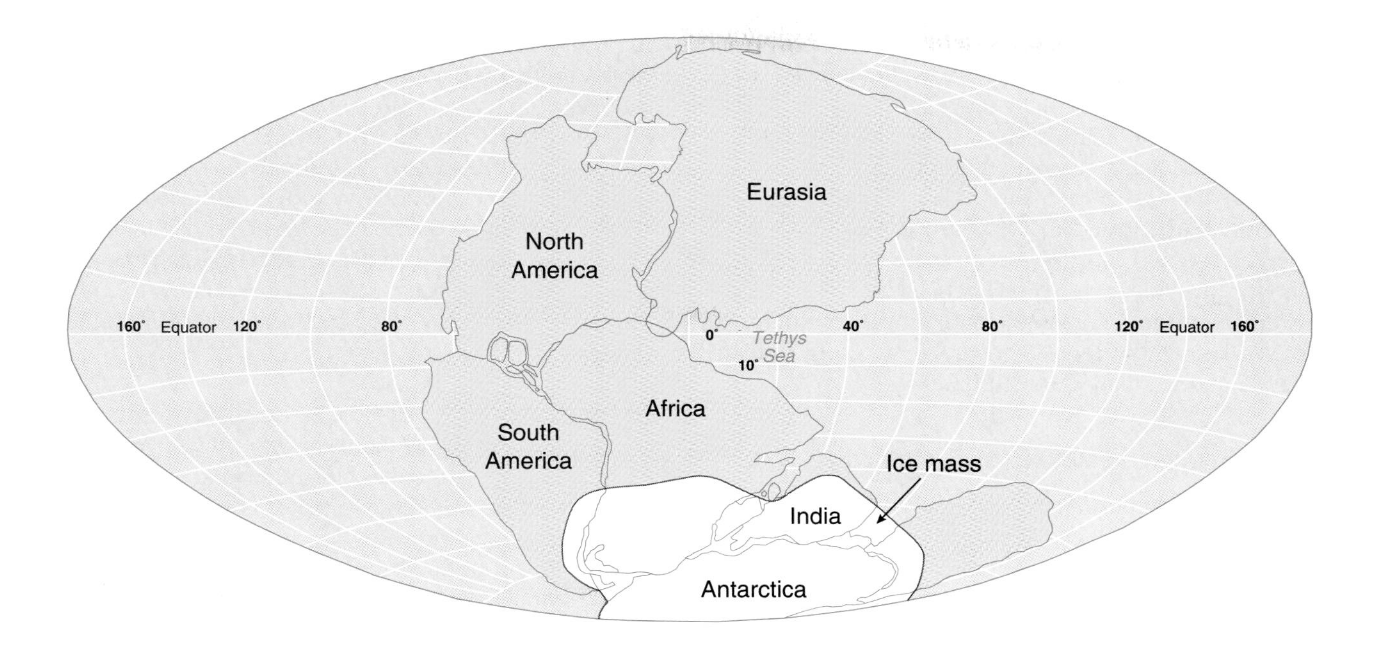

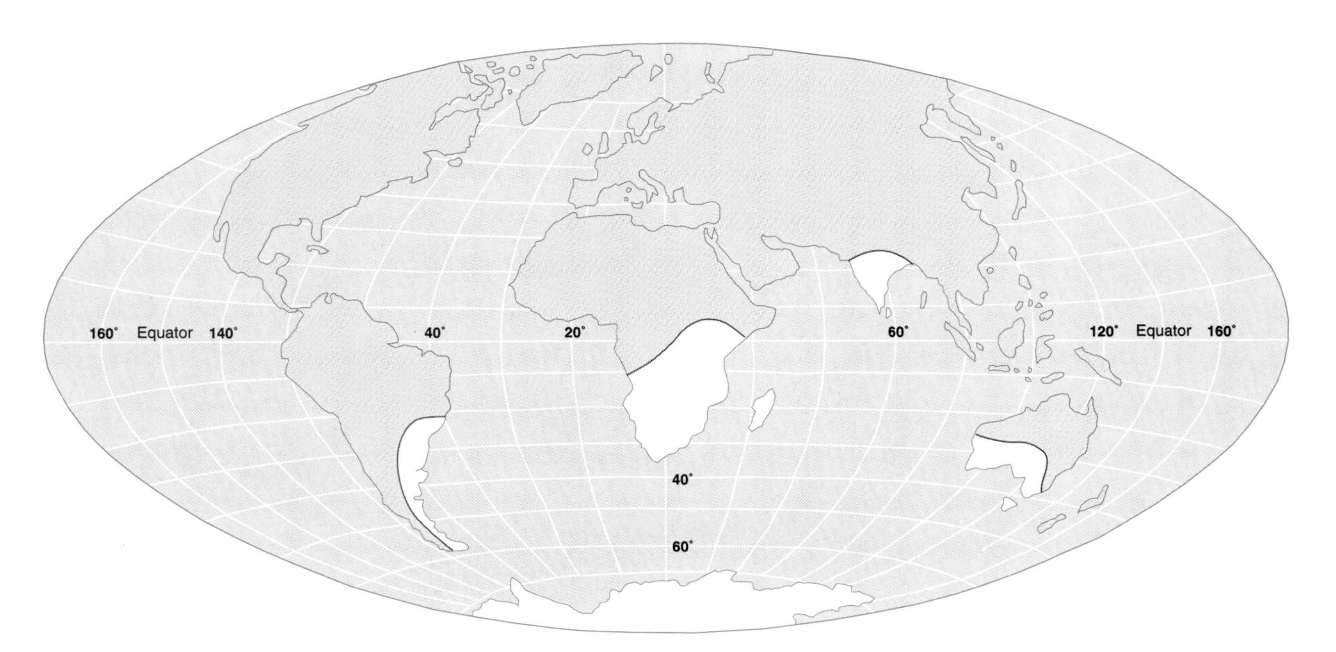

FIGURE 4-2
Earth as we believe it appeared 300 million years ago (a) and as it appears today (b). In (a), note the "fit" of the continental outlines. Evidence for this fit is impressive, including glacial evidence (shown in white), plus distinct intercontinental patterns of plant and animal fossils, coal seams, and ancient deserts. The single, nearly continuous continental mass in (a) is called *Pangaea* ("all Earth"). (From E. J. Tarbuck and F. K. Lutgens, *The Earth,* 4th ed., Macmillan Publishing Co., 1993, after R. F. Flint and B. J. Skinner, *Physical Geology,* 2d ed., New York: Wiley, 1977.)

and that Earth was only a few thousand years old. This notion of a "fixed Earth" was challenged early in this century by German scientist Alfred Wegener. He argued that Earth's land areas once had been joined in a single "supercontinent," now known as *Pangaea,* and that over thousands of years the continents had moved apart (Figure 4–2). Because Wegener could not explain *how* the continents had moved, his ideas were rejected at the time, especially by American geologists. But researchers in the 1960s vindicated Wegener by working out an explanation, called plate tectonics theory.

Earth's Moving Crust

Earth resembles an egg with a cracked shell. Its thickly fluid interior is surrounded by a thin, rigid crust, averaging 45 kilometers (25 miles) thickness. The rock in the interior, known as the **mantle,** moves very slowly. As it moves, the mantle carries heat from Earth's core toward the surface in incredibly slow *convection currents,* similar to those that stir the atmosphere and oceans (Chapter 2). Geologists believe that this motion of the mantle causes Earth's rigid crust to move in pieces, called **tectonic plates**—hence the image of a cracked eggshell. This is the **plate tectonics theory.** Movement of the plates causes earthquakes to rumble, volcanoes to erupt, and mountains to build (Figure 4–3).

Earthquakes. Thousands of **earthquakes** occur every day, as plates of Earth's crust move. Figure 4–4 shows major earthquakes zones around the world, notably clustered where two plates adjoin. Within the United States, earthquakes are more likely to occur on the West Coast, along the boundary between the North Atlantic and Pacific plates (Figure 4–5).

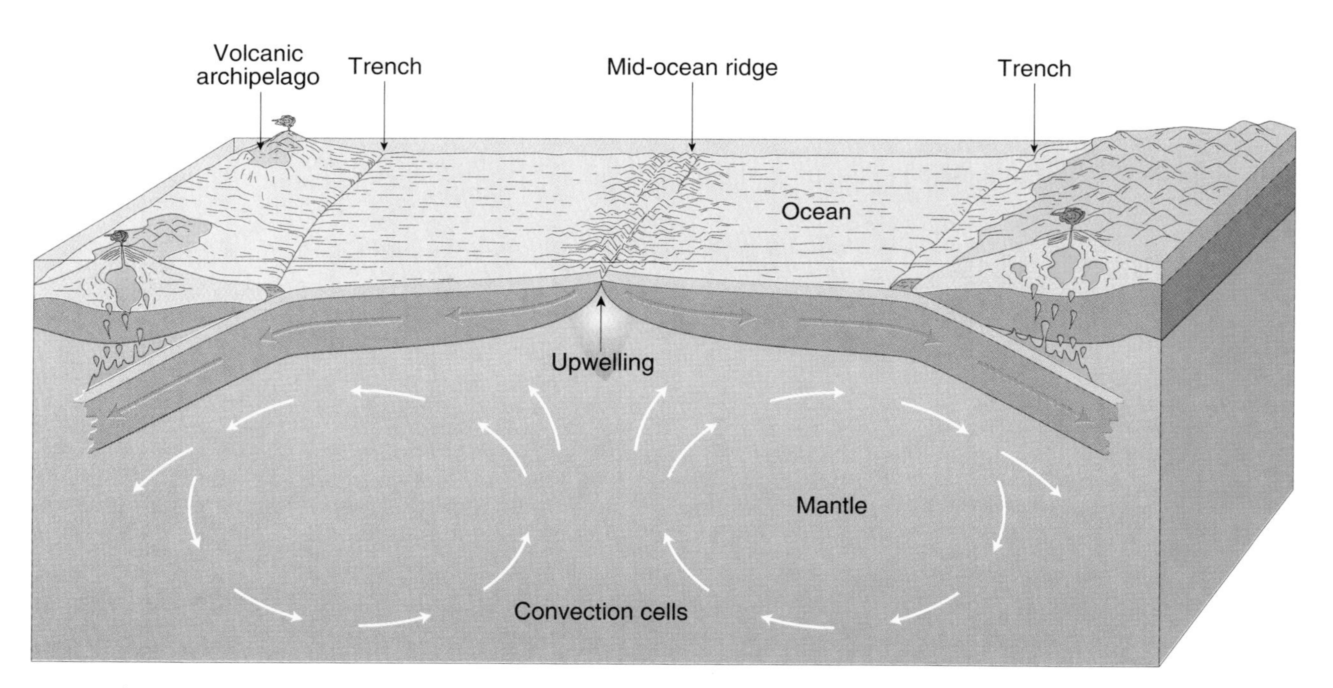

FIGURE 4–3
Plate tectonics theory explains the occurrence of moving continents, mid-ocean ridges, deep-sea trenches, volcanoes, and earthquakes. At center, rock magma melted in Earth's interior rises along mid-ocean ridges, emerging to chill into new oceanic crust. Along this ridge, plates of oceanic crust are spreading apart. Where crustal plates collide (at left and right), crust is forced downward (subducted) and recycled (melted) back into the interior. This generates new magma, which migrates toward the surface, sometimes emerging as a volcano. Less dense continental crust (the continents) ride on moving plates of basaltic crust. This plate-tectonic process is believed to be driven by convection currents in Earth's mantle, shown by arrows. (From E. J. Tarbuck and F. K. Lutgens, *The Earth,* 4th ed., Macmillan Publishing Co., 1993)

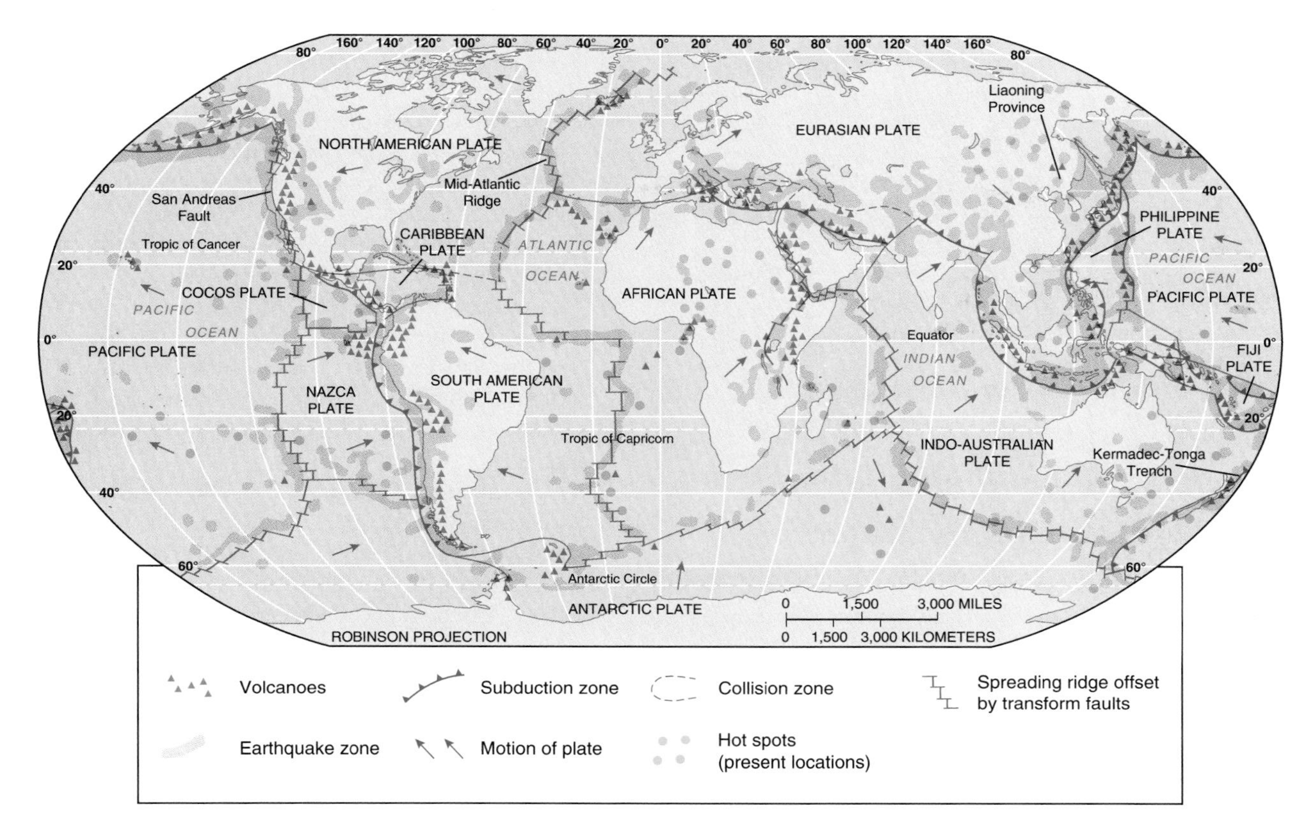

FIGURE 4–4
Earth's plates. Earth's crust is like a cracked eggshell, with the pieces—called plates—slowly moving relative to one another. Arrows show the direction of plate movement. Plates are generally spreading apart in the ocean basins, and colliding around the continents. Most of Earth's active areas of mountain-building, volcanic eruptions, and earthquakes lie along plate boundaries. (From R. W. Christopherson, *Geosystems,* 2d ed., Macmillan Publishing Co., 1994.)

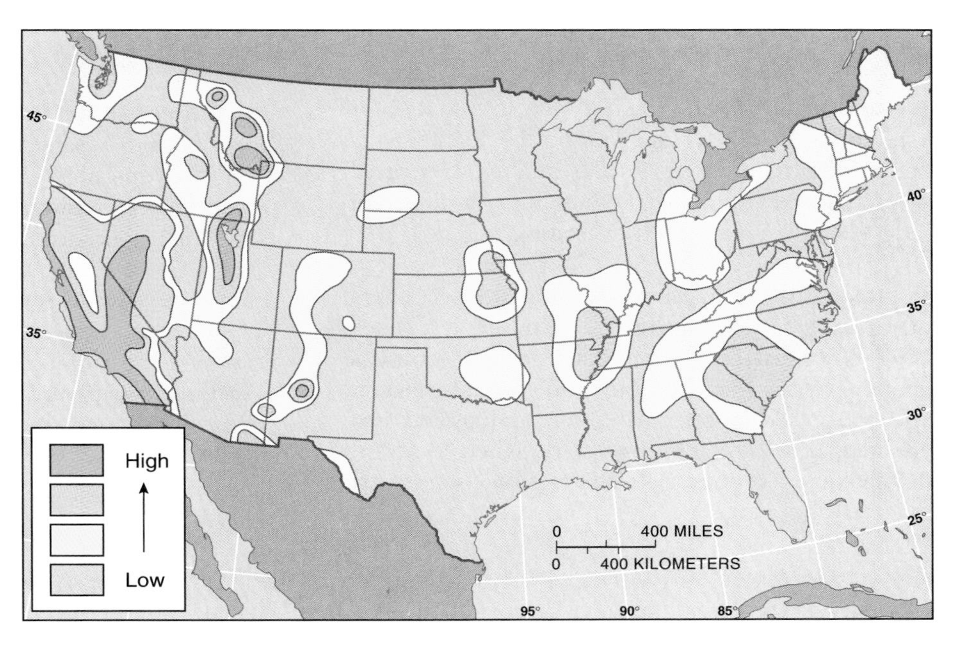

FIGURE 4–5
Earthquake risk in the United States, showing the *likelihood of damage* during a 50-year period. The risk in California is common knowledge, but you may be surprised at other risk areas, such as Utah, Missouri, and South Carolina. New Madrid, Missouri experienced the strongest quakes in U.S. history during 1811 and 1812, and Charleston, South Carolina, experienced the East Coast's strongest tremor, in 1886. These areas are fairly quiet—for now. (From E. J. Tarbuck and F. K. Lutgens, *The Earth,* 4th ed., Macmillan Publishing Co., 1993, p. 411; data from Environmental Science Service Administration.)

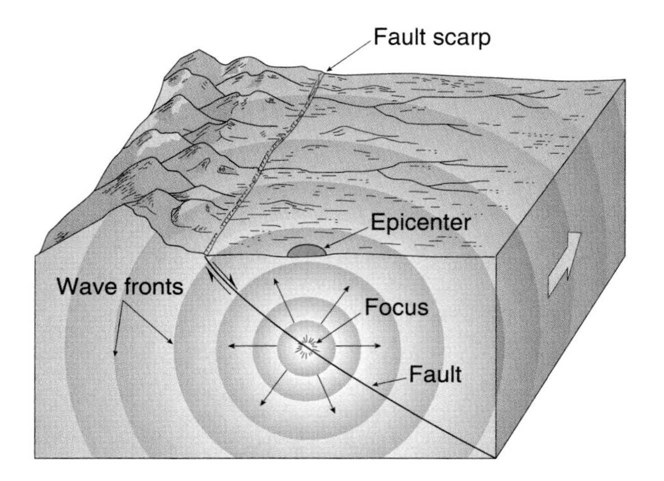

FIGURE 4–6
Earthquake focus and epicenter. Where Earth's crust actually moves is the *focus* of an earthquake. The point on the surface directly above is the *epicenter*. (From E. J. Tarbuck and F. K. Lutgens, *Earth*, 7th ed., Macmillan Publishing Co., 1994)

The place where Earth's crust actually moves is the **focus** of an earthquake. The focus is generally near the surface, but can be as deep as 600 kilometers (400 miles). The point on the surface directly above the focus is the **epicenter** (Figure 4–6). The tremendous energy released at the focus travels worldwide in all directions, at various speeds through different layers of rock.

Most earthquakes are too small for people to feel, and they are detectable only with a **seismograph,** which records the quake's **seismic waves,** or vibrations. Earthquake intensity is measured on a 0-to-9 scale developed by Charles F. Richter. Magnitude 3 to 4 earthquakes on the Richter scale are minor; magnitude 5 to 6 quakes can break windows and topple weak buildings; and magnitude 7 to 8 quakes are devastating killers if they affect populated areas.

In addition to an earthquake's intensity, several other factors determine the damage it causes. Generally, damage is greater at places closer to the epicenter and at places built on unstable ground and loose sediment. Earthquake damage is also greater where buildings are not designed to absorb the shaking. To appreciate this, let us contrast two major earthquakes:

- *Loma Prieta, California*. This earthquake struck the San Francisco Bay Area in 1989, minutes before the start of a World Series game. It was being played, ironically, between the two Bay Area teams, the San Francisco Giants and the Oakland A's. The Bay Bridge buckled, freeways crumbled, and 67 perished. Some 100,000 buildings were damaged or destroyed, many in the Marina District, a neighborhood built on unstable fill in the San Francisco Bay. But the death and destruction was limited, considering the quake's magnitude of 7.1 on the Richter scale.
- *Northwestern Armenia*. This 1988 earthquake killed nearly 55,000, injured 15,000, and left at least 400,000 homeless. Although registering 6.9 on the Richter scale—lower than the Loma Prieta quake—thousands were trapped inside collapsing buildings not designed to withstand earthquakes. People in less economically developed societies cannot finance the cost of earthquake-proofing their structures, as is routinely done with new buildings in quake-prone communities in relatively wealthy societies like the United States.

Earthquake prediction is unreliable. Suspected faults are closely monitored, and computer models attempt to replicate conditions along plate boundaries. But predicting an earthquake is like trying to forecast the weather several months from now—it just cannot be done accurately with current technology.

Volcanoes. Like earthquakes, volcanoes cluster along boundaries between tectonic plates (Figure 4–4). Movement within Earth and between the plates generates **magma** (molten rock). Being less dense than the surrounding rock, magma migrates toward the surface. Some reaches the surface and erupts, and it then is called **lava.** A **volcano** is the surface vent where lava emerges. The magma may flow over the surface, forming a *basalt plain,* or it may build up to form a volcanic mountain. The chemistry of the magma/lava determines its texture, and therefore the type of landform it builds.

Shield volcanoes erupt runny, basaltic lava. They are called shields because of their shape (Figure 4–7). Each of the Hawaiian Islands is a large shield volcano, although the only currently active ones

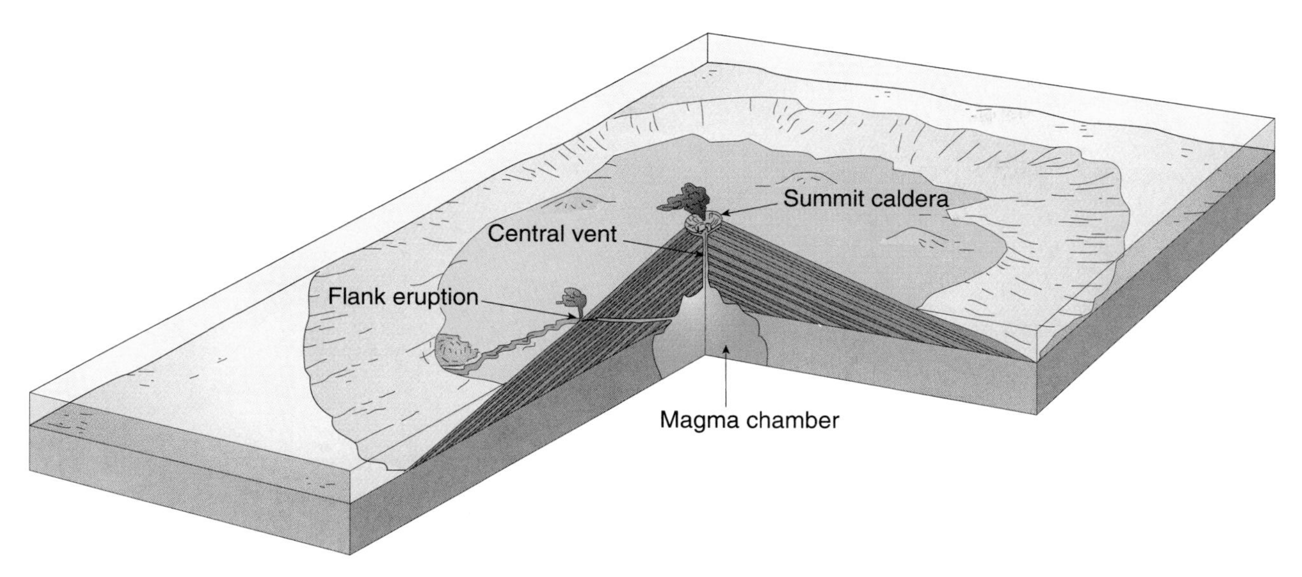

FIGURE 4–7
Shield volcanoes, exemplified by the Hawaiian Islands, are the largest volcanoes on Earth. Their eruptions are comparatively "gentle"; they flow, toss, and bubble lava instead of exploding. (From E. J. Tarbuck and F. K. Lutgens, *Earth,* 7th ed., Macmillan Publishing Co., 1994.)

are Kilanea and Mauna Loa, on the island of Hawaii (the Big Island). These generally sedate volcanoes make news on the rare occasions when they grow more active, and flows of lava threaten settlements. The mid-ocean ridges are formed of similar basaltic lava.

Explosive volcanoes that cause death and destruction are more likely to be **composite cone volcanoes** (Figure 4–8). Their magma is thick and gassy, and it may erupt explosively through a vent. This sends ash, glassy cinders (called *pyroclasts)*, and clouds of sulfurous gas high into the atmosphere. It may also pour lethal gas clouds and dangerous mudflows down the volcano's slopes. Repeated eruptions build a cone-shaped mountain, composited of a mixture of lava and ash layers.

Composite cone volcanoes have killed tens of thousands of people at a time, but such disasters are much less frequent than severe earthquakes. One of the greatest volcanic explosions in recorded history was the 1883 eruption of the island of Krakatau in present-day Indonesia. Two-thirds of the island was destroyed, and the event killed about 36,000, most of whom died in a flood triggered by the eruption. Ash discharged into the atmosphere by Krakatau significantly blocked sunlight and caused noticeable cooling of Earth's climate for a couple of years.

The deadliest volcanic eruption in the twentieth century, in 1902, was Mt. Pelée on the island of Martinique in the Caribbean, which wiped out the city of St. Pierre in minutes, leaving 29,000 dead and only one survivor, a man protected from the holocaust in his prison cell. This catastrophe exemplifies the problematic relationship that will forever exist between the physical world and its people: Mt. Pelée steamed and rumbled, giving months of warning before erupting. The fearful citizens of St. Pierre wanted to leave, but the politicians forced them to stay in the city to vote in an election—one that did not happen.

The 1985 eruption of Nevado del Ruiz, in Colombia, killed 23,000, and triggered giant mud slides that buried most of a town. In the United States, Washington state's Mount Saint Helens erupted in 1980, leaving 54 dead.

Thousands of volcanoes stand dormant around the world. About 600 are actively spewing lava, ash, and gas—some daily—but they rarely cause damage. Fortunately, predicting volcanic eruptions is more accurate than predicting earthquakes, because volcanoes give many warnings before erupting.

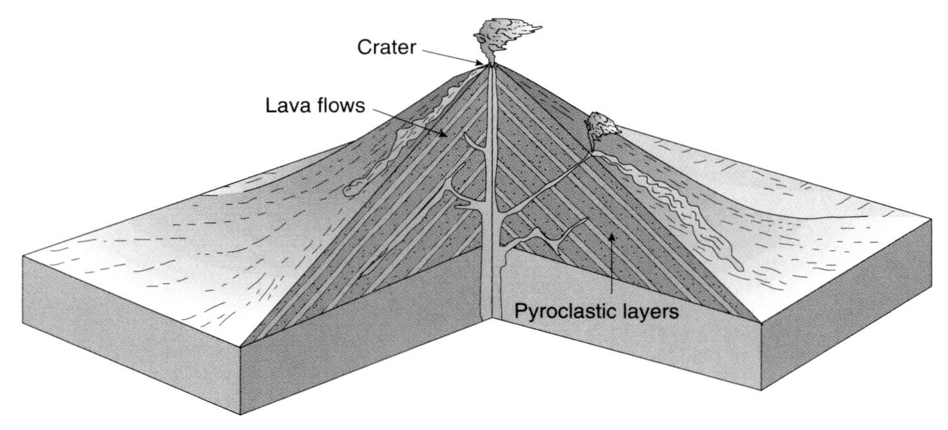

FIGURE 4–8
Composite cone volcanoes, composed of alternating lava and ash, are relatively explosive. (From E. J. Tarbuck and F. K. Lutgens, *Earth Science,* 7th ed., Macmillan Publishing Co., 1994.)

Mt. Pinatubo, a composite cone volcano in the Philippines, erupted in June 1991, spewing ash over a large area and causing a brief cooling period in the Earth's climate. (Gamma Liaison)

Types of Boundaries Between Plates

Three types of boundaries form between moving plates of Earth's crust. The type of boundary depends on whether the plates are spreading apart, pushing into each other, or grinding past each other.

Divergent plate boundaries. A boundary where plates are spreading apart is a **divergent plate boundary.** People seldom are aware of plates spreading apart, first because it happens at only centimeters per year, and second because it happens mostly deep in mid-ocean. European explorers who ventured across the Atlantic several hundred years ago did not realize that they were passing over spreading plates and active volcanoes along the Mid-Atlantic Ridge, deep beneath them. Divergent plate boundaries occur on land, too. The rift valleys of East Africa are an example, visible on pages 126 and 133.

Where two plates are diverging on the seafloor, a phenomenon called **seafloor spreading,** lava continually erupts. Rapidly chilled by seawater, it solidifies to forms new seafloor crust. In fact, this is how all the seafloor crust visible in the chapter-opening illustration formed, very slowly spreading from mid-ocean ridges (Figure 4–3).

Convergent plate boundaries. A boundary where plates push together is a **convergent plate boundary.** Material from the crust is very slowly forced downward by the collision, back into the mantle. Seafloor crust is denser than continental crust. Because of this difference in density, when a plate of continental crust collides with a plate of oceanic crust, the denser oceanic plate is forced beneath the lighter continental crust. The oceanic plate is carried into Earth's mantle, where some of it is remelted. This magma then migrates toward the surface, causing volcanic eruptions at sites above the plunging plate (Figure 4–3).

Transform plate boundaries. A boundary where the plates neither converge nor diverge, but grind past each other, is a **transform plate boundary.** California's San Andreas Fault is a famous example. Along this fault, the Pacific plate is moving northwest relative to the North American plate (Figure 4–4). The boundary between these plates is not a smooth one, and ridges and mountains are built as the two plates grind against one another. The plates bind for long periods and then abruptly slip, causing the earthquakes for which California has become noted.

Vertical movements of Earth's crust. Parts of the crust move vertically as well as horizontally. As two plates collide, material may be forced downward, as in the case of converging plates, or upward. Over millions of years, vertical movements along plate boundaries produce mountain ranges thousands of meters high.

Vertical movement of crust also occurs because of **isostatic adjustments,** which are caused by addition or removal of large volumes of material from the crust. The crust "floats" on the underlying mantle, like a boat floating in water. If material is added to the crust it will sink, and if material is removed it will rise. Deposition of sediment or accumulation of glacial ice can cause the crust to sink. Removal of this material by erosion or melting of glaciers then allows the crust to isostatically adjust, or "rebound." Crust that was buried under continental ice sheets has risen vertically over 100 meters (300 feet) in the last 15,000 years because of glacial melting.

Rock Formation

Although by human standards Earth's surface moves very slowly—by at most a few centimeters per year—this movement produces Earth's great diversity of rocks. As Earth's crust moves, its materials are eroded and deposited, heated and cooled, buried and exposed.

Types of rocks. To help understand rocks, we group them into three basic categories that reflect how they form:

- **Igneous rocks** are formed when molten crustal material cools and solidifies. The name derives from the Greek word for *fire*—the same root as the English word *ignite.* Examples of igneous rocks are basalt (of which the seafloor is made) and granite (of which much continental crust is made).
- **Sedimentary rocks** result when rocks eroded from higher elevations (mountains, hills, plains) accumulate at lower elevations (like swamps and ocean bottoms). When subjected to high pressure and the presence of cementing materials to bind their grains together, rocks like sandstone, shale, and some limestones result.
- **Metamorphic rocks** have been exposed to great pressure and heat, altering them into more compact, crystalline rocks. In Greek, the name means *to change form.* Examples include marble (which metamorphosed from limestone) and slate (which metamorphosed from shale).

Minerals. Earth's rocks are also diverse because the crust contains thousands of minerals. Depending on the kind of minerals present, rocks can be more dense or less dense:

- Denser rocks are dominated by compounds of silicon, magnesium, and iron minerals; they are called **sima** (for **si**licon-**ma**gnesium).
- Less dense rocks are dominated by compounds of silicon and aluminum minerals; they are called **sial** (for **si**licon-**al**uminum).

Denser sima rocks make up much of the oceanic crust. Less dense sial rocks make up much of the continental crust. The lower density and greater thickness of sial rocks cause the continents to have higher surface elevations than the oceanic crust, just as a less-dense dry log will float higher in water than a denser wet one.

The formation and distribution of many minerals is caused by the movements of Earth's crust. Vast areas of the continental crust, known as **shields,** have not been eroded or changed for millions of years. Shield areas usually contain rich concentrations of minerals, such as metal ores and fossil fuels. Shields are located in the core of large continents, such as Africa, Asia, and North America. Many of the world's mining districts operate where these continental shields are exposed at the surface.

Stress on rocks. Crustal movements along plate boundaries exert tremendous stress on rocks. Despite their rigidity, the rocks bend and fold (Figure 4–9). When stressed far enough, they fracture. The fractured pieces may then be transported to new locations. Fracturing takes place in different ways, depending on the type of boundary:

- Near a divergent plate boundary, rocks break apart because they are stretched; the fracture is called a *normal fault* (Figure 4–10). The Wasatch Fault of Utah is an example.
- Near a convergent plate boundary, rocks fracture because they are compressed; such fractures are called *reverse faults* (because they are the opposite of normal faults), or *thrust faults* if there is large horizontal movement (Figure 4–11). Alternatively, the crust may rumple like a rug, creating folds. The Appalachians, the European Alps, and the Himalayas are examples of mountain ranges created by faulting and folding along convergent plate boundaries.
- Faulting also occurs along transform boundaries. Rock movement near a transform boundary is mostly horizontal rather than the vertical movement typical near the other two types of boundaries.

Fractures in Earth's crust provide openings for magma to penetrate, sometimes reaching the surface (Figure 4–12). Most magma cools and hardens before it reaches the surface, forming igneous rock. But some magma reaches the surface as lava erupted from volcanoes. Magma reaching the surface has formed lava flows in the U.S. northwest and India's Deccan Plateau; volcanic mountains in the ocean (Hawaii, Iceland, West Indies); and chains of volcanoes like that running from British Columbia to California.

Slopes and Streams

Landforms created through endogenic processes are attacked by exogenic processes as they are being formed. The wearing down of Earth's crust through exogenic processes reshapes Earth's crust into very different landforms. Exogenic processes shape Earth's surface in two principal steps. Rocks

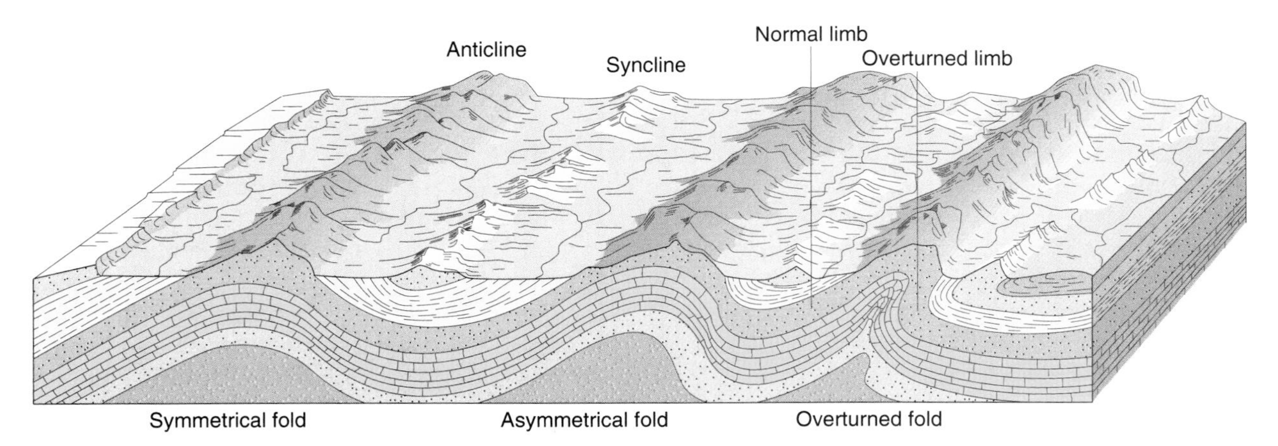

FIGURE 4–9
In many places, the rocks of Earth's crust are warped, folded, and broken (faulted). A good example is the Appalachian Mountains of the eastern United States. Subsequent erosion and covering with sediment, soil, and vegetation masks these contortions from easy view in most places, but these diagrams illustrate what has happened underground. The result is complex patterns of different rock types exposed at Earth's surface, and distinctive landforms. (From E. J. Tarbuck and F. K. Lutgens, *The Earth,* 4th ed., Macmillan Publishing Co., 1993.)

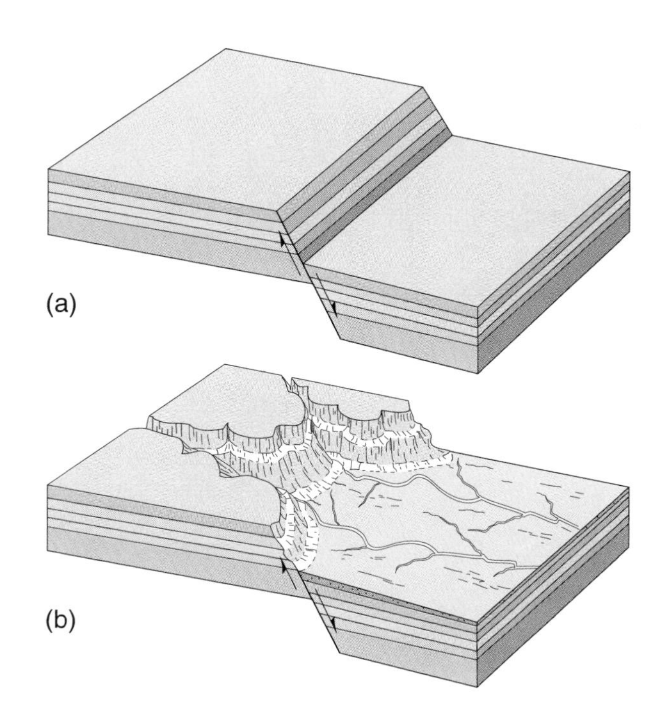

FIGURE 4–10
Cut-away views of what happens beneath Earth's surface along a *normal fault.* (a) Rocks break when stressed by forces that pull them apart. (b) How erosion alters the surface of rocks along a normal fault. (From E. J. Tarbuck and F. K. Lutgens, *The Earth,* 4th ed., Macmillan Publishing Co., 1993.)

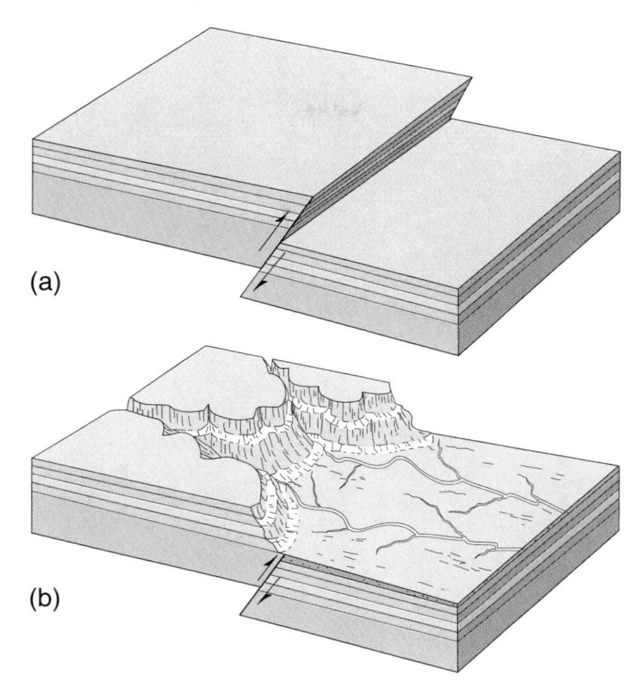

FIGURE 4–11
Cut-away views of what happens beneath Earth's surface along a *reverse fault.* (a) Rocks break when stressed by forces that compress them. (b) How erosion alters the surface of rocks that have been uplifted along a reverse fault. (From E. J. Tarbuck and F. K. Lutgens, *The Earth,* 4th ed., Macmillan Publishing Co., 1993.)

The Wasatch Mountains in Utah rise abruptly from the Salt Lake valley along the Wasatch Fault, an active normal fault just east of Salt Lake City. (Tom McHugh/Photo Researchers, Inc.)

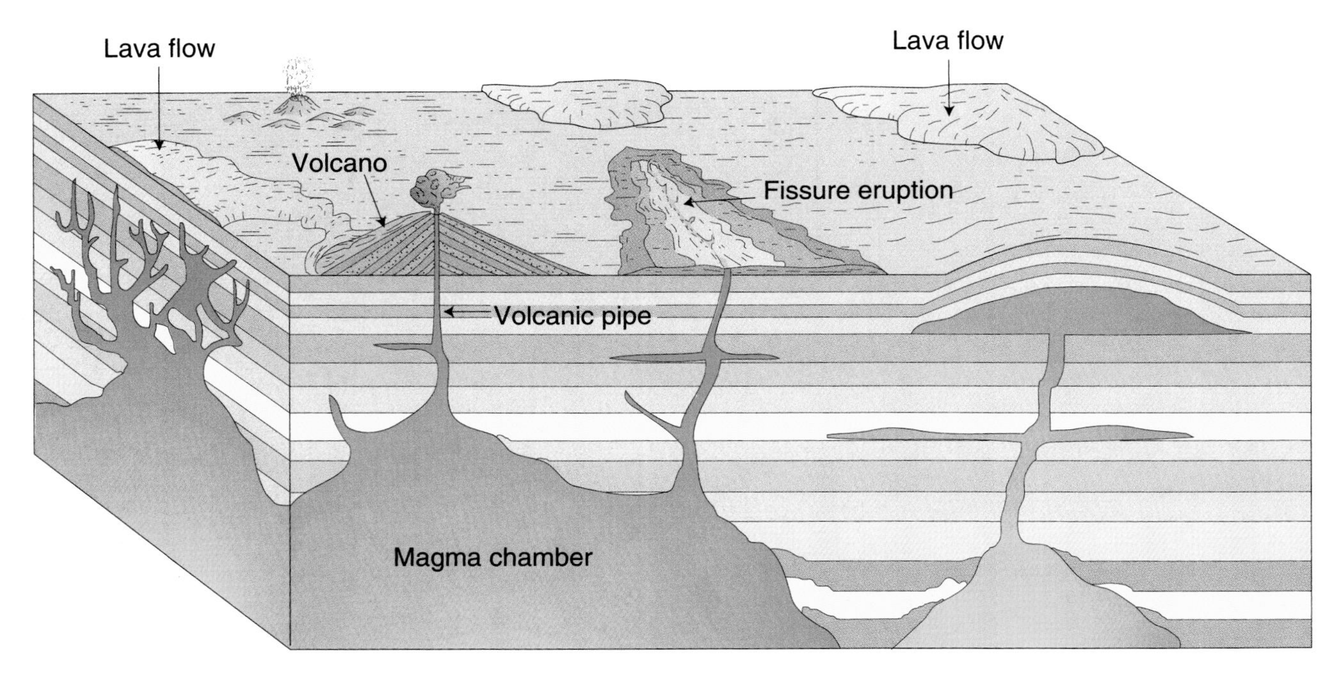

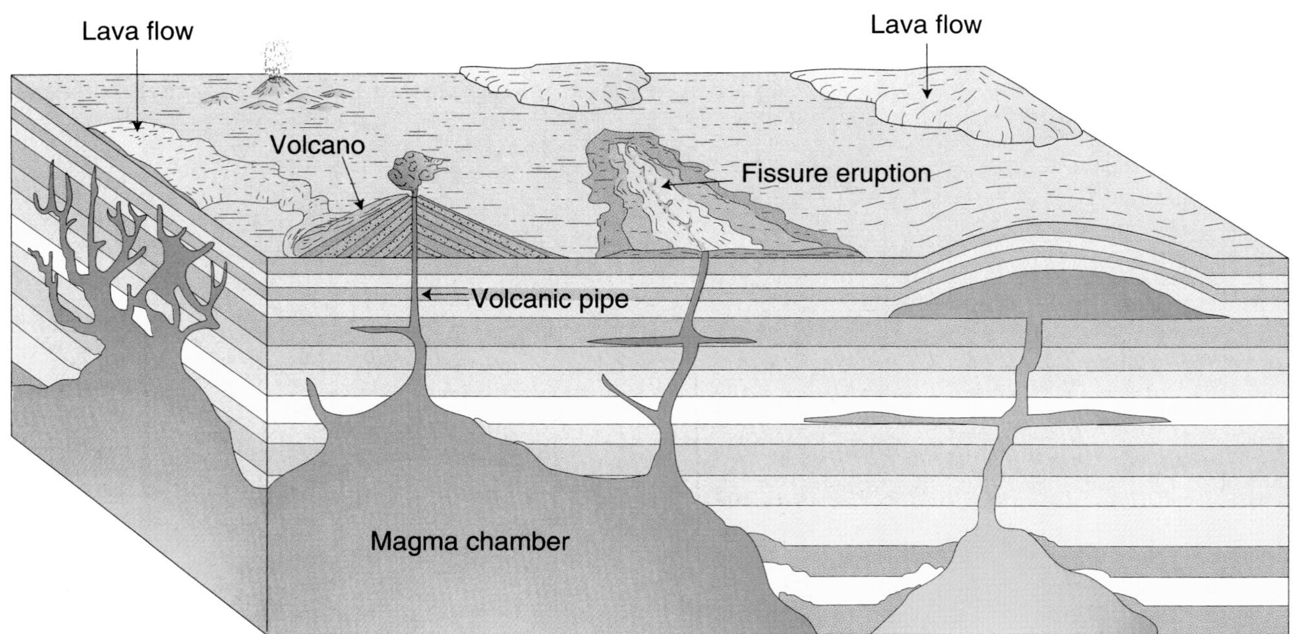

FIGURE 4–12
Igneous rock structures. In (a), molten rock (*magma*) slowly rises, following fractures and weak areas in rock. Most magma stops short of the surface, cooling and hardening to form *igneous rocks*. Magma that reaches the surface may either flow as *lava* or erupt more violently. Illustration (b) shows how igneous rock structures erode over time. (From E. J. Tarbuck and F. K. Lutgens, *The Earth,* 4th ed., Macmillan Publishing Co., 1993.)

shape Earth's surface in two principal steps. Rocks are first broken down into smaller pieces through weathering. Then they are carried by gravity down the slopes of hills or are transported by water, wind, or ice from one place to another.

On most of Earth's land surface, water is the most important agent in moving weathered rock fragments downhill. Streams are powerful transporters of sediment, carving deep valleys. The slopes on the sides of these valleys become steeper as streams erode into their beds, and erosion on slopes then moves material to the streams. In this way streams exert a strong influence on the shape of the land.

Weathering

Through weathering, rocks are broken into pieces ranging in size from boulders to pebbles, sand grains, silt, down to microscopic clay particles. Without weathering, the force of gravity and the agents of water, wind, and ice would have nothing to move. Rocks begin to break down the moment they are exposed to the weather at Earth's surface. They are attacked by water, oxygen, carbon dioxide, and temperature fluctuations. Weathering takes place in two ways: chemical weathering and mechanical weathering.

Chemical weathering. Rocks may be broken down as a result of **chemical weathering,** which is a change in the minerals that compose rocks when they are exposed to air and water. Chemical weathering occurs faster in places with warm temperatures and abundant water. Acids released by decaying vegetation also chemically weather rocks.

One example of chemical weathering is oxidation. Iron is a common element in rocks, and it combines with oxygen in the air to form iron oxide, or rust. Iron oxide has very different properties from the original iron—it is physically weaker and more easily eroded. You can see the effects of oxidation on a steel can: the rusty oxide easily flakes away.

Another example of chemical weathering is the decomposition of calcium carbonate. It is a major component of limestone and other sedimentary rocks. Calcium carbonate dissolves in water, in which it separates into ions of calcium and carbonate that are carried by streams into the sea.

Mechanical weathering. Rocks are also broken down by physical force, called **mechanical weathering.** Frequent changes in temperature can expand and contract rocks, causing them to break apart. Freezing and thawing in cold environments are an example. Rainwater seeps into cracks in rocks and freezes into ice crystals when the temperature turns colder. The water, which expands about 9 percent when frozen, pushes apart the rocks, opening them further, in a phenomenon called *frost-wedging*. Plant roots growing in cracks

A boulder of volcanic rock reveals the effects of weathering. Water has penetrated the rock and weakened the bonds between grains, allowing the rock to be easily broken open. The outer layers are discolored by chemical reactions, indicating greater weathering near the surface. The outer surface has a thin crust of white salt crystals that formed when water dried.

Death Valley, California was formed by faulting that raised the land in the foreground and the distance, while lowering the valley floor. The valley is filled with sediments eroded from the adjacent mountains. The surface in the foreground is covered with desert pavement. On the far side of the valley is a large alluvial fan. (Roy Morsch/The Stock Market)

between rocks also contribute to mechanical weathering; you probably have observed sidewalks that have been heaved by tree roots.

Mechanical and chemical weathering work together to break down rocks. Often, mechanical forces open cracks, and water seeps in to chemically weather the rock.

Moving Weathered Material

Once rocks are weathered, they may be carried from one place to another. Material most commonly moves downhill by gravity. This happens in two ways: by mass movement or by surface erosion. In **mass movement,** rocks roll, slide, or free-fall downhill under the steady pull of gravity. In **surface erosion,** water—which flows downhill because of gravity—carries material with it. Surface erosion may also result when the wind or ice carries material from one place to another.

Slope. Material moves faster down steeper hills. The steepness of a hill is measured through *slope,* which is the difference in the elevation between two points (known as the *relief* or rise) divided by the horizontal distance between the two points (known as the run). The greater the rise and the shorter the run between the two points, the faster will be the movement of materials down the hill. Wherever slopes occur, gravity is available to move material. Even the gentlest slope provides the potential energy necessary to move at least some material downward, either through mass movement or surface erosion. But erosion is usually much more rapid on steeply sloping land than on gentle slopes.

Mass movement. The most common form of mass movement is **soil creep.** As the name suggests, creep is a very slow, gradual movement of material down the slope of a hill (Figure 4–13). A tiny movement can cause creep—a rodent digging, a worm burrowing, an insect pushing aside soil. Creep occurs near the surface, in the top 1 to 3 meters (3 to 10 feet) of soil.

More dangerous and dramatic mass movements, such as rock slides and earth flows, can occur on steep slopes, especially during wet conditions. Steep slopes are prone to rock slides because the force of gravity pushing down on the rocks is likely to exceed the strength of the crust to hold onto them. Earthslides down steep slopes can follow intense rains, because material with a high water content is weaker and less able to resist the force of gravity. The sliding material may break down into fluid mud, which flows downhill. Houses built on very steep slopes, such as along the U.S. West Coast, risk damage from landslides and mud flows.

Surface erosion. The most common form of **surface erosion** is caused by rainfall. Intense rain sometimes falls faster than soil can absorb it. Water that can't soak into the ground must run off the

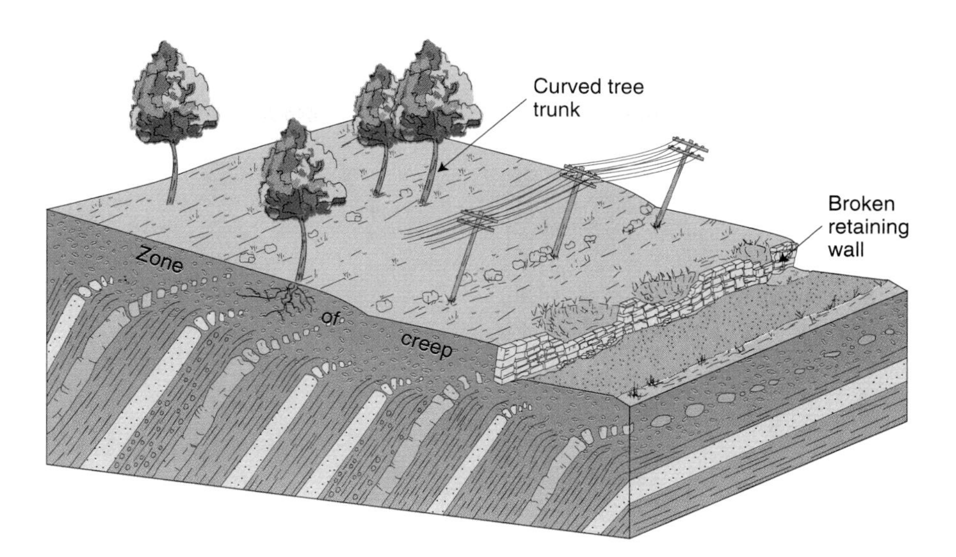

FIGURE 4–13
Soil creep is the most common form of mass movement. Creep, as the name suggests, is a very slow downslope movement of the uppermost part of the soil. (From E. J. Tarbuck and F. K. Lutgens, *The Earth,* 4th ed., Macmillan Publishing Co., 1993.)

surface. As it runs off the surface, water picks up soil particles and carries them down the slope. With enough runoff, water can carve channels into the landforms.

The smallest channels eroded by the flow of water—only a few centimeters deep—are called **rills.** Rills are so small that soil creep or a farmer's plow can obliterate them. But if channels gather enough water, they become larger and permanent carriers of water. As these channels deepen, they gather water and eroded soil from adjacent slopes, and when they gather enough water they form ravines, valleys, or canyons.

Except in deserts and coastal areas, surface erosion by water is slow where the ground is covered by grass and trees. But on much of Earth's surface, humans have removed vegetation by plowing fields or building settlements. Once the vegetative ground cover is removed, slow surface erosion can suddenly become severe erosion. Ground where vegetation has been removed can suffer more surface erosion in a few months than it experienced during the previous several thousand years. The eroded soil can pollute places downstream, and the remaining soil may be less productive for agriculture.

This landslide occurred on a steep slope on Santa Cruz Island, California, following heavy rain. The rain weakened the rocks, triggering the landslide. Small scars near the top of the slide reveal fractured surfaces in the rock. Once the weathered rock began to slide downhill, it was jostled and broken apart by the motion, and turned to mud. By the time it reached the bottom of the slope, the landslide had become a mudflow.

Stream drainage. Streams collect water from two sources: groundwater and overland flow (Figure 4–14). When rain falls on the land surface, most of it soaks into the soil and accumulates as **groundwater.** Groundwater migrates slowly through the soil and underlying rocks. During dry periods, most of the water flowing into streams is supplied not by rainfall, but by groundwater. If rain falls intensely, the soil may not be able to absorb it as fast as it falls. Water then may run directly to streams through **overland flow.**

In some areas of limestone bedrock, underground streams fed from groundwater are capable of eroding rocks beneath the surface. Regions of limestone rock may contain few streams at the surface, but underground water movement may dissolve passageways and even large caverns in the limestone. In other regions, streams carve landforms on the surface.

A stream drains groundwater and overland flow from an area called its **drainage basin.** The greater the area of its drainage basin, the more water a stream must carry. A basin with plentiful runoff from groundwater and overland flow may carve a complex network of many channels to remove the water and sediment. Small rills deliver water and material to larger streams, which join others to form rivers, which carry increasing amounts of water and sediment to the sea.

The volume of water that a stream carries per unit time is its **discharge.** Discharge ranges from a few cubic centimeters per second in rills to an average of more than 200,000 cubic meters per second at the mouth of the Amazon, the world's largest river. Discharge of any stream usually increases after storms and decreases during dry spells.

Drainage density is the combined length of all of the stream channels in a basin, divided by the area of the drainage basin. A basin that has soil capable of absorbing and storing most of the rain that falls will usually have a low drainage density. Landscapes with soils that cannot absorb rainfall very rapidly are easily eroded to form channels, and they have high drainage densities.

Streams shape their channels by alternately eroding and depositing material on their beds and banks. The turbulent, swirling motion of the water erodes particles from the channel. The water transports these particles, along with loose sediment in the channel and minerals dissolved in the water. This movement of material in a stream is called

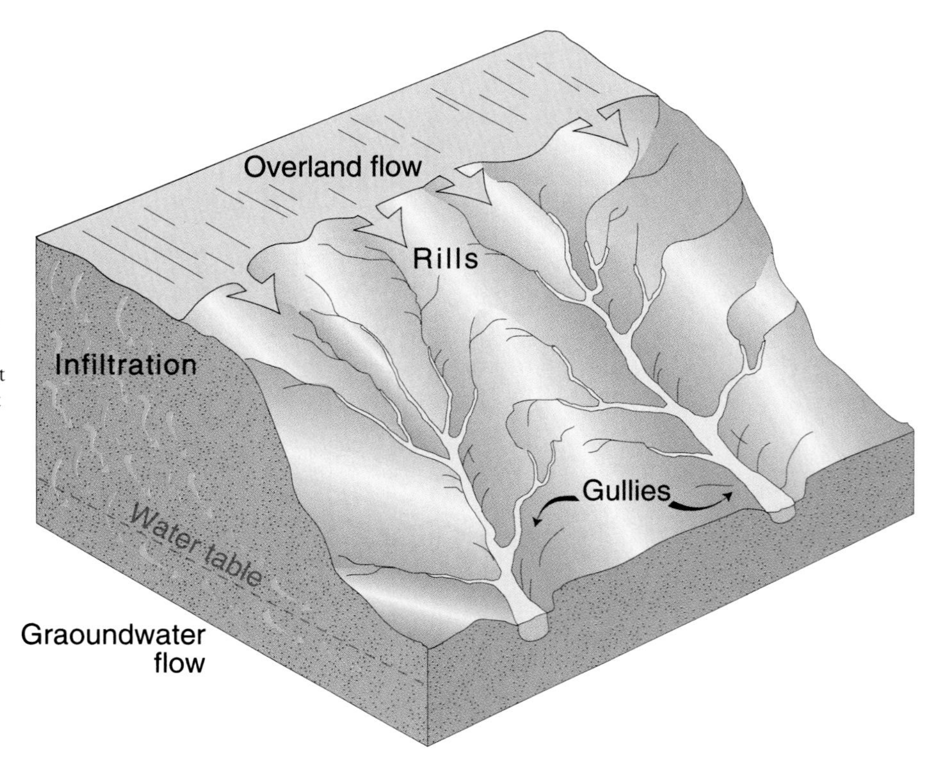

FIGURE 4–14
Surface water (overland flow) and *groundwater* are the two water sources for streams. Under the pull of gravity, rainwater soaks into the soil. This groundwater slowly migrates through the soil and porous rock. The surface flow follows the lowest path available, eventually running in streams. Streams merge into other streams and rivers, so most water ultimately ends up in the ocean. If rainfall is intense and the soil cannot absorb it all, overland flow can become great enough to cause flooding.

sediment transport. Transport increases greatly after a heavy rain. During dry periods, the stream's channel may partially fill with sediment, so less is transported downstream.

A stream is also responsible for shaping its floodplain, which is a nearly level surface at the bottom of the valley through which the stream is flowing. The surface of the floodplain is formed from deposits of sediment when the stream periodically floods (Figure 4–15). Streams develop a **meandering** course through their floodplains, as shown in (b). The channel continually shifts from side to side as the stream erodes material from one side of the channel, where current is swifter, and deposits it on the other side, where the slower current has less energy.

By continually eroding and depositing material in channels and floodplains, streams tend toward a stable condition, known as **grade.** A graded stream transports exactly as much sediment as it has collected (Figure 4–16). Streams rarely operate at a condition of grade for long, because daily changes in weather and disturbances from erosion and human activities continually upset the balance. As the stream's stable condition is upset—and the transport of sediment increases or decreases—the shape of the stream channel may change. An espe-

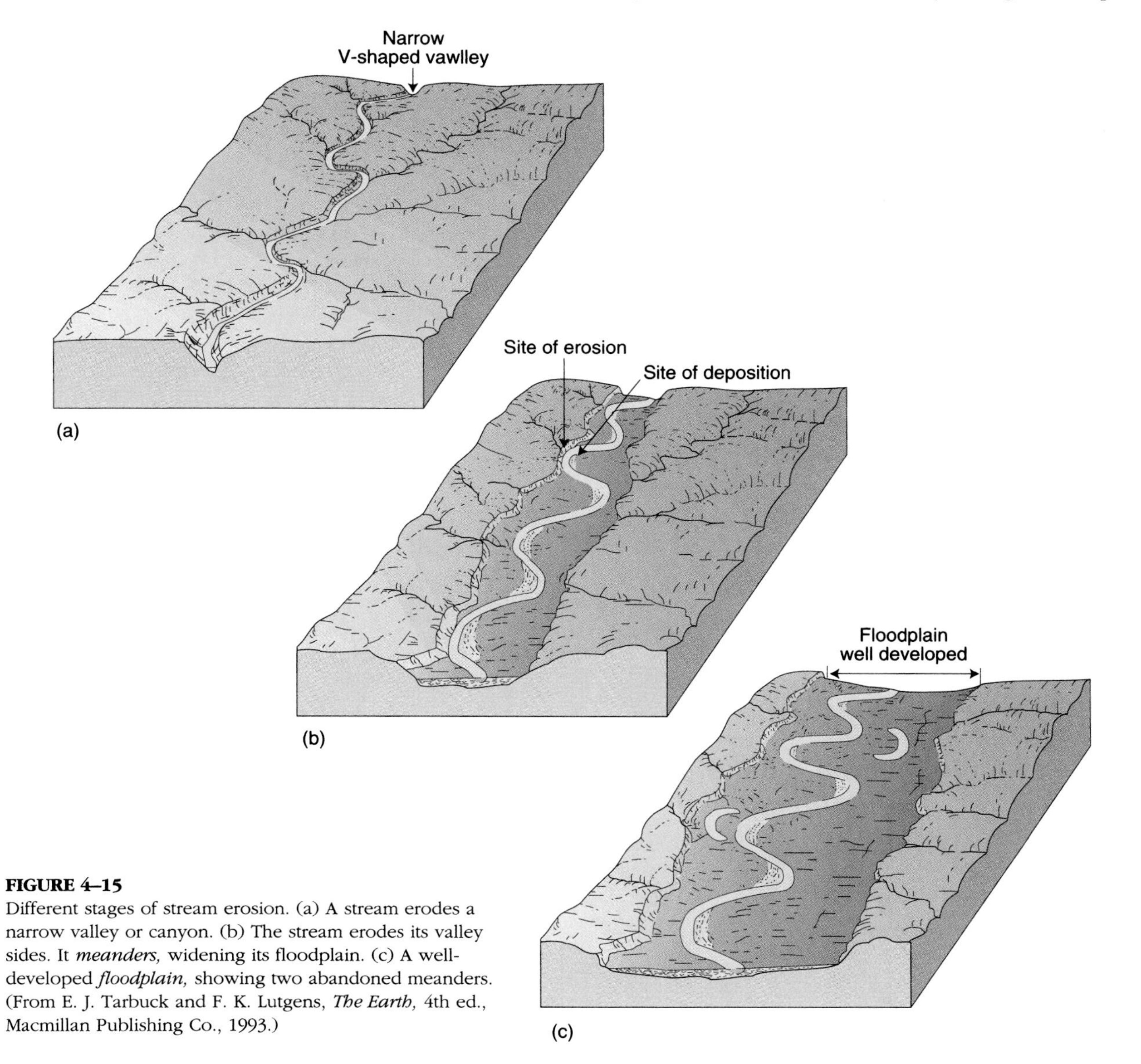

FIGURE 4–15
Different stages of stream erosion. (a) A stream erodes a narrow valley or canyon. (b) The stream erodes its valley sides. It *meanders,* widening its floodplain. (c) A well-developed *floodplain,* showing two abandoned meanders. (From E. J. Tarbuck and F. K. Lutgens, *The Earth,* 4th ed., Macmillan Publishing Co., 1993.)

FIGURE 4–16
An idealized profile of a stream and its numerous tributaries. Note the concave-upward shape of the profile, with steeper gradient toward the headwaters of streams and gentler gradient downstream. As the stream flows downhill, it accumulates more and more water, and its erosive power is greatly increased. The gentler gradient downstream slows the flow, balancing the erosive power of the stream with the amount of sediment it must carry. (From E. J. Tarbuck and F. K. Lutgens, *Earth Science,* 7th ed., Macmillan Publishing Co., 1994.)

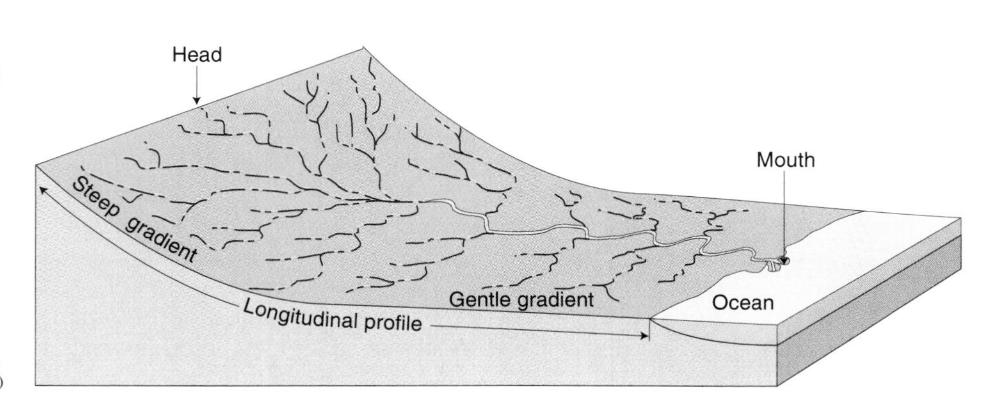

cially heavy flow following a storm may cause the channel to shift.

A stream adjusts to conditions by regulating its transport of sediment. If increased erosion upstream generates more sediment than it can carry, the stream deposits the excess in the channel or on the floodplain (Figure 4–17). By depositing the additional sediment, the stream transports only what it is capable of carrying.

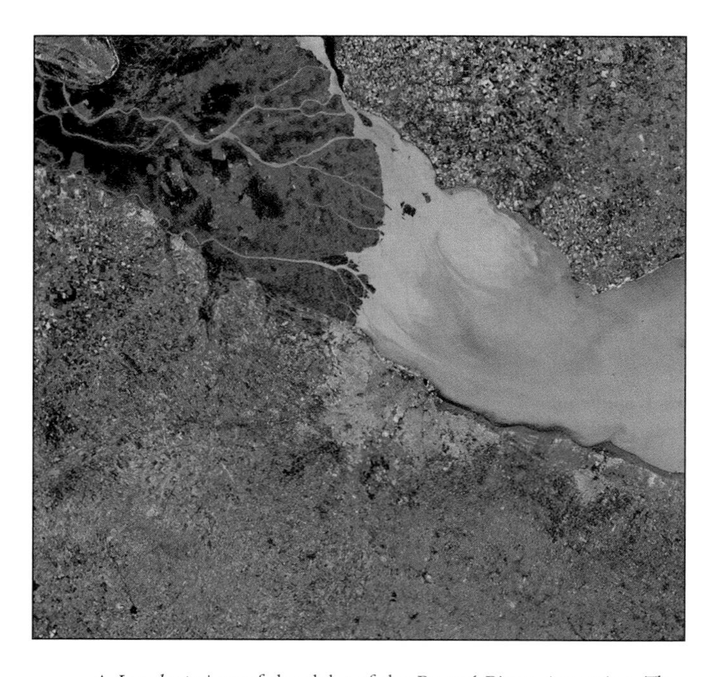

A *Landsat* view of the delta of the Paraná River, Argentina. The Paraná enters from the northwest; the city of Buenos Aires is at the center of the image. The river discharges about 10,000 tons of sediment per hour to the River Plate estuary. (Photo Researchers, Inc.)

Depositing excess sediment slowly raises the elevation of a stream. This reduces the difference in elevation between the deposit and places upstream. This reduction in relief also reduces the slope. Lowering the slope reduces the amount of sediment arriving from upstream.

Most sediment carried by a stream does not move to its final resting place in a single step. Typically, sediment becomes temporarily stored in a floodplain, then eroded-transported-deposited a second time, then a third time, and so on many times along its journey. Eventually, most sediment reaches the sea. Where a river enters the sea, the water velocity drops abruptly, and the sediment may form a delta.

Increased erosion from human activity. Human activities sharply increase the amount of sediment being eroded into streams. Major contributors include deforestation, agriculture, and urban development.

People clear forests because they want to use trees for fuel, lumber, and paper, or they want to use the land for another purpose, especially agriculture and urban development. Erosion has increased as a result of elimination of the vegetation cover for both reasons, but it is particularly severe where agriculture has replaced the forest.

For much of Earth's agricultural land, erosion of soil into streams is a major loss. To meet the needs of a growing population, food production expands in two principal ways—by opening up new land for agriculture and by using existing farmland

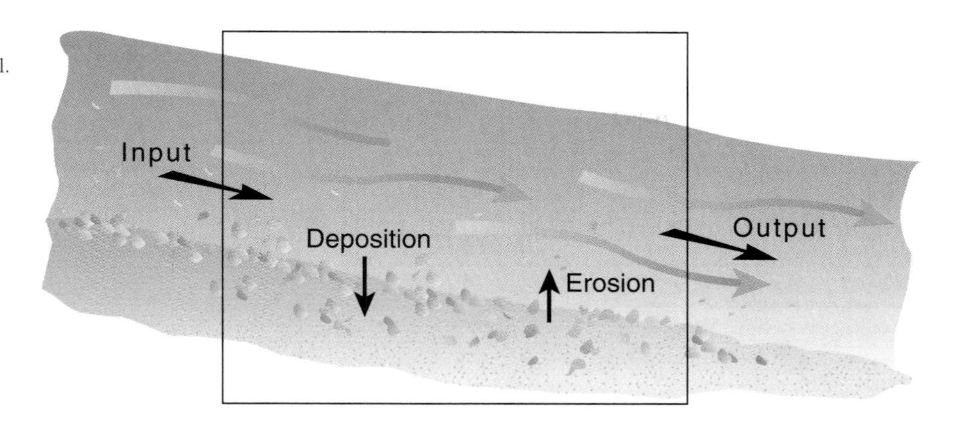

FIGURE 4–17
A systems view of a stream channel. In the segment of channel selected for study, sediment flows in and out. If more sediment enters than leaves, the excess is deposited and the channel shrinks or the floodplain accumulates sediment. On the other hand, if more sediment leaves than enters this portion of the channel, it means that the channel is being eroded, and therefore enlarged.

more intensively. But both strategies can result in erosion of the rich topsoil necessary for productive agriculture.

In places with rapid population growth, new lands may be opened for agriculture that may not be suited for intensive farming. For example, farmers in East Africa are clearing, tilling, and planting extremely steep mountain slopes. The rate of erosion in such areas can reach a few millimeters per year. Within a few years, the rich topsoil may be completely eroded in such areas.

Opening up new land for agriculture was a major contributor to increased erosion in the United States during the eighteenth and nineteenth centuries. More than 2 million square kilometers (800,000 square miles) of forest were cleared and replaced with plowed fields and pastures in the eastern and midwestern United States. This deforestation probably increased the rate of soil erosion by 10 to 100 times.

Erosion of agricultural land is further increasing as farmers use existing fields more intensively. The soil on about one-fifth of U.S. cropland is being lost faster than it can be replaced by natural soil formation. Also, to obtain higher yields, farmers plant profitable crops like corn, soybeans, and wheat every year, instead of periodically planting cover crops such as clover or alfalfa to restore nutrients to the soil. By failing to restore nutrients to the soil, farmers can exhaust its productivity.

Soil erosion lessened east of the Appalachians during the twentieth century because farmland was abandoned. Its soil was depleted of nutrients, and more productive land opened in the Midwest. In

Floodplain development in a valley in northern England. The high rainfall of this marine west-coast climate provides ample water for stream erosion. Glaciers occupied this area until about 12,000 years ago; since that time the stream has eroded the valley bottom and continues to widen its floodplain by meandering and lateral erosion.

A buried soil profile in southwestern Wisconsin. The dark band is the A-horizon of a floodplain soil that has been buried by more than 1.5 meters of sediment that was eroded from upland farmland since the area was first cleared and plowed in the 1830s. Beginning in the 1930s, soil erosion rates in this area have been reduced, and streams are now cutting down through these valley deposits. (James C. Knox/James C. Knox)

fact, much of the former farmland in the eastern United States has returned to forest. The soil erosion problem simply moved west of the Mississippi River during the early twentieth century, culminating in the "Dust Bowl" of the 1930s. Since then, government-sponsored soil conservation measures have significantly reduced erosion on U.S. farms, but the problem remains serious in some areas.

A heavy rainfall or rapid snow melt may dump an overload of eroded sediment into a stream, and it may be deposited on the flood plain downstream. Such sediment can bury nutrient-filled soils already in the floodplain, as well as roads and buildings constructed on the floodplain. Flooding may increase if a sediment-choked stream overflows its banks.

Even after the rate of soil erosion into a stream declines, a large quantity of sediment from the valley bottom may continue to be sent downstream. Many U.S. rivers have higher sediment loads than we would expect, because they are currently excavating from their floodplains sediment deposited during a past time of severe agricultural erosion.

Urban development also increases erosion. Land is cleared of vegetation for building new houses, factories, and shops. Erosion rates during the con-

Steep slopes being cultivated in Rwanda. Soil erosion rates under these conditions are high, and soil deterioration can be rapid. (Betty Press/ Woodfin Camp and Associates)

The Rhône Glacier, Switzerland. Crevasses on the glacier's surface are caused by flow. The budget of this alpine glacier has been out of balance over most of the past 200 years, with melting exceeding the accumulation rate. As a result, the glacier is much smaller than it was in the 1700s, probably as a result of climatic warming over that period. (Gabe Palmer/The Stock Market)

struction phase often are hundreds of times greater than for undisturbed land. If construction proceeds slowly, the land may be subject to high rates of erosion for several years.

Once land has been developed, surfaces once covered with vegetation are covered instead with roofs, streets, and parking lots. Because rain does not soak into these nonporous surfaces, it must run off into stream channels. Sewers collect storm water in cities and often discharge it into a stream at a rapid rate. If the streams are unable to handle the flow during a heavy rainfall, low-lying areas in the city may flood. To handle increased discharge, streams in urban areas may enlarge their channels through erosion of their banks.

Ice, Wind, And Waves

In some parts of the world, running water is a less powerful erosional agent than other forces. In some cold places, for example, the land is covered with flowing ice that grinds away at rocks. The wind can cause surface erosion in places with sparse vegetation, especially in the world's extensive deserts. In coastal areas, a combination of water and wind can cause surface erosion. Waves—which are driven by winds—pound the shore with turbulent water. Even the strongest rocks and vegetation are eventually broken down by the continual rushing forth and back of the waves.

Glaciers

Greenland, the South Pole, and many high mountain areas are currently covered with thick layers of moving ice, or **glaciers. Alpine glaciers** form on the peaks of individual mountains, wherever snow accumulates year after year without melting. **Continental glaciers** exceeding 3 kilometers (almost 2 miles) thickness cover vast areas of Greenland and Antarctica.

Glaciers are rivers of ice that flow from places where snow accumulates year-to-year to places that are warmer where the ice melts. A **glacial budget** can account for the water entering a glacier and leaving it over time. Water enters the head of the glacier as snow and leaves at its terminus as meltwater. Some snow also *sublimates,* a term for evaporating directly from the frozen state to vapor. Reasonably accurate glacial budgets can be calculated for individual alpine glaciers, but are much harder to calculate for the large continental ice masses covering Antarctica and Greenland.

Glaciers flow very slowly. They grow when they receive more snowfall at their head, and they shrink if warm temperatures increase the melting at the terminus. Glaciers can also act erratically. For many years, a glacier may move only a few meters per year; then for a few years, it may suddenly surge forward at several hundred meters per year, before slowing again.

A glacier is like a conveyor belt because it hauls sediment from areas of erosion and deposits it elsewhere. As ice accumulates and begins to flow, the glacier picks up material, and where it melts the sediment is deposited. Glacial deposits play an especially important role in shaping landforms in places with rapid melting, because large quantities of material are dropped in these places, forming **moraines** (Figure 4–18). A **terminal moraine** is a ridge of material dumped at the end of the glacier.

Meltwater leaving a glacier deposits some of the debris close to the glacier in a broad, gently sloping plain, known as an **outwash plain.** The outwash plain contains larger rocks deposited close to the glacier in a layer of sand and gravel that can exceed a thickness of 100 meters (300-350 feet). The finer silt and clay materials are usually carried much farther, and may be deposited in lakes, seas, or distant valleys.

Impact of past glaciations. Only 18,000 years ago—a very short time in Earth's history—glaciers covered much of North America, Europe, and northern Asia (Figure 4–19). These glaciers acted like bulldozers, shaping landforms as they advanced. When Earth's atmosphere warmed, these rivers of ice melted away, but they left behind debris that shapes the landforms we see today in many regions.

As they advanced and retreated, alpine glaciers created distinctive landforms. In an unglaciated mountainous area, a stream may carve a V-shaped valley (Figure 4–20a). As an alpine glacier flows through a V-shaped stream channel, it scours away the rock and rounds the valley bottom into a U-shape (b). When the ice melts, a U-shaped valley remains, surrounded by knife-edged ridges (c).

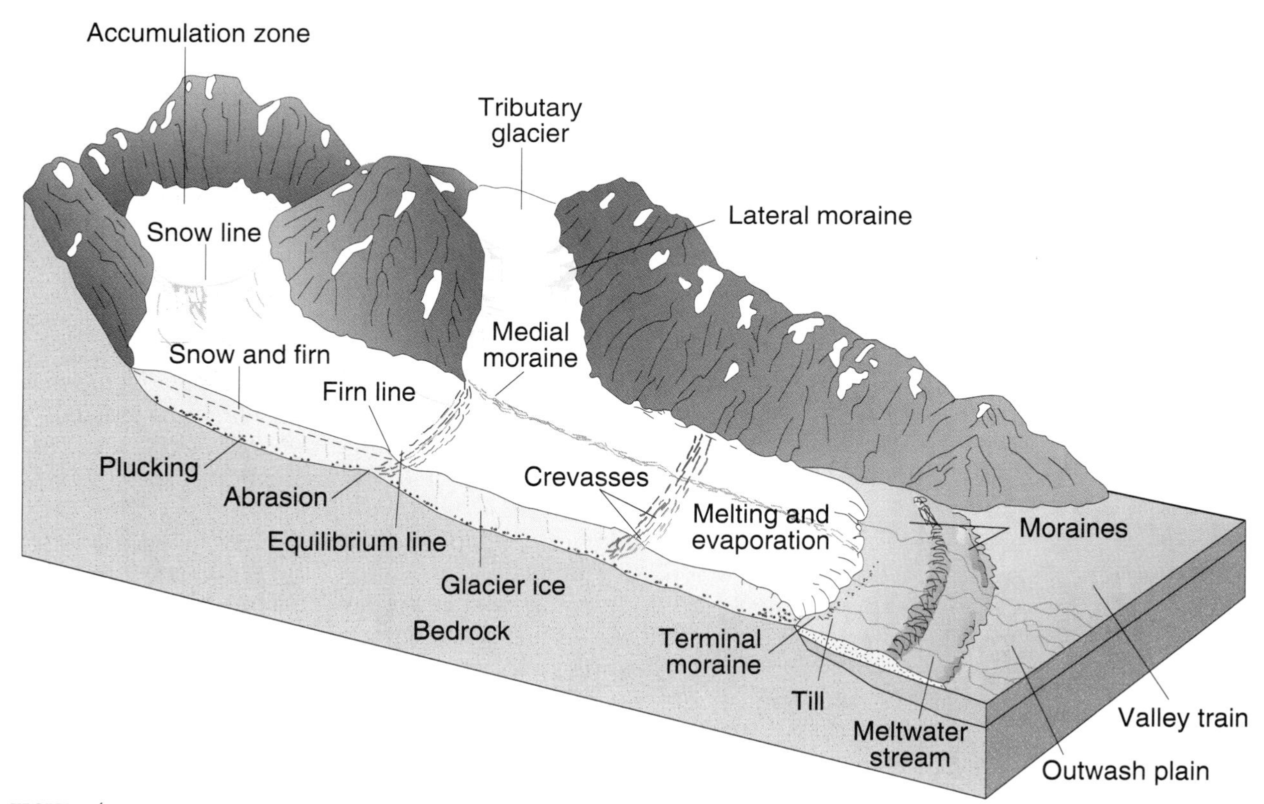

FIGURE 4–18
Alpine glaciers are fed by snow in a *zone of accumulation* and flow downhill under the influence of gravity. They melt at lower elevations, leaving debris to accumulate in *moraines* or to be carried away in *meltwater streams*. (From R. W. Christopherson, *Geosystems,* 2d ed., Macmillan Publishing Co., 1994.)

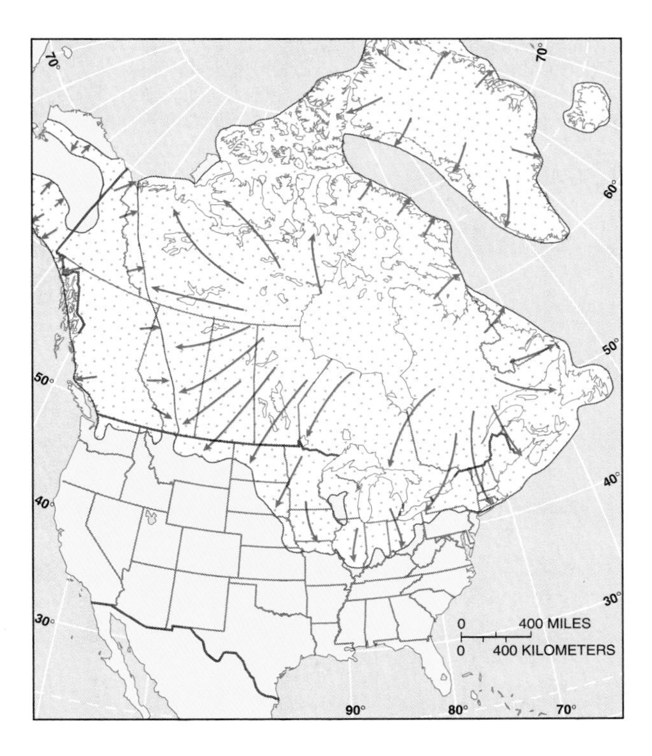

FIGURE 4–19
The extent of continental glaciation in North America during the last Ice Age, about 18,000 years ago. The glaciers flowed from areas where accumulation of snow was greatest, such as the Rockies and the Hudson Bay area. Continental glaciers are large enough to flow over some hills and mountains, such as the uplands of northern New England and southeastern Canada.

The advance and retreat of continental glaciers have influenced the distribution of many human activities. The following are examples of how glaciation has affected three activities in the United States: agriculture, drinking water, and transportation.

- *Agriculture.* The advance and retreat of glaciers in the U.S. Midwest broke up bedrock and brought it to the surface. Weathering releases from this rock nutrients that are important for modern agriculture, such as calcium and magnesium. Today, the U.S. Midwest is the world's most important center for growing corn. Soybeans, wheat, and hay are also grown in large quantities (see Chapter 11).
- *Water supply.* Retreating glaciers left deposits of sand and gravel capable of yielding a large supply of high-quality groundwater in places such as Long Island, New York (Figure 4–21). Long Island, which is more than 100 miles long and about 20 miles wide, is the top of a terminal moraine marking the southern extent of the most recent glaciation. Much of Long Island consists of outwash deposits of sand and gravel which absorb precipitation readily. This groundwater is the primary source of water for the 2.5 million inhabitants of Long Island.

  With rapid population growth in recent years, demand for water has exceeded the rate at which the groundwater is recharging in Long Island. As more of the island is paved with concrete for roads and buildings, precipitation is blocked from entering the ground. As a result, the level of groundwater has lowered, and saltwater from the Atlantic Ocean has migrated in to replace freshwater in some places. The quality of the groundwater has also been hurt by pollution from landfills and industry. Similar problems are occurring in other urbanized areas of the northeastern United States and Europe that depend on groundwater left behind by retreating glaciers.
- *Transportation routes.* When the glaciers retreated, they left behind the rivers and lakes seen on contemporary maps, such as the Great Lakes and the Ohio, Missouri, and Mississippi rivers. The lakes and rivers of North America have been important transportation routes for migrants heading west to settle and for farmers sending their output to the large cities in the east. Of equal importance in the development of transportation routes in the United States are the dry channels formed by glaciers, known as **meltwater channels.** The dry meltwater channels were used in the nineteenth century as the location of major canals, including the Champlain, Erie, Trent-Severn, and Illinois Barge, which encouraged the movement of people and goods between the interior of North America and the East Coast (see Box 4-1).

Effects of Wind on Landforms

Wind is an important shaper of Earth's landforms, especially in the dry regions (shown on the world climate map, Figure 3–1). Wind acts with water to reshape coastal landforms as well.

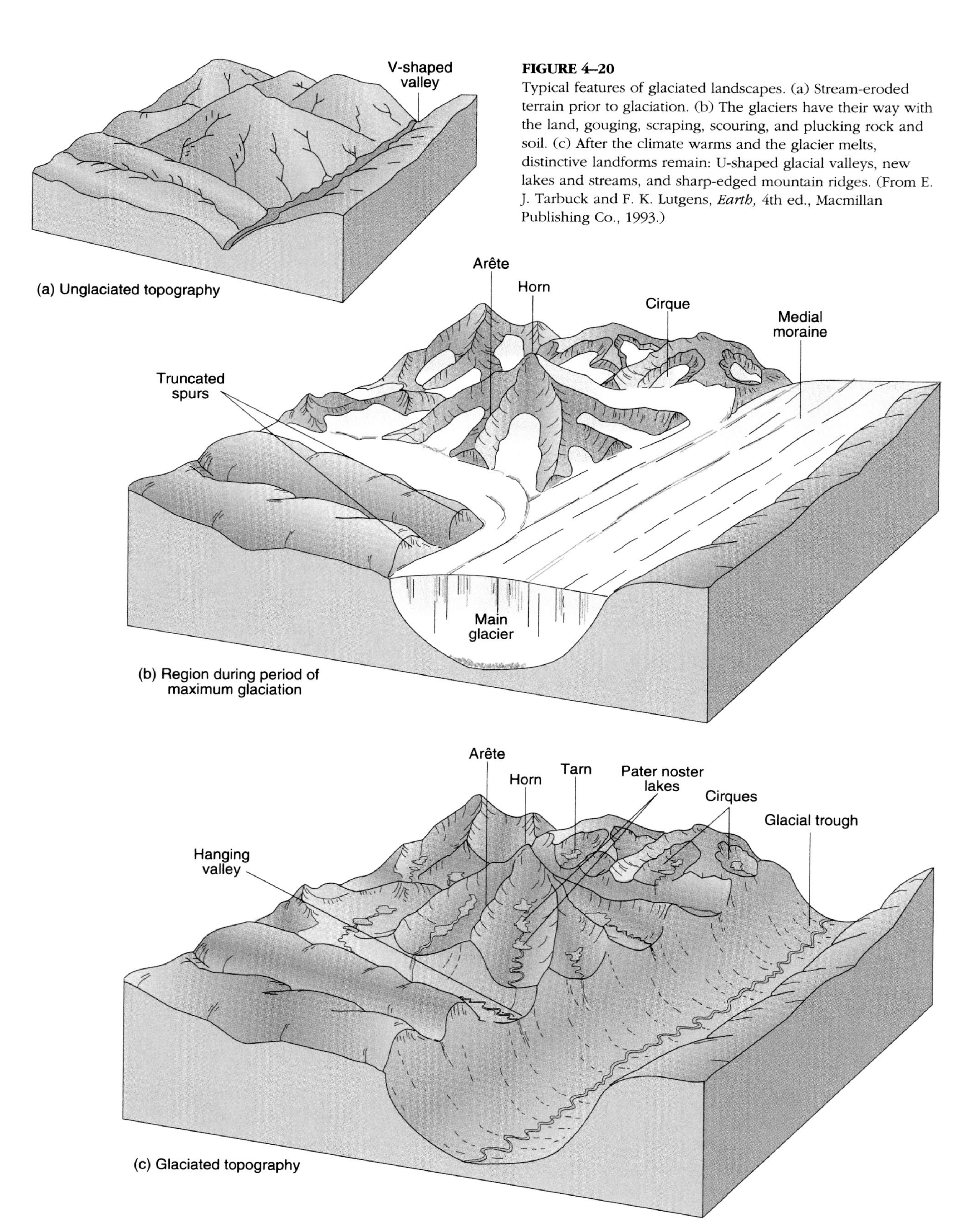

FIGURE 4–20
Typical features of glaciated landscapes. (a) Stream-eroded terrain prior to glaciation. (b) The glaciers have their way with the land, gouging, scraping, scouring, and plucking rock and soil. (c) After the climate warms and the glacier melts, distinctive landforms remain: U-shaped glacial valleys, new lakes and streams, and sharp-edged mountain ridges. (From E. J. Tarbuck and F. K. Lutgens, *Earth,* 4th ed., Macmillan Publishing Co., 1993.)

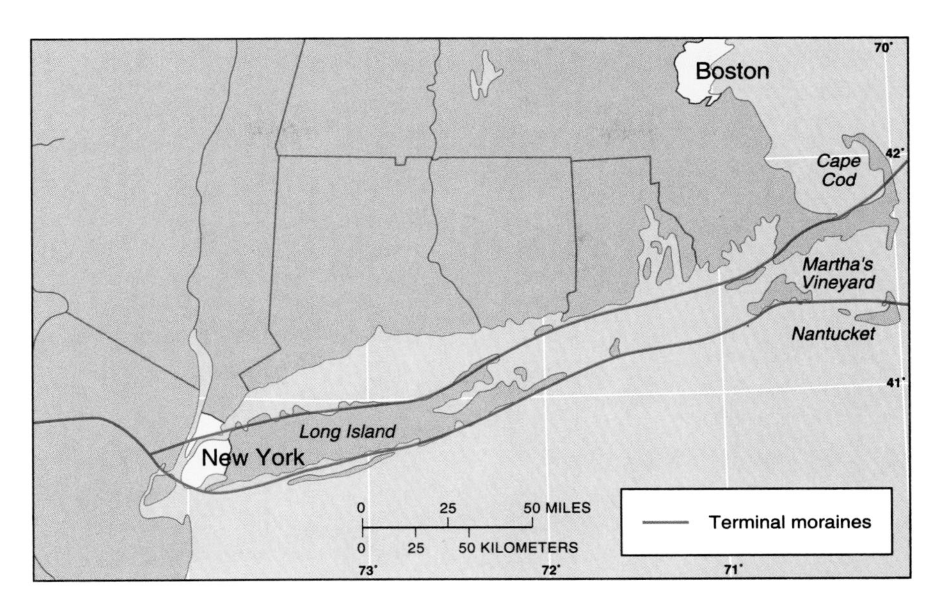

FIGURE 4–21
Long Island, New York, is completely covered with glacial deposits. The island is actually the top of a string of terminal moraines that also includes Martha's Vineyard, Nantucket, and Cape Cod, Massachusetts. The sandy soils of Long Island absorb water readily, and groundwater resources once were abundant. Today, however, these resources are threatened by overuse, intrusion of saltwater from the ocean, and pollution.

Dry climate erosion and landforms. Vegetation is crucial in protecting the soil from erosion. Recall from Chapter 3 that vegetation is sparse in deserts, because potential evapotranspiration greatly exceeds precipitation. The lack of a complete vegetation cover leaves dry lands vulnerable to erosion from the wind. Wind erosion can also occur in semiarid areas if the vegetation is removed.

Wind is not capable of moving large particles like gravel, but it can carry great amounts of fine-grained sediment such as sand. Where wind velocities are lower, or where topography encourages deposition, the sand accumulates in **dunes.** These accumulations of shifting sand are difficult places for vegetation to become established.

When the fine sand is eroded from the soil surface, larger rocks, pebbles, and gravel are left behind. The landforms in the dry lands can form a hard, armored surface, called **desert pavement.** The most popular image of a desert landform is a sand dune, but about 90 percent of Earth's desert areas are covered with desert pavement rather than sand. Wind erosion is minimized once desert pavement has formed because the remaining rocks are too heavy to be moved by the wind.

Wind also can affect landforms in humid regions, because very fine-grained sediment eroded by the wind can be carried some distance before being deposited. Thick layers of wind-blown silt, called **loess,** blanket many areas, including central China and the Mississippi River valley of the United States. Originally, this material was carried by meltwater from glaciers to nearby valley bottoms. Wind then carried the sediment to adjoining areas, forming loess, a fine agricultural soil (Chapter 5). Running water subsequently has carved deep gullies and ravines in these loess deposits.

Rainfall, though rare in dry areas, also is an important erosion agent. Because dry lands lack vegetation, the occasional rainfalls cause rapid erosion. Bare soils are much more prone to erosion than vegetated ones—look at any construction site after a heavy storm. Thus, a dry landscape often is covered with gullies and dry stream channels, even though only a few centimeters of rain may fall per year.

A prominent landform in many deserts is the **alluvial fan,** which is formed from sand and gravel deposited where a fast-moving stream emerges from a narrow canyon onto a broad valley floor. When the stream emerges, the water quickly spreads out and loses velocity. Given the limited amount and frequency of rainfall, a stream in dry lands is unlikely to carry sediment very far. Instead, sediment that the stream can no longer carry is deposited in a fan-shaped pattern. Occasional floods crossing the alluvial fan can disturb the pattern and encourage wind erosion. Large depositional areas in desert valleys form important sources of wind-eroded sediment.

FOCUS BOX 4-1

Glaciers and Canals

Humans have long relied on water transport for communication. European explorers of the North American continent during the seventeenth and eighteenth centuries were no exceptions. Settlement of the North American interior was heavily influenced by the location of the major inland rivers feeding into the Mississippi, such as the Ohio, Missouri, and Illinois.

Before the last continental ice sheet formed tens of thousands of years ago, the Missouri River flowed the opposite direction: northward into Hudson Bay. Ohio, Indiana, and Illinois were drained by a river that flowed west well north of the current path of the Ohio River. But as the glaciers melted and began to recede, the meltwater from the Ohio and Missouri was blocked by the glacier and could not take its normal northward route to the sea. Instead, the meltwater was forced to flow along the edge of the glacier until it found a new route. In the case of the Ohio and Missouri Rivers, the meltwater flowed south into the Mississippi River. The glaciers also left behind the Great Lakes, which explorers used to reach the interior of the continent.

Travel from the East Coast to the interior was difficult, because no convenient direct water route existed from the Atlantic to the Great Lakes and the Mississippi River system. In the years immediately after U.S. independence, the solution was to build canals. Many were built in dry meltwater channels left behind by receding glaciers:

- The Champlain Canal followed a meltwater channel to connect the Hudson River (which flows into the Atlantic at New York City) with Lake Champlain (which empties into the St. Lawrence River).
- The Erie Canal followed a meltwater channel across New York State to connect the Mohawk River with lakes Erie and Ontario.
- The Trent-Severn canal system in Canada followed a meltwater channel to connect Lake Ontario with Lake Huron, while avoiding Niagara Falls and a long journey through Lake Erie.
- The Illinois and Michigan Canal connected the Mississippi River with Lake Michigan at Chicago, in part by following a meltwater spillway occupied by the Illinois River.

Some of these canals are used only for recreational traffic today, but others remain important routes for bulk cargo. The Mississippi itself was greatly enlarged by glacial meltdown, especially in its northern reaches. This enlargement has enhanced the navigability of the Mississippi.

Erosion by the San Juan River has exposed folded sedimentary rocks in southeastern Utah. Even though wind erosion is an important process in this region, water erosion has also made a prominent mark on the landscape. (Tom Bean/The Stock Market)

Land degradation in semiarid areas. Human action can greatly accelerate wind erosion in semiarid areas. This process is called **desertification,** because the semiarid land becomes more like a desert. Desertification removes millions of hectares of land from agricultural production each year.

During the 1930s, wind-generated erosion was widespread in the U.S. Great Plains, and the region was known then as the Dust Bowl. Faced with declining prices for their crops, farmers in the Dust Bowl tried to plant more crops than the region's soil could handle. Several unusually hot and dry summers exacerbated the erosion problem. Winds carried the very-fine grained silt and clay dust hundreds of miles, even as far east as Washington, DC—a visual warning to government officials about the severity of the problem.

In recent years, erosion has been widespread in the Sahel region of northern Africa. Population growth, the breakdown of traditional farming and grazing patterns, and periodic drought have led to overgrazing of animals and excessive cultivation of crops. Removal of vegetation to feed people and animals has left soil unprotected and prone to wind erosion. Wind-blown dust carried from the Sahel out over the Atlantic Ocean can be seen in satellite images. Soil fertility may be reduced for many generations.

Coastal Erosion

Wind-blown sand is common in coastal zones as well as dry climates, because of the lack of vegetation. But vegetation is scarce along a coast because of the pounding of waves rather than an arid climate. A coast is an especially active area of erosion because an enormous amount of energy is concentrated on the shorelines from pounding waves. Land may be lost to erosion or gained through deposition at rates up to several meters per year. At this rate, over several decades, a house built near the shore may end up much farther from the beach, or much closer, even washed away.

Waves. Winds blow across the sea surface, transferring their energy to the water by generating waves. Waves are a form of energy, traveling horizontally along the boundary between water and air. As the wind blows harder and longer, it transfers more energy and generates bigger waves. Waves also grow larger with an increase in the expanse of water across which the wind blows.

The speed at which a wave travels is affected by its **wavelength.** A small ripple travels very slowly, whereas an ocean swell wave may travel 10 to 50 kilometers per hour (6 to 30 miles per hour). A **tsunami,** which is an extremely long wave created by an underwater earthquake, may travel hundreds of kilometers per hour (Figure 4–22).

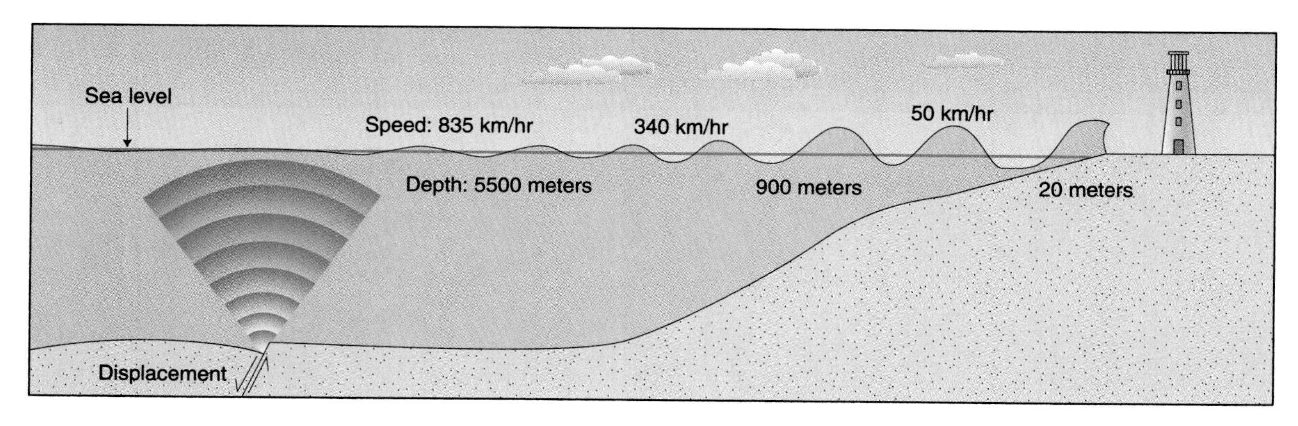

FIGURE 4–22
When a fault in the seafloor slips, it violently displaces the water above, creating a very low, fast-moving wave at the surface, called a Tsunami. As the wave comes ashore, it can build to a height of tens of meters, especially in a confined harbor (*tsunami* is a Japanese word meaning *harbor wave*). Historically, tsunamis have killed thousands in coastal settlements. After losing 150 people in a 1946 tsunami, Hawaii established the Pacific Tsunami Warning Center, which alerts nations around the Pacific when a tsunami is detected. (From E. J. Tarbuck and F. K. Lutgens, *The Earth,* 4th ed., Macmillan Publishing Co., 1993.)

A wave can travel thousands of kilometers across deep ocean water relatively unchanged, but when it nears the shore, the shallow bottom restricts water motion and distorts the wave's shape. The wave is slowed, causing the top part of it to rush forward and break (Figure 4–23).

The energy of a wave is released as a tremendous erosive force of rushing water on the beach. Beach pebbles and granules of sand are rolled back and forth, constantly being ground ever-finer. The finest particles are carried into deep water and settle on the seafloor. The larger sand- and gravel-sized particles are left behind on the shore to form a **beach,** which is a surface on which waves break and constantly move sand up and down.

A beach reflects the characteristics of the waves that form it. A coastal area pounded by larger waves will likely form a beach of very coarse mate-

Satellite view of a dust storm over the Algeria-Nigeria border in the Sahara Desert. Small cumulus clouds have formed over parts of the dust front. (NASA/Science Source Photo Researchers, Inc.)

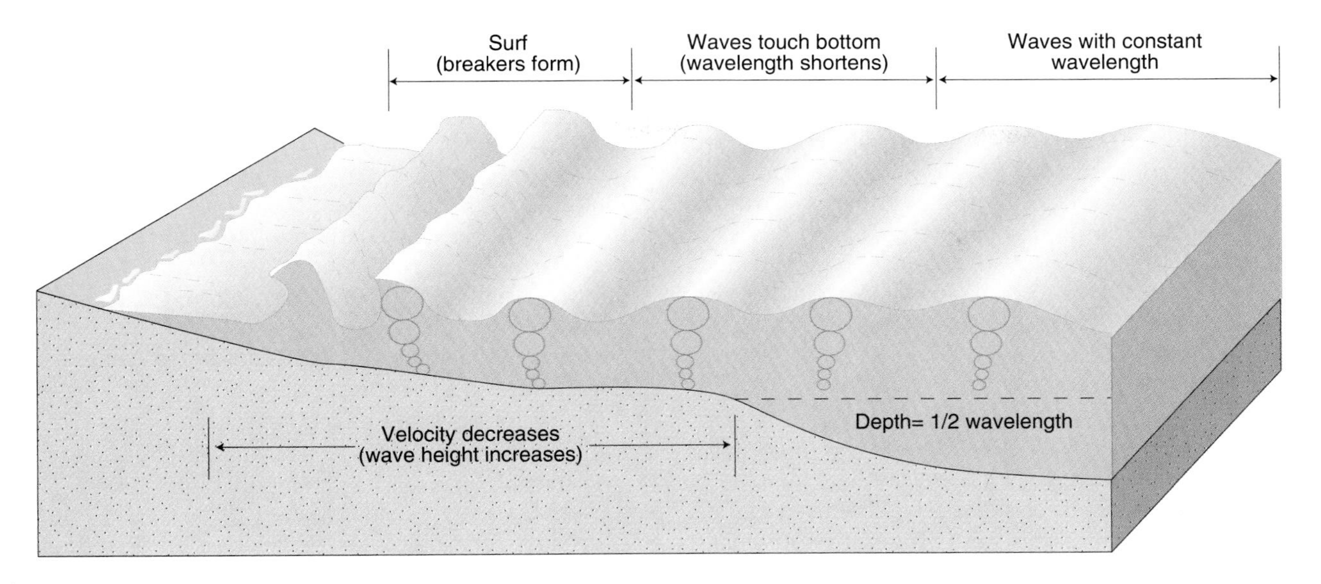

FIGURE 4–23
As waves reach shallow water they change shape, with long, low waves becoming short and tall, eventually breaking and releasing their energy on a beach. (From E. J. Tarbuck and F. K. Lutgens, *The Earth,* 4th ed., Macmillan Publishing Co., 1993.)

rial, because the waves carry the finer sand offshore and deposit it in quieter waters. The shape and size of the beach can vary if storms hit during some seasons, while other seasons are quiet.

Longshore current. When you watch waves break on a beach, the most obvious motion of the water is perpendicular to the shoreline—the waves move up the beach and then recede. When the waves break, the energy also gives a push to the water in a direction parallel to the shore, and the repeated breaking of many waves generates a **longshore current** traveling parallel to the shore (Figure 4–24). The longshore current is like a river, carrying sediment from areas where it is eroded by waves and depositing it where breaking waves

Beach-front homes in Fort Lauderdale, Florida. The strip of land in the foreground is a barrier island, built by longshore transport of sand along the shoreline. Most of the beaches in this part of Florida are eroding, and sand must be imported to replace what is lost to the waves. If sea level rises, erosion rates will increase. (Alese and Mort Pechter/The Stock Market)

lose the energy to carry it—usually in deep water. Longshore currents can carry enormous amounts of sediment great distances.

Like rivers, landforms along shorelines are shaped by the balance between sediment arriving in a portion of the shore and being removed from it. Distinctive landforms develop, such as beaches, bars, and spits. If more sediment is removed than arrives, the coast is eroded—which is the condition of most of the world's shorelines. But in some areas, more sediment arrives than is removed, and the land area grows.

Sea-level change. The edge of the land—the shoreline—is defined by the elevation of the sea, or **sea level.** We think of sea level as being fixed, but on most shorelines it is not constant: it continually rises or falls relative to the adjacent land. In the short term, the sea can rise or fall several meters because of tides and storms. But two long-term factors can cause sea level to rise or fall: climate change and movements of Earth's crust.

Over the past few hundred years, sea level has risen on Earth as a whole, at about 1 millimeter per year along coasts that are stable otherwise. Sea level has risen because the volume of seawater has increased. The increase probably results from the melting of glaciers, due to the overall warming trend since the eighteenth century. This change in sea level is small compared to the sea-level rise of about 85 meters (280 feet) that occurred at the end of the most recent ice age.

Sea-level changes are significant at two different time scales. In the short term—a few decades—the direction of sea-level change affects how a shoreline erodes. If sea level rises, the water offshore becomes deeper and waves will break closer to the land, causing more erosion. If sea level falls, the shallower water causes waves to break further offshore, dissipating their energy and reducing shoreline erosion. Because many shorelines have gentle slopes, a minor increase in sea level can translate into a much larger landward migration of the shoreline (Figure 4–25).

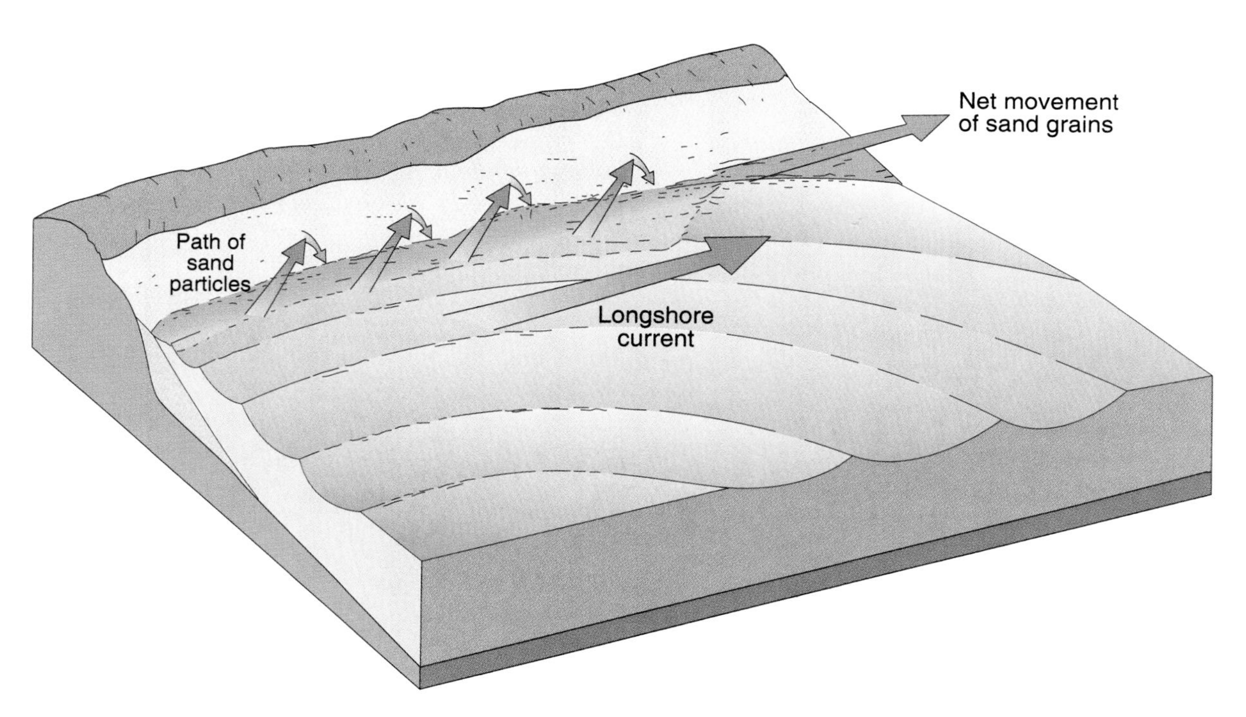

FIGURE 4–24
The *longshore current* is driven by waves breaking at an angle to the coastline. The current flows parallel to the shore. Breaking waves stir beach sand, causing it to be carried along with the current. (From E. J. Tarbuck and F. K. Lutgens, *The Earth,* 4th ed., Macmillan Publishing Co., 1993.)

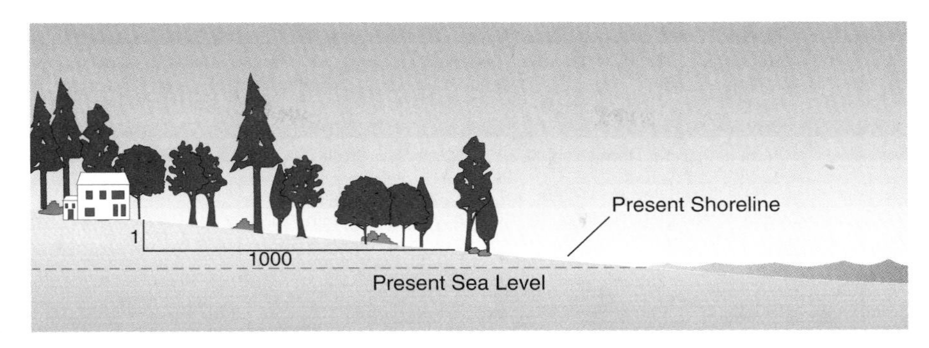

FIGURE 4–25
Sea level rise and the position of a shoreline are directly related. On gently sloping shorelines, such as coastal plains, a relatively small increase in sea level can translate into significant horizontal relocation of the shoreline. For example, if the land elevation increases 1 meter for each kilometer of distance inland (a ratio of 1:1,000), a 1 millimeter per year sea level rise could mean horizontal shoreline displacement of 1 meter per year. Such rates of shoreline erosion usually create major problems in developed coastal areas.

Over thousands of years, large sea-level change can reshape shorelines. During continental glaciation, the sea was substantially lower, and rivers in coastal areas cut deep valleys as they approached the sea. When the sea level rose worldwide, about 15,000 to 10,000 years ago, the river valleys "drowned," became estuaries such as Chesapeake Bay and Delaware Bay.

Sea level has fallen rather than risen in some places. This has left inland soils and vegetation that are typical of beaches. For example, much of the U.S. west coast has been tectonically uplifted in the last few million years as the North American and Pacific plates have ground together. As a result, the region lacks the deep river mouths of the U.S. East Coast, but it does have exposed marine terraces.

Human impact on coastal processes. The sea has been important for thousands of years for trade, communication, and food. This long tradition has established most of the world's densest populations near the sea (see Figure 7-1). The shoreline is an attractive place to build houses and recreational facilities, so coastal areas worldwide have become focal points for settlement and investment.

A drawback of coastal living is erosion from shifting shorelines. In wealthy societies like the United States, the typical response is to control coastal processes by modifying the coastline. These

The broad green area in this photo of California's Big Sur coast is a marine terrace, a former beach raised above the sea by tectonic uplift. Such terraces are common on the U.S. West Coast. (Pat and Tom Leeson/Photo Researchers, Inc.)

structures achieve the desired erosion control and stabilize the shoreline—but only temporarily. For example, people build *groins* perpendicular to the shore to slow **longshore transport** of sediment (Figure 4–26). A groin constructed in one location along the shore to interrupt the movement of sand will cause less sand to be deposited on a beach farther along the shore, worsening the erosion problem there.

People also build *seawalls* parallel to the shore, but the constant pounding of waves removes sand from around the seawalls and ultimately undermines them. People may feel confident in building beachfront homes "protected" by seawalls, but a major storm can wipe out the seawall and destroy the properties.

The shoreline is one of the most dynamic features of Earth's surface. On an eroding barrier island of the eastern United States, erosion over the past few decades could average between 10 centimeters and 2 meters (between 4 inches and 6 feet) per year. At that rate, someone who takes out a 30-year mortgage to buy a house that is 30 meters (100 feet) from the beach may see waves lapping at its edge by the time the last payment is made. And if a hurricane hits, the house may be washed away even sooner.

Homebuilders may not know about the threat of erosion when they build new houses on an attractive beach. A coastal development is usually well-established before its residents discover the dynamic nature of the coast. By then it is too late to abandon the substantial investment, and more money is spent to protect the structures already built.

Most of the erosion and damage to these communities takes place during major storms such as hurricanes. After such a disaster, sympathy for the victims induces the government to help rebuild communities at considerable public expense. Over the long run, though, buying up the properties and relocating the families would probably be cheaper than constantly rebuilding shorefront communities.

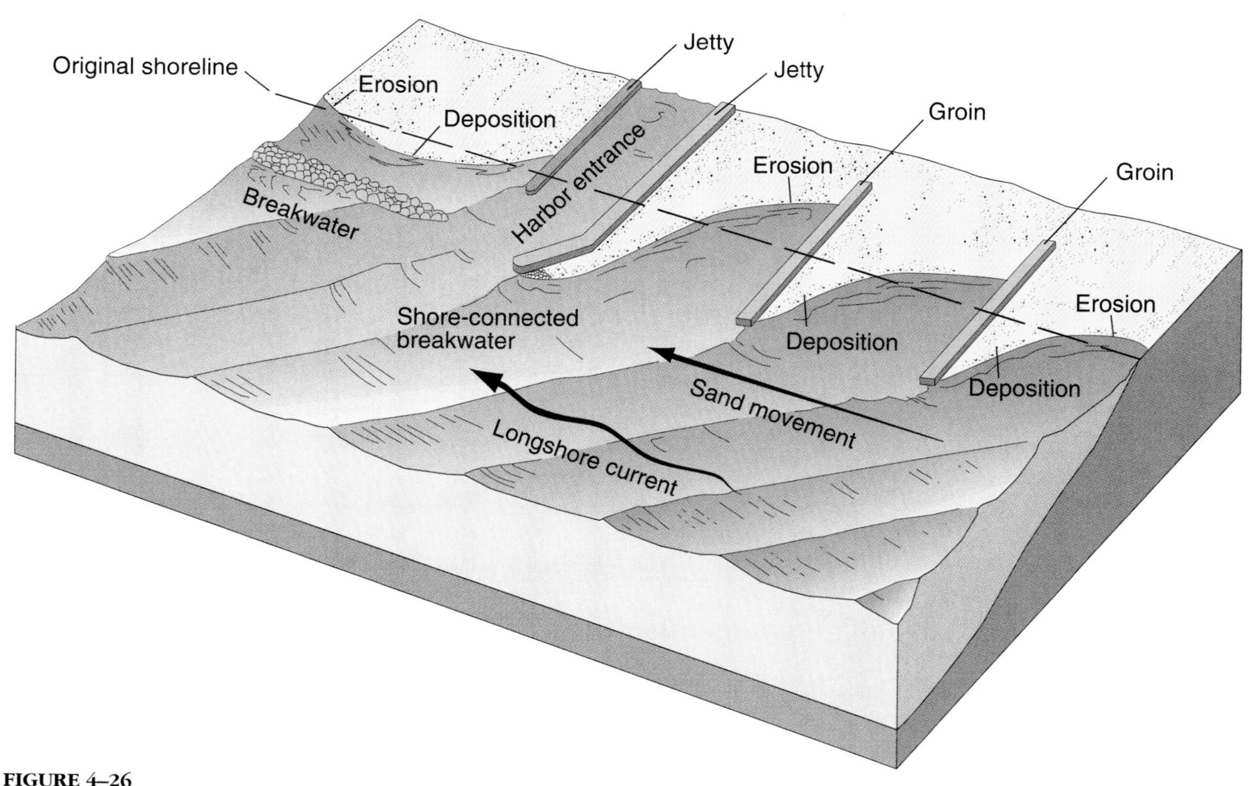

FIGURE 4–26
People's efforts to modify a shoreline only change the pattern and cause problems elsewhere. (From R. W. Christopherson, *Geosystems,* 2d ed., Macmillan Publishing Co., 1994.)

Our Dynamic Planet

The preceding sections have shown how Earth's surface is continually being shaped by both endogenic and exogenic processes, and landforms reflect the interaction between these two processes. Tectonic uplift and lowering by weathering and erosion occur simultaneously, so to fully understand how they interact we must consider the rates at which these changes occur.

Rates of Landform Change

Most landforms change so slowly that they provide metaphors for permanence, such as *everlasting hills* and *rock of ages*. Rates of horizontal movement of continents relative to one another are typically millimeters per year to centimeters per year. Vertical movements of the land are somewhat slower: generally less than a few millimeters per year. Rates of landscape lowering by weathering and erosion are even slower: the surface of a mountain may erode at 5 to 50 millimeters per century, and a gentle slope at only 0.1 to 1 millimeter per century.

Change may be somewhat faster where processes are strong and materials are weak. For example, massive granite usually erodes very slowly, but a steep slope of fractured shale may erode rapidly. Similarly, waves erode a shoreline of glacial sand and gravel much faster than one of solid rock. Much of the Massachusetts shoreline, which consists of glacial deposits, is eroding horizontally at a rate of 0.5 to 2 meters per year (2 to 7 feet). But in Maine, resistant rocky coasts are eroding at less than 10 millimeters a year (0.4 inch).

One way to assess the significance of a rate of change is to determine how long it takes to create a landform that is in balance with its environment. For example, how long does a river need to create a channel of suitable size to carry its runoff? How long do waves need to shape a coastline into a smooth beach? How long does volcanic rock need to develop into mature soil? The time required to shape the land varies greatly, depending on the processes and the materials involved (Figure 4–27).

In virtually every landscape, features that take a long time to develop can retain the fossil marks of past events, such as climate and sea-level changes tens of thousands of years ago. These same landscapes also contain more dynamic features that respond to the weather or changes in land use, and so reflect the environmental conditions of the present or very recent past.

Another way to view rates of change is to see their effects on human activity in particularly sensitive areas, such as along coastlines. In the Mississippi Delta, land is sinking relative to the sea at 5 to 10 millimeters per year (a quarter to half an inch). The city of New Orleans was founded about 200 years ago, at a site close to sea level. If the land sinks at 5 millimeters per year for 200 years the total subsidence is 1 meter (3 feet), a significant amount for a low-lying city. Today parts of the city are below sea level, and pumps operate continuously to keep the streets dry.

Environmental managers must understand natural rates of landform change so they can predict and manage human impacts on the land. To maintain naturally productive soils, for example, we cannot allow soil to erode severely and then expect it to recover in a few years. On the other hand, if we pave an area and thus alter the amount of water entering a stream, we should not be surprised that the stream responds quickly with flooding and increased erosion.

Environmental Hazards

Some environmental change, such as volcanoes, earthquakes, landslides, tornadoes, and hurricanes can occur so rapidly as to cause death and destruction. From a human perspective, these changes are momentous, but from a geologic perspective they are routine. As you walk down the street you pay little attention to individual steps you take, but for an ant on the sidewalk, one of your steps might be the event of a lifetime. Abrupt environmental changes are much the same for humans.

In some areas, rapid change is frequent enough for people to recognize the threat and avoid it. But a coastal lowland may experience a severe hurricane only once in several decades, so people feel they can safely build homes there. Because many environmental hazards are once-in-a-lifetime events, people often rebuild after a disaster rather than move to a safer area. They believe that another similar event is unlikely in the near future, and the risk is too small to justify giving up a liveli-

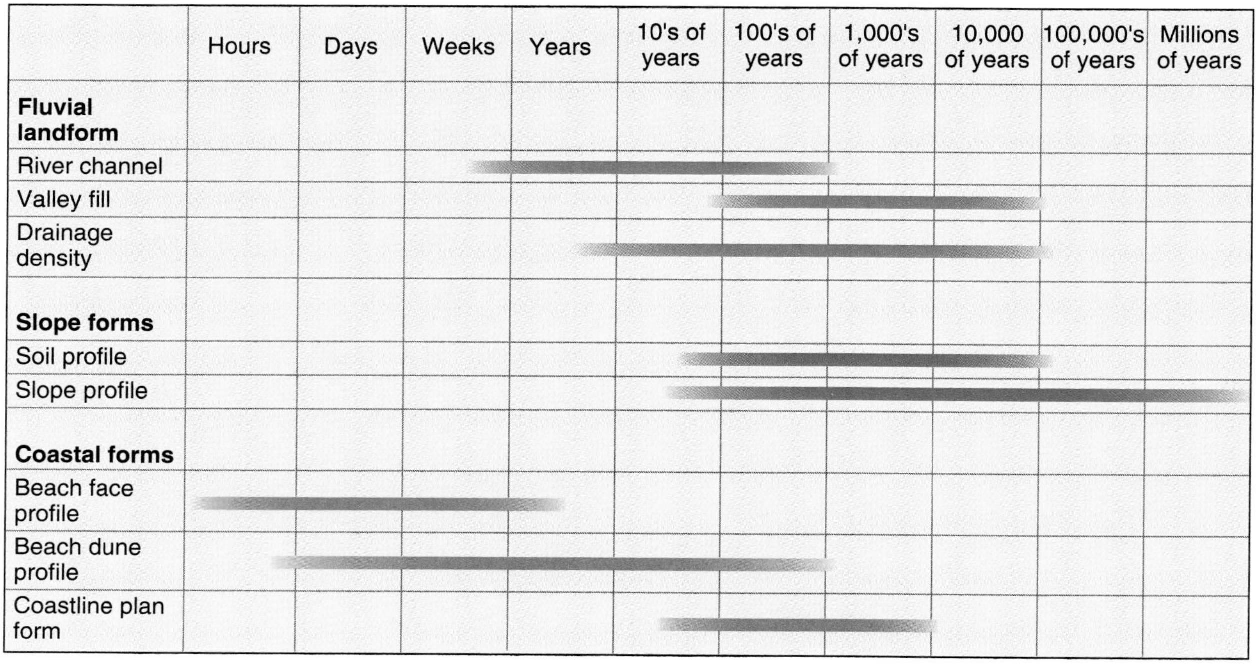

	Hours	Days	Weeks	Years	10's of years	100's of years	1,000's of years	10,000 of years	100,000's of years	Millions of years
Fluvial landform										
River channel										
Valley fill										
Drainage density										
Slope forms										
Soil profile										
Slope profile										
Coastal forms										
Beach face profile										
Beach dune profile										
Coastline plan form										

FIGURE 4–27
Length of time to form selected landscape features. The length of time needed to create a landform is short where processes are strong and materials are weak. If large amounts of material must be moved, landforms can take very long periods to be shaped by weak processes.

hood or a home. Rebuilding often is accompanied by public pressure to build new structures to protect the community, such as a levee or seawall.

Unfortunately, structures such as seawalls on the shoreline, levees along rivers, or fire-breaks in chaparral woodland cannot protect people and property against the most extreme events in a given environment. And in many cases, these structures worsen the problem. For example, levees confine floods within a narrow channel instead of letting the water spread out across a floodplain, so the water height during a flood increases. Protective structures also worsen the problem by giving people a false sense of security, leading them to make greater investments and so increase their vulnerability. When the next flood, storm, or wildfire strikes, the damage is even greater.

Geographers studying these problems recognize that natural disasters are caused by a combination of natural events and human vulnerability. In a flood, a house is destroyed for a simple reason: someone chose to build where a flood can occur. In most cases, the best approach is not to control the threat, but instead to change human behavior.

In the wake of the 1993 floods on the Missouri and Mississippi Rivers, some levees were neither repaired nor rebuilt. Instead, flood-prone land was bought or leased by the government, so that future floods can inundate this land instead of damaging valuable properties again. Allowing land to be inundated instead of keeping out floodwaters has two benefits: damage is reduced to structures in the path of the flood and downstream communities are protected.

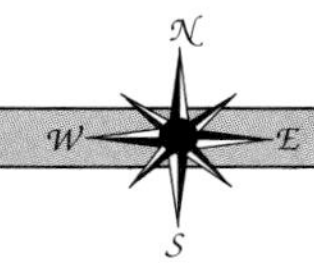

CONCLUSIONS: CRITICAL ISSUES FOR THE FUTURE

Vegetation reflects soil patterns, which in turn are controlled by erosion, deposition, and climate. A floodplain has soils that have developed from river deposits, tree species that are adapted to the moist conditions there, and a vegetation pattern that reflects the position of the channel. It is a geographer's task to understand these connections among Earth's dynamic processes and landforms, vegetation, soils, and climate.

With modern technology, humans now rival earthquakes, rivers, waves, and glaciers as significant agents in shaping Earth's landforms. By understanding the relationship between environmental processes and landforms, we gain insight into how our activities can affect the environment. By recognizing that vegetation and soils on a floodplain are linked to erosion and deposition in a river channel, we see that accelerated soil erosion upstream may alter floodplain ecosystems downstream. Because we can estimate the rates of these processes, we can predict the impact of our actions. These human impacts are greater than ever before, and we must manage them.

Human activity can destabilize landforms, meaning danger for people living in that environment. A destabilized drainage basin might now have more frequent floods that erode river banks. An unstable coastline might now experience severe erosion. A deforested hillslope now might be prone to catastrophic landsliding.

People must learn to live with instability, rather than trying to control it. When Earth is moving, either in the sudden jolts of an earthquake or through gradual erosion, human structures are likely to fail. Future urban growth and land development must be sensitive to Earth's dynamic nature, by avoiding hazardous areas. Floodplains and other low-lying areas should be left undeveloped or used for parks. Construction of housing on landslide-prone slopes should be discouraged, and soils that are easily eroded should be protected with special conservation measures.

Geographers' training and experience in both the natural and social sciences places them in a unique position to help solve problems of human occupance of hazardous environments. Geographers have been leaders in identifying natural hazards and developing strategies for coping with them. Such skills are increasingly important on our densely populated and dynamic Earth.

Chapter Summary

1. Plate tectonics

Major landforms of the world are created by a combination of endogenic and exogenic landforming processes. Endogenic mechanisms are forces which cause movement of Earth's crustal plates, raising some portions and lowering others. This motion can cause earthquakes, volcanoes, and mountains, depending on whether the boundaries between plates are convergent, divergent, or transform.

2. Slopes and streams
Exogenic processes are erosional forces that wear down Earth's crust and reshape it into new landforms. Rocks are first broken into smaller pieces through weathering. Then they are eroded by the agents of water, wind, and ice working together with gravity. Streams collect groundwater and overland flow and transport it to the sea. Streams are conveyor belts for sediment from hillsides, and they play a major role in shaping landforms.

3. Ice, wind, and waves
In many mountain and poleward environments, glaciers flow across the land, eroding rock and depositing it where higher temperatures melt the ice. Wind plays a major role in moving material in deserts, and the absence of vegetation allows even infrequent rainfall to be a major event. Along coastlines, waves caused by wind blowing across the ocean surface cause intensive erosion and rapidly change landforms.

4. The dynamic planet
Most change on Earth's surface is slow in human terms, usually taking thousands of years to significantly reshape the land. But in some areas change may be dramatic. Geographers have learned to identify areas that are subject to rapid or sudden change and to better understand how natural processes occur and how they affect human settlements. In general, we have learned that in dynamic environments, it is better to live with nature and avoid hazards than to attempt to control them.

Key Terms

Alluvial fan, 153
Alpine glacier, 149
Beach, 156
Chemical weathering, 141
Composite cone volcano, 135
Continental glacier, 149
Convergent plate boundary, 136
Desert Pavement, 153
Desertification, 155
Discharge, 144
Divergent plate boundary, 136
Drainage basin, 144
Drainage density, 144
Dune, 153
Earthquake, 132
Endogenic processes, 130
Epicenter, 134
Exogenic processes, 130
Focus (of an earthquake), 134
Geomorphology, 129
Glacial budget, 149
Glacier, 149
Grade, 145
Groundwater, 144
Igneous rocks, 137
Isostatic adjustment, 137
Lava, 134
Loess, 153
Longshore current, 157
Longshore transport, 160
Magma, 134
Mantle, 132
Mass movement, 142
Meandering, 145
Mechanical weathering, 141
Meltwater channel, 151
Metamorphic rocks, 137
Moraine, 150
Outwash plain, 150
Overland flow, 144

Questions for Study and Discussion

1. What are endogenic and exogenic processes? Make a list of mechanisms that affect the shape of Earth's surface, and classify each as endogenic or exogenic.
2. What three types of relative motion occur at plate boundaries? What are examples of boundaries where each of these three types of motion is occurring?
3. What is isostatic adjustment, and what are its causes?
4. What environments favor rapid rates of mass movements?
5. What is a drainage basin? What measurable characteristics of drainage basins are important to understanding river processes?
6. What landforms provide evidence of sea-level rise? Sea-level fall?
7. How do glacial budgets help us to understand geographic variations in glacial activity, and expansion and contraction of glaciers over time?
8. How has agriculturally accelerated soil erosion modified sediment transport, erosion, and deposition in rivers?
9. What are some examples of landforms whose shape can be understood in terms of the processes acting on them? How long does it take for these landforms to develop?

Thinking Geographically

1. What were the environmental conditions where you live (or go to college) at the time of the last glacial maximum, 20,000 years ago? Were glaciers present? If not, how was the climate different? How was the vegetation different? What is the evidence that reveals these differences?
2. In your library, obtain a copy of the Soil Survey for your area. Read the sections pertaining to the development of soils in the area. Select two different sites that exemplify different kinds of landforms or deposits, and visit these sites, comparing what you see with what is written in the soil survey.
3. Look at a topographic map of an area that includes a river. Can you identify the meanders of the channel? Measure the width of the river. Now measure a distance along the channel that is long enough to include several bends of the river. Count the number of right-hand bends in the measured portion of the river, and divide into the distance. The result is the meander wavelength, normally 10 to 15 times the river width. Try these measurements on other maps to see how consistent the relation is.

4. Mount Saint Helens in Washington State erupted violently in 1980. The Loma Prieta earthquake in California occurred in 1989. How far apart are these two points on a map? Was there a possible connection between these two events? List reasons why there might be a connection, and why there might not.
5. On Figure 4–5 (earthquake risk map), mark dots locating your school, home, and locations where you have close friends and relatives. What level of earthquake risk does each live under? Pick any of these locations and research the geology of that location to discover why it has the risk level shown on the map.

Suggestions for Further Reading

Abrahams, Athol D., and R. A. Marston. "Drainage Basin Sediment Budgets: An Introduction." *Physical Geography* 14 (May–June, 1993): 221–24.

Alexander, David. "Applied Geomorphology and the Impact of Natural Hazards on the Built Environment." *Natural Hazards* 4 (1991): 57–80.

Balling, Robert C., and S. G. Wells. "Historical Rainfall Patterns and Arroyo Activity Within the Zuni River Drainage Basin, New Mexico." *Annals of the Association of American Geographers* 80 (December 1990): 603–17.

Beach, Timothy. "Estimating Soil Loss from Medium-Size Drainage Basins." *Physical Geography* 13 (July–September, 1992): 206–24.

Birkeland, Peter W. *Soils and Geomorphology*. New York: Oxford, 1984.

Boardman, John. "Periglacial Geomorphology." *Progress in Physical Geography* 15 (1991, no. 1): 77–82.

Bull, William B. *Geomorphic Responses to Climatic Change*. New York: Oxford, 1990.

Cooke, R. U., and J. C. Doornkamp. *Geomorphology in Environmental Management: A New Introduction*. Oxford: Clarendon, 1990.

Cooke, R. U., A. Warren, and A. Goudie. *Desert Geomorphology*. London: UCL Press, 1993.

Costa, John R. and V. Baker. *Surficial Geology*. New York: John Wiley & Sons, 1981.

Dawson, A. G. *Ice Age Earth: Late Quaternary Geology and Climate*. London: Routledge, 1992.

Dobson, Jerome E. "Spatial Logic in Paleogeography and the Explanation of Continental Drift." *Annals of the Association of American Geographers* 82 (June 1992): 187–206.

Dolan, Robert, M. Fenster, and S. Holme. "Erosion of U.S. Shorelines." *Geotimes* 35 (1990): 22–24.

Douglas, Ian. *The Urban Environment*. London: Edward Arnold, 1983.

Dregne, H. E. "Erosion and Soil Productivity in Africa." *Journal of Soil and Water Conservation* 45 (July–August 1990): 431–36.

Dunne, Thomas, and L. B. Leopold. *Water in Environmental Planning*. San Francisco: Freeman, 1978.

FitzGerald, Duncan M., and P. S. Rosen. *Glaciated Coasts*. San Diego: Academic Press, 1987.

Francek, Mark A. "A Spatial Perspective on the New York Drumlin Field." *Physical Geography* 12 (January–March, 1991): 1–18.

Gerrard, John. *Soil Geomorphology*. New York: Wiley, 1992.

Goudie, Andrew. "Human Influence in Geomorphology." *Geomorphology* 7 (July 1993): 37–60.

Graf, William L. "Mercury Transport in Stream Sediments of the Colorado Plateau." *Annals of the Association of American Geographers* 75 (December 1985): 552–65.

Harbor, J. M. "Glacial Geomorphology: Modeling Processes and Landforms." *Geomorphology* 7 (July 1993): 129–40.

Higgitt, D. L. "Soil Erosion and Soil Problems." *Progress in Physical Geography* 17 (1993, no. 4): 461–72.

James, L. Allan. "Sustained Storage and Transport of Hydraulic Gold Mining Sediment in the Bear River, California." *Annals of the Association of American Geographers* 79 (December, 1989): 570–92.

Knox, James C. "Historical Valley Floor Sedimentation in the Upper Mississippi Valley." *Annals of the Association of American Geographers* 77 (June 1987): 224–44.

Lee, Jeffrey A., K. A. Wigner, and J. M. Gregory. "Drought, Wind and Blowing Dust on the Southern High Plains of the United States." *Physical Geography* 14 (January–February, 1993): 56–67.

Leopold, Luna B. *A View of the River*. Cambridge: Harvard, 1994.

Martin, Charles W. "The Response of Fluvial Systems to Climate Change: An Example from the Central Great Plains." *Physical Geography* 13 (April–June, 1992): 101–14.

McIntyre, S. W. "Reservoir Sedimentation Rates Linked to Long-Term Changes in Agricultural Land Use." *Water Resources Bulletin* 29 (June, 1993): 487–95.

McManus, John, and R. W. Duck. *Geomorphology and Sedimentology of Lakes and Reservoirs*. New York: Wiley, 1993.

Meyer, Grant A., S. G. Wells, R. G. Balling, and A. J. Timothy. "Response of Alluvial Systems to Fire and Climate Change in Yellowstone National Park." *Nature* 357 (May 14, 1992): 147–50.

Millman, John D., and R. H. Meade. "World-Wide Delivery of River Sediment to the Oceans." *Journal of Geology* 91 (January 1983), 1–21.

Nuhfer, Edward B. *The Citizen's Guide to Geologic Hazards*. Arvada, CO: American Institute of Professional Geologists, 1993.

O'Connor, Jim E. *Hydrology, Hydraulics, and Sedimentology of the Bonneville Flood*. Boulder: Geological Society of America, 1993.

Phillips, Jonathan D. "Fluvial Sediment Budgets in the North Carolina Piedmont." *Geomorphology* 4 (1991): 231–41.

Rhoads, Bruce L. "Mutual Adjustment Between Process and Form in a Desert Mountain Fluvial System." *Annals of the Association of American Geographers* 78 (June 1988):271–78.

Rhoads, Bruce L. "The Impact of Stream Channelization on the Geomorphic Stability of an Arid-Region River." *National Geographic Research* 6 (Spring 1990): 157–77.

Ritter, Dale F. *Process Geomorphology*. 2nd ed. Dubuque, IA: WCBrown, 1986.

Thornes, John B. *Vegetation and Erosion: Processes and Environments*. New York: John Wiley & Sons.

Trenhaile, Alan S. *The Geomorphology of Canada: An Introduction*. Toronto: Oxford, 1990.

Trimble, Stanley W. "The Distributed Sediment Budget Model and Watershed Management in the Paleozoic Plateau of the Upper Midwestern United States." *Physical Geography* 14 (May–June, 1993): 285–303.

_____ and S. W. Lund. *Soil Conservation and the Reduction of Erosion and Sedimentation in the Coon Creek Basin, Wisconsin*. U.S. Geological Survey Professional Paper 1234. Washington: U.S. Government Printing Office, 1982.

Wells, N. A., and B. Andriamihaja. "The Initiation and Growth of Gullies in Madagascar: Are Humans to Blame?" *Geomorphology* 8 (September 1983): 1–46.

Woo, Ming-ko, A. G. Lewkowick, and W. R. Rouse. "Response of the Canadian Permafrost Environment to Climatic Change." *Physical Geography* 13 (October–December, 1992): 287–317.

We also recommend these journals: *American Journal of Science; Arctic & Alpine Research; Catena; Earth Surface Processes & Landforms; Environmental Geology & Water Sciences; Geografiska Annaler Series B; Geology; Geomorphology; Journal of Soil & Water Conservation; Physical Geography; Progress in Physical Geography; Soil Science; Soil Science Society of America Journal; Water Resources Research*.

5
The Biosphere

- The biosphere
- The hydrosphere
- Carbon and oxygen flows in the biosphere
- Soil
- Biomes: Global patterns in the biosphere

Ripening wheat in Washington

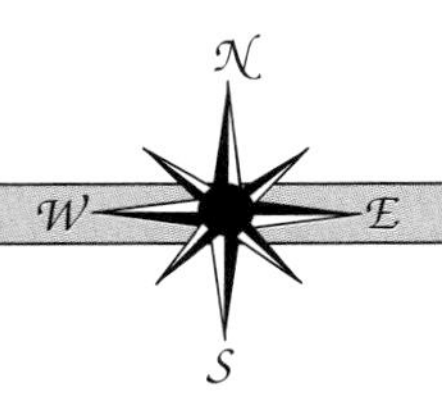

1. The biosphere
Earth's surface is divided into several biomes, which are major vegetation regions, where plants, animals, and their physical environments interact. Primary production is the basis of food chains and supports animal life. Vegetation varies in relation to climate and may change over time through succession, especially following disturbance.

2. The hydrosphere
The hydrologic cycle is a continual worldwide flow of water among the four spheres. Water budgets are useful in analyzing water movements in the environment and evaluating water resources for human use.

3. Carbon and oxygen flows in the biosphere
The carbon cycle is a fundamental link among the atmosphere, oceans, biosphere, and soil. Atmospheric carbon is absorbed by the biosphere during photosynthesis, stored in living organisms and in the soil, and released through respiration. Fossil fuel combustion releases bountiful carbon. A net flow of carbon to the atmosphere may be causing global warming.

4. Soil
Soil is a mixture of mineral particles, organic matter, water, and air. Soil is formed through a combination of physical, chemical, and biological processes. Soil characteristics vary in relation to climate, and the world map of soils has patterns similar to the world climate map. In many areas, people have reduced the ability of soil to support plant life by removing nutrients and exposing soil to erosion.

5. Biomes: Global patterns in the biosphere
World maps of soils, vegetation, and climate show strong similarities because climate is a powerful control of soil formation and plant growth. Throughout the populated world, vegetation and soils show extensive human impact. The biosphere is a system in which humans play a crucial role.

INTRODUCTION

Traditionally, studies of diverse topics like vegetation, weather, human population, and industrial activity have been carried out in isolation from each other. But in recent decades, scientists have recognized the *intimate interaction among all Earth's systems,* including natural and human-dominated phenomena. Although studies of these connections are conducted in many disciplines, geographers are in a unique position to understand them spatially.

In this chapter, we examine the biosphere and the hydrosphere. The biosphere is the thin layer of living things, of which we are an inseparable part. The biosphere is intimately connected to the solid Earth, ocean, and air: plants depend on solar energy for photosynthesis, and on atmospheric circulation for temperature regulation, carbon supply, and water supply that make growth possible. At the same time, they exchange oxygen and carbon dioxide with the atmosphere, and in so doing powerfully influence atmospheric composition and climate. Plants and animals help in the slow conversion of rock into soil, the dynamic medium that stores water and nutrients and supports life. Because of these interactions among climate, soil, living things, and rocks, patterns of plant and animal life correspond closely with patterns of climate, topography, and geology.

The hydrosphere consists of the water on Earth's surface and in its atmosphere. Water constantly circulates among the lithosphere, biosphere, and atmosphere. In each of these subsystems, water plays a critical role in regulating environmental processes and in determining spatial patterns. The hydrosphere is the most important link connecting all of Earth's subsystems.

The flow of water among the lithosphere, biosphere, and atmosphere is an example of a **biogeochemical cycle.** Biogeochemical cycles are recycling processes that supply essential substances such as carbon, nitrogen, and other nutrients to the biosphere. Biogeochemical cycles interconnect Earth's subsystems.

In Chapter 4, we described environmental changes that occurred over the past few million years. These changes have been driven mostly by natural climatic variability. But during the last 10,000 years, and especially the last 1,000 years, our population has expanded so rapidly that we now profoundly influence global environmental patterns as much as the most dramatic climatic variations (glaciations) in the Pleistocene Epoch.

Just a few thousand years ago, the human species had only a minor environmental impact. But our large numbers and our enormous consumption of natural resources and discharge of waste give us a significant role in global biospheric processes like energy exchange, food webs, and movements of materials on the surface of Earth. As a consequence, we can understand Earth's environments today only by recognizing how humans modify and regulate them.

The Biosphere

Some areas have trees, while others are dominated by grasses or isolated shrubs. Some trees are conifers and stay green all the year, while others are deciduous and lose their leaves every autumn. We take for granted this variety of vegetation and its distribution, but it reflects the enormous complexity of the biosphere. Two essential concepts help us understand the biosphere: the biome and the ecosystem.

A **biome** is a large region of Earth's surface characterized by particular plant and animal types. Biomes may be marine or terrestrial. A *marine biome* is a large area of broadly similar plants and animals found in the sea. A *terrestrial biome* is a similar area on land. This chapter will focus on terrestrial biomes.

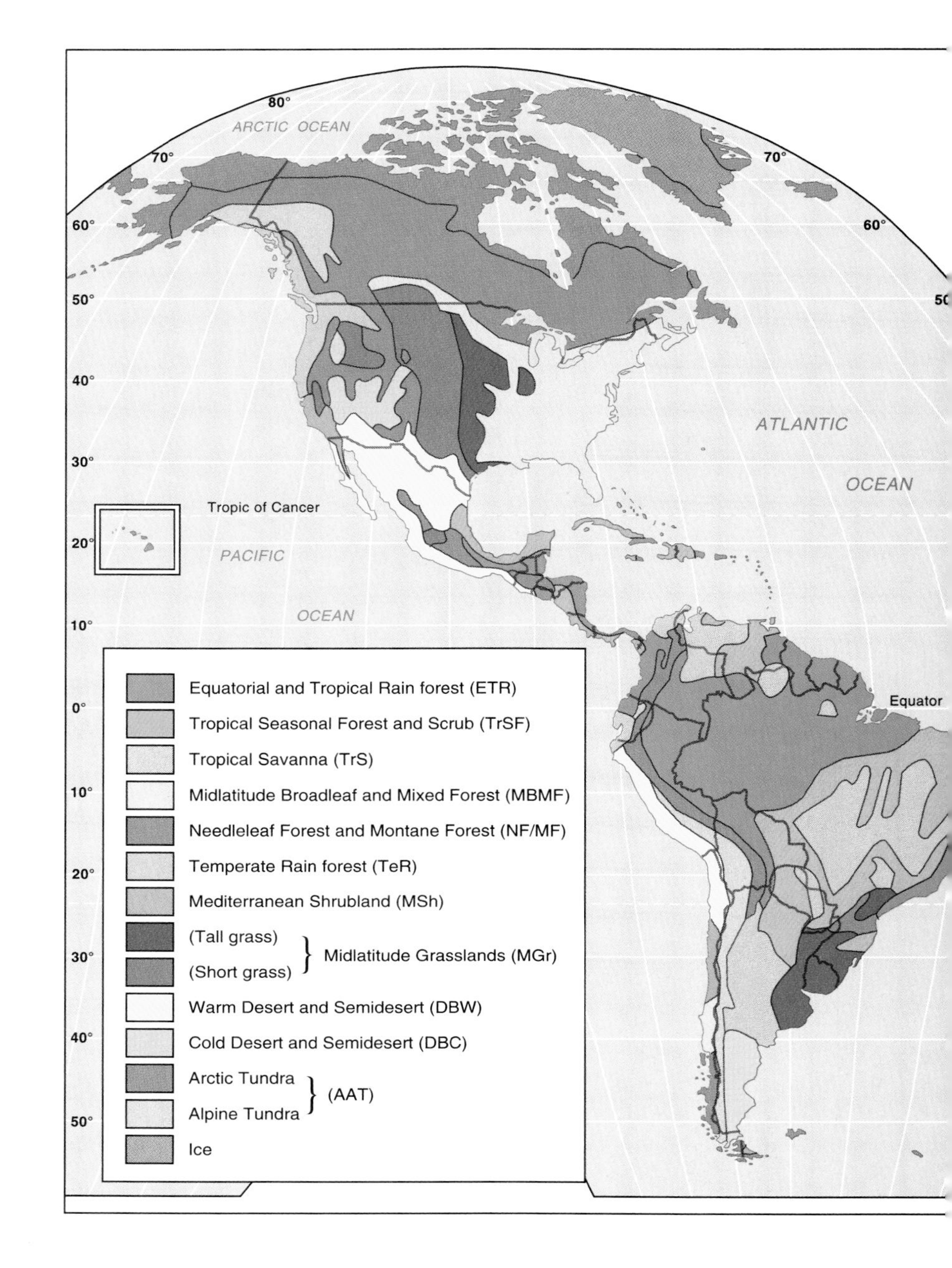

FIGURE 5–1
The thirteen major terrestrial biomes. The map also shows ice-covered regions that are not part of any biome. (From R. W. Christopherson, *Geosystems,* 2d ed., Macmillan Publishing Co., 1994.)

Terrestrial biomes reflect two especially visible features: climate and vegetation type. These generally give a biome its name, such as "tropical rainforest" or "artic tundra." Underlying each biome's simple name, however, is a great diversity of plants and animals.

Note that the global distribution of biomes (Figure 5–1) imitates very closely the distribution of climate regions described in Chapter 3 (Figure 3–1). To grow, plants depend on proper levels of the two most important elements of a climate: temperature and water availability.

A biome may contain many ecosystems. An ecosystem is an interrelated collection of plants and animals and the physical environment with which they interact (Figure 5–2). An ecosystems can cover a much smaller area than a biome, such as a field or a pond. Regardless of whether the

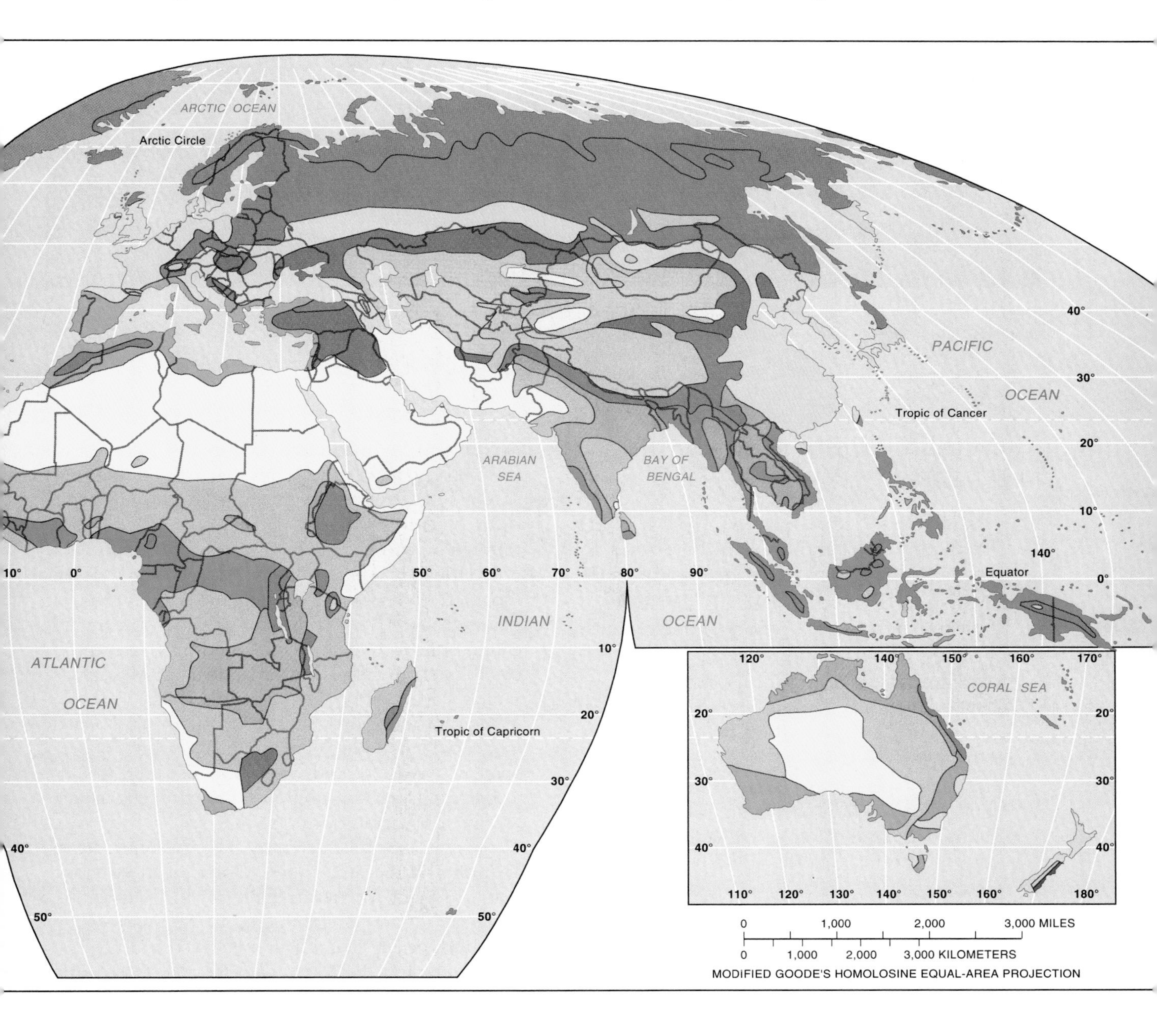

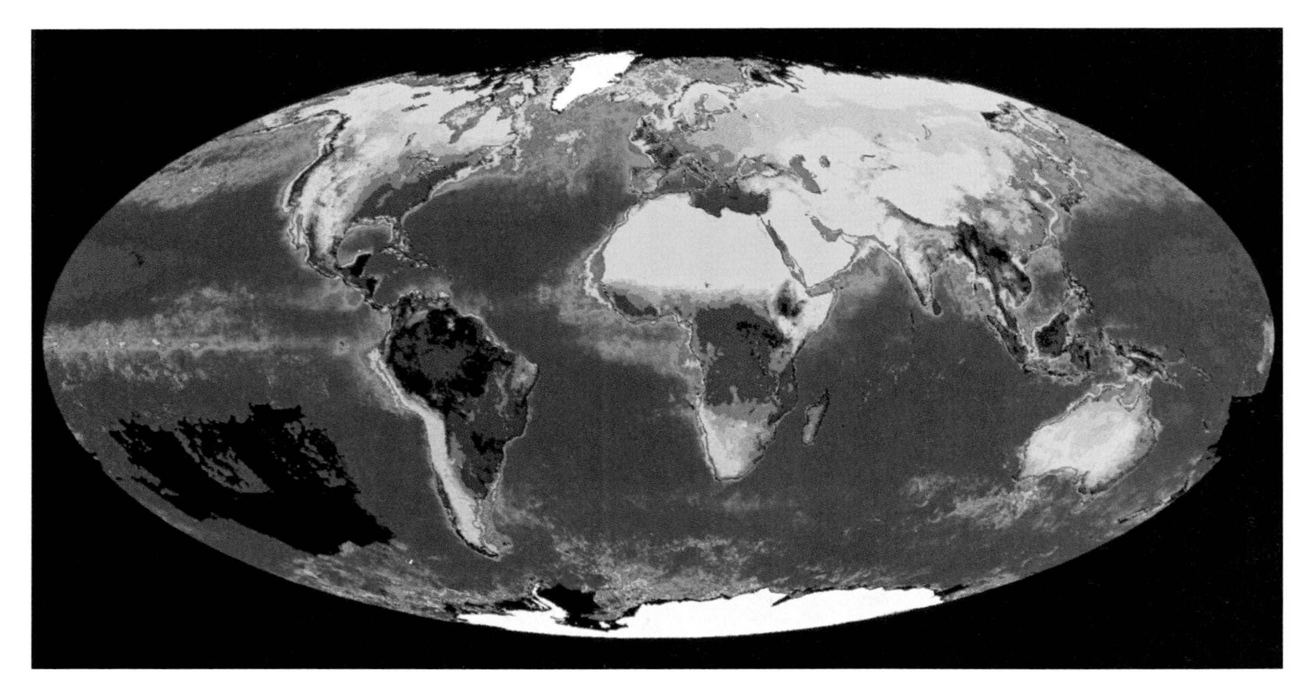

A global map of plant growth, produced from satellite data. On land areas, dense vegetation is shown in dark green with sparse vegetation and desert shown in pale green and tan. In oceans, areas of high plant growth are shown in red, yellow, and pale green, while less productive areas are shown in blue and magenta. Data are lacking in black areas. (NASA/Science/Photo Researchers, Inc.)

ecosystem is small or large, the four same elements and processes drive its operation:

1. **Producers** of food: green plants, which through photosynthesis produce food for themselves and for consumers that eat them.
2. **Consumers** of food: animals that eat producers (herbivores), other animals (carnivores), or both (omnivores).
3. **Decomposers** of producers and consumers: small organisms that digest and recycle dead plants and animals; they include bacteria, fungi, insects, and worms.
4. Materials and energy essential for production and consumption to occur: water, mineral nutrients, gases such as oxygen and carbon dioxide, and energy (light and heat).

Ecosystem Function

Energy from the sun is the starting point for understanding the operation of an ecosystem. Sunlight makes possible **photosynthesis,** which is the formation of food in green plants as a result of exposure to light. Green plants produce food in the form of carbohydrates. This food is distributed through an ecosystem by way of a **food chain.**

Plant-eating animals, known as *herbivores,* begin the food chain by consuming the food stored in plants (Figure 5–3). *Carnivores,* which are meat-eating animals, eat herbivores and may in turn be eaten by other carnivores or by *omnivores,* such as humans, who eat both plants and animals.

Most of the food that animals consume is used to keep their bodies functioning. Animals excrete some of the food, and when they die their bodies are rich in stored-up nutrients. Decomposers attack animal excretions and bodies when they die. These small organisms return chemical nutrients to the soil and the atmosphere. Some of these nutrients are used for new plant growth, completing the cycle. In any environment, the dry mass of living and formerly living matter is called **biomass.**

Each step in the food chain is called a **trophic level.** At each trophic level, food is passed from one

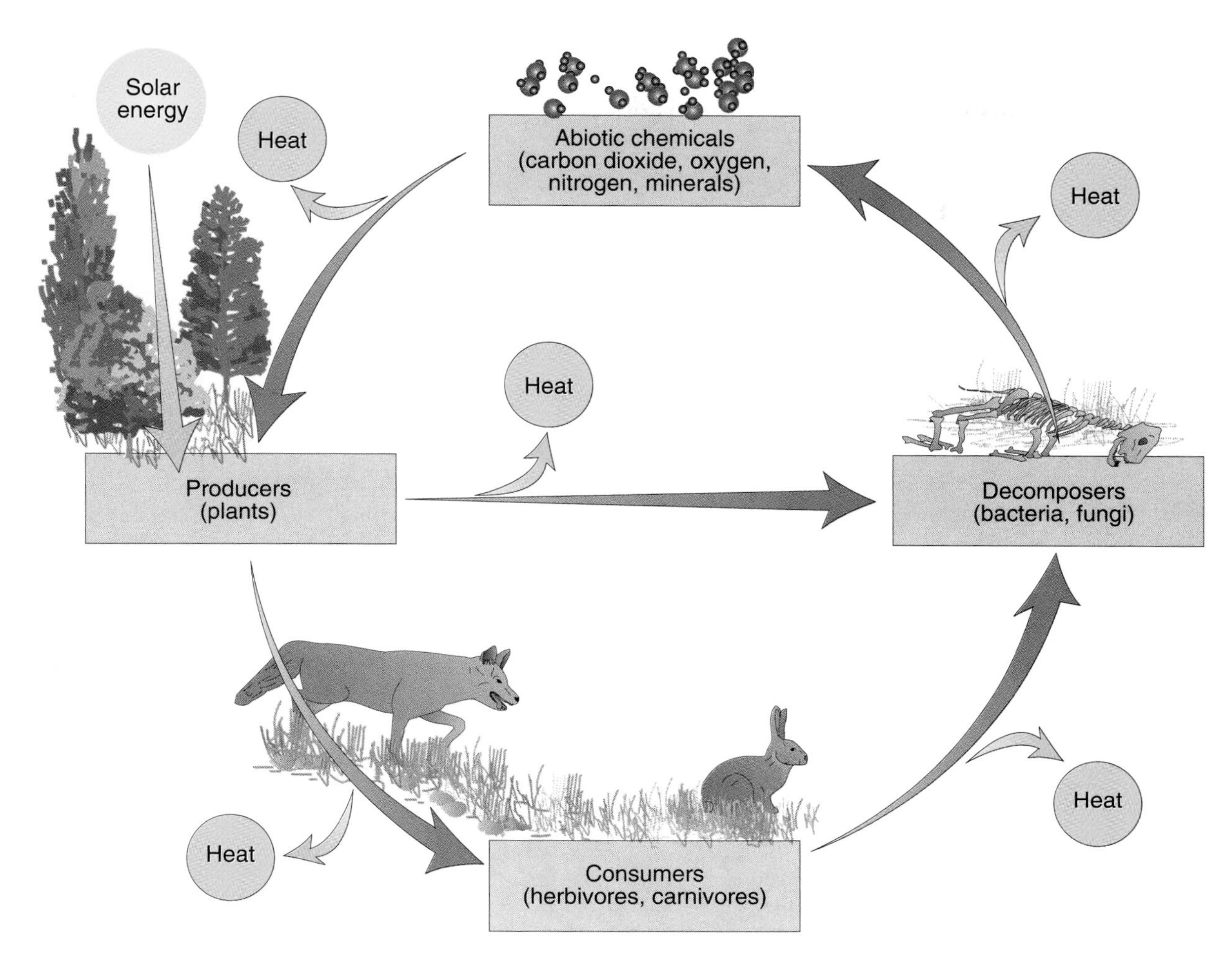

FIGURE 5–2
Ecosystems consist of plants, animals, decomposers, and the physical environment with which they interact. Energy is supplied to an ecosystem from the sun, passed through it via a food chain, and dissipated as heat.

level to the next, but most of the energy is lost. As a rule of thumb, only 10 percent of the energy consumed as food at a given trophic level is converted to new biomass, and the remaining 90 percent is dissipated as heat. Because of this, the volume of biomass decreases as we go from the first trophic level, the green plants, to higher levels. This is why a natural system has vast numbers of plants, numerous herbivores such as mice and rabbits, but very few large carnivores like wolves and lions.

Passing food energy through a food chain also passes along any poisons in the plants, especially long-lived pesticides like DDT. Such chemicals, which break down very slowly in the environment, tend to accumulate in animal tissues. If a pesticide accumulates in an animal rather than being excreted, then at each step of the food chain the concentration of that pesticide increases, a process called **biomagnification.**

Consider this example: If a falcon weighing 1 kilogram consumes in its lifetime 50 smaller birds, and if each of these smaller birds consumes in its lifetime 2 kilograms of plant-eating insects, then the falcon indirectly consumes 100 kilograms of insects. If each of these insects consumed, during its lifetime, ten times its weight in plants, then the falcon has indirectly consumed 1,000 kilograms of plants!

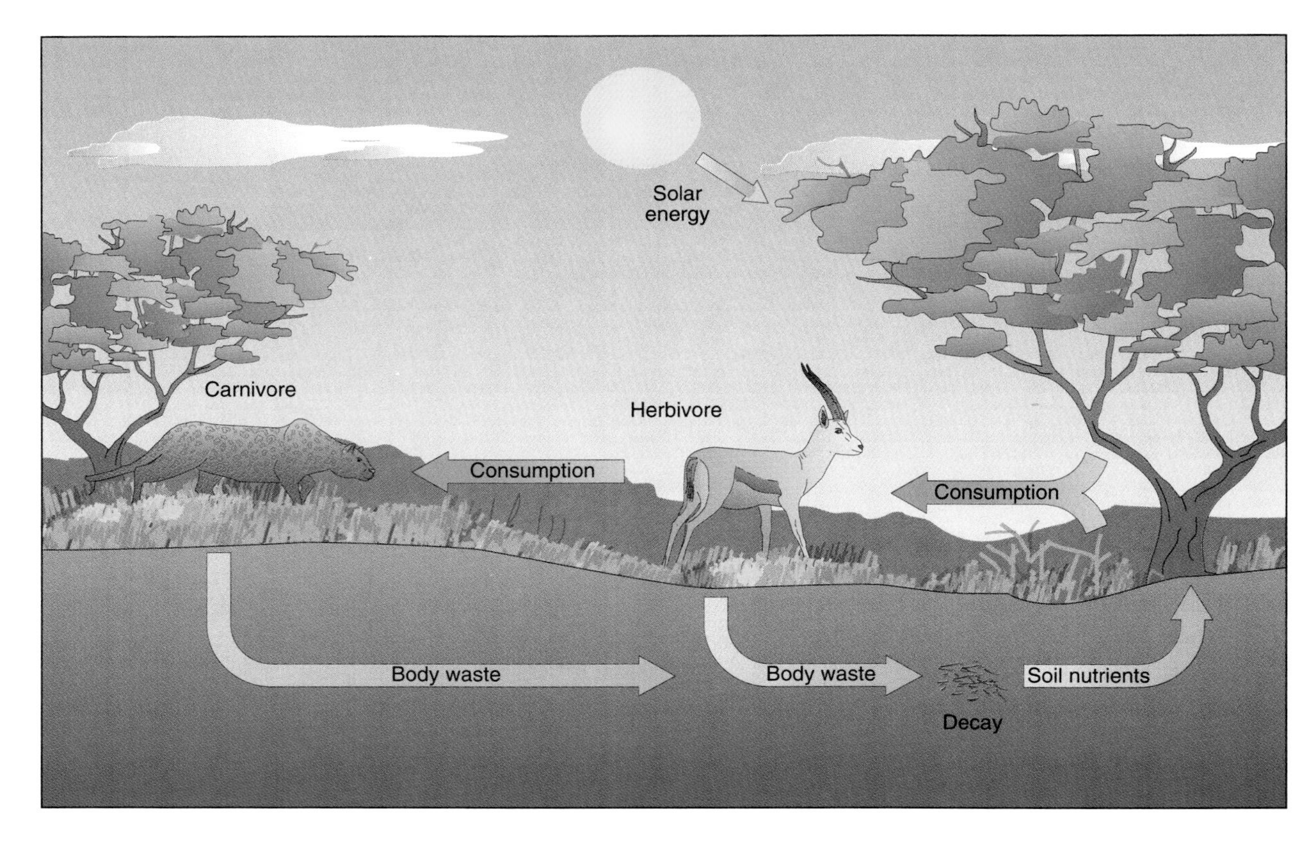

FIGURE 5–3
A food chain. Green plants are the primary producers of food, which is eaten by the primary consumers: herbivores and omnivores such as humans. Carnivores, or secondary consumers, derive their energy from other animals rather than directly from plants. At each step of the food chain energy is lost as heat, and waste materials are broken down by decomposers. (From T. L. McKnight, *Physical Geography: A Landscape Appreciation,* 4th ed., Macmillan Publishing Co., 1993.)

Even a very small concentration of pesticides at a low trophic level, such as an insect can become a high concentration in large carnivores. Because of these problems, DDT and similar pesticides have been banned or their use severely restricted, and pesticides used today are designed to decay relatively rapidly to reduce this problem.

Plant and Animal Success in Ecosystems

Within any particular ecosystem, living things compete for resources such as food, water, and space. The more successful plants and animals in this competition will dominate their environment. *Success* of a species means that it survives, thrives, or even comes to dominate an ecosystem for a long period. Humans are an example: we survived for the first 2 or 3 million years of our existence as a species, then thrived over the last few thousand years, and today dominate Earth, for better or worse.

The success of one species over another results from competition. In the case of plants, this competition is for light, water, nutrients, and space. Plants require all of these factors to grow, but in any specific ecosystem, one factor usually is restricted, and this forces competition and adaptation. For example:

- In an arid environment, plants compete for scant water but do not need to compete for the abundant sunlight.
- In a humid environment, water is abundant but the great number of plants must compete for sunlight.

- An area with adequate water and light may have poor soils, so plants must compete for nutrients.

Through evolution, plants have survived by adapting life-forms, physiological characteristics, and reproductive mechanisms to achieve success in particular environments. The plants that best compete in an environment dominate the vegetation there.

Plants display many strategies for success. In a humid area, for example, the strategy of height is effective: the tallest plants not only are fully exposed to solar energy, but also cast shadows on their competition, slowing the growth of lower species. Short growing seasons, drought, and fire favor organisms that have adapted means to survive or even benefit from the stress. For example, some species have adapted the ability to take advantage of occasional fires. After a fire has felled other vegetation and heated the soil, long-dormant seeds respond to the flash heating and germinate quickly. They occupy the burned site, completing their life cycle and spreading new seed before the other vegetation recovers from the fire. Once the other vegetation regains dominance, the fire-resistent seeds lie dormant, awaiting the next fire opportunity.

Community Succession

Plant success is the basis of **community succession,** in which plants in a community are supplanted by more dominant, longer-term species. Plants are not as passive as they seem. As they grow and multiply, plants alter their environments, allowing other plants to compete. A common example of community succession occurs when agricultural land is abandoned in eastern North America (Figure 5–4).

- In the first couple of growing seasons, the fields sprout fast-growing herbaceous weeds and other pioneer species. These species tolerate bright sunshine and spread seeds widely.
- Over the next few growing seasons, the weeds and pioneer species are replaced by slower-growing perennial shrubs and trees, such as pines, which prefer bright sunshine and gradually shade out the smaller plants beneath them.
- Hardwoods that tolerate shade then grow beneath the pines, eventually overtaking them in height. As the pines die away in the competition, a **climax** forest of shade-tolerant species comes to dominate the canopy. They are replaced by their own seedlings as they die, thus leading to a stable mix of vegetation.

The length of time necessary for succession to reach a climax varies, depending on the growth rates and lifespans of plants, rates of seed introduction and establishment, and whether soil development proceeds along with plant growth. If the seeds of the climax species are present at the time succession begins and growth rates are fast, a well-developed forest may be established in a few decades. The illustration shows a more common situation, where forest climax is achieved in 150–200 years. But if succession also involves soil development, through weathering of rock and accumulation of organic matter, it may take thousands of years for a site to be converted from bare rock to climax forest.

In environments subject to frequent disturbance, such as fire, windstorm, disease outbreaks, floods, and volcanic eruptions, the time to establish a climax community can be long. Under these conditions, vegetation may continually change, always tending toward a climax forest, but frequently interrupted.

Earth's recent history has been one of such frequent, dramatic change. Between 10,000 and 20,000 years ago, major climatic changes covered and uncovered vast areas with ice or water, and even where these changes did not occur there were large shifts in temperature, water availability, or both. In the last few hundred years, agriculture has dramatically disturbed many places that were untouched for millennia. These environmental changes have profoundly influenced vegetation distribution in many areas, favoring plants that are adapted to dynamic conditions.

Biodiversity

Biodiversity is the diversity of species present in any environment. Biodiversity is important, for it brings multiple food options to living things, improving each one's odds for survival. Thus, biodiversity can improve the stability of an entire community.

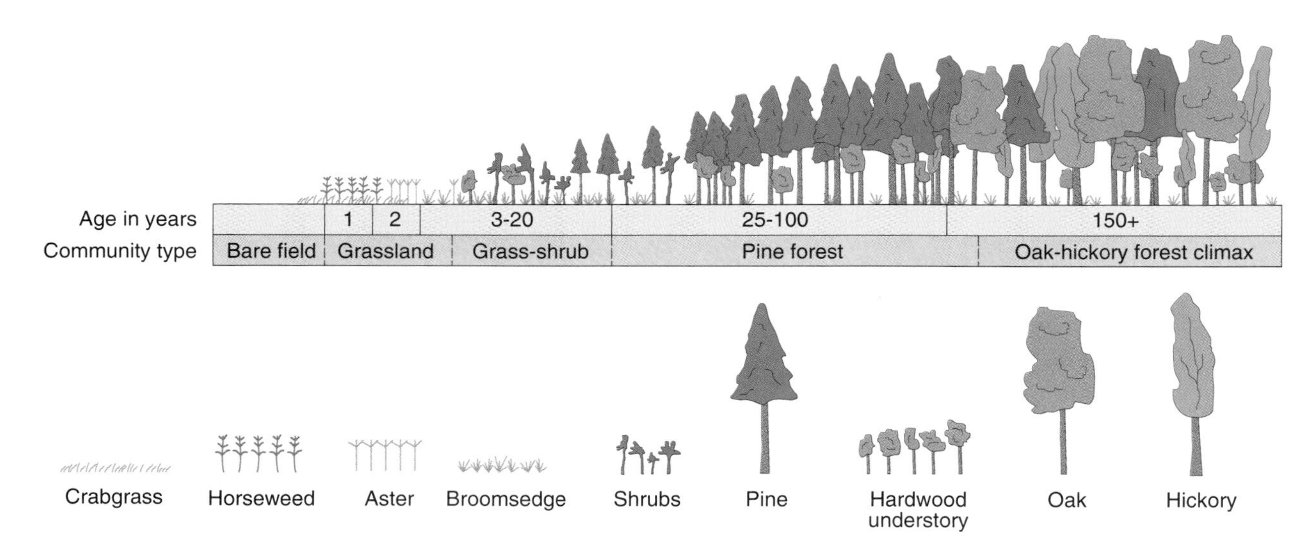

FIGURE 5–4
Clearing a field sets the stage for the succession sequence shown here, from left to right. This example is typical of the southeastern United States. (From R. W. Christopherson, *Geosystems,* 2d ed., Macmillan Publishing Co., 1994.) After Figure 9–4 from *Fundamentals of Ecology* by Eugene P. Odum, © 1971 by Saunders College Publishing, a division of Holt, Rinehart and Winston, Inc., adapted by permission of the publisher.)

Earth's biodiversity is vast, encompassing an estimated 10 million plant and animal species. Each of these has specific habitat needs. For plants, critical needs include water, light, and nutrient availability, the absence of soil disturbance, effective seed dispersal, and good germination conditions. For animals, the most critical requirement is food supply, meaning a sufficient quantity of specific plants or animals to feed upon. In a world of rapidly changing environments, the critical needs of species increasingly are not met.

Habitat loss is so widespread that a significant portion of the world's species may be threatened with extinction. The largest single cause of extinctions is land use change, resulting from increasing human settlement worldwide. About 36 percent of Earth's land area (excluding Antarctica) now is in cropland or permanent pasture. Between 1978 and 1988, the land area in such agricultural uses increased about 2.2 percent, whereas the world's forest area decreased 1.8 percent. Hunting and pollution also contribute to extinctions.

In Europe and North America, radical changes in the landscape have taken place over centuries of human habitation. While a few extinctions were noted, little attention was paid to the broader, unknown impacts of that landscape change. Rapid change is now occurring in the Amazon Basin at the same time that scientists are beginning to explore tropical environments. Biologists working in the Amazon discover new species virtually every time they look, raising concern that deforestation is causing extinctions of species even before we discover them.

Biosphere Reserves. To reduce the impact of land-use change on species diversity, protected areas are being established by national governments and international agencies such as the United Nations Biosphere Reserve Program. In South America, for example, about 5.7 percent of the entire continent (490,000 square kilometers or 190,000 square miles) now is nationally protected, including 24 Biosphere Reserves (some overlap nationally protected areas) that cover about 119,000 square kilometers (46,000 square miles). Worldwide, approximately 4.8 percent of all land area and 1.5 percent of water area are nationally protected, and 283 Biosphere Reserves cover about 0.8 percent of Earth's surface. However, the degree of protection (restrictions on land use and disturbance) varies considerably.

The size of individual protected areas is important. A single animal may range over only a few square kilometers, but that small area cannot

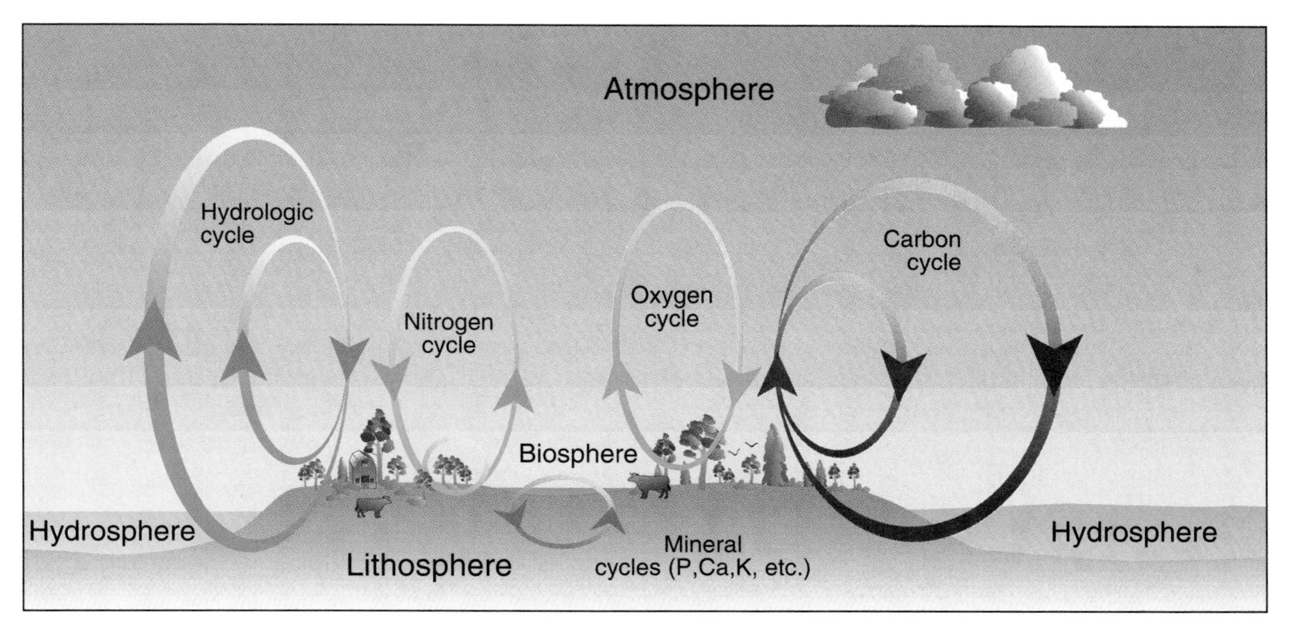

FIGURE 5–5
Biogeochemical cycles transfer matter between the atmosphere, biosphere, hydrosphere, and lithosphere. The cycles represented here are shown greatly simplified.

ensure species survival. A reserve must be large enough to sustain enough individuals to allow genetic diversity in the breeding population. Deciding how large a protected area should be is difficult, and becomes part of the political tug-of-war among conflicting land users.

Biodiversity is a barometer of the consequences of human modification of the environment. A commitment to biodiversity means a commitment to a broad range of environmental policies—not just preserving wilderness areas, but maintaining diversity everywhere.

Biogeochemical Cycles

The operation of an ecosystem in the biosphere—the growth and decay of plants, eating and decay of animals, and every other ecosystem process—depends on exchanges of energy and matter. Energy is continually directed to Earth from the sun. Matter, including the essential substances of water and carbon, is available in the atmosphere, hydrosphere, and lithosphere, as well as in the biosphere itself.

The law of *conservation of energy* states that energy can neither be created nor destroyed under ordinary conditions, but it may be changed from one form to another. Similarly, the law of *conservation of matter* states that matter can neither be created nor destroyed under ordinary conditions, but it may be changed from one form to another. The exception to "ordinary conditions" is nuclear reactions, in which matter is converted to energy.

Although energy and matter are neither created nor destroyed, they are constantly being cycled through pathways in Earth's systems (Figure 5–5). Energy and matter can be stored in any of Earth's spheres—the atmosphere, hydrosphere, biosphere and lithosphere—and exchanges occur between any pair of spheres. Each of the biogeochemical cycles involves storage of energy or matter, and flows or exchanges of energy and matter from one place to another.

The following sections focus on two of the most important biogeochemical cycles, the hydrologic cycle and the carbon/oxygen cycle.

The Hydrosphere

Water is central to every ecosystem in the biosphere. Some of the distinctive features of the hydrosphere that make it so important include:

- Earth has the largest supply of water known on any planet.
- Earth's moderate temperatures, compared to other planets, allow water to exist in all three states—solid, liquid, and gas (vapor)—and to change readily from one to another. This makes water incredibly mobile, for it can permeate rock as groundwater, exist in frozen reservoirs we call glaciers, flow worldwide in the oceans, and drift worldwide through the atmosphere. Water is the only common substance that exists in all three of these states at normal environmental temperatures.
- Relatively large amounts of heat energy are involved in the changes among solid, liquid, and vapor.
- Water is an excellent solvent, readily dissolving many substances, making them more mobile and available for chemical reactions with whatever the water contacts.
- Most important, all living things are primarily made of water (roughly 70 percent in the case of humans).

Geographical variation in water availability is the key to understanding environmental variations at the global scale.

Water flows from its storage places in the atmosphere, lithosphere, and hydrosphere by means of evaporation, condensation, precipitation, and runoff. Water falls from the atmosphere to the ground and ocean as **precipitation. Runoff** carries this water over and through the land to the sea. **Evaporation** converts liquid water in lakes, oceans, and streams into vapor, delivering it into the atmosphere. This flow is the **hydrologic cycle,** or water cycle (Figure 5–6).

Hydrologic Cycle and Water Budgets

Water is stored in the atmosphere, hydrosphere, and lithosphere in solid, gaseous, and liquid forms (Figure 5–7). The salty oceans are by far the largest reservoirs of water—over 97 percent of all water on Earth. Over 2 percent is stored in glacial ice. The remaining 0.6 percent is available as freshwater, mostly groundwater. About 0.2 percent of all water (one-third of all freshwater) is in rivers and lakes. Although powerful in controlling weather and climate, the atmosphere contains a mere 0.001 percent of all the world's water.

These proportions change little from year to year for Earth as a whole. But over hundreds or thousands of years, climate substantially alters the size of glaciers, and this changes the volume of water in the oceans. As explained in Chapter 4, a much greater amount of water was stored in glacial ice about 20,000 years ago, making sea level substantially lower than today.

Estimates of the global average rates of evaporation, condensation, precipitation, and runoff are shown in the global **water budget** (Figure 5–6). Each process varies geographically with the amount of water available. Excess water evaporated into the atmosphere over the oceans is carried by wind over land areas, where it condenses into clouds and falls as precipitation. About two-thirds of the water that falls on land areas evaporates there, and the remaining one-third drains into rivers, which return the excess to the sea as runoff.

The global water budget shown in Figure 5–6 does not indicate the variability of water from place to place and season to season. Evaporation from the sea depends on the insolation available to heat water. Precipitation and evaporation occur in both land and ocean areas, but not in equal amounts; over the oceans, more evaporation than precipitation occurs.

Evapotranspiration. When we focus on local water budgets, we recognize the essential role of plants in evaporating water from the soil. In well-vegetated areas, much of the conversion of water from liquid to vapor takes place in the leaves of plants. This water is replaced by water drawn from the soil through plant roots. This plant mechanism is called **transpiration,** and when combined with evaporation we call the process **evapotranspiration (ET).**

ET occurs when enough water is available for plants to transpire, when the air is not already saturated with moisture, and when energy is available to evaporate the water. Recall that, when water vaporizes from a liquid to a gas, energy must be absorbed from the environment. In fact, *the rate of evapotranspiration depends mostly on energy availability.* ET occurs fastest under warm conditions and virtually halts below freezing. Atmospheric humidity and wind speed also are

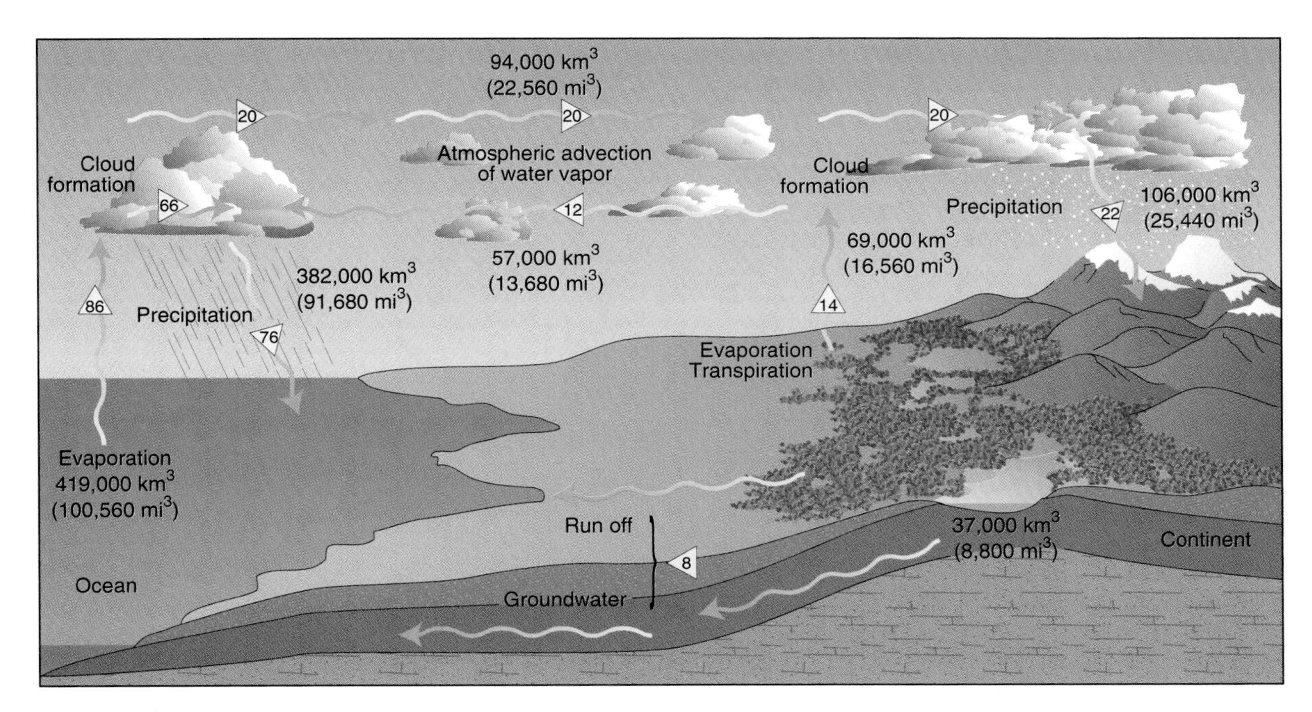

FIGURE 5–6
The hydrologic cycle and global water budget. Water is transferred continually among land, sea, and air. The quantities of water transferred in an entire year are shown. Precipitation exceeds evapotranspiration over land areas, while over the oceans there is more evaporation than precipitation. (From R. W. Christopherson, *Elemental Geosystems—A Foundation in Physical Geography,* Prentice Hall, 1995)

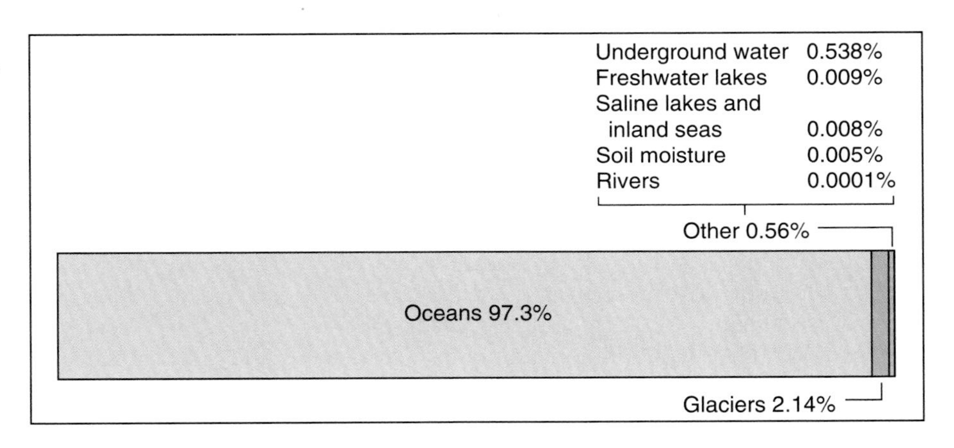

FIGURE 5–7
Quantities of water in storage. The oceans hold most water; glaciers hold the next largest amount, and the remainder is in streams, lakes, groundwater, and the atmosphere. (From N. Coch and A. Ludman, *Physical Geology,* Macmillan Publishing Co., 1991.)

significant; ET is faster on dry, windy days than on humid, calm days. As you can see, ET rates vary tremendously over space and time.

In warm weather, conditions may favor high ET, but the soil may be too dry to supply this demand. Therefore, we distinguish between potential ET and actual ET. Potential evapotranspiration (POTET) is the amount of water that would be evaporated *if it were available.* **Actual evapotranspiration (ACTET)** is the amount that *actually is evaporated* under existing conditions. If water is plentiful, then ACTET equals POTET. But if water is in short supply, ACTET is less than POTET. (ACTET can never exceed POTET.)

Comparing POTET with precipitation is a useful way to describe water availability in various climates. For example, if precipitation always is greater than POTET, plants have plenty of water and the climate is quite humid. If precipitation is less than POTET most of the time, plants do not get as much water as they need, and so the natural vegetation is adapted to dryness, and the climate is arid. Farmers using irrigation to supply water to plants can calculate the difference between precipitation and POTET and thereby know how much water they must apply to crops. POTET can be calculated from mean monthly temperature data.

Local water budgets. A water budget compares precipitation, POTET, and ACTET. For a humid mid-latitude site such as Chicago, a graph of a water budget reveals that precipitation occurs fairly consistently throughout the year (Figure 5–8). POTET is low in winter and high in summer, reflecting seasonal transpiration by plants. In the winter, when POTET is low, precipitation exceeds POTET. The surplus water either runs off or becomes stored as groundwater.

In springtime, plants develop leaves and POTET rises rapidly, so that by early summer it exceeds precipitation. Initially, this excess water demand is met by water stored in the soil, but by late summer, soil moisture is exhausted and ACTET is limited. At this time shallow-rooted grasses may wither, while deep-rooted shrubs and trees remain green. In autumn, POTET falls with the temperature, and once again precipitation exceeds evapotranspiration. At this time, excess precipitation restores soil moisture to higher levels, and runoff increases again.

Water budgets vary tremendously from one climate to another (Figure 5–9). In some humid climates, precipitation almost always meets vegetation demand (POTET), and stored soil moisture provides the small additional amount of water needed during brief dry spells. In semiarid and arid climates, POTET considerably exceeds precipitation during the year, so soil moisture is never fully replenished, and plants must withstand severe moisture deficits.

Soil's role. Soil is critical in the water budget, for it stores water and makes it available for evapotranspiration; soil is a water "bank." How effectively soil does this depends on how fast it can absorb precipitation, how much water it can

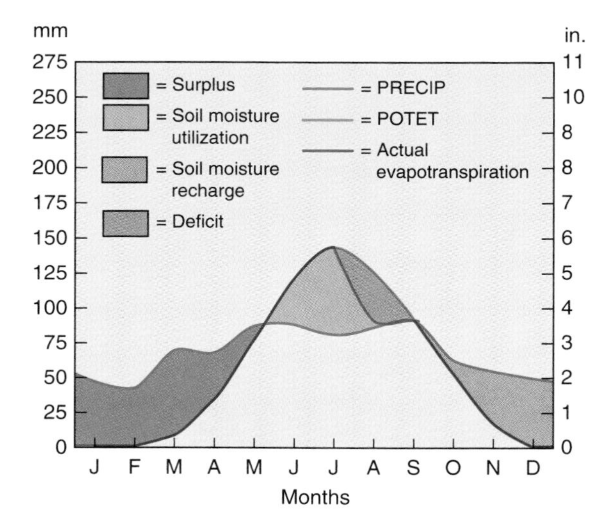

FIGURE 5–8
Water budget diagram for Chicago, Illinois, a typical humid mid-latitude site. Precipitation is shown as a blue line, and POTET is shown in red. In the cool months, precipitation is greater than POTET and the excess becomes runoff (dark blue shaded area). In the warm season, POTET is greater than precipitation. Part of this excess demand is made up by soil moisture use (green shading), but not all. The difference between POTET and ACTET is a deficit (brown shading). The soil water that was used in early summer is recharged in autumn (light blue).

store in the root zone (the depth of soil used by plant roots), and the ability of local vegetation to transpire water into the atmosphere.

Soil's **infiltration capacity** determines how well rainfall infiltrates (soaks into) the soil or runs off. Infiltration is good and most precipitation will soak in when dense vegetation exists at ground level, a layer of organic litter covers the soil, and the soil is kept porous by worms and insects. If the soil is bare or compacted, infiltration is poor, less water is stored, and runoff is greater.

Soil texture, the size of mineral particles in the soil, affects how much water soil can hold. Sandy soils drain quickly and hold little water (pour water on sand and watched how quickly it soaks in). Soils with high contents of finer particles—silt and clay—drain slowly (pour water on modeling clay and see how slowly it soaks in). Deep-rooted plants (several meters) have a longer opportunity to absorb water as it drains through the soil; shallow-rooted plants (half a meter or less) must rely

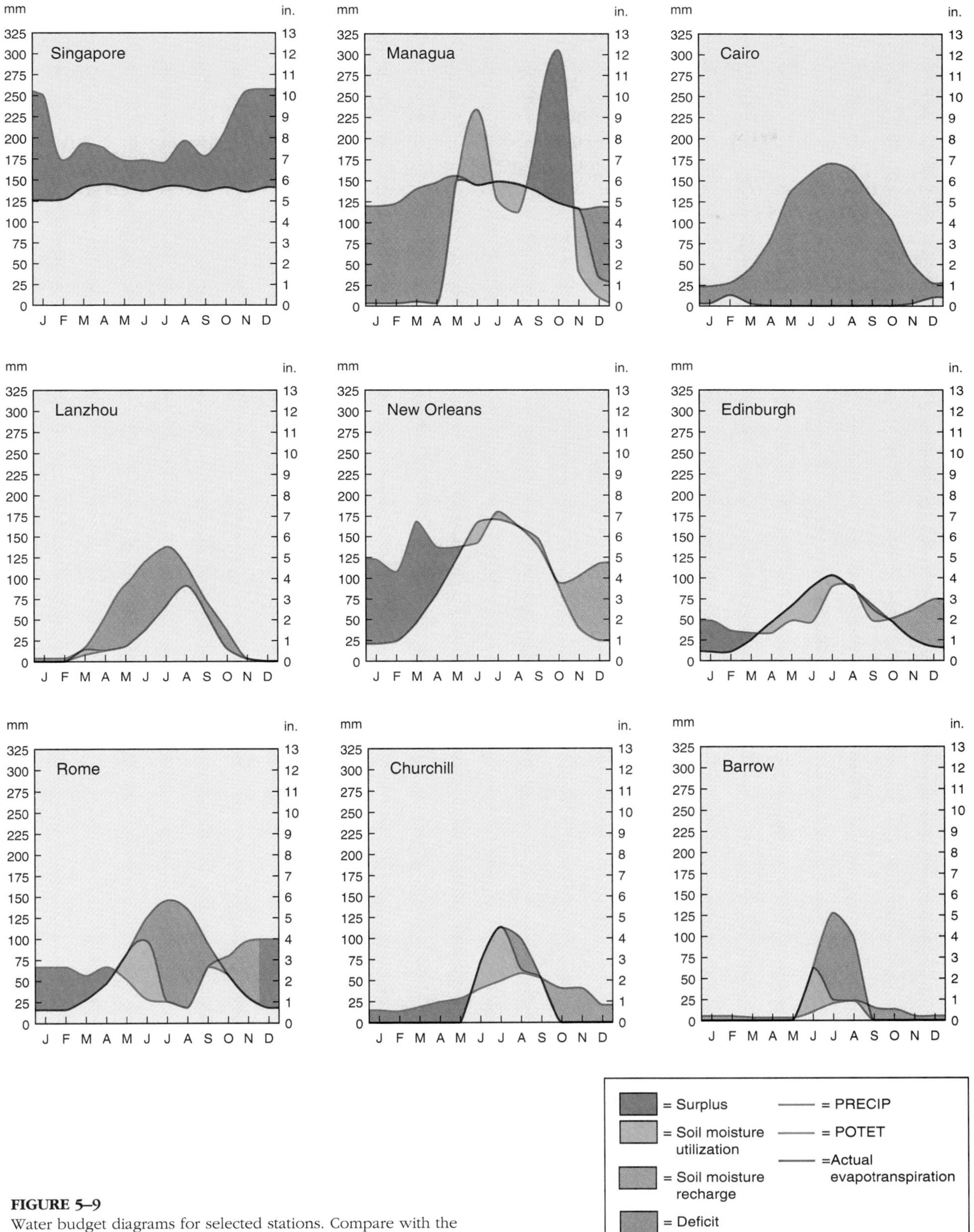

FIGURE 5–9
Water budget diagrams for selected stations. Compare with the climatic data in Chapter 3, Boxes 1 through 7, 9, and 10.

on moisture contained in the upper soil. Thus, shallow-rooted grasses and herbs generally use less water (and potentially allow more runoff) than deep-rooted trees. Soil and vegetation management is critical for maintaining adequate water resources for agriculture and other human needs.

Water budgets and resource management. River water is used for multiple purposes, including domestic and commercial supplies, as cooling water for power plants and industries, for hydroelectric power generation, irrigation, fish and wildlife habitat, navigation, recreation, and waste removal. Knowing how much water is available for these purposes is important, so water resource analysts study drainage basins. A **drainage basin** is the entire land area that contributes runoff to a particular stream. For example, the drainage basin of South America's Amazon River has an area of 6.15 million square kilometers (2.37 million square miles); Europe's Rhine River drains 143,360 square kilometers (55,351 square miles); New York's Hudson River drains 20,000 square kilometers (7,720 square miles); and Colorado's Two Buck Gulch drains 4.2 square kilometers (1.6 square miles).

The amount of water flowing in a stream is measured at sites called *gaging stations,* which have been established along most rivers. However, an analyst often needs to know the flow rate of smaller, ungaged streams. These flows can be predicted from measured precipitation and calculated evapotranspiration. For example, if we wanted to know a stream's average flow, we could estimate it from the water budget, as shown in Figure 5–10. This procedure works reasonably well in humid areas where precipitation exceeds POTET. In arid regions, more careful consideration is necessary for storms, seasonal variations, and water storage in the ground.

Water budgets help us understand why river flows are so much greater in humid regions than arid regions. For example, Canada's MacKenzie River has a drainage area of 1.8 million square kilometers and an average flow of 9,600 cubic meters per second. The Nile has a much larger

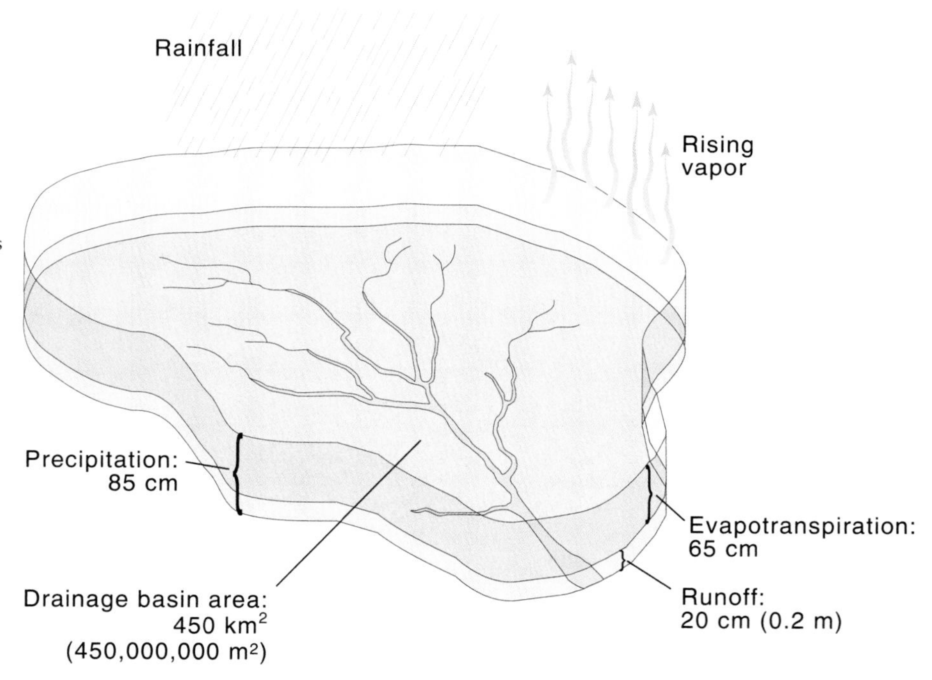

FIGURE 5–10
Average streamflow can be calculated from the water budget for a drainage basin, showing the close connection between water resources and climate. The average annual evapotranspiration is subtracted from the annual precipitation to determine the average annual runoff, expressed as a depth of water on the land. This value, when multiplied by the drainage basin area (in comparable units) gives the volume of runoff carried by the stream in a year.

drainage area of 3 million square kilometers but discharges only 950 cubic meters per second because so much of the basin is arid. The MacKenzie's annual flow is equivalent to a depth of about 17 centimeters of water spread over the entire basin area; the Nile's annual flow is equivalent to an average depth of only 1 centimeter.

Water budgets are controlled partly by soil moisture storage. Thus, land use affects evapotranspiration and runoff. Under natural vegetation cover, water soaks into the soil and becomes available for plants to transpire back into the atmosphere. But urban areas replace vegetation and porous soil with streets, roofs, and parking lots. Water cannot soak into these surfaces, and fewer plants exist. The result is less evapotranspiration and more runoff. In addition, urban runoff occurs much faster than runoff from vegetated surfaces, increasing the flood hazard.

Vegetation and the Hydrologic Cycle

Forests demonstrate the close association between vegetation and the water budget. Forests occur where ample moisture is available for most of the year. Trees demand large volumes of water; a single large tree can transpire 1,000 liters per day in warm weather. They gather water through extensive and usually deep root systems and transpire through leaves that often are tens of meters above the ground surface. Trees use so much water that they are the single greatest medium for returning rainwater to the atmosphere in large forested regions like eastern North America and the Amazon River basin.

This role of forests has water management implications. In the southeastern United States, areas formerly covered with slow-growing deciduous trees (that shed their leaves in autumn) have been replanted with faster-growing evergreens (conifers), to feed the timber industry. Because conifers do not lose their leaves in the winter, they keep growing and transpiring water year-round, when deciduous trees cannot. Studies in an experimental watershed in North Carolina showed that this measurably increased transpiration and correspondingly decreased streamflow.

In contrast to trees, grasses are relatively shallow-rooted, and so experience significant variations in moisture availability. When soil water is plentiful, grasses grow quickly and transpire at rates similar to trees. But during periods of limited soil moisture, grasses become dormant and transpiration virtually ceases.

Deforestation, or clear-cutting of forest, is proceeding at the rate of about 15,000 square kilometers per year (5,770 square miles per year) in the Amazon River Basin. By 1995, over 450,000 square kilometers (173,000 square miles) had been deforested, an area about the size of Montana. Deforestation may be critical to the water balance of this region, for a very interesting reason. In this tropical rainforest, rainfall averages 200 to 300 centimeters (80 to 120 inches) per year, but ACTET is only 110 to 120 centimeters (44 to 48 inches) per year. The excess precipitation runs off to the Atlantic Ocean via the Amazon River, which has the world's greatest discharge. The Atlantic is the original source of water that falls as precipitation on the Amazon Basin, so this a fine example of the hydrologic cycle at work.

The Amazon River in Peru. The rainwater that feeds the western portion of the Amazon Basin is derived from the Atlantic Ocean, but most of the water falls as precipitation, evaporates, and is returned to the atmosphere more than once on its westward journey across the continent. (Jack Fields/Photo Researchers, Inc.)

Harvesting peat for fuel in Ireland. The Irish climate provides ample moisture for plant growth, but because of the cool temperatures organic matter does not decay rapidly. Plant remains accumulate to form peat, a precursor of coal. Substantial amounts of carbon are stored in high-latitude peat deposits like these. (Tom Bean/The Stock Market)

But the Amazon Basin's interior is more than 2,000 kilometers (1,200 miles) from the Atlantic, and most of the atmospheric water falls closer to the coast. This precipitated water infiltrates the soil, trees transpire it back into the air, and easterly winds carry it further inland. This cycle repeats, moving water step-by-step westward. Thus, rainfall in the Amazon Basin's interior may have been precipitated and transpired several times in its westward journey.

Forests clearly are integral to this process. If the forests continue to be cut over very large areas, and if the grasses that replaces them transpire much less than the trees, this could reduce atmospheric moisture available for precipitation in the interior. As yet, such precipitation changes are only the predictions of computer models, but they help to show the important role that vegetation plays in the hydrologic cycle.

Carbon and Oxygen Flows in the Biosphere

As noted, biogeochemical cycles are recycling processes that supply essential substances such as carbon, nitrogen, and other nutrients to the biosphere. Biogeochemical cycles interconnect Earth's subsystems. Critical to life on Earth is the carbon/oxygen cycle.

The Carbon/Oxygen Cycle

Carbon is not the most abundant element on Earth but, in combination with hydrogen, it is the most important for sustaining life. Compounds of carbon and hydrogen are the major component of the foods that plants produce (carbohydrates) and that

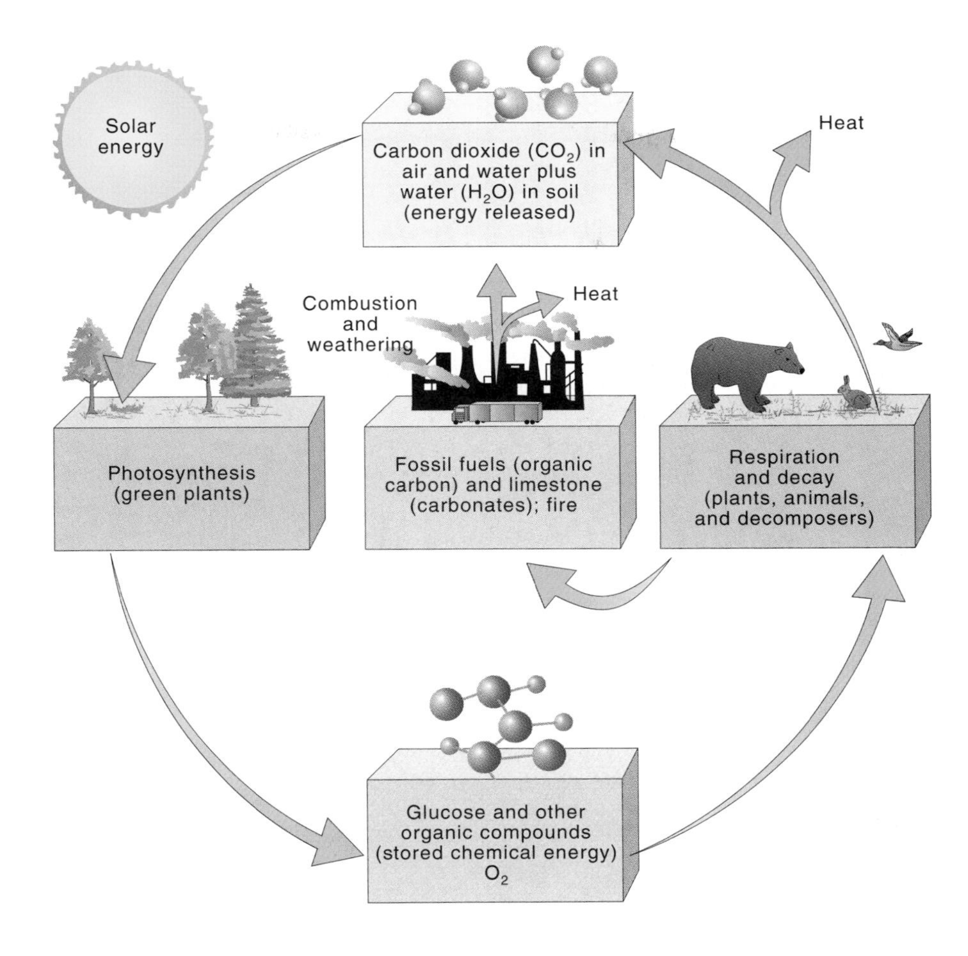

FIGURE 5–11
The carbon and oxygen cycles. (From R. W. Christopherson, *Geosystems,* 2d ed., Macmillan Publishing Co., 1994.)

animals consume, and in fossil fuels (hydrocarbons) that are our most important sources of power. Living things constantly exchange carbon with the environment, by photosynthesis, respiration, eating, and disposing of waste.

Exchanges of carbon among the biosphere, atmosphere, oceans, and rocks are collectively called the **carbon cycle.** Because carbon and oxygen often bond to each other and to other elements, the *oxygen cycle* is inextricably involved with the carbon cycle, so we will consider them together (Figure 5–11).

Carbon in the lithosphere. Of the four "spheres," the lithosphere is the greatest storehouse of carbon. Through geologic time, carbon has entered the lithosphere slowly through rock formation, principally from oceanic sediments. A widespread example is limestone, or calcium carbonate, which combines calcium, carbon, and oxygen—$CaCO_3$.

Carbon in the hydrosphere. Carbon also is exchanged between the atmosphere and ocean. This occurs primarily when carbon dioxide dissolves in seawater. Ocean circulation carries atmospheric carbon downward, where it is stored in deep ocean waters.

Carbon also enters the ocean through biological activity. In the upper 200 meters (650 feet) of the ocean, tiny floating plants called *phytoplankton* absorb carbon dioxide as they photosynthesize their food. Phytoplankton are the basis of the marine food chain, so carbon from their tissues passes into the tissues of consumers, many of which incorporate the carbon into shells or bones. As sea plants and animals die, much of the debris is deposited in seafloor sediments, where the carbon is stored. The oceanic component of the carbon cycle is the least understood, and estimates of ocean-stored carbon vary widely, but it is clearly a major absorber of carbon.

Carbon in the biosphere and atmosphere. When plants photosynthesize their food (carbohydrates), they bind carbon into their tissues. The carbon comes from atmospheric carbon dioxide, CO_2. By consuming plants, animals acquire this carbon. Then, through respiration, animals and decomposers return the carbon to the atmosphere as CO_2. Together, photosynthesis and respiration move carbon back and forth between the biosphere and atmosphere.

Photosynthesis in green plants is a chemical reaction, described in this equation:

Carbon dioxide + water + light energy —>
carbohydrates + oxygen

For this reaction, terrestrial plants obtain carbon dioxide from the air, water from the soil, and light energy from the sun. They store carbohydrates in tissue for later use, and release oxygen to the atmosphere. Plants are the source of atmospheric oxygen, without which animals—including humans—could not exist.

Respiration, which occurs in both plants and animals, is the opposite reaction:

Carbohydrates + oxygen —>
carbon dioxide + water + energy (heat)

In respiration, plants combine carbohydrates with oxygen from the atmosphere to produce CO_2, water, and energy, which is released in the form of heat.

Net primary productivity is the dried weight of biomass produced by plants in an area. The two most important factors in primary productivity are energy and water. Both must be present for plant production (photosynthesis), but their availability to plants varies widely (Figure 5–12). Net primary productivity is greatest in warm, wet environments: the humid tropics and warm, wet areas such as swamps and estuaries. In mid-latitude humid environments, productivity is lower because photosynthesis is restricted to part of the year. In dry areas, productivity is reduced by lack of water. In extremely cold environments and in the nutrient-poor open ocean, productivity is low.

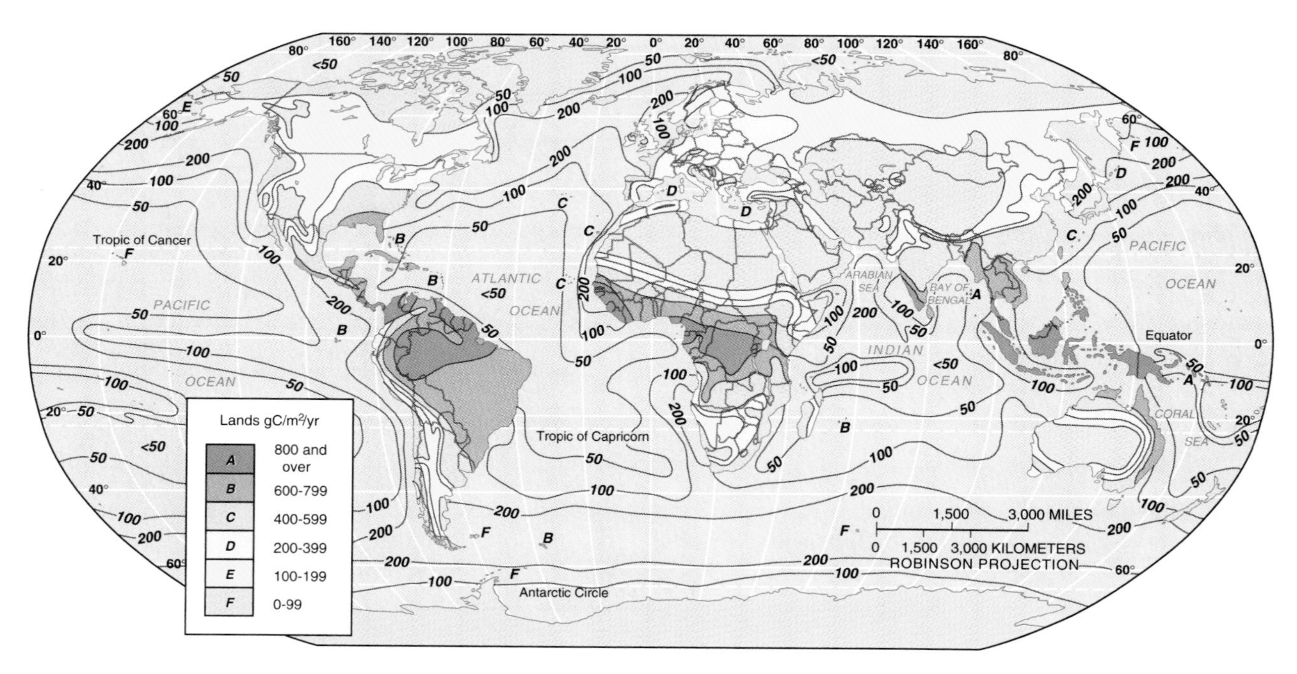

FIGURE 5–12
Map of net primary productivity. Units are grams of carbon produced by plants, per square meter, per year. Generally, productivity is at maximum along the equator and decreases poleward. Compare this map to the satellite data displayed in the satellite image on page 174. (From R. W. Christopherson, *Geosystems,* 2d ed., Macmillan Publishing Co., 1994. After D. E. Reichle, *Analysis of Temperate Forest Ecosystems,* Heidelberg, Germany: Springer-Verlag, 1970. Adapted by permission.)

Because photosynthesis is driven by solar energy, and because solar energy input varies with Earth's seasons, photosynthesis increases and decreases in an annual cycle. These seasonal cycles are reflected directly in the CO_2 content of the atmosphere. The air we breathe has been monitored for carbon dioxide concentration since the late 1950s at the Mauna Loa observatory in Hawaii (Figure 3–9):

- As expected, CO_2 varies annually with the seasons, by a few parts per million.
- The greatest concentrations in the Northern Hemisphere are during spring, because during winter, photosynthesis is reduced (or ceases completely in cold climates) while respiration continues, returning CO_2 to the atmosphere.
- The lowest concentrations are during autumn. CO_2 concentration drops during the summer because plants are actively photosynthesizing, removing CO_2 from the atmosphere and storing it in biomass.

The Mauna Loa data also clearly reveal the steady increase in atmospheric CO_2 levels that are resulting from fossil fuel combustion.

The Global Carbon Budget

Suppose we know how much carbon is stored in the wood and soil of a mature forest. Also suppose we know the rate at which carbon flows from atmospheric storage into storage within the trees, as the forest grows. With these facts, we can calculate the impact of forest growth on the amount of atmospheric carbon dioxide. However, our ability to do so depends on fully understanding the entire carbon cycle. For example, if we predict changing atmospheric CO_2 but ignore the great carbon exchanges that occur continuously between the atmosphere and ocean, our estimate will be extremely inaccurate and useless.

Table 5–1 presents a crude global carbon budget. Fossil fuel combustion dominates, contributing at least three-fourths of the carbon to the atmosphere. (This should erase any doubt that humans powerfully influence the environment.) Ocean uptake of carbon from the atmosphere also is important, accounting for perhaps half the carbon discharged to the atmosphere by humans. If we add up all the carbon discharged to the atmosphere, it should equal the amount added to the

TABLE 5–1

The carbon budget. Values are in billions of tons (gigatons) per year; a positive value indicates flow to the atmosphere.

Inputs of carbon to the atmosphere:	
+5.9 ±0.1	Fossil fuel combustion
+0.9 ±1.0	Net deforestation and soil carbon loss
Removal of carbon from the atmosphere:	
−2.0 ±0.8	Oceanic uptake
−3.4 ±0.2	Increased storage of carbon in the atmosphere as CO_2
"Missing" carbon:	
1.4 ±2.1	

A *Landsat* satellite image of deforestation in Rondônia, Brazil. Deforestation follows the road patterns. Forest is shown in dark green; grass areas are pale green, and newly cleared sites are in red and white tones. This part of Brazil experienced very rapid deforestation in the 1980s; in the early 1990s the rate of clearance slowed significantly. (NRSCLTD/Science Photo Library/Photo Reseachers, Inc.)

oceans and the atmosphere. However, most attempts to balance the carbon budget cannot account for all the carbon discharged to the atmosphere via fossil fuel combustion. Hence, some of the carbon is "missing." Scientists are working hard to improve their accounting of the carbon cycle and more closely balance this budget.

Use of coal, oil, and natural gas since the Industrial Revolution began in the late 1700s is a new factor affecting the carbon cycle. We have been removing these fossil fuels and their stored carbon from the lithosphere, and burning them, which injects their stored carbon into the atmosphere. This is something new on Earth. Rates of carbon release into the atmosphere through this pathway have increased steadily over the past 200 years, and are expected to increase well into the twenty-first century. The long-term CO_2 increase so evident in Figures 2-15 and 3-9 illustrates this. Uncertainty about future fossil fuel use makes it difficult to predict future atmospheric CO_2 concentrations.

Deforestation. Deforestation is a significant factor in the global carbon budget. Most notably in tropical rainforests, trees are being harvested for lumber or fuel and to make way for agriculture. In some areas trees are growing back, but at a global level, more trees are being cut than are growing. This is important to the global carbon budget because relatively large amounts of carbon are stored in forest biomass, and when trees are cut and burned, the carbon that they contain is released to the atmosphere.

The most rapid deforestation, and hence the largest release of carbon, is in tropical Central and South America, West Africa, and Southeast Asia (Figure 5–13). Deforestation rates in most mid-latitude forests are much lower, and in many areas, such as the eastern United States, forests are regrowing and are thus a "sink," or storage facility, for carbon. In Arctic regions, large amounts of carbon are stored in soils as peat deposits—accumulations of plant matter in swamps and bogs. How these areas might change in a global warming is a point of con-

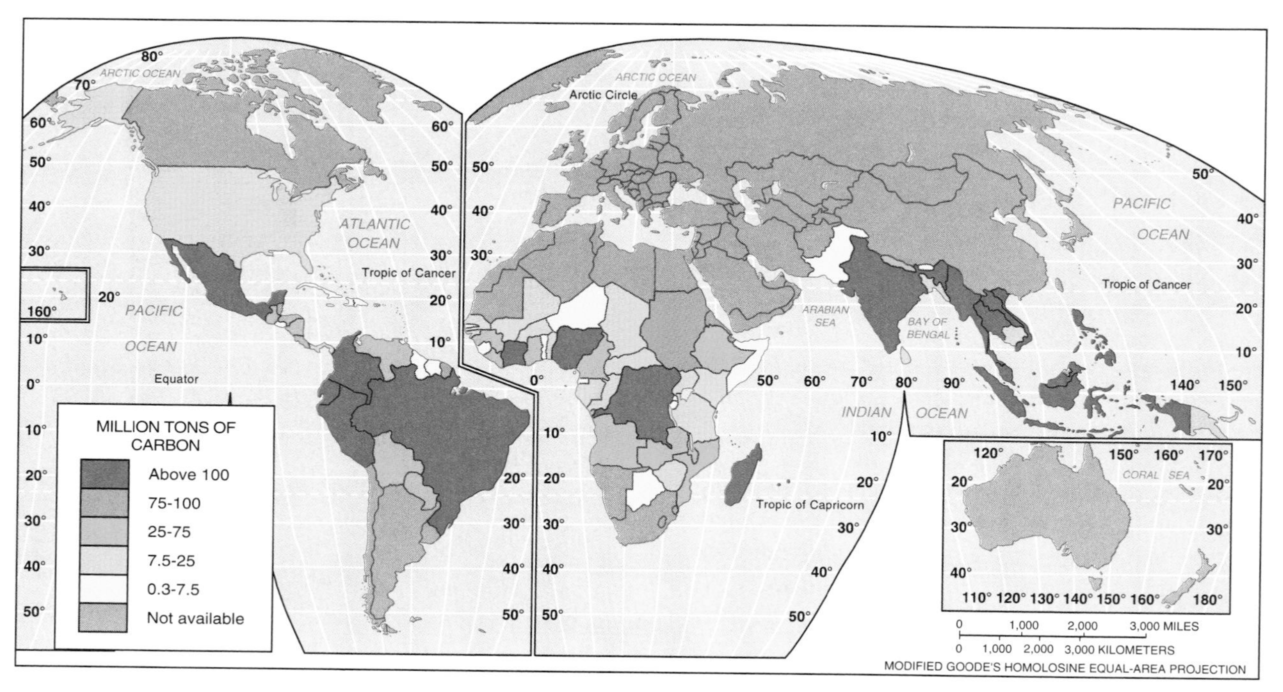

FIGURE 5–13
Net emissions of carbon from deforestation, 1989. Values are in millions of tons of carbon. The largest emissions from this source are in the humid tropics, especially tropical South America, Africa, and southeast Asia. However, carbon emissions from fossil fuel combustion are much greater than these from deforestation. (Data from World Resources Institute, *World Resources 1992-93*. Oxford University Press, 1992.)

cern. Warming would increase both respiration and photosynthesis, but we do not know whether the net change would be toward storage of carbon in biomass or in the atmosphere.

It is sobering to think about the relationships among the sun, atmosphere, and biosphere. If all of Earth's vegetation were abruptly removed, the same thing would happen. Our lives depend entirely on the carbon cycle, driven by the sun.

Soil

Like the atmosphere and hydrosphere, the lithosphere is a critical part of biogeochemical cycles. The interface between the lithosphere and the biosphere is the *soil*. We can view soil as the uppermost part of the lithosphere, and at the same time as the lowermost part of the biosphere. Soil moderates flows among the biosphere, lithosphere, atmosphere and hydrosphere. It is a storage site for water, carbon, and plant nutrients.

Soil properties are attributable to five major factors:

1. *Parent material* (rock) from which soil is formed.
2. *Climate,* which regulates both water movements and biologic activity.
3. *Biological activity* (plants and animals), which moves minerals and adds organic matter to the soil.
4. *Topography,* controls water movement and erosion rates.
5. *Time,* during which all of these factors work, typically requiring many thousands of years to create a mature soil.

Soil Formation

The first step in the soil-formation process is weathering, which is the breakdown of rock into smaller particles and new chemical forms. Chapter 4 explained the two ways that rocks weather: mechanical weathering (such as ice expansion or tree roots growing in the cracks between rocks) and chemical weathering (such as oxidation of iron and aluminum and decay from organic acids in the soil). The **parent material** (rocks) from which a soil forms is important because it influences the chemical and physical characteristics of the soil, especially in young soils.

Water plays an important role in rock weathering and soil formation. A water budget indicates how much water moves through the soil in a place, and measures the significance of water in local weathering. For example, if mean annual precipitation is 100 centimeters and POTET is 70 centimeters, then 30 centimeters of water would percolate down through the soil in an average year. This water is a powerful weathering agent. However, if an area has 70 centimeters of precipitation and 150 centimeters of POTET, virtually all water is evapotranspired by plants, so none would percolate down through the soil to help weather the rocks.

In a very humid climate, much water passes through the soil and dissolves soluble minerals on its way. Because of this, soils in humid climates generally have lower amounts of remaining soluble minerals, such as sodium and calcium, compared to soils in dry climates. In dry areas, water enters the soil and dissolves soluble minerals, which are drawn toward the surface as water is evapotranspired. Consequently, soils of semiarid and arid climates often have a layer rich in calcium near the surface. Topography affects the amount of water present in the soil, largely through controlling drainage. Steeply sloping areas generally have better drainage than flat or low-lying areas.

Plant and animal activity is also critical to soil formation. Plants produce organic matter that accumulates on the soil surface, and animals like worms and moles mix this organic matter through the soil. Soil formation is a slow process that takes place very gradually over thousands of years. Soils that have been forming only for a few hundred years, or even a few thousand, have very different characteristics from those that have been modified by chemical and biological processes for tens of thousands of years.

Soil horizons. Soil is a complex medium, containing six principal components:

- *Rocks and rock particles,* which constitute the greatest portion of the soil (rock pieces, sand, silt, and clay). They may weather to release nutrients.
- *Humus,* which is dead and decaying plant and animal matter. Humus holds water, sup-

ports soil organisms, and supplies nutrients.

- *Dissolved substances* including phosphorus, potassium, calcium, and other nutrients needed for plant growth.
- *Organisms,* including animals such as insects and worms, bacteria, and fungi.
- *Water from rainfall and snowmelt,* which is necessary for plant growth and which helps to distribute other substances through the soil.
- *Air,* which shares soil pore spaces with water and which is necessary for respiration by plant roots and soil organisms.

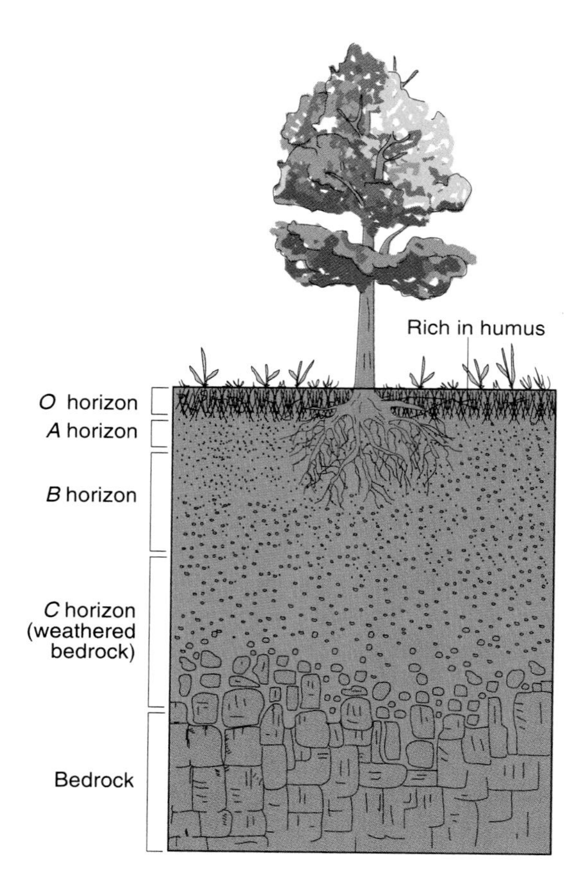

FIGURE 5–14
A soil profile. The sequence from the bottom up shows the typical steps of soil formation. Solid rock at the bottom weathers into large pieces, which weather further into ever-smaller particles. At top is the organic layer, which contains organic matter from decayed plants and animals. Designations like O and A are given to horizons in well-developed soils. Poorly developed soils lack some of these layers. (From E. J. Tarbuck and F. K. Lutgens, *Earth Science,* 7th ed., Macmillan Publishing Co., 1994.)

These substances are not uniformly distributed in soils, but found in layers called **soil horizons** (Figure 5–14). Horizons are different from layers in sedimentary rocks, which are deposited in sequence from the bottom up. Soil horizons are formed through the movement of water, minerals, and organic matter vertically in the soil, and by variations in biological and chemical activity at different depths in the soil.

For example, litter—leaves, twigs, dead insects and other organic matter—accumulates to form a horizon at the surface known as the O (organic) horizon. As this litter decays, decomposers such as insects, worms, and bacteria consume it and carry it underground, where it helps form the A horizon. Waste from these burrowing animals, as well as their own dead carcasses, adds more organic matter to the A horizon. In many soils, the A horizon contains much of the nutrients that support plant life.

Water may carry some substances from the A horizon to the B horizon. Clay minerals formed from chemical weathering often accumulate in the B horizon. In dry regions, soluble minerals such as calcium will accumulate in the B horizon. The C horizon, beneath the B horizon, contains weathered parent material that has not been altered as completely by soil-forming processes as materials above it.

Thousands of Soils

Soils are divided into 11 broad clusters, called **soil orders** (Figure 5–15). These 11 soil orders are further divided into 47 suborders, about 230 great groups, 1,200 subgroups, 6,000 families, and thousands of soil series. In other words, Earth's dynamic lithosphere, hydrosphere, atmosphere, and biosphere have worked as a team to create an enormous diversity of soil. There is nothing simple about "ordinary dirt"!

Detailed maps developed by the U.S. Department of Agriculture's Soil Conservation Service are used by farmers and land-use planners because they show the great local variability of soils (Figure 5–16). Often, two or three different great groups of soils occur within a few kilometers of each other, and these groups are subdivided into the subgroups, families, and series of more specific characteristics. Generalizations about soils must be made with caution because of this great local variability.

Oxisol, Hawaii (USDA Soil Conservation Service)

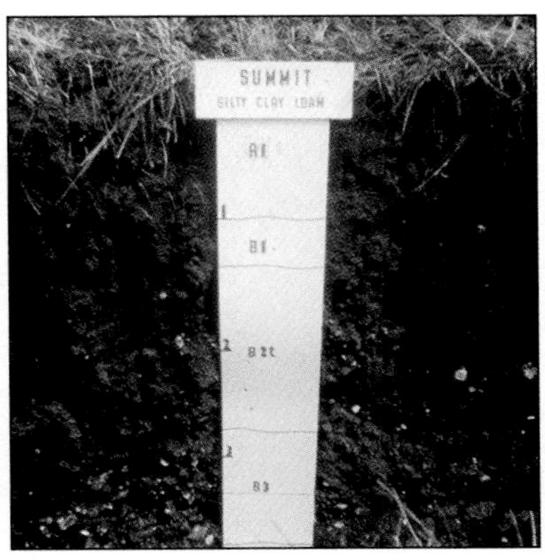

Mollisol, Montana (USDA Soil Conservation Service)

Spodosol, New York (Soil Science Society of America)

Inceptisol, New Zealand (USDA Soil Conservation Service)

Aridisol, New Mexico (USDA Soil Conservation Service)

Entisol, Thailand (USDA Soil Conservation Service)

Soil profiles

Climate, Vegetation, Soil, and the Landscape

At a global scale, distinctive environmental regions have been created by the interactions among climate, vegetation, and soil:

- *Humid tropical and subtropical areas* have highly weathered soils. They have lost soluble minerals due to leaching by heavy precipitation. This leaching depletes the soils of essential nutrients, so fertilizer must be added to increase crop yields. Tropical soils are usually oxidized and red in color.
- *Arid regions* have soils with high contents of soluble minerals, because little leaching occurs due to the scant rainfall. These soils are low in organic matter, but can be very productive if sufficiently irrigated.
- *Mid-latitude humid soils* are moderately leached where they are dominated by deciduous forest vegetation, and heavily leached where they are developed beneath the acidic litter of coniferous forests.
- *Mid-latitude subhumid soils* are fertile and are the resource base for many important grain-producing regions.

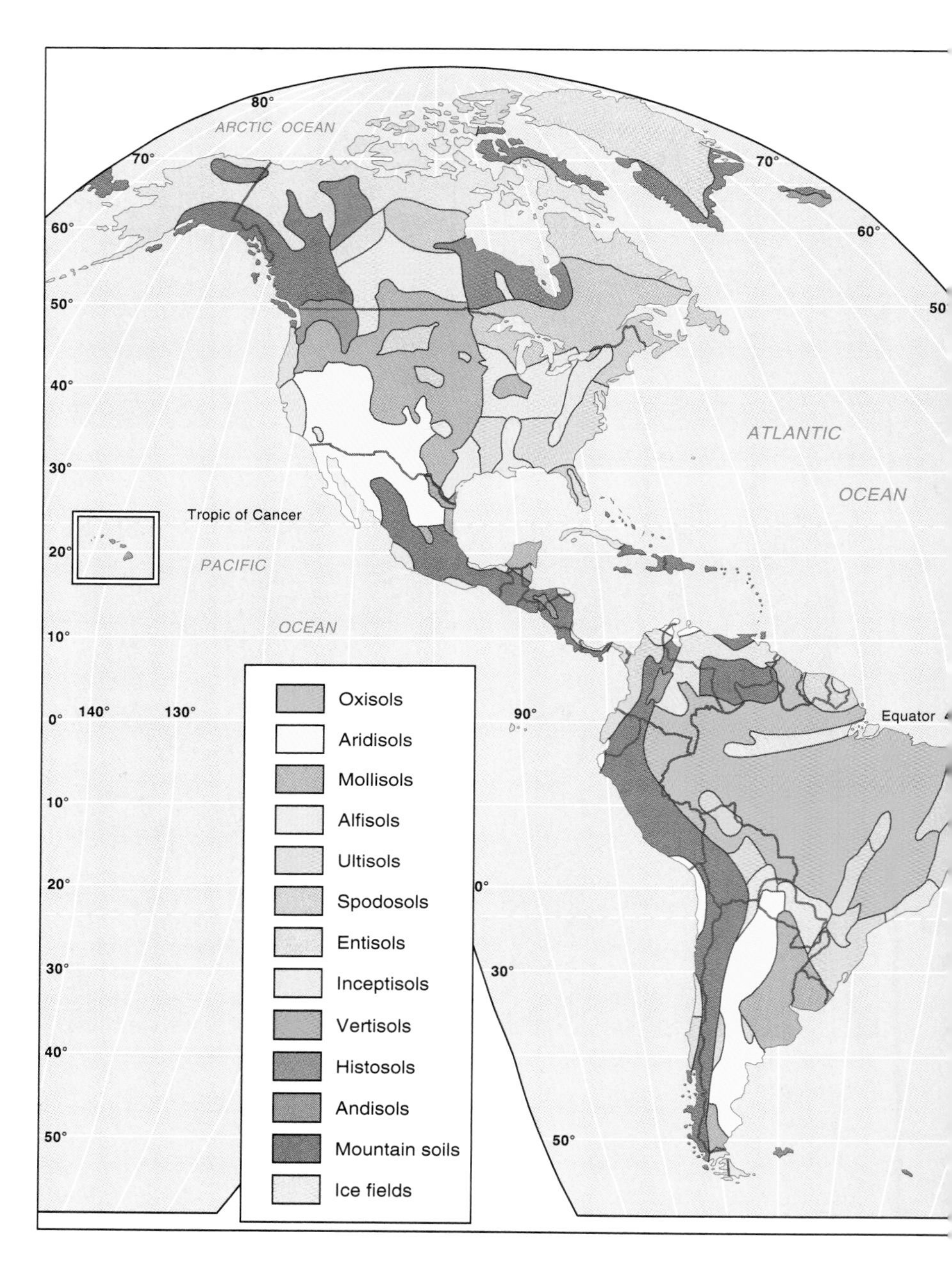

FIGURE 5–15
World soils map. (From R. W. Christopherson, *Geosystems*, 2d ed., Macmillan Publishing Co., 1994.)

Oxisol—deeply weathered, heavily oxidized soil of the humid tropics, red in color.
Aridisol—dry desert soil with limited organic matter, limited chemical weathering, and accumulations of soluble minerals.
Mollisol—grassland soil of subhumid/semiarid land, dark with plentiful organic material. Mollisols are an important agricultural soil in U.S. Midwest.
Alfisol—A moderately leached soil of humid subtropical and mid-latitude forests, with clayey B horizon.
Ultisol—red-to-yellow-colored, well-weathered soil with clayey B horizon, typically occurring in warm, humid climates.
Spodosol—acidic soil formed under coniferous forest, with a light-colored, sandy A horizon and red-brown B horizon in which iron accumulates.
Entisol—Poorly developed soil, usually as a result of low weathering rates, with characteristics dominated by parent material.
Inceptisol—weakly developed soils, often found in mountainous areas where erosion limits soil development.
Vertisol—soils with a high content of clay minerals that swell upon wetting and shrink upon drying. Deep cracks open in the dry season.
Histosol—Organic soils found in peat bogs and other areas of organic matter accumulation.
Andisol—young volcanic soils developing on ash and lava deposits.

These climatic soil regions are interrupted by major mountain chains of the world. In mountain areas, poorly developed soils predominate because erosion of the steep topography removes soil as rapidly as it is formed.

In Figure 5–16, we "zoom in" on a small area of the world soils map shown in Figure 5–15. This lets us examine vegetation and soil at a much larger map scale. Note that the same interaction among climate, vegetation, and soil that exists worldwide is also obvious at the scale of the local ecosystems shown in this small area of northwestern Colorado. The vegetation and soils are far more diverse than is visible on the world map.

This semiarid area is a sandstone and shale upland, incised with narrow valleys. It is an open woodland dominated by Pinyon Pine and Utah Juniper, and a shrub/grass community that is most-

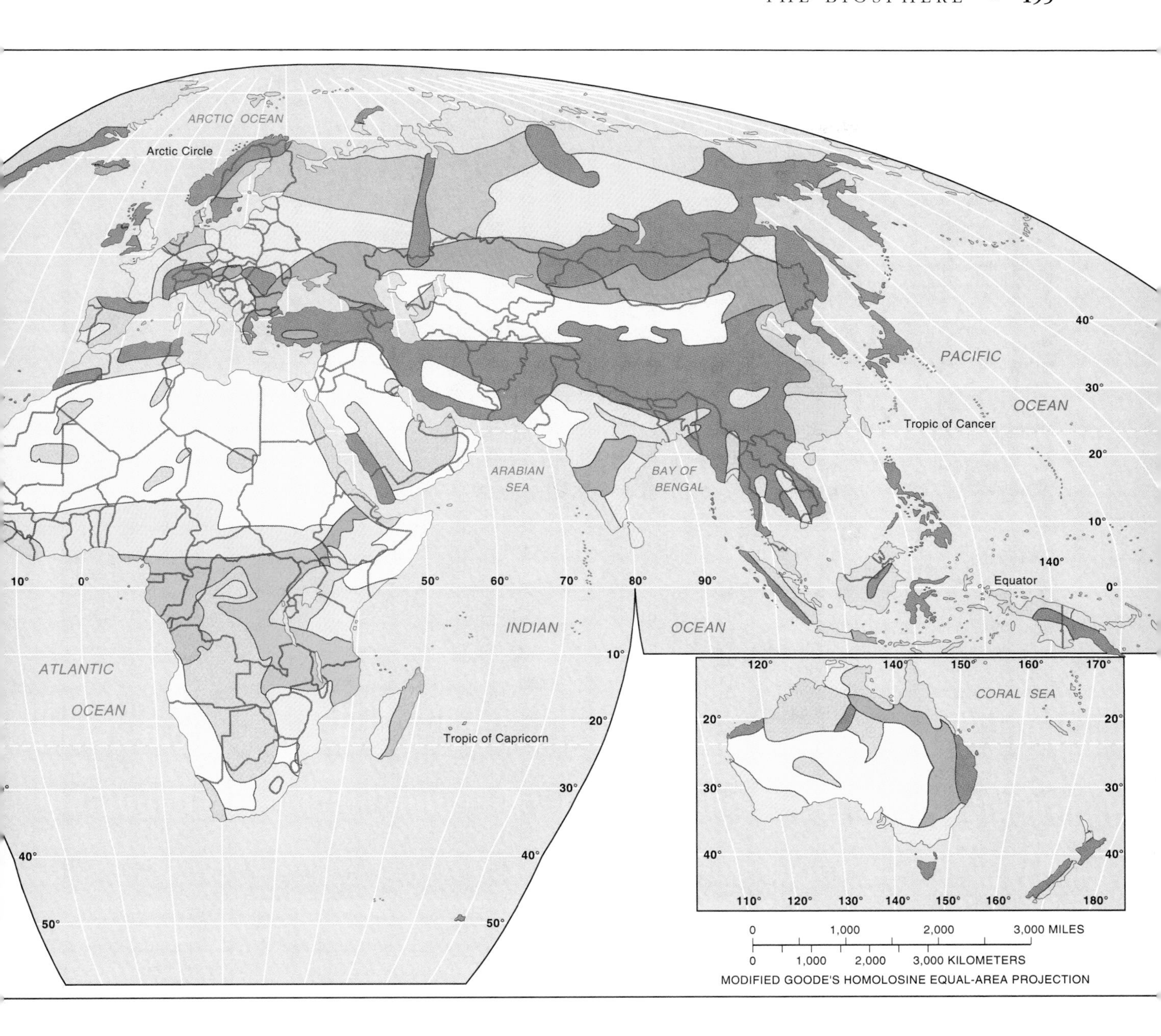

ly sagebrush. Within a few square kilometers, you may see forest of one or more types, disturbed areas, shrublands, and grasslands.

Soil Problems

In modern industrial agriculture, and in many traditional agricultural systems, nutrients are added artificially to the soil. These nutrients may be delivered in the form of natural fertilizer (animal manure), or by plowing into the soil a soil-enriching crop, or by adding manufactured inorganic fertilizers. The more intensively a soil is worked, the more important such additions become. Unfortunately, increasingly intensive agriculture has meant that nutrients, especially organic matter, are not being replaced as rapidly as they are withdrawn. Soil fertility usually declines over decades, so farmers are less aware of

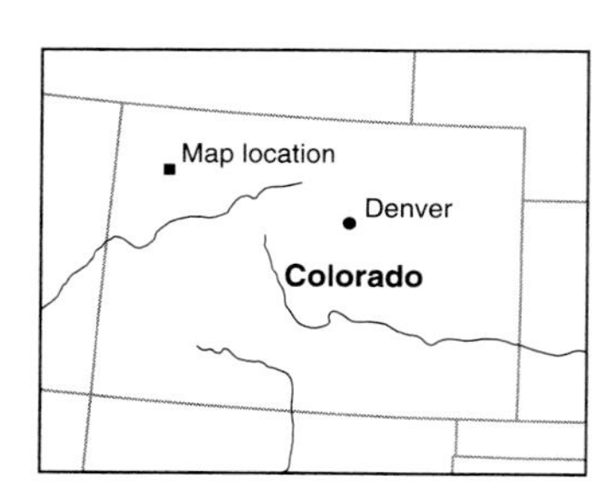

FIGURE 5–16
(Right) A soil map for part of Rio Blanco County, Colorado, shows the close correspondence among landforms, soils, and vegetation. Sagebrush prefers deep, well-drained soils (36); Pinyon-Juniper woodland prefers thin, rocky soils (73). On gently sloping uplands, moderately thick soils have developed from sandstone bedrock; vegetation is dense-to-open woodland of Pinyon Pine and Juniper, with scattered sagebrush and grasses (70); small patches of drier, wind-blown material show a distinctive fine-textured soil with sagebrush (33). (Map from U.S. Department of Agriculture, Soil Conservation Service.) (Below) The landscape of Rio Blanco County, Colorado. View is taken from lower center of map, looking north.

this problem than if it occurred over just a few years. Variations in yield from year to year caused by weather, insects, plant diseases, and changing technology mask the effects of long-term soil degradation. Therefore, a long-term decline in soil quality is in progress worldwide (Figure 5–17).

To this problem we must add the erosion caused by cultivation. Plowing a field rips apart the soil-holding system of plant roots and lays bare the soil to attack by rain, wind, freezing, and thawing. Erosion removes the uppermost part of the soil—the O or organic horizon, usually its most fertile part—and with it go both nutrients and the ability to store them. In many agricultural regions, topsoil

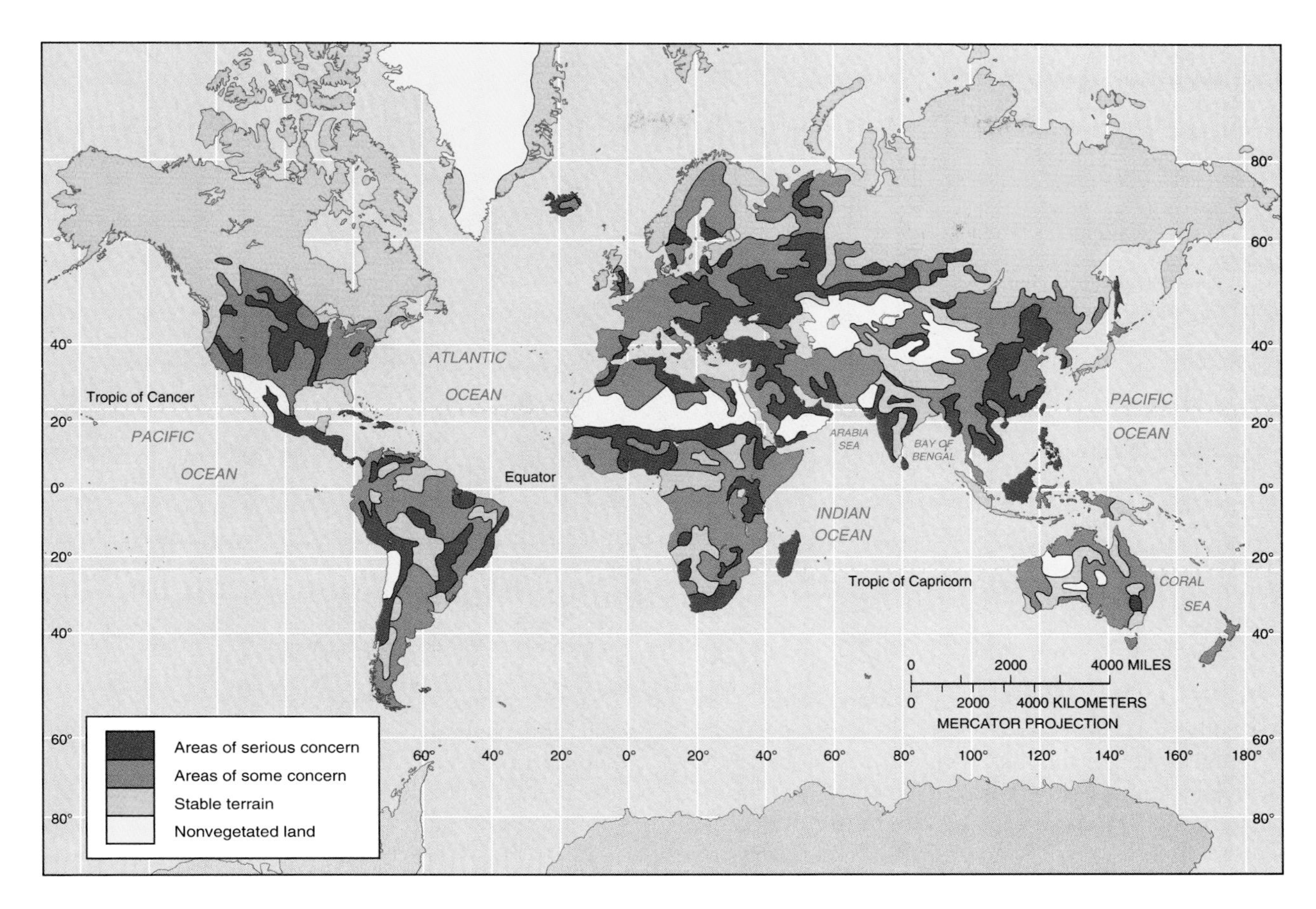

FIGURE 5–17
Map of soil degradation. About 1.2 billion hectares (3 billion acres) of Earth's soils suffer degradation through erosion due to human misuse. (From R. W. Christopherson, *Geosystems,* 2d ed., Macmillan Publishing Co., 1994. Adapted from United Nations Environment Programme, International Soil Reference and Information Center, "Map of Status of Human-Induced Soil Degradation," Sheet 2, Nairobi, Kenya, 1990.)

loss has ranged from several centimeters (1 to 2 inches) to tens of centimeters (4 inches to a foot or more). Often, the depth of erosion amounts to a significant part of the entire A horizon.

Soil Fertility: Natural and Synthetic

Plants need sunlight, carbon dioxide, oxygen, and water. But they also require nutrients for growth, including nitrogen, phosphorus, potassium, calcium, magnesium, sulfur, and others. **Soil fertility** refers to the ability of a soil to support plant growth though making nutrients available. Nutrients are present in many different forms—dissolved in soil water, as part of organic matter, attached to the surfaces of mineral grains, and in the crystalline structure of minerals. Some forms are more available to plants than others.

Organic matter, both living and dead, stores nutrients and makes them available to plants. In most soils, depletion of organic matter reduces fertility. (This explains why a standard soil amendment is to add manure.) Most available nutrients occur in the uppermost soil horizons: the O, A, and upper B horizons. Cultivation of the soil and harvesting agricultural crops accelerates the breakdown of organic matter and losses of nutrients stored in the soil. Sustaining soil fertility requires maintaining soil nutrients and maintaining the soil's ability to store and release them to plants.

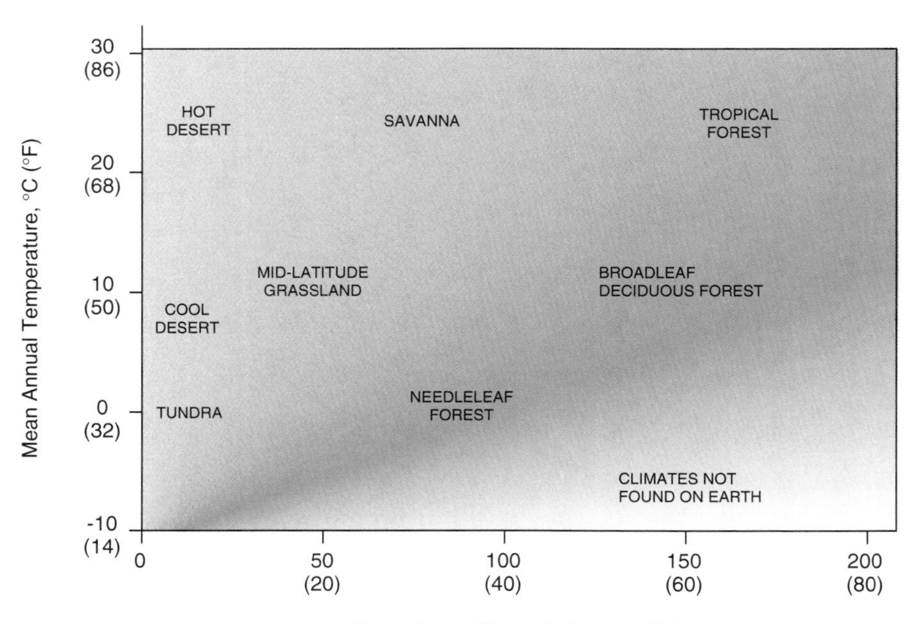

FIGURE 5–18
Generalized relation between vegetation type and climate. The amount of precipitation necessary to support a given amount of vegetation (such as forest relative to grassland) is greater in warm climates, where evapotranspiration is greater, than in cold climates. (Cold climates with high rainfall amounts do not exist on Earth.)

Farmers long have known that measures must be taken to restore soil fertility. In traditional agricultural systems, this has meant a *fallow period,* a season when a field is allowed to rest and nothing is harvested from it. During this rest period, the natural processes of plant growth, organic matter accumulation, mineral weathering, and biological activity can restore fertility (see Chapter 11). Earth's burgeoning population demands more agricultural products, so fallow periods have been reduced or eliminated. Many areas that once were fallowed now are cultivated every year, reducing their opportunity to restore natural soil fertility.

The low cost, availability, and easy application of inorganic (synthetic) fertilizers has led them to replace organic (natural) fertilizers. Although manufactured fertilizers contain valuable nutrients, they lack the organic matter that soils need to store those nutrients. In the short run (a single crop season), nutrients like nitrogen and phosphorus can be added to soil to supply plants' needs. But in the long run, the ability of the soil to support plant growth is diminished.

Farmers and resource managers around the world increasingly worry about the impact of this soil degradation on the world's food-producing capacity. As discussed in Chapter 4, about one fifth of U.S. cropland is being eroded faster than it can be replaced. American agricultural policy is oriented toward helping farmers to better conserve the soil. The challenge is to develop and adopt agricultural practices that sustain the land, not only in relatively developed nations, but in developing countries as well.

A study of landscape patterns at local and regional levels, using vegetation and soil maps, almost always reveals human disturbance. As settlement increases, environmental managers worry about human impact on ecosystems. Although global generalizations are useful, such as reports of the total amount of forest area lost in the tropics each year, impacts are best understood at local scale.

Biomes: Global Patterns in the Biosphere

We can identify thirteen major biomes: four forest, three savanna and woodland, four dryland, and two tundra. Differences in vegetation and climates distinguish major biomes. Figure 5–18 shows important associations among temperature, water availability, and vegetation types.

Forest Biomes

Forest biomes are dominated by trees: tall, woody-stemmed perennials with spreading

Tropical forest, Amazon River Basin. The density of the species is evident in the foreground. (Claudia Parks/The Stock Market)

canopies. Forests occur over a wide range of humid and subhumid climates, where ample water is available most of the year. We discuss three of the four forest biomes: tropical rainforest, mid-latitude broadleaf deciduous forest, and needleleaf forest. The fourth forest biome (the temperate rainforests of western North America) are also needleleaf forests.

Tropical rainforest biome. The **tropical rainforest** has tall, broad-leaved trees that retain their leaves all year, and the forest is characteristic of humid equatorial environments of Central America, the Amazon River Basin, equatorial Africa, and southeast Asia. Most biomes are complex horizontally, but the tropical rainforest is also complex vertically, with a canopy layer and two more layers beneath the canopy (Figure 5–19). Each layer has different dominant species and associated animal communities.

The highest canopy is intermittent, with very tall trees—heights of 40 to 60 meters (130 to 200 feet). Underneath is the dense middle canopy. Most photosynthesis occurs in the two upper layers, where sunlight is most intense. The middle canopy blocks most light, keeping the forest floor in twilight, so plants at the lowest levels must tolerate deep shade to survive. Lianas (vines) are common in many tropical rainforests, as are epiphytes (plants that grow on the surfaces of other plants, to gain height advantage without having long trunks to the ground). The rainforest and the ocean are analogous, with the greatest productivity (photosynthesis) at the top, and life forms decreasing toward the bottom.

Tropical rainforests are noted for biodiversity, with hundreds of tree species within a single hectare. In contrast, mid-latitude forests may have only 5 to 15 tree species per hectare, and high-latitude coniferous forests may have only 3 to 5. This diversity is paralleled by a great variety of animals. The vertical structure adds to diversity of habitats and species.

A walk through a zoo illustrates the vast number of large animals species that are native to tropical climates. Primates exist almost exclusively in the tropics, especially tropical forests. The biodiversity of the tropical rainforest places this biome at the center of controversy over deforestation, species extinctions, and biodiversity.

In the rainforest biome, much nutrient storage is in living biomass rather than in the soil. This is yet another reason why deforestation is a major concern. Organic matter decays rapidly in warm, wet conditions, and nutrients are not easily retained in the mineral soil. When forests are removed, their store of nutrients may be lost quickly to leaching and runoff, leaving the soil impoverished.

Mid-latitude broadleaf deciduous forest biome. This forest exists in subtropical and mid-latitude humid environments where seasonally cold

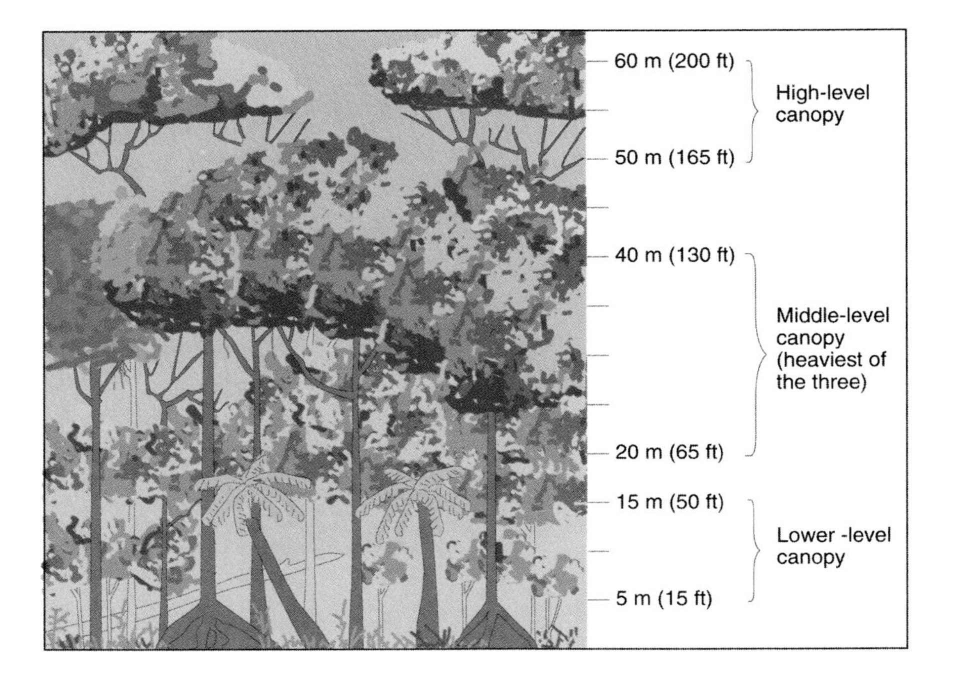

FIGURE 5–19
Three levels of rainforest canopy. (From R. W. Christopherson, *Geosystems,* 2d ed., Macmillan Publishing Co., 1994)

conditions limit plant growth. The eastern United States and much of China share this biome, which is why some Chinese ornamental species are popular on eastern U.S. lawns. On the map you can see that this biome occurs mostly in the Northern Hemisphere.

During the summer growing season, long days and high solar angle promote rapid growth, so that annual net primary productivity may attain 60 to 75 percent of that in the tropics, even though the growing season is only 5 to 7 months long. These plants have evolved wide, flat leaves to capture as much sunlight as possible. Broadleaf deciduous forests are much less diverse than tropical rainforests, but still support many species. Because of the mid-latitude location, many animal species either hibernate in winter or migrate to warmer locations.

Soils of this region vary with the underlying geology, but generally are leached less than soils of the humid tropics. This is because annual precipitation usually is less than in the rainforest, and because lower temperatures slow chemical weathering. Lower temperatures also contribute to reduced decay rates of organic matter. Consequently, undisturbed soils typically have a substantial O horizon ranging in thickness from a few centimeters (1 to 2 inches) to over 10 centimeters (4 or more inches). When these forests are cleared and the soil is brought under cultivation, they generally are quite fertile, but that fertility declines once the organic matter decays and is not replaced.

Needleleaf forest biomes. In poleward portions of the mid-latitudes, a needleleaf (coniferous) evergreen forest flourishes. The name boreal forest comes from the northern location of this vegetation (Latin *borealis* = northern). The boreal forest thrives in cold mid-latitude climates, and because these climates are rare in the Southern Hemisphere, extensive boreal forests are restricted to the Northern Hemisphere. Needleleaf forests also occur in some areas of seasonally dry soils.

During cold winters, low humidity and frozen ground cause moisture stress. However, the leaves of needle-leaved trees have low surface area in relation to their volume, plus a waxy coating to reduce water loss. This adaptation allows them to survive in colder climates that would kill broad-leaved trees. Because they are evergreen, these trees photosynthesize whenever temperatures are warm enough, instead of having to wait until springtime to produce leaves.

Coniferous forests also occur in the southeastern United States, where sandy soils drain quickly and have limited water storage. During warm summers, when ACTET exceeds precipitation, these soils dry out, and the moisture-retaining properties of coniferous trees allow them to compete well against broadleaved trees. Fire-adaptation is an added advantage of many coniferous species in areas of summer aridity.

Pine needles accumulate beneath coniferous forests, producing strong acids in the water that percolates through the soil. This acidic water leaches the upper soil of all but the most insoluble minerals, leaving a light-colored, sandy upper soil layer. Some of the iron and aluminum leached from the upper soil accumulates lower in the soil profile. This produces a distinctive soil associated with coniferous forests called a *spodosol* (see Figure 5-15).

Savanna, Open Woodland, and Scrubland Biomes

Savannas and open woodlands have trees spread widely enough for sunlight to support dense grasses and shrubs beneath them. This vegetation is common in climates that have a pronounced dry season.

Tropical savanna and scrub biomes. The three major biomes in this group are tropical savanna, tropical seasonal forest and scrub, and Mediterranean shrubland. The term **savanna** refers especially to vegetation characteristic of large seasonally dry areas in tropical Africa and South America. The rainy season brings green, lush grass but during the dry season the grass dries out and only deeper-rooted trees continue to photosynthesize. If the dry season is pronounced, even the trees may lose their leaves, and in some areas dry-season deciduous trees are common.

In more arid margins of tropical savannas, a thorn forest or thorn-scrub vegetation occurs. Drought-resistant trees and shrubs dominate, as well as cacti, and grasses are less important. Savanna-like open woodlands (also called parklands) occur in some semiarid mid-latitude areas.

Fire is common in the savanna, aided by the seasonal aridity and probably by humans. Fire does not destroy most grasses, which grow back quickly. Slower-growing trees might in time become dominant, shading out the grasses, but they are more susceptible to fire damage, so the savannas remain. Some expanses of African savanna occur in climates that support forests elsewhere, which supports a hypothesis that human-caused fires have created the extensive African savannas.

Soils of tropical savannas usually are deeply weathered, as in the humid tropics, but they may be less leached of nutrients as a consequence of the reduced rainfall. High temperatures and reduced primary productivity combine to lower the organic content. Many soils of these regions are quite productive if sufficient irrigation water is available.

A forest in the transition zone between mid-latitude deciduous forest and northern coniferous forest, in Northern Michigan. (Stan Osolinski/The Stock Market)

FOCUS BOX 5-1

Fire and Chaparral in California

In the chaparral woodlands of California, fire is an important natural part of ecosystems. Unfortunately, people have built homes and businesses close to the chaparral. Conflagrations like the 1993 fires near Los Angeles destroy entire communities, causing vast property damage, and occasionally taking lives.

When a fire starts near a populated area, the natural reaction is to extinguish it at once. However, this actually makes conditions worse. In the chaparral, plant growth produces flammable biomass that does not decay as rapidly as it is produced and thus accumulates over time. The longer it accumulates, the more fuel is available for severe fires. If a fire starts in a recently burned area, little fuel is available and the fire burns slowly at cooler temperatures and does not spread widely. But if a fire starts where one has not burned for a long time, a hot, dangerous fire is likely.

Consequently, many areas that routinely would burn have not, and fuel accumulations—dry chaparral brush—are unusually large. When a fire starts during dry, windy conditions, it becomes very dangerous and uncontrollable. Its high temperatures make it much more destructive than fires that occur during moist conditions.

Decades of fire-fighting in coastal California and the western United States have not eliminated fires, and have only allowed fuel to accumulate. Normal, "natural" fires, low-temperature and slow-burning, have been eliminated and replaced by infrequent but catastrophic conflagrations. Fire managers now use controlled burns in some areas to reduce fuel accumulations.

Mediterranean shrubland biome. Mediterranean climates promote a distinctive vegetation type called **chaparral.** This is a shrub woodland dominated by hard-leaved trees and shrubs that withstand the severe summer aridity. Fire is common in this area because of summer aridity and flammable waxy leaf coatings. Many plant species in the chaparral are adapted to frequent fire, and some even require fire for their continued presence in the landscape (see Focus box 5–1).

Dryland Biomes

The dryland biomes include two mid-latitude grasslands—tall grass and short grass—and two desert biomes—warm desert and cold desert.

Mid-latitude grassland biomes. In Figure 5–1, extensive mid-latitude grasslands are shown in deep red and brown. Major areas include interior Asia, the Great Plains of Canada and the United States, and central Argentina. The mid-latitude semiarid climate of these areas features hot summers, cold winters, and moderate rainfall. In moister portions, the grass is usually taller (up to 2 meters or 5 to 6 feet) and the biome is called **prairie.** In drier areas, such as Asia, the biome is called *short-grass prairie* or **steppe.**

Grasses are well-suited to this climate because they grow rapidly in the short season when temperature and moisture are favorable (generally spring and early summer). During dry or cold periods, above-ground parts of these plants die back, but the roots become dormant and survive. This trait also allows grasses to survive fire and grow back rapidly, using available moisture at the expense of trees or shrubs that might invade. Many biogeographers believe that occasional fires, either natural or deliberately set by humans, are at least partly responsible for establishing and maintaining extensive grasslands.

The soils of these mid-latitude grasslands have a dark upper horizon with high nutrient content and are very fertile. Because of this fertility, many of the world's mid-latitude grasslands are called "bread baskets" because they are primary production areas for wheat and other cereal grains. In North America, virtually all original *tall-grass prairies* are farmed to produce wheat, corn, soybeans, and other small grains. Much of the *short-grass prairie* present at the time of European settlement remains, although it has been heavily grazed.

Desert biomes. In deserts, moisture is so scarce that large areas of bare ground exist and the sparse vegetation is entirely adapted to moisture stress. Some desert plants are drought-tolerant varieties of those in more humid areas, such as grasses. Others, such as cacti, are almost exclusive to deserts. Most desert plants are adapted to limit evaporative losses, with thick, wax-coated leaves or the replacement of leaves with needles, as in cacti.

Another common adaptation is water-collecting root systems. These are extensive shallow roots to gather occasional rainwater or long tap roots to reach deep groundwater. Many desert plants survive long periods without moisture by means of *dormancy,* either as a mature plant or as seed. When occasional rains occur, these dormant plants spring to life and flower in a few weeks before becoming dormant again. Animal life in the desert is widely dispersed because of the reduced plant production, yet it is diverse. Many desert animals are nocturnal, foraging and hunting at night to avoid dehydration during the hot daytime hours.

Scant moisture and consequent slow chemical activity leave many desert soils poorly developed. Leaching of soluble nutrients is limited, and usually the horizons in desert soils are developed through vertical movements of soluble minerals, controlled by rainfall and evaporation. When water enters the soil, it may dissolve some minerals, but as evaporation draws this water back to the soil surface, these minerals are redeposited. As a result, high concentrations of soluble minerals near the surface (especially salts) are common. Desert soils may be fertile if irrigated, but irrigation usually introduces more minerals to the soil, and the soil can become salty (salinization).

Tundra Biomes

The two major tundra biomes are arctic tundra and alpine tundra. In cold, high-latitude environments or in high mountains above the treeline, freezing and short growing seasons severely limit plant growth. During winter, water is locked up in the soil as ice. Dehydration and abrasion by blowing snow damage exposed upper parts of plants. The

tundra thus is a high-latitude, high-altitude, often snow-covered biome, dominated by low, slow-growing herbaceous plants and a few low shrubs. These survive the cold by lying dormant below the wind, often buried in snow, growing only in the short, cool summer. Animal life in this biome is very limited in winter, but in summer the tundra comes alive with insects, migratory birds, and grazers such as reindeer and caribou.

As in deserts, tundra soils often are poorly developed because of slow chemical activity. Permafrost—soil that does not thaw, even in the summer—occurs where the mean annual temperature is below 0°C (32°F). Soil horizons are weakly developed and often disrupted by frost action. In poorly drained areas, organic matter often accumulates rather than decaying, and *peat* is formed. Peat is an accumulation of plant debris, and a precursor of coal.

Because of slow plant growth in this environment, recovery from disturbances can be very slow. This, combined with the disruption of permafrost caused by construction activities and vegetation disturbance, has led to concern about the long-term effects of oil exploration and related activities in Alaska, Siberia, and similar areas.

(Left) Tall grass prairie in the Flint Hills of Kansas. (Richard Parker/Photo Researchers, Inc.)
(Right) Today, most of the tall grass prairie in Kansas has been converted to row crops.
(Russell Munson/The Stock Market)

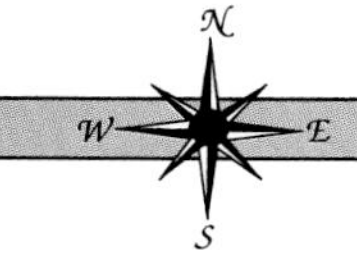

CONCLUSIONS: CRITICAL ISSUES FOR THE FUTURE

Plants, soils, climate, and human activity—each reflects the influence of the other. Overall, climate is the strongest control: climate controls soil formation and plant growth so strongly that, under natural conditions, world vegetation and soil patterns correspond closely to world climates. Locally, topography and geology may be stronger controls over soil development than climate, creating complex vegetation patterns.

Regionally and globally, climate in turn is affected by ecological processes and soil-vegetation linkages:

- Plants regulate evapotranspiration from land surfaces, and thus moderate the hydrologic cycle. If vegetation is cut back, reducing the number of plants available to draw water from the soil and transpire it to the atmosphere, moisture flow to the atmosphere may be reduced, and the excess water runs off to the sea as stream flow. Thus, vegetation can modify climate by affecting humidity.
- Soil eroded by wind from semiarid and arid areas may cloud the atmosphere with dust and modify energy exchanges there, thus modifying climate.
- The biosphere stores and regulates carbon, helping to control the global carbon budget. If vegetation grows (absorbing atmospheric CO_2) faster than organic matter can decay (returning CO_2 to the air), there is net storage of carbon in biomass. If the reverse is true, there is net release of carbon as CO_2 into the atmosphere. Thus, vegetation partly controls the level of an important greenhouse gas—CO_2—in the atmosphere.

These intimate relations among the soil, water, air, and living things lead some scientists to view Earth as a single interconnected dynamic system—almost a functioning organism—rather than just a set of cause-and-effect linkages. We can see all of these subsystems interacting with each other in much the same way that different parts of a single living organism interact. In fact, the **Gaia hypothesis** (*Gaia* is Greek for Earth) argues that Earth is best seen as a single self-regulating "organism."

Whether or not Earth is an organism, it certainly has been disturbed by another one, humankind. Beginning thousands of years ago, but especially within the past two centuries, we have disrupted ecosystems worldwide, largely through population increase and resource demand. By disturbing environments, we have unbalanced them, and they probably now are much more dynamic than they are naturally, laboring to restore their own equilibrium. We have increased biomass stores here and reduced them there; we have introduced species where they did not exist; and we have changed the sediment and nutrient flows of rivers and lakes, forcing them to adjust.

Only a few thousand years ago, Earth was regulated by nonhuman processes, and people were only negligible players, analogous to a few fleas on an elephant. Today, we are major participants in Earth's environmental processes, sometimes even dominant. Environments in which human impacts are minor are increasingly scarce, and the future of Earth's environmental systems depends on how we manipulate biological and physical processes at the surface of Earth.

The scientific study of human interactions with the biosphere is still young, and there is much to be learned. Until recently most studies of human-environment relations have been carried out at relatively local scales. But today we recognize that human impacts are truly global, and we need to increase our understanding of Earth's systems at that scale.

Chapter Summary

1. The biosphere

Photosynthesis is the basis of food chains and ecological systems. Plants compete for water, sunlight, and nutrients, and plants that are best adapted to compete for limiting factors will dominate a given environment. Such adaptations help explain the world distribution of major vegetation types. Disturbances modify ecological communities, and succession following disturbance is a fundamental process of landscape change. Very large portions of the world's land surface have been modified by human activity, and humans are major players in most of the world's ecosystems.

2. The hydrosphere

Climate is the dominant control on local environmental processes, primarily through its influence on water availability and movement. Because the water budget is a critical regulator of ecosystem activity, knowledge of the water budget is essential to understanding both physical and biological processes. This influence extends to human use of the landscape, through determining the natural vegetation cover to agricultural potential.

3. Carbon and oxygen flows in the biosphere

Carbon is the basis of life on Earth. Flows of carbon among the atmosphere, oceans, biosphere, and soil are regulated by biologic processes. Photosynthesis transfers carbon from the atmosphere to the biosphere, and respiration returns it to the atmosphere. Net productivity is highest in warm, humid climates where water and energy are plentiful. Productivity is less in dry or cool climates. Large amounts of carbon are exchanged between the atmosphere and the oceans, and the oceans are a major sink for carbon dioxide added to the atmosphere by fossil fuel combustion.

4. Soil

The outermost layer of Earth's surface in land areas is soil, a mixture of mineral and organic matter formed by physical, chemical, and biological processes. Climate is a major regulator of soil development, through its control on water movement. Plant and animal activity in the soil produces organic chemicals and mixes the upper soil layers. In humid regions, water moves downward through the soil, carrying dissolved substances lower in the profile or removing them altogether. In arid regions, these substances are not so easily removed. Soil characteristics reflect these processes, and there is close correspondence between the world climate map and the world soil map.

5. Biomes: Global Patterns in the biosphere

The world vegetation map closely mirrors the world climate map. Ecologically diverse and complex forests occupy humid environments, storing most nutrients in their biomass. In arid and semiarid regions, sparse vegetation is adapted to moisture stress. Forests adapted to winter cold are in humid mid-latitude climates, with broadleaf forests in warmer areas and coniferous forests in subarctic latitudes. In cold, high-latitude climates, cold-tolerant short vegetation occupies areas that have a mild summer season. Vegetation is absent in ice-bound polar climates.

Key Terms

Actual evapotranspiration (ACTET), 181
Biodiversity, 177
Biogeochemical cycles, 171
Biomagnification, 175
Biomass, 174
Biome, 172
Carbon cycle, 187
Chaparral, 203
Climax, 177
Community succession, 177
Consumer, 174
Decomposer, 174
Drainage basin, 184
Evaporation, 180
Evapotranspiration, 180
Food chain, 174
Gaia hypothesis, 205
Hydrologic cycle, 180
Infiltration capacity, 182
Net primary productivity, 188
Parent material, 191
Photosynthesis, 174
Prairie, 203
Precipitation, 180
Producer, 174
Respiration, 188
Runoff, 180
Savanna, 201
Soil fertility, 197
Soil horizon, 192
Soil order, 192
Steppe, 203
Transpiration, 180
Trophic level, 174
Tropical rainforest, 199
Tundra, 204
Water budget, 180

Questions for Study and Discussion

1. Diagram the hydrologic cycle, using boxes and arrows. Proportion the box sizes to indicate the relative amounts of water stored at each point in the cycle. Label the arrows linking the boxes with the appropriate hydrologic process (evapotranspiration, condensation, precipitation, runoff, and so on).
2. Sketch a water-budget diagram for the climate where you live, including precipitation, POTET, ACTET, soil moisture use and recharge, runoff, and moisture deficit.
3. What are photosynthesis, respiration, and net primary productivity? How are they related? How do they vary seasonally in relation to temperature and water availability? How do they vary in relation to major climate types?
4. What is soil texture? How does it affect the movement and storage of water in the soil?
5. How are soil-forming processes affected by water availability?
6. For each major climate type described in Chapter 3, describe the major vegetation type generally associated with that climate.

Thinking Geographically

1. In your library, locate flow data for a river near you. (Look for a book entitled *Water Resources Data for [your state],* published by the U.S. Geological Survey). Make a bar graph showing mean monthly flow. Explain the variations in river flow in terms of seasonal variations in precipitation and evapotranspiration.
2. In your library, find the Soil Survey for a place near you where excavation is occurring, such as for a highway or building foundation. Discover the soil characteristics in the area. Then visit the excavation site and view the soils in the field, comparing what you see with the descriptions in the soil survey.
3. Consider a typical water budget for a mid-latitude continental site in which soil moisture use is important for part of the year. How is ACTET affected by the amount of water stored in the soil, and the rooting depth of the vegetation?
4. World maps of climate and vegetation have many similarities. If climate were to change significantly, do you think the vegetation map would also change? Why or why not?
5. How has the vegetation cover where you live changed in the past 200 years? How do you think this change may have affected the local water budget?

Suggestions for Further Reading

Abrams, Marc D. "Fire and the Development of Oak Forests." *Bioscience* 42 (May 1992): 346–53.

Arno, Stephen F., and J. K. Brown. "Managing Fire in Our Forests: Time for a New Initiative." *Journal of Forestry* 87 (December 1989): 44–46.

Bahre, Conrad J., and M. L. Shelton. "Historic Vegetation Change, Mesquite Increases, and Climate in Southeastern Arizona." *Journal of Biogeography* 20 (September 1993): 489–504.

Barrett, Linda R., and R. J. Schaetzl "Soil Development and Spatial Variability on Geomorphic Surfaces of Different Age." *Physical Geography* 14 (January–February 1993): 39–55.

Beatty, Susan W. "Mass Movement Effects on Grassland Vegetation and Soils on Santa Cruz Island, California." *Annals of the Association of American Geographers* 78 (September 1988): 491–504.

Brandon, Katrina E. "The Principles and Practice of Buffer Zones and Local Participation in Biodiversity Conservation." *Ambio* 22 (May 1993): 157–62.

Briggs, John C. *Biogeography and Plate Tectonics*. Amsterdam: Elsevier, 1987.

Brown, Dwight A. "Early Nineteenth-Century Grasslands in the Midcontinent Plains." *Annals of the Association of American Geographers* 83 (December 1993): 589–612.

Burnett, Howard. "In Hugo's Wake." *American Forests* 96 (January/February 1990): 17–20.

Chahine, Moustafa T. "The Hydrological Cycle and its Influence on Climate." *Nature* 359 (October 1, 1992): 373–80.

Cook, Edward R., and A. H. Johnson. "Climate Change and Forest Decline: A Review of the Red Spruce Case." *Air, Water and Soil Pollution* 48 (November 1989): 127–40.

Corlett, Richard T. "The Ecological Transformation of Singapore, 1819–1990. *Journal of Biogeography* 19 (July 1992): 411–20.

Cowell, C. Mark. "Environmental Gradients in Secondary Forests of the Georgia Piedmont, U.S.A." *Journal of Biogeography* 20 (March 1993): 199–208.

Cox, C. Barry. *Biogeography: An Ecological and Evolutionary Approach*. 5th ed. Oxford: Blackwell, 1993.

Curley, Anne, and Randy Urich. "The Flood of '93: An Ecological Perspective." *Journal of Forestry* 91 (September 1993): 28–30.

Flannagan, M. D., and C. E. Van Wagner. "Climate Change and Wildfire in Canada." *Canadian Journal of Forest Research* 21 (1991, no. 1): 66–72.

Gillis, Anna M. "Why Can't We Balance the Globe's Carbon Budget?" *BioScience* 41 (July/August 1991): 442–47.

Goldblum, David, and T. T. Veblen. "Fire History of a Ponderosa Pine/Douglas Fir Forest in the Colorado Front Range. *Physical Geography* 13 (April–June 1992): 133–48.

Haines-Young, Roy. "Biogeography." *Progress in Physical Geography* 15 (1991, no. 1): 101–13.

Hartley, Suzanne, and S. L. Dingman. "Effects of Climatic Variability on Winter-Spring Runoff in New England River Basins." *Physical Geography* 14 (July–August 1993): 379–403.

Houghton, R. A., R. D. Boone, J. R. Fruci, J. E. Hobbie, J. M. Melillo, C. A. Palm, B. J. Peterson, G. R. Shaver, G. M. Woodwell, B. Moore, D. L. Skole, and N. Myers. "The Flux of Carbon From Terrestrial Ecosystems to the Atmosphere in 1980 Due to Changes in Land Use: Geographic Distribution of the Global Flux." *Tellus* 39B (1987): 122–39.

Kupfer, John A., and G. P. Malanson. "Structure and Composition of a Riparian Forest Edge." *Physical Geography* 14 (March–April 1993): 154–70.

MacArthur, Robert W. *Geographical Ecology*. New York: Harper & Row, 1972.

McLean, Herbert E. "Paradise Burning: How to Live with Wildfire." *American Forests* 98 (January–February 1992): 22–26.

Meentemeyer, Vernon. "The Geography of Organic Decomposition Rates." *Annals of the Association of American Geographers* 74 (December 1984): 551–60.

Merrens, Edward J., and D. R. Peart. "Effects of Hurricane Damage on Individual Growth and Stand Structure in a Hardwood Forest in New Hampshire, USA." *Journal of Ecology* 80 (No. 4, 1992): 787–95.

Milly, P. C. D., and K. A. Dunne. "Sensitivity of the Global Water Cycle to the Water-Holding Capacity of the Land." *Journal of Climate* 7 (April 1994): 506–37.

Minnich, Richard A. "Fire Mosaics in Southern California and Northern Baja California." *Science* 219 (1983): 1,287–94.

Myers, Norman. "Tropical Forests: The Main Deforestation Fronts." *Environmental Conservation* 20 (Spring 1993): 9–16.

Paoletti, M. G., D. Pimentel, B. R. Stinner, and D. Stinner. "Agroecosystem Biodiversity: Matching Production and Conservation Biology. *Agricultural Ecosystems and Environment* 40 (May 1992): 3–23.

Parsons, Peter A. "Conservation and Global Warming: A Problem in Biological Adaptation to Stress." *Ambio* 18 (no. 6, 1989): 322–35.

Phillips, Jonathan D. "Biophysical Feedbacks and the Risks of Desertification." *Annals of the Association of American Geographers* 83 (December 1993): 630–40.

Pimm, Stuart L., and J. L. Gittlemen. "Biological Diversity: Where Is It"? *Science* 255 (February 21, 1992): 940.

Post, Wilfrid. M., T-H. Peng, W. R. Emanual, A. W. King, V. H. Dale, and D. L. DeAngeles. "The Global Carbon Cycle." *American Scientist* 78 (July–August 1990): 310–26.

Prentice, I. Colin, W. Cramer, S. P. Harrison, R. Leemans, R. A. Monserud, and A. M. Solomon. "A Global Biome Model Based on Plant Physiology and Dominance, Soil Properties and Climate." *Journal of Biogeography* 19 (March 1992): 117–34.

Reid, Walter V., and M. C. Trexler. *Drowning the National Heritage: Climate Change and U.S. Coastal Biodiversity*. Washington: World Resources Institute, 1991.

Robinson, George R., R. D. Holt, M. S. Gaines, S. P. Hamburg, M. L. Johnson, H. S. Fitch, and E. A. Martinko. "Diverse and Contrasting Effects of Habitat Fragmentation." *Science* 257 (July 24, 1992): 524–26.

Romme, William H., and D. G. Despain. "The Yellowstone Fires." *Scientific American* 261 (November 1989): 36–45.

Savage, Melissa, M. Reed, and T. T. Veblen. "Diversity and Disturbance in a Colorado Subalpine Forest." *Physical Geography* 13 (July–September 1992): 240–49.

Schaetzl, Randall J., and S. A. Isard. The Distribution of Spodosol Soils in Southern Michigan: A Climatic Interpretation." *Annals of the Association of American Geographers* 81 (September 1991): 425–42.

Schlesinger, William H., J. F. Reynolds, G. L. Cunningham, L. F. Huenneke, W. M. Jarrell, R. A. Virginia, and W. G. Whitford. "Biological Feedbacks in Global Desertification." *Science* 247 (March 2, 1990): 1,043–48.

Sedjo, Roger A. "Climate, Forests and Fire: A North American Perspective." *Environment International* 17 (1991, No. 2–3): 163–68.

Simmons, Ian G. *Biogeographical Processes*. London: Allen and Unwin, 1982.

Sioli, Harald. "The Effects of Deforestation in Amazonia." *The Ecologist* 17 (No. 4/5, 1987): 134–38.

Stickney, Peter F. "Early Development of Vegetation Following Holocaustic Fire in Northern Rocky Mountain Forests." *Northwest Science* 64 (1990, No. 5): 243–46.

Tchebakova, Nadja M., R. A. Monserud, R. Leemans, and S. Golovanov. "A Global Vegetation Model Based on the Climatological Approach of Budyko." *Journal of Biogeography* 20 (March 1993): 129–44.

Thompson, Stephen A. "Simulation of Climate Change Impacts on Water Balance in the Central United States." *Physical Geography* 13 (January–March 1992): 31–52.

Turner, Billie Lee. *Earth as Transformed by Human Action*. New York: Cambridge University Press, 1990.

Vitousek, Peter M., P. R. Ehrlich, A. H. Ehrlich, and P. A. Matson. "Human Appropriation of the Products of Photosynthesis." *BioScience* 36 (June 1986): 368–73.

Yool, Stephen R., D. W. Eckhardt, J. E. Estes, and M. J. Cosentino. "Describing the Brushfire Hazard in Southern California." *Annals of the Association of American Geographers* 75 (September 1985): 417–30.

We also recommend consulting these journals: *Agriculture, Ecosystems & Environment; American Scientist; Conservation Biology; Ecological Monographs; Ecology; Geografiska Annaler Series B; Journal of Biogeography; Journal of Ecology; Journal of Forestry; Landscape Ecology; Paleogeography, Paleoclimatology, Paleoecology; Progress in Physical Geography; Sierra.*

6
Natural Resources

- What is a natural resource?
- Mineral and energy resources
- Air and water resources
- Natural resources and land use

Soybeans, corn, and oil in Michigan.

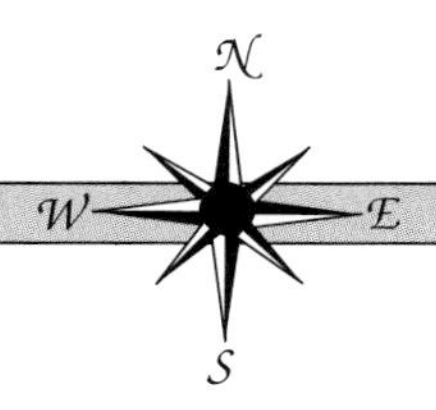

1. What is a natural resource?
A natural resource is something that is useful to people. Cultural, technological, and economic factors determine whether something is a resource.

2. Mineral and energy resources
Mineral resources include metals and nonmetals. Principal energy resources come from three fossil fuels—oil, coal, and natural gas.

3. Air and water resources
Pollution results when a substance is discharged into the air or water faster than it can be dispersed or removed by natural processes. Recycling and pollution prevention are growing in importance as methods for solving pollution problems.

4. Natural resources and land use
People often must balance competing uses of natural resources. The use of a resource for one purpose may be incompatible with its use for another purpose. The use of a natural resource may have a positive impact on some environmental systems and a negative impact on others.

INTRODUCTION

Everything we consume is both extracted from our planet and returned to it. As we use Earth's resources of air, water, minerals, energy, plants, and animals, we simultaneously discharge our waste back into the environment. As Earth's human population of nearly 6 billion increases to 10 billion in the twenty-first century, consumption of resources will increase. This expanded population will place tremendous stress on Earth's remaining resources, and the ability of the planet's air, water, and land to accommodate human waste.

Consumption and waste vary among cultures, and over time. For example, different countries at different times have obtained energy from fuelwood, coal, petroleum, natural gas, running water, nuclear fission, wind, and the sun. Issues of natural resource use and environmental quality must be understood in both their physical and human dimensions.

Resource management is exceedingly complex, because each resource varies geographically and physically. Each resource also varies in value, depending on the human factors: culture, technology, beliefs, politics, economics, and style of government. Because many resources are publicly controlled, resource management is a political process.

In this chapter, we explore the factors that affect the value of resources. These factors include the physical characteristics of resources and the natural systems in which they exist, the changing technology of resource use, and human value systems. We then consider how changing resource values influence which resources we use, and how much. Finally, we examine environmental pollution, resource conflicts, and their management.

What Is a Natural Resource?

A **natural resource** is anything that people use and value, that has been created through natural processes. Examples include plants, animals, coal, water, air, land, metals, rock, and energy. Natural resources are especially important to geographers, because they are the specific elements of the atmosphere, biosphere, hydrosphere, and lithosphere with which people interact. Natural resources can be distinguished from **human-made resources,** which are human creations or inventions, such as money, weapons, computers, information, and labor.

We often use natural resources without considering the environmental consequences of doing so. For example, the burning of oil to generate heat or to power an automobile engine can pollute the atmosphere, streams, and groundwater.

Characteristics of Resources

Any substance is merely a part of nature until a society finds a use for it. Consequently, a natural resource is defined by the three elements of a people's culture, as defined in Chapter 1:

- A society's *cultural values* influence its determination that a commodity is desirable and acceptable to use.
- A society's level of *technology* must be high enough to use the resource.
- A society's *economic system* affects whether a resource is affordable and accessible.

Let us apply these three elements of culture to petroleum as an example of a natural resource in North America:

- *Cultural values:* North Americans want to drive private automobiles and enjoy the convenience of petroleum to heat homes and generate electricity.
- *Technology:* Petroleum extraction and use is highly developed, making it the preferred fuel in North American transportation systems, power plants, and other forms of technology.
- *Economic system:* North Americans have the economic system and willingness to pay high

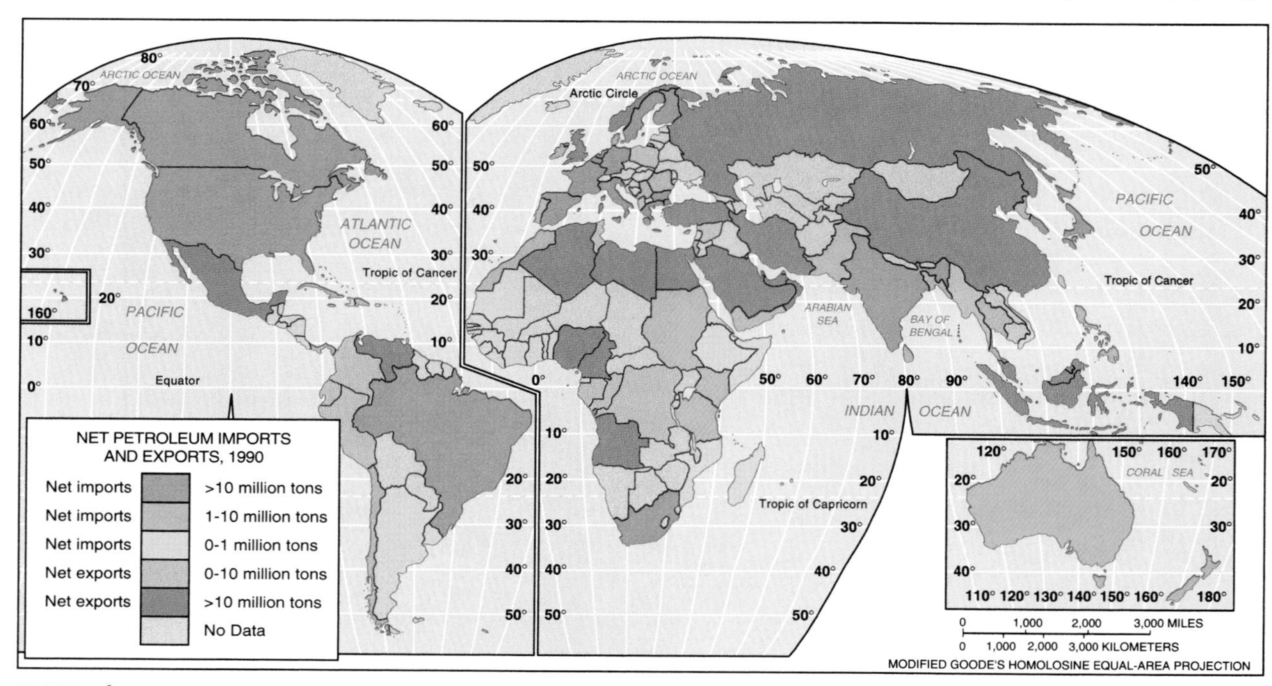

FIGURE 6–1
Net imports and exports of petroleum and petroleum products, 1990, in metric tons. The largest importers are wealthy nations of North America and Europe. Most developing nations are also net importers, although the quantities are much smaller.

enough prices to justify removing petroleum from beneath the land surface and sea floor, and to import it from distant places (Figure 6–1).

The same elements of culture apply to any example of a natural resource: rice to the Japanese, diamonds to South Africans, forests to Brazilians, and air quality to residents of Los Angeles. In every case, a combination of the three factors is necessary for a substance to be valued as a natural resource. Differences among societies in cultural values, level of technology, and economic systems help geographers to understand why a resource may be valuable in one place and ignored elsewhere.

Cultural values and natural resources. To survive, humans need shelter, food, and clothing, and we make use of a variety of resources to meet these needs. We can build homes of grass, wood, mud, stone, or brick. We can eat the meat of cattle, pigs, fish, or mice—or we can consume grains, fruit, and vegetables. We can make clothing from animal skins, cotton, silk, or rayon. Cultural values guide a process of identifying substances as resources to sustain life.

A swamp is a good example of how shifting cultural values can turn an unused feature into a resource. A century ago, swamps were seen in the United States as noxious, humid, buggy places where diseases thrived and agriculture did not. Swamps were valued only as places to dump waste or that could be converted into agricultural land. Eliminating swamps was good, because it removed the breeding ground for mosquitoes while simultaneously creating productive and valuable land.

During the twentieth century, however, cultural values changed in the United States. Scientists and environmentalists began to praise the value of swamps and documented their importance in controlling floods, providing habitat for wildlife, and reducing water pollution. Philosophers increasingly regarded nature as beautiful and praiseworthy. Artists painted wilderness scenes instead of events drawn from the Bible. The result has been a changing public attitude toward swamps, which is reflected in our vocabulary: instead of *swamps,* we now use a more positive term, *wetlands.*

As a consequence of cultural change, wetlands now are a valued land resource, protected by law. We restore damaged wetlands, create new ones,

Wetlands in the Sacramento Valley, California. These wetlands provide wildlife habitat, remove pollutants from water, and help to control floods. (Renee Lynn/Photo Researchers, Inc.)

and restrict activities that might harm them. As described in Chapter 1, in the Netherlands, a long-standing policy of draining wetlands to increase farm land has been reversed: at considerable expense, agricultural lands are being converted back to wetlands to improve water quality and enhance species diversity.

Technology and natural resources. The utility of a natural substance depends on the technological ability of a society to obtain it and to adapt it to their purposes. A metal ore is not a resource if the society lacks knowledge of how to recover its metal content and how to shape the metal into a useful object, such as a tool, structural beam, coin, or automobile fender.

Earth has many substances that we do not use today because we lack the means to extract them or the knowledge of how to use them. Things that might become resources in the near future are called **potential resources.** For example, as a result of its high level of biodiversity, the tropical rainforest is brimming with plants and animals that North Americans regard as potential resources. New medicines, pesticides, and foods might be developed from these substances. To the indigenous peoples of the Amazon rainforest, some of these plants and animals are already resources. Deforestation threatens the availability of these resources to indigenous peoples and others. By destroying the rainforest, we are diminishing Earth's pool of both current and potential resources.

Human need can drive technological advances. Desire to live in cold climates drove people to invent insulated homes and heating technology. The need to increase the supply of food drove people to develop new agricultural technology.

Because human need drives technological advance, new technologies may emerge when a resource becomes scarce. New technology for reusing materials is being developed in part because space in landfills has become a scarce resource, especially in large urban areas of relatively developed countries, where consumption is highest (see Focus Box 6–1). This scarcity is stimulating development of new methods for reusing and recycling materials. Most waste currently is not reusable. But as we deplete the resource of landfill space, we will make more things recyclable and manufacture new products from waste, and waste materials themselves will become resources.

Economics and natural resources. Natural resources acquire a dollar value through exchange in a marketplace. The price of a substance in the marketplace, as well as the quantity that is bought and sold, is determined by supply and demand. Common sense tells us some principles of supply and demand:

- A commodity that requires less labor, machinery, and raw material to produce (bicycle) will sell for less than a commodity that is harder to produce (automobile).
- The greater the supply, the lower the price (corn). The greater the demand, the higher the price (Super Bowl tickets).
- Consumers will pay more for a commodity if they strongly desire it (stereo) than if they have only a moderate desire (textbook).
- If a product's price is low (beer and pizza), consumers will demand more than if the cost is high (champagne and prime rib).

In general, natural resources are produced, allocated, and consumed according to these rules of supply and demand. Water is a good example. In areas where water is plentiful because of high rainfall and low demand—such as northern Minnesota—consumers pay a low price, essentially the cost of pumping it to their homes. But in the arid southwestern United States, water rights must be purchased and scarce water must be carried hundreds of kilometers through pipelines and aqueducts, so prices are much higher.

Uranium. In every society, *cultural values, level of technology,* and *economic system* interact to determine which elements of the physical environment are resources and which are not. Uranium is a good example:

- *Cultural values:* Until the 1930s, uranium was a resource only because its salts made a pretty yellow glaze for pottery. After German physicists realized that great energy stored in uranium atoms might be released by "splitting" their nuclei, uranium gained value as a powerful weapon during World War II.

FOCUS BOX 6-1

Landfills: An Example of Changing Resource Values

Sanitary landfills across the country are building mountains of trash. New York City's Fresh Kills landfill is an extreme example. It contains 60 million cubic meters (80 million cubic yards) of trash, stands taller than the Statue of Liberty, and has a volume about 25 times greater than the Great Pyramid at Giza, Egypt.

Garbage dumps are nothing new, as excavations of ancient cities reveal, but they are growing much more rapidly than in the past. Traditionally, food waste was fed to livestock, iron was used in durable goods and not just thrown away, packaging was scant, and plastic products were unknown. Far fewer goods were manufactured. Our "throwaway culture" is a modern invention.

Garbage dumps have been sited in swamps or other low-value land. For years, garbage simply was dumped and left uncovered. Rubber tires, deliberately burned in landfills to keep fires going, emitted offensive black smoke. Fires smoldered, rats thrived, flies buzzed, and homes situated downwind rarely enjoyed cookouts. Many coastal cities dumped their garbage at sea.

As people became more numerous, so did their garbage dumps. Technical innovations have allowed us to produce goods more cheaply. With increased wealth, we are less likely to hold onto older goods—we discard them instead. Paper consumption has risen steadily, and therefore we produce more paper trash. Products that were once recyclable have been replaced by disposable ones; examples include beverage bottles, hand towels, styrofoam cups, cigarette lighters, and medical supplies. Disposable diapers have become a significant component of solid waste.

To reduce fire, vermin, and odor, cities have converted open dumps to **sanitary landfills,** in which a layer of earth is bulldozed over the garbage each day. They are considered *sanitary* because burying the garbage reduces emissions of gases and odors from the decaying trash, prevents fires, and discourages vermin. Sanitary landfills have indeed reduced some environmental problems, but have aggravated others. Landfilled toxics are leached by groundwater, polluting it. This, plus much-publicized problems like the chemical landfill at Love Canal near Buffalo, New York, have heightened public concern.

As landfills run out of space, new ones are difficult to site because of the NIMBY phenomenon—Not In *My* Back Yard. Stricter regulations have forced some landfills to close and have prevented others from opening. The number of operating landfills in the United States declined from about 30,000 in 1976 to just over 6,000 in 1987. Since that time the few remaining landfills have grown deeper and higher as they accept the waste that formerly went to other sites.

Today landfills are a scarce resource that must be used carefully. To slow the rate of filling, many landfills no longer accept grass clippings and leaves, which can be composted (decomposed harmlessly) at home. Many cities now require residents to separate recyclable waste (newspapers, metals, glass, plastic), and send the remainder to a landfill.

Near Lancaster, Pennsylvania, the reverse process is occurring. The landfill is being dug up, and the wastes are being burned to generate electricity for 15,000 homes. This is encouraging: an industry that disposes of unwanted things also sells a commodity that people need. Of course, digging up the landfill is driven mainly by need for space, rather than the desire to reuse solid waste. But it shows how we can adapt to meet multiple needs.

- *Technology:* The United States and its allies, fearing that Germany might develop a nuclear bomb, conducted a crash program to develop the bomb technology first, known as the Manhattan Project. Germany surrendered before the "Manhattan Project" succeeded, but the technology was used to end the war against Japan.
- *Economic system:* After World War II, nuclear technology was applied to generating electrical power. Nuclear-powered electricity was slow to gain acceptance because it was more expensive than alternative sources, but after Middle East petroleum supplies were threatened during the 1970s, more nuclear power plants were built.
- *Cultural values:* With the construction of more nuclear power plants, the public became increasingly concerned about the risks. Following power plant accidents at Three Mile Island in Pennsylvania (1979) and Chernobyl in Ukraine (1986), government agencies regulated nuclear plant safety much more closely. Higher safety standards increased the cost of nuclear power, at a time when conservation efforts had succeeded in reducing demand for electricity. Fear that plutonium, which is generated by nuclear power plants, might be used by hostile countries to make bombs also dampened support for nuclear power. Orders for new power plants ceased during the 1980s.

Wood is used in house construction in the United States mostly because it is relatively inexpensive and easy to work with. Here, oriented strand board, a manufactured wood product, is being used in new construction. This board is much cheaper than sawn lumber, but has ample strength for covering exterior walls, floors, and roofs. This wood product costs about the same as a foam board made from petroleum that can be used as a substitute on exterior walls. (John Blaustein/Woodfin Camp & Associates)

Substitutability and Renewability

Many natural resources are valued for *specific properties*—coal for the heat it releases when burned, wood for its strength and beauty as a building material, fish as a source of protein, clean water for its healthiness. In most cases, several substances may serve the same purpose, so if one is scarce or expensive another can be substituted. Copper is an excellent conductor of electricity, but is expensive. Wire made of cheaper aluminum can be substituted, or information can be transmitted using light in a fiber-optic cable instead of electrons in a copper wire.

Substitutability. The **substitutability** of one substance for another is important in stabilizing resource prices and limiting problems caused by resource scarcity. If one commodity becomes scarce and expensive, cheaper alternatives usually are found. Such substitution is central to our ability to use resources over extended periods without exhausting them and without declines in our standard of living.

However, many resources have no substitutes. There is only one Old Faithful Geyser and only one species of sperm whale, so if we destroy Old Faithful or force extinction of the sperm whale, we have no substitutes waiting in the wings. Other geysers and whales exist, but they are not the same as those we would have destroyed. The uniqueness of these resources is the essence of their value.

Renewable and nonrenewable resources. To manage Earth's resources, we distinguish between those that are renewable and those that are not:

- **Nonrenewable resources** form so slowly that for practical purposes they cannot be replaced when used. Examples include coal, oil, gas, and ores of uranium, aluminum, lead, copper, and iron.
- **Renewable resources** are replaced continually within a human lifespan. Examples include solar energy, air, wind, water, trees, grain, livestock, and medicines made from plants.

Even a renewable resource can be *depleted,* or consumed to the point where it is no longer economically available. The only resources we cannot deplete are solar energy and its derivatives, wind and flowing water.

Mineral and Energy Resources

Mineral resources are substances that we derive from the lithosphere. They are basic materials that we use to construct roads and buildings, manufacture goods, and power transportation. Without minerals, modern industrial societies could not function. Our use of mineral resources in industry and commerce is governed primarily by technology and economics.

We value most minerals for their properties of strength, malleability (ability to be shaped), weight, and chemical reactivity, rather than for their aesthetic characteristics. Few car owners care if the engine is made from aluminum or iron—what matters is that it is powerful and durable. Few people care whether the roof of their house is made from slate or asphalt shingles, as long as the roof keeps out the rain. Gold has been a rare exception, for it is a mineral valued mostly for its beauty in jewelry. However, even gold is increasingly demanded for the manufacture of electronics.

Because we value a mineral primarily for its mechanical or chemical properties, our use of mineral resources is continually changing as our technology and economy change. As new technological processes and products are invented, demand can suddenly increase for materials that had little use in the past. When these new processes and products replace older ones, demand may diminish for minerals that were important in the past. Changing consumer demand makes one mineral more favored while another becomes less desired.

This section distinguishes between two types of mineral resources: metals and nonmetallic minerals, and energy resources. Metals and nonmetallic minerals are valued primarily for their properties of strength and malleability. Energy sources are valued primarily for their ability to power machines and provide heat.

Mineral Resources

The terms *Stone Age, Iron Age,* and *Bronze Age* capture the importance of particular minerals at various times in the past. Minerals are as essential to civilization as plant, animal, water, and energy resources. Everyday objects made from minerals include cars, buildings, sidewalks, highways, coins, gemstones, staples, weapons, salt, electrical wiring, glass, pipe, cookware, home appliances, light bulbs, computer chips, bicycles, railroad tracks, oil barges, cameras, farm machinery, abrasives, hand tools, paint pigments, wedding rings, and pencil "lead" (Figure 6–2).

Earth has ninety-two natural elements, but most of the crust comprises only eight—oxygen, silicon, aluminum, iron, calcium, sodium, magnesium, and potassium. These elements, as well as rarer ones, combine to form thousands of minerals, each with its own properties and spatial distribution. Each mineral potentially is a resource, if people find a use for it.

Metallic minerals, such as copper, lead, silicon, tin, aluminum, and iron, usually occur in ores. Nonmetallic minerals include building stone, graphite, gemstones, sulfur, slate, and quartz. These minerals must be discovered, mined, transported, refined, and manufactured into useful goods.

Variations in mineral use. Historically, the use of particular metals and nonmetals has fluctuated between periods of high demand and price and periods of low demand and price. Discovery of a new resource could create a "rush" of people to the area of discovery. The "Gold Rushes" to California, Colorado, and Alaska are nineteenth-century examples. The period between 1970 and 1985 featured especially volatile mineral prices, as a

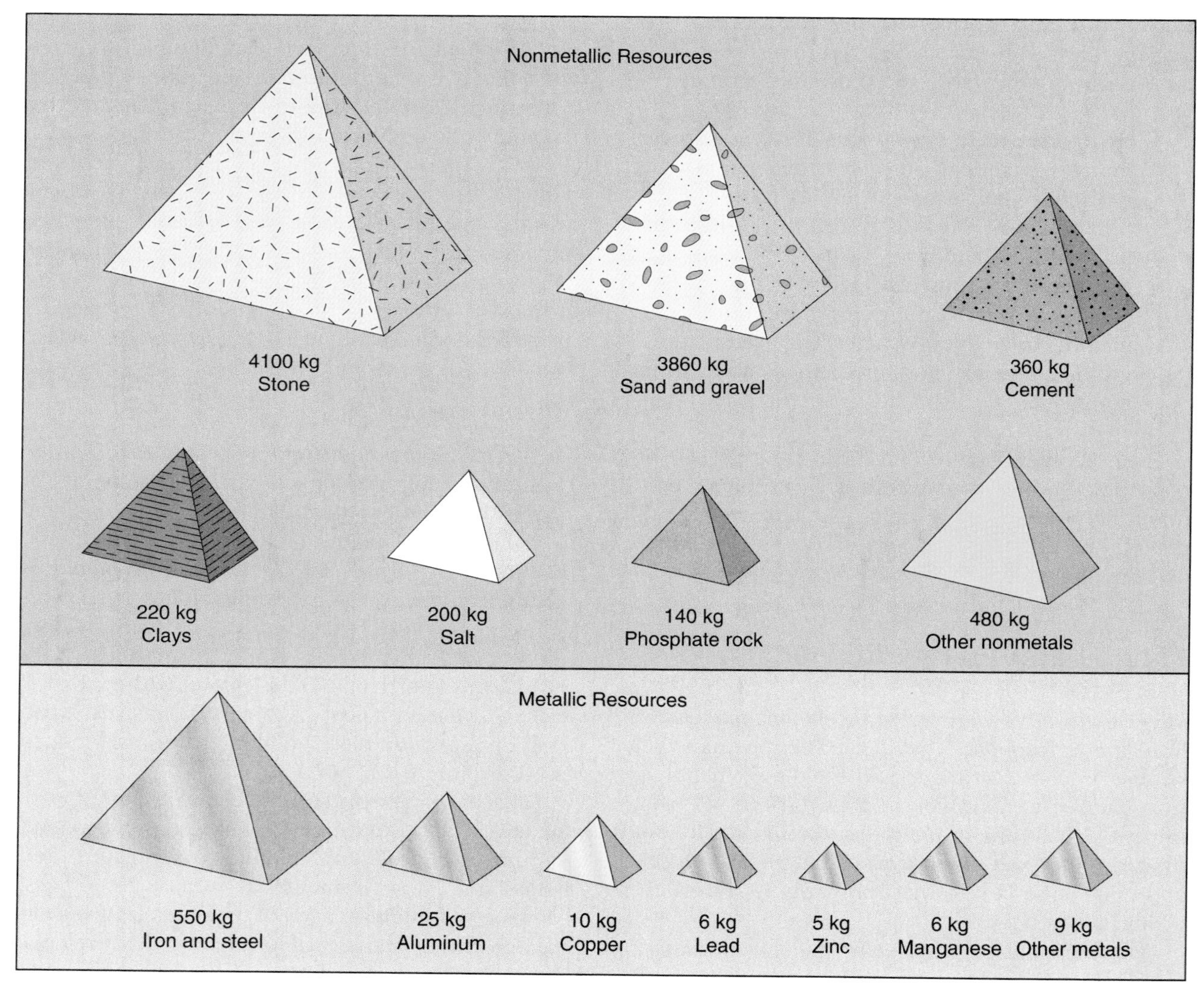

FIGURE 6–2
The annual per capita consumption of non-energy mineral resources in the United States is nearly 10,000 kilograms (10 metric tons or 11 short tons). Over 90 percent is nonmetallic. (From E. J. Tarbuck and F. K. Lutgens, *The Earth,* 4th ed., Macmillan Publishing Co., 1993, after U.S. Bureau of Mines.)

result of rises and declines in industrial output and inflation. For example, the price of copper doubled between 1973 and 1980, then fell nearly 40 percent between 1980 and 1985. Prices of most minerals have been relatively stable since the mid-1980s.

Mineral deposits are not uniformly distributed around the world. A handful of countries produces most of the world's supply of particular minerals. For example, four countries—Australia, Guinea, Jamaica, and Brazil—produce nearly three-fourths of the world's total bauxite (aluminum ore). Russia, Canada, and New Caledonia together produce about two-thirds of the world's nickel. Australia, the United States, China, Canada,and Kazakhstan together produce three-fourths of the world's lead. The United States, Russia, Canada, China, and Australia are especially rich in metallic and nonmetallic mineral resources.

The concentration of mineral resources and production in a few countries favors the establishment of **cartels.** A cartel is a group of countries that agree to control the market by limiting production, in order to drive up prices. During the 1970s, when world demand for minerals was strong, a few cartels were able to control world markets for brief periods. But weak demand, falling prices, and political instability have limited the strength of cartels in recent years. In addition, the United States—the largest consumer in most cases—has accumulated a stockpile of important minerals to protect against short-term reductions in supply caused by high prices, political instability, or hostile foreign governments.

Depletion and substitution. Fluctuations in the price of a mineral as a result of actions by a cartel, a political dispute, or a limited supply rarely continue for a long period. If high prices persist for several decades, technological innovations usually result in substitution of cheaper minerals for more expensive ones. For example, as the price of copper rose rapidly in the 1970s, plumbers began to substitute pipe made of polyvinyl chloride (PVC). Today, PVC has largely replaced copper pipe for plumbing in new buildings.

The constant substitution of one mineral for another has an important consequence: *Even though the world supply of a mineral resource may be limited, we will never actually run out of it.* The reason is that, if the supply of a resource dwindles, its price will rise. The increase in price has three important consequences.

- Demand for the mineral will decrease, slowing its rate of depletion.
- Mining companies will have added incentive to locate and extract new deposits of the mineral, especially poorer deposits that might have been neglected when prices were lower. Recycling becomes more feasible.
- Research to find feasible substitutes will intensify, and as use of the substitute increases, demand for the scarce mineral ceases before the supply is exhausted.

For these reasons, although we will exhaust Earth's remaining supply of lead in eighteen years, zinc in twenty years, and copper in thirty-three years at current rates of use, you should not worry about running out of these materials in your lifetime. On the other hand, if you know of a company that has just invented a product that will replace one of these minerals cheaply, consider investing in it!

Energy Resources

Earth has bountiful and varied sources of renewable energy that humans have been able to harness:

- Solar energy comes from the sun.
- Hydroelectric power and wind power come from natural movements of water and air caused by solar energy and gravity.
- Hydrothermal energy comes from Earth's internal heat in volcanic areas, including California, Iceland, and New Zealand.

However, most of the world's energy comes from chemical energy stored in such substances as wood, coal, oil, natural gas, alcohol, and manure. Energy is released by burning these substances. People do so to heat their homes, run factories, generate electricity, and operate motor vehicles. Burning these substances reduces their supply, for they are nonrenewable. We can continue to burn them for another few decades, but eventually we must switch to other energy resources if we are to maintain our current standard of living.

Energy from fossil fuels. Oil, natural gas, and coal are known as **fossil fuels** because they come from the residue of small plants and animals buried millions of years ago. In essence, fossil fuels are stored solar energy. Through photosynthesis, plants convert solar energy to the chemical energy stored in their tissues. When plants die, this energy remains in their tissues, waiting to be released. The energy may be released promptly if animals or decomposers consume the plants, or if a fire burns the plants in their field or forest. However, the plants may become fossilized as coal, and the stored energy may wait millions of years to be released when the coal is mined and burned.

Oil and gas are also stored solar energy. When sea animals eat plant plankton in the ocean, they store the solar energy incorporated into the plant tissues. Countless sea creatures have died over millions of years, their bodies sinking to the bottom, creating an organic sediment. Over time, this became fossilized in the form of oil, accompanied by natural gas.

When we burn fossil fuels today, we are releasing the solar energy originally stored in plants millions of years ago. Coal, oil, and natural gas were created over millions of years, and still are being created, but the processes are so slow that from a human perspective, fossil fuels are nonrenewable resources: once burned, they are gone as useful sources of energy.

From the time that humans first lived in North America—probably at the end of the last ice age about 18,000 years ago—until the mid-1800s, wood was the most important source of energy. Prior to the arrival of European colonists, indigenous North Americans used wood almost exclusively, but because their total population was small, they did not deplete the resource. But Europeans arrived in North America beginning in the seventeenth century, and they harvested the forests for fuel and lumber and cleared the land for agriculture. By the nineteenth century, most of the forests in the eastern United States had been cut, and fuelwood became very expensive. Similarly, in developing countries today, wood provides a large portion of energy, though supplies are dwindling in many areas.

Coal became a substitute for fuelwood during the nineteenth century. Although large amounts of coal have been consumed, abundant supplies still remain in the United States and several other countries. Another fossil fuel, oil, was a minor resource until the diffusion of motor vehicles early in the 1900s. Today it is the world's most important energy resource. A third fossil fuel, natural gas, once was burned off during oil drilling as a waste product, but in recent years it has become an important energy source. Today, these three fossil fuels provide more than 80 percent of the world's energy and more than 90 percent in relatively developed countries (Figure 6–3).

For U.S. industry, the main energy resource is coal, followed by natural gas and oil. Some businesses directly burn coal in their operations, whereas others rely on electricity generated primarily at coal-burning power plants. At home, we use energy primarily to produce heat and hot water. Natural gas is the most common source for home use, followed by petroleum in the forms of heating oil and kerosene. Nearly all transportation

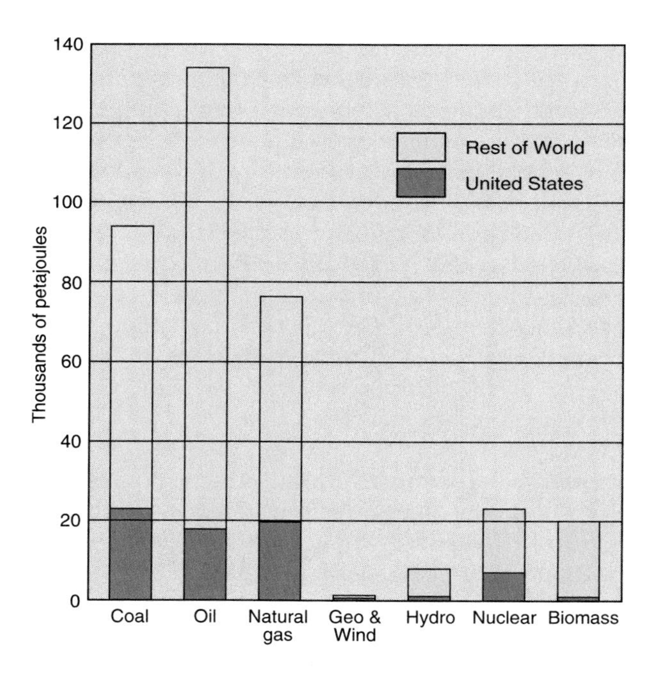

FIGURE 6–3
Energy production in the U.S. and the world. Fossil fuels clearly are the major share of current global energy production; renewable and nuclear energy are relatively minor.

systems operate on petroleum products, including automobiles, trucks, buses, aircraft, and most railroads. Only subways, streetcars, some trains, and a few electric cars run on electricity.

Distribution of fossil fuels. Fossil fuels are not uniformly distributed beneath Earth's surface (petroleum, Figure 6–4; coal, Figure 6–5). Some regions have abundant reserves, whereas others have none. This distribution reflects how fossil fuels are formed in Earth. Coal forms in swampy areas, rich in plants. The lush tropical wetlands of 250 million years ago are today the coal beds of the world, relocated to mid-latitudes by the ponderous movement of Earth's tectonic plates (see Chapter 4). Oil and natural gas form in seafloor sediment, but Earth's tectonic movements eventually elevate some areas of seafloor above sea level to become land. Today we drill for petroleum on both land and the seafloor.

Relatively developed countries—which comprise about one-fourth of the world's population—possess nearly 90 percent of the world's coal and more than 40 percent of its natural gas. The United

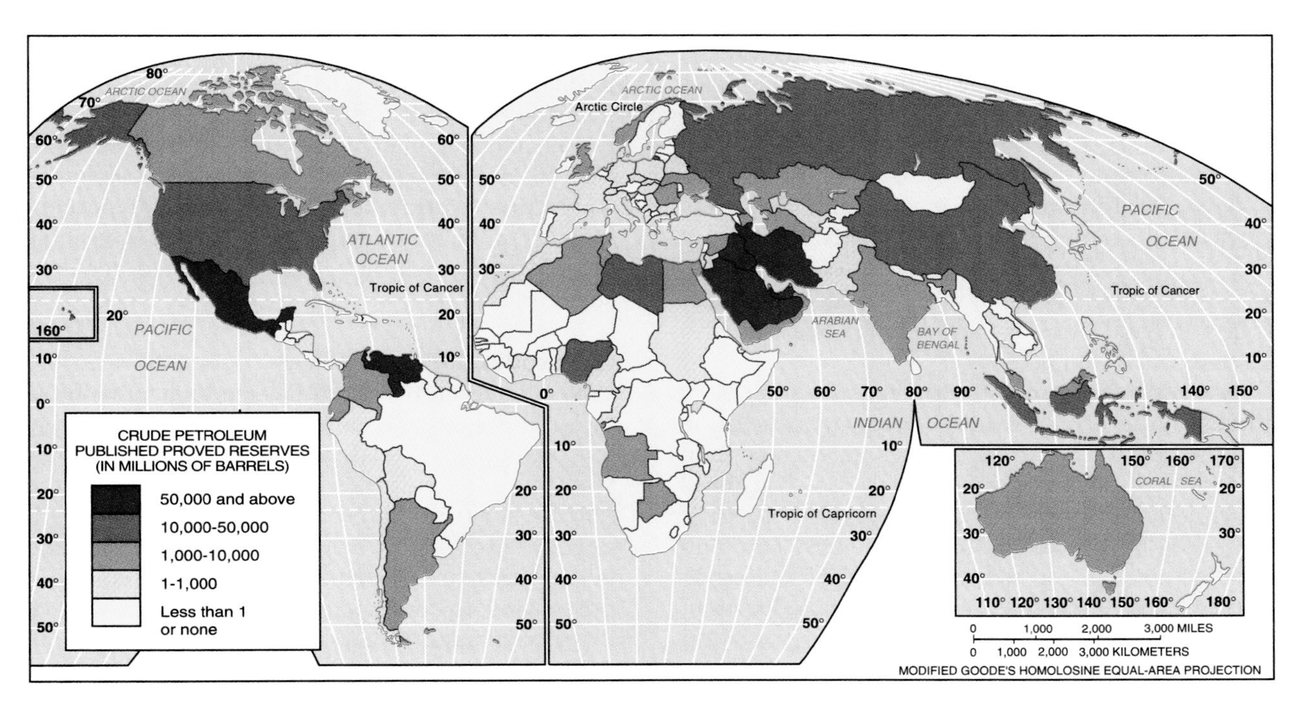

FIGURE 6–4
Substantial petroleum reserves are found in many parts of the world. The ratio of reserves to production is much higher in the middle east than in most other oil-producing regions.

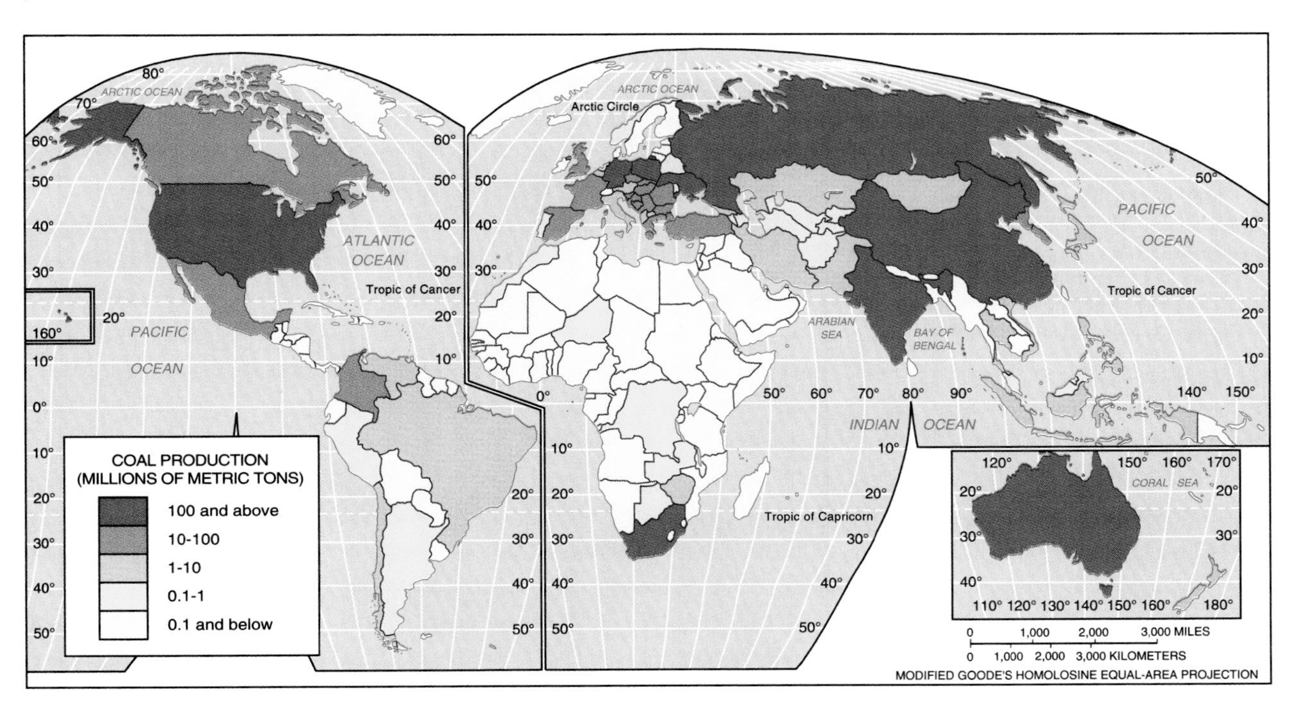

FIGURE 6–5
Coal production is more highly clustered than petroleum production. Four countries—China, the United States, Russia, and Germany—account for nearly two-thirds of all coal production.

States is the leading producer of coal, while Russia is the leading producer of natural gas. China has extensive reserves of coal, but most other countries in Africa, Asia, and Latin America have few reserves of either fuel.

The distribution of oil is different. Relatively developed countries possess only 12 percent of the world's oil reserves. Two-thirds of the world's oil reserves are in the Middle East, including one-fourth in Saudi Arabia. Mexico and Venezuela also have extensive oil fields.

North America and Europe account for nearly three-fourths of the world's energy consumption (Figure 6–6). The United States, with less than five percent of the world's population, consumes nearly one-fourth of the world's energy; Russia is the second largest fossil fuel consumer. The high level of energy consumption supports a lifestyle rich in food, goods, services, comfort, education, and travel.

Because relatively developed countries consume more energy than they produce, they must import energy, especially oil, from developing countries. The United States imports about half of its oil, Western European countries more than half, and Japan more than 90 percent. U.S. dependency on foreign oil began in the 1950s, when oil companies determined that the cost of extracting domestic oil had become higher than foreign sources. U.S. oil imports have increased from 14 percent of total consumption in 1954 to over 50 percent today. European countries and Japan increasingly depend on foreign oil because of limited domestic supplies.

Oil production and prices. Companies in the United States and Western Europe originally drilled for oil in the Middle East and sold it inexpensively to consumers in relatively developed countries. Western companies set oil prices and paid the Middle Eastern governments only a small percentage of their oil profits. To reduce dependency on Western companies, the countries possessing the oil created the Organization of Petroleum Exporting Countries (OPEC) in 1960. OPEC members include eight countries in the Middle East (Algeria, Iran, Iraq, Kuwait, Libya, Qatar, Saudi Arabia, and United Arab Emirates), plus Venezuela in South America, Gabon and Nigeria in Africa, and Indonesia in Asia.

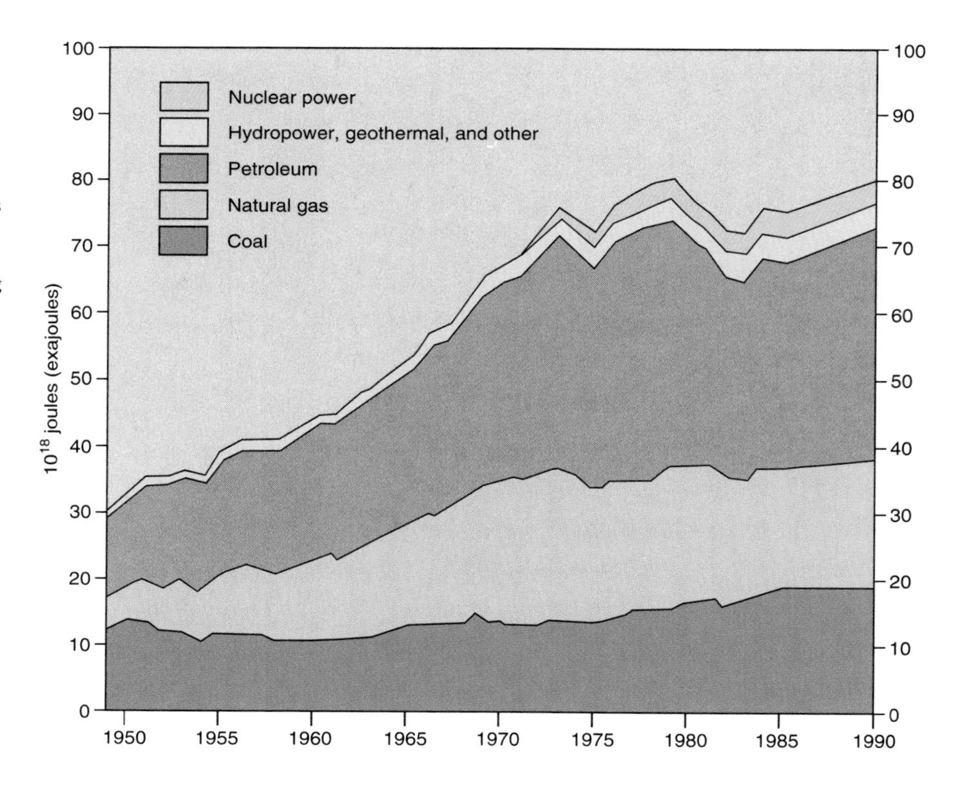

FIGURE 6–6
Fossil fuels provide the lion's share of energy consumed in the United States. Use of oil and natural gas increased rapidly from the 1950s to the 1970s, but growth has been much slower since the energy crises of 1974 and 1979. (From E. J. Tarbuck and F. K. Lutgens, *The Earth,* 4th ed., Macmillan Publishing Co., 1993. Data from Energy Information Administration.)

Angry at the United States and Western Europe for supporting Israel during its 1973 war with the Arab states of Egypt, Jordan, and Syria, OPEC's Arab members organized an oil boycott during the winter of 1973–1974 (see Chapter 9). OPEC states refused to sell oil to countries that had supported Israel. Soon, gasoline supplies dwindled in relatively developed countries, and prices at U.S. gas pumps soared from 30 cents per gallon to more than one dollar.

Each U.S. gasoline station was rationed a small quantity of fuel, which ran out early in the day. Long lines formed at gas stations and some motorists waited all night for fuel. Gasoline was rationed by license plate number; cars with licenses ending in an odd number could buy only on odd-numbered days. Some countries took more drastic action; the Netherlands banned all but emergency motor vehicle travel on Sundays.

OPEC lifted the boycott in 1974 but raised oil prices from $3 per barrel to more than $35 by 1981. To import oil, U.S. consumers spent $3 billion in 1970, but $80 billion in 1980. This rapid price increase caused severe economic problems in relatively developed countries during the 1970s (see Chapter 10).

Developing countries were especially hurt by the price rises. They depended on low-cost oil imports to spur industrial growth, and could not afford the higher oil prices. Relatively developed countries somewhat lessened the impact of higher oil prices on themselves by encouraging OPEC countries to invest some of their wealth in real estate, banks, and other assets in North America and Western Europe. However, poorer countries could not offer this opportunity for reinvesting oil wealth.

To protect against another boycott, the United States has created a strategic petroleum reserve: about one-half billion barrels of oil are stored in several caverns in thick salt beds, located in Texas and Louisiana along the Gulf Coast. This reserve would allow the United States to maintain current levels of oil consumption for about two months if all imports were halted.

The price of oil plummeted from more than $30 to less than $10 per barrel during the 1980s. Internal conflicts weakened OPEC's influence: Iraq and Iran fought an eight-year war, and Libya was isolated by other countries because it supported several terrorist activities.

Petroleum exporting countries then flooded the world market with more oil in an unsuccessful attempt to maintain the same level of revenues they had received during the 1970s. But, as supplies increased, demand in relatively developed countries remained lower than before the boycott. Oil prices briefly rose to about $40 during the 1991 Gulf War, but then quickly dropped back to about $20.

Even if the oil-producing countries of the Middle East remain politically stable in the future, world oil prices may rise because of changes in supply and demand. Demand for oil has risen in recent years in the developed world, while production is now decreasing in many countries, including Russia and the United States.

Future of fossil fuels. How much of the fossil fuels remain? Oil, natural gas, and coal occur beneath Earth's surface, so we cannot see them. Geologists can estimate fairly accurately the *proven reserves* of fossil fuels available in fields that have been explored. But analysts disagree sharply on the potential reserve of fossil fuels—the amount in fields not yet discovered and explored.

If we divide Earth's current proven oil reserves (about 120 billion metric tons) by current annual consumption (about 3 billion metric tons), we get 40 years of oil supply remaining. Rates of consumption will change and new reserves will be discovered, but proven oil reserves will probably last only a few decades. Thus, unless large potential reserves are discovered, Earth's oil reserves will be depleted during the twenty-first century, possibly in your lifetime. (Remember, depleted does not mean completely gone; it means that the commodity has become too scarce and expensive to continue to use.)

Every discovery of new oil deposits extends the life of the resource. But extracting oil is becoming harder, and therefore more expensive. When geologists seek oil, they look first to easily accessible areas where geologic conditions favor accumulation. The largest, most accessible deposits already have been exploited. Newly discovered reserves generally are smaller and more remote, such as beneath the seafloor, where extraction is costly. Exploration cost also has increased because methods are more elaborate and the probability of finding new reserves is less.

Unconventional sources of oil are being studied and developed, such as *oil shale* and *tar sands*

An oil field in China. China has large petroleum reserves, which it can use to fuel economic development into the twenty-first century. (Xinhua/Gamma-Liaison, Inc.)

(sandstones). Oil shale is a "rock that burns" because of its tarlike content. Tar sandstones are saturated with a thick petroleum. The states of Utah, Wyoming, and Colorado contain large reserves of oil in shale, more than ten times the conventional oil reserves found in Saudi Arabia.

Unconventional energy sources such as oil shale are not being used today because no extraction technology is available that is economically feasible and environmentally sound. The cost of conventional oil resources must increase dramatically before unconventional sources could become profitable. By the time prices have risen that high, different sources of energy may be less expensive to obtain than unconventional sources of oil.

Will we deplete fossil fuels and abandon them, or will we gradually switch to other energy sources? Historically, the United States has taken both routes:

- Americans switched to coal when our wood resource grew depleted in the 1800s. In this case, a depleted resource was abandoned for a more abundant one.
- Today, coal, oil, and natural gas coexist as important energy sources. Coal is smoky, polluting, bulky, dirty, expensive to transport, and hard to burn in automobiles, yet oil and natural gas have not replaced it. The plentiful supply of inexpensive coal has kept this resource low in cost and in use in parallel with petroleum. Coal still supplies about a quarter of total U.S. energy consumption.

If we increase our rate of fossil fuel consumption without finding more reserves, we will run out sooner. However, since the 1970s, we have made more efficient use of energy. For example, motor vehicles have lighter engines, fewer metal parts, and more efficient transmissions than before the oil crises of the 1970s. The typical vehicle now travels 12 kilometers on a liter of gasoline (30 miles on a gallon) instead of 6 to 8 kilometers (15 or 20 miles) as in the 1970s.

If progress in efficiency continues, and alternative energy sources are developed, we probably will not deplete Earth's oil reserves. We may be able to continue using oil for those few tasks for which it is best suited. Thus, Earth is not "running out" of oil, but our dependency on this nonrenewable fossil fuel will prove a remarkably short period of human history.

Natural gas and coal: Short-term oil substitutes. In searching for alternatives to oil, we look first to the other two major fossil fuels—natural gas and coal. Natural gas is important in the United States as the clean-burning fuel of choice to heat more than half of the country's homes. A 1.6 million-kilometer (1-million mile) pipeline network efficiently distributes gas from the production areas in Gulf Coast, Oklahoma, and Appalachia to the rest of the country.

Natural gas does not offer much relief from oil, because the global distribution of the two resources is similar. At current rates of consumption, natural gas reserves would last for 60 years, although potential reserves may be greater than for oil. Russia and the Middle East have about two-thirds of the world's natural gas reserves; the United States has only a few years of proven natural gas reserves.

Coal reserves are more abundant than oil or natural gas; at current rates of use, the world has nearly 400 years of proven reserves. Coal can play an especially important role in providing the United States with energy, because the country has a large percentage of the world's proven reserves of coal. But several problems hinder expanded use of coal:

- *Air pollution.* Uncontrolled burning of coal releases sulfur oxides (SO_x), hydrocarbons, carbon dioxide, and particulates into the atmosphere. The sulfur oxides are a major component of acid rain. Many communities suffered from coal-polluted air earlier in this century and encouraged their industries to switch to cleaner-burning natural gas and oil. The Clean Air Act now requires utilities to use low-sulfur coals or to install "scrubbers" on smokestacks. Pittsburgh, Pennsylvania, once noted for severe air pollution caused by the burning of coal in its steel mills and glass factories, today has air relatively clean of particulates and SO_x. Despite this progress, coal-fired power plants still pump large amounts of CO_2 into the atmosphere.
- *Land and water impacts.* Both surface mining and underground mining of coal cause great environmental damage, part of the "hidden" cost of burning this fuel. The damage from surface mining is visible: vegetation, soil, and rock are stripped away to expose the coal. Today, surface-mined land must be restored after mining, but restoration may leave the land less productive than it was before. Underground mining causes surface subsidence, and acidic groundwater may also be released.
- *Economics.* Coal is expensive to mine—as much as $30 per ton. This heavy, bulky rock is expensive to transport long distances. The coal market carries low profit margins, because supply is abundant while demand has not increased.
- *Limited uses.* Coal is a bulky solid, hard to ship through a pipeline and incapable of directly fueling the internal combustion engines used in cars, trucks, and buses. However, coal-generated power plants can provide the power to operate electric vehicles, which will become more common in the next few years, but more than 60 percent of the energy is lost in the conversion of coal to electricity.

Renewable Energy Resources

Where can we turn for energy that is safe, economical, nonpolluting, widely available, and not controlled by a handful of countries? In the long run, we must look to energy sources that are renewable—or at least to resources so abundant that they are in effect renewable. The two most promising energy sources are nuclear and solar. Other alternatives at present include biomass and hydroelectric power. Let us now look at each of these resources.

Nuclear energy. The big advantage of nuclear power is the tremendous energy that is available from a small amount of material. A kilogram of nuclear fuel contains more than 2,000,000 times the energy in a kilogram of coal.

The peaceful use of nuclear reactors to generate electricity began in the 1950s, and today about 400 operate around the world. Nuclear power provides about one-third of all electricity in Europe. Japan (which has virtually no fossil fuels), South Korea, and Taiwan also rely on nuclear-generated electricity (Figure 6–7). Nuclear power generates approximately 20 percent of North American electricity (Figure 6–8). The United States uses less nuclear energy than other developed countries, in part because of its abundant coal reserves.

Like coal, nuclear power presents serious problems. These include potential accidents, generation of radioactive waste, public opposition, and high cost.

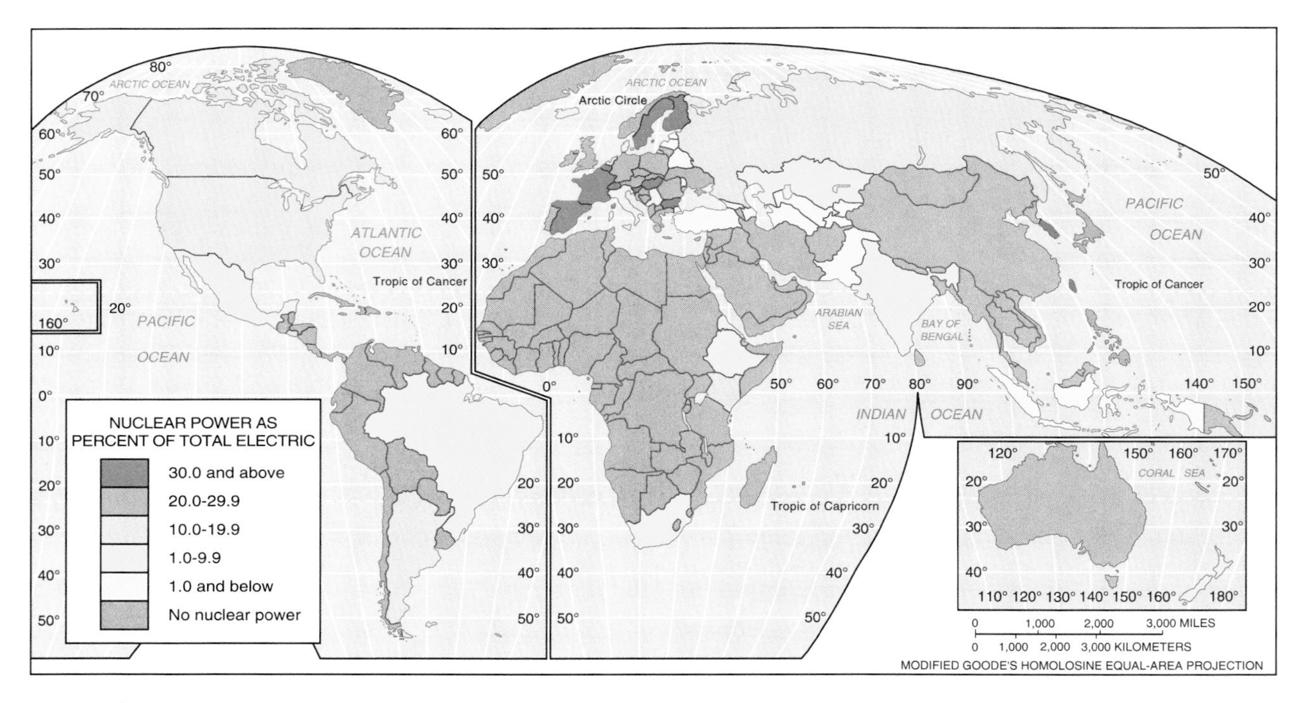

FIGURE 6–7
Nuclear power plants are clustered in more developed regions, including Western Europe, North America, and Japan. Nuclear power has been especially attractive to developed European nations that lack abundant reserves of petroleum or coal.

Potential Accidents—A nuclear power plant cannot explode like a nuclear bomb, because the quantity of uranium in it is too small. It is possible to have a runaway chain reaction, though. The reactor can overheat, causing a meltdown, possible steam explosions, and scattering of radioactive material into the atmosphere.

Such a meltdown occurred in 1986 in a nuclear power plant at Chernobyl, Ukraine, near the border with Belarus (both countries at the time were part of the Soviet Union). Soviet officials reported that the accident caused 31 deaths, including 2 at the accident site and 29 elsewhere who were exposed to severe radiation. Most scientists dismiss these figures as much too low. Hundreds of thousands of residents and cleanup workers were exposed to high levels of radiation. For example, officials in Belarus, where 70 percent of the fallout landed, have reported that, within a decade of the accident, cancer cases increased 45 percent in the districts closest to the plant.

The impact of this accident extended through Europe: most European governments temporarily banned the sale of milk and fresh vegetables, which were contaminated with radioactive fallout. Half of the eventual victims may be residents of other European countries.

In recent years, nuclear power plants constructed by the Soviet Union in Eastern Europe, especially during the 1960s, have reported problems as a result of defective parts and secrecy about operations. At a Soviet-built nuclear power plant in East Germany, eleven of the twelve cooling pumps were knocked out by a fire and a power failure in 1975. Had the twelfth pump failed, the reactor core would have begun an unstoppable meltdown, and about 50,000 inhabitants of the nearby city of Greifswald would have been exposed to high levels of radioactivity. The Soviet Union and East Germany kept this incident secret for fifteen years.

Radioactive Waste—Waste materials in "spent" (exhausted) nuclear fuel remain highly radioactive for thousands of years. Spent fuel can also be reprocessed to extract plutonium for use in nuclear bombs. Pipes, concrete, and water near the fission-

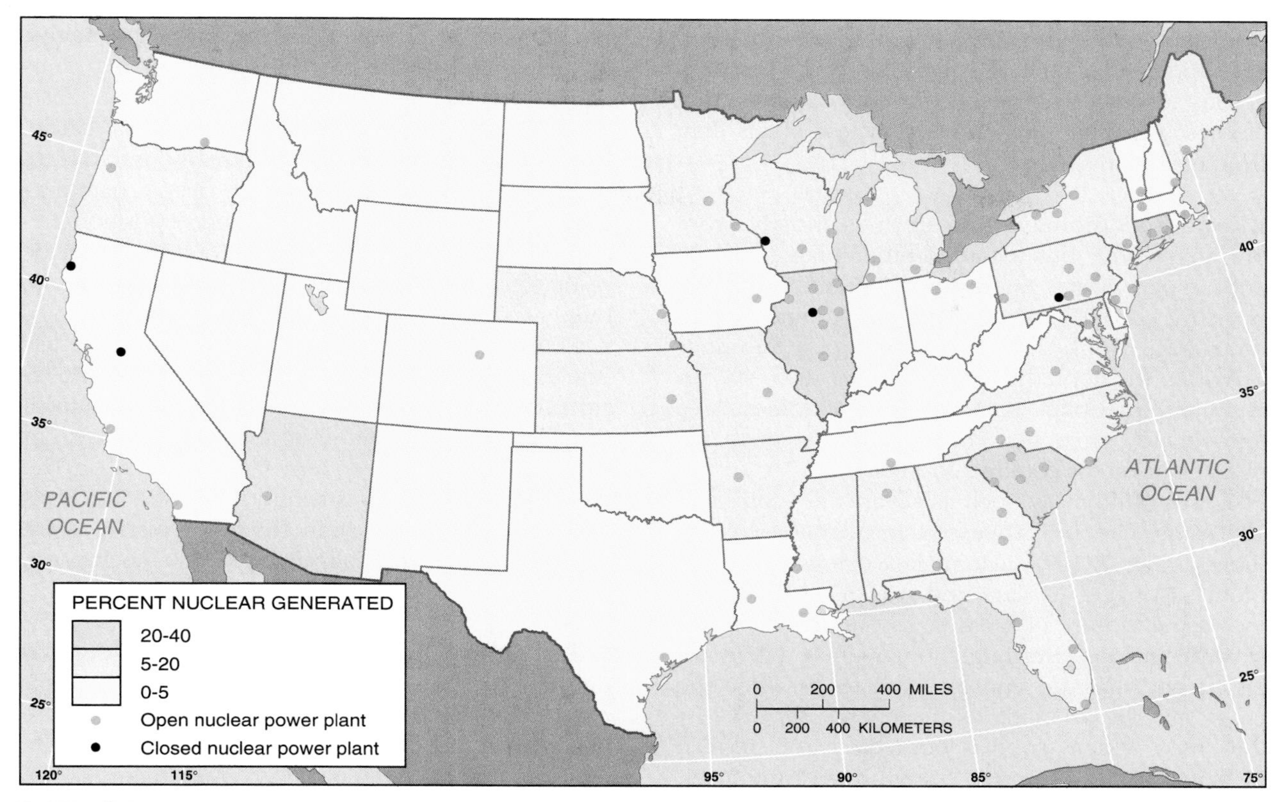

FIGURE 6–8
Percentage of electricity generated by nuclear reactors, by state. Nuclear power is an important source of electricity in several northeastern and midwestern states. Some stations have more than one nuclear reactor.

ing fuel also become "hot" with radioactivity. This waste cannot be burned or chemically treated: it must be isolated for several thousand years until it loses its radioactivity. Currently, spent fuel is stored in cooling tanks at nuclear power plants, but these are nearly full. The United States is Earth's third largest country in land area, yet it has failed to find a suitable underground storage site. Locations in Nevada and New Mexico have been rejected because of risks from volcanic activity and groundwater contamination.

No one has yet devised permanent storage for radioactive waste. Proposals abound: rocketing it into the sun or burying it at sea, in abandoned mines, or in deep layers of rock. But cultural values are critical to selecting the disposal method. No one wants a storage facility near their community, and the time required for radioactive waste to decay to a safe level is far longer than any country has yet existed.

Public opposition—Public concern about safety has been an obstacle to the diffusion of nuclear power since the technology first emerged. The accident at Chernobyl, as well as less damaging incidents in the United States and other countries, has dramatically increased public concern about nuclear power. At Shoreham, Long Island, near New York City, a nuclear power plant was ready to operate and begin generating electricity when, under pressure from worried citizens, the government decided not to allow the plant to open and compensated the electric company for the loss of the plant. In 1992, Italian voters rejected future nuclear power development in that country, and public opposition is similarly strong in Germany and Scandinavia. Even in France, where over 70 percent of electricity is generated from nuclear power, public opposition is a major barrier to new development.

High cost—Nuclear power plants cost several billion dollars to build, primarily because of elaborate safety measures. Without double and triple backup systems, nuclear energy would be too dangerous to use. As a result, the cost of generating electricity is much higher from nuclear plants than from coal plants.

Biomass. Biomass fuel derives mostly from burning wood, but includes other plant material and animal waste. Energy is generated either by burning directly or converting substances to charcoal, alcohol, or methane gas. Biomass provides most of the energy in several African countries. In China, some individual homes have fermentation tanks that convert waste to methane, which is used for cooking and heating.

Forms of biomass, such as sugar cane, corn, and soybeans, can be processed into motor vehicle fuels; Brazil in particular makes extensive use of biomass to fuel its cars and trucks. Potential for increasing the use of biomass for fuel is limited for several reasons. Burning biomass may be inefficient, because the energy used to *produce* the crops may be as much as the energy *supplied* by the crops. When wood is burned for fuel instead of being left in the forest, the fertility of the forest soil may be reduced. But the most important limitation on using biomass for energy is that it already serves other important purposes: providing much of Earth's food, clothing, and shelter.

Hydroelectric power. Flowing water has been a source of mechanical power since before recorded history. In the past, water was used to rotate a wheel; the turning of the wheel could operate machines capable of grinding grain, sawing timber, and pumping water. During the twentieth century, the energy of moving water has been used primarily to generate electricity, called **hydroelectric power.** It supplies about one-fourth of the world's electricity, more than any other source except for coal.

To generate hydroelectric power, water must abruptly change height at a waterfall. The falling water turns turbines which power electrical generators. Usually, the required abrupt change of height is created by damming a river. A hydroelectric plant produces clean, inexpensive electricity, and a reservoir behind the dam can be used for flood control, drinking water, irrigation, and recreation.

Hydroelectric power has several drawbacks, however. Opposition to construction of big dams and reservoirs is strong among environmentalists who fear the environmental damage they cause. Hydroelectric dams may flood formerly usable land. By breaking the natural slope of a river, a dam may disturb its flow, erosion pattern, and deposition pattern (see Chapter 4). Construction of a hydroelectric plant can alter aquatic life. And few good sites to build new dams remain in the United States.

Many good sites for generating hydroelectric power remain outside the United States, but political considerations restrict their use, especially if the river flows through more than one country. For example, Turkey's recently built dam on the Euphrates River was strongly opposed by Syria and Iraq, through which the river also passes. They argue that the dam diverted too much water from the river and increased its salinity.

Solar energy. Solar energy—energy derived directly from the sun—offers the best potential for providing the world's energy needs in future centuries. The sun is a very safe, essentially limitless, and renewable source of energy.

At present, solar energy is used in two principal ways: for thermal energy and photovoltaic electricity production. Solar thermal energy is heat collected directly from sunshine. Collection may be achieved by designing buildings to capture the maximum amount of solar energy. Alternatively, special collectors may be placed near a building or on the roof to gather sunlight. The heat absorbed by these collectors is then carried in water or other liquids to the places where it is needed.

Photovoltaic electric production is a direct conversion of solar energy to electricity in **photovoltaic cells.** Each cell generates a small electric current, but banks of them wired together collectively produce a large amount of electricity. Solar-generated electricity is now used in pocket calculators and in places where conventional power is unavailable, such as spacecraft and remote places on Earth. As more photovoltaic cells are produced, and as the technology and efficiency are improved, the cost of solar power will decline. Photovoltaic cells may become competitive with conventional energy sources in many new residential and commercial installations early in the twenty-first century.

Solar collectors on apartment roofs in Portland, Oregon. Solar energy is economically competitive with conventional heating and electricity in new construction today, and relatively small price changes could make solar energy significantly cheaper than fossil fuels. (John V.A.F. Neal/Photo Researchers, Inc.)

Solar energy can be generated either at a central power station or in individual homes. Many countries are wired for central distribution, so central generation by utilities makes sense. But solar power now makes feasible individual home systems. An installation costing several thousand dollars provides a solar energy system that provides virtually all household heat and electricity. Because the high installation price is offset by low monthly operating costs, home-based solar energy is economical for consumers who remain in the same house for many years. Individual solar energy users do not face rising electric bills from utilities, which pass on their cost of purchasing fossil fuels and constructing facilities.

The United States, Israel, and Japan lead in solar use at home, mostly for heating water. Solar energy may become more attractive as other energy sources become more expensive. One indication of solar energy's bright future is that the major U.S. manufacturers of photovoltaic cells are owned by petroleum companies, which are concerned about the world's limited oil reserves.

Transition to new energy sources. The world offers a variety of energy sources. You have just seen that, in addition to fossil fuels, people use hydroelectric, biomass, solar, and nuclear power. Other technologies are also in commercial use today, including geothermal, wind, and tidal power.

The emergence of new energy technologies suggests that we are beginning the transition from a fossil-based energy system to a new mix of energy supplies. Oil and gas are likely to remain important for several decades. But they will likely diminish in importance relative to other resources, especially renewable alternatives.

At present, the emerging energy sources are less versatile than oil and gas, and are likely to find only specialized uses. Solar thermal energy might be used to heat buildings, whereas photovoltaic cells can power small electrical appliances. Centralized electric power plants—coal, nuclear, or hydroelectric—will probably remain the major source of energy for heavy users, such as large factories and shopping malls.

If history is a reliable guide, the mechanism that will drive this transition to new energy supplies is the market. If the price of oil rises significantly, alternative technologies that are currently uncompetitive would become attractive. As these new technologies become more extensively used, their prices will decrease, further encouraging a shift away from oil.

Perhaps the greatest source of energy reserves will be conservation. More efficient use of energy means that we can produce more goods, operate more motor vehicles, and heat more buildings with the same amount of energy. Further conservation is equivalent to discovering a new energy source.

Industrial growth in China is fueled by coal and, more recently, oil. Until recently, there had been little concern about the environmental consequences of industrial growth, so pollution problems are widespread. (Novosti/Gamma-Liaison, Inc.)

Air and Water Resources

Consumption of resources is half of the resource equation, but waste disposal is the other half. The burning of fossil fuels pumps carbon dioxide into the atmosphere. The sulfur from burning oil and coal also enters the atmosphere and returns as acid deposition. Thus, the geography of resources also is the geography of environmental pollution. In this section, we consider air and water pollution, the consequences of our resource use.

Air and water share two important properties: they are critical to human and other life on Earth, and they are important places to deposit waste. An environmental extremist might argue that we should not discharge *any* waste into air and water, but in practice we can rely on air and water to remove and disperse some waste. Not all human actions harm the environment, for the air and water can accept some waste. When we wash chemicals into a river, the river may dilute them until their concentration grows insignificant. **Pollution** results when more waste is added than a resource can accommodate.

Pollution levels generally are greater where people are concentrated, especially in urban areas. When many people are clustered in a small area, the amount of waste they generate is more likely to exceed the capacity of the air and water to accommodate it.

When we use something, that product cannot disappear from Earth. Natural processes may transport the waste from one part of the environment to another: waste discharged into the air may turn up in a river, and organic waste dumped in a landfill can produce gases that leak to the atmosphere. The following section looks at pollution of air and water.

Air Pollution

Air is our most immediate resource, used with each breath. Because we are immersed for our lifetimes in this invisible mixture of gases, the purity of air is paramount to life on Earth. As we discussed in Chapter 5, the atmosphere is a central part of Earth's biogeochemical cycles.

Air can become polluted through natural processes unrelated to human actions, such as dust, forest fire smoke, and volcanic discharges. Humans add to this natural pollution by discharging into the atmosphere smoke and gas from burning fossil fuels, incinerators, evaporating solvents, and industrial processes.

The atmosphere is constantly stirred by temperature and pressure differences, mixing both vertically and horizontally (see Chapter 2). As air moves from one place to another, it carries various wastes. The more waste we discharge into the atmosphere, and the less the air circulates, the greater is the concentration of pollution.

Average air at the surface contains about 78 percent nitrogen, 21 percent oxygen, and less than 1 percent argon. The remaining 0.04 percent of air's composition includes several *trace gases.* **Air pollution** is a human-caused concentration of trace substances at a greater level than occurs in average air. The most common air pollutants include **carbon monoxide, sulfur dioxides (SO_x), nitrogen oxides (NO_x), hydrocarbons,** and **particulates.** Concentrations of these pollutants in the air can damage property and affect the

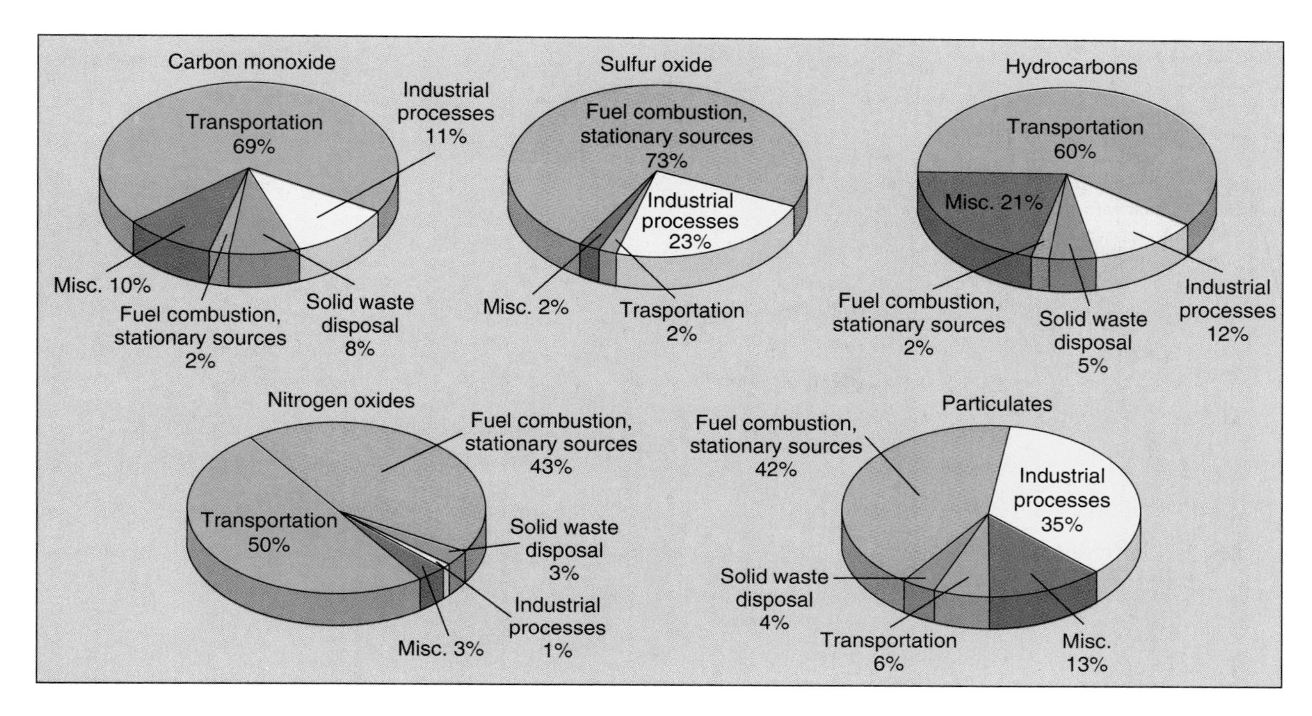

FIGURE 6–9
Major human-caused air pollution sources in the United States. (From R. W. Christopherson, *Geosystems,* 2d ed., Macmillan Publishing Co., 1994.)

health of people, other animals, and plants (Figure 6–9).

Each pollutant entering the atmosphere is affected differently. For example:

- SO_x and NO_x combine with water and fall to Earth as acid precipitation.
- *Particulates* in smoke are quickly cleansed from the atmosphere by gravity and precipitation.
- Smog—the product of *hydrocarbons,* as well as NO_x and sunlight—is created by chemical reactions that occur in the atmosphere itself.
- Chlorofluorocarbons (CFCs) remain in the air long enough to be widely dispersed and carried into the upper atmosphere.

We are polluting our air in many ways. We will now focus on two important air pollution issues: acid deposition and urban air pollution.

Acid deposition. Acid deposition occurs when sulfur oxides and nitrogen oxides, produced mainly from burning fossil fuels, are discharged into the atmosphere. The sulfur and nitrogen oxides combine with oxygen and water in the atmosphere to produce sulfuric acid and nitric acid. When dissolved in water, the acids may fall as *acid precipitation,* or they may be deposited in dust. We use the term **acid deposition** to include both types of pollution (Figure 6–10).

Acid deposition seriously damages lakes and kills fish and plants. But the most severe damage from acid deposition occurs when it makes soil too acidic. High concentrations of acid in the soil may slow plant growth, increasing their susceptibility to disease and ultimately killing them. Some of the acid deposited in soil is neutralized by calcium, magnesium, and other naturally present chemicals, but the amount of acid deposited can exceed the capacity of the soil's chemicals to neutralize the acid. If soil water grows too acidic, plant nutrients are leached away, and thus become unavailable to plants. Acid deposition can dissolve aluminum in the soil, which can be toxic to plants and interfere with their nutrient uptake. Acids may also harm the worms and insects in soil that decompose organic matter.

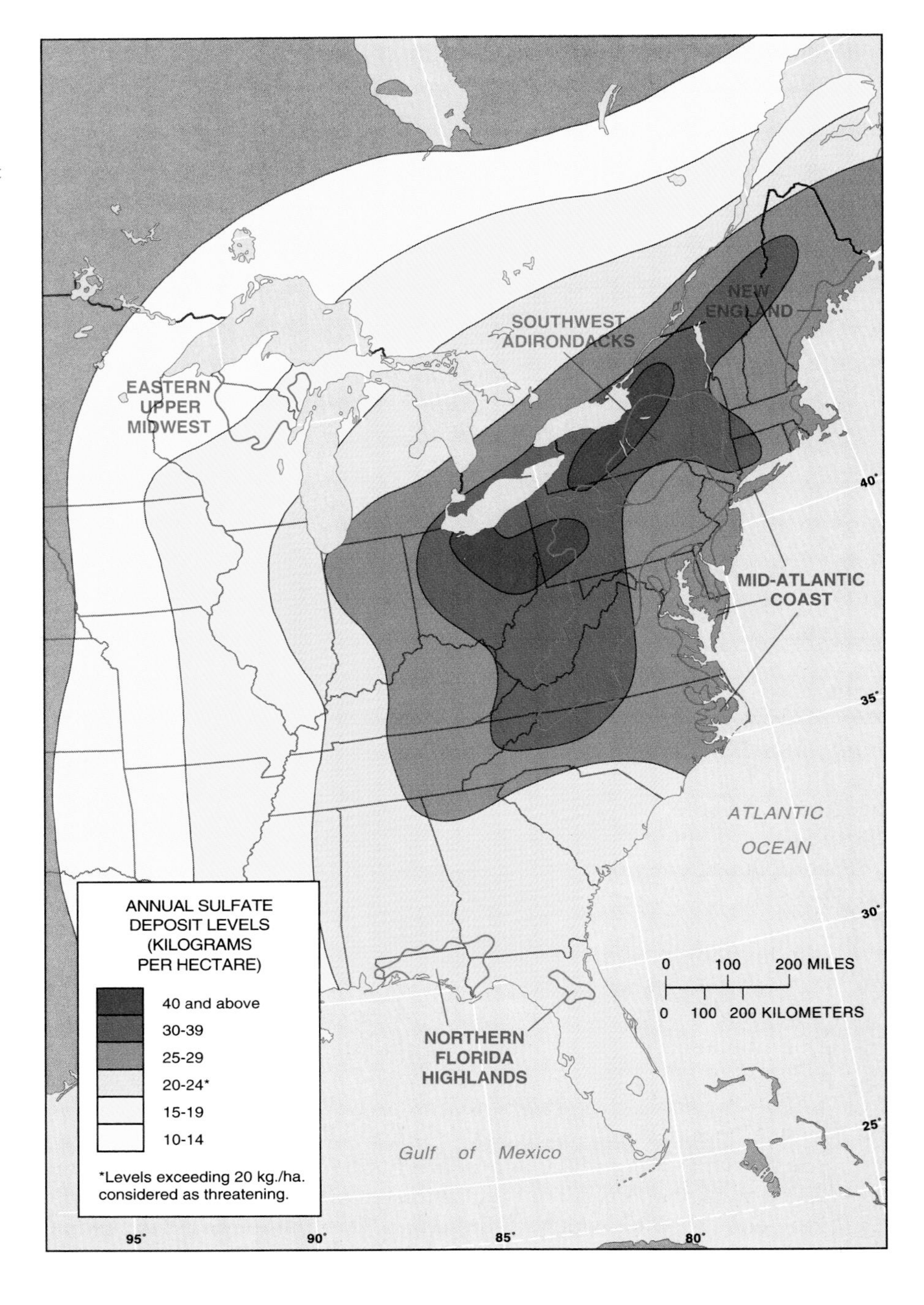

FIGURE 6–10
The most severe effects from acid precipitation do not occur where the pollutants are emitted into the atmosphere. Prevailing westerly winds across North America export the problem eastward to other states and into Canada. Regions of particular sensitivity to acid deposition because of lack of buffering capacity in soils are outlined in red.

Acid deposition is a regional problem, most severe in the densely populated industrial regions in Europe and eastern North America. Acid deposition is the major source of damage to forests, especially in the highly industrialized and densely populated regions of eastern United States and central Europe, where discharges are greatest. Damage to individual forests varies widely, depending on its age, tree species, the buffering capacity of the soil, and interactions between trees and other organisms in the forest. Because the relationship between tree damage and high dis-

Trees in the Black Forest, Germany, suffering from forest decline probably caused by acid deposition. The Black Forest is in the heart of the western European industrial region, and has been a focus of concern about the effects of acid deposition. (Ken Hayman/Woodfin Camp & Associates)

charges of SO_x and NO_x has not been documented precisely, strong efforts have not been made to control emissions. Few governments are willing to impose the cost of controlling emissions on their industries and consumers.

During the past quarter-century, the United States has reduced SO_x emissions about 25 percent. Over this same period, though, emissions have been cut by larger percentages in other relatively developed countries, including 50 percent in Canada, 60 percent in France, 75 percent in Germany, and 86 percent in Sweden.

Despite progress in reducing SO_x emissions in relatively developed countries, acid deposition continues to be a problem. Although precise figures are not available, SO_x emissions have probably increased in developing countries, especially China, which is responsible for burning nearly one-fourth of the world's coal, the major source of SO_x emissions. And NO_x emissions, which are more difficult than SO_x to control, have remained at about the same level in the United States during the past quarter-century.

Urban air pollution. Urban air pollution results when a large volume of emissions are discharged into a small area. The problem is aggravated in cities when the wind cannot disperse them. Urban air pollution has three basic components:

- *Carbon monoxide.* Proper burning in power plants and vehicles produces carbon dioxide (CO_2), but incomplete combustion produces *carbon monoxide (CO).* Breathing CO reduces the oxygen level in blood, impairing vision and alertness, and threatening those with breathing problems.
- *Hydrocarbons* also result from improper fuel combustion, as well as from evaporation of paint solvents. Hydrocarbons and NO_x in the presence of sunlight form **photochemical smog,** which causes respiratory problems and stinging in the eyes.
- *Particulates* include dust and smoke particles. You can see particulates as a dark plume of smoke emitted from a smoke stack or a diesel truck.

Three weather factors are critical to urban air pollution: wind, temperature, and sunlight:

- When the *wind* blows, it disperses pollutants. When it is calm, pollutants build.
- Air *temperature* normally drops rapidly with increasing altitude. But over cities, conditions sometimes cause **temperature inversions,** in which warmer air lies above cooler air. This limits vertical circulation, trapping pollutants near the surface (Figure 6–11).
- *Sunlight* is the catalyst for smog formation.

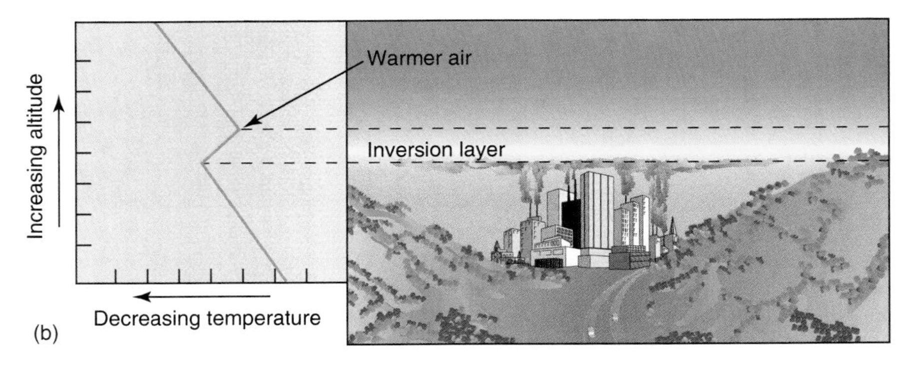

FIGURE 6–11
A normal temperature profile (a) and a temperature inversion (b) above a city. A temperature inversion traps pollutants. (From R. W. Christopherson, *Geosystems,* 2d ed., Macmillan Publishing Co., 1994.)

Mexico City suffers from a combination of circumstances that all lead to significant air pollution problems. The city lies in a mountain basin that limits dispersion of pollutants. Automobiles are numerous, traffic is heavy, and there are few emission controls. In the early 1990s the problem reached crisis proportions, and a pollution-control effort is now underway. (Tom McHugh/Photo Researchers, Inc.)

As a result of these three factors, you should recognize that the worst urban air pollution occurs under a *stationary high-pressure cell:*

- Winds are slight.
- Descending air encourages a *temperature* inversion.
- Clear skies assure that *sunlight* is present.

A city that experiences frequent stationary highs, such as Denver, Colorado, has frequent pollution problems.

Mexico City is notorious for severe air pollution, especially in winter, when high pressure often dominates, and the surrounding mountains discourage dispersal of pollutants by wind. In the eastern United States, pollution problems are worst in summer and autumn, because stationary highs are most common then. In West Coast cities such as Los Angeles and San Francisco, the pollution "season" is also summer and autumn because inversions are more persistent then.

Progress in controlling urban air pollution is mixed. In relatively developed countries with strict regulations, air quality has improved. Reduced use of coal and improvements in automobile engines, manufacturing processes, and generation of electricity have all contributed to higher-quality urban air.

To reduce auto emissions in the United States, for example, catalytic converters have been attached to exhaust systems since the 1970s. As a result, carbon monoxide emissions have declined by more than three-fourths and nitrogen oxide and hydrocarbon emissions by more than 95 percent. Gains in relatively developed countries have been offset somewhat, though, by increased use of cars and trucks in recent years.

In many developing countries, urban air pollution is getting worse. Although ownership of motor vehicles is less common than in wealthy countries, the cars and trucks are older and lack pollution controls found on newer vehicles in relatively developed countries. Instead of relying on gas or electricity, many urban residents in developing countries burn wood, coal, and dung for cooking and heating. These smoky fires can create acute air pollution problems in poorly ventilated areas, even in the country. In developing countries, an estimated 4 million children die each year from acute respiratory problems, for the most part caused or aggravated by air pollution.

Water Pollution

Water is our most immediate resource other than air—it is consumed daily and comprises about 70 percent of our bodies. Oceans occupy 71 percent of

Women washing and collecting water at a river in India. Less than 25% of the wastewater generated in India's 12 largest cities is collected by sewers. Most smaller cities have no wastewater treatment. It is estimated that between 5 and 10 million persons die each year from waterborne diarrheas in Asia, Africa, and Latin America. (Forrest Anderson/Gamma-Liaison, Inc.)

The bed of the Aral Sea, Kazakhstan. The Aral is an inland sea in an arid environment where water is scarce. Water has been diverted from streams that formerly flowed to the Aral Sea. The loss of inflow has caused the lake level to fall, stranding these ships. (Virginia Moos/The Stock Market)

Earth's surface. From the sea we obtain fish, shellfish, oil, gas, sand, gravel, salt, and sulfur. The seafloor may someday yield manganese and cobalt, and become a burial site for nuclear waste. Countries with inadequate freshwater supplies, such as Saudi Arabia, desalinate seawater. It is difficult to imagine an Earth without its plentiful resource of water.

In the United States, about 1.3 billion cubic meters (350 billion gallons) of fresh water are pumped from the ground or from rivers and lakes each year. About 13 percent of the water is used in homes, 45 percent in industry, and 42 percent in agriculture. About two thirds of the water we use is returned to rivers and lakes, while the remaining one third is evapotranspired, mostly from irrigated fields.

Pure water is essential to human survival, yet we allow waste to run off into our streams and seep into our groundwater. We dump trash and spill oil into the sea and wash fertilizers, pesticides, chemicals, and wastewater into our rivers and seas.

Water is a "universal solvent," the great cleanser of Earth. It can dissolve a wide range of substances, and it can transport bacteria, plants, fish, dead trees, sediment, toxic chemicals, and trash of all kinds. As with air pollution, water pollution results when substances enter the water faster than they can be carried off, diluted, or decomposed. Water pollution is measured as the amount of waste being discharged in relation to a body of water's ability to handle the waste.

We usually can tell when a stream or lake is polluted. Sediment pollution is visible, and an odd color, smell, or taste can betray trouble. Pollutants may be detected with simple chemical tests. For example, pollution from sewage may be revealed if tests show high concentrations of coliform bacteria, which are common in the digestive tracts of mammals. However, some toxics, such as pesticides, are harder to detect and require costly testing.

Pollutants have diverse sources. Some come from a *point source,* meaning that they enter a stream at a specific location, such as a wastewater discharge pipe. Others may come from a *nonpoint source*—they come from a large, diffuse area, such as organic matter or fertilizer washed from a field by a storm. Point-source pollutants are usually smaller in quantity and much easier to control. Nonpoint sources usually pollute in greater quantities and are much harder to control.

The *concentration* of pollution, such as sewage effluent, usually declines downstream from where the waste is discharged (Figure 6–12). This reduction occurs because the waste is diluted, and natural processes decompose pollutants and remove them from the water.

Reduction of oxygen in water. Because aquatic plants and animals need oxygen, the **dissolved oxygen concentration** in water indicates the health of a stream and lake. The oxygen consumed by decomposing organic waste constitutes the **biochemical oxygen demand.** If too much waste is discharged to a lake or stream, the water becomes oxygen-starved, and fish die.

This condition is typical when water becomes loaded with municipal sewage or industrial waste. The sewage and industrial pollutants consume so

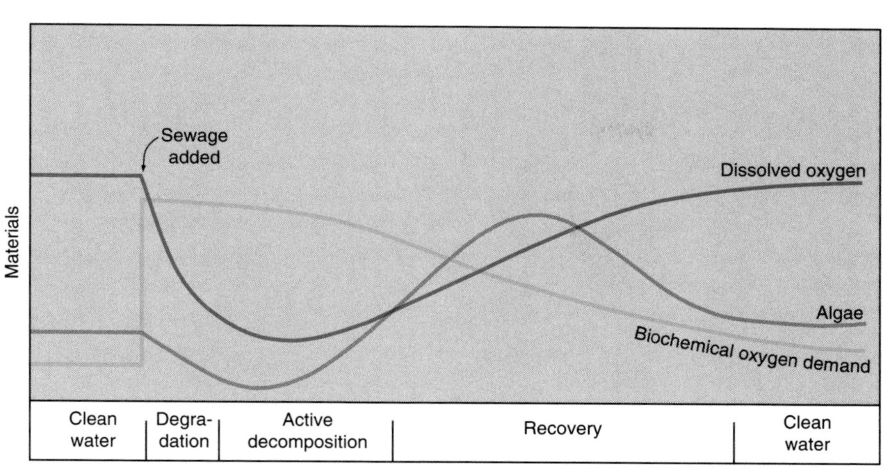

FIGURE 6–12
Water quality varies downstream from a point where sewage is added. Dissolved oxygen becomes depleted as organic matter is consumed, but recovers further downstream. Nutrients are released to the water and algae increase downstream as a result. (Adapted from Hynes, *The Biology of Polluted Waters*, University of Liverpool Press, 1978.)

much oxygen that the water can become unlivable for normal plants and animals, creating a "dead" stream or lake. Similarly, when runoff carries fertilizer from a farm field into a stream or lake, the fertilizer nourishes excessive aquatic plant production (a "pond scum" of algae), which consumes oxygen when it decomposes.

Agriculture is the leading consumer of water, primarily for irrigation, and it also is the leading water polluter. Runoff from agricultural land discharges sediment, fertilizers, animal waste, and small quantities of chemicals into nearby streams. The fertilizer and animal waste, which are rich in nitrogen compounds, can nourish aquatic plants to the point of overproduction.

Wastewater and disease. About 13 percent of the total flow in U.S. rivers is wastewater. Improved wastewater treatment is critical, especially in a world with a rapidly growing human population. Relatively developed countries like the United States generate more wastewater than developing countries, but they also have greater capacity to treat this wastewater. Thanks to stricter legislation, treatment facilities have been upgraded over the last half century in relatively developed countries, and their rivers are cleaner than a few decades ago. Yet, many treatment plants in the United States still do not meet the standards of the country's Clean Water Act.

In the developing world, untreated sewage usually goes directly into rivers, which also supply drinking water. Untreated water—combined with poor sanitation, nutrition, and medical care—can make drinking of this water deadly. Waterborne diseases, such as cholera, typhoid, and dysentery, are major causes of death in developing countries. Because of improper sanitation, millions of people in Asia, Africa, and South America die each year from diarrhea. As people in these rapidly growing regions crowd into urban areas, drinking water becomes less safe, and waterborne diseases flourish.

Chemical and Toxic Pollutants. Any waste discharged onto the ground, or into it, may pollute groundwater and streams. Pesticides applied to lawns and golf courses find their way into streams and groundwater. Landfills and underground tanks at gasoline stations can leak pollutants into groundwater, contaminating nearby wells, soil, and streams. Petroleum spilled from ocean tankers contaminates seawater (Figure 6–13). During times of flooding, normally secure tanks of chemicals may be dislodged and broken open, and their contents mixed with the floodwaters.

Toxic substances are chemicals that are harmful even in very low concentrations. Contact with them may cause mutations, cancer, chronic ailments, and even immediate death. Major toxic substances include PCB oils from electrical equipment, cyanides, strong solvents, acids, caustics, and heavy metals such as mercury, cadmium, and zinc.

During the 1950s and 1960s, toxic wastes were often buried, but by the 1970s this method of isolating toxic wastes had proved inadequate because many waste sites were leaking. One of the most notorious is Love Canal, near Niagara Falls, New York, where several hundred families were exposed

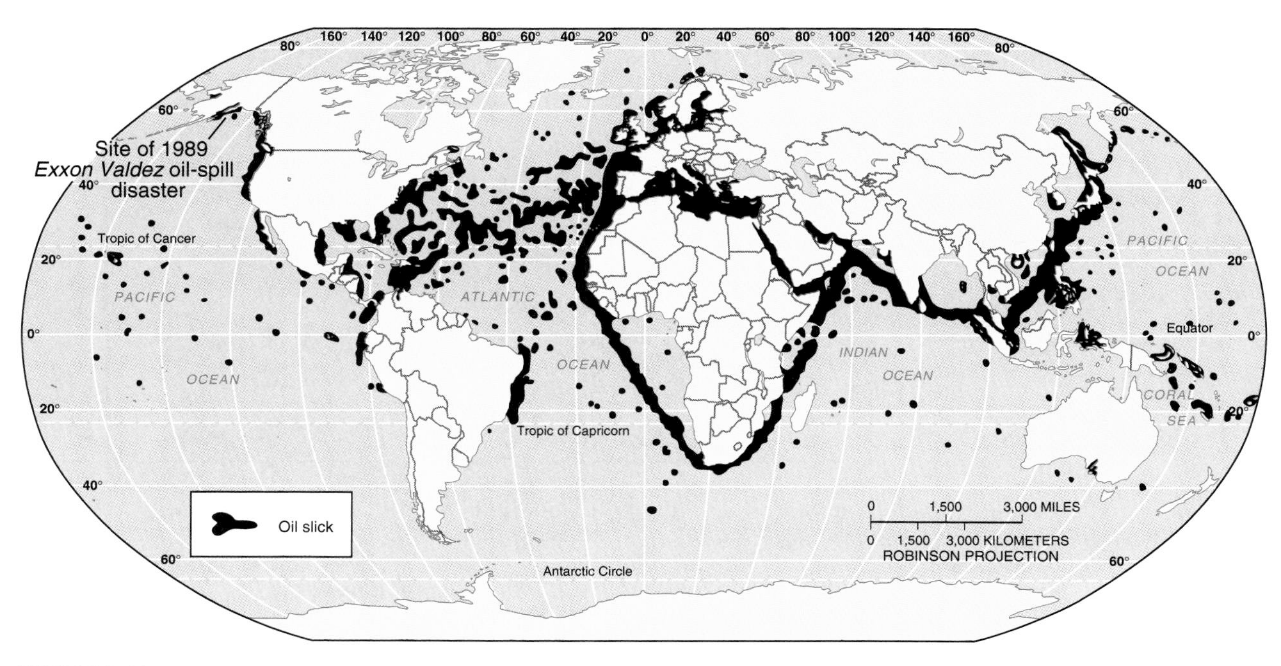

FIGURE 6–13
There are far more oil spills than most people realize. Most of the spillage is along the main shipping routes, especially those used by oil tankers. (From R. W. Christopherson, *Geosystems,* 2d ed., Macmillan Publishing Co., 1994. Data from Organization for Economic Cooperation and Development, *The State of the Environment,* Paris, 1985.)

to chemicals released from a waste disposal site used by the Hooker Chemicals and Plastic Company.

During the 1930s, Hooker buried toxic wastes in metal drums. In 1953, a school and several hundred homes were built on the site. Eventually, erosion exposed the metal drums. Beginning in 1976, residents noticed a strong stench and slime oozing from the drums, and they began to suffer a high incidence of health problems, such as liver ailments and nervous disorders. After four babies on the same block were born with birth defects, New York State officials relocated most of the families and began an expensive clean-up effort.

Love Canal is not unique. Toxic wastes have been improperly dumped into thousands of landfills. The U.S. Environmental Protection Agency maintains a list of the sites most in need of cleanup (Figure 6–14). As sites available for disposal of toxic waste grow scarce, some European and North American firms have tried to transport toxic waste to West Africa.

Since the Federal Clean Water Act became law over twenty years ago, about 30 percent of Americans have benefited from new sewage-treatment facilities. Industry has spent heavily to meet water-quality standards. Many streams and lakes that had become open sewers now are suitable for recreation. But the job is unfinished. Serious efforts are made to control chemical waste in relatively developed countries, but developing countries often give little attention to this problem, and wastes are freely released into streams.

Alternatives For Reducing Air And Water Pollution

Recognizing that pollution results from wastes being discharged to the environment faster than they can be dispersed, two alternatives are obvious: we can either increase the capacity of the environment to accept waste, or we can reduce the amount of waste. Both strategies are currently used, but in the future reducing waste must become the preferred alternative.

Increase environmental capacity. We can increase environmental capacity to accept waste by using two strategies:

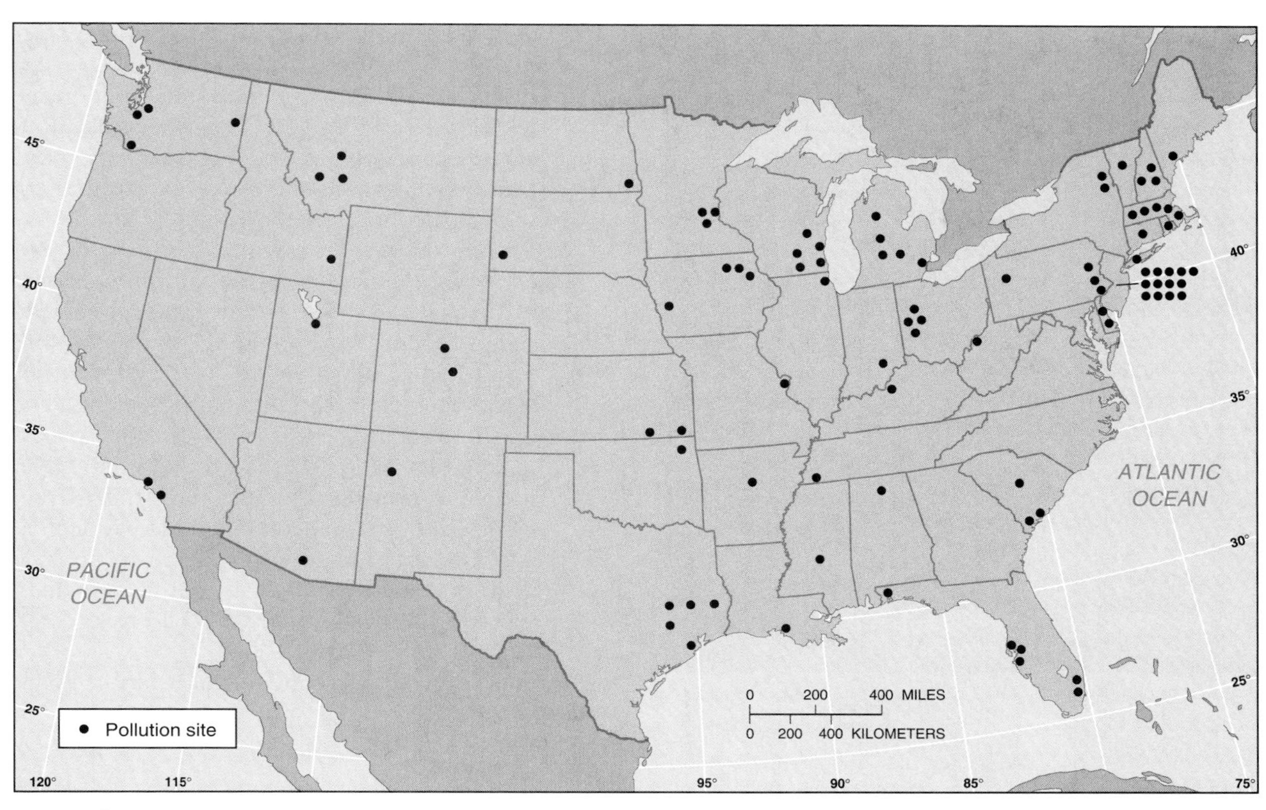

FIGURE 6–14
The U.S. Environmental Protection Agency maintains a national priorities list of "Superfund" hazardous waste sites, ranked according to the hazard severity. This map shows only the 100 worst sites.

1. *Use more efficiently the air, water, or land that receives the waste.* The capacity of air, water, or land to accept waste is widely variable. In the case of streams, adding wastewater may or may not constitute pollution, depending on streamflow. A deep, fast-flowing river has a greater capacity to absorb wastewater than a shallow, slow-moving one. Wastewater can be stored when the river level is low and released when the river is high.

   The same is true for air: exhaust released into a brisk wind is quickly dispersed, but exhaust released into stagnant air during a temperature inversion quickly accumulates to irritating levels. Industries and utilities reduce local air pollution by building taller smokestacks, which better-disperse the gases at greater heights. Air quality may also be improved by staggering workers' hours so pollution from cars is spread more evenly through the day.
2. *Transform waste so it can be discharged into a different part of the environment.* Coal-burning power plants pollute the air, so *wet scrubbers* are installed to wash particulates from the gas before venting to the atmosphere. Wet scrubbers capture the airborne particulates in water. The water then can be discharged into a stream, or placed in a settling basin where the particulates drop out. This transforms the residue into a solid waste for disposal on land.

Both strategies have limitations. Because we do not always know the environment's capacity to assimilate a particular waste, we are likely to exceed it at times. Recent history is filled with examples of people discharging wastes to the environment in the belief that they would be dispersed or isolated safely: toxic wastes at Love Canal, CFCs in the stratosphere, and garbage dumped offshore from barges that has washed up onto beaches.

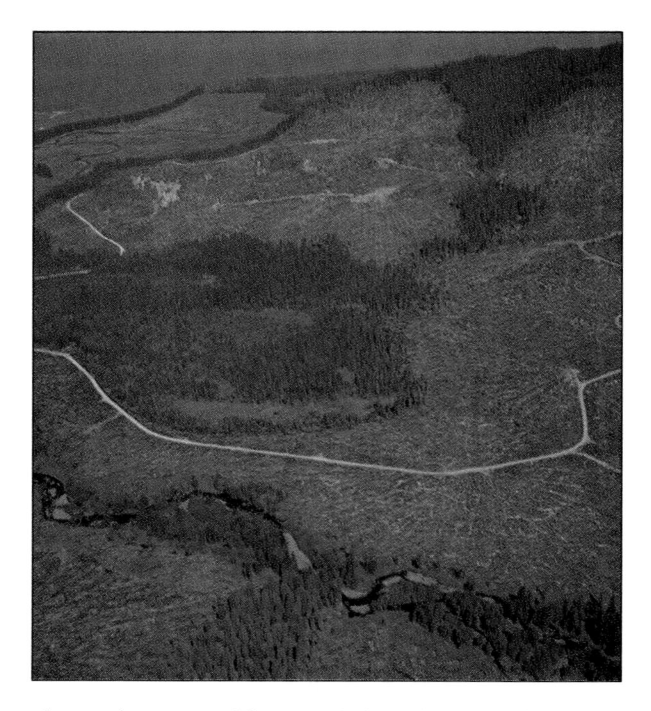

Clear-cut harvesting of forest in Alaska. While most of the original forest area of North America has already been cleared, large areas remain in Canada and Alaska. Lumber from these forests is shipped to Asia and Europe, as well as supplying the vast North American market. (Tom Bean/The Stock Market)

Although wastes may be dispersed, they can remain harmful. Early efforts to control SO_x pollution around coal-burning industries and metal smelters included building tall smokestacks that would disperse the sulfur over a larger area. The sulfur was better-dispersed, but it caused acid rain hundreds of miles away, instead of locally.

Finally, many pollutants are very mobile. They often travel from air to soil, or soil to water. A pollutant like SO_x might exist at tolerable levels in the air, but when it accumulates in the soil, forest decline results.

Reducing the waste discharged. The preferred approach to reducing pollution is to reduce the amount of waste discharged. We can use two strategies to achieve this:

1. *Reduce or eliminate polluting substances.* Many opportunities exist for reducing or eliminating industrial pollution without sacrificing quality or increasing cost. For example, toxic cleaning products are used in manufacturing to remove oil, grease, soldering residues, dust, coatings, and metal fragments. Trichloroethylene (TCE) is an example. It is an excellent solvent, but it is also very toxic. Less toxic cleaning agents can be substituted, such as alcohol or detergents.
2. *Recycle polluting substances instead of discharging them into the environment.* Some industrial processes generate recyclable scraps, such as the molding of plastics and stamping of sheet metal. Normally, larger scraps are recycled, but small fragments, like dust from polishing or sanding, may be placed in landfills as solid waste or discharged in water. Capturing and recycling these fragments has dual benefits: it reduces pollution and increases the supply of scrap for making new materials. Health benefits also may accrue if the dust presents a health hazard.

In the early 1990s, a fundamental shift in pollution management began, from pollution *control* to pollution *prevention*. This change was spurred by several factors:

1. Despite pollution controls, the environment still was not clean enough.
2. Polluters faced legal liability if they discharged substances that *later* were found to cause harm, so reducing discharges to the environment reduces the potential for liability later.
3. Pollution prevention often is cheaper than pollution control.
4. Publicizing its efforts to reduce pollution can improve a company's image and help it to market products, to recruit employees, and to negotiate with government regulators.

Pollution prevention has quickly been adopted by many industries, especially in relatively developed countries of Europe and North America. Large corporations—which are especially visible to governments, consumers, and the general public—have led the way in reducing discharges into the environment. Companies increasingly recognize that the best way to prevent pollution is to design a product that can be manufactured with a minimum of toxic chemicals, used without generating pollution, and recycled easily when it is no longer wanted.

Although the relatively developed world has been successful in reducing pollution during the past few decades, pollution problems are increasing in the developing world. Most of the differences in pollution between rich and poor countries result from differences in wealth and level of industrial development. Where electricity, clean-burning fuels, and pollution-control devices are widely available—as in Europe, North America, and Japan—urban air pollution has largely been controlled. But problems are much more acute in places where people cook and heat with wood and drive old vehicles kept running with cleverness rather than scheduled maintenance.

As manufacturing expands in the developing world, new facilities could be built with pollution control in mind. But pollution controls can be costly. Reducing the threat of large-scale pollution in developing countries requires development of pollution prevention methods that reduce—not increase—costs.

Natural Resources and Land Use

As world population grows and resources are used for more purposes, conflicts over resource use are inevitable. In some cases, the conflict concerns *who* has access to a resource in short supply, such as water in a desert region. But increasingly, conflicts concern *how* a resource should be used. Should a valley be dammed to generate hydroelectricity, or should people continue to live there and practice their traditional way of life? Should a wilderness area be protected as a habitat for endangered species, or should it be opened for development?

One resource may have competing uses, some of which may be incompatible with others. Balancing competing and incompatible uses requires careful management, and difficult choices must be made. In this section, we look at two important conflicts in resource management: forests and solid waste.

Forests

A forest is part of an ecosystem in which vegetation plays a major role in biogeochemical cycles. By absorbing carbon dioxide and releasing oxygen, a forest is an essential part of a local climate; a forest also stores carbon that otherwise would be in the atmosphere as carbon dioxide. A forest reduces erosion, aids in flood prevention, and provides a place for recreation. A forest is drained by rivers that provide drinking water, irrigation, habitat, and waste removal. People may use a forest as a habitat and a place to obtain food, fuel, shelter, and medicine.

Conflicting uses. Despite the many important benefits of maintaining forests, societies harvest them for lumber, paper, and fuel. Without the forest cover, water quality in rivers is substantially reduced, and eroded sediment clogs reservoirs and kills fish.

The most severe long-term damage from harvesting forests may be the release of stored carbon into the atmosphere, through decay and burning of unusable parts of trees. Deforestation in the tropics may be responsible for as much as 10 percent of the increase in CO_2 in the atmosphere.

In much of the world, especially in the tropics, the rate of deforestation exceeds the rate of reforestation. Net deforestation is occurring, especially in the tropical rainforest, because people are cutting trees faster than they can be replanted or regrow naturally. The United States, Russia, China, India, and Brazil are major timber-harvesting regions because their citizens demand wood prod-

Environmentalists like these students at the University of California might argue that a tree is more valuable standing in a forest than sawn into lumber. (Andy Levin/Photo Researchers, Inc.)

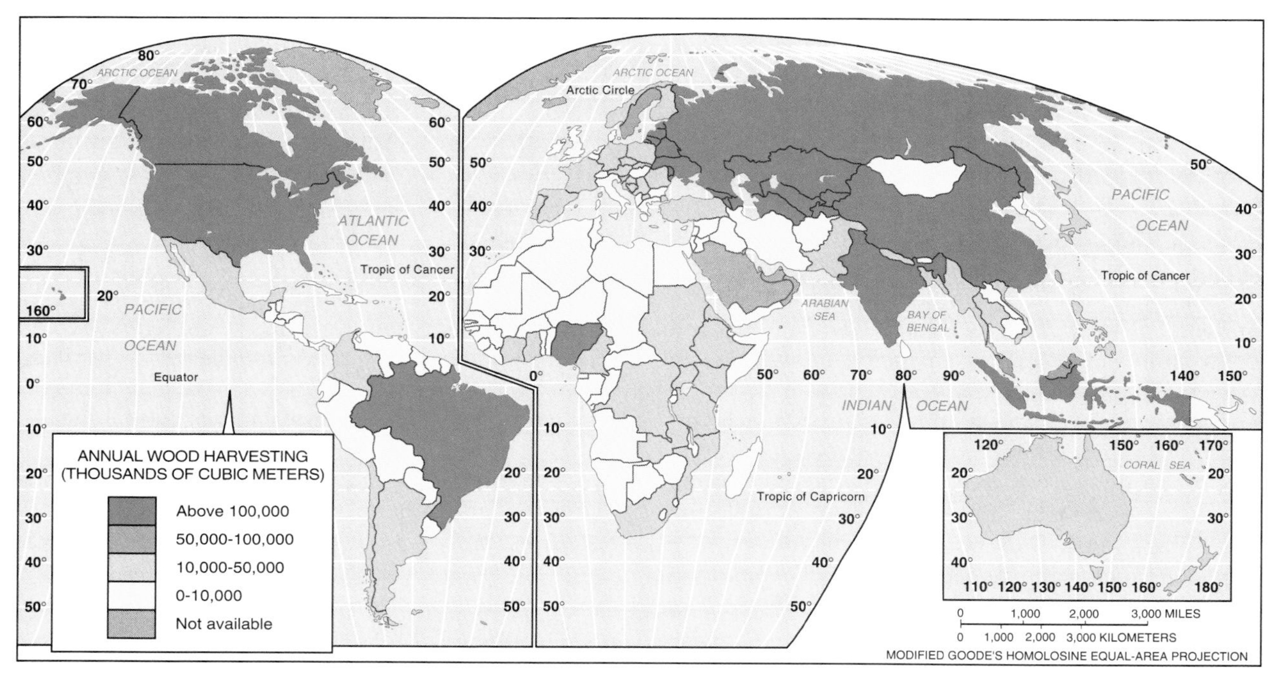

FIGURE 6–15
World timber harvesting. Data for the Soviet Union are lumped; virtually all of this harvesting takes place in Russia. Data from World Resources Institute, 1994.

ucts (Figure 6–15). Only about 4 percent of forests worldwide, and 14 percent in the United States, are protected from cutting. The timber industry does replant forests, especially in the mid-latitudes, but forests can require more than a century to regrow.

Managing forest resources. Forests, the wildlife they support, and the water that flows through them are all renewable resources. Conflicts occur between those who wish to harvest timber and those who wish to leave the trees standing.

Nearly all forests in the United States (except in Alaska) have been cut at least once during the past 300 years, so very little original forest remains. The timber industry wants to cut remaining original forests because their large, straight trees yield high-quality lumber. Harvesting provides jobs and benefits U.S. companies by keeping the price of timber low. Restrictions on cutting, as well as competition from Canada and other countries, would threaten severe unemployment in regions with extensive original-growth forests, especially in Oregon and Washington.

Environmentalists argue that, when trees are cut, the forest no longer supports the wildlife or maintains clean water in the way it did before. Cutting a forest places its inhabitants at risk. During the late 1980s, the spotted owl was endangered in the original-growth forests of Oregon and Washington when timber interests wanted to cut in the habitat. Environmentalists point out that U.S. laws prohibit the government from actions that could extinguish an endangered species. U.S. government attempts to find a compromise between the timber industry and environmentalists have proved elusive, because the two uses of the original-growth forest resource—commercial timbering and the spotted owl habitat—ultimately are incompatible.

Sustained yield management is a strategy that can maintain the productivity of a resource even as it is being used. In forestry, sustained yield management means that the number of trees harvested should not exceed the number replaced by new growth. The strategy also emphasizes harvesting in a manner that minimizes soil erosion, and therefore enhances the ability of the forest to regenerate. Sustained yield management is the official policy of the U.S. Forest Service, which manages the national forests, although the agency seldom practices it.

Similar conflicts are taking place in developing countries. For example, during the 1980s, tropical deforestation became a global issue, primarily because the rate of deforestation in the Amazon Basin was increasing rapidly. Environmentalists argued that, once the forest was cut, it could never be restored to its original condition. Species would become extinct, the soil would be ruined, and indigenous cultures would be wiped out. But those involved in clearing the forest responded that most of the forest land in the United States and Western Europe had already been cleared, some of it many centuries ago. By what right can relatively developed countries tell Brazilians to save their trees?

Ideally, the solution to conflicting uses of forests is sustainable management. Trees can be cut, but not to the extent that the future productivity of the forest is reduced. Some original forest should be protected, to provide habitat for native species and to prevent extinctions. Resource managers in northwestern North America and in the Amazon are working toward such compromise solutions.

Solid Waste

Today, average Americans discard 2 kilograms (4 pounds) of solid waste per day, nearly double what their parents threw out in 1960. Paper accounts for one-third of all solid waste in the United States. Discarded food and yard waste accounts for another one-third (Figure 6–16). Relatively developed societies generate large quantities of packaging and containers made of paper, plastic, glass, and metal. We dispose of this solid waste in three ways: landfills, incineration, and recycling. Each of these methods poses significant problems, either in environmental degradation or in costs of disposal. Normally, the choice of one disposal method over another means that costs are shifted from one group to another, and conflicts are inevitable.

Landfill Disposal. About 80 percent of solid waste generated in the United States is trucked to landfills and buried under soil in sanitary landfills. Unlike air and water pollution, which are reduced by dispersal into the atmosphere and rivers, solid waste pollution is minimized by concentrating the waste in thousands of landfills. However, landfills have been closed in many communities because they can contaminate groundwater, devalue property, and eventually run out of space. Opening new landfills is difficult because environmental regulations are more stringent today, and local opposition to new landfills is usually overwhelming. The result has been a solid waste crisis. Disposal sites are few and costly. Some communities pay to use landfills elsewhere. San Francisco trucks solid waste to Altamont, California, 100 kilometers (60 miles) away. Passaic County, New Jersey, hauls waste 400 kilometers (250 miles) west to Johnstown, Pennsylvania.

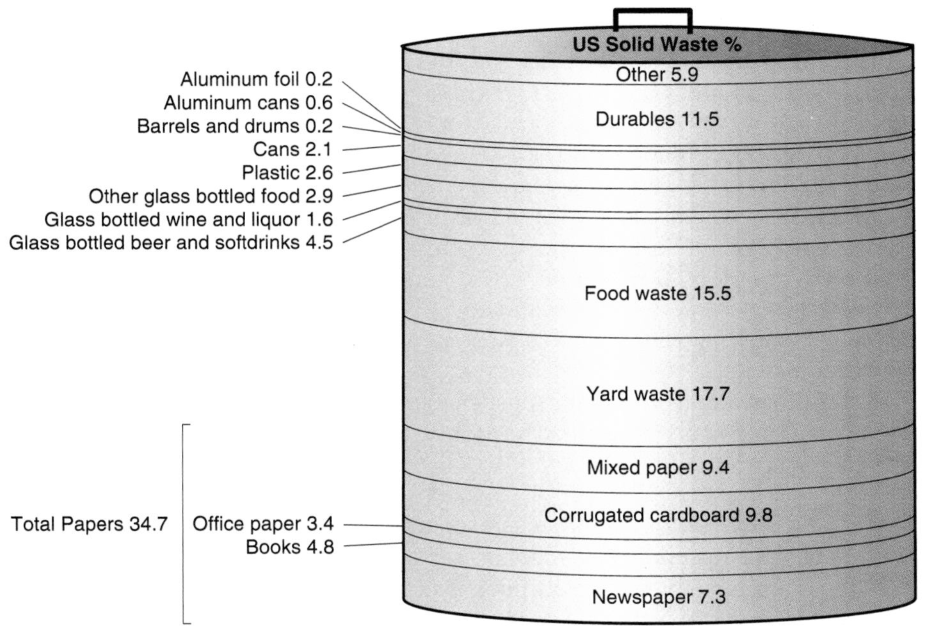

FIGURE 6–16
Paper products account for the largest percentage of solid waste in the United States, followed by food products and yard rubbish.

Solid waste being deposited in a sanitary landfill on the outskirts of Chicago. Despite objections to such techniques, Americans find it both easier and cheaper to discard wastes than to recycle them. (Roger A. Clark, Jr./Photo Researchers, Inc.)

Incineration. One alternative to landfills is **incineration.** This method reduces the bulk of trash by about three-fourths, and the remaining ash requires far less landfill space. Incineration also releases heat energy, which can be used to boil water, producing steam that can heat homes or generate electricity by operating a turbine. More than a hundred incinerators now burn about 15 percent of the trash generated in the United States. However, because solid waste is a mixture of many materials, it does not burn efficiently. Burning also releases some toxics into the air, and some remain in the ash.

Recycling. Recycling solid waste reduces the need for landfills and incinerators and reuses natural resources that already have been extracted. Recycling simultaneously addresses both pollution and resource depletion. Most U.S. communities have instituted some form of mandatory recycling. Table 6–1 shows savings possible with recycling of different materials.

Several barriers to recycling must be surmounted: separating waste, consumer resistance, lack of market, cost, and indirect losses.

- *Waste separation.* Solid waste comprises a variety of materials that must be separated to be recycled. Metals containing iron can be pulled from the waste magnetically, but paper, yard waste, food waste, plastics, and glass can not be separated easily from each other. Consequently, many communities require consumers to practice **source separation** of solid waste. Typically, consumers must place newspaper, glass, plastic, and aluminum in separate containers for pickup. Each type may be collected on a different day by a different truck, and it is shipped to a specialized processor. The procedure is generally more expensive for the community than picking up all of the trash together.
- *Consumer resistance.* Separation for recycling is a nuisance for people in relatively developed countries used to throwing things away. To encourage recycling, some communities charge high fees to pick up nonrecyclables, but take recyclables at little or no charge. Citizens may be fined for failing to comply. Bottle and can laws requiring a deposit on beverage containers have been enacted in many states to encourage recycling and reduce roadside litter.
- *Lack of market.* To succeed, recycled products must have a market. The lack of an assured market for many recycled products is perhaps the most difficult obstacle to increased recycling. Demand for recycled goods among industries and consumers is uncertain. For example, mixed plastics can be used as a sub-

TABLE 6–1

Savings possible with recycling of different materials.

Product	Recycled Waste Used	Natural Resource Saved	Other Savings
Plastics	Discarded plastic containers	Oil	Reduces oil exploration, drilling, transportation, refining, oil spill risk, energy consumption
Metals	Discarded cans, large appliances, cars	Metal ores (iron, aluminum)	Reduces environmental damage and energy consumption from mining and refining
Glass	Discarded glass bottles	Glassmaking sand	Reduces sand mining, energy consumption
Paper	Discarded paper	Trees	Reduces logging, destruction of forests, water pollution from pulp-making

stitute for wood in picnic tables and playground equipment, but few consumers are willing to purchase these products.

- *Hidden costs.* Recycling involves far more than just "melting down and reshaping." For example, to recycle paper, ink must be removed, an additional step compared to conventional papermaking, and therefore an added cost. Because it is difficult to remove all of the ink, recycled paper is unacceptable for some uses because it is too grey or speckled. The relatively high cost of processing and the lower value of the recycled products has prevented some recyclers from finding markets.
- *Indirect losses.* Trash burns only if it contains enough combustibles. If paper is recycled and yard waste is composted instead of being thrown in the trash, the trash may be difficult to burn. Because recycling has increased during the 1990s, some communities have not had enough combustible waste to operate their incinerators.

Many problems have resulted from the fact that recycling has grown faster than consumer demand for recycled materials. Some recyclable material is being placed in landfills, because more is being generated than the market can absorb. Recycling will require changes in manufacturing technology and product design.

Resource recovery from solid waste requires sorting and separating different types of materials. One approach is to collect them together and separate them at a centralized facility like this one. Alternatively they can be separated at the source (home, school or work) and transported separately to reprocessing facilities. (Palmer/Kane/The Stock Market)

To increase consumer acceptance, the quality of recycled material must also improve. Poor quality has resulted because recycling became mandatory before proper manufacturing methods had been developed.

Many companies are developing manufacturing methods and packaging that facilitate recycling. To reduce packaging volume, detergent is being sold in concentrated form. Refillable containers are available for more products. The 3M Corporation is producing a desk organizer designed for recycling; when thrown away, its snap-together parts can be easily separated for recycling. In Germany, manufacturers of some goods—including motor vehicles—are required to accept their products back when their useful lives are over. Such regulations are costly, but they encourage manufacturers to design recyclability into their products.

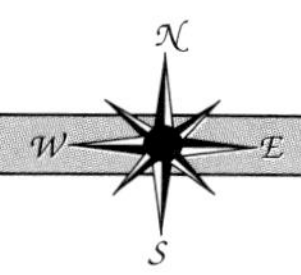

CONCLUSION: CRITICAL ISSUES FOR THE FUTURE

During the 1970s, a book titled *The Limits to Growth* argued that depletion of Earth's natural resources had placed us on a disastrous course. The report predicted that natural resource depletion, combined with population growth, would disrupt the world's ecosystems and economies and lead to mass starvation. If natural resource protection were not in place within twenty years, the report claimed, environmental systems would be permanently damaged, and everyone's standard of living would decline.

Few contemporary geographers accept the pessimistic predictions made by *The Limits to Growth* a quarter-century ago. Use of many resources has declined significantly since then. We may still deplete some resources, but substitutes and other strategies are available. Pollution continues to degrade natural resources, but industrial development has been made compatible with environmental protection in some locations. However, many natural resource problems are likely to worsen in the coming decades because of continued population growth, as discussed in the next chapter.

The developing countries face the greatest challenges. In places suffering from extreme poverty, sound management of resources for the future is difficult. Careful management of Earth's natural resources is even more difficult where population is increasing rapidly, as the next chapter demonstrates. Some regions, especially in Africa, already contain more people than natural resources can sustain with existing technology, at even minimal standards of human existence.

Chapter Summary

1. What is a natural resource?
A natural resource is an element of the physical environment that is useful to people. Cultural values determine how resources are used. Technological factors limit our use of some resources, whereas economic factors influence whether a resource is used, and how much. Natural resources include air, water, soil, plants, animals, fuels, and non-energy minerals.

2. Mineral and energy resources
Our actions deplete mineral resources. A notable example is our dependence on fossil fuels for energy. We continually deplete deposits of oil and natural gas, and economically recoverable deposits are increasingly difficult to find. These resources are nonrenewable because millions of years are required to form them. Over time, the increasing cost of obtaining resources can lead to the adoption of alternative resources.

3. Air and water resources
Air pollution is a concentration of trace substances at a greater level than occurs in average air. Acid deposition and pollution of urban areas are particularly serious forms of contemporary air pollution. Water pollution results from both point and nonpoint sources. Chemical and toxic discharges comprise an especially critical water pollution issue. Air and water pollution can be reduced by increasing environmental capacity, but the preferred alternative is to reduce discharges.

4. Natural resources and land use
Conflicts over the use of a natural resource arise because some people prefer to preserve the resource while others wish to use it for economic purposes. Forests are a renewable resource that provide many useful products, but harvesting trees has undesired impacts. Sustained yield management is a strategy that attempts to balance the productive use of a resource while not depleting its supply. Solid waste disposal is a continuing problem.

Key Terms

Acid deposition, 235
Acid pollution, 234
Biochemical oxygen demand, 240
Carbon monoxide, 234
Cartel, 223
Dissolved oxygen concentration, 240
Fossil fuel, 223
Human-made resource, 216
Hydrocarbon, 234
Hydroelectric power, 232
Incineration, 248
Natural resource, 216
Nitrogen oxide, 234
Nonrenewable resource, 221
Particulate, 234
Photochemical smog, 237
Photovoltaic cell, 232
Pollution, 234
Potential resource, 218
Renewable resource, 221
Sanitary landfill, 219
Source separation, 248
Substitutability, 220
Sulfur dioxide, 234
Sustained yield management, 246
Temperature inversion, 237
Toxic substance, 241

Questions for Study and Discussion

1. What is a resource? How do political/cultural, technological, and economic factors determine whether something in the environment becomes a valuable resource, or does not?
2. What is a renewable resource? What is a nonrenewable resource? Give examples of each. Can any resources be considered either renewable or nonrenewable, depending on their use?
3. How has human use of coal, oil, and natural gas changed over the past 300 years? What changes are likely in use of these fuels over the next few decades? What are some energy resources that are likely to be substituted for fossil fuels?
4. How do emission rates and rates of pollution dispersal interact to determine the severity of air and water pollution? What weather conditions aggravate air pollution?
5. What are the major causes and consequences of acid deposition? Of urban air pollution?
6. How has water quality in U.S. rivers changed in the past few decades? What have been the major factors in this change? How does water pollution in wealthy countries differ from that in poor ones?

Thinking Geographically

1. Where does the drinking water come from where you live or attend school? What are the most significant potential pollution sources affecting your water supply? Your air supply?
2. If you drive, estimate the number of gallons of gasoline your automobile consumes in a year, by dividing the number of miles driven annually by your fuel efficiency in miles per gallon. If each gallon of gasoline contains about 2 kilograms of carbon, how many kilograms of carbon were emitted to the atmosphere by your automobile? Compare this with the number of acres of growing forest it would take to store this carbon in biomass, if an acre of forest accumulates carbon at the rate of 300 kilograms per year.
3. Of the approximately 90 million new people added to the world's population each year, over 80 million are in developing countries. But the per capita resource consumption of people in developing countries is a small fraction of that in rich countries. If population control is important for limiting global use of resources, should population control efforts be focused in developing countries or industrial countries? Why?
4. What are the most important issues of environmental quality where you live? Has environmental quality there improved or deteriorated in the past thirty years? What is your evidence?
5. The next time you have a full bag of trash, sort the contents of the bag before you throw it away. Measure the amount in each of the four categories in Table 6-1. For each category, estimate the quantity of waste that could be eliminated by not acquiring it in the first place (such as buying products with less packaging). Where would the remaining materials need to go if they were to be recycled? How would they get there? How much would it cost?

Suggestions for Further Reading

Adams, R. M., C. Rosenzweig, R. M. Peart, J. T. Ritchie, B. A. McCarl, J. D. Glyer, R. B. Curry, J. W. Jones, K. J. Boote, and L. H. Allen, Jr. "Global Climate Change and U.S. Agriculture." *Nature* 345 (1990): 219–24.

Addiscott, T. M. *Farming, Fertilizers and the Nitrate Problem*. Wallingford, CT: C.A.B. International, 1991.

Beach, Timothy, and P. Gershmehl. "Soil Erosion, T Values, and Sustainability: A Review and Exercise." *Journal of Geography* 92 (January–February 1993): 16–22.

Becker, C. Dale. *Water Quality in North American River Systems*. Columbus: Batelle Press, 1992.

Bolgiano, Chris. "Yellowstone and the Let-Burn Policy." *American Forests* 95 (January–February 1989): 21–30.

Brower, Michael. *Cool Energy: The Renewable Solution to Global Warming*. Cambridge: Union of Concerned Scientists, 1990.

Brown, Lester R., A. Durning, C. Flavin, H. French, N. Lenssen, M. Lowe, A. Misch, S. Postel, M. Renner, L. Starke, P. Weber, and J. Young. *State of the World 1994*. New York: Norton (see most current edition of this annual book).

Brown, Lester R., H. Kane, and D. M. Roodman. *Vital Signs 1994*. New York: Norton, 1994 (see most current edition of this annual book).

Burnett, Howard. "Report on Our Stressed-Out Forests." *American Forests* 95 (March–April 1989): 21–26.

Charbonneau, Robert, and G. M. Kondolf. "Land Use Change in California, USA: Nonpoint Source Water Quality Impacts." *Environmental Management* 17 (July–August 1993): 453–60.

Cherfas, J. "The Fringe of the Ocean—Under Siege From Land." *Science* 248 (1990): 161–65.

Costanza, Robert, and H. E. Daly. "Natural Capital and Sustainable Development." *Conservation Biology* 6 (March 1992): 37–46.

Crosby, Alfred W. *Ecological Imperialism: The Biological Expansion of Europe, 900–1900*. Cambridge: Cambridge University Press, 1986.

Cutter, Susan L., Renwick, H. L., and Renwick, W. H. *Exploitation, Conservation, Preservation*. 2d ed. New York: Wiley, 1991.

Dakers, Sonya. *Sustainable Agriculture: Future Dimensions*. Ottawa: Library of Parliament, Research Branch, 1992.

Dodd, Jerrold L. "Desertification and Degradation in Sub-Saharan Africa: The Role of Livestock." *BioScience* 44 (January 1994): 28–34.

Drew, David. *Man-Environment Processes*. London: Allen & Unwin, 1983.

Flavin, Christopher. *Building on Success: The Age of Energy Efficiency*. Washington: Worldwatch Institute, 1988.

Francis, Charles A. (ed.) *Sustainable Agriculture in Temperate Zones*. New York: Wiley, 1990.

Franklin, Jerry F. "Lessons from Old Growth." *Journal of Forestry* 91 (December 1993): 10–13.

Glantz, Michael H. "Floods, Fires and Famine: Is El Niño to Blame?" *Oceanus* 27 (Summer 1984): 14–19.

Goldstein, Joan. *Demanding Clean Food and Water: The Fight for a Basic Human Right*. New York: Plenum Press, 1990.

Gulliford, Andrew. *Boomtown Blues*. Niwot, Colorado: University Press of Colorado, 1989.

Hammer, Monica, A. Jansson, and B-O Jansson. "Diversity Change and Sustainability: Implications for Fisheries." *Ambio* 22 (May 1993): 97–105.

Hinrichsen, Don. *Our Common Seas: Coasts in Crisis*. London: Earthscan, 1990.

Hollander, Jack M. (ed.) *The Energy-Environment Connection*. Washington: Island Press, 1992.

Hulme, Mike, and M. Kelly. "Exploring the Links Between Desertification and Climate Change." *Environment* 35 (July-August 1993): 4–11, 39–45.

Hynes, H. B. N. *The Biology of Polluted Waters*. Liverpool: University of Liverpool Press, 1978.

Jones, John G. *Agriculture and the Environment*. New York: Horwood, 1993.

Lewandrowski, J. K., and R. J. Brazee. "Farm Programs and Climate Change." *Climate Change* 23 (January 1993): 1–20.

Maybeck, Michel, D. Chapman, and R. Helmer. *Global Freshwater Quality: A First Assessment*. Oxford: Blackwell, 1989.

Meadows, D. H., and others. *The Limits to Growth*. New York: Universe Books, 1972.

Myers, Mary F., and G. F. White. "The Challenge of the Mississippi Flood." *Environment* 35 (December 1993): 6–9, 25–35.

Nemerow, Nelson L. *Stream, Lake, Estuary and Ocean Pollution*. 2d ed. New York: Van Nostrand Reinhold, 1991.

Omernik, James M., and G. E. Griffith. "Ecological Regions Versus Hydrologic Units: Frameworks for Managing Water Quality." *Journal of Soil and Water Conservation* 46 (September-October 1991): 334–40.

Orians, Gordon H. "Ecological Concepts of Sustainability." *Environment* 32 (November 1990): 10–15, 34–39

Pickup, G. and D. M. Stafford Smith. "Problems, Prospects and Procedures for Assessing the Sustainability of Pastoral Land Management in Arid Australia." *Journal of Biogeography* 20 (September 1993): 471–88.

Pimentel, David, J. Allen, A. Beers, L. Guinand, R. Linder, P. McLaughlin, B. Meer, D. Musonda, D. Perdue, S. Poisson, S. Siebert, K. Stoner, R. Salazar, and A. Hawkins. "World Agriculture and Soil Erosion." *BioScience* 37 (April 1987): 277–83.

Pimentel, David, L. McLaughlin, A. Zepp, B. Lakitan, T. Kraus, P. Kleinman, F. Bancini, W. J. Roach, E. Graap, W. S. Keeton, and G. Selig. "Environmental and Economic Effects of Reducing Pesticide Use." *BioScience* 41 (June 1991): 402–09.

Pimentel, David, H. Acquay, M. Biltonen, P. Rice, M. Silva, J. Nelson, V. Lipner, S. Giordano, A. Horowitz, and M. D'Amore. "Environmental and Economic Costs of Pesticide Use." *BioScience* 42 (November 1992): 750–60.

Redford, Kent H. "The Empty Forest." *BioScience* 42 (June 1992): 412–22.

Reisner, Marc. *Cadillac Desert.* New York: Viking, 1986.

Rhodes, Richard. *The Making of the Atomic Bomb.* New York: Simon & Schuster, 1986.

Smith, Nigel J. H., and R. E. Schultes. "Deforestation and Shrinking Crop Gene-Pools in Amazonia. *Environmental Conservation* 17 (Autumn 1990): 227–34.

Turner, Billie Lee, and K. W. Butzer. "The Columbian Encounter and Land-Use Change" *Environment* 34 (October 1992): 16–20, 37–44.

Turner, R. Eugene and N. N. Rabelais. "Changes in Mississippi River Water Quality this Century." *BioScience* 41 (March 1991): 140–47.

World Commission on Environment and Development. *Food 2000: Global Policies for Sustainable Agriculture.* London: Zed Books, 1987.

World Resources Institute. *World Resources 1994–95.* New York: Oxford University Press, 1994 (see most current edition of this annual book).

We also recommend these journals: *Ambio; Ecological Economics; Ecologist; Environment; Environmental Management; Environmental Pollution; Journal of Environmental Management; Oil & Gas Journal; Water, Air & Soil Pollution; Water Resources Bulletin; Worldwatch.*

7 Population

- Population distribution
- Population growth
- The demographic transition
- Migration
- Threat of overpopulation

Family in Iraq

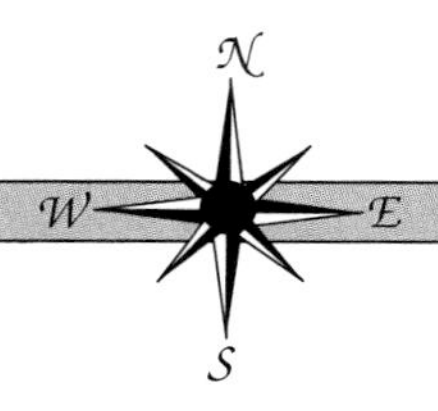

1. Population distribution

People are not distributed uniformly around the world; some regions support clusters of people, while other regions are sparsely inhabited. People tend to live in climates that support agriculture and avoid regions that are especially hot, cold, wet, dry, or mountainous.

2. Population growth

The population of a region changes according to the rate of natural increase, which is the difference between the numbers of births and deaths. Natural increase rates are much higher in relatively poor, less economically developed countries than in relatively wealthy, more developed ones.

3. The demographic transition

The process of change in a country's population is known as the demographic transition. Global population is increasing rapidly at this time because most of the world's countries are in a stage of the demographic transition characterized by high population growth.

4. Migration

A region's population also changes because of migration from other regions. People in countries with high rates of natural increase may migrate from rural to urban areas or to other countries in search of better economic conditions.

5. Threat of overpopulation

Overpopulation results when the number of people in an area exceeds the capacity of the environment to support life at a decent standard of living. The capacity of Earth as a whole to support human life may be high, but some regions have a favorable balance between people and available resources, while others do not.

INTRODUCTION

Throughout most of history, humanity's most critical resource in growing food and producing goods was human labor itself. A family with many children thus had an important resource in its daily work to obtain food and other necessities. Children were also an important resource to couples as they aged and could no longer produce their own food. Because infants and children often died at young ages, couples hoped to have enough babies to assure that some would live long enough to be family resources.

Earth's human population has increased rapidly in recent decades: with modern medicine and preventive health care, infants are more likely than in the past to survive. At the same time, humans have harnessed as important resources many elements of the physical environment, such as fossil fuels. But rapid population growth has resulted in the depletion and degradation of some of these resources.

Geographers and other social scientists study the population problems of countries from a variety of viewpoints. The scientific study of population characteristics is called **demography**.

The study of population is critically important for three reasons:

- More people are alive at this time than at any point in the past.
- The world's population has been increasing at a more rapid rate since the end of World War II than ever before in history.
- Virtually all global population growth is concentrated in the poor, developing countries.

These facts lend an urgency to the task of understanding the diversity of population problems in the world today.

Some demographers worry that the world may contain too many people in the future. Will the world's population exceed the capacity of Earth to support people without widespread starvation, poverty, and war? Geographers cannot offer a simple yes or no answer to the issue of global overpopulation. However, geographic concepts help explain the nature of global population problems—and suggest possible solutions.

Using concepts from analysis of physical and human systems, some geographers argue that the so-called overpopulation problem is not simply a matter of the total number of people in the world, but relationships between the distribution of people and the distribution of resources. **Overpopulation** results when the number of people in an area exceeds the capacity of the environment to support life at a given level of technology and decent standard of living.

Using spatial analysis and area analysis concepts, geographers find that overpopulation is a threat in some regions of the world but not in others. The capacity of Earth as a whole to support human life may be high, but some regions have adequate resources to support the population, while others do not. In addition, people have greater control over the use and allocation of resources in some regions than they do in others. The regions with the most people are not necessarily the same as the regions with an unfavorable balance between population and resources.

Population Distribution

Population is not distributed uniformly across Earth's surface (Figure 7–1). People are clustered in some regions, while other regions contain few inhabitants. Approximately three-fourths of the world's population lives on 5 percent of Earth's surface, while the balance of Earth's surface consists of oceans and less intensively inhabited land. The portion of Earth's surface occupied by permanent human settlement is called the **ecumene.**

Population Concentrations

The world's population is clustered in five regions: East Asia, South Asia, Southeast Asia, Western Europe, and Eastern North America. These five

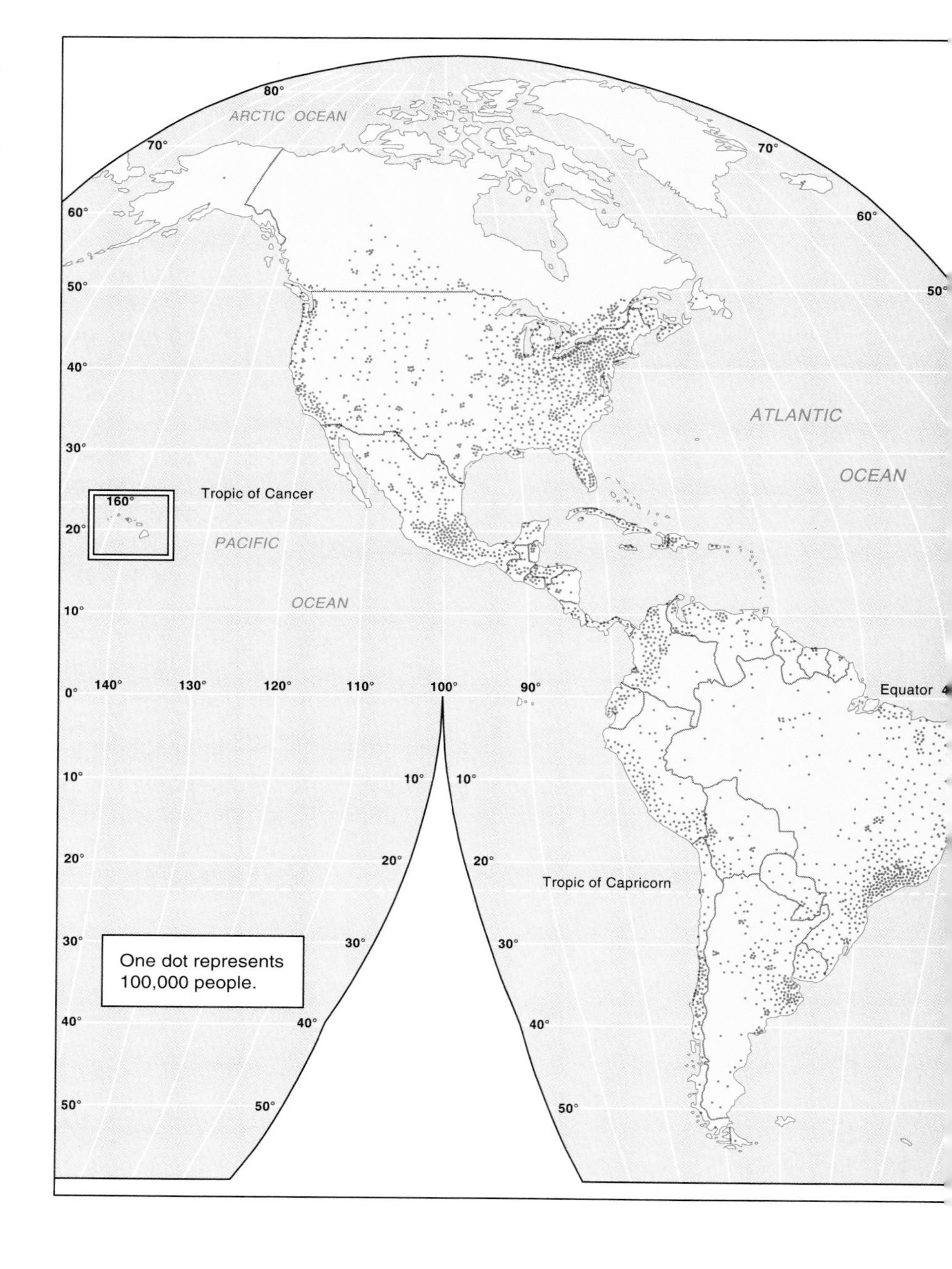

FIGURE 7–1
People are not distributed uniformly across Earth. This map illustrates the use of both density and concentration to describe the regularities in the distribution of population.
In Chapter 1, we defined density as the frequency of occurrence of something in a given unit of area. On this map, the phenomenon is people and the unit of area is square kilometers (or square miles). The density of population in much of Asia and Europe exceeds 50 persons per square kilometer (125 persons per square mile). In contrast, the density in most of the Western Hemisphere is fewer than 30 persons per square kilometer (75 persons per square mile).
Concentration is the extent of spread of something, such as people, over a given area. More than three-fourths of the world's population is clustered in these five regions which are listed according to number of people: (1) East Asia; (2) South Asia; (3) Western Europe; (4) Southeast Asia; and (5) Northeastern United States and southeastern Canada.

regions display some similarities. Most people live near an ocean, or a river with easy access to an ocean, rather than in the interior of major land masses. In fact, approximately two-thirds of the world's population lives within 500 kilometers (300 miles) of an ocean, and 80 percent lives within 800 kilometers (500 miles). Access to a river and ocean is important for facilitating transport of people's goods and diffusion of their customs and ideas.

The five population clusters occupy generally low-lying areas that have fertile soil and climates that support agriculture. With the exception of part of the Southeast Asia concentration, the regions are located in the Northern Hemisphere, between 10° and 55° north latitude. Despite these similarities, we can see significant differences in the pattern of occupancy of the land in the five concentrations.

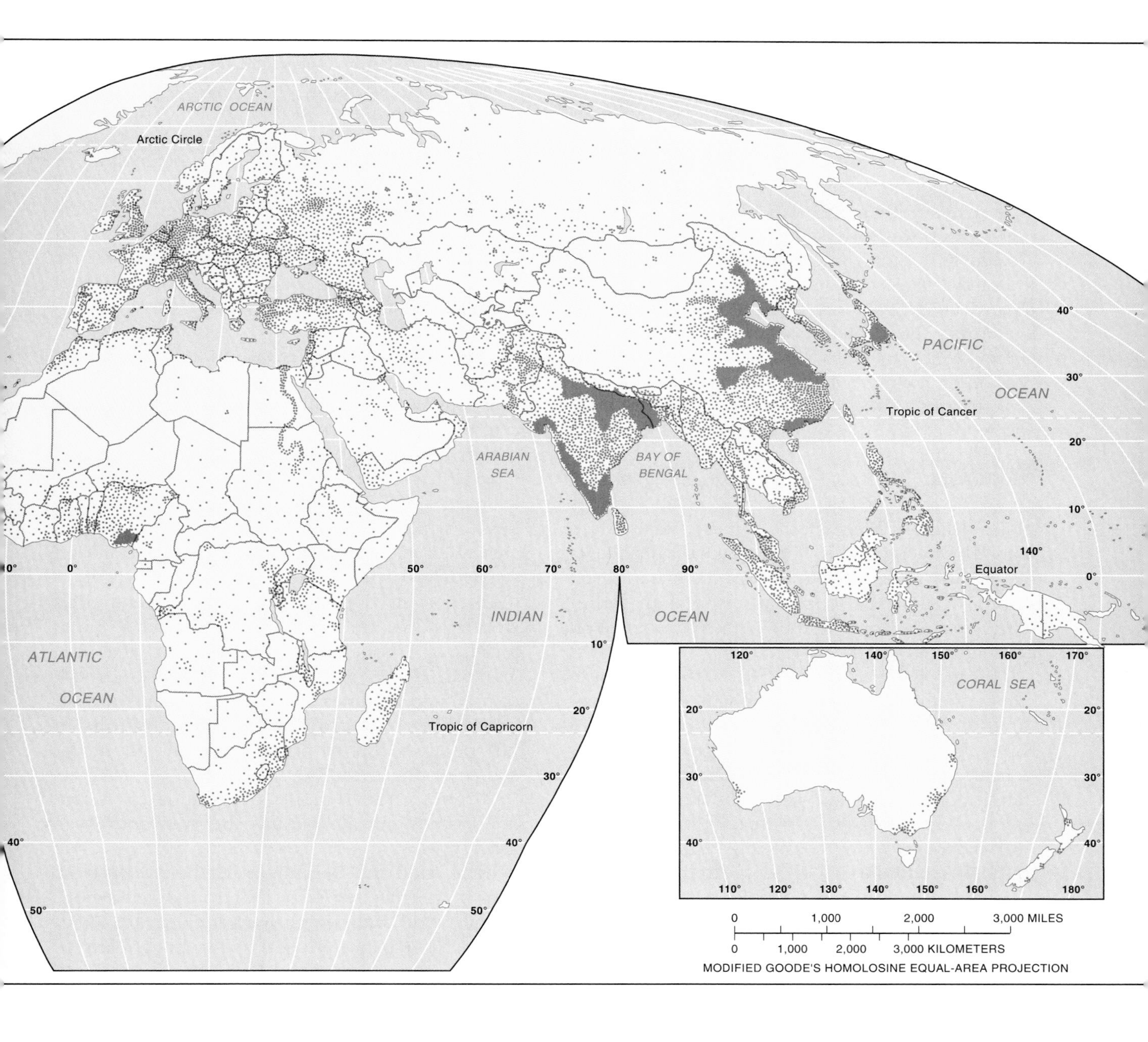

East Asia. Approximately one-fourth of the world's people live in East Asia, the largest cluster of human inhabitants. The region, bordering the Pacific Ocean, includes the eastern part of China, Japan, the Korean peninsula, and the island of Taiwan.

Five-sixths of the people in this concentration live in the People's Republic of China, the world's most populous country. China is the world's third largest country in land area, but much of its interior consists of sparsely inhabited mountains and deserts. Instead of spreading out over this land area, the Chinese population is clustered in mid-latitude warm (C) climate regions near the Pacific coast and in several fertile river valleys that extend inland, such as the Huang and the Yangtze. Although China has eight cities with more than two million inhabitants, three-fourths of the people live in rural areas, where they work as farmers.

Population is not distributed uniformly within Japan, either. More than one-fourth of the people live in two large metropolitan areas, Tokyo and Osaka, which together comprise less than 3 percent of the country's land area. Overall, more than three-fourths of the Japanese live in urban areas and work at industrial or service jobs.

South Asia. The second largest concentration of people is in South Asia, a region which includes India, Pakistan, Bangladesh, and Sri Lanka. More than 20 percent of the world's human inhabitants live in South Asia. India, the world's second most populous country, contains more than three-fourths of the people in the South Asia population concentration.

The most important concentration of people within South Asia is situated along a 1,500-kilometer (900-mile) corridor from Lahore, Pakistan, through India and Bangladesh to the Bay of Bengal. Much of this area's population is concentrated along the plains of the Indus and Ganges Rivers. Population is also heavily concentrated near India's two long coastlines—the Arabian Sea to the west and the Bay of Bengal to the east.

Like the Chinese, most people in South Asia are farmers living in rural areas. The region contains 10 cities with more than two million inhabitants, but only one-fourth of the total population lives in urban areas.

Southeast Asia. A third important Asian population cluster, and the world's fourth largest, is in Southeast Asia. Nearly 500 million people live in Southeast Asia, mostly on a series of islands that lie between the Indian and Pacific oceans. The largest population concentration is on the island of Java, inhabited by more than 100 million people. Indonesia, which consists of 13,677 islands and includes Java, is the world's fourth most populous country. A number of islands that are part of the Philippines contain high population concentrations as well. Population is also clustered along several river valleys and deltas at the southeastern tip of the Asian mainland, known as Indochina. Like China and South Asia, the Southeast Asia concentration is characterized by a high percentage of people working as farmers in rural areas.

The three Asian population concentrations together comprise over half of the world's total population, in less than 10 percent of the world's land area. Asia held a similar percentage of global population 2,000 years ago.

Europe. The world's third largest population cluster encompasses much of Europe, from the United Kingdom to western Russia. Approximately 15 percent of the world's people live in this cluster. The region includes more than two dozen countries, ranging from Monaco, which has only 1.9 square kilometers (465 acres) and a population of 30,000, to Russia, which is the world's largest country in land area (17.1 million square kilometers or 6.6 million square miles), when its Asian portion is included.

In contrast with the three Asian concentrations whose people are largely rural inhabitants, three-fourths of Europe's inhabitants live in cities and less than 20 percent are farmers. A dense network of road and rail lines links the settlements together. The highest population concentrations in Europe are near coal fields, historically the major source of power for industry.

Although the region's predominantly warm mid-latitude climate permits the cultivation of a variety of crops, Europeans do not produce enough food for themselves to survive. Instead, they import needed resources from other regions of the world. The search for additional resources has been a major incentive for Europeans to explore and colonize other parts of the world during the past six centuries. Today they continue to turn many of these resources into manufactured products.

North America. The largest concentration of population in the Western Hemisphere is in the northeastern United States and southeastern Canada. The population cluster extends along the Atlantic coast from Boston, Massachusetts, to Newport News, Virginia, and west along the Great Lakes to Chicago, Illinois. Approximately 150 million people live in the area. Like the Europeans, most Americans are urban dwellers, and fewer than 5 percent are farmers.

Sparsely Populated Regions

Human beings avoid clustering in certain physical environments. In general, people prefer not to live in regions that are too dry, too wet, too cold, or too mountainous for activities such as agriculture and industry. Instead, they seek regions that allow for a safe and at least somewhat comfortable life.

Dry lands. Areas that are too dry for farming cover approximately 20 percent of Earth's land surface. Two large desert regions exist in the world, mostly between 15° and 50° north latitude and between 20° and 50° south latitude caused by the subtropical high-pressure zones (Chapter 2). The largest desert region extends from North Africa to Southwest and Central Asia and is known by several names, including the Sahara, Arabian, Thar, Takla Makan, and Gobi. The largest desert region in the Southern Hemisphere comprises much of Australia.

Because potential evapotranspiration exceeds precipitation, the deserts generally lack sufficient water to grow crops that could feed a large population. Some people survive in the deserts by raising animals, such as camels, that are adapted to the climate. By constructing irrigation systems, people can grow crops in some parts of the desert. While dry lands are generally inhospitable to intensive agriculture, they contain natural resources useful to people—notably, much of the world's petroleum reserves. The increasing demand for this resource has led to a growth in settlements in or near deserts.

Wet lands. Tropical climate regions that receive very high levels of precipitation may also be inhos-

Well-water irrigates crops near Timbuktu, Mali. Although arithmetic density is low, the desert land cannot provide enough food for the country's people at the level of technology currently prevailing in the country. (Contact: Larry C. Price/The Stock Market)

pitable for intensive human occupation. These lands are located primarily near the equator between 20° north and south latitudes in the interiors of South America, Central Africa, and Southeast Asia. Rainfall averages more than 125 centimeters (50 inches) per year, with most areas receiving more than 225 centimeters (90 inches) per year. The combination of rain and heat hinders agriculture, because nutrients are rapidly depleted from the soil.

Precipitation may be concentrated into specific times of the year or spread through the year. In seasonally wet tropics, such as Southeast Asia, enough food can be grown to support a large population.

Cold lands. Few human beings live in very cold climates. Territory near the North and South Poles is not suitable for planting crops, and few animals can survive the extremely cold climate. Much of the land is perpetually covered by an ice cap, as you will recall from Chapter 3.

High lands. Finally, relatively few people live at high elevations. The highest mountains in the world are sparsely settled. For example, approximately half of Switzerland's land is more than 1,000 meters (3,300 feet) above sea level, but only 5 percent of the country's people live on this high land. We can find some significant exceptions, especially in Latin America and Africa. People may prefer to occupy higher lands if temperatures and precipitation are uncomfortably high at lower elevations. In fact, Mexico City, one of the world's largest cities, is located at an elevation of 2,243 meters (7,360 feet).

Population Density

The concept of density helps geographers measure the relationship between population and available resources. Recall in Chapter 1 that we defined density as the number of objects in a unit of area. Several measures of density are helpful in understanding the distribution of population.

Arithmetic density. Social scientists most frequently use the **arithmetic density** (or population density), which is the total number of people

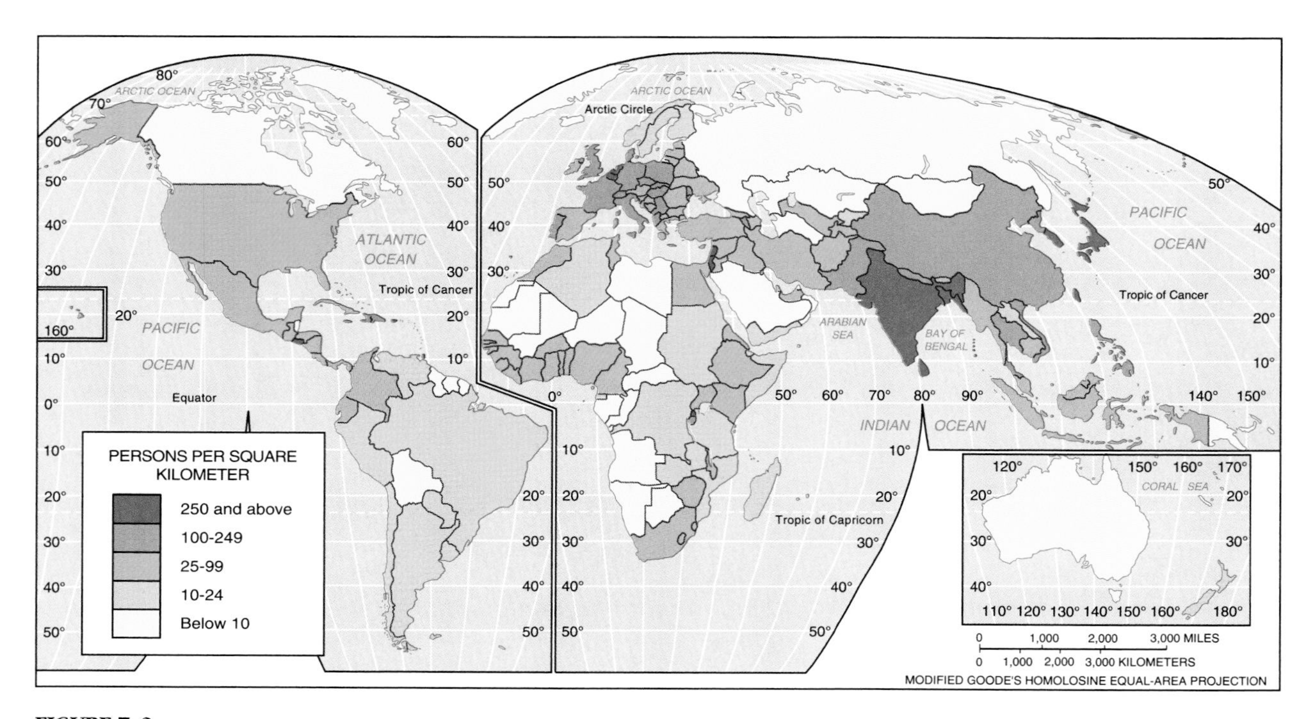

FIGURE 7–2
Arithmetic population density is the total number of people divided by the total land area. The highest population densities are found in Asia, Europe, and Central America, while the lowest are in North and South America and Australia.

divided by total land area. The arithmetic density of different countries can be easily compared, because the two pieces of information needed to calculate the measure—total population and total land area—are readily available.

For example, to compute the arithmetic or population density for the United States, we can divide the population (approximately 260 million people) by the land area (approximately 9.2 million square kilometers or 3.5 million square miles). The result shows that the United States has an arithmetic density of 28 persons per square kilometer (74 persons per square mile). By comparison, the arithmetic density is much higher in South Asia; it is approximately 810 persons per square kilometer (2,100 persons per square mile) in Bangladesh and 286 persons per square kilometer (741 persons per square mile) in India. On the other hand, the arithmetic density is only 3 persons per square kilometer (8 persons per square mile) in Canada and 2 persons per square kilometer (6 persons per square mile) in Australia (Figure 7–2).

Arithmetic density varies even more within individual countries. In the United States, for example, New York County (Manhattan Island) has a population density of approximately 21,000 persons per square kilometer (53,000 persons per square mile), while Esmeralda County, Nevada, has a population density of approximately 0.15 persons per square kilometer (0.38 persons per square mile). In Egypt, the arithmetic density is nearly 2,000 persons per square kilometer (5,000 persons per square mile) in the delta and valley of the Nile River, compared to 3 persons per square kilometer (8 persons per square mile) in the rest of the country.

Physiological density. Measures other than arithmetic density more fully explain why people are not uniformly distributed across Earth's surface. Physiological density and agricultural density provide insights into the relationship between the size of a population and the availability of resources to support life in a region. **Physiological density** is the number of people per unit area of *arable land*, which is land suitable for agriculture. The higher the physiological density, the greater the potential pressure people may place on the land to produce enough food (Figure 7–3).

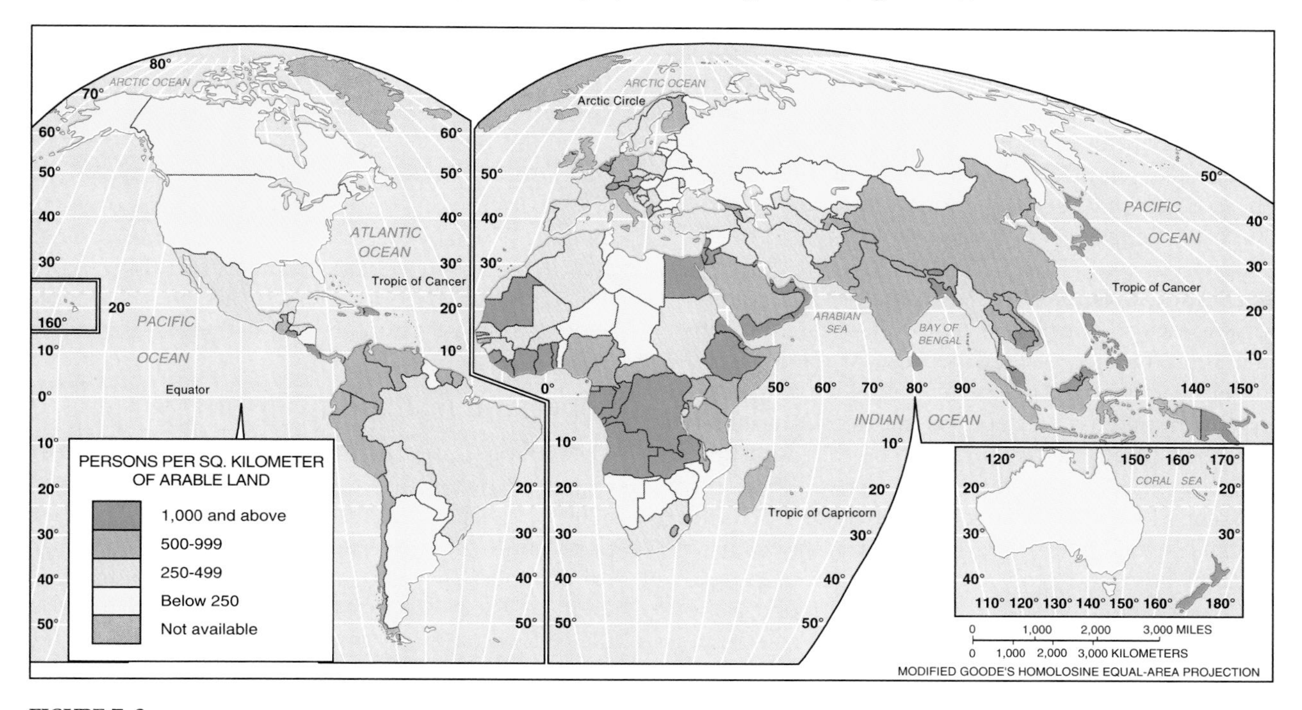

FIGURE 7–3
Physiological density is the number of people per unit of area of arable land, which is land suitable for agriculture. Physiological density is a better measure than arithmetic density of the relationship between population and the availability of resources in a society.

TABLE 7–1

Measures of density in selected countries, expressed as population per square kilometer.

	Arithmetic Density	Physiological Density	Agricultural Density	Percent Farmers	Percent Arable
Canada	3	61	2	4	5
United States	28	148	4	3	19
Egypt	59	1,952	703	36	3
India	286	596	393	66	48
Japan	370	3,083	216	7	12
Netherlands	371	1,766	71	4	21
Bangladesh	810	1,473	840	57	55
United Kingdom	239	889	18	2	27

For example, in the United States the physiological density is 148 persons per square kilometer (384 persons per square mile) of land suitable for agriculture, compared to 1,952 (5,056) in Egypt. The large difference in physiological densities demonstrates that crops grown on a given unit of land in Egypt must feed more people than in the United States.

Comparing physiological and arithmetic densities helps geographers understand the capacity of the land in different countries to yield enough food for the needs of the people. If the physiological density is much larger than the arithmetic density, as it is in Egypt, most of the country's land area may be unsuitable for intensive agriculture. In fact, all but 5 percent of the Egyptian people live in the Nile River valley and delta, because it is the only area in the country that receives enough moisture to allow intensive cultivation of crops (Table 7–1).

Agricultural density. Two countries with similar physiological densities may produce significantly different amounts of food because of different economic conditions. **Agricultural density** is the ratio of the number of farmers to the total amount of land suitable for agriculture; use of this ratio helps explain these economic differences. Relatively developed societies have lower agricultural densities because a few people are able to farm an extensive area of land and feed a large number of people. Most people are therefore available to work in factories, offices, or shops rather than in the fields. Having a large percentage of people employed in industry and services rather than in agriculture is a major element of economic development.

To develop a picture of the relationship between population and resources in a country, geographers examine its physiological and agricultural densities in combination. As Table 7–1 shows, the Netherlands has a higher physiological density that Bangladesh, but the Dutch have a much lower agricultural density than the Bangladeshis. Geographers conclude that both the Dutch and Bangladeshis put heavy pressure on the land to produce, but the more efficient Dutch agricultural system requires a much smaller number of farmers than the Bangladeshi system. Dutch farmers can generate a relatively large agricultural output from a limited resource.

Population Growth

For most of human history the size of the population was virtually unchanged at perhaps one-half million. The species multiplied in some regions and declined in others, while remaining sparse throughout the world. During this period people lived as nomadic hunters and gatherers of food.

Population Increases

The global rate of population growth sharply increased during three periods, beginning around 8000 B.C., A.D. 1750, and 1950 (Table 7–2). Each population spurt was accompanied by technological advances that gave people greater control over their physical and social environments. In turn, these technological improvements increased the capacity of Earth to support larger populations.

First period of population increase. For several hundred thousand years prior to approximately 8000 B.C., global population had increased very

TABLE 7–2

World population and growth rates.

Date	Estimated Population (millions)	Percent Average Yearly Growth in Prior Period	Number of Years in Which Population Doubles at Period Growth Rate
400,000 B.C.	0.5	—	—
8000 B.C.	5	.001	59,007
1 A.D.	300	.05	1,354
1750	791	.06	1,250
1800	978	.43	163
1850	1,262	.51	136
1900	1,650	.54	129
1950	2,517	.85	82
1994	5,607	1.78	38

modestly, at an average of only a few dozen people per year. Then around the year 8000 B.C., the annual growth rate surged to 50 times higher than in the past, and world population grew by several thousand per year. Between 8000 B.C. and A.D. 1750, global population increased from approximately 5 million to 800 million.

What caused the burst of population growth around 8000 B.C.? Scientists point to the **agricultural revolution**, which is the time when human beings first domesticated plants and animals and no longer relied entirely on hunting and gathering for food. By growing plants and raising animals, human beings created larger and more predictable sources of food, and more people could survive.

The agricultural revolution involved a series of accidents and experiments over a period of several hundred years. Major advances in agriculture often are credited to people in the Fertile Crescent, an area that extends from the eastern edge of the Mediterranean Sea to present-day Iran. However, anthropologists have discovered that plant and animal domestication apparently originated independently in several locations.

Second period of population increase. For nearly 10,000 years after the agricultural revolution, the world population grew at a fairly steady pace. After around 1750, the world's population suddenly began to grow ten times faster than in the past. The average annual increase jumped from several thousand in the early eighteenth century to several hundred thousand in the late eighteenth century. Global population rose from approximately 800 million in 1750 to 2.5 billion in 1950, an average annual increase of 0.6 percent per year.

The second spurt in the rate of population increase resulted from the **industrial revolution**, which began in England in the late eighteenth century and spread to the European continent and North America during the nineteenth century. The industrial revolution involved a series of improvements in industrial technology that transformed the process of manufacturing goods. The result of this industrial transformation was the production of unprecedented levels of wealth, some of which was used to make the community a healthier place to live.

New machines helped farmers to increase agricultural production and feed the rapidly growing population. More efficient agriculture opened up the opportunity for people to work in factories that produced other goods and generated enough food for a growing number of industrial workers and their families. The wealth produced by the industrial revolution was also used to improve sanitation and personal hygiene. Sewer systems were installed in cities, and food and water supplies were protected. As a result of these public improvements, people were healthier and lived longer.

Third period of population increase. The third dramatic increase in global population began in the late 1940s after World War II. The average annual population increase jumped from approximately one-half of one percent early in the twentieth century to nearly two percent by the middle of the twentieth century. Instead of adding a few million people per year, as was the case at the beginning of the twentieth century, the world has grown by more than 70 million per year since the late 1940s. During the 1990s, world population is increasing by about 90 million people per year.

The recent era of population growth has been caused by the **medical revolution**. Medical and health care technology that was invented in Europe and North America has diffused to the poorer countries of Latin America, Asia, and Africa. Improved medical and health care practices have eliminated many of the traditional causes of death in poorer countries and enabled more people to have longer and healthier lives. Penicillin, vaccines, and insecticides such as DDT effectively and inexpensively have controlled many infectious diseases, such as tuberculosis, smallpox, and malaria. Current death rates are 60 to 80 percent lower in Africa, Asia, and Latin America than in the late 1940s.

Components of Population Change

The rate of **natural increase** is the percentage by which a population grows in a year; it is computed by subtracting the crude death rate from the crude birth rate. The term *natural* means that a country's growth rate excludes migration.

During the past decade, the world rate of natural increase was 1.7, meaning that world population grew in one year by 1.7 percent. For many things in life—such as an examination grade—the difference between 1 percent and 2 percent is not important. In population studies, the difference is critical.

The rate of natural increase affects the **doubling time**, which is the number of years needed to double a population, assuming a constant rate of natural increase. At the current rate of natural increase, the population of the world will double in approximately 40 years. Should the natural increase rate decline to 1.0, global population will double in approximately 70 years. A century from now, people will notice the difference between these two natural increase rates. The current rate would place global population in the year 2100 at 35 billion, but if the natural increase rate during the twenty-first century declines to 1.0, the world's population would be less than 20 billion in 2100.

Very small changes in the natural increase rate dramatically affect the size of the population, because the base from which we derive the percentage is so high. When we multiply a natural increase rate of 1.7 by the current global population base of more than 5 billion, the result is an annual increase of about 90 million people. If the natural increase rate immediately dropped to 1.0, then the annual population increase would decline to approximately 55 million. As the base continues to grow in the twenty-first century, a change of one-tenth of 1 percent will produce very large swings in population growth.

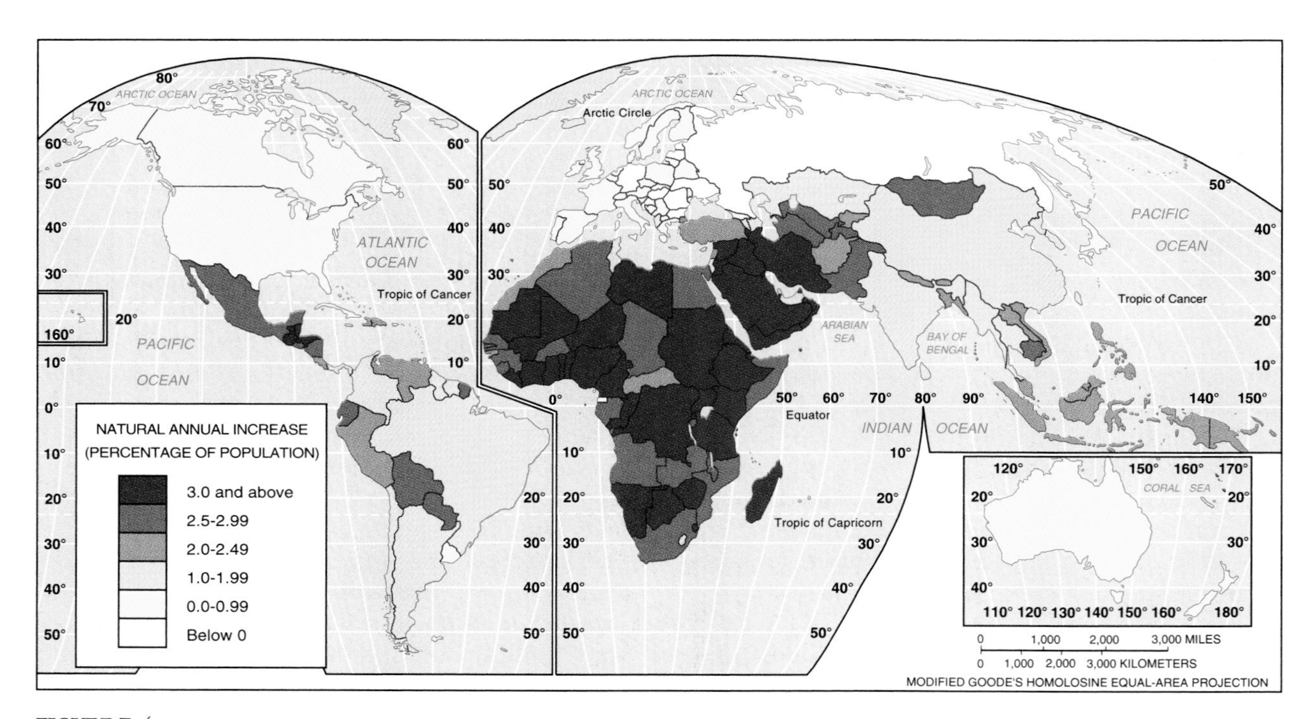

FIGURE 7–4
The natural increase is the percentage by which the population of a country grows in a year. World average in recent years has been about 1.7 percent per year. Countries with the highest natural increase rates are concentrated in Africa and Southwest Asia.

Distribution of natural increase rates. The distribution of natural increase rates shows significant regional differences (Figure 7–4). The natural increase rate exceeds 3.0 in much of Africa and southwestern Asia, while at the other extreme, the United States, Canada, and every European country with the exception of Albania and Iceland have natural increase rates below 1.0. Japan, Australia, New Zealand, and a handful of smaller countries also have natural increase rates below 1.0. Several European countries have negative natural increase rates, meaning that in the absence of any migration into the country, population actually would be declining.

Not only is world population increasing faster than ever before, but virtually all of the growth is concentrated in poorer countries. This year, approximately 64 percent of the world's population growth is in Asia, 23 percent in Africa, and 10 percent in Latin America. Europe and North America account for only 3 percent of global population growth. Regional differences in natural increase rates mean that virtually all of the world's additional people live in countries whose economic and social systems are the least able to support them. To explain the differences in growth rates, geographers point to regional differences in fertility and mortality rates.

Fertility. To study the number of births at the national or global scale, geographers most frequently refer to the **crude birth rate**, which is the total number of live births in a year for every 1,000 people alive in the society. A crude birth rate (or CBR) of 20, for example, means that for every 1,000 people in a country, 20 babies are born over a one-year period.

The word *crude* means that we are concerned with society as a whole, rather than with particular individuals. In communities with an unusually large number of people of a certain age—such as a college town—we may study separate birth rates for women or men of each age. These numbers, for example, are age-specific birth rates rather than crude birth rates.

Demographers also use the **total fertility rate** (TFR) to measure the number of births in a society. The total fertility rate is the average number of children a woman will have throughout her child-bearing years (roughly age 15 through 49). To compute the total fertility rate, scientists must assume that, as a woman reaches a particular age in the future, she will be just as likely to have a child as are women of that age today. The crude birth rate provides a picture of a society as a whole in a given year, while the total fertility rate attempts to predict the future behavior of individual women.

The world map of crude birth rates mirrors the distribution of natural increase rates. As was the case with natural increase rates, the highest crude birth rates are in Africa and the lowest are in Europe and North America (Figure 7–5). Most African countries have a crude birth rate over 40, and several exceed 50. Crude birth rates over 30 are common in Asia and Central America. On the other hand, the United States, Canada, and every European country with the exception of Albania have crude birth rates below 20. Japan, Australia, New Zealand, and a handful of smaller countries also have crude birth rates below 20.

Mortality. The **crude death rate** compares the total number of deaths and the total number of people living in a country in one year. Comparable to the crude birth rate, the crude death rate is expressed as the annual number of deaths per 1,000 population. Geographers also compute different death rates for specific age groups or for males and females.

The **infant mortality rate** is the annual number of deaths of infants under age one year compared to the total number of live births. The infant mortality rate is usually expressed as the number of deaths among infants per 1,000 births.

The global distribution of infant mortality rates follows the familiar pattern. The highest rates are in the poorer countries of Africa and Asia, while the lowest rates are in Europe, North America, and other wealthier societies. Infant mortality rates exceed 100 in many African and Asian countries, meaning that over 10 percent of all babies die before reaching their first birthday. Infant mortality rates are less than 10 in most European countries, the United States, Canada, Japan, Australia, and New Zealand (Figure 7–6).

In general, the infant mortality rate reflects a country's health care system. We find lower infant mortality rates in countries with well-trained doctors and nurses and large supplies of hospitals and medicine. Although the United States is well-endowed with medical facilities, it suffers from

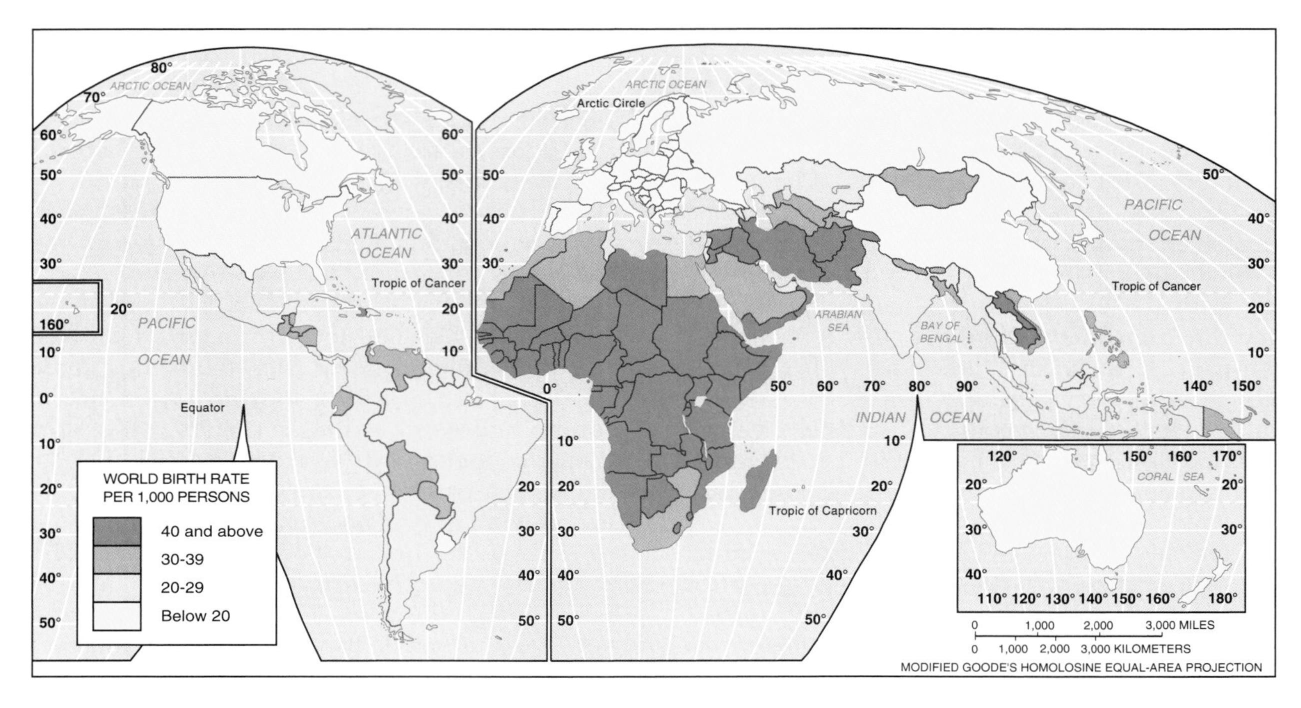

FIGURE 7–5
The crude birth rate is the total number of live births in a year for every 1,000 people alive in the society. The global distribution of crude birth rates parallels that of natural increase rates. Again, the highest crude birth rates are found in Africa and Southwest Asia, whereas the lowest are in Europe.

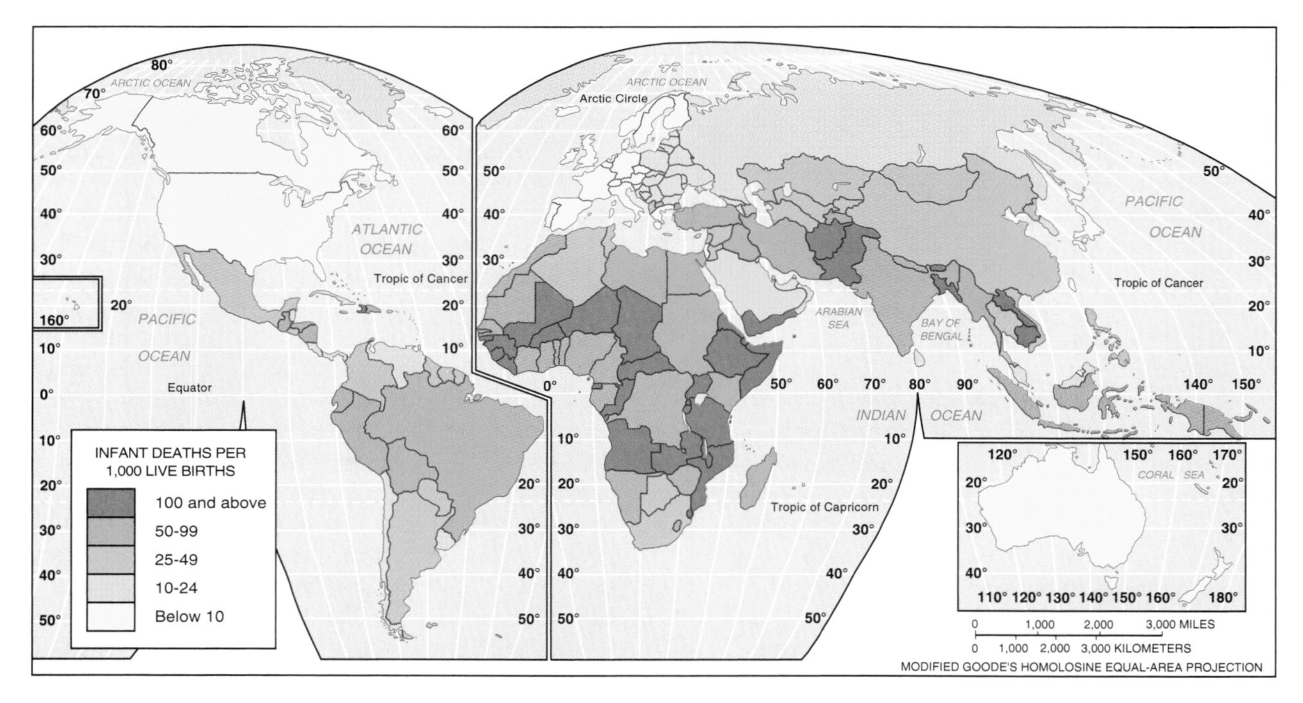

FIGURE 7–6
The infant mortality rate is the number of deaths of infants under age one per 1,000 live births in a year. European and North American countries generally have infant mortality rates of under 10 per 1,000, whereas rates of more than 100 per 1,000 are common in Africa.

somewhat higher infant mortality rates than Canada and most Western European countries. This is because in the United States health care is applied somewhat inconsistently. African-Americans and other minorities have infant mortality rates that are twice as high as the national average. Many health experts attribute this to the fact that many poor people in the United States, especially minorities, cannot afford good health care for their infants, and free health care is not as easily accessible to the general population as it is in Canada and some Western European countries.

Life expectancy at birth measures the average number of years a newborn infant can expect to live under current mortality levels. Like the infant mortality, crude birth, and natural increase rates, life expectancy varies sharply among different regions. Babies born today can expect to live into their early fifties if they are African and into their late seventies if they are Western European or North American (Figure 7–7).

The global distribution of crude death rates does not follow the pattern set by the other mortality and fertility variables. Consistent with the other demographic characteristics, the highest crude death rates are in Africa. But perhaps unexpectedly, the lowest crude death rates are in Latin America and Asia, rather than in the wealthy countries of North America and Europe (Figure 7–8).

Furthermore, the spread among different countries between the highest and lowest crude death rates is relatively low. Crude birth rates for individual countries range from around 10 to 50, a spread of more than 40. However, the highest crude death rate is only in the low 20s, and the difference between the highest and lowest rates is around 20.

Why does Sweden, one of the world's wealthiest countries, have a crude death rate higher than Thailand, one of the poorest? Why does the United States, with its extensive system of hospitals and physicians, have a higher crude death rate than Costa Rica or Panama? The answer is that the populations of different countries are at various stages in a process discussed in the following section, called the demographic transition.

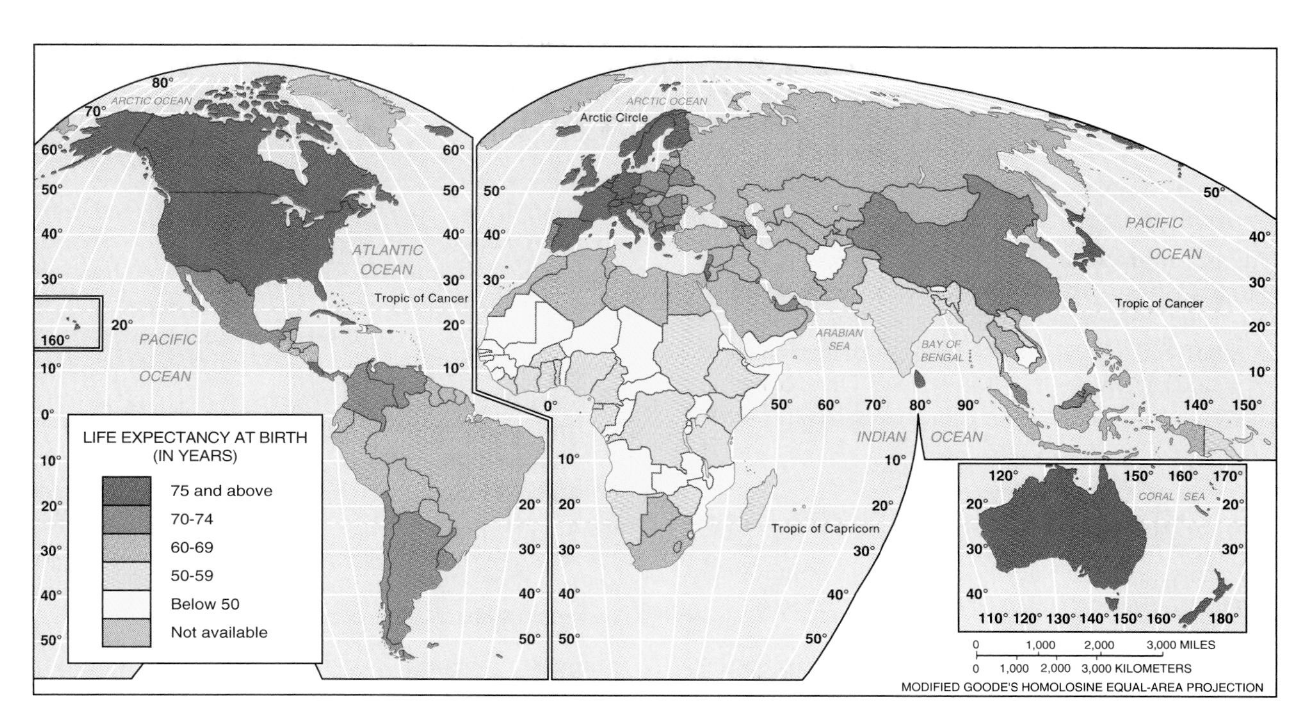

FIGURE 7–7
Babies born this year are expected to live until their mid-sixties. Life expectancy for babies, however, ranges from the early forties in several African countries to the late seventies in much of Europe, Australia, North America, and Japan.

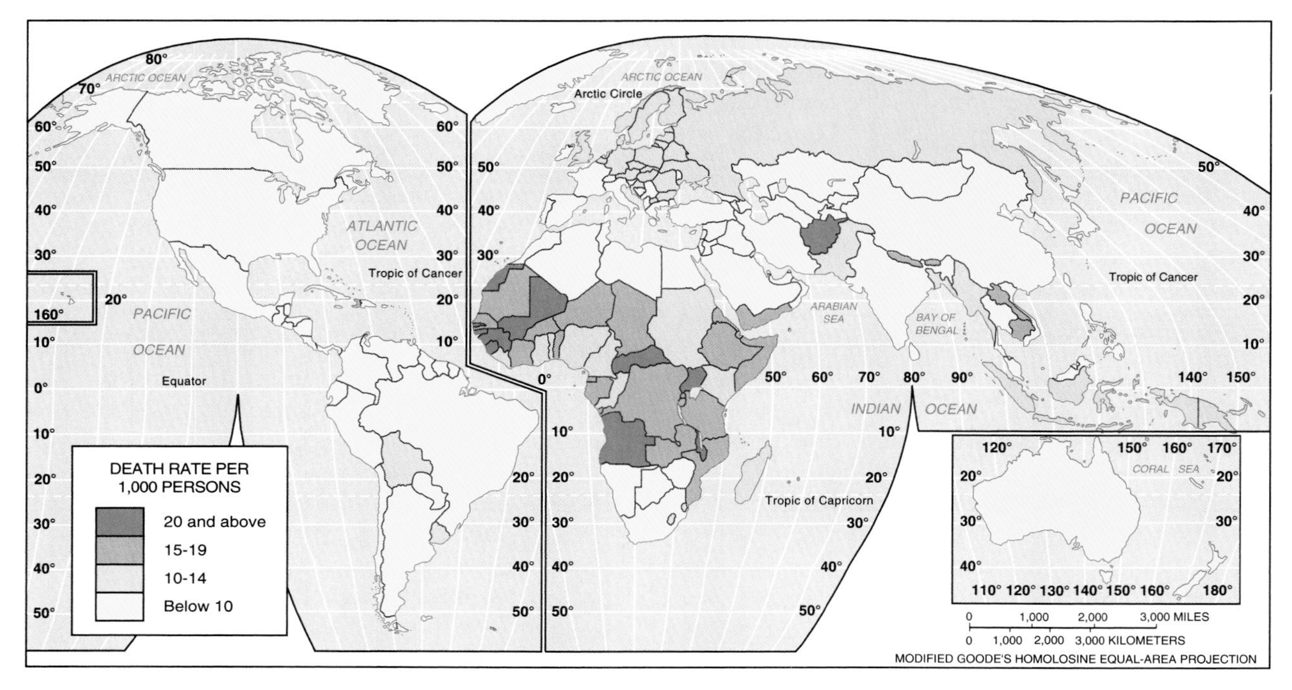

FIGURE 7–8
Crude death rate is the total number of deaths in a year for every 1,000 people alive in the society. The global pattern of crude death rates differs from those for the other demographic variables already mapped in this chapter. First, while Europe has the lowest natural increase, crude birth, and infant mortality rates, it does not have low crude death rates. Second, the variance between the highest and lowest crude death rates is much lower than was the case for the crude birth rates. The concept of the demographic transition helps to explain the distinctive distribution of crude death rates.

Demographic Transition

The **demographic transition** is a process of population change with several stages, and every country is in one of the stages. The process has a beginning, middle, and end, and—barring a catastrophe such as a nuclear war—it is irreversible. Once a country moves from one stage of the process to the next it does not revert to an earlier stage.

Stages in the Demographic Transition

The demographic transition consists of four stages (Figure 7–9). The stages are described below.

Stage 1. The first stage of the demographic transition consists of generally high crude birth and death rates. These rates may vary considerably from one year to the next, but over the long term the rates are roughly comparable. As a result, the natural increase rate is very low.

Survival is unpredictable in a Stage 1 society. The population may depend on hunting and gathering for food. When food is easily obtained, the population increases; but it declines in times of shortage. People who practice settled farming, or farming in one place, prosper during abundant harvests and suffer when unfavorable climatic conditions result in low output. Wars and diseases also have their effects on the death rate in a Stage 1 society.

Most of human history saw every existing society in Stage 1 of the demographic transition, but changes have come rapidly, and today no Stage 1 country remains. Every country has moved on to at least Stage 2 of the demographic transition—and with that movement has experienced profound changes.

Stage 2. In Stage 2 of the demographic transition the crude death rate suddenly plummets, while the crude birth rate remains roughly the same as in Stage 1. Because the difference between the crude birth rate and crude death rate is very high, the natural increase rate is also very high.

Countries in Europe and North America entered Stage 2 of the demographic transition in the late eighteenth or nineteenth century. The change has come in the twentieth century to countries in Africa, Asia, and Latin America. New technology that permits increases in the permanent food supply and control of diseases accounts for the rapid decline in the crude death rate.

Stage 3. A country moves from Stage 2 to Stage 3 of the demographic transition when the crude birth rate begins to drop sharply. The crude death rate continues to fall in Stage 3 but at a much slower rate than in Stage 2. The population continues to grow because the crude birth rate is still higher than the crude death rate. However, the rate of natural increase is more modest in Stage 3 countries than in Stage 2 countries, because the gap between the crude birth and death rates narrows.

European and North American countries generally moved from Stage 2 to Stage 3 of the demographic transition during the first half of the twentieth century. Some countries in Africa, Asia, and Latin America have moved to Stage 3 in recent years, but others remain in Stage 2.

The sudden drop in the crude birth rate during Stage 3 results from a different reason than that for the rapid decline of the crude death rate of Stage 2. The crude death rate declines in Stage 2 following introduction of new technology into the society, but the crude birth rate declines in Stage 3 because of changes in social customs.

A society enters Stage 3 of the demographic transition when people choose to have fewer children. The decision is partly a delayed reaction to a decline in mortality, especially the infant mortality rate. In Stage 1 societies, the survival of any one infant could not be confidently predicted, and families typically had a large number of babies to improve the chances of some living to adulthood. Medical and health care practices introduced in Stage 2 societies greatly improve the probability of an infant surviving but many years elapse before families react by conceiving fewer babies.

Economic changes in Stage 3 societies also induce people to have fewer babies. People in Stage 3 societies are more likely to live in cities rather than the countryside and work in offices, shops, or factories rather than on farms. Farmers often consider a large family to be an asset because children can do some farm work. In contrast, children living in cities are generally not economic assets to their parents, because they do not work in most types of urban jobs. Children are either unable to perform the work, or they are prohibited from working. In addition, urban homes are relatively small and may not have enough space to accommodate large families.

Stage 4. A country reaches Stage 4 of the demographic transition when the crude birth rate declines to the point where it equals the crude death rate, and the natural increase rate approaches 0. **Zero population growth (ZPG)**, a term which is often applied to Stage 4 countries, may occur when the crude birth rate is still slightly higher than the crude death rate. Some females die before reaching child-bearing years, and the num-

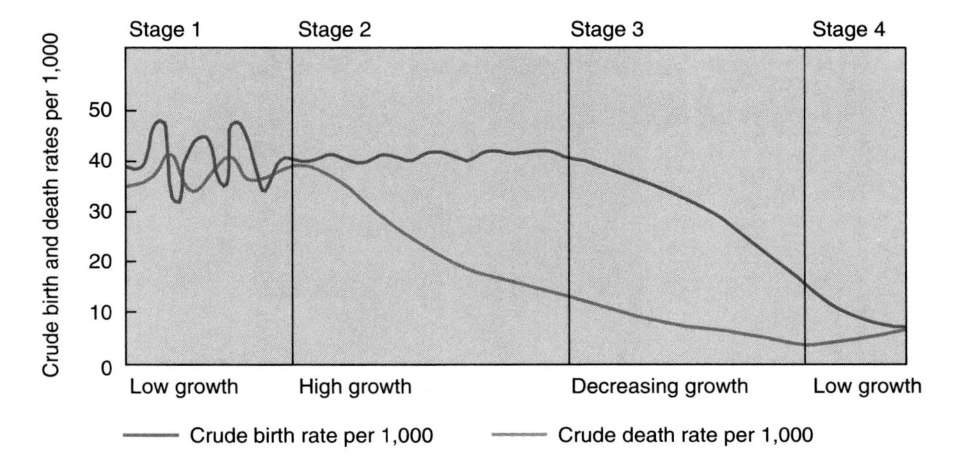

FIGURE 7–9
The demographic transition model consists of four stages. *Stage 1*—very high birth and death rates produce virtually no long-term natural increase. *Stage 2*—rapidly declining death rates combined with very high birth rates produce very high natural increase. *Stage 3*—birth rates rapidly decline, while death rates continue to decline; natural increase rates begin to moderate. *Stage 4*—very low birth and death rates produce virtually no long-term natural increase.

ber of females in their child-bearing years can vary. To account for these discrepancies, zero population growth is frequently expressed as the total fertility rate that results in a lack of change in the total population over a long term. At this time, a TFR of approximately 2.1 produces ZPG. However, a country experiencing a high level of migration into the country may need a lower total fertility rate to achieve a stable population.

Several countries in Western and northern Europe have reached Stage 4 of the demographic transition, including Sweden, Germany, and the United Kingdom. The United States has not completely moved into Stage 4 because birth rates remain higher among some groups, such as recent immigrants from Latin America. In several European countries, including Denmark, Germany, and Hungary, the crude birth rate slips below the crude death rate in some years. However, total population has not been declining, at least not in Denmark and Germany, because people have been immigrating from other countries.

Social customs again explain the movement from one stage of the demographic transition to the next. Increasingly, women in Stage 4 societies work in the labor force, rather than work at home with children. When most families lived on farms, economic activities and child rearing were conducted at the same place, but in urban societies most parents must leave the home to work in an office, shop, or factory. Employed parents must arrange for child care during their working hours.

Changes in lifestyle also encourage smaller families. People who have access to a wider variety of birth-control methods are more likely to use some of them. When people devote a great deal of time to their jobs or careers, they tend to limit their family size, often because of the time away from the family or the energy their jobs can take. More people start families in their thirties or even in their forties due to career or lifestyle decisions, and these families are generally smaller as well. Also, with increased income and leisure time, more people choose lifestyles with entertainment and recreation activities that tend to exclude young children, such as attending cultural events.

Natural increase rates have been negative in Russia during the 1990s as a result of rapidly falling crude birth rates. Russians may be choosing to bear fewer children, because of widespread uncertainty about the future of the country following the breakup of the Soviet Union and the dismantling of the communist economic system.

One of India's 1.5 million census takers questions a family in a poor section of Delhi, India, as the government tries to find out how many people actually live in the country. (Reuters/Bettmann)

A country that passes through all four stages of the demographic transition has in some ways completed a cycle—from little or no natural increase in Stage 1 to little or no natural increase in Stage 4. Two crucial demographic differences underlie this process, however. First, at the beginning of the demographic transition the crude birth and death rates are high—35 to 40 per 1,000—while at the end of the process the rates are approximately 10 per 1,000. Second, the total population of the country is much higher in Stage 4 than in Stage 1.

Population Pyramids

Countries display distinctive population structures, depending on their stage in the demographic transition. The population of countries varies in two ways: first, the percentage of people in each age group, and second, distribution between males and females.

We can display the distribution by gender and age groups of a country's population on a bar graph called a **population pyramid**. A population pyramid normally shows the percentage of the total population in each five-year age group, with the youngest group (zero to four years old) at the base of pyramid and the oldest group at the top. The length of the bar represents the percentage of the total population contained in that group. Males are usually shown on the left side of the pyramid and females on the right (Figure 7–10).

The shape of a pyramid is determined primarily by the crude birth rate in the country. A country with a high crude birth rate has a relatively large number of young children. Consequently, the base of the population pyramid is broad. On the other hand, if the country has a relatively large number of older people, the top of the pyramid is wider, and the graph looks more like a rectangle than a pyramid.

Age distribution. The age structure of the population is extremely important in understanding similarities and differences among different countries in the world. The most important factor is the **dependency ratio**, which is the percentage of people who are too young or too old to work compared to the number of people in their productive years. The larger the percentage of dependents, the greater is the financial burden on those who are working to support those who cannot.

To compare the dependency rates of different countries, we can divide the population into three age groups: zero to fourteen, fifteen to sixty-four, and sixty-five and over. People under fifteen and over sixty-four are normally classified as dependents. Approximately one-half of all people living in countries in Stage 2 of the demographic transition are dependents, compared to only one-third in Stage 4 countries. Consequently, the dependency ratio is nearly 1:1 in Stage 2 countries, whereas in Stage 4 countries the ratio is 1:2 (1 dependent for every 2 workers). Young dependents outnumber elderly ones by ten to one in Stage 2 countries, but the numbers of young and elderly dependents are roughly equal in Stage 4 countries.

In nearly every African country and in many Asian and Latin American countries, more than 40 percent of the people are under age fifteen. This high percentage follows from the high crude birth rates in these regions. In contrast, the percentage of children under fifteen is around 20 percent in the European and North American countries at or near Stage 4 of the demographic transition.

A large percentage of children strains the ability of a poorer country to develop needed services, such as schools, hospitals, and day-care centers. When the children reach the age of leaving school they must find jobs, but the community must continue to allocate scarce resources to meeting the needs of the still growing number of young people.

As countries pass through the stages of the demographic transition, the percentage of elderly people increases. The higher percentage partly reflects the lower percentage of young people produced by declining crude birth rates. Older people also benefit in Stage 4 countries from improved medical care and higher incomes. People over age sixty-five exceed 15 percent of the population in several European countries, such as Denmark, Sweden, the United Kingdom, and Germany, compared to less than 5 percent in most African countries.

Older people must receive adequate levels of income and medical care after they retire from their jobs. The "graying" of the population places a burden on European and North American social services and governments to meet these needs.

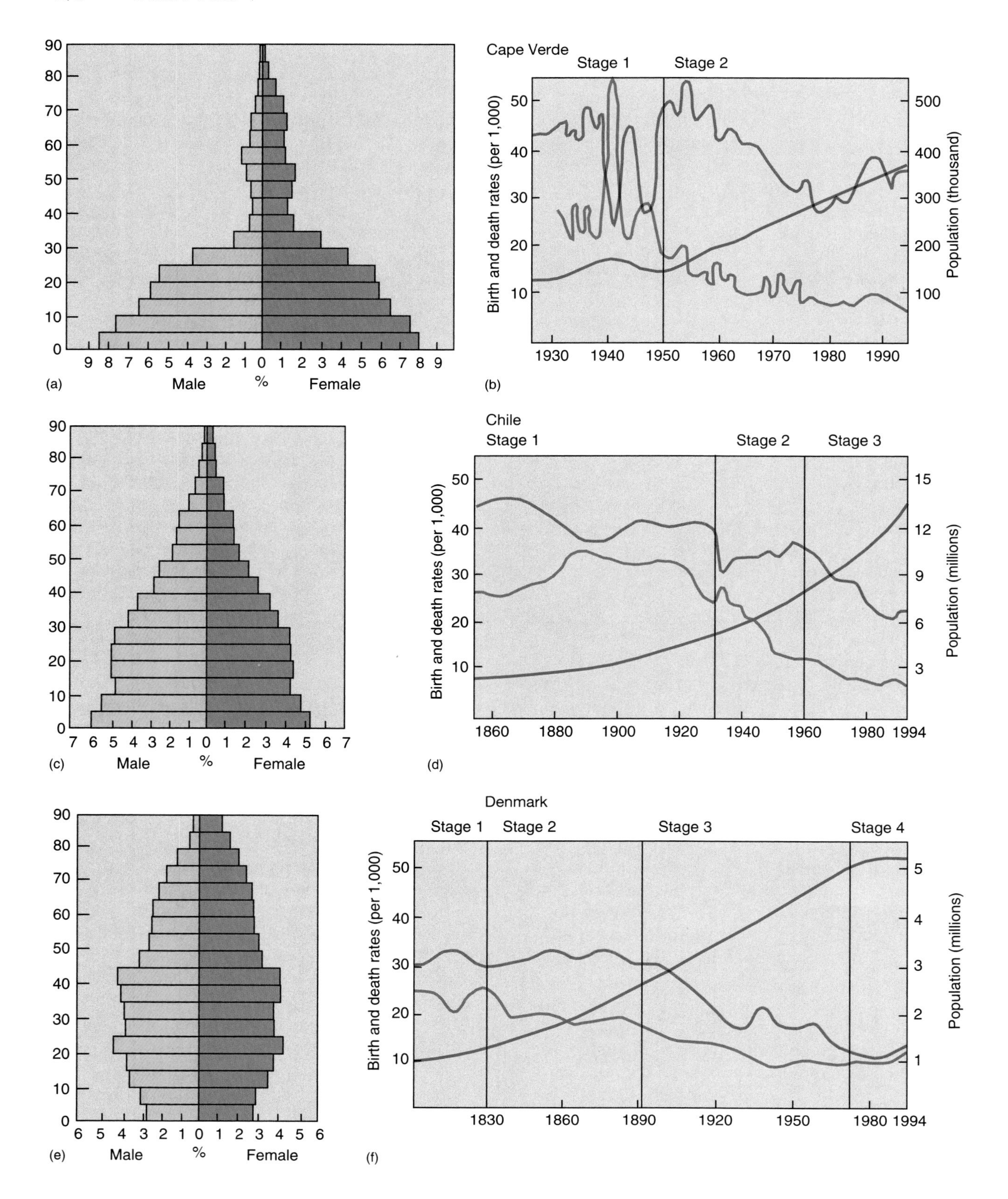

Cape Verde
Stage 1
Stage 2
Birth and death rates (per 1,000)
Population (thousand)
Chile
Stage 1
Stage 2
Stage 3
Birth and death rates (per 1,000)
Population (millions)
Denmark
Stage 1
Stage 2
Stage 3
Stage 4
Birth and death rates (per 1,000)
Population (millions)
Male
%
Female
(a)
(b)
(c)
(d)
(e)
(f)

More than one-fourth of all government expenditures in many European and North American countries go to social security, health care, and other programs for the elderly population. Because of the larger percentage of elderly people, countries in Stage 3 or 4 of the demographic transition like the United States and Sweden have higher crude death rates than Stage 2 countries.

Sex ratio. The number of males per 100 females in the population, which is called the **sex ratio**, varies among countries, depending on the particular birth and death rates. In general, slightly more males than females are born, but males have higher death rates. The ratio of men to women is about 95:100 (that is, 95 men for each 100 women) in Europe and North America, compared to 102:100 in the rest of the world.

In the United States, males under 19 exceed females by a ratio of 105 to 100. Women first outnumber men at about age 30 and comprise 60 percent of the population over age 65. High mortality rates during childbirth partly explains the lower percentage of women in poorer countries. The difference also relates to the age structure because poorer countries have a larger percentage of young people, where males generally outnumber females, and a lower percentage of elderly people, where females are much more numerous.

Societies with a high rate of immigration typically have more males than females because males are more likely to undertake long-distance migration. Frontier areas and boom towns typically have more men than women.

Migration

In addition to natural increase, a region's rate of population growth is influenced by **migration**, which is a permanent move to a new location. Migration has two forms, emigration and immigration. **Emigration** is migration *from* a location while **immigration** is migration *to* a location. Given two locations, A and B, some people emigrate from A to B while at the same time others immigrate to A from B. The difference between the number of immigrants and the number of emigrants is the **net migration**.

If the number of immigrants exceeds the number of emigrants, the net migration is positive, and the region has what is called net in-migration. If the number of emigrants exceeds the number of immigrants, the region has net out-migration. A region's population rises because of births and in-migration of people from other regions, while population declines as a result of deaths and out-migration.

Net in-migration currently accounts for nearly one-third of the total U.S. population growth, and by 2020 it will account for more than one-half. Were it not for net in-migration, total population already would be declining in several European countries. The impact of migration on population change is even greater in regions within countries. Two-thirds of Florida's population growth results from net in-migration, while several northern states have experienced population declines due to large-scale emigration.

If a series of global population distribution maps similar to Figure 7–1 were created for several points in time and shown one after the other like frames in a movie film, the pattern would change constantly. Some regions would increase relatively rapidly, while other regions would decline. Much of the change results from patterns of migration.

Migrants head for one of two types of destinations—either other countries or other locations in the same country. **International migration** refers to permanent movement from one country to another, whereas **internal migration** refers to permanent movement within a country. Regardless of destination, individuals migrate for a variety of push and pull factors. **Push factors** induce people to move away from their old residence, whereas **pull factors** attract people to a particular new location. Most people migrate because of economic push and pull factors, but political and environmental push and pull factors influence some migration decisions. Political push factors are discussed in Chapter 9.

FIGURE 7–10
Population pyramid and demographic transition for Cape Verde (now in Stage 2 of the demographic transition), Chile (now in Stage 3), and Denmark (now in Stage 4).
(a) Population pyramid for Cape Verde
(b) Demographic transition for Cape Verde
(c) Population pyramid for Chile
(d) Demographic transition for Chile
(e) Population pyramid for Denmark
(f) Demographic transition for Denmark

Many people emigrate from countries that are in Stage 2 of the demographic transition. A country with rapid rates of natural increase may offer limited prospects for economic advancement. Most inhabitants are farmers in countries entering Stage 2, but rural areas may be unable to absorb large increases in the number of people who wish to farm.

In the past, emigrants from countries in Stage 2 of the demographic transition could head for sparsely inhabited frontier regions where land could be cleared for farming. Today, emigrants from countries with high natural increase rates are more likely to reach highly urbanized, densely populated regions in Stage 3 or 4 of the demographic transition.

Immigration to the United States

Historically, the largest stream of international migrants for economic reasons has been from Europe. Rapid population growth pushed emigrants from Europe, especially after 1800. Application of new technology spawned by the industrial revolution—such as public health measures, medicine, and food—produced a rapid decline in the death rate and pushed much of Europe into Stage 2 of the demographic transition.

As the population increased, many Europeans found opportunities for economic advancement limited. Family farms, for example, often had to be divided among a great number of relatives, and the average farm was becoming too small to be profitable. In other cases, farms were inherited by the oldest son, leaving younger siblings without land.

To promote more efficient agriculture, European governments forced the consolidation of several small farms into larger units. In England the consolidation policy was known as the *enclosure movement* (Chapter 13). The enclosure movement forced millions of people to migrate from rural areas. Displaced farmers could choose between moving to cities and working in factories or migrating to another part of the world where farmland was plentiful.

At the same time, many Europeans were pulled to other countries by the prospect of economic improvement. European migrants were most attracted to the mid-latitude climates of North America, Australia, New Zealand, southern Africa, and southern South America, where farming methods used in Europe could be most easily transplanted.

Emigration from Europe. The most popular destination for European emigrants has been the United States. Of the 60 million European migrants since 1500, 37 million have come to the United States. Between 1776 and 1840, approximately 15,000 migrants came each year to the United States, 90 percent of them from Britain. During the 1840s and 1850s, the level of immigration to the United States surged to an annual level of more than 200,000. More than 90 percent of these immigrants came from northern and western Europe, including more than 40 percent from Ireland and more than 30 percent from Germany.

Immigration to the United States declined somewhat during the 1860s as a result of the Civil War but began to climb again in the 1870s. A second peak was reached during the 1880s, when more than one-half million people per year immigrated to the United States. More than three-fourths of the immigrants during the late 1800s again came from northern and western Europe. Germans accounted for one-third of all immigrants to the United States

This Italian family immigrated through Ellis Island, in New York harbor, in 1905. (Lewis W. Hine)

during this period, and large numbers of Irish emigrated. However, other countries in northern and western Europe sent increasing numbers of migrants, especially the Scandinavian countries of Norway and Sweden, which had entered Stage 2 of the demographic transition.

Economic problems in the U.S. discouraged immigration during the early 1890s, but by the end of the decade the level reached a third peak. One million people per year migrated to the United States during the first 15 years of the twentieth century. Nearly one-fourth each came from Italy, Russia, and Austria-Hungary, which encompassed portions of present-day Austria, Belarus, Bosnia and Herzegovina, Croatia, Czech Republic, Estonia, Hungary, Italy, Latvia, Lithuania, Moldova, Poland, Romania, Slovakia, Slovenia, Ukraine, and Yugoslavia.

The era of massive European migration to the United States ended with the start of World War I in 1914, and the level has steadily declined since then. Europeans accounted for one-third of all immigrants to the United States in the 1960s and only 10 percent since 1980.

Current origin of U.S. immigrants. The national origin of immigrants to the United States has changed sharply since the 1960s. Today, more than three-fourths of all immigrants come from Latin America and Asia. Increased desire to emigrate from Latin America came a bit earlier. Latin Americans have comprised more than 40 percent of all immigrants since the 1960s. In recent years, one-fourth of all immigrants to the United States have come from Mexico. Jamaica, Haiti, the Dominican Republic, and several other Caribbean islands have also become major points of origin for immigrants to the United States (Figure 7–11).

Several Latin American countries currently in Stage 2 of the demographic transition siphon off a large percentage of their population growth to the United States. For example, Jamaica, a country of approximately 2.5 million, has a crude birth rate of 24 per 1,000, a crude death rate of 5, and a natural increase rate of 1.8 percent per year. Otherwise stated, in one year Jamaica has approximately 60,000 births, approximately 13,000 deaths, and therefore a natural increase of approximately 47,000. However, about 20,000 Jamaicans have migrated to the United States annually over the past decade. As a result of emigration to the United States, Jamaica's annual population increase is actually only about 1.1 percent, less than two-thirds of the natural increase rate.

Asia was the leading source of immigrants to the United States between the late 1970s and late 1980s until supplanted again by Latin America in 1988. The largest numbers of Asians have come from the Philippines, followed by South Korea, China, India, and Vietnam. Asians continue to constitute by far the largest source of immigrants to Canada.

Quota laws. All Americans or their ancestors migrated to the North American continent at some point in the recent or distant past—from Europe perhaps a few decades ago or from Asia many thousands of years ago. People already living in America have always regarded newer arrivals with suspicion.

Americans tempered their dislike for recent immigrants during the nineteenth century, because the new arrivals helped to settle the frontier and extend U.S. control across the continent. By the early twentieth century, most U.S. citizens believed that the frontier had closed, and the country no longer had the space to accommodate an unlimited number of immigrants. Opposition to immigration also intensified in the United States when the majority of immigrants no longer came from northern and western Europe.

The Quota Act in 1921 and the National Origins Act in 1924 established **quotas**, or maximum limits, on the number of people who could immigrate to the United States from each country during a one-year period. The total number of immigrants from the Eastern Hemisphere was limited to 150,000 per year, virtually all of whom had to be from Europe. The system continued with minor modifications until the 1960s. The United States adopted quotas for each hemisphere rather than individual country in 1968 and replaced the hemisphere quotas with a global quota in 1978.

Because the number of applicants far exceeds the quotas, Congress has determined that preference should go to immigrants who have family members in the United States or who have distinctive employment skills. Scientists, researchers, doctors, and other professionals are migrating to the United States so that they can make better use of their abilities. Large-scale emigration by talented people is known as a **brain drain.** Other countries fear that U.S. immigration policy now contributes to

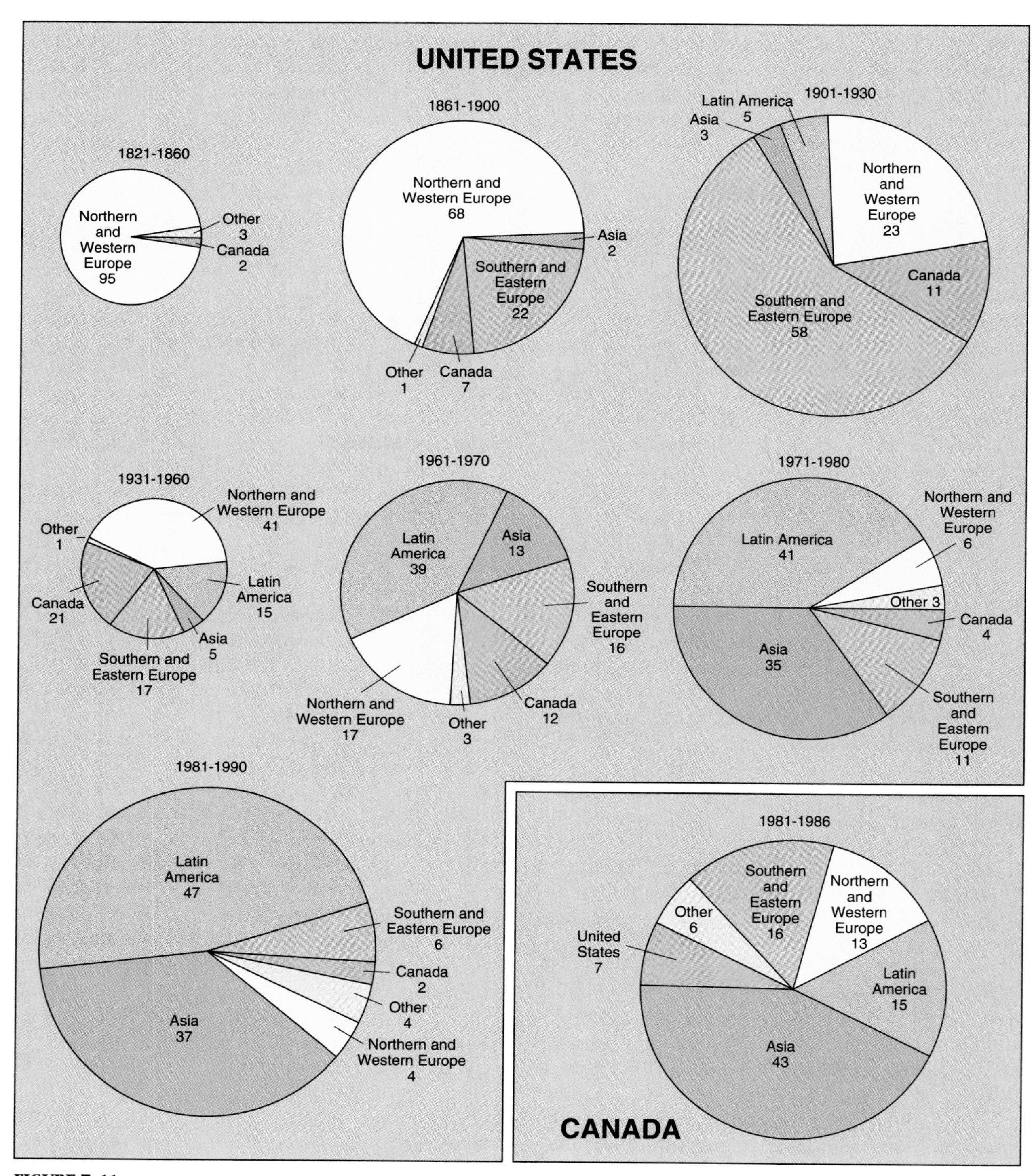

FIGURE 7–11
Europeans comprised more than 90 percent of immigrants to the United States during the nineteenth century. Since the 1960s, Latin America has replaced Europe as the most important source of immigrants to the United States, although Asia was the most important region during the early 1980s. The area of the circles is proportional to the relative number of immigrants during each period.

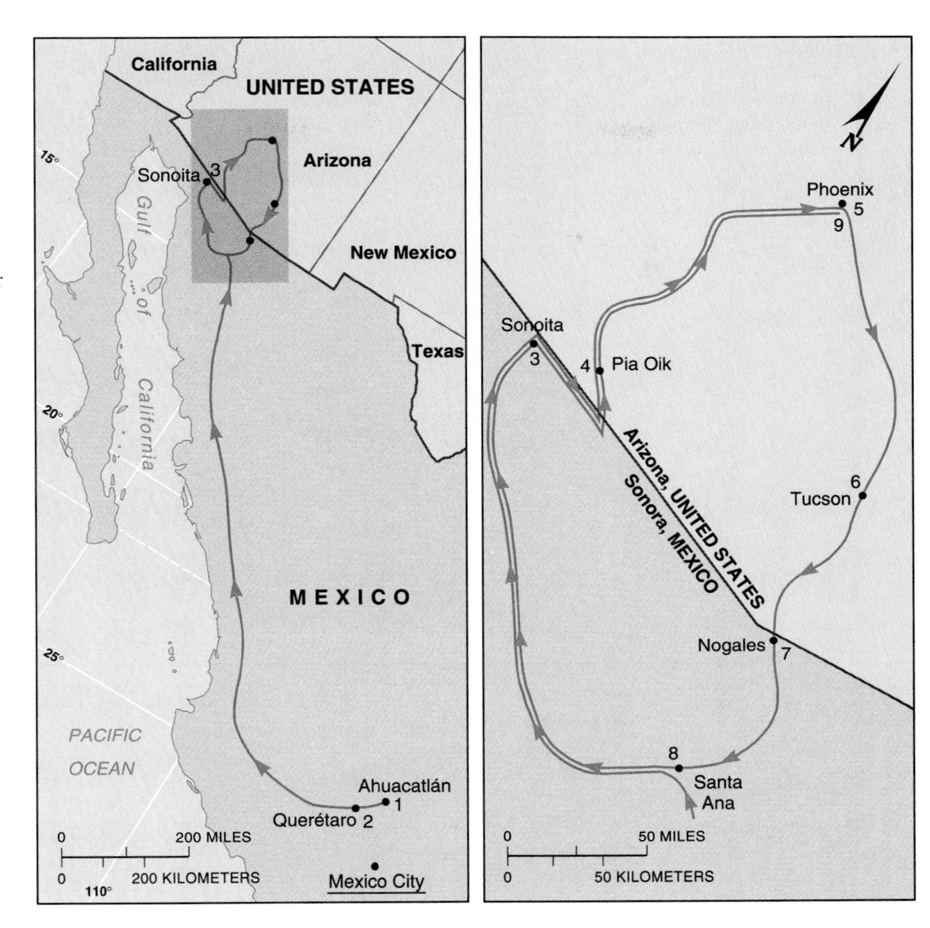

FIGURE 7–12
The route of one group of undocumented immigrants from Mexico to the United States began in Ahuacatlán (1), a village of 1,000 inhabitants in Querétaro State. The immigrants took buses to Querétaro (2) and Sonoita (3), hired a driver to take them to a remote location, crossed the U.S. border on foot near Pia Oik, Arizona (4), and paid a driver to take them to Phoenix (5). Arrested in Phoenix by the Border Patrol, they were driven to Tucson (6) and then to the Mexican border at Nogales (7), where they took buses to Santa Ana (8) and Sonoita and repeated the same route back to Phoenix (9).

a brain drain by giving preference to skilled workers. For example, in recent years Russia has lost many scientists and scholars who had been prevented from emigrating by the Communists.

Once admitted to the United States as skilled workers, many immigrants bring in relatives under the family reunification provisions of the quota. Eventually, these relatives can bring in a wider range of other relatives, a process known as **chain migration**. Asians have made especially good use of preferences for family reunification in the U.S. immigration laws.

Undocumented immigration to the United States. Many people who cannot legally be admitted into the United States immigrate illegally, that is, without proper documents. The U.S. Immigration and Naturalization Service, which apprehends more than 1 million persons per year trying to cross the border, estimates that for every person caught at least two are successful. More than half of the illegal immigrants—sometimes called undocumented immigrants—are estimated to be Mexicans, and another one-fourth are from other Latin American countries (Figure 7–12).

Most undocumented residents have no difficulty finding jobs in the United States. Some work in agriculture, while others obtain jobs in factories. Allowing undocumented residents to stay in the United States could encourage more to come and raise the country's unemployment rate. On the other hand, undocumented residents take low-paying jobs that many American citizens won't accept.

Guest Workers

Western Europe, once the world's principal place of origin for international migrants, has in recent years joined North America as one of the world's leading destinations. Most immigrants to Western

Turkish guest workers provide services in Hamburg, Germany, including opening grocery stores on Sundays when most German-owned stores are closed. (Ergun Cagatay/ Gamma Liaison, Inc.)

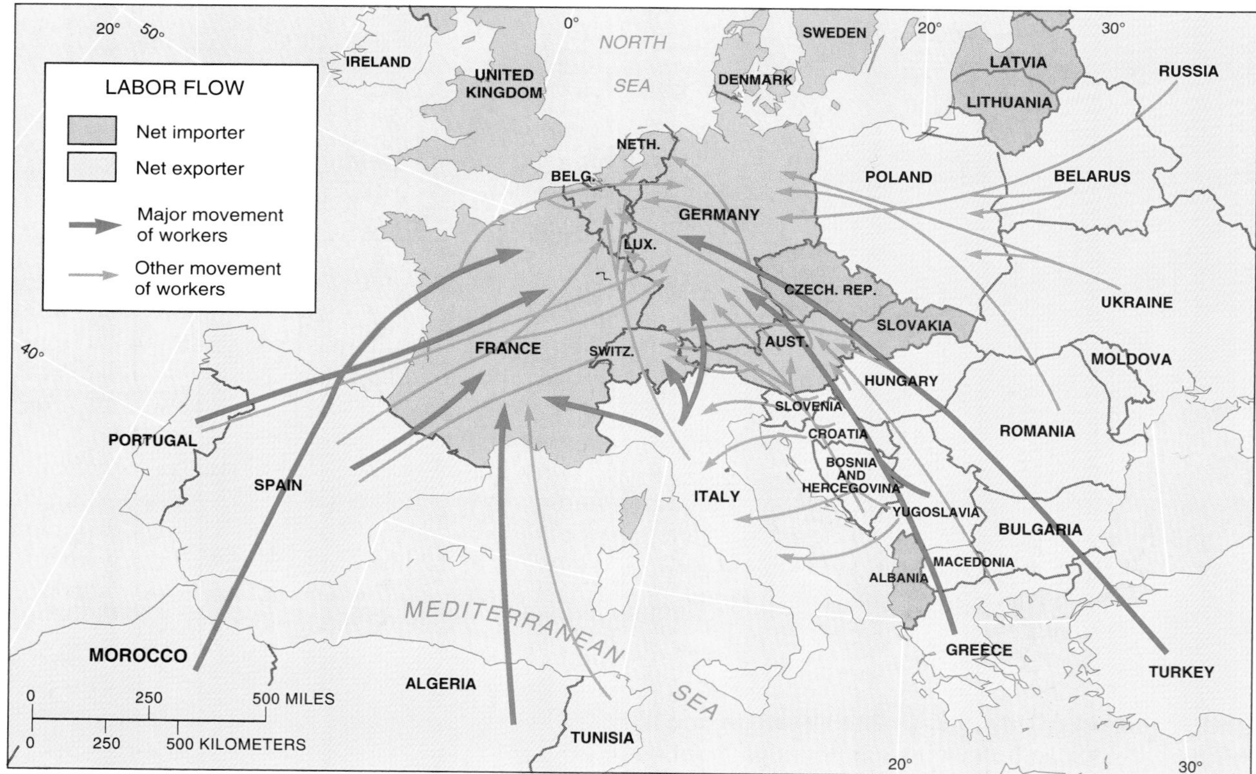

FIGURE 7–13
Guest workers emigrate primarily from southern Europe and northern Africa to work in the relatively developed countries of northern and western Europe. Guest workers follow distinctive migration routes. The selected country may be a former colonial ruler, have a similar language, or have an argeement with the exporting country.

Europe come from Africa and Asia, as well as from southern and Eastern Europe, rather than from Latin America, as is the case in the United States. Foreigners who work in Western European countries are known as **guest workers**.

Guest workers serve a useful role in Western Europe because they take low-status and low-skilled jobs that local residents won't accept, such as driving buses and taxis, collecting garbage, repairing streets, and washing dishes. While relatively low-paid by European standards, guest workers earn far more than they would at home. Unlike most undocumented workers in the United States, European guest workers are protected by minimum wage laws and labor union contracts.

The economy of the guest worker's native country also gains from the arrangement. By letting people work elsewhere, poorer countries reduce their own unemployment problems. Guest workers also help their native country by sending a large percentage of their earnings back home to their families. The injection of foreign currency then stimulates the local economy.

Origin and destination of guest workers in Europe. Guest workers exceed 10 percent of the population in Switzerland and Luxembourg and 5 percent in Belgium, France, and Germany. Most of the guest workers in Europe come from southern and Eastern Europe, northern Africa, the Middle East, and Asia (Figure 7–13).

Distinctive migration routes have emerged between exporting and importing countries. Italy and Turkey send the largest number of guest workers to Western Europe, especially to Germany as a result of government agreements. Switzerland attracts a large number of Italians, while Luxembourg receives primarily Portuguese. Many guest workers in France come from Algeria, Morocco, and Tunisia, which are former French colonies in North Africa.

Middle East guest workers. International migration of guest workers also is widespread in the Middle East. Two-thirds of the workers in Middle East petroleum-exporting states such as Kuwait, Qatar, Saudi Arabia, and United Arab Emirates are foreign. Guest workers emigrate primarily from poorer Middle Eastern countries and from Asia. One-fourth of the labor force in Jordan, Lebanon, Syria, and Yemen immigrate to petroleum-exporting states to seek employment. India, Pakistan, Thailand, and South Korea also send several million guest workers to the Middle East.

Petroleum-exporting countries fear that the increasing numbers of guest workers will spark political unrest and abandonment of traditional Islamic customs. After the 1991 Gulf War, Kuwaiti officials expelled hundreds of thousands of Palestinian guest workers who had sympathized with Iraq's invasion of Kuwait in 1990. To minimize long-term stays, other host countries in the Middle East force migrants—primarily men—to return home if they wish to marry and prevent them from returning once they have families.

Migration in Asia. Many Asians have migrated voluntarily to other countries for economic reasons. During the late nineteenth and early twentieth centuries, millions of Asians migrated to other countries for economic advancement. A proportion of the voluntary migrants comprised so-called time-contract laborers, recruited for a fixed period of time to work in mines or on plantations. When their contracts expired, many laborers settled permanently in the new country.

Indians went as time-contract laborers to Burma (now Myanmar), Malaysia, British Guiana (present-day Guyana), eastern and southern Africa, Fiji, Mauritius, and Trinidad. Japanese and Filipinos went to Hawaii, and Japanese also went to Brazil. More than 29 million ethnic Chinese live permanently in other countries, for the most part in Asia. Chinese comprise three-fourths of the population in Singapore, one-third in Malaysia, and one-tenth in Thailand (Figure 7–14).

Threat of Overpopulation

Global population is increasing rapidly because most countries are either in Stage 2 or Stage 3 of the demographic transition—the stages with high natural increase. No country remains in Stage 1, only a handful have reached Stage 4, and few others are likely to reach Stage 4 in the near future.

The demographic transition involves two big breaks with the past: first, the sudden drop in the death rate, and second, the sudden drop in the birth rate. The first break, which comes from technological innovation, has been accomplished throughout

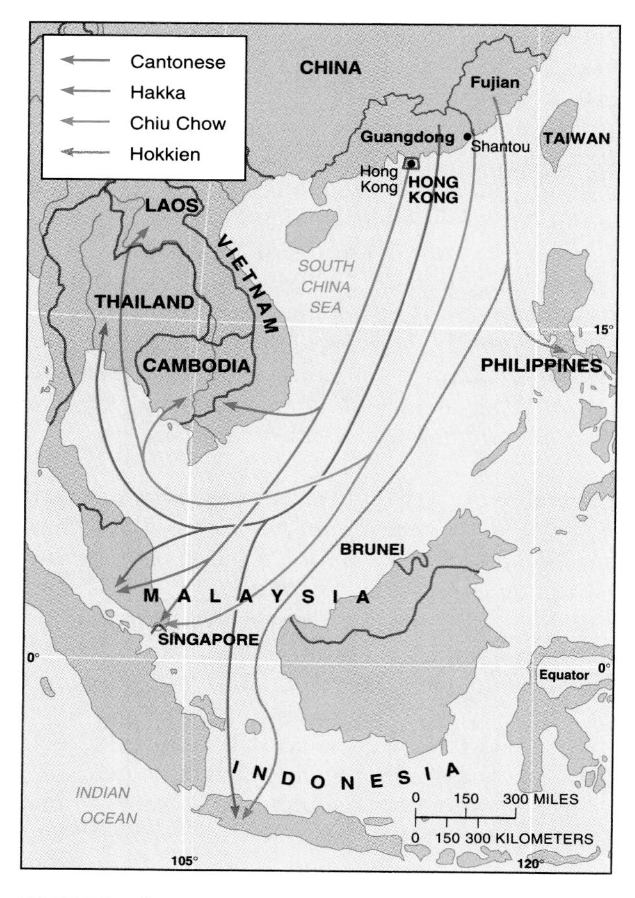

FIGURE 7–14
Various ethnic Chinese peoples have distinctive streams of migration to other Asian countries. Most migrate to communities where other members of the ethnic group have already established businesses. Most emigrate from Guangdong and Hokkien (Fujian) provinces.

the world. The second break, which comes from changing social customs, has not yet been achieved in most countries. If most countries in Europe and North America can reach—or at least approach—Stage 4 of the demographic transition, why can't countries elsewhere in the world? Fundamental problems prevent other countries from replicating the experience in Europe and North America.

The first demographic change—the sudden decline in the crude death rate—occurred for different reasons in the past. The nineteenth century decline in the crude death rate in Europe and North America took place in conjunction with the industrial revolution. The unprecedented level of wealth generated by the industrial revolution was used in part to stimulate research by European and North American scientists into the causes of diseases, as well as their cures. These studies ultimately led to medical advances, such as pasteurization, X-rays, penicillin, and antibiotics.

In contrast, the sudden drop in the crude death rate in Africa, Asia, and Latin America in the twentieth century has been accomplished without a parallel economic transformation. For example, the crude death rate in Sri Lanka (then known as Ceylon), plummeted 43 percent between 1946 and 1947. The most important reason for the sharp drop was the use of the insecticide DDT to control malaria.

European and North American countries were responsible for inventing and manufacturing the DDT and training the experts to supervise its use. Furthermore, the houses of the Sri Lankans were sprayed. The spraying process and other medical services, which cost only $2 per person per year, were paid for primarily by international organizations.

Thus, Sri Lanka's crude death rate was reduced by nearly one-half in a single year with no change in the country's economy or social system. Medical technology was imported from Europe and North America instead of arising within the country as part of an economic adaptation. This pattern has been repeated in dozens of countries in Africa, Asia, and Latin America.

Having caused the first break with the past through diffusion of medical technology, European and North American countries now urge countries elsewhere in the world to complete the second break with the past, the reduction of the birth rate. However, reducing the crude birth rate is more difficult. A decline in the crude death rate can be induced through introduction of new technology by outsiders, but the crude birth rate will drop only when people decide for themselves to have fewer children.

While many people in Africa, Asia, and Latin America may not be prepared for the second break with the past, they are being urged to move through the demographic transition rapidly. In Europe and North America, Stage 2 of the demographic transition lasted for approximately 100 years. During that time, global population increased by about 1 billion. If Stage 2 of the demographic revolution in Africa, Asia, and Latin America also lasts for 100 years—from around 1950 to 2050—15 billion people may be added to the world during that time.

Thomas Malthus on Population

In view of the current size and natural increase rate of the world's population, many wonder whether there will soon be too many people on Earth. Will continued population growth lead to global starvation, war, and lower quality of life?

The English economist and theologian Thomas Malthus was one of the first to argue that the world's rate of population increase was far outrunning the development of food supplies. In his *Essay on Population*, published in 1798, Malthus claimed that population was growing much more rapidly than Earth's food supply, because population increased geometrically, while food supply increased arithmetically. According to Malthus, these growth rates would produce the following relationships between people and food in the future:

Today	1 person	1 unit of food
25 years from now	2 persons	2 units of food
50 years from now	4 persons	3 units of food
75 years from now	8 persons	4 units of food
100 years from now	16 persons	5 units of food

Malthus was writing during the second era of global population increase, which had begun around 1750 in association with the industrial revolution. He concluded that population growth would press against available resources in every country, unless what he termed moral restraint produced lower crude birth rates, or unless disease, famine, war, or other disaster produced higher crude death rates.

In Malthus's time, only a few relatively wealthy countries had entered Stage 2 of the demographic transition, which, again, is characterized by rapid population increase. Malthus did not foresee critical social, economic, and technological changes that would induce relatively wealthy societies to move into Stages 3 and 4 of the demographic transition. He also failed to anticipate that relatively poor countries would have the most rapid population growth, because medical technology would transfer from relatively developed countries, but not wealth.

Malthus's views influence contemporary scientists who believe that the continued high rate of population growth, combined with depletion of resources and unrestricted use of industrial technology, will disrupt global environmental and economic systems and lead to mass starvation and widespread suffering. While many geographers may not agree with Malthus, they may still recognize the contribution to contemporary geographic analysis of his view that overpopulation represents an imbalance between the number of people and the amount of resources.

Debate Over Solutions to the Population Problem

Overpopulation results when an area's population exceeds its physical, social, and economic resources. From this definition of overpopulation, two types of solutions emerge: Either reduce the size of the population or increase the level of an area's resources. Experts sharply disagree on appropriate solutions to avoid global overpopulation at some point in the future. The debate is between giving priority to the population side or the resources side of the relationship.

Proponents of avoiding overpopulation by increasing resources argue that the world can accommodate additional people through further economic development and redistribution of wealth. Those who favor the solution of reduced population growth argue that global resources are limited and must be conserved. Let's look at both sides of the argument.

Increasing resources. One approach to meeting the threat of overpopulation emphasizes economic development. According to this point of view, economic growth can generate sufficient resources to eliminate global hunger and poverty. The world can support further population growth as long as the economy continues to expand, because the world's resource base is not fixed and limited but is constantly being increased by scientific inventions. New manufacturing processes, agricultural practices, energy sources, and communications systems all help expand the world's resources. Can we not expect further technological improvements in the future?

The biggest flaw in these arguments is that, in recent years, many developing countries have expanded their economies significantly and yet have more poor people than ever before. For example, income in Eastern African countries rose during the past decade by approximately 2 percent per year above inflation, but the population grew by approximately 3 percent per year. Because population growth outpaced economic development, all the economic growth was absorbed simply in accommodating the additional population. Economic growth notwithstanding, the average resident of East Africa is worse off today than a decade ago.

Advocates of reducing birth rates through economic development also point out the importance of improving the condition of women in society. As a country develops, women have more access to education, jobs, and health care. Women will choose on their own to curb the size of their families if they are better educated, hold more responsible jobs, and receive intensive health-care counseling. And men will take increased responsibility for using birth control and preventing sexually transmitted diseases.

Reducing population growth. Adherents of this solution argue that humanity can avoid mass starvation primarily by a drastic reduction in the current rate of population increase. Only two approaches can reduce the current natural increase rate: increase the crude death rate or decrease the crude birth rate.

An increase in the crude death rate may halt the growth of the human population, but few people wish to see population growth curbed through an increase in the death rate. However, unless humanity acts, population growth could result in widespread famine. Some fear that sending food to starving Africans now may not prevent mass starvation in the future. Others fear that millions could die in wars—especially in an era of nuclear weapons—or from a natural disaster, such as a major earthquake.

The crude death rate may also rise as a result of the spread of diseases. One-third of all childhood deaths in developing countries other than China result from diarrhea; another one-third die from one of six infectious diseases: polio, measles, diphtheria, tetanus, whooping cough, and tuberculosis. These diseases have been virtually eliminated in relatively developed countries through inoculation, improved nutrition, and hygiene. However, only a minority of children in developing countries have been immunized against these diseases, and water supplies remain unsafe in many places. Even where programs have been implemented to fight these preventable diseases, they have not always been successful because of a lack of qualified medical staff.

The only demographic alternative to seeing the world suffer from higher death rates is to reduce birth rates. A variety of reliable birth control devices can help people limit the size of a family. Many governments have undertaken extensive programs to educate citizens about the benefits and risks of the various methods of birth control.

The percentage of people using so-called "modern" contraception, including the pill, IUD, condom, and sterilization, is especially low in Africa. A United Nations study of women found that only 11 percent of women in Sub-Saharan Africa were employing modern contraceptive devices in the late 1980s, compared to 48 percent in Latin America and over 54 percent in Asia.

Many people in the world oppose birth control programs for religious and political reasons. Adherents of several religions, including Roman Catholics, fundamentalist Protestants, Muslims, and Hindus, consider that their religious convictions prevent them from using some birth control devices. Opposition is strong within the United States to terminating pregnancies through abortions, and the U.S. government has at times withheld aid to countries and family-planning organizations that counsel abortions, even as a small part of overall activities.

A health center in Ethiopia distributes free birth control pills. (*Source:* Marta Sentis/Photo Researchers, Inc.)

Very high birth rates in Africa and southwestern Asia reflect the relatively low status of women. In societies where women receive less formal education and hold fewer legal rights than men, women regard having a large number of children as a measure of status, while men regard it as a sign of virility.

A cycle of overpopulation already may be irreversible in Africa. Rapid population growth has led to the overuse of land. As the land declines in quality, more effort is needed to yield the same amount of crops, extending the working day of women, who have primary responsibility for growing food for their families. Women then regard having another child as a means of securing additional help in growing food.

Many advocates of giving priority to reducing population growth also recognize the importance of expanding economic resources. However, people who are opposed to birth control techniques frequently argue exclusively for the economic growth solution. Compromise between the two positions is therefore difficult.

Population Policies in India and China

The world's two most populous countries, India and China, will heavily influence future prospects for global overpopulation. These two countries, together encompassing more than one-third of the world's population, have adopted different policies to control population growth. In the absence of strong family-planning programs, India adds about 4 million more people each year than China. At current rates of natural increase, India will surpass China as the world's most populous country by the middle of the twenty-first century.

India's population policies. India, like most other countries in Africa, Asia, and Latin America, remained in Stage 1 of the demographic transition until the late 1940s. Immediately following independence from England in 1947, India's crude death rate declined sharply, while the crude birth rate remained high, and natural increase jumped to 2 percent per year. The demographic pattern has not changed much in India during the past half-century.

The government of India has launched various programs to encourage family planning, but none have been very successful. The government has established clinics and distributes educational information about alternative methods of birth control. Birth control devices are distributed free of charge or at subsidized prices. Abortions have been performed at a rate of several million per year since they were legalized in 1972.

India's most controversial family-planning program was the establishment of camps in 1971 to perform sterilizations, surgical procedures by which people were made incapable of reproduction. A sterilized person was entitled to a payment, generally equivalent to the average monthly income in India. At the height of the program, 8.3 million sterilizations were performed during a six-month period in 1976.

Widespread opposition to the sterilization program grew in India, because people feared that they would be forcibly sterilized. A government-sponsored family-planning program continues, but with emphasis on education, including advertisements on national radio and television networks and informational material distributed through local health centers. Given the cultural diversity of the Indian people, a national campaign has had only limited success. The dominant form of birth control continues to be sterilization of women, many of whom have already borne several children. Effective methods have not been devised to induce recently married couples to have fewer children—nor to involve men in birth control and sterilization.

China's population policies. In contrast with India, China has made substantial progress in reducing its rate of natural increase. The government of the People's Republic of China has acted strongly to reduce the number of children. The core of the government's policy is limiting families to one child. Couples receive financial subsidies, a long maternity leave, better housing, and in rural areas more land if they agree to have just one child. The government prohibits marriage for men until they are twenty-two and women until they are twenty. To discourage births further, people receive free contraceptives, abortions, and sterilizations. A family with more than one child must pay a fine, amounting to 5 or 10 percent of its income for ten years, and job promotions may be denied. Some officials in rural villages maintain records of women's menstrual cycles to assure that no unplanned babies are born.

China's crude birth rate has increased somewhat since the mid-1980s. The increase has resulted partly from looser enforcement of the "one-child" rules, especially in rural villages, where families may receive permission to have a second child. The Chinese government has also relaxed enforcement because of international criticism that the one-child policy encouraged the killing of baby girls. If they are limited to one child, most Chinese families prefer to have a boy, in part because of cultural tradition and in part because a boy is regarded as stronger and better able to take care of aging parents. The crude birth rate is also rising in China because of greater wealth. The average Chinese family increasingly can afford the fine for the opportunity of having a second child.

Despite recent increases, China is likely to maintain a much lower natural increase rate than India into the twenty-first century. Following years of intensive educational programs, as well as coercion, the Chinese people have accepted to a greater degree than the Indian people the benefits of family planning. As China moves closer to a market economy, especially in rural areas, Chinese women increasingly recognize that having fewer children opens greater opportunities to obtain a job and earn more money.

The government of China aggressively promotes a one-child policy on public billboards. (Anderson/Gamma-Liaison, Inc.)

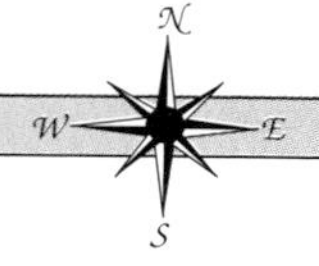

CONCLUSIONS: CRITICAL ISSUES FOR THE FUTURE

Birth rates have declined sharply during the past decade in some countries, but not in others. Delegates to the International Conference on Population and Development, held in Cairo, Egypt, in late 1994, disagreed sharply on the reasons. Some point to progress in economic development, especially improved health care and education of women. Others claim that birth rates have declined in places because of diffusion of contraceptives. Underlying this debate is disagreement concerning the best way to spend scarce family planning funds—educating women and expanding health-care services or distributing condoms and implanting intrauterine devices (IUDs).

Geographers caution that population growth in a region is not by itself an indication of overpopulation. Some densely populated regions are not overpopulated, while some sparsely inhabited areas may be. Instead, overpopulation is a relationship between the size of the population and a region's level of resources. The capacity of the land to support life derives partly from features of the physical environment and in part from human actions to modify the environment through agriculture, industry, and exploitation of raw materials. Further, a region's overpopulation may result from unequal distribution of resources rather than from an inadequate supply. The world as a whole does not immediately face overpopulation, but current trends must be reversed to prevent a crisis in some regions.

Chapter Summary

1. Population distribution

Global population is concentrated in a few places. Humans tend to avoid those parts of Earth's surface considered too wet, too dry, too cold, or too mountainous. The capacity of Earth to support a much larger population depends heavily on people's ability to make use of sparsely settled lands more effectively.

2. Population growth

The natural increase rate of the world's population rose dramatically at three points in history: the agricultural revolution around 8000 B.C., the industrial revolution around A.D. 1750, and the medical revolution around 1950. On each occasion significant changes in technology enabled more people to survive. Today, virtually all of the world's population increase is concentrated in the relatively poor countries of Africa, Asia, and Latin America. Many of these countries do not have the resources to meet the needs of their rapidly growing populations. In contrast, most European and North American countries now have low population growth rates, and some countries are now experiencing population declines.

3. The demographic transition

The demographic transition is a process of change in a country's population. A country moves from a condition of high birth and high death rates and little growth in the size of the population to a condition of low birth and low death rates and low population growth. During the process, the total number of people increases, because the death rate starts to decline some years before the birth rate. The relatively developed countries of Europe and North America have reached or neared the end of the demographic transition. African, Asian, and Latin American countries are at the stages of the demographic transition characterized by rapid population growth. Death rates have declined sharply, but birth rates remain relatively high.

4. Migration

Overpopulation is defined not in terms of numbers of people on Earth but by the capacity of a given region to support the population that it has. Historically, humans reduced the threat of overpopulation by migrating from a region with limited resources to another region with abundant resources. Migration cannot alter the overall balance between people and resources on Earth, but given the uneven distribution of global population, migration remains an important means of reducing environmental pressure in some regions.

5. Threat of overpopulation
The rate at which global population has been growing in the last three decades is unprecedented in history. A dramatic decline in the death rate has produced the increase. With death rates controlled, for the first time in history the most critical factor determining the size of the world's population is the birth rate. Scientists agree that the current rate of natural increase must be reduced, but they disagree on the appropriate methods for achieving the goal.

Key Terms

Agricultural density, 266
Agricultural revolution, 267
Arithmetic density, 264
Brain drain, 279
Chain migration, 281
Crude birth rate (CBR), 269
Crude death rate (CDR), 269
Demographic transition, 272
Demography, 259
Dependency ratio, 275
Doubling time, 268
Ecumene, 260
Emigration, 277
Guest workers, 283
Immigration, 277
Industrial revolution, 267
Infant mortality rate, 269
Internal migration, 277
International migration, 277
Life expectancy, 271
Medical revolution, 267
Migration, 277
Natural increase, 268
Net migration, 277
Overpopulation, 259
Physiological density, 265
Population pyramid, 275
Pull factors, 277
Push factors, 277
Quota, 279
Sex ratio, 277
Total fertility rate (TFR), 269
Zero population growth (ZPG), 273

Questions for Study and Discussion

1. List and describe differences among the world's five major population concentrations.
2. Compare climate conditions in sparsely inhabited regions with population concentrations.
3. Explain differences among population density, physiological density, and agricultural density.
4. Describe why population increased during each of the three revolutions—agricultural revolution, industrial revolution, and medical revolution.
5. Define crude birth rate and crude death rate.
6. Outline the births, deaths, and natural growth rates typical of each stage in the demographic transition, and discuss reasons that a society moves from one stage to the next.
7. Explain the information presented in a population pyramid.
8. Discuss typical patterns of international migration during each stage of the demographic transition.
9. Describe where most immigrants to the United States have originated during the nineteenth century, the early twentieth century, and the past two decades.
10. Why did the United States enact quota laws?

11. Outline Malthus's argument that population grows more rapidly than food supply.
12. Compare policies in China and India for reducing population growth.

Thinking Geographically

1. Paul and Anne Ehrlich argue in *The Population Explosion* that a baby born in a relatively developed country such as the United States poses a graver threat to global overpopulation than a baby born in a developing country. The reason is that people in a relatively developed country place much higher demands on the world's supply of energy, food, and other limited resources. Do you agree with this view? Why?
2. What policies should governments in relatively developed countries pursue with regard to reducing global population growth? If a relatively developed country provides funds and advice to promote family planning, does it gain the right to tell developing countries how to spend the funds and utilize the expertise? Explain your answer.
3. In a recent *Scientific American* article, Bryant Robey, Shea O. Rutstein, and Leo Morris stated "Contraceptives are the best contraceptives." They argued that "If fertility rates are to fall, don't wait around for the forces of modernization to slowly work their way. Make family planning available, and you might be surprised by the number of couples that use them." Do you agree with this approach to lowering birth rates? Explain your answer.
4. Russia's population is declining, in part because of declining birth rates, as in other countries. But more critical in Russia is a rapid increase in death rates during the 1990s, following the fall of the Communist government. What connections can you find between the end of communism and an increase in death rates?
5. According to the concept of chain migration, current migrants tend to follow the paths of relatives and friends who have moved earlier. Can you find evidence of chain migration in your community? Does chain migration apply primarily to the relocation of people from one community in a developing country to one community in a relatively developed country, or is chain migration more applicable to movement within a relatively developed country? Explain.
6. What is the impact of large-scale emigration on the places from which migrants depart? On balance, do these places suffer because of the loss of young, upwardly mobile workers, or do these places benefit from the draining away of surplus labor? In the communities from which migrants depart, is the quality of life on balance improved through reduced pressures on local resources or damaged through the deterioration of social structures and institutions? At the same time that some people are migrating from developing to relatively developed countries in search of employment, transnational corporations have relocated some low-skilled jobs to developing countries to take advantage of low wage rates. Should developing countries care whether their surplus workers emigrate or remain as employees of foreign companies? Why?

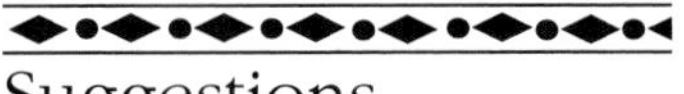

Suggestions for Further Reading

Bennett, D. Gordon. *World Population Problems: An Introduction to Population Geography*. Champaign, IL: Park Press, 1984.

Bouvier, Leon F., and Robert W. Gardner. "Immigration to the U.S.: The Unfinished Story." *Population Bulletin* 41(4). Washington, DC: Population Reference Bureau, 1986.

Brown, Lester R., and Jodi L. Jacobson. "Our Demographically Divided World." *Worldwatch Paper* 74. Washington, DC: Worldwatch Institute, December 1986.

Brown, Lawrence A., and Victoria A. Lawson. "Migration in Third World Settings, Uneven Development, and Conventional Modeling: A Case Study of Costa Rica." *Annals of the Association of American Geographers* 75 (March 1985): 29–47.

Brown, Larry A., and R. L. Sanders. "Toward a Development Paradigm of Migration: With Particular Reference To Third World Settings." In *Migration Decision Making: Multidisciplinary Approaches to Micro-level Studies in Developed and Developing Countries*, edited by G. F. DeJong and R. W. Gardner. New York: Pergamon Press, 1981.

Cadwallader, M. *Migration and Residential Mobility: Macro and Micro Approaches.* Madison: University of Wisconsin Press, 1992.

Chant, Sylvia, ed. *Gender and Migration in Developing Countries.* London: Belhaven Press, 1992.

Clark, W. A. V. *Human Migration.* Beverly Hills: Sage Publications, 1986.

Clarke, John I. *Geography and Population: Approaches and Applications.* Oxford and New York: Pergamon Press, 1984.

Cole, H. S. D., Christopher Freeman, Marie Jahoda, and K. L. R. Pavitt. *Models of Doom: A Critique of the Limits to Growth.* New York: Universe Books, 1972.

Coleman, David, and Roger Schofield, eds. *The State of Population Theory: Forward from Malthus.* Oxford and New York: Basil Blackwell, 1986.

Demko, George, George Schnell, and Harold Rose. *Population Geography: A Reader.* New York: McGraw-Hill, 1970.

Donaldson, Peter J., and Amy Ong Tsui. "The International Family Planning Movement." *Population Bulletin* 45. Washington, DC: Population Reference Bureau, November 1990.

Du Toit, Brian M. and Helen I. Safa, eds. *Migration and Development.* The Hague, Netherlands: Mouton, 1975.

Ehrlich, Paul, and Anne Ehrlich. *The Population Explosion.* New York: Simon and Schuster, 1990.

Freedman, Ronald. "Family Planning Programs in the Third World," *The Annals of the American Academy of Political and Social Science* 510 (July 1990): 33–43.

Gober, Patricia. "Americans on the Move." *Population Bulletin* 48. Washington, DC: Population Reference Bureau, 1993.

Goliber, Thomas J. "Africa's Expanding Population: Old Problems, New Policies." *Population Bulletin* 44. Washington, DC: Population Reference Bureau, 1989.

Jacobsen, Judith. "Promoting Population Stabilization: Incentives for Small Families." *Worldwatch Paper* 54. Washington, DC: Worldwatch Institute, June 1983.

Jensen, Leif. *The New Immigration: Implications for Poverty.* Westport, CT: Greenwood Press, 1989.

Jones, Richard C. "Undocumented Migration from Mexico: Some Geographical Questions." *Annals of the Association of American Geographers* 72 (March 1982): 77–78.

Kosinski, Leszek A., and R. Mansell Prothero. *People on the Move.* London: Methuen, 1975.

Kritz, Mary M., Charles B. Keely, and Silvano M. Tomasi. *Global Trends in Migrations: Theory and Research on International Population Movements.* New York: Center for Migration Studies, 1981.

Lee, Everett. "A Theory of Migration." *Demography* 3 (no. 1, 1966): 47–57.

Lutz, Wolfgang. "The Future of World Population." *Population Bulletin* 49. Washington, DC: Population Reference Bureau, 1994.

Malthus, Thomas. *An Essay on the Principles of Population.* 1978 Reprint; London: Royal Economic Society, 1926 (first published 1798).

McFalls, Joseph A., Jr. "Population: A Lively Introduction." *Population Bulletin* 46. Washington, D.C.: Population Reference Bureau, 1991.

McNeill, William, and Ruth S. Adams. *Human Migration: Patterns and Policies.* Bloomington: Indiana University Press, 1978.

Meadows, Donnella H., Dennis L. Meadows, Jorgen Randers, and William W. Behrens III. *The Limits to Growth*, 2d ed. New York: Universe Books, 1973.

Meadows, Donnella H., Dennis L. Meadows, and Jorgen Randers. *Beyond the Limits*. Post Mills, VT: Chelsea Green Publishing Co., 1992.

Merrick, Thomas W. "World Population in Transition." *Population Bulletin* 41. Washington, DC: Population Reference Bureau, 1986.

Newland, Kathleen. "International Migration: The Search for Work." *Worldwatch Paper* 33. Washington, DC: Worldwatch Institute, November 1979.

_____. "Refugees: The New International Politics of Displacement." *Worldwatch Paper* 43. Washington, DC: Worldwatch Institute, March 1981.

Papademetrion, Demetrios G. "International Migration in a Changing World." *International Social Science Journal* 36, No. 3 (1984): 409–24.

Peters, Gary L., and Robert P. Larkin. *Population Geography: Problems, Concepts, and Prospects*, 3d ed. Dubuque, IA: Kendall-Hunt, 1989.

Ravenstein, Ernest George. "The Laws of Migration." *Journal of the Royal Statistical Society* 48 (1885): 167–227.

Robert, Godfrey. *Population Policy, Contemporary Issues*. New York: Praeger, 1990.

Rogge, John R., ed. *Refugees: A Third World Dilemma*. Totowa, NJ: Rowman and Littlefield, 1987.

Scientific American. *The Human Population*. San Francisco: W. H. Freeman, 1974.

Simon, Rita J., and Caroline B. Brettell, eds. *International Migration: The Female Experience*. Totowa, NJ: Rowman and Allanheld, 1986.

Stephenson, George M. *A History of American Immigration*. New York: Russell and Russell, 1964.

Tien, H. Tuan. "China: Demographic Billionaire." *Population Bulletin* 38. Washington, DC: Population Reference Bureau, 1983.

Trewartha, Glenn T. *A Geography of Population*. New York: Wiley, 1969.

United Nations. *Demographic Yearbook*. New York: United Nations, published annually.

_____. *Statistical Yearbook*. New York: United Nations, published annually.

United States Department of Commerce, Bureau of the Census. *Statistical Abstract of the United States*. Washington, DC: Government Printing Office, published annually.

Van der Kaa, Dirk J. "Europe's Second Demographic Transition." *Population Bulletin* 42. Washington, DC: Population Reference Bureau, 1987.

Visaria, Pravin, and Leela Visaria. "India's Population: Second and Growing." *Population Bulletin* 36. Washington, DC: Population Reference Bureau, 1981.

Weeks, John R. *Population: An Introduction to Concepts and Issues*. 4th ed. Belmont, CA: Wadsworth, 1989.

Wolpert, Julian. "Behavioral Aspects of the Decision to Migrate," *Papers, Regional Science Association* 15 (1965): 159–69.

World Bank. *World Bank Development Report*. New York: Oxford University Press, published annually.

Wrigley, E. A. *Population and History*. New York: World University Library, 1969.

Zelinsky, Wilbur. "A Bibliographic Guide to Population Geography." Research Paper Number 80. Chicago: University of Chicago Department of Geography, 1962.

_____. "The hypothesis of the mobility transition." *Geographical Review* 61 (July 1971): 219–49.

_____. *A Prologue to Population Geography*. Englewood Cliffs, NJ: Prentice-Hall, 1966.

_____, Leszek A. Kosinski, and R. Mansell Prothero, eds. *Geography and a Crowding World*. New York: Oxford University Press, 1970.

We also recommend these journals: *American Demographics; Demography; Intercom; Population; Population and Development Review; Population Bulletin; Population Studies*. In addition, the Population Reference Bureau publishes a World Population Data Sheet every year.

$25,000 TODAY!
DAILY NEWS
Shark takes
of the water!
Moore llegó
Y comenzó
A trabajar
DEPORTES
UTUADO
BOXING85

8
Culture

- Culture regions
- Languages
- Religions
- Social customs
- Problems of cultural diversity

Spanish New York

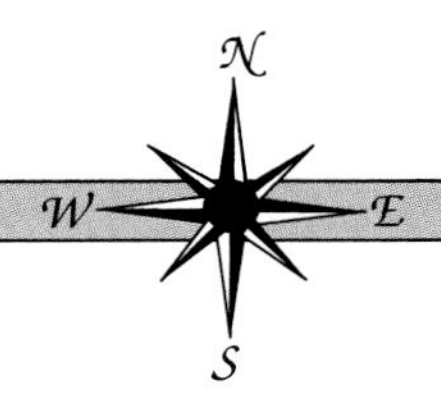

1. Culture regions
Countries can be grouped into nine major culture regions, including Anglo-America, Latin America, Western Europe, Eastern Europe, East Asia, South Asia, Southeast Asia, the Middle East, and Sub-Saharan Africa.

2. Languages
Most of the thousands of languages spoken on Earth's surface can be grouped into a handful of language families. Nearly three-fourths of the people in the world speak languages belonging to the Indo-European and Sino-Tibetan language families. The spatial distribution of languages is primarily a function of migration over thousands of years.

3. Religions
Nearly 90 percent of the world's people who profess belief in a religion are Christian, Muslim, Buddhist, or Hindu. Hinduism is an ethnic religion, whereas the other three are universalizing religions. As with languages, global distribution of religions is heavily influenced by migration.

4. Social customs
Social customs can be categorized as either folk or popular. Geographers document differences in the origin, process of diffusion, and current distribution of folk and popular customs. Folk customs vary widely from one place to another, whereas popular customs vary more in time than in place.

5. Problems of cultural diversity
Adherents of one religion, speakers of one language, and practitioners of one set of social customs often have difficulty living peacefully with other cultural groups. Such conflicts reflect the intense meaning people draw from their cultural characteristics and the lengths to which they will go to defend their cultural identity against real or perceived threats from other groups.

INTRODUCTION

Why does everyone in the world not speak the same language and practice the same religion? The diversity of languages, religions, and other social customs in the world are cultural characteristics largely taken for granted. The heterogeneous collections of languages and religions are clear and obvious characteristics of cultural diversity.

Geographers are interested in two aspects of culture: regions and landscapes. First, geographers study cultural characteristics such as languages and religions, because, along with, for example, climate and vegetation, they are major features of a region. Culture is a source of pride to a people, a symbol of unity. Studying the distribution of cultural traits across Earth's surface helps geographers identify the regions that various cultural groups occupy.

Geographers observe the origin of languages, religions, and other cultural features, study how in the past they diffused from one region to another, and examine their current distribution. The process of spatial diffusion and the interactions of people around the world distribute cultural traits. Culture is like a piece of luggage: people carry it with them when they move from place to place. They add elements of their culture to the new places they reach.

Second, geographers study the relationships between culture and the landscape. Different social groups take particular elements from the physical environment and in turn leave different impacts on the landscape. Culture is influenced by natural events and by features of the physical environment. People incorporate features of the landscape—such as mountains, rivers, and trees—into their cultural traditions and attach meaning to these physical features. At the same time, people place objects on the landscape, from fast-food restaurants and gas stations to vineyards and golf courses. In doing so, they alter the physical environment in accordance with their cultural values.

The distribution of cultural traits is essential for understanding the causes of political unrest and ethnic conflict around the world. Conflicts arise when the territory occupied by one cultural or ethnic group does not coincide with the political boundaries of a country. Wars have broken out in countries containing more than one widely spoken language, practiced religion, or ethnic group.

Culture Regions

The previous chapter introduced important global demographic patterns. We have seen that birth, death, and natural increase rates vary from one region or country to another. This chapter examines geographic differences in languages, religions, and other social customs, all of which are grouped under the concept of culture.

Countries fall into nine major regions according to a given set of demographic and cultural characteristics (Figure 8–1). We call these areas *culture regions*. In the Western Hemisphere, Anglo-America (Canada and the United States) and Latin America can be distinguished on the basis of predominant languages, religions, and natural increase rates. Although there is considerable diversity within these regions, at a global scale the individual countries within these regions display cultural similarities.

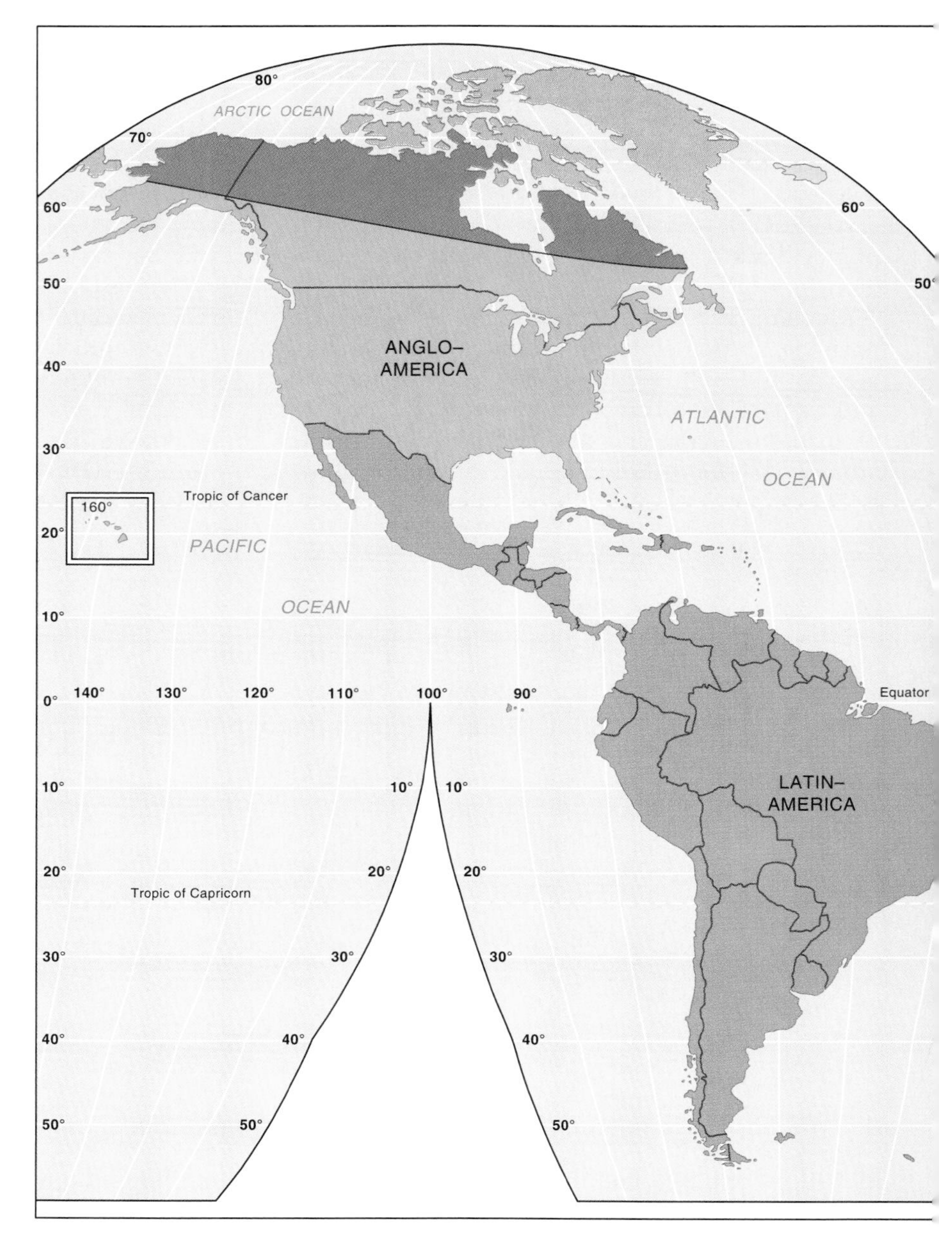

FIGURE 8–1
Culture regions. The nine regions are Anglo-America, Western Europe, Eastern Europe, the Middle East, Latin America, Sub-Saharan Africa, East Asia, Southeast Asia, and South Asia. As we will see in Chapter 10, the first three rank among the world's relatively developed regions, whereas the other six are developing regions. Japan, the South Pacific, and South Africa are other relatively developed areas surrounded by developing regions.

Europe can be divided into two regions: Western and Eastern. Although the two regions share cultural characteristics, distinctive political experiences have produced different levels of economic development during the twentieth century.

Asia includes four regions: East, South, Southeast, and Southwest. Major cultural, demographic, and political differences distinguish these four regions. Southwest Asia can be combined with Northern Africa to form a region known as the Middle East, because of similarities in language, religion, and population growth. Africa south of the Sahara comprises the ninth major region.

In addition to those nine major regions, three other important areas can be identified: Japan, South Africa, and the South Pacific. Japan and South Africa are populous countries with cultural and demographic characteristics that contrast sharply with most of their neighboring states. The

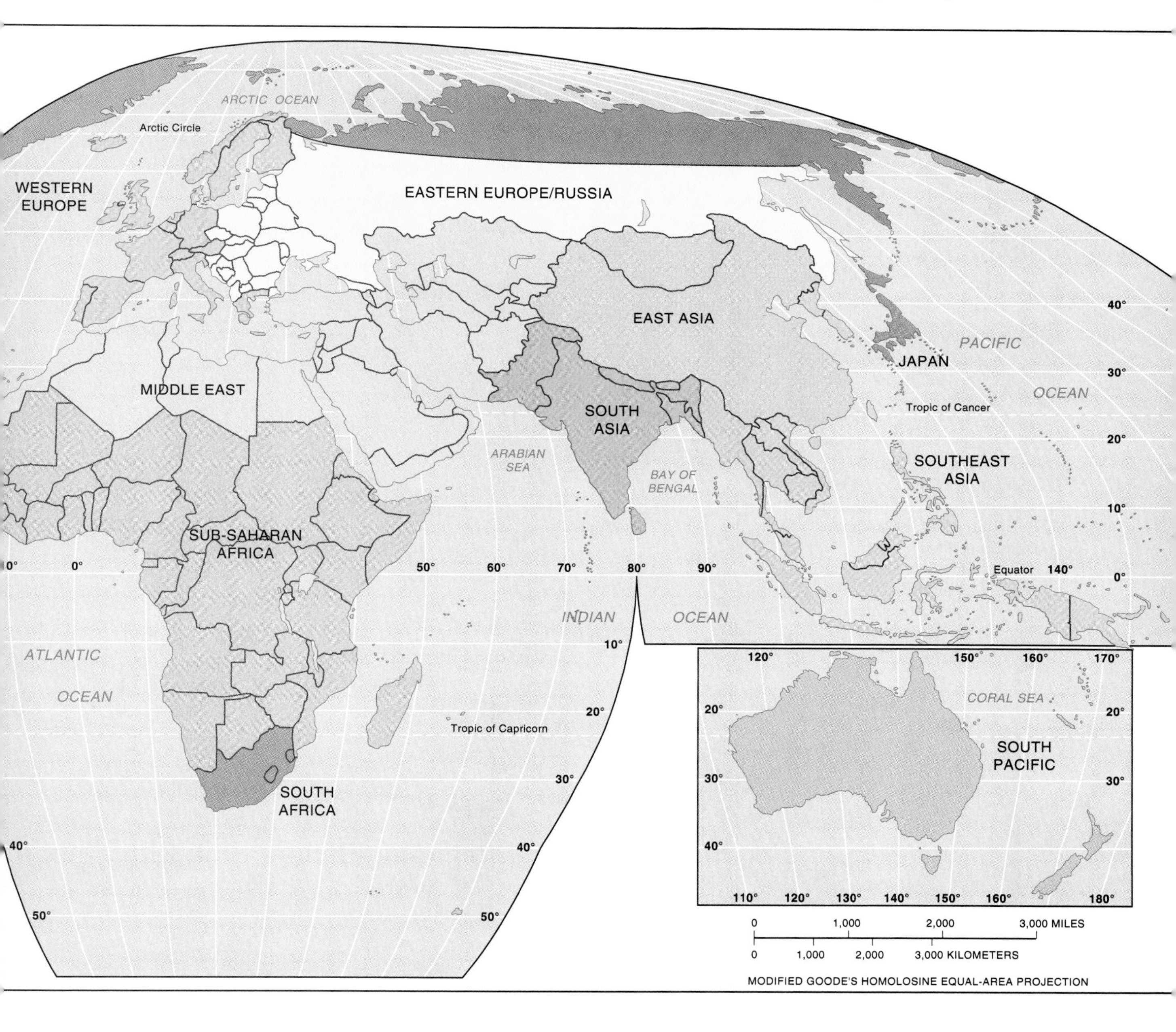

South Pacific, primarily Australia and New Zealand, covers an extensive area of Earth's surface but is much less populous than the nine major regions.

Countries within these nine regions differ sharply in the types of political systems they follow, as discussed in the next chapter. The regions also differ from each other in their distinctive types of economic activities, their methods of production, how the societies use their wealth, and other economic characteristics. As we move toward a global economy, geographers increasingly study the similarities and differences in the economic patterns of the various regions. Economic patterns are emphasized beginning with Chapter 10. This chapter concentrates on similarities and differences in cultural patterns, including languages, religions, and social customs.

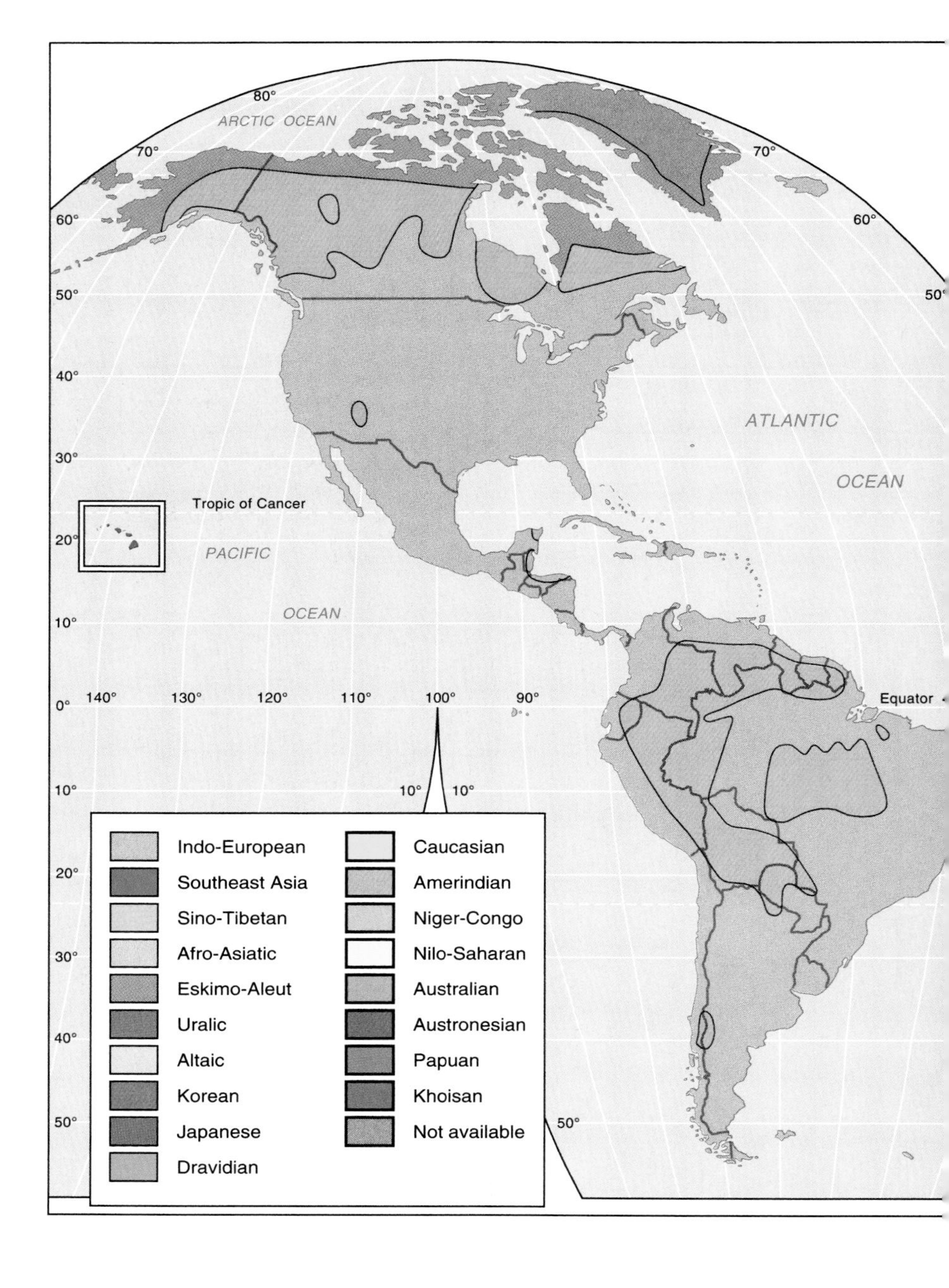

FIGURE 8–2
Language families. Most people speak a language that can be classified into one of a handful of language families. Languages in the following families are spoken by the largest number of people in the world. *Indo-European* is spoken by about 50 percent of the world's people, especially in Europe, the Western Hemisphere, and southern and western Asia. *Sino-Tibetan* is spoken by about 20 percent of the world's people, primarily in East Asia.
Languages belonging to one of these four families are each spoken by about 5 percent of the world's people. *Austronesian* (Malayo-Polynesian) is spoken primarily in Southeast Asia. *Afro-Asiatic* (Semito-Hamitic) is spoken primarily in northern African and southwestern Asia. *Niger-Congo* is spoken primarily in central Africa. *Dravidian* is spoken primarily in southern Asia. (Adapted from A. Meillet and M. Cohen, *Les Languages du Monde,* 1952 (Paris: Centre National de La Recheiche Scientifique), carte XIB)

Languages

Language is a system of communication through the use of speech, a collection of vocal sounds that are understood by a group of people to have the same meaning. Experts disagree on the total number of language families and individual languages in the world. Estimates of the total number of languages range from 2,000 to 4,000, but they can be grouped into a small number of language families.

A **language family** is a collection of individual languages related to each other by virtue of having a common ancestral language before recorded history (Figure 8–2). About 50 percent of the world's people speak a language belonging to the Indo-European family, while another 20 percent speak a Sino-Tibetan language. Four other language families—Austronesian (once known as Malay-Polynesian),

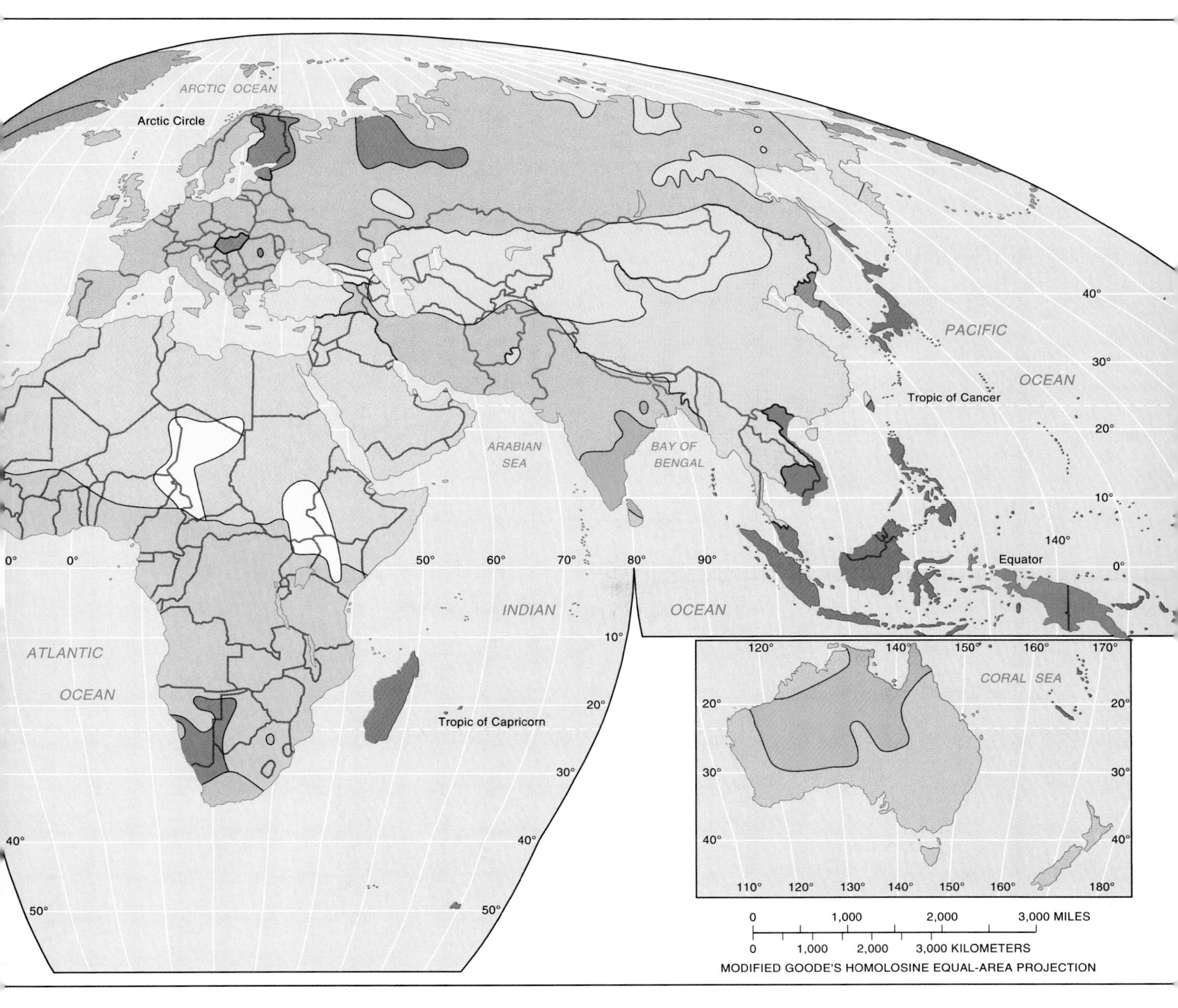

TABLE 8–1

Language with more than 5 million speakers (in millions of speakers)

Language	Speakers
Indo-European Family	2,971
Germanic branch	633
English*	463
German	320
Netherlandish (Dutch)	21
Afrikaans	10
Swedish	9
Danish	5
Norwegian	5
Romance branch	780
Spanish*	371
Portuguese	179
French*	124
Italian	63
Romanian	26
Catalan	9
Other	8
Baltic Slavic Branch	446
Russian*	291
Ukrainian	46
Polish	44
Serbo-Croatian	20
Czech	12
Belorusian	10
Bulgarian	9
Slovak	5
Other	9
Indo-Iranian Branch	1,090
Hindi	400
Bengali	192
Urdu	98
Punjabi	92
Oriya	31
Assamese	23
Pashto	21
Sindhi	18
Nepali	16
Sinhalese	13
Kurdish	11
Baluchi	5
Sylhetti	5
Tajiki	5
Other	23
Greek	12
Albanian	5
Armenian	5
Sino-Tibetan Family	1,279
Sinitic branch	1,144
Mandarin*	930
Cantonese (Yue)	65
Wu	65
Min	50
Hakka (Kijia)	34
Austro-Thai branch	86
Thai (& Lao)	53
Zhuang	15
Miao	6
Other	12
Tibeto-Burman branch	49
Burmese	31
Yi	7
Tibetan	5
Other	6
Austronesian (Malayo-Polynesian) Family	378
Malay-Indonesian	152
Javanese	63
Tagalog	43
Sundanese	26
Cebuano	13
Malagasy	12
Madurese	10
Ilocano	7
Panay-Hilgaynon	7
Minangkabau	6
Other	9
Afro-Asiatic (Semito-Hamtic) Family	309
Semitic branch	241
Arabic*	214
Amharic	19
Other	8
Chadic (Hamitic) branch (Hausa)	37
Cushitic branch	19
Oromo (Galla)	10
Somali	6
Other	3
Berber branch	10
Other branches	2
Dravidian Family	224
Telugu	72
Tamil	68
Kannada	43
Malayalam	35
Other	6
Uralic Family	22
Finnic branch	8
Finnish	6
Other	2
Ugric branch (Magyar)	14
Altaic Family	124
Turkish	58
Azerbaijani	15
Uzbek	14
Kazakh	8
Tatar	8
Uighur	8
Mongolian (Khalkha)	6
Other	8
Austro-Asiatic (Southeast Asian) Family	78
Vietnamese	63
Khmer	8
Santali	5
Other	2
Japanese	126
Korean	74
Niger-Congo Family	284
Benue-Congo branch	184
Swahili	47
Ruanda	8
Shona	8
Xhosa	8
Zulu	8
Lingala	7
Luba-Lulua	7
Rundi	6
Kikuyu (Gekoyo)	5
Nyamwezi-Sokuma	5
Nyanja	5
Other	64
Kwa branch	52
Yoruba	19
Igbo (Ibo)	17
Akan	7
Other	9
West Atlantic branch	27
Fula	13
Wolof	7
Other	7
Mande branch	12
Malinke-Bambara-Dyula	9
Other	3
Other branhes	9
Nilo-Saharan Family	13
Amerindian Family	14
Quechua	8
Other	6
Caucasian Family (Georgian)	4
Total	5,907

Source: Adapted from Sidney S. Culbert, "The Principal Languages of the World," *The World Almanac 1994*, pp. 578–579.

Afro-Asiatic (once known as Semito-Hamitic), Niger-Congo, and Dravidian—are the language families of another 20 percent of the world's speakers. The remaining 10 percent of the world's population speak a wide variety of other languages (Table 8–1).

The distribution of languages is a function of the interplay between interaction and isolation. People living in different regions of the world speak similar languages because of diffusion, especially through migration. When people migrate, they add new words to the language they speak that they take from the new place, and they contribute words to the existing stock of language at the new location. On the other hand, different groups of people speak different languages because of lack of interaction between the groups. Geographers look at the similarities and differences among languages to understand the diffusion and interaction of people around the world.

The map of world languages shows one particularly striking oddity in Africa: The people of Madagascar, the large island situated in the Indian Ocean off the east coast of Africa, are classified as speaking an Austronesian language, even though they are separated by 3,000 kilometers from the world's principal concentration of Austronesian speakers in such Southeast Asian countries as Indonesia. That Malagasy, the language of Madagascar, is classified as an Austronesian language is evidence of migration from Southeast Asia. Malayo-Polynesian people apparently set sail in small boats and reached Madagascar approximately 2,000 years ago.

Indo-European Languages

Within a language family, a **language branch** is a collection of languages that share a common origin but that have evolved into individual languages. The Indo-European language family includes eight branches. More than 400 million people each speak languages in four of its branches, which are Germanic, Romance, Balto-Slavic, and Indo-Iranian. Four language branches within the Indo-European family that are used less extensively are Albanian, Armenian, Celtic, and Greek (Figure 8–3). Differences among individual languages within a language branch are not as extensive as those among languages belonging to different branches of a single language family.

A **language group** comprises several individual languages within a language branch that share a common origin in the relatively recent past and display comparatively few differences in grammar and vocabulary. English is a language in the West Germanic group of the Germanic language branch (Figure 8–4).

Diffusion of Indo-European languages. The development of individual languages within branches of the Indo-European language family is a product of migration in the recent past. English is now the principal language of the United States and Canada (other than Québec) because people from England migrated to the region and gained control of the territory beginning in the seventeenth century. English is the principal language of England as a result of migration as well. In the fifth century A.D., England was invaded by several Germanic peoples, known as the Angles, Jutes, and Saxons, and the name England is derived from *Angles' land* (Figure 8–5).

Like English, Spanish and Portuguese have achieved worldwide importance because of the colonial activities of their European speakers over the past few centuries. Spanish is the official language of eighteen Latin American states, while Portuguese is spoken in Brazil, which has as many people as all the other South American countries combined and fifteen times more than Portugal itself. These two Romance languages were diffused to the Americas by Spanish and Portuguese colonists.

The spatial distribution of a language is a measure of the fate of a distinctive cultural group. Before the conquest of England by the Angles, Jutes, and Saxons, the British Isles were inhabited primarily by various groups who spoke Celtic languages. Two thousand years ago, Celtic languages were spoken in much of present-day Germany, France, and northern Italy, as well as in the British Isles. In the wake of the Germanic invasions, the Celts fled to the more mountainous and remote northern and western parts of the British Isles, including Ireland, Scotland, Wales, and Cornwall, as well as to the region of Brittany in France. Today, Celtic languages survive only in these relatively isolated regions.

The survival of a language depends on the political and military strength of its speakers. The

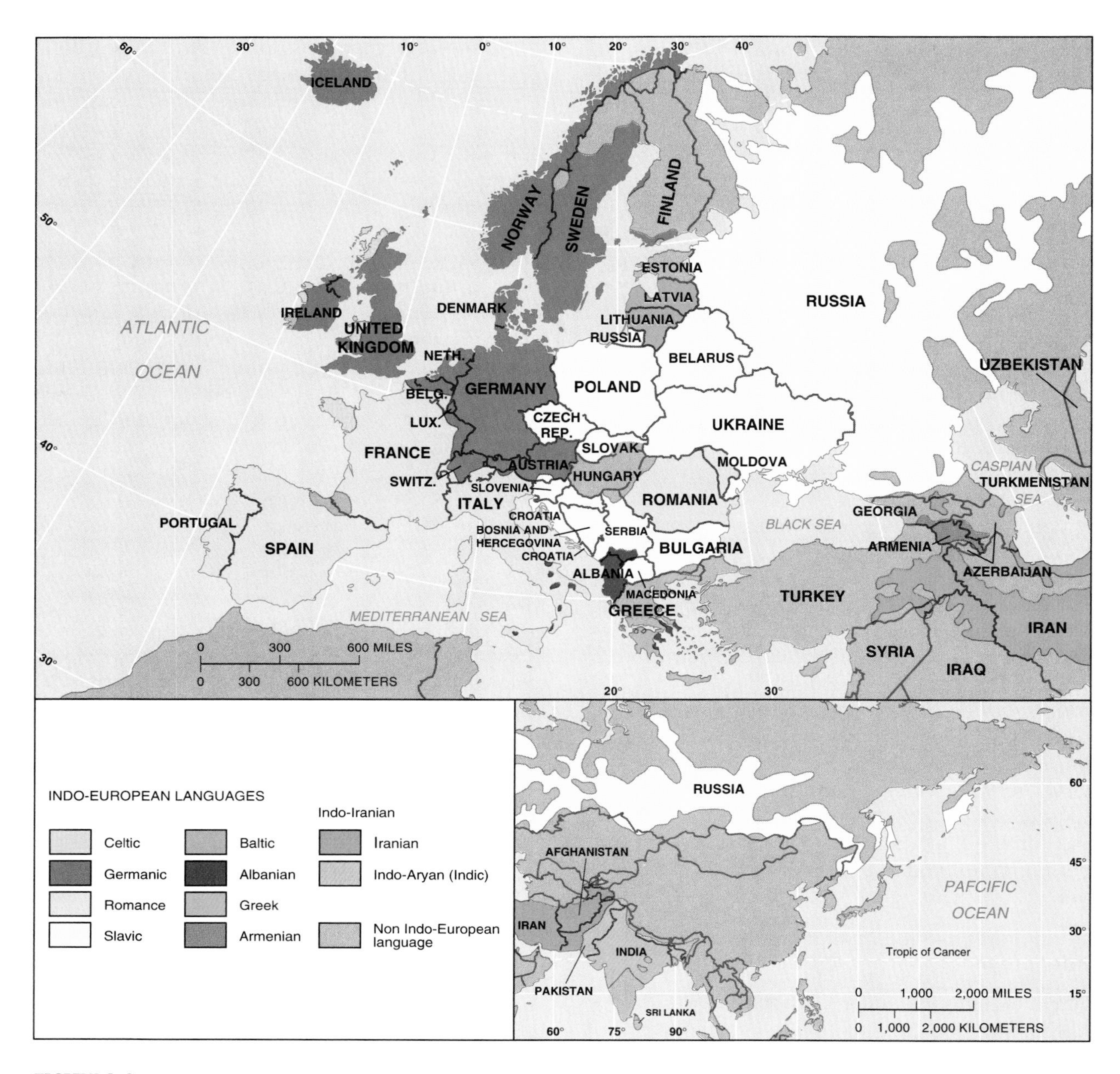

FIGURE 8–3
Indo-European languages. The three most important branches of the Indo-European family in Europe are Germanic in the north and west, Romance in the south and west, and Slavic in the east. Speakers of Indo-Iranian, the fourth important branch of Indo-European, are clustered in southern and western Asia. (Adapted from Antoine Meillet and Marcel Cohen, *Les langues du monde,* 1952.)

Celtic languages declined in importance because the Celts lost most of the territory they once controlled to speakers of other languages. Few Celtic speakers remain who do not also know the language of their English or French conquerors.

The European Union operates the European Bureau of Lesser Used Languages, based in Dublin, Ireland, to provide financial support for the preservation of a couple of dozen languages. However, the long-term decline of languages such

FIGURE 8–4
Germanic languages. The Germanic language branch of Indo-European includes North Germanic and West Germanic languages. The main North Germanc languages include Swedish, Danish, Norwegian, and Icelandic. The main West Germanic languages are English, German, and Dutch (spoken in the Netherlands and northern Belgium). (Adapted from *Encyclopaedia Britannica,* 15th edition (1987), 22: 660)

as Celtic provides an excellent example of the precarious struggle for survival many languages experience. Faced with the diffusion of alternatives used by people with greater political and economic strength, speakers of Celtic and other languages must take pains to preserve their cultural identity.

Common origin of Indo-European language family. If different languages are part of the same family, they must ultimately be descended from the same language. Just as all languages in the Romance and Germanic branches can be traced back to a common language, so can all languages in the Indo-European family. The ancestral language of all Indo-European speakers predates the invention of writing or recorded history, but recent discoveries by linguists and archaeologists confirm that all Indo-European languages can be traced to one source, known as Proto-Indo-European.

Scholars disagree on where and when the first speakers of Proto-Indo-European lived. An influential theory, espoused by Marija Gimbutas, is that the first speakers of Proto-Indo-European were the Kurgan people, whose homeland was in the steppes near the Volga River north of the Caspian Sea, east of the Don River, north of the Caucasus, and west of the Ural Mountains, near the border between present-day Russia and Kazakhstan. The earliest archaeological evidence of the Kurgans dates to around 4300 B.C.

The Kurgans, who lived primarily by herding animals, were among the first people to domesticate horses and cattle. Migrating in search of grasslands for their animals took the Kurgans westward through Europe, eastward to Siberia, and southeast to Iran and South Asia. Between 3500 and 2500 B.C., Kurgan warriors, using their domesticated horses as effective transportation, conquered much of Europe and South Asia (Figure 8–6).

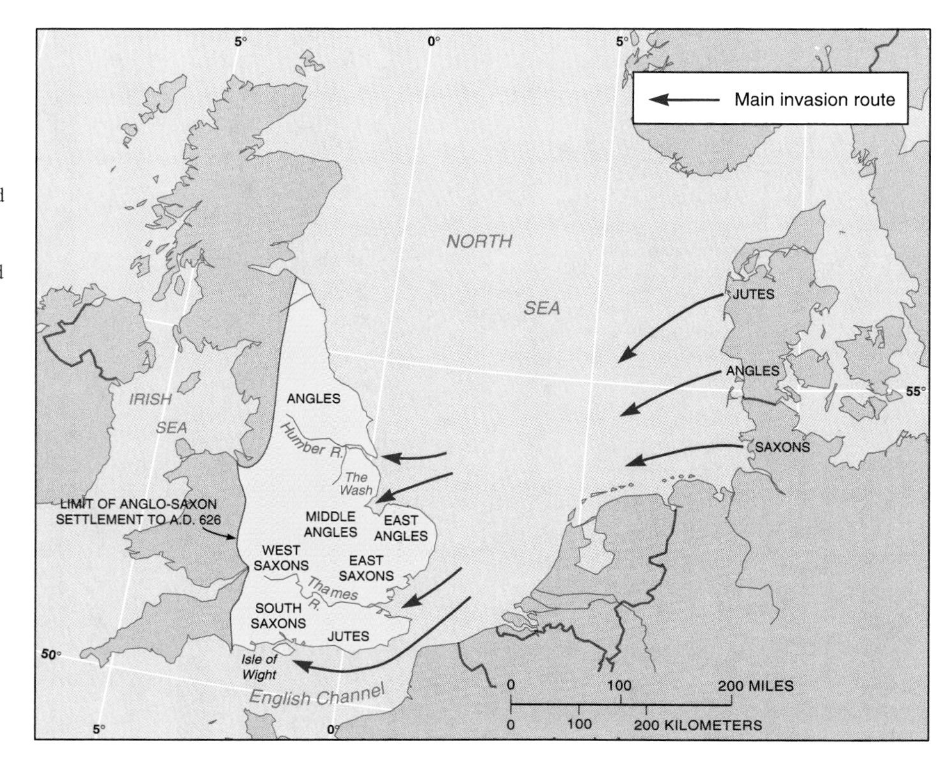

FIGURE 8–5
Migration of Germanic tribes to England. The first speakers of the language that became known as English were tribes that lived in present-day Germany and Denmark before invading England in the fifth century. The Jutes settled in southeastern England, the Saxons in the south and west, and the Angles in the north. From this original spatial separation the first major regional dialect differences developed in England. (Source: From Albert C. Baugh and Thomas Cable, *A History of the English Language*, 3d ed., 1978, p. 47. Reprinted by permission of Prentice Hall, Inc., Englewood Cliffs, NJ.)

Other scholars, including Colin Renfrew, argue that the first speakers of Proto-Indo-European lived 2,000 years before the Kurgans, and that the language originated in eastern Anatolia, part of present-day Turkey. According to this theory, the speakers of Indo-European diffused from Anatolia westward into Europe and eastward into South Asia along with the diffusion of agricultural practices. The language triumphed because its speakers became more numerous and prosperous through growing their own food instead of relying on hunting.

Regardless of whether Proto-Indo-European diffused across Europe and South Asia through military conquest or agricultural innovation, on-going communications between different groups of warriors or farmers were poor. After many generations of complete isolation, various groups spoke increasingly distinct languages.

Dialects

Just as languages evolve from a common ancestor, so can several dialects derive from one language. A **dialect** is a form of a language spoken in a local area. One dialect of a language is normally recognized as the **standard language**, which is the form used for official government business, education, and mass communications.

Dialects, like language families, acquire distinctive distributions across the landscape through various social processes, such as migration, interaction, and isolation. Dialects vary in three ways: pronunciations, spellings, and meanings of particular words.

Major dialect differences have originated in the United States because of the differences in dialects among the various groups of original English settlers. Two-thirds of the New England colonists were Puritans from the East Anglia section of Southeast England, and only a few came from the north of England. Around half of the southeastern settlers came from Southeast England. The dialect spoken in the Middle Atlantic colonies differed significantly from those spoken further north and south, because most of the settlers came from the north rather than the south of England, or from countries other than England. Today, major dialect differences within the United States continue to exist primarily on the East Coast, although some distinctions can be found elsewhere in the country (Figure 8–7).

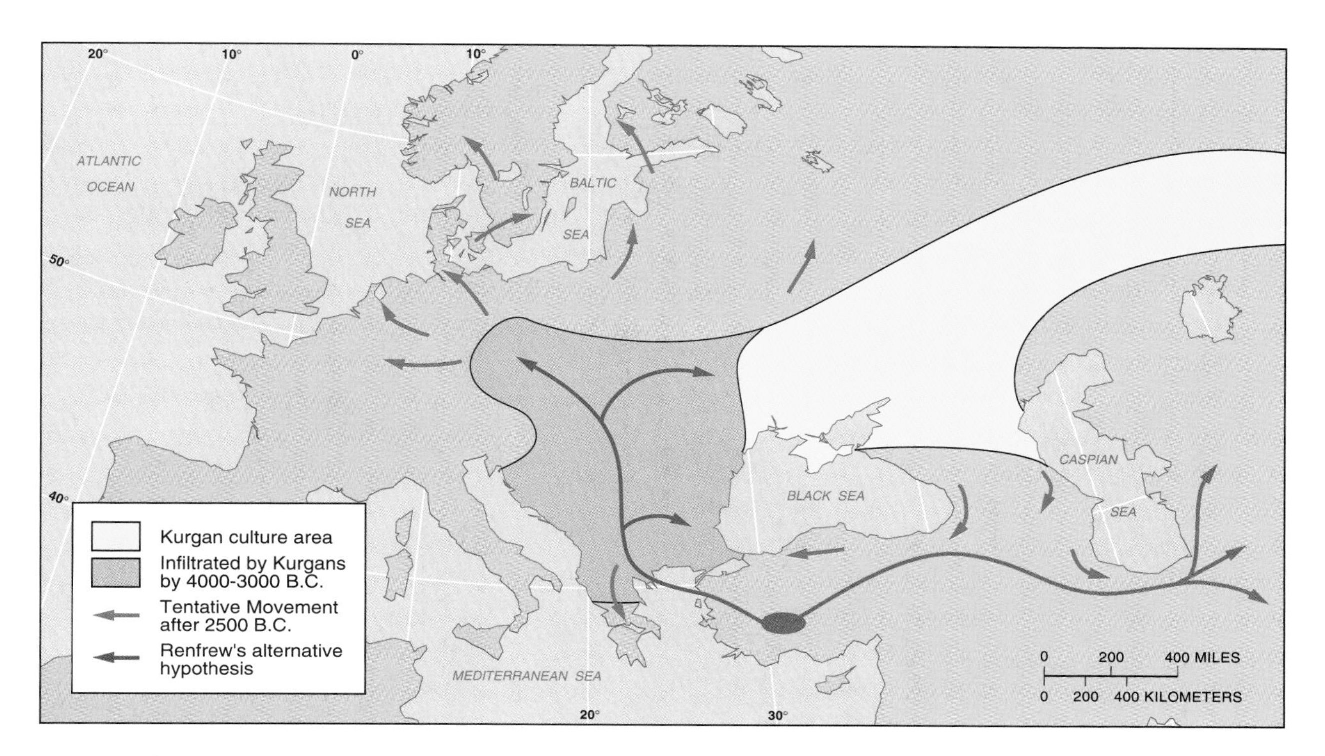

FIGURE 8–6
The first speakers of an Indo-European language, called Proto-Indo-European, may have been the Kurgans, who lived in the steppes north of the Caspian Sea. Beginning about 4000 B.C., the Kurgans may have expanded into much of Europe and southwestern Asia. Centuries of isolation and the lingering impact of languages spoken in localities prior to the Kurgan invasion may have led to the separation of Proto-Indo-European into many branches. According to Renfrew's alternate theory, Proto-Indo-European may have originated in present-day Turkey and diffused along with agricultural innovations. (Reprinted from George Cardona, Henry M. Hoenigswald, and Alfred Senn, "Indo-European and Indo-Europeans", in *Pronto-Indo-European Culture* by Marija Gimbutes, University of Pennsylvania Press, Philadelphia, 1970.)

Religions

Geographers explain the spatial distribution of religions, like languages, in large measure as a consequence of diffusion. The diffusion of religions can be reconstructed through patterns of communications, decision-making, and migration. Differences among religions in rituals, holidays, and other practices also explain distinctive distributions.

Distribution of Religions

The global distribution of religions is less complex than that of languages, because only a few religions can claim the adherence of large numbers of people (Table 8–2). Approximately 80 percent of the people of the world are considered followers of any religion; and nearly 90 percent of these people adhere to one of four religions—Christianity, Islam, Buddhism, and Hinduism (Figure 8–8).

Some religions can be divided into branches, denominations, and sects. A **branch** refers to a large and fundamental division within a religion, a **denomination** to a division within a branch, and a **sect** to a relatively small denominational group that has broken away from an established church.

Christianity. Christianity has 1.9 billion adherents, far more than any other religion, and it has the most widespread distribution. Christianity is the predominant religion in North America, South America, Europe, and Australia, and countries with a Christian majority exist in Africa and Asia, as well.

Early diffusion of Christianity from its point of origin in Palestine was facilitated by the Roman Empire (Figure 8–9). The followers of Jesus Christ carried the religion to people in other locations

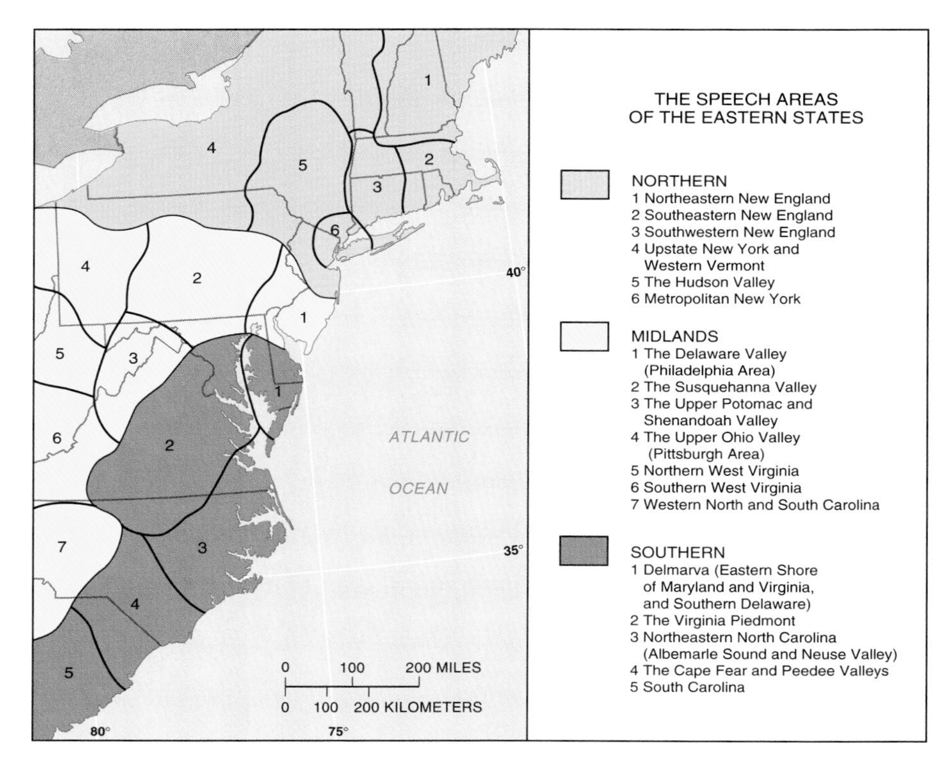

FIGURE 8–7
Dialects in Eastern United States. The most comprehensive classification of U.S. dialects was made by Hans Kurath in 1949. He found the greatest diversity of dialects in the eastern part of the country, especially in vocabulary used by farmers. He divided the eastern United States into three major dialect regions—Northern, Midlands, and Southern—each of which contained a number of important subareas. (Adapted from Hans Kurath, *A Word Geography of the Eastern United States,* Ann Arbor: University of Michigan Press, 1949.)

TABLE 8–2.

Estimated membership of the principal religions of the world (in millions)

	Africa	Asia	Europe	Latin America	North America	Oceania	Eurasia (Former USSR)	All
Christians	341	300	410	443	241	23	112	1,870
Roman Catholics	128	130	260	412	98	8	6	1,043
Protestants	119	87	106	19	105	13	10	458
Orthodox	30	4	36	2	6	1	96	174
Other	64	80	8	10	33	1	9	195
Muslims	285	668	14	1	3	a	43	1,014
Hindus	2	747	1	1	1	a	c	751
Buddhists	b	332	a	1	1	b	a	334
Chinese folk	b	141	a	a	a	b	c	141
New-Religionists	b	122	a	1	1	b	c	124
Animists	70	29	c	1	b	a	d	100
Sikhs	b	19	a	b	a	b	c	20
Jews	a	6	1	1	7	a	2	18
Shamanists	c	11	c	c	c	c	a	11
Confucians	c	6	c	c	b	c	c	6
Baha'is	2	3	a	1	a	a	b	6
Jains	a	4	b	c	c	c	d	4
Shintoists	d	3	c	c	c	c	d	3
Other	a	13	1	4	a	c	a	19
Nonreligious[1]	3	888	74	22	26	4	137	1,155
Total	703	3,292	50	475	282	28	295	5,576

a=<500,000; b=<50,000; c=<5,000; d=<500
[1]Includes atheists
Sum of columns may not match "total," and sum of rows may not match "all" because of rounding.

Source: *Encyclopaedia Britannica Book of the Year, 1994.*

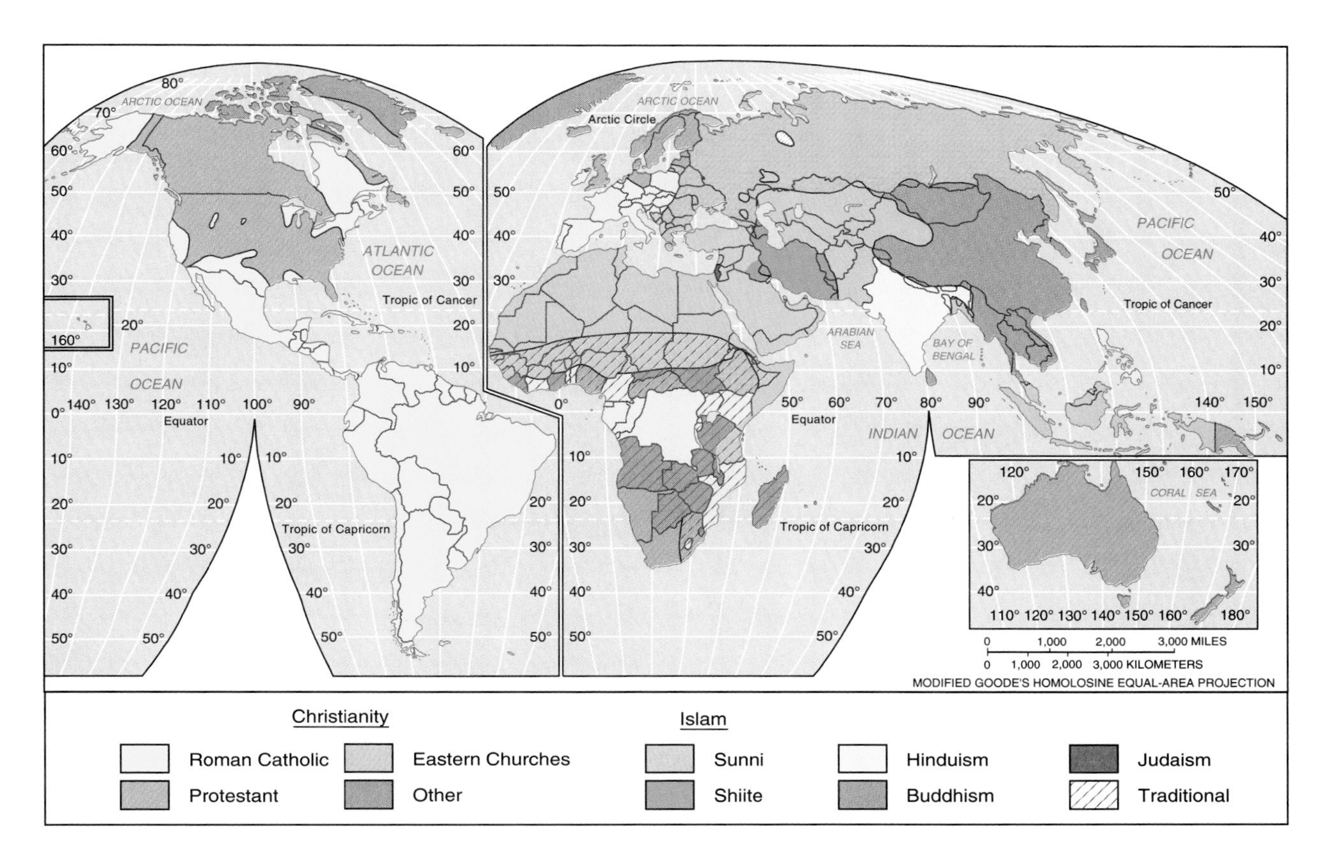

FIGURE 8–8
More than two-thirds of the world's people are adherents of one of these four religions—Christianity, Islam, Hinduism, and Buddhism. (Adapted from D. Sopher, *Geography of Religions,* Englewood Cliffs, N.J.: Prentice Hall, 1967)

along the empire's protected sea routes and excellent road network. Thus, people in the commercial and military settlements directly linked by the communications network received the message first. When the Romans adopted Christianity as their official religion during the fourth century, the empire's administrative organization assured further diffusion of the religion over a larger region.

Migration and missionary activity by Europeans has spread Christianity worldwide. Through permanent resettlement of Europeans, Christianity became the dominant religion in North and South America, Australia, and New Zealand. Christianity's dominance was further achieved by conversion of the indigenous populations in the lands to which Europeans migrated, as well as by intermarriage.

Christianity is divided into three major branches—Roman Catholic, Protestant, and Eastern Orthodox—each with a distinctive spatial distribution. Roman Catholics comprise approximately 56 percent of the world's Christians, Protestants 24 percent, and Eastern Orthodox 9 percent. The remaining 10 percent include Catholics other than Roman and followers of isolated African, Asian, and Latin American Christian churches.

Within Europe, Roman Catholicism is the predominant Christian branch in the southwest and northeast, Protestantism in the northwest, and Eastern Orthodoxy in the east and southeast (Figure 8–10). A fairly sharp boundary between Roman Catholic and Protestant branches exists in the Western Hemisphere as well. Roman Catholics comprise 87 percent of the population in Latin America compared to 35 percent in the United States and Canada.

Migration patterns explain the distribution of Christian branches and denominations. In the Western Hemisphere, for example, Latin Americans are predominantly Roman Catholic because their

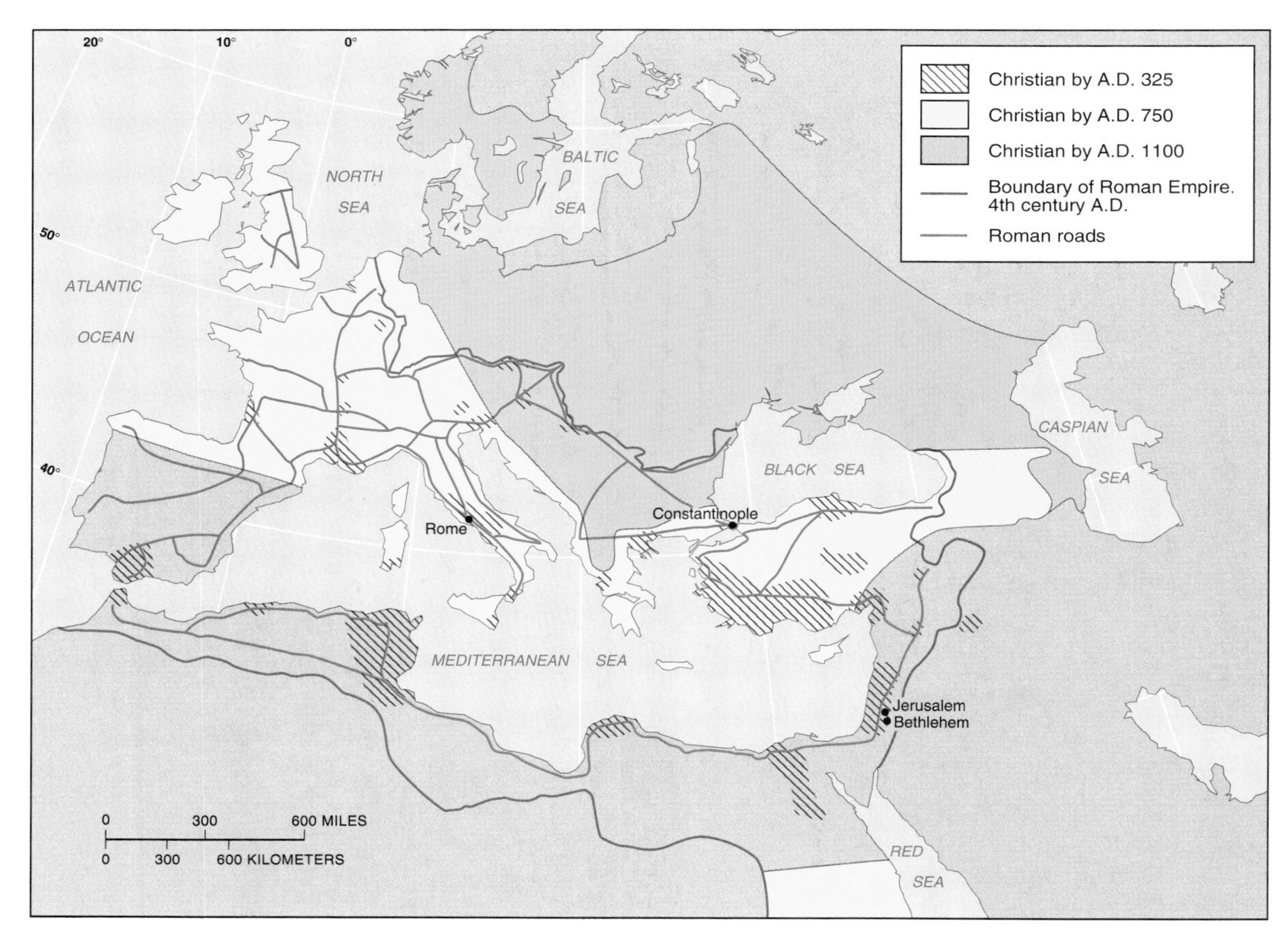

FIGURE 8–9
The diffusion of Christianity from Palestine through Europe began during the time of the Roman Empire and continued after its collapse. Muslims controlled portions of modern-day Spain for more than 700 years, until 1492. Much of southwestern Asia was predominantly Christian at one time but today is predominantly Muslim. (From W. Shepherd, *Historical Atlas,* by permission of Barnes and Noble Books.)

territory was colonized by the Spanish and Portuguese, who brought with them to the Western Hemisphere their religion as well as their languages. The United States and Canada (other than Québec) have Protestant, English-speaking majorities because their early colonists came primarily from Protestant, English-speaking England.

Some regions and localities within the United States and Canada are predominantly Roman Catholic because of immigration from Roman Catholic countries (Figure 8–11). New England and large midwestern cities such as Cleveland, Chicago, Detroit, and Milwaukee have concentrations of Roman Catholics because of immigration from Ireland, Italy, and Eastern Europe, especially in the late nineteenth and early twentieth centuries. Immigration from Mexico and other Latin American countries has produced concentrations of Roman Catholics in the Southwest, while immigration from France has produced a predominantly Roman Catholic Québec.

Nearly half of all Africans presently are classified as Christians, split evenly among Roman Catholic, Protestant, and other religions (see Figure 8–8). In some African countries, traditional religious ideas and practices have been merged with Christianity, while in other cases distinct Christian churches developed independently from the three main branches.

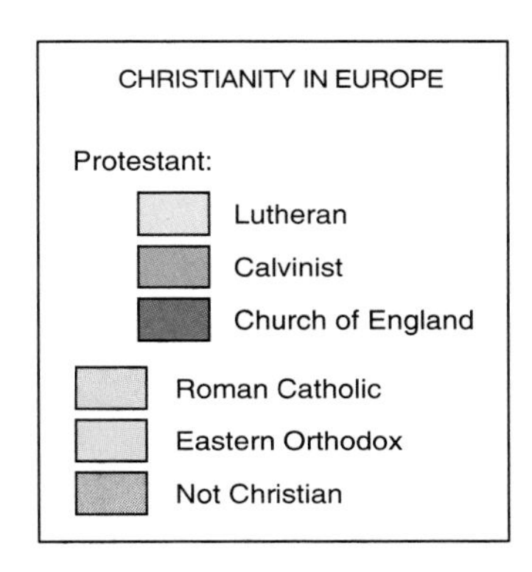

FIGURE 8–10
Branches of Christianity in Europe. Europe is divided into three major regions according to predominant Christian branch. The majority of people adhere to a Protestant denomination in northwestern Europe, while Eastern Orthodoxy is the most important branch in eastern and southeastern Europe. Roman Catholicism predominates in southern, central, and southwestern Europe.

Islam. Islam is the religion of approximately 1 billion people, located predominantly in a region that extends from North Africa to Central Asia, from Morocco to Pakistan. The two most important concentrations of Muslims outside this region are in Bangladesh and Indonesia.

Islam traces its origin from the same narrative as Judaism and Christianity. All three religions consider Adam to have been the first man and Abraham one of his descendants. According to legend, Abraham married Sarah, who did not bear children. Abraham then married Hagar, who bore a son, Ishmael. Sarah then had a child named Isaac, and she prevailed upon Abraham to have Hagar and Ishmael banished. Jews and Christians trace their story through Isaac, Muslims through Ishmael. After their banishment, Ishmael and Hagar wandered through the Arabian desert, eventually reaching Makkah (Mecca), in present-day Saudi Arabia. Centuries later, one of the Ishmael's descendants, Muhammad (A.D. 570?–632), became the Prophet of Islam.

Muhammad's successors extended the religion of Islam over an extensive area of Africa, Asia, and Europe. Within a century of Muhammad's death, Muslim armies conquered Palestine, the Persian Empire, and much of India, resulting in the conversion of many non-Arabs to Islam. To the west, the Muslims captured North Africa, crossed the Strait of Gibraltar, and retained part of Western Europe until 1492 (Figure 8–12). During the same century that the Christians regained all of Western Europe, Muslims took control of much of southeastern Europe and Turkey.

Islam was carried to portions of sub-Saharan Africa and Southeast Asia by **missionaries**—individuals who help to diffuse a religion through practices of conversion rather than war. Indonesia, the world's fourth most populous country, is predominantly Muslim, although it is spatially isolated from the Islamic heartland in Southwest Asia, because Arab traders introduced Islam to the islands of Indonesia in the thirteenth century.

Islam is divided into two important branches: Sunni and Shiite. Sunnis comprise 83 percent of Muslims and are the largest branch in most Muslim countries. Approximately 70 percent of all Shiites live in Iran, where they constitute about 90 percent

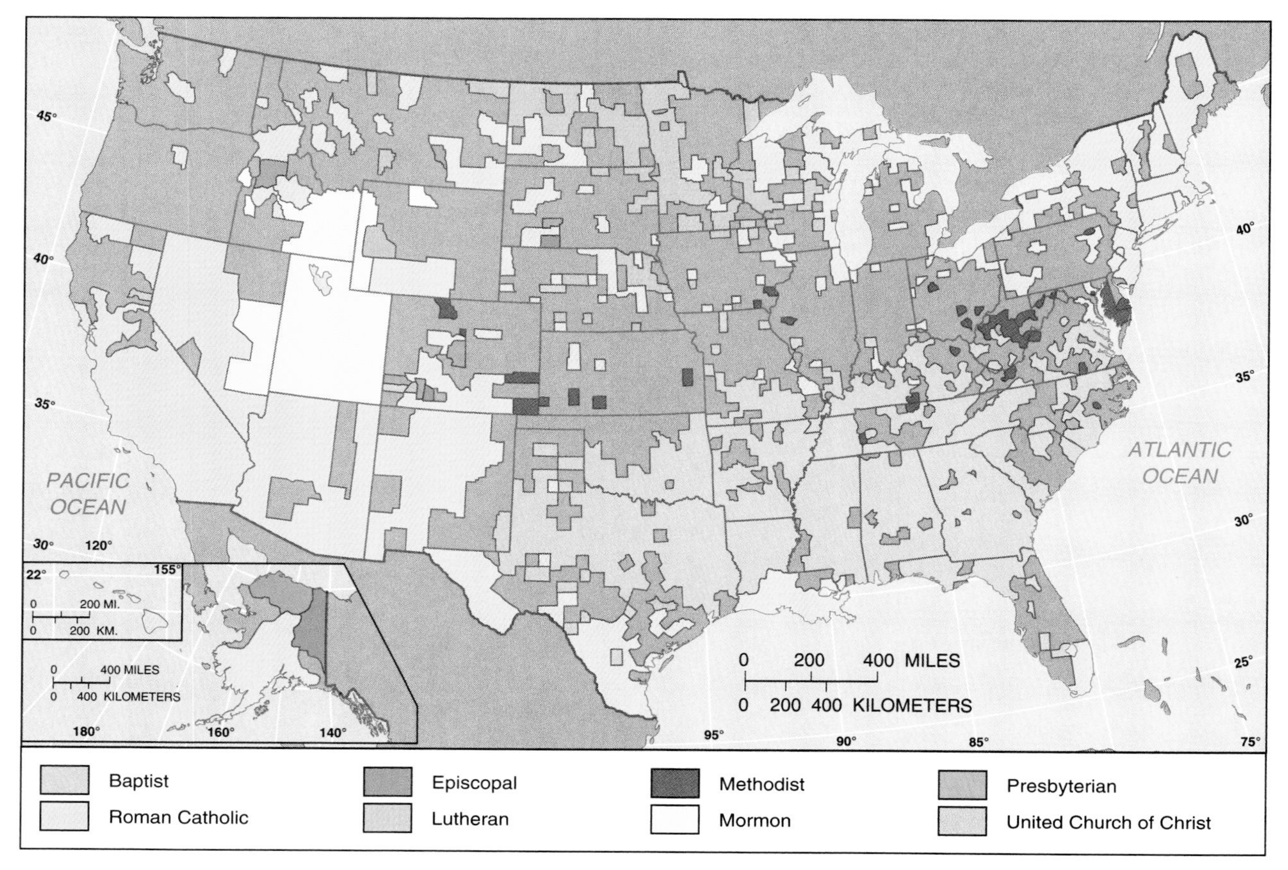

FIGURE 8–11
Christianity in the United States. The shaded areas are U.S. counties in which more than 50 percent of church membership is concentrated in either Roman Catholicism or one Protestant denomination. The distinctive distribution of religious groups with the United States results from patterns of migration, especially from Europe in the nineteenth century and from Latin America in recent years. (Adapted from Douglas W. Johnson, Paul R. Picard, and Bernard Quinn, *Churches and Church Membership in the United States,* 1971)

of the country's population, and another 15 percent of the world's Shiites are in Iraq, where they are twice as numerous as Sunnis. Shiites also outnumber Sunnis in Azerbaijan, Lebanon, and Bahrain. Most of the remaining Shiites live in Yemen and Afghanistan, where they constitute important minorities.

Buddhism. The founder of Buddhism, Siddhartha Gautama, was born in about 563 B.C. in Lumbini, in present-day Nepal near the border with India. The son of a lord, he led a privileged existence sheltered from life's hardships. Gautama had a beautiful wife, palaces, and servants.

According to Buddhist legend, Gautama's life changed after a series of four trips. He encountered a decrepit old man on the first trip, a disease-ridden man on the second trip, and a corpse on the third trip. After witnessing these scenes of pain and suffering, Gautama began to feel he could no longer enjoy his life of comfort and security. Then, on a fourth trip, Gautama saw a monk, who taught him about withdrawal from the world.

At age twenty-nine, Gautama left his palace one night and lived in a forest for the next six years, thinking and experimenting with forms of meditation. Gautama emerged as the Buddha, the awakened or enlightened one, and spent forty-five years preaching his views across India. In the process, he trained monks, established religious orders, and preached to the public.

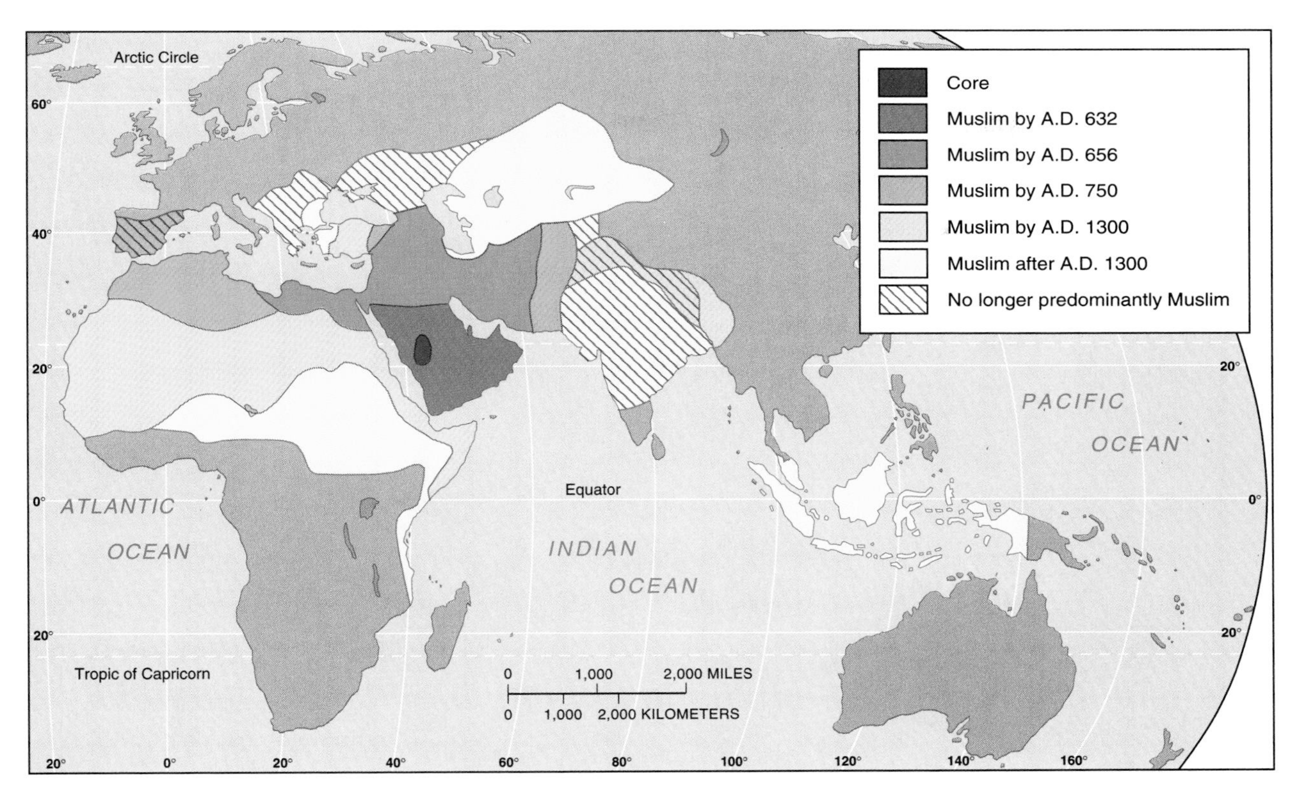

FIGURE 8–12
Islam diffused rapidly from its point of origin in present-day Saudi Arabia. Within 200 years, followers of Islam controlled much of North Africa, southwestern Europe, and southwestern Asia. Subsequently, Islam became the predominant religion as far east as Indonesia. (From Ismail Ragi al Farugi and David E. Sopher, *Historical Atlas of the Religions of the World*, New York: Macmillan, 1974.)

Buddhism did not diffuse rapidly from its point of origin in southern Nepal (Figure 8–13). The individual most responsible for the spread of Buddhism was Asoka, emperor of the Magadhan Empire from about 273 to 232 B.C. Around 257 B.C., at the height of the Magadhan Empire's power, Asoka became a Buddhist and thereafter attempted to put into practice Buddha's social principles.

A council organized by Asoka at Pataliputra decided to send missionaries to territories neighboring the Magadhan Empire. Emperor Asoka's son, Mahinda, led a mission to the island of Ceylon (now Sri Lanka), where the king and his subjects were converted to Buddhism. As a result, Sri Lanka is the country that claims the longest continuous tradition of practicing Buddhism. Missionaries were also sent in the third century B.C. to the Kashmir, Himalayas, Burma (Myanmar), and elsewhere in India.

In the first century A.D., merchants along the trading routes from northeastern India introduced Buddhism to China. Many Chinese were receptive to the ideas brought by Buddhist missionaries, and Buddhist texts were translated into Chinese languages. Chinese rulers allowed their people to become Buddhist monks during the fourth century A.D., and in the following centuries, Buddhism evolved into a genuinely Chinese religion. Buddhism further diffused from China to Korea in the fourth century and from Korea to Japan two centuries later. During the same era, Buddhism lost its original base of support in India.

Buddhism is also split into more than one branch, as followers disagreed on interpretation of the founder's statements. The two main branches are Mahayana and Theravada. Mahayana Buddhism predominates in Central and East Asia, including Tibet, Mongolia, China, Japan, and Vietnam, as well

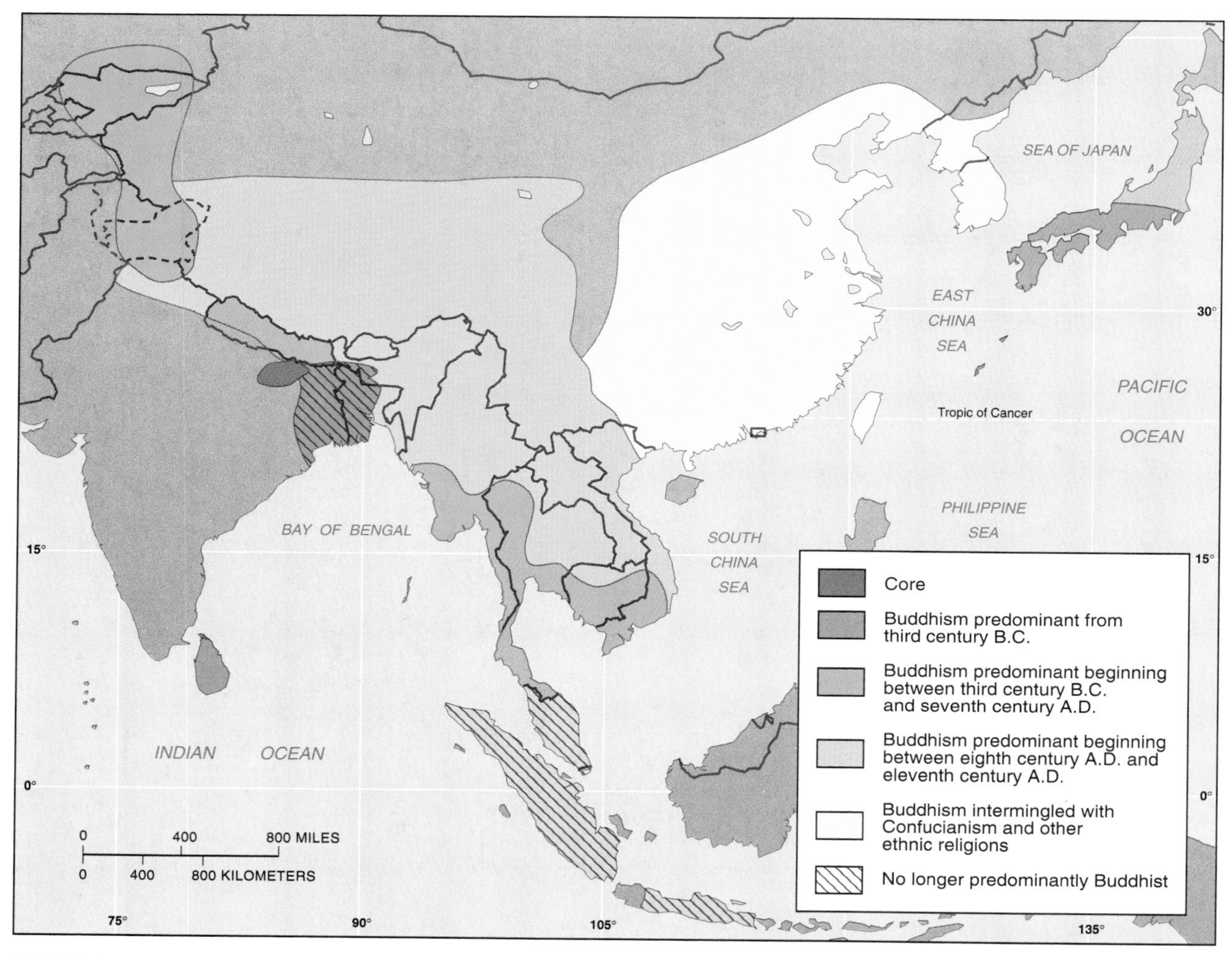

FIGURE 8–13
Buddhism diffused slowly from its core in Nepal and northeastern India. Buddhism was not well established in China until 800 years after Buddha's death. (From Ismail Ragi al Farugi and David E. Sopher, *Historical Atlas of the Religions of the World,* New York: Macmillan, 1974)

as Sri Lanka. Theravada Buddhism is most prevalent in Southeast Asia, especially Thailand, Myanmar, Cambodia, and Laos.

Theravadists believe that the practice of Buddhism is a full-time occupation. Therefore, to become a good Buddhist, one must renounce worldly goods and become a monk. Theravada means "the way of the elders," which indicates the Theravada Buddhists' belief that they are closer to Buddha's original approach.

Mahayanists claim that their approach to Buddhism can help more people because it is less demanding and all-encompassing. While the Theravadists emphasize Buddha's life of self-help and years of solitary introspection, the Mahayanists emphasize Buddha's later years of teaching and helping others. The Theravadists cite Buddha's wisdom, the Mahayanists his compassion. The word *mahayana* is translated as "the bigger ferry" or "raft," and Mahayanists call Theravada Buddhism by the name *hinayana*, or "the little raft."

Mahayana Buddhism is divided into at least six distinct denominations. Geographers cannot draw a map showing the spatial distribution of Mahayana denominations, such as the one for Christian branches within Europe, because the different groups do not occupy distinct geographical areas.

Buddhism currently has more than 300 million adherents, although an accurate count is impossible to obtain. Only a few people in Buddhist countries participate in Buddhist institutions, and religious functions are performed primarily by monks rather than by the general public. The number of Buddhists is also difficult to count, because someone can be both a Buddhist and a believer in other Eastern religions. Most Buddhists in China and Japan, in particular, simultaneously believe in other religions, such as Confucianism and Daoism (also spelled Taoism) in China and Shintoism in Japan.

Hinduism. The spatial distribution of Hindus differs from that of the other three major religions already described. Whereas followers of Christianity, Islam, and Buddhism are spread among many countries, more than 99 percent of the world's Hindus are concentrated in one country, India.

Hindus adhere to the belief that more than one path exists to reach the divinity, or God. Because people start from different backgrounds and experiences, the appropriate form of worship for any two persons may not be the same. Hinduism does not have a central authority or a single holy book, so each individual selects suitable rituals. If one person practices Hinduism in a particular approach, other Hindus will not think that the individual has made a mistake or strayed from orthodox doctrine.

According to Hinduism, because everyone is different, it is natural that each individual should belong to a particular position in the social order, known as a **caste,** which is the class or distinct hereditary order into which a Hindu is assigned according to religious law. The caste system apparently originated when Aryans invaded from the west about 1400 B.C. The Aryans were divided into four groups: Brahman, Kshatriya, Vaisya, and Shudra.

Considerable differences in social and economic positions developed among the four castes. The Brahmans were the priests and top administrators, the Kshatriyas the warriors, and the Vaisyas the merchants, and the Shudras were the agricultural workers and artisans. Over the centuries, the original castes split into thousands of subcastes.

Below the four castes were the so-called untouchables, or outcasts, who did the work considered too dirty for other castes. The untouchables theoretically were descended from the people living in India prior to the Aryan conquest. Until recently, social relations among the five groups were limited, and the rights of non-Brahmans, especially untouchables, were restricted. The rigid caste system has been considerably relaxed in recent years. The government of India legally abolished the untouchable caste, and the people formerly in that caste now have equal rights with other Indians.

The type of Hinduism practiced will depend in part on the individual's caste. A high-caste Brahman may practice a form of Hinduism based on knowledge of relatively obscure historical texts. At the other end of the caste system, an illiterate person in a rural village may perform religious rituals without a highly developed set of written explanations for them. The average Hindu has allegiance to a particular deity or concept within a broad range of possibilities. The three approaches that have the largest number of followers are probably Saivism, Vaishnavism, and Shaktism.

Although most Indians are Hindus, other religions have large numbers of adherents as well. More than 10 percent of the Indian people are Muslims, and India is also home to nearly all of the world's 16 million Sikhs, who adhere to Sikhism, an ethnic religion.

Sikhs have clashed with Hindus several times during the 1980s and 1990s over control of Punjab state in northwestern India, where the Sikhs are clustered. In 1984, India's Prime Minister Indira Gandhi was assassinated by two Sikhs who had been her bodyguards. The Sikhs were protesting the prime minister's order to seize the religion's holiest shrine, the Golden Temple in Amritsar. Sikh holy men and gunmen used the shrine as a base for launching violent raids as part of a campaign to gain greater independence for Punjab.

Differences Between Universalizing and Ethnic Religions

Hinduism's more clustered distribution in comparison to Christianity, Islam, and Buddhism, is due to differences in the structures of these religions. Geographers distinguish two kinds of religions:

universalizing and *ethnic*. Christianity, Islam, and Buddhism are the three universalizing religions with the largest numbers of adherents, while Hinduism is the ethnic religion with the most followers.

A **universalizing religion** attempts to appeal to all people, not just to residents of one cultural background or location. The three main universalizing religions—Christianity, Islam, and Buddhism—each began with an individual founder who preached a message accepted initially only by immediate followers. These followers in turn transmitted the message to people in other places on Earth. Today, these three universalizing religions have hundreds of millions of adherents distributed across wide areas of the world.

In contrast, the religious principles of an **ethnic religion** are more likely to be based on the physical characteristics of a particular location on Earth's surface. Consequently, an ethnic religion carries meaning primarily either for people living in the particular environment or for those attracted to a particular environment or belief. An ethnic religion such as Hinduism differs from a universalizing religion in that it typically has a more clustered geographic distribution. The word *Hinduism*, in fact, is simply a term which means "the religion of India."

Because its social forms are rooted in a specific location, an ethnic religion is harder to transmit to people elsewhere in the world. An ethnic religion may change as social, economic, and physical conditions in the homeland change, but the region of the religion's followers is unlikely to expand extensively.

Many Africans follow traditional ethnic religions, sometimes called **animism**. Animists believe that such inanimate objects as plants and stones or such natural events as thunderstorms and earthquakes have discrete spirits and conscious life.

The universalizing religions, especially Christianity and Islam, have sent missionaries to regions of Africa where traditional ethnic religions once predominated. Nearly 50 percent of all Africans are now classified as Christians—split about evenly among Roman Catholic, Protestant, and other—and another 40 percent are Muslims. Followers of traditional African religions now constitute a majority in only a handful of small West African countries, including Benin, Côte d'Ivoire, Guinea-Bissau, Sierra Leone, and Togo.

One ethnic religion that has not remained clustered in a single location is Judaism. It developed as a local religion in the territory of the Middle East called Canaan in the Bible, Palestine by the Romans, and the state of Israel since 1948. However, since A.D. 70 most Jews have not lived in this territory but instead were forced to disperse throughout the world, a process known as a **diaspora**. Unlike other ethnic religions, Judaism is practiced in many countries throughout the world as a result of a combination of forced and voluntary migration. Having been exiled from the home of their ethnic religion, Jews have lived among other nationalities, retaining their separate religious practices but adopting other cultural characteristics of the host country, such as language.

Historically, the Jews of many European countries were forced to live in a **ghetto**, which strictly speaking was defined as a neighborhood in a city set up by law to be inhabited only by Jews. The origin of the word *ghetto* is unknown, but it may have originated during the sixteenth century in Venice, Italy, as a reference to the city's foundry or metal casting district, where Jews were forced to live. Ghettos were frequently surrounded by walls, and the gates were locked at night to prevent people from leaving.

Holidays. Holidays in an ethnic religion typically derive from characteristics of the physical environment where the religion is clustered. The most significant element of the natural environment incorporated into many ethnic religions is the calendar—the annual cycle of variation in climatic conditions. Knowledge of the calendar is critical to successful agriculture, whether for sedentary crop farmers or nomadic animal herders. The seasonal variations of temperature and precipitation help farmers to select the appropriate times for planting and harvesting and to make best choice of crops.

The major religious events of the Bontok people of the Philippines, for example, revolve around the agricultural calendar. Sacred moments, known as *obaya*, include the times when the rice field is initially prepared, the seeds are planted, the seedlings are transplanted, the harvest is begun, and the harvest is complete.

Because universalizing religions are practiced in a variety of locations around the world, the role of the landscape in forming rituals is different than in ethnic religions. The main holidays in the Buddhist and Christian calendars relate to events in their founders' lives. All Buddhists observe Buddha's birth, enlightenment, and death, although not all of them on the same days. For example, Japanese Buddhists celebrate Buddha's birth on April 8, his enlightenment on December 8, and his death on February 15; Theravadist Buddhists observe all three events on the same day, usually in April.

Christians may relate Easter to the agricultural cycle, but that relationship differs among Christians, depending on where they live. In southern Europe, Easter is a joyous time of harvest, while in northern Europe it is a time of anxiety over planting new crops, as well as a celebration of spring's arrival after a harsh winter. Christians outside the Mediterranean countries lack a specific harvest holiday which would be placed in the fall, although Thanksgiving in the United States and Canada has been endowed with Christian prayers to play that role. Holidays are even less related to events in the physical environment for Christians in the Southern Hemisphere, where Easter comes in the fall and Christmas in the summer.

Holy places. As an ethnic religion of India, Hinduism is closely tied to the Indian landscape. According to a survey conducted by the geographer Surinder Bhardwaj, the natural features most likely to rank among the holiest shrines in India are riverbanks or coastlines.

Hindu holy places are organized into a hierarchy. Some prominent shrines attract pilgrims from the entire country, while other shrines are important to a local community but are seldom visited by people from other regions (Figure 8–14). Because Hinduism has no central authority, the relative importance of shrines is established by tradition, not by doctrine. For example, many Hindus make long-distance pilgrimages to Mt. Kailās, located at the source of the Ganges in the Himalayas, holy because the deity Siva lives there. At the same time, other mountains may attract only local pilgrims. Throughout India, local residents may consider a nearby mountain to be holy if Siva is thought to have visited it at one time.

As a universalizing religion, Buddhism holds places to be holy because important events in Buddha's life occurred there. The most important places include Lumbinī, where Buddha was born, around 563 B.C.; Bodh Gayā, where he reached perfect wisdom; Deer Park in Sarnath, where he gave his first sermon; and Kuśinagara, where he passed into Nirvana, a state of peaceful extinction. Four other sites in northeastern India are also sacred because they were the locations of Buddha's principal miracles (Figure 8–15).

The holiest locations in Islam are places in cities associated with events in Muhammad's life. The holiest cities for Muslims include Makkah (Mecca), Muhammad's birthplace; Madinah, where Muhammad received his first support; and Jerusalem, where he ascended to heaven.

Pilgrimages. Both ethnic and universalizing religions that have a collection of important holy shrines may also have an organized procedure by which adherents from around the world visit them. Hindus and Muslims are especially encouraged to make **pilgrimages** to visit holy places in accordance with recommended itineraries, and Shintoists are encouraged to visit holy places in Japan.

Every healthy Muslim who has adequate financial resources is expected to undertake a pilgrimage to Makkah (Mecca), known as a *hajj*. The word *mecca* now has a general meaning in the English language: a place sought as a goal or a center of activity. Regardless of nationality and economic background, all pilgrims dress alike in plain white robes to emphasize common loyalty to Islam and the equality of people in the eyes of Allah. A precise set of rituals is practiced, culminating in a visit to the Ka'ba (Figure 8–16).

Hindus consider a pilgrimage, known as a *tirtha*, to be an act of purification. Although not a substitute for meditation, the pilgrimage is an important act in achieving redemption. Particularly sacred places attract Hindus from all over India, despite the relatively remote locations of some, while less important shrines attract primarily local pilgrims.

The remoteness of holy places from population clusters once meant hardship. However, recent improvements in transportation have increased the accessibility of shrines. Hindus can now reach holy places in the Himalaya Mountains by bus or car, and Muslims from all over the world can reach Makkah by airplane.

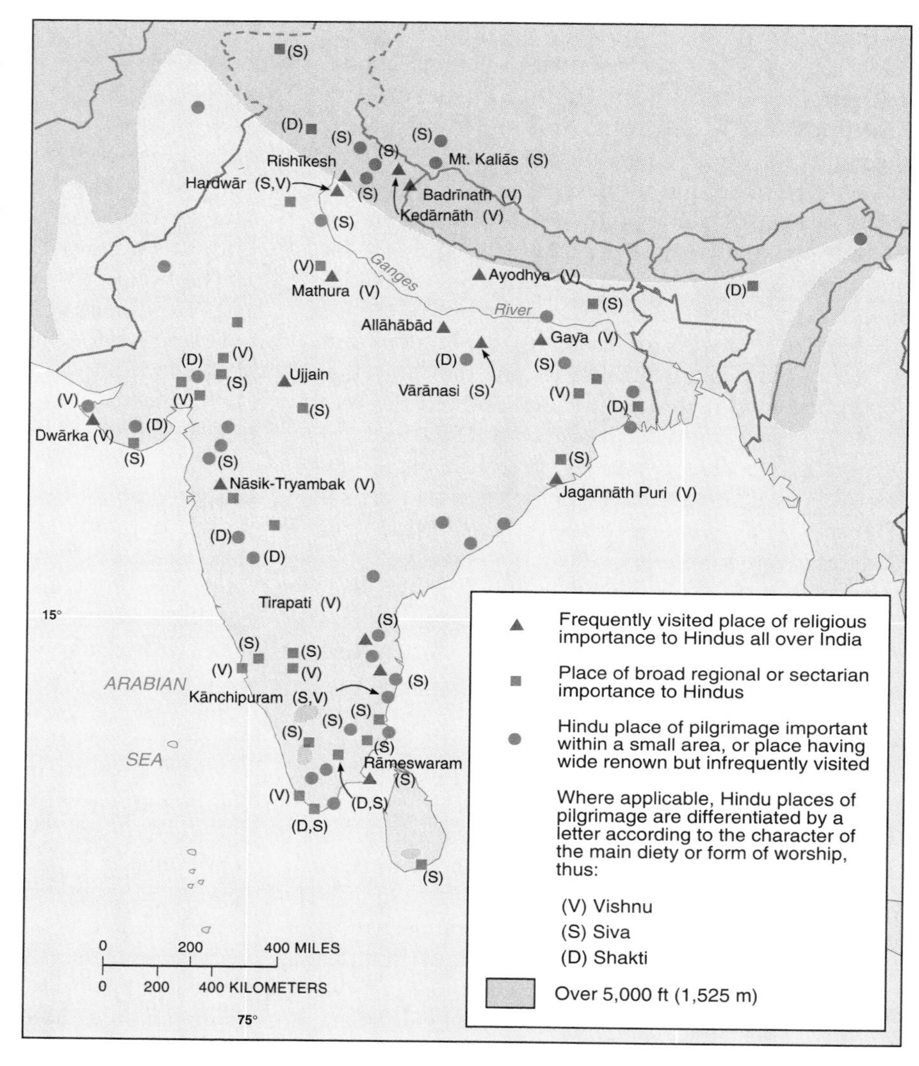

FIGURE 8–14
Hindu holy places can be organized into a hierarchy. Some places are important to Hindus all over India and are visited frequently, while others have importance only to nearby residents. The map also shows that holy places for particular deities are somewhat clustered in different regions of the country—Shakti in the east, Vishnu in the west, and Siva in the north and south. (From Ismail Ragi al Farugi and David E. Sopher, *Historical Atlas of the Religions of the World*, New York: Macmillan, 1974.)

Social Customs

A **custom** is an act frequently repeated, to the extent that it becomes characteristic of the group of people performing the act. Custom is a more precise word than culture, which is one of the hardest words to define in English. In Chapter 1 culture was defined as the body of customary beliefs, social forms, and material traits of a group of people. **Habit**, a word similar to custom, is a repetitive act that a particular individual performs. Unlike custom, habit does not imply that the act has been adopted by most of the society's population. A custom is therefore a habit that a group of people has widely adopted.

Geographers are interested in two aspects of social customs. First, each social custom has its own spatial distribution, and geographers study the origin, process of diffusion, and integration of different characteristics. Second, geographers study the relationships between social customs and the physical environment. Different social groups take particular elements from the physical environment in their culture and in turn construct landscapes (built environments) that modify nature in distinctive ways.

Geographers examine two kinds of social customs. First are customs deriving from necessary activities of daily life: all human beings must have food, clothing, and shelter to survive, but social groups provide these needs in their own ways.

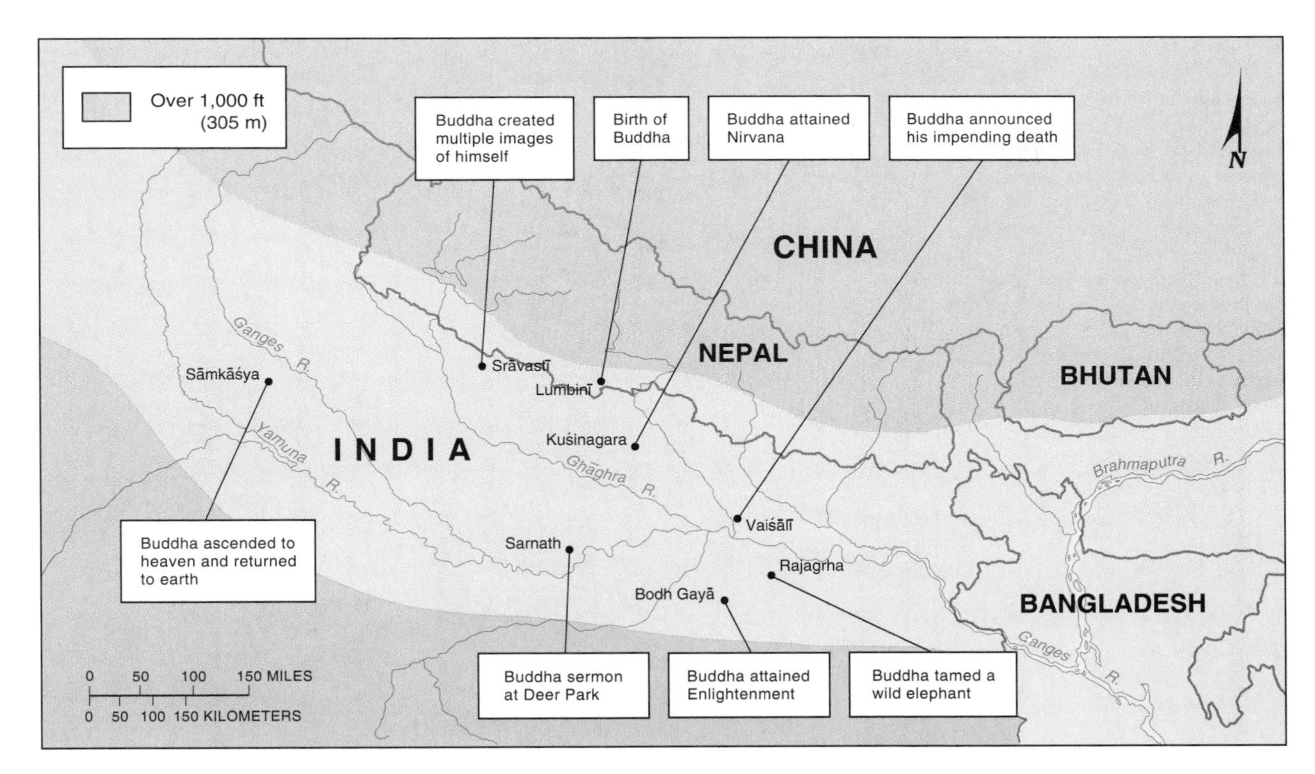

FIGURE 8–15
Holy places in Buddhism are clustered in northeastern India and Nepal, because they were the locations of important events in Buddha's life. Most of the sites are in ruins today. (From Ismail Ragi al Farugi and David E. Sopher, *Historical Atlas of the Religions of the World,* New York: Macmillan, 1974.)

Second are customs involving leisure activities, especially the arts and recreation. Each social group has its own definition of meaningful art and stimulating recreation.

Differences Between Folk and Popular Customs

Social customs fall into two basic categories: folk and popular. **Folk customs** are traditionally practiced primarily by small, homogeneous groups living in isolated rural areas. **Popular customs** are found in large, heterogeneous societies that share certain habits despite differences in other personal characteristics.

Popular customs generally have more extensive distributions around the world than do folk customs. Two basic factors help to explain geographic differences between popular and folk customs: the process of origin and the pattern of diffusion.

Hindus believe that the Ganges River springs from the hair of Siva, one of the main deities. The river attracts pilgrims from all over India, who achieve purification by bathing in it. Bodies of the dead are washed with water from the Ganges before being cremated. (Benjamin Rondel/The Stock Market)

FIGURE 8–16
Makkah, in Saudi Arabia, is the holiest city for Muslims, because Muhammad was born there. Thousands of Muslims make a pilgrimage to Makkah each year and gather at al-Haram al-Sharīf, a mosque located in the center of the city.
(Below) The black cubelike structure in the center of the mosque, called the Ka'ba, once had been a shrine to tribal idols until Muhammad rededicated it to Allah. Muslims believe that Abraham and his son Ishmael originally built the Ka'ba.

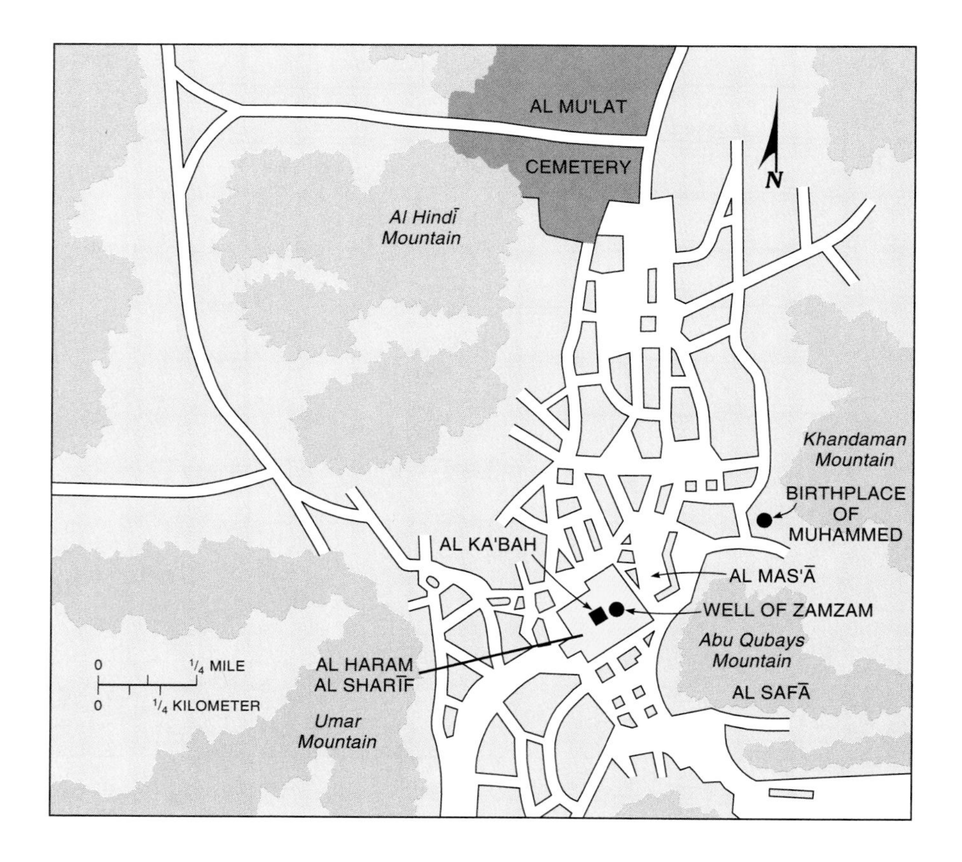

Mohamed Lounes. (Gamma-Liaison, Inc.)

Origin of social customs. Social customs originate at a hearth, defined in Chapter 1 as a center of innovation. Folk customs often have anonymous or multiple hearths, originating from anonymous sources, at an unknown date, through unidentified originators.

In contrast to folk customs, popular customs are most often a product of the economically developed countries, especially in North America, Western Europe, and Japan. They arise from a combination of advances in industrial technology and increased leisure time. Industrial technology permits the uniform reproduction of objects in large quantities. Many of these objects help people enjoy leisure time, which has increased as a result of the widespread change in the labor force from predominantly agricultural to predominantly service and manufacturing jobs.

Folk customs, such as folk music, may have multiple origins, as a result of a lack of communication among groups of people living in different places. Country music in the United States provides a recent example of the process by which folk customs originate independently at multiple locations. The geographer George Carney identified four major hearths of country music in the southeastern United States during the late nineteenth and early twentieth centuries: southern Appalachia, central Tennessee and Kentucky, the Ozark and Ouachita mountains of western Arkansas and eastern Oklahoma, and north-central Texas (Figure 8–17). Carney documented these hearths on the basis of the birthplaces of performers and other individuals active in the field.

Popular music has different origins than folk music. Popular music is written by specific individuals and recorded for the purpose of being sold to a large number of people. Recorded popular music displays a high degree of technical skill and is frequently capable of being performed only in a studio with electronic equipment.

Diffusion of social customs. The most significant geographic process that distinguishes popular from folk customs is interaction. Popular customs are based on the ease of simultaneous global interaction through communications systems, transportation networks, and other modern technology. The spread of popular customs typically follows the process of hierarchical diffusion from *nodes of innovation*. Prominent examples of such nodes for innovation of popular customs in the United States include Hollywood, California, for the film industry and Madison Avenue in New York City for advertising agencies. Paris, France, and Milan, Italy are nodes of innovation for the apparel industry. Rapid diffusion facilitates frequent changes in dominant popular customs.

In contrast, folk customs are transmitted from one location to another more slowly and on a smaller scale, primarily through migration rather than electronic communications. Even groups living in proximity to each other may produce a variety of folk customs in a limited geographic area

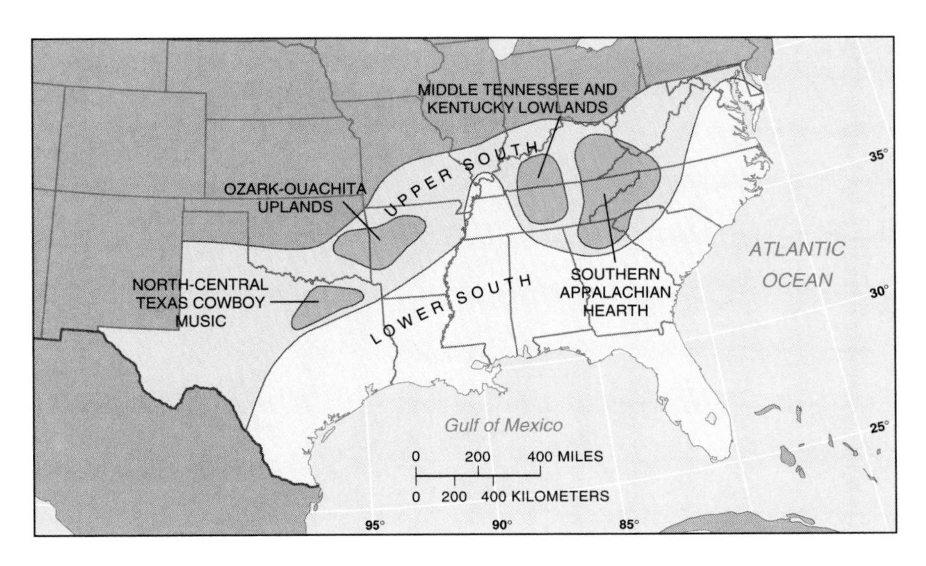

FIGURE 8–17
U.S. country music has four major hearths, or regions of origin. These include southern Appalachia, central Tennessee and Kentucky, the Ozark and Ouachita mountains of western Arkansas and eastern Oklahoma, and north-central Texas. (Adapted from John F. Rooney, Jr., Wilbur Zelinsky, and Dean R. Louder eds., *This Remarkable Continent: An Atlas of United States and Canadian Society and Culture,* College Station, TX, Texas A & M University Press.)

because of lack of communication. The spread of folk customs is an example of relocation diffusion, the spread of a characteristic through migration. Folk customs are more likely than popular customs to vary from one place to another at a given point in time, while popular customs are more likely than folk customs to vary from one point in time to another at a given place.

Distribution of Folk Customs

Folk customs observed at a given point in time vary widely from one place to another, even among nearby places. In a study of artistic customs in the Himalaya Mountains, geographers P. Karan and Cotton Mather demonstrate that distinctive views of the physical environment emerge among cultural groups who live near each other, yet are isolated. The study area—a narrow, 2,500-kilometer (1,500-mile) corridor in the Himalaya Mountains of Bhutan, Nepal, and northern India—contains four religious groups: Tibetan Buddhists in the north, Hindus in the south, Muslims in the west, and Southeast Asian animists in the east (Figure 8–18). Despite their spatial proximity, limited interaction among the groups produces distinctive folk customs.

Physical environmental processes, such as climate, landforms, and vegetation, also play a significant role in the development of unique folk customs. Environmental determinists theorized that processes in the physical environment cause social customs, but most contemporary geographers reject this notion. There are many examples of different peoples living in similar physical environments who adopt different social customs and, conversely, of peoples living in different environmental conditions who adopt similar social customs. However, geographers generally recognize that people do not ignore their physical environment.

Social customs such as food, clothing, and shelter practices are partially influenced by the prevailing climate, soil, and vegetation. For example, residents of arctic climates may wear fur-lined boots, which protect against the cold, and snowshoes to walk on soft, deep snow without sinking in. On the other hand, people living in warm and humid climates may not need any footwear at all due to the heavy rainfall, and time they spend in water.

Environmental conditions can limit the variety of human actions anywhere, but folk societies are particularly responsive to the physical environment because of the level of technology and the predominant form of economy. People living in folk societies are likely to be farmers growing their own food utilizing hand tools and animal power.

Yet folk customs may ignore the physical environment. Not all arctic residents wear snowshoes, nor do all people in moist temperate climates wear wooden shoes. Geographers merely observe that broad differences in folk customs arise with some consideration of physical conditions, even though those conditions may produce a variety of customs.

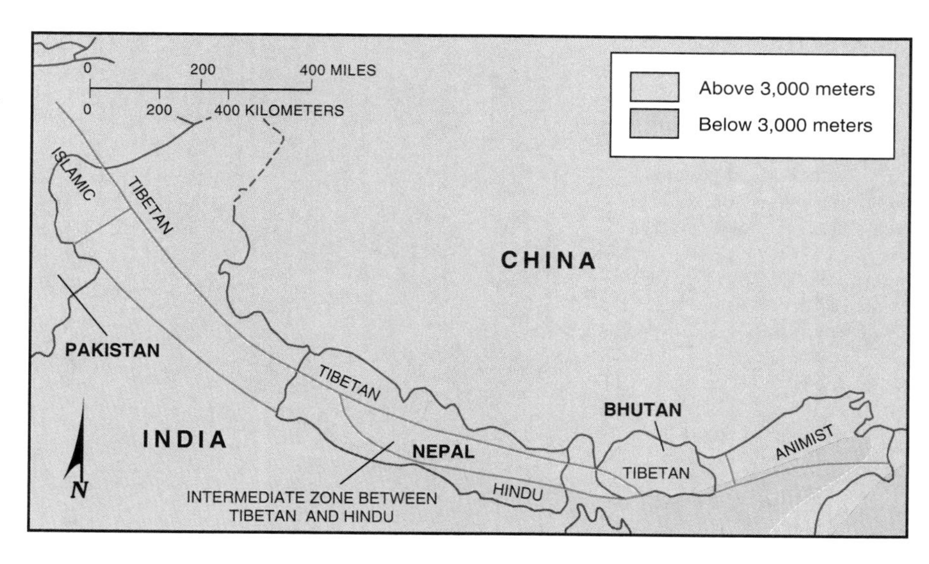

FIGURE 8–18
Karan and Mather found four culture regions in the rugged Himalaya mountains of Bhutan, Nepal, and northern India. Variations among the four groups were found in religion, painting, dance, and other folk customs. (Reproduced by permission from the *Annals of the Association of American Geographers,* Volume 66, 1976.)

Food customs and taboos. Two necessities of daily life, food and shelter, demonstrate the influence of both distinctive cultural values and physical environmental patterns on the development of unique folk customs. People in different folk societies display distinctive preferences for food and styles of house construction, based in part on cultural values and in part on characteristics of the local physical environment.

Food is composed for the most part of living things that spring from the soil and water of a region in the form of either plants or animals. Inhabitants of a region must consider the soil, climate, and other characteristics of the physical environment in deciding to produce particular foods. Rice must be grown in a relatively mild, moist climate, while wheat thrives in colder, dryer regions.

People adapt their food preferences to conditions in the physical environment. Soybeans are widely grown in Asia, but in a raw state they are toxic and indigestible. To make them edible, soybeans must be cooked for a long time, but sources of fuel are in short supply in Asia. Asians have adapted to this environmental problem by consuming foods derived from soybeans that do not require extensive cooking, including bean sprouts (germinated seeds), soy sauce (fermented soybeans), and bean curd (steamed soybeans).

According to many folk customs, everything in nature carries a signature, or distinctive characteristic, based on its appearance and natural properties. Consequently, people may desire or avoid certain foods in response to perceived beneficial or harmful natural traits.

Certain foods are eaten because their natural properties are perceived to enhance qualities considered desirable by the society, such as strength, fierceness, or love-making ability. The Abipone Indians of Paraguay eat jaguars and bulls to make them strong, brave, and swift. The mandrake, a plant native to Mediterranean climates, was thought to enhance an individual's love-making abilities. The smell of the plant's orange-colored berries is attractive, but the mandrake's association with sexual prowess comes primarily from the appearance of the root, which is thick, fleshy, and forked, suggestive of a man's torso. In parts of Africa and the Middle East, the mandrake's root is administered as a drug, and several references to its powers are found in the Bible.

People refuse to eat particular plants or animals that are thought to embody negative forces in the physical environment. Such a restriction on behavior imposed by social custom is known as a **taboo**. The Ainus in Japan avoid eating otters, because they are believed to be forgetful animals, and consuming them could cause loss of memory. Europeans blamed the potato, the first edible plant they had encountered that grew from tubers rather than seeds, for a variety of problems during the seventeenth and eighteenth centuries, including typhoid, scrofula, and famine. Initially, Europeans also resisted eating the potato because it resembled human deformities caused by leprosy.

Before becoming pregnant, Mbum Kpau women of Chad do not eat chicken or goat. Abstaining from consumption of these animals is thought to help escape pain in childbirth and prevent birth of an abnormal child. During pregnancy, Mbum Kpau women avoid meat from antelopes with twisted horns, which could cause them to bear deformed offspring. In the Trobriand Islands off the eastern tip of Papua New Guinea, couples are prohibited from eating meals together before marriage, while premarital sexual relations are an accepted feature of social life.

Some folk customs may establish food taboos because of concern for the natural environment. These taboos may help to protect endangered animals or conserve scarce natural resources. For example, to preserve scarce animal species, only a few high-ranking people in some tropical regions are permitted to hunt, while the majority of the population cultivate crops. However, most food avoidance customs arise from cultural values.

Relatively well-known taboos against consumption of certain foods can be found in the Bible. The ancient Hebrews were prohibited from eating a wide variety of foods, including animals that didn't chew their cud or have cloven feet and fish without fins or scales. These taboos arose partially from concern for the physical environment by the Hebrews, who lived as pastoral nomads in lands bordering the eastern Mediterranean. The pig, for example, is prohibited in part because it is more suited to sedentary farming than pastoral nomadism and in part because its meat spoils relatively quickly in hot climates, such as the Mediterranean.

Similarly, Muslims embrace the taboo against pork, because pigs are unsuited for the dry lands of the Arabian peninsula (Figure 8–19). Pigs would compete with humans for food and water, without offering compensating benefits, such as being able to pull a plow, carry loads, or provide milk and wool. Widespread raising of pigs would be an ecological disaster in Islam's hearth.

Hindu taboos against consuming cows can also be explained partly on environmental reasons. Cows give birth to oxen (castrated male bovine), the traditional animal of choice for pulling plows and carts. A large supply of oxen must be maintained in India, because every field has to be plowed at approximately the same time, when the monsoon rains arrived. Religious sanctions kept India's cow population large as a form of insurance against the loss of oxen in the face of increasing population.

But the taboo against consumption of meat among many people, including Muslims, Hindus, Buddhists, and Jews, cannot be explained primarily by physical environment factors. Social values must influence the choice of diet, since people in similar climates and with similar levels of income consume different foods. The biblical food taboos were established in part to set the Hebrew people apart from others. That Christians ignore the biblical food injunctions reflects their desire to distinguish themselves from Jews. Furthermore, as a universalizing religion, Christianity was less tied to taboos that originated in the Middle East.

Folk housing. The type of building materials used to construct folk houses is influenced in part by the availability of particular resources in the environment. The two most common building materials in the world are wood and bricks, although stone, grass, sod, and skins are also used. If available, wood is generally preferred for house construction. In the past, pioneers who settled in forested regions could build log cabins for themselves.

The form of houses in some societies may reflect religious values. For example, houses may have sacred walls or corners. The east wall of a house is considered sacred in Fiji, and the northwest in parts

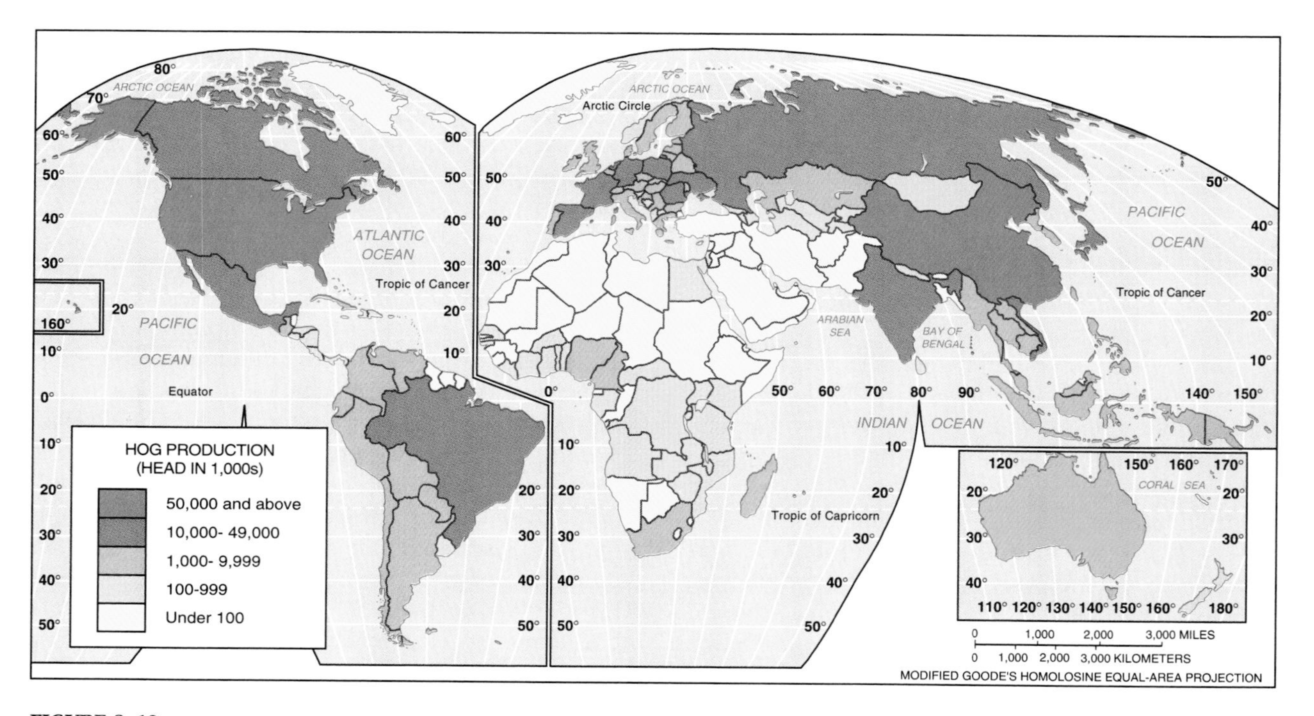

FIGURE 8–19
Annual hog production. The distribution is heavily influenced by religious taboos against consuming pork. Hog production is virtually nonexistent in predominantly Muslim regions, such as northern Africa and southwestern Asia, while the level is high in predominantly Buddhist China and predominantly Christian Europe and Western Hemisphere.

of China. Sacred walls or corners are also found in parts of the Middle East, India, and Africa. In the south-central part of the island of Java, the front door always faces south, the direction of the South Sea Goddess, who holds the key to Earth.

In northern Laos, the Lao people arrange beds perpendicular to the center ridge pole of the house. Because the head is considered high and noble and the feet low and vulgar, people sleep so that their heads will be opposite their neighbor's heads and their feet opposite their neighbor's feet (Figure 8–20). The principal exception to this arrangement: a child who builds a house next door to the parents sleeps with his or her head towards the parents' feet as a sign of obeying the customary hierarchy.

Although they speak similar Southeast Asian languages and adhere to Buddhism, the Lao people do not orient their houses in the same manner as the Yuan and Shan peoples in nearby northern Thailand. The Yuan and Shan ignore the position of neighbors and all sleep with their heads towards the east, which Buddhists consider the most auspicious direction. Staircases must not face west, the least auspicious direction, the direction of death and evil spirits.

The form of housing is related in part to environmental as well as social conditions. The con-

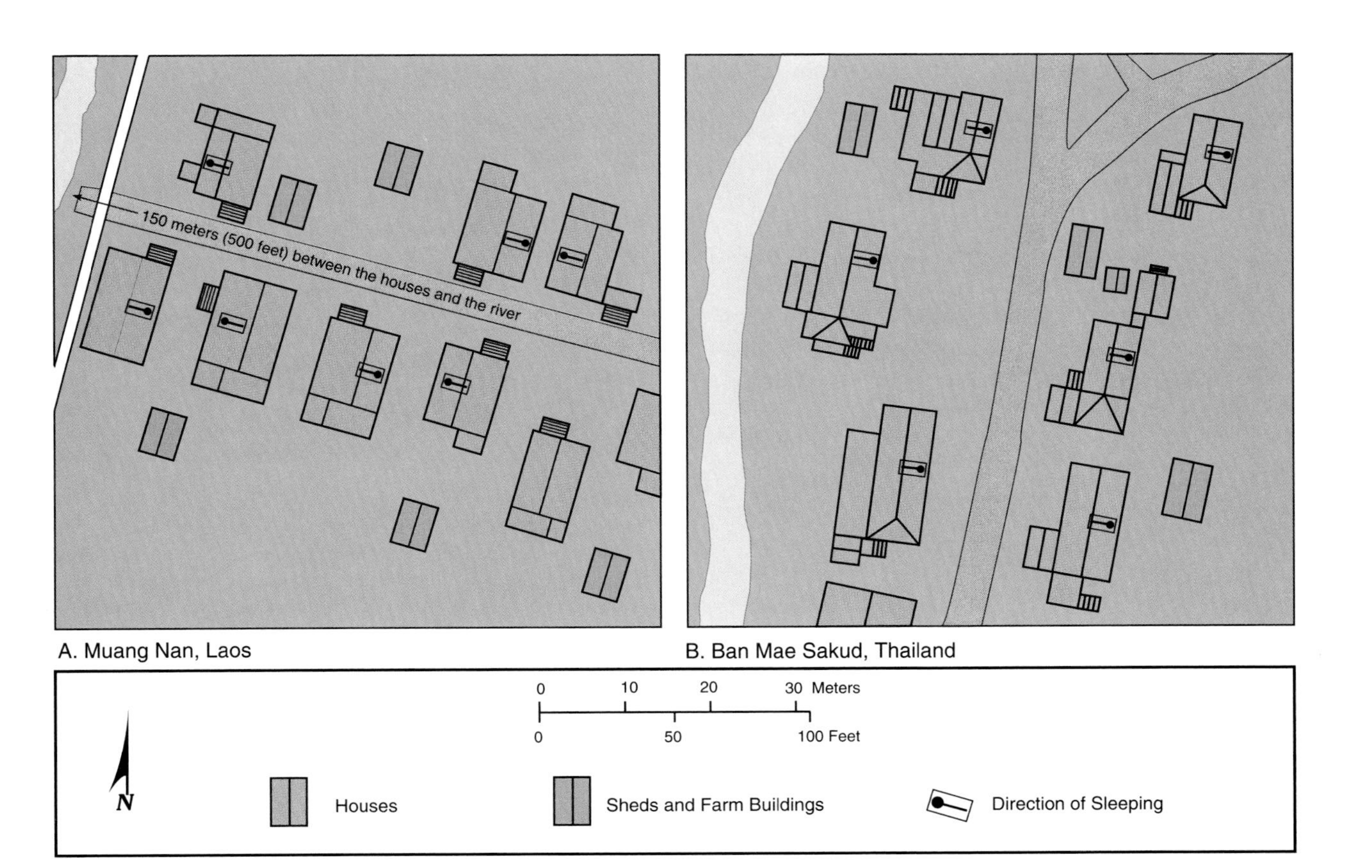

FIGURE 8–20
Arrangement of houses in two Southeast Asian communities.
(a) The front gables of Lao houses, such as those in Muan Nan, Laos, face one another across a path, and the backs face each other at the rear. Ridgepoles are set perpendicular to the path and parallel to a stream if one is nearby. Inside the house, the head of a sleeping person is in the opposite direction to that of the neighbor, so that heads and feet are always together.
(b) In Ban Mae Sakud, Thailand, houses are not set in a straight line because of a belief that evil spirits move in straight lines. Ridgepoles are set parallel to the path, and the heads of all sleeping persons face east.

struction of a pitched roof is important in wet or snowy climates to facilitate runoff. Windows may be aligned to the south in temperate climates to take advantage of the sun's heat and light. In hot climates, on the other hand, window openings may be smaller in order to protect the interior from the full heat of the sun.

U.S. folk house forms. In the United States, when a family migrated to the west in the eighteenth or nineteenth century, it would cut down the trees to clear fields for planting and use the wood to build the house, barn, and fences. The shape and layout of the house that the pioneering family built was influenced by the style associated with economic attainment that prevailed at the time in the region of the east coast from which they migrated.

Kniffen identified three major hearths or nodes of house forms in the United States: New England, Middle Atlantic, and Lower Chesapeake (Figure 8–21). Migrants carried house types from New England northward to Upper New England and westward across the southern Great Lakes region, from the Middle Atlantic westward across the Ohio Valley and southwestward along the Appalachian trails, and from the Lower Chesapeake southward along the Atlantic coast (Figure 8–22).

Today, such distinctions are relatively difficult to observe in the United States. The style of housing does not display the same degree of regional distinctiveness, because rapid communication and transportation systems provide people throughout the country with knowledge of alternative styles. Furthermore, most people do not build the houses in which they live. Instead, houses are usually mass-produced by construction companies.

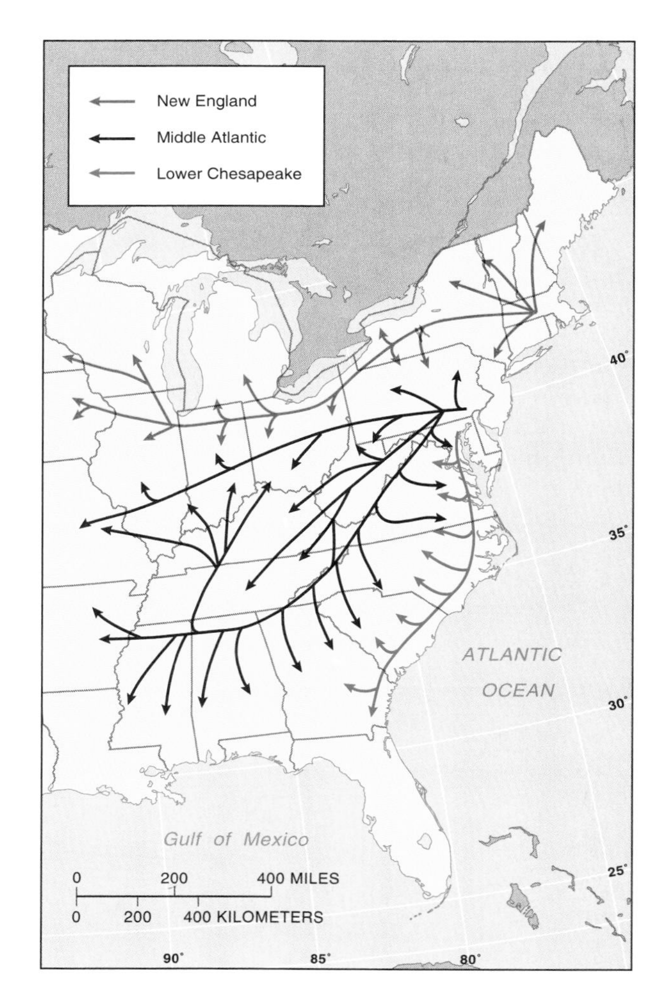

FIGURE 8–21
Source areas of U.S. folk house types. According to Kniffen, house types in the United States originated in three main source areas and diffused to the west along different paths. These paths coincided with the predominant routes taken by migrants from the East Coast to the interior of the country. (Reproduced by permission from the *Annals of the Association of American Geographers*, Volume 55, 1965.)

Distribution of Popular Customs

Popular customs vary more in time than in space. They may originate in one location, within the context of a particular social and physical environment, but in contrast to folk customs, they diffuse rapidly throughout the world to locations that vary in physical conditions. Rapid diffusion depends on whether groups of people have a sufficiently high level of economic development to acquire the material possessions associated with the popular custom.

Housing style. The housing built in the United States since the 1940s demonstrates how popular customs vary more in time than in place. In contrast with folk housing characteristic of the early nineteenth century, newer housing in the United States has been built with an attempt to reflect rapidly changing fashion concerning the most suitable house form.

Houses show the influence of shapes, materials, detailing, and other features of architectural style in vogue at any one point in time. In the years immediately after World War II, most houses in the

FIGURE 8–22
Diffusion of folk house types from southern New England. Kniffen suggests that these four houses, in turn, were popular in southern New England at the indicated dates. As settlers migrated, they carried memories of familiar house types with them and built similar structures on the frontier. Thus, New Englanders were most likely to build houses such as the one shown here in yellow when they began to migrate to upstate New York in the 1790s, because this was the predominant type of house constructed in southern New England at the time they began to migrate. During the nineteenth century, when New Englanders began to migrate farther west to Ohio and Michigan, they built the type of house typical in New England at that time, shown here in blue.

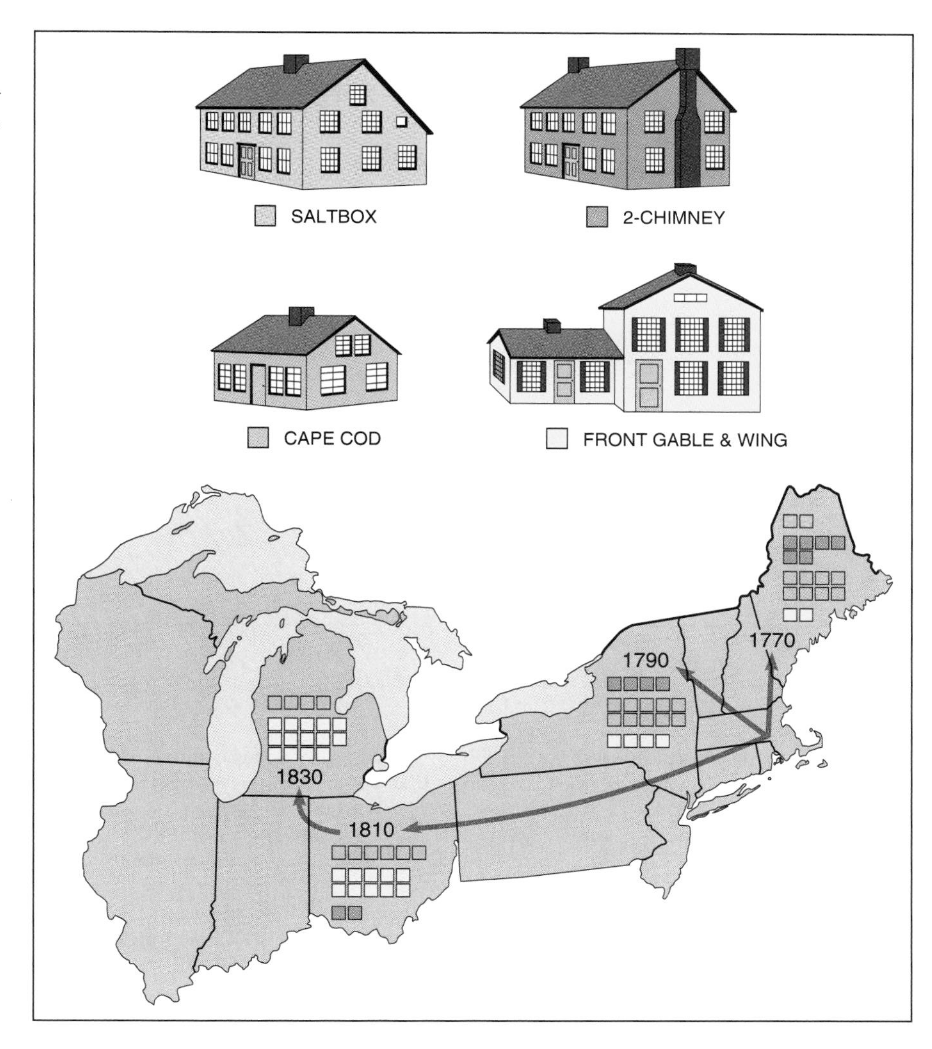

United States were built in a modern style—boxy, low to the ground, and unadorned with ornamental trim. Since the 1960s, styles that architects call neoeclectic have predominated (Figure 8–23).

Specific types of modern-style houses were popular at different times. In the late 1940s and early 1950s, the dominant type was known as minimal traditional, usually one-story, with a dominant front gable and few decorative details. They were small modest houses designed to house large numbers of young families and returning veterans immediately after World War II. The ranch house replaced minimal traditional as the dominant style of housing in the 1950s and into the 1960s. With all of the rooms on one level rather than two or three, the ranch house took up larger lot sizes and encouraged the sprawl of urban areas (see Chapter 13). The split-level house was a popular variant of the ranch house between the 1950s and 1970s. The lower level of the typical split-level house contained the garage and the newly invented family room, where the television set was placed. In the late 1960s, neoeclectic styles became popular and by the 1970s had become much more popular than so-called modern styles.

Differences in housing among communities in the United States derive largely from differences in the time period in which the houses were built. The ranch house was more common in the southeast than in the northeast primarily because the southeast grew much more rapidly during the 1950s and 1960s, the period when the ranch house

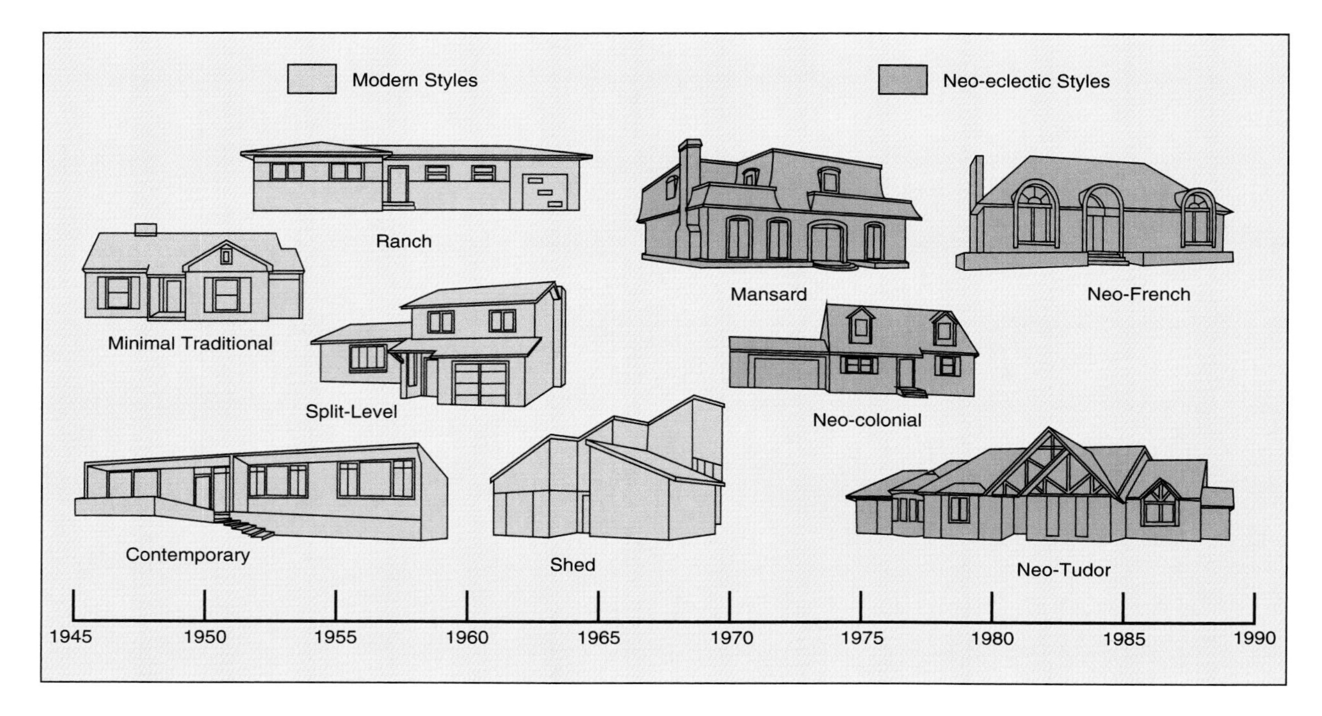

FIGURE 8–23
Popular house types in the United States since 1945. *Modern* styles dominated until around 1970. The dominant type of modern construction in the United States was *minimal traditional* during the late 1940s and early 1950s, followed by *ranch houses* during the late 1950s and 1960s. The *split-level* house was a popular variant of the ranch house between the 1950s and 1970s. The *contemporary* style was popular for architect-designed houses between the 1950s and 1970s. The *shed* style was popular in the late 1960s.
Neo-eclectic styles became popular in the late 1960s, beginning with the *mansard*. The *neo-Tudor* style was popular in the late 1970s and the *neo-French* in the 1980s. The *neoclassical* style has been popular since the 1950s but never dominant.

was especially popular. A housing development built in one region will resemble more closely developments built at the same time elsewhere in the country than developments built in the same region at other points in time (Figure 8–24).

Importance of television. Watching television is an especially significant popular custom for two reasons. First, it is the most popular leisure activity in relatively developed countries throughout the world. Second, television is the most important mechanism by which knowledge of popular customs, such as professional sports, is rapidly diffused across Earth.

Currently, the level of television service falls into four categories. The first category consists of countries where nearly every household owns a television set. This category includes the relatively developed countries of North America and Europe as well as Australia, New Zealand, and Japan. A second category consists of countries in which ownership of a television is common but by no means universal. These are primarily Latin American countries and the poorer European states, such as Turkey and Portugal.

The third category consists of countries in which television exists but has not yet been widely diffused to the population as a whole because of the high cost of the sets. This category includes some countries in Africa, Asia, and Latin America. Finally, about 20 countries, most of which are in Africa and Asia, have very few television sets. Some of these countries do not have operating television stations, although programs from neighboring countries may be received with antennas (Figure 8–25).

FIGURE 8–24
Distribution of popular house types within the United States. Jakle, Bastian, and Meyer identified types of single-family houses in twenty small towns. Houses were grouped into five types: bungalow, double-pile, irregularly massed, ranch, and single-pile. Ranch houses were more common in the southeastern towns, while double-pile predominated in the northeast.

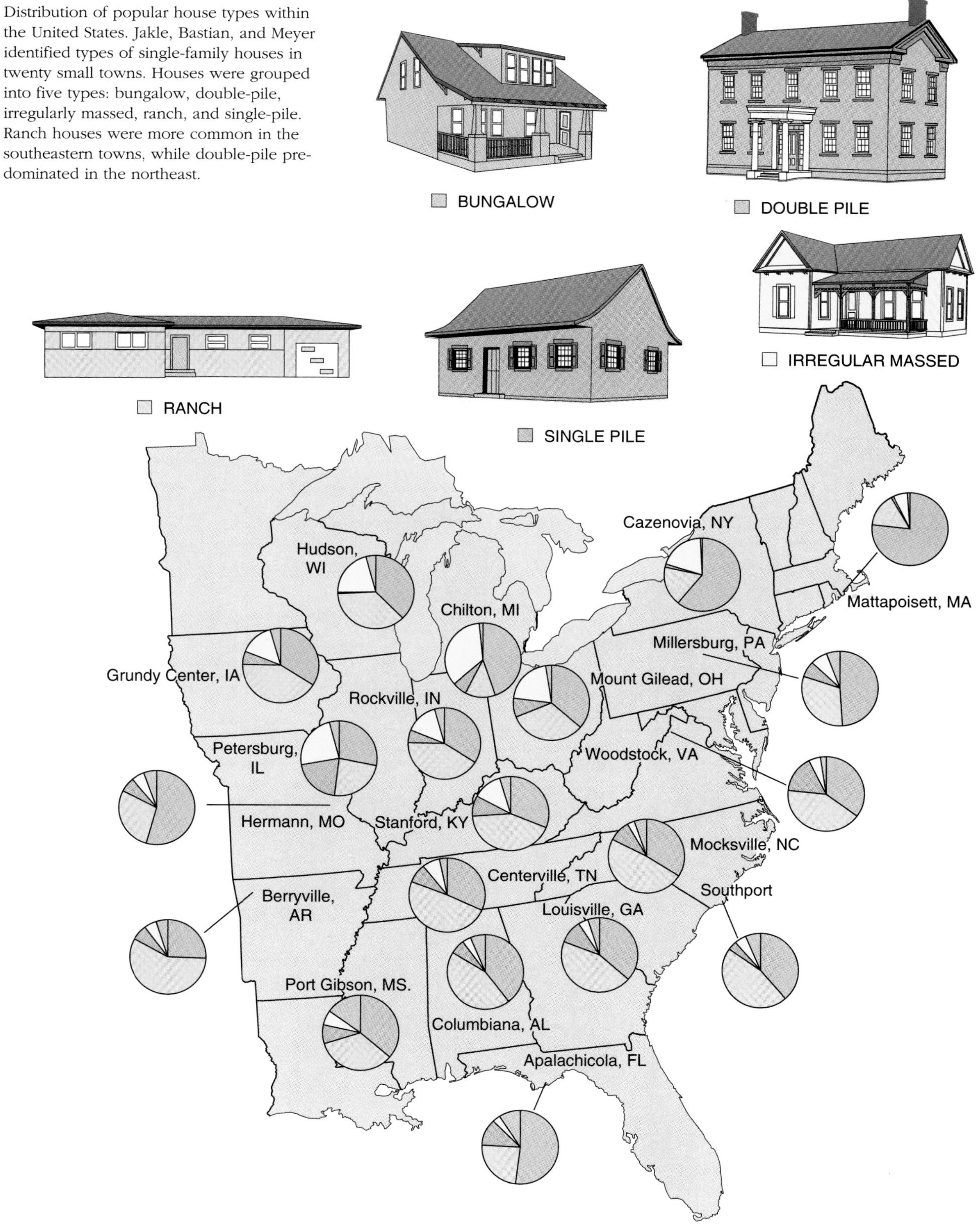

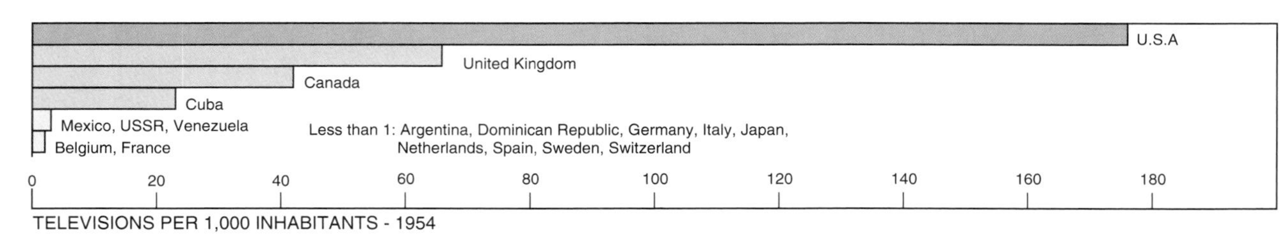
U.S.A
United Kingdom
Canada
Cuba
Mexico, USSR, Venezuela
Belgium, France
Less than 1: Argentina, Dominican Republic, Germany, Italy, Japan, Netherlands, Spain, Sweden, Switzerland
0
20
40
60
80
100
120
140
160
180
TELEVISIONS PER 1,000 INHABITANTS - 1954

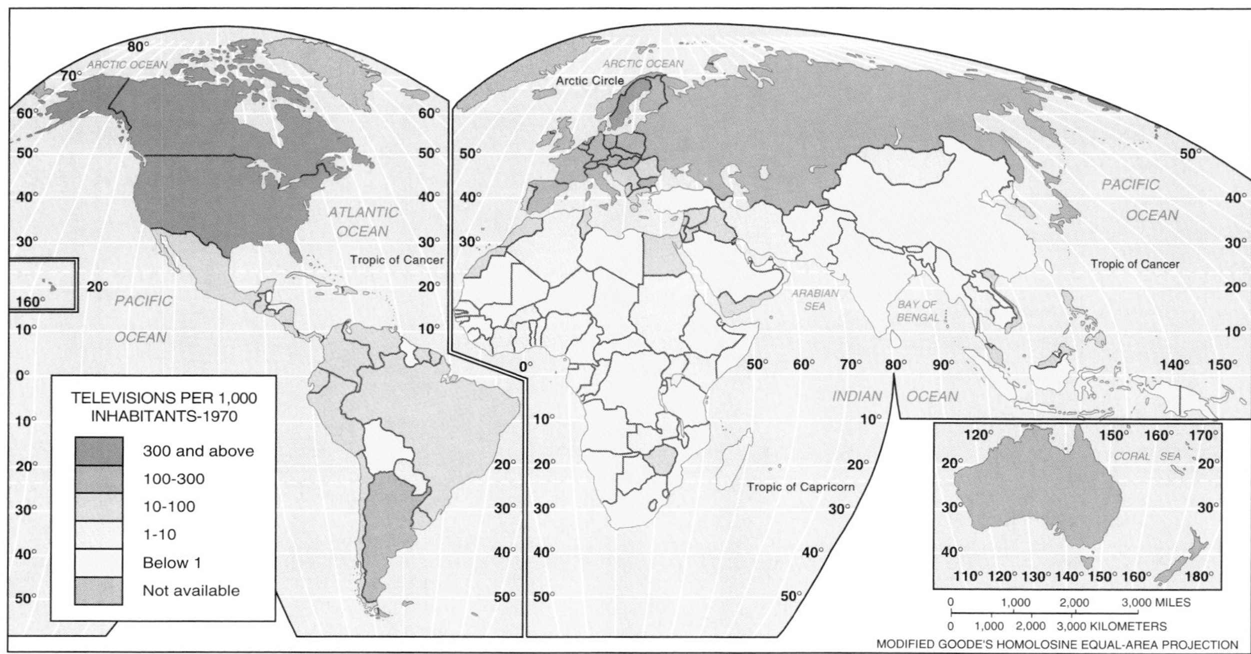
TELEVISIONS PER 1,000 INHABITANTS-1970
300 and above
100-300
10-100
1-10
Below 1
Not available
ARCTIC OCEAN
Arctic Circle
ATLANTIC OCEAN
PACIFIC OCEAN
INDIAN OCEAN
ARABIAN SEA
BAY OF BENGAL
CORAL SEA
Tropic of Cancer
Tropic of Capricorn
0 1,000 2,000 3,000 MILES
0 1,000 2,000 3,000 KILOMETERS
MODIFIED GOODE'S HOMOLOSINE EQUAL-AREA PROJECTION

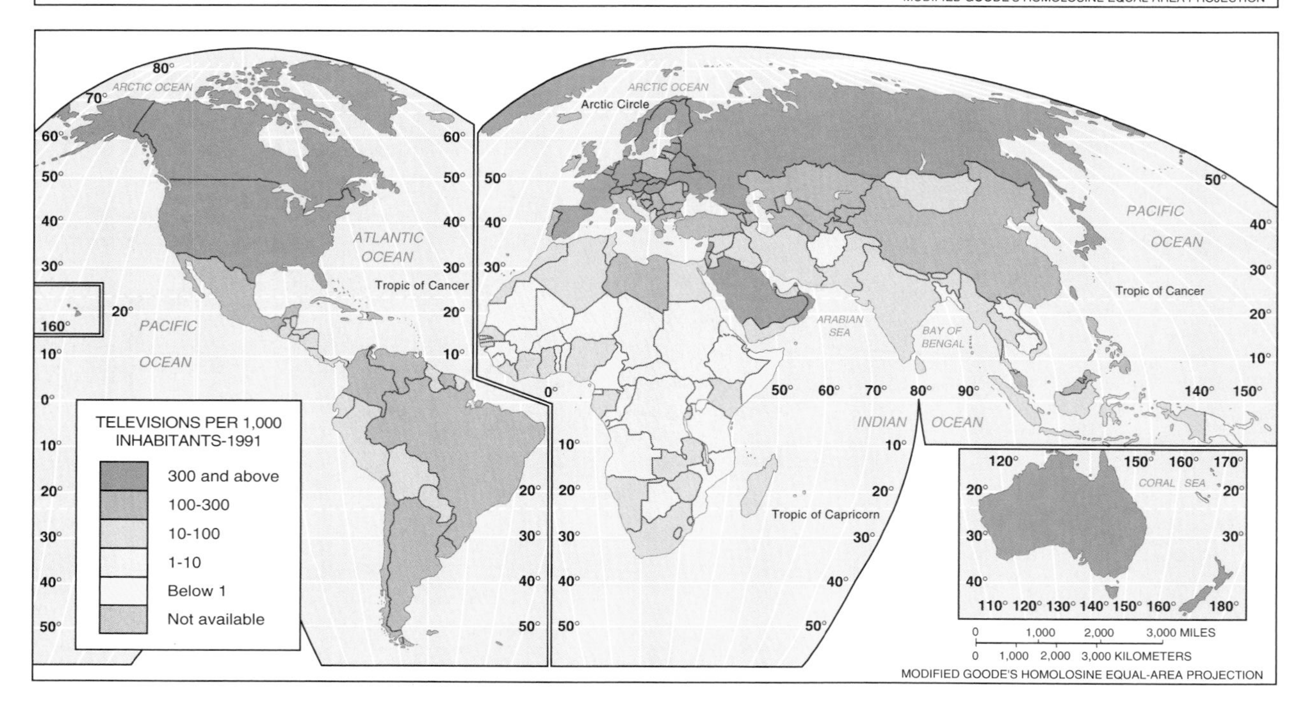
TELEVISIONS PER 1,000 INHABITANTS-1991
300 and above
100-300
10-100
1-10
Below 1
Not available
ARCTIC OCEAN
Arctic Circle
ATLANTIC OCEAN
PACIFIC OCEAN
INDIAN OCEAN
ARABIAN SEA
BAY OF BENGAL
CORAL SEA
Tropic of Cancer
Tropic of Capricorn
0 1,000 2,000 3,000 MILES
0 1,000 2,000 3,000 KILOMETERS
MODIFIED GOODE'S HOMOLOSINE EQUAL-AREA PROJECTION

Muslim woman at keyboard. Exposure to modern technology does not necessarily change all traditional social customs. Women in predominantly Muslim countries have been urged to wear the *chador*, a combination head covering and veil, rather than Western clothes. Wearing traditional clothes is a sign of adherence to traditional religious principles. (Esaias Baitel/Gamma-Liaison, Inc.)

Problems of Cultural Diversity

A person's culture, including language, religion, and other social customs, can be a source of pride and a means of identification. People adhering to one religion, speaking one language, and practicing one set of social customs frequently fight with other cultural groups for control of land across Earth's surface. These struggles reflect the importance placed on cultural identification. A group of people strongly committed to their language, religion, and social customs may feel compelled to fight to expand—or at least to preserve— the area of the world dominated by their culture.

FIGURE 8–25
Diffusion of television. Television has diffused from North America and Europe to other regions of the world, but the number of television sets per capita still varies considerably among countries. The number of television sets per capita is also an important indicator of a society's level of development, as shown in Chapter 10.

Global Dominance of Popular Customs

The international diffusion of popular customs has led to two problems, both of which can be understood in a geographic perspective. First, the diffusion of popular customs may threaten the survival of traditional folk customs in many countries. Second, popular customs may be less responsive to the diversity of local conditions in the physical environment than folk customs and consequently may generate adverse environmental impacts.

Loss of traditional values. People in many countries fear the loss of folk customs for two reasons. First, the disappearance of folk customs may be symbolic of the loss of traditional values in society. Second, the diffusion of popular customs from relatively developed countries can lead to dominance of Western perspectives in their countries.

The global diffusion of popular customs threatens the subservience of women to men embedded in many folk customs. Women were traditionally relegated to performing household chores, such as cooking, cleaning, and bearing large numbers of children. Those women who were working outside the home were likely to be obtaining food for the family, either through agricultural work or by trading handicrafts. Advancement was limited by low levels of education and high rates of victimization from violence, often inflicted by husbands. The concepts of legal equality and availability of economic and social opportunities outside the home have become widely accepted concepts in relatively developed countries, even where women in reality continue to suffer from discriminatory practices.

Developing countries do not fear the incursion of popular customs only for symbolic reasons. Leaders of some developing countries consider the dominance of popular customs by developed countries to be a threat to their independence. The threat is posed primarily by the media, especially news-gathering organizations and television.

Three countries—the United States, the United Kingdom, and Japan—dominate the television industry in developing countries. Leaders of many developing countries view the spread of television as a new method of economic and cultural imperialism on the part of the relatively developed countries, especially the United States. American television, like other media, presents characteristically American beliefs and social forms, such as upward social mobility, relative freedom for women, glorification of youth, and stylized violence. These themes may conflict with and drive out traditional social customs.

Developing countries fear the effects of the news-gathering capability of the media even more than the entertainment function. The diffusion of information to newspapers around the world is dominated by the Associated Press (AP) and Reuters, owned by American and British companies, respectively. Many African and Asian government officials criticize the Western concept of freedom of the press. They argue that U.S. news-gathering organizations are more interested in covering earthquakes, hurricanes, or other sensational disasters than more meaningful but less visual and dramatic domestic stories, such as birth control programs, health care innovations, or construction of new roads.

The emergence of English as an international language has facilitated the diffusion of popular customs, as well as science and trade. In the twentieth century, English has become the world's most important language of international communication, or **lingua franca,** a language mutually understood and commonly used in trade by people who have different native languages. When well-educated speakers of two different languages wish to communicate with each other in countries such as India or Nigeria, they frequently use English. A Polish airline pilot who flies over France speaks to the traffic controller on the ground in English. The number of people in the world who speak English as a second language is unknown, but it is at least as large as the number who use English as a first language.

People in smaller countries perceive the need to learn English to participate more fully in a global economy and culture. Nearly all schoolchildren learn English in Northern European countries, such as the Netherlands and Sweden, in order to facilitate communication with people of other countries. Obviously, it is more likely that several million Dutch people will learn English than that several hundred million English speakers around the world will learn Dutch.

Traditionally and conversely, language has been a very important source of national pride and identity in France. The French government attempted to prohibit use of English words, such as *cheeseburger* and *weekend,* but the country's highest court ruled that the ban violated freedom of thought and expression. Many French speakers believe that people who forsake their native language must weigh the benefits of using English against the cost of losing a fundamental element of local cultural identity.

The French-speaking people of Québec live in a region dominated by English speakers (Figure 8–26). Until recently, Québec suffered from cultural isolation and lack of French-speaking leaders. It was one of Canada's poorest and least developed provinces, and its economic and political activities were dominated by an English-speaking minority.

Québec has restricted the use of languages other than French. Québec's Commission de Toponyme is renaming towns, rivers, and mountains that have English-language origins. The word *Stop* has been replaced by *Arrêt* on the red octagonal street signs, even though *Stop* is used throughout the world, even in other French-speaking countries. French must be the predominant language on all commercial signs, and the legislature passed a law, ultimately ruled unconstitutional by the Supreme Court, which banned all non-French outdoor signs altogether.

To assure preservation of their distinctive culture, many Québecois are willing to become independent from Canada, although in a 1980 referendum, a majority of voters in Québec opposed separation. The continuing challenge for the citizens of Québec, as well as the rest of Canada, is to maintain a country that encourages cultural diversity.

Impact on physical environment. Folk customs are more likely to be responsive to the physical environment, whereas popular customs may be distributed with relatively little regard for physical features. Instead, the spatial organization of popular customs is more dependent on the distribution of social and economic characteristics. Popular customs can significantly modify or control the physical environment and may appear to be imposed on the physical environment, rather than springing forth from the local physical environment, as with many folk customs. With changing technology, popular customs acquire an increasingly uniform appearance.

Golf courses, because of their large size (80 hectares, or 200 acres), provide a particularly prominent example of the imposition of popular customs on the physical environment (Figure 8–27). Golf courses are designed partially in response to local physical conditions. Species of grasses are selected depending on what will thrive in the local climate and still be suitable for the distinctive needs of greens, fairways, and roughs. Yet, like other popular customs, golf courses require remaking the physical environment: creating or flattening hills, cutting grass or letting it grow tall, carting in or digging up sand for traps, and draining or expanding bodies of water for hazards.

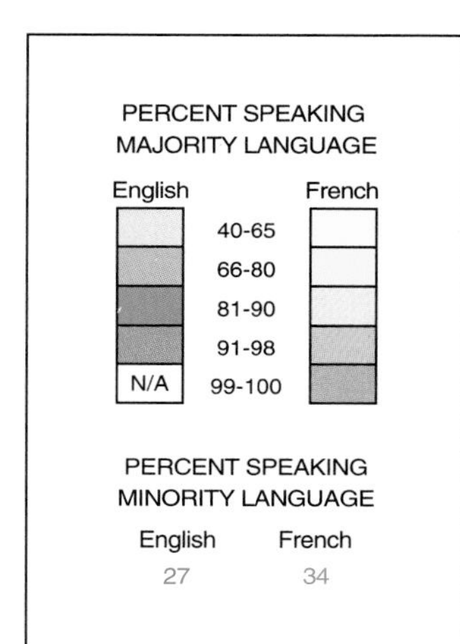

FIGURE 8–26
French speakers in eastern Canada. More than 80 percent of the residents of Québec speak French, compared to about 6 percent for the rest of Canada.
(Right) Laws require that signs in public places such as restaurants be in French.

(Steve Liss/Gamma-Liaison, Inc.)

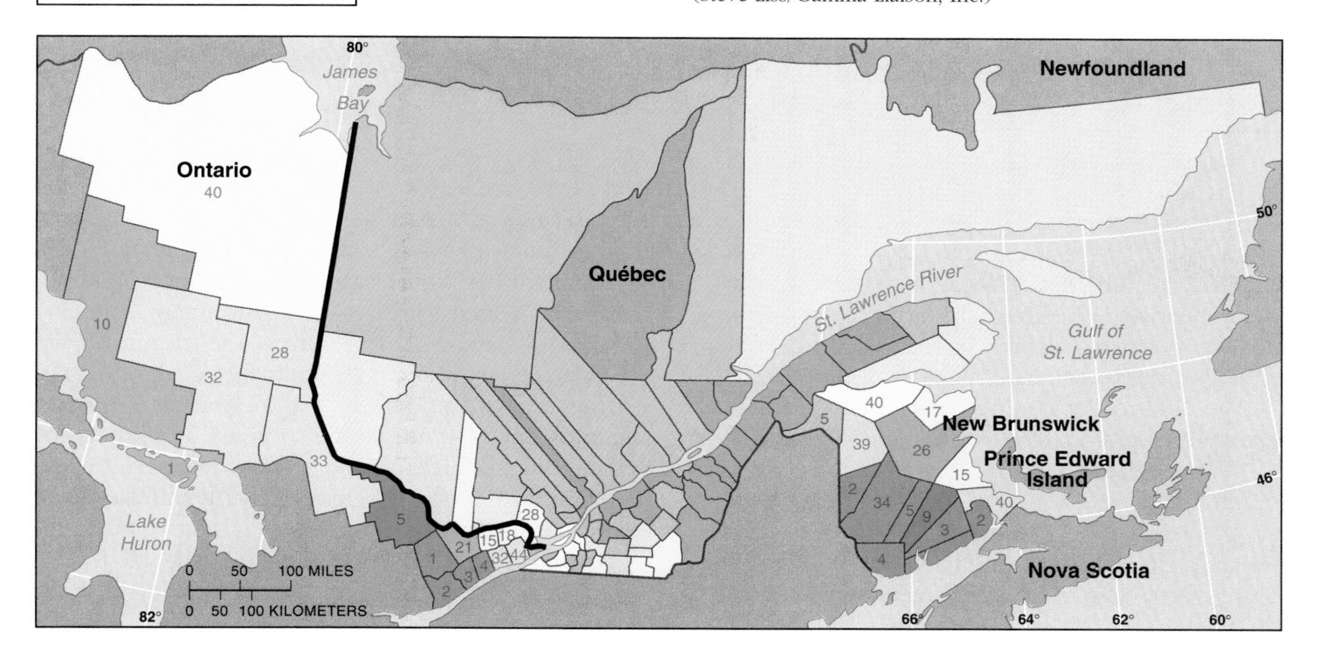

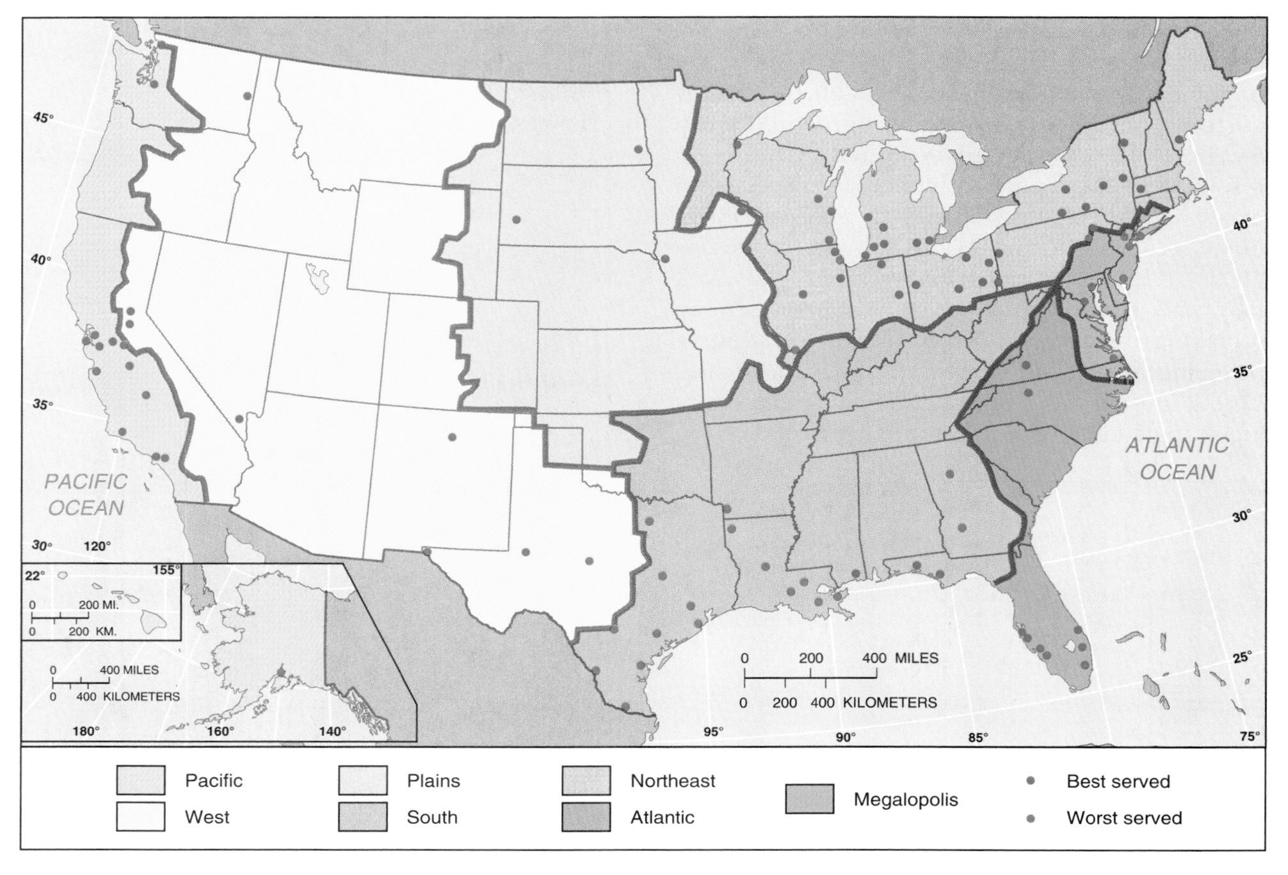

FIGURE 8–27
The fifty best-served and fifty worst-served metropolitan areas in terms of the number of golf holes per capita. In the North Central states, people have a long tradition of playing golf, even if it is confined to summer months. The ratio is less favorable for golfers in the large urban areas of the East Coast, as well as in the rapidly growing areas of the South and West. (Adapted from John F. Rooney, Jr., "American Golf Courses: A Regional Analysis of Supply," Sport Place International, Vol. 3, 1989)

The distribution of popular customs across Earth's surface tends to produce more uniform landscapes. The spatial expressions of a particular popular custom in one location will be similar to the appearance elsewhere. In fact, the promoters of various popular customs want a uniform appearance in order to generate higher consumption.

The diffusion of fast-food restaurants provides an example of uniformity. The buildings are designed to be immediately recognizable as part of a national or multinational company to both local residents and travelers. The success of fast-food restaurants depends on large-scale mobility: people who travel or move to another city immediately recognize a familiar place. Newcomers to a particular place know what to expect in the restaurant, because the establishment does not reflect strange and unfamiliar local customs that could be uncomfortable.

The diffusion of some popular customs can have an adverse impact on environmental quality. Popular customs may increase demand for resources such as certain animals, resulting in the extinction of some species and imbalance in other ecological systems. Some animals are killed for their skins, which can be shaped into fashionable clothing and sold to people living thousands of kilometers from the animals' habitat. Many folk customs may also encourage the use of animal products, but the need is usually smaller than for popular customs.

Increased demand for consumption of some products can strain the capacity of the environment to support popular customs. Increased consumption of meat, for example, has not resulted in the elimination of cattle and poultry. However, animal consumption is a relatively inefficient method for people to acquire needed calories because the animals are normally fed grain, which could otherwise be used to feed people directly. On average, nearly 10 kilograms (20 pounds) of grain are used to produce 1 kilogram of beef sold in the supermarket and nearly 3 kilograms (6 pounds) of grain are used for every kilogram of chicken. With a large percentage of the world's population undernourished, some people question the use of grain to feed animals for eventual human consumption.

Tokyo MacDonalds. U.S. fast-food chains have diffused to other countries, including Japan. Corporate logos enable customers to instantly identify the establishment, such as this one in Tokyo, regardless of whether they know the language. (Greg Davis/ The Stock Market)

Popular customs can also cause adverse environmental impacts through pollution. The physical environment has the capacity to accept some discharges from human activities. However, by virtue of their lack of sympathy with the environment as well as their widespread nature, popular customs generate a relatively high amount of residuals—in the form of either solid, liquid, gas, heat, or noise—that must be disposed of in the environment. With an increasing number of people in the world adopting popular customs, the adverse effects are greater.

Religious Conflicts

Religious beliefs run deeply in many people and societies; many activities of daily life are influenced by religious principles and organized around religious practices. Conflicts may break out between followers of different religions, as in the Middle East, or between followers of different denominations or sects within the same religion, as in Ireland.

Ireland. Roman Catholics and Protestants have long fought for control of the island of Ireland, also known as Eire. The Republic of Ireland is approximately 95 percent Roman Catholic, but the six northern counties of the island, which are part of the United Kingdom rather than Ireland, are divided between Roman Catholic and Protestant.

All of Eire was an English colony for many centuries. The English owned most of the land and did little to develop the island's economy. In 1801, Ireland was made part of the United Kingdom, although agitation for independence was extremely strong among the Irish. Following a series of bloody confrontations, Ireland became a self-governing dominion within the British Empire in 1921. Complete independence was declared in 1937, and a republic was created in 1949.

Twenty-six of the twenty-nine districts in the Ulster region of northeastern Ireland were grouped into six counties and became known as Northern Ireland. The people of Northern Ireland voted to remain in the United Kingdom rather than join the Republic of Ireland because a majority were Protestant—as is the case elsewhere in the United Kingdom—while 95 percent of the people in the Irish Republic were Roman Catholic (Figure 8–28). Today, most Protestants in Northern Ireland wish to remain part of the United Kingdom, even

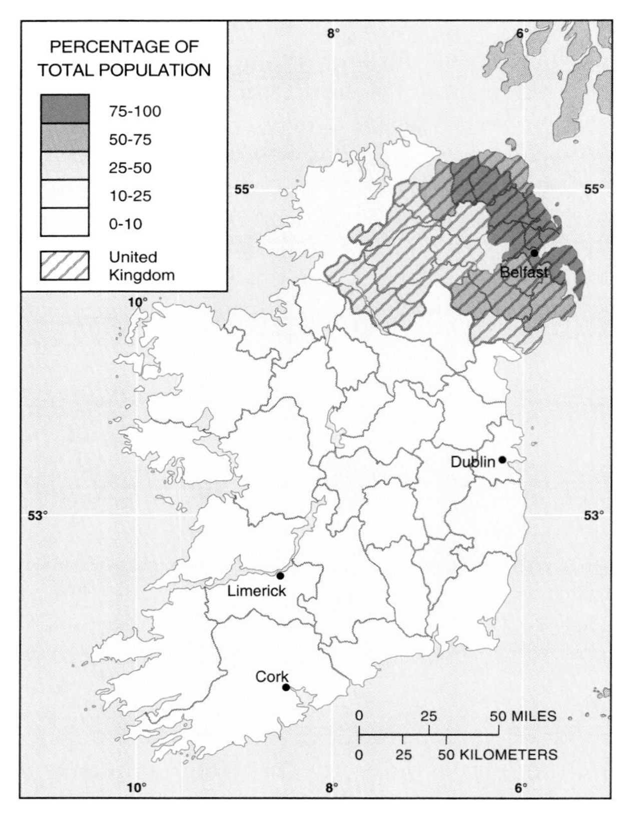

FIGURE 8–28
Distribution of Protestants in Ireland, 1911. Ireland, long a colony of England, became a self-governing dominion within the British Empire in 1921 and a completely independent country in 1937. However, twenty-six districts in northeastern Ireland were detached from the rest of Ireland and remained part of the United Kingdom. The Republic of Ireland is more than 95 percent Roman Catholic, while Northern Ireland has a Protestant majority. The boundary between Roman Catholics and Protestants does not coincide precisely with the international border, and Northern Ireland includes some communities that are predominantly Roman Catholic.

though the rest of the country is separated from Northern Ireland by 50 kilometers (30 miles) of water. Many Roman Catholics in Northern Ireland would prefer the six counties to be unified with the Republic of Ireland immediately to the south.

Roman Catholics living in Northern Ireland have been victimized by discriminatory practices, such as exclusion from higher paying jobs and better schools. A small number of Irish Catholics in both Northern Ireland and the Republic have joined the Irish Republican Army (IRA), a militant organization dedicated to achieving Irish national unity by whatever means are available. Similarly, a scattering of Protestants have created extremist organizations to fight the IRA, including the Ulster Defense Force. While the overwhelming majority of Northern Ireland's citizens prefer peace, extremists disrupt daily life. As long as some Protestants are firmly committed to remaining in the United Kingdom and some Roman Catholics are equally committed to union with the Republic of Ireland, peaceful settlement is difficult.

Religious wars in the Middle East. Competing claims by Muslims, Christians, and Jews frequently have led to battles for territorial control in the Middle East territory known since the time of the Roman Empire as Palestine. Christians consider Palestine holy because the major events in Jesus's life were concentrated there. Muslims regard Jerusalem as their third holiest city, after Makkah and Madinah, because Muhammad ascended to heaven from there. Between the eleventh and thirteenth centuries, Christians and Muslims fought a series of campaigns, known as Crusades, for control of the territory.

As an ethnic religion, Judaism considers Palestine important because the major events in the Hebrew Bible took place there, and the religion's customs and rituals acquired meaning from the agricultural life of the ancient Hebrew tribe. Jews, however, have often been dispersed from Palestine by conquering armies and restricted or prohibited altogether from visiting holy places.

When the Muslim Ottoman Empire, which controlled Palestine for most of the period between 1516 and 1917, was defeated in World War I, Great Britain took over Palestine under a mandate from the League of Nations and later the United Nations. For a few years, the British allowed some Jews to return to Palestine, but emigration was restricted again during the 1930s in response to intense pressure by Arabs in the region.

As violence initiated by both Jewish and Arab settlers escalated after World War II, the British announced their intention to withdraw from Palestine. The United Nations voted to partition Palestine into two independent states, one Jewish and one Arab Muslim. Jerusalem was to be an international city, open to all religions, and run by the United Nations (Figure 8–29).

When the British withdrew on May 15, 1948, Jews declared an independent state of Israel, in the

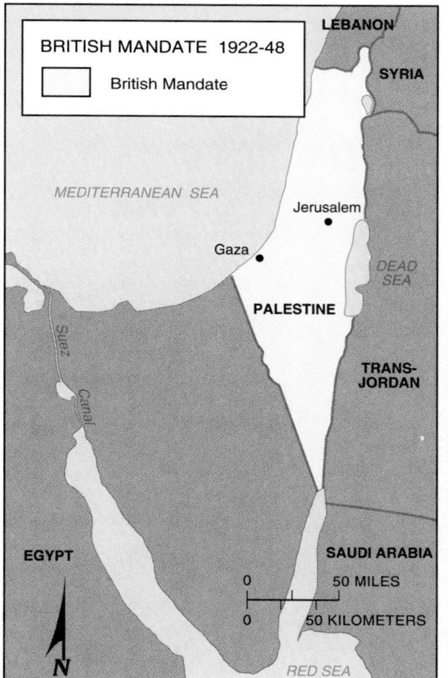

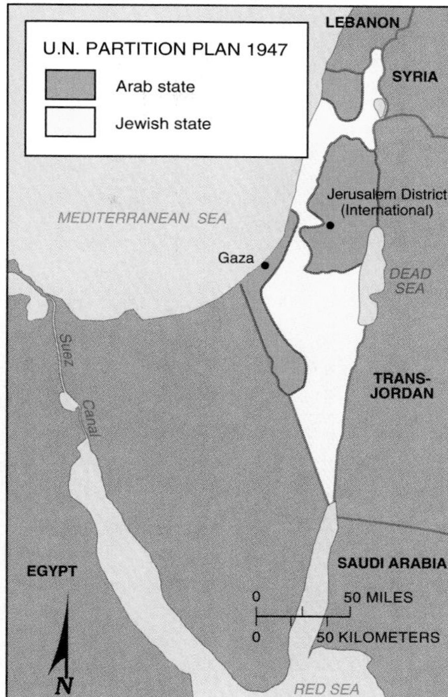

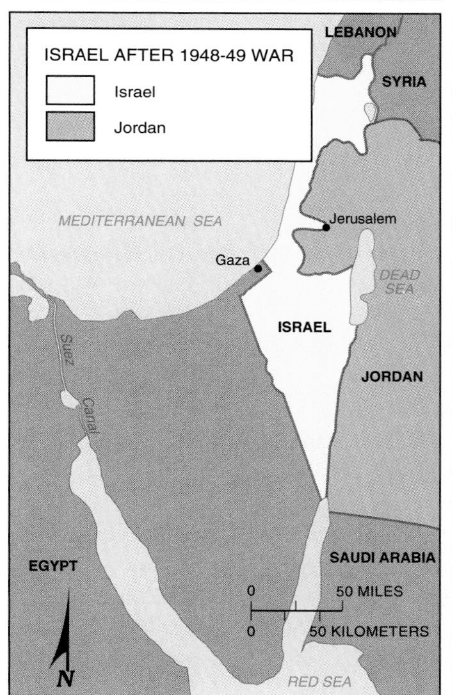

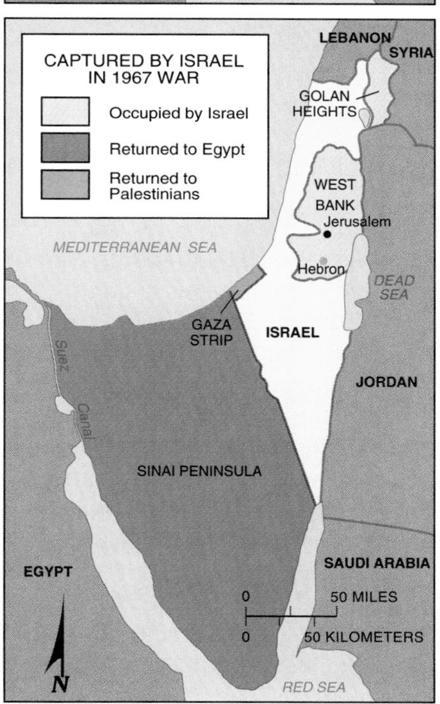

FIGURE 8–29
Upper left: Palestine under British control, between 1922 and 1948.
Upper right: The 1947 United Nations plan to partition Palestine. The plan was to create two countries, with the boundaries drawn to separate the predominantly Jewish areas from the predominantly Arab Muslim areas. Jerusalem was intended to be an international city, run by the United Nations.
Lower left: Israel after the 1948–1949 War. The day after Israel declared its independence, several neighboring states began a war, which ended in an armistice. Israel's boundaries were extended beyond the U.N. partition to include the western suburbs of Jerusalem. Jordan gained control of the West Bank and East Jerusalem, including the Old City.
Lower right: The Middle East since the 1967 War. Israel captured the Golan Heights from Syria, the West Bank and East Jerusalem from Jordan, and the Sinai Peninsula and Gaza Strip from Egypt. Israel returned the Sinai to Egypt in 1979 and turned over control of the Gaza Strip and a portion of the West Bank to the Palestinians in 1994. Israel is negotiating with the Palestinians and Syria to return other occupied territories in exchange for peace treaties. Jordan has renounced its claim on the West Bank and signed a peace treaty with Israel in 1994.

Route 66. The diffusion of popular customs can result in an ugly landscape, as well as an environmentally unsound one. Route 66 was once a well-known symbol of an especially prominent American popular custom—the freedom to drive a car across the country's wide open spaces. Route 66, which once connected Chicago and Los Angeles, has been largely replaced by interstate highways. Remaining stretches, such as this one, are often cluttered by unattractive strip development, dominated by large signs for national motel, gasoline, and restaurant chains. (Bryan F. Peterson/The Stock Market)

boundaries prescribed by the U.N. resolution. The next day, its neighboring Arab Muslim states declared war, but after failing to defeat Israel, they signed an armistice in 1949. Israel fought three more wars with its neighbors in 1956, 1967, and 1973. Each time, Israel won.

After the 1949 armistice, Jerusalem became a divided city. The Old City of Jerusalem, which contained the famous religious shrines, was part of the Muslim country of Jordan, while the newer western area was in Israel. During the 1967 Six-Day War, Israel captured the entire city and removed the barriers that had prevented Jews from visiting and living in the Old City of Jerusalem. Israel captured the Sinai Peninsula, but later returned it to Egypt in exchange for Egypt recognizing Israel's right to exist.

Three decades after the Six-Day War, the status of the other territories occupied by Israel has not been settled. In 1981, Israel annexed the Golan Heights, a sparsely inhabited, mountainous area where Syria launched attacks against Jewish settlements in the valley. In 1994, Israel turned over to the Palestinians control of the Gaza Strip and a portion of the West Bank. The future of the remainder of the West Bank and Jerusalem has been especially difficult to resolve—both Israelis and Palestinians make strong claims to these territories.

Israeli Jews are divided between those who wish to retain most of the occupied territories and those who wish to return most of them in exchange for peace treaties with neighboring states. Palestinians are divided between those who wish to continue fighting against Israel for control of the entire territory of Palestine and those who are willing to settle for control of a portion of the West Bank and the Gaza Strip.

People draw identity from their religion, as well as language and other social customs. The conflicts in Ireland and the Middle East show that people are willing to take extreme measures—including going to war—to preserve their culture from real or imagined threats.

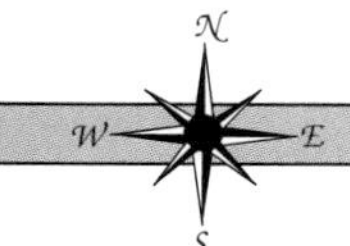

CONCLUSIONS: CRITICAL ISSUES FOR THE FUTURE

During the 1960s and 1970s, Yugoslavs liked to repeat a refrain that roughly translated as follows: "Yugoslavia has seven neighbors, six republics, five nationalities, four languages, three religions, two alphabets, and one dinar." Specifically:

- Yugoslavia's seven neighbors included three long-time democracies (Austria, Greece, and Italy) and four states then governed by Communists (Albania, Bulgaria, Hungary, and Romania).
- Yugoslavia's six republics—Bosnia-Herzegovina, Croatia, Macedonia, Montenegro, Serbia, and Slovenia—had more autonomy from the national government to run their own affairs than was the case in most other countries.
- Five of the republics were named for the country's five recognized nationalities—Croats, Macedonians, Montenegrens, Serbs, and Slovenes. Bosnia-Herzegovina comprised a mix of Serbs, Croats, and Muslims.
- Yugoslavia had four official languages—Croatian, Macedonian, Serbian, and Slovene. Montenegrens spoke Serbian.
- The three major religions included Roman Catholic in the north, Eastern Orthodox in the east, and Islam in the south. Croats and Slovenes were predominantly Roman Catholic, Serbs and Macedonians predominantly Eastern Orthodox, and the Bosnians and Montenegrens predominantly Muslim.
- Two of the four official languages—Croatian and Slovene—were written in the Western alphabet, while Macedonian and Serbian were written in Cyrillic. Most linguists outside Yugoslavia considered Serbian and Croatian to be the same language except for different alphabets.
- The refrain concluded that Yugoslavia had one dinar, the national unit of currency. Despite cultural diversity, according to the refrain, common economic interests kept Yugoslavia's cultural groups unified.

Common economic interests were not sufficiently strong to keep Yugoslavia's diverse cultural groups united. Yugoslavia splintered into many pieces in the early 1990s; small countries were created under the control of the various cultural groups, and several wars broke out among cultural groups fighting for possession of territory. Bosnia-Herzegovina, Croatia, Slovenia, and Macedonia all broke away, leaving only Serbia and Montenegro remaining in Yugoslavia. Bosnia-Herzegovina has suffered from a civil war among its Serb, Croat, and Muslim populations.

In the modern world, as the next chapter shows, countries in addition to Yugoslavia have suffered political unrest because cultural groups could not live together peacefully. The political unrest in ethnically diverse countries shows the importance to people of controlling a portion of Earth.

Chapter Summary

1. Culture regions
Countries can be grouped into nine major culture regions, including Anglo-America, Latin America, Western Europe, Eastern Europe, East Asia, South Asia, Southeast Asia, the Middle East, and Sub-Saharan Africa. While considerable cultural diversity exists within each of these regions, at a global scale each displays distinctive patterns of languages, religions, and other social customs.

2. Languages
Indo-European languages predominate in Europe because of prehistoric migrations of Indo-European speakers, although analysts disagree on whether the migrations were for peaceful or military reasons. English is the language of the British Isles because Germanic tribes migrated there about 1,500 years ago. Similarly, migrations of English-speaking peoples during the past 500 years have diffused the language outside Europe.

3. Religions
When people migrate they take with them their religion, as well as their language and other cultural characteristics. The distribution of religions across Earth' surface is a function of the migrations of adherents in the relatively recent past. Ethnic religions such as Hinduism, which are rooted in the physical geography and culture of particular places, are less likely to diffuse to other places than are universalizing religions, such as Christianity, Islam, and Buddhism.

4. Social customs
Folk customs are more likely to have anonymous origins and to diffuse slowly through migration, whereas popular customs are more likely to be invented and diffused rapidly with the use of modern communications. Differences in folk customs may be observed among different places at one point in time, whereas differences in popular customs are more likely to be observed in one place at different points in time.

5. Problems of cultural diversity
Popular customs, generally originating in Western economically developed countries, may cause elimination of some folk customs and adversely affect the physical environment. Adherents of different religions have battled each other repeatedly for control of particular places, especially in the Middle East, regarded as holy by Christians, Muslims, and Jews.

Key Terms

Animism, 316
Branch, 307
Caste, 315
Custom, 318
Denomination, 307
Dialect, 306
Diaspora, 316
Ethnic religion, 316
Folk custom, 319
Ghetto, 316
Habit, 318
Language, 301
Language branch, 303
Language family, 301
Language group, 303
Lingua franca, 332
Missionary, 311
Pilgrimage, 317
Popular custom, 319
Sect, 307
Standard language, 306
Taboo, 323
Universalizing religion, 316

Questions for Study and Discussion

1. List the nine main culture regions. Identify the principal religions and language families in each.
2. Name the language family with the largest number of speakers. Name the four main branches of this language family. Name the language family with the second largest number of speakers.
3. Describe differences between the two main theories for the origin and diffusion of Indo-European languages.
4. Name the universalizing religion with the largest number of adherents. Name the three major branches of this religion.
5. Name the universalizing religion with the second largest number of adherents. Name the two major branches of this religion.
6. Name the universalizing religion with the third largest number of adherents. Name the two major branches of this religion.
7. Name the ethnic religion with the largest number of adherents. In what country are the followers of this religion clustered?
8. Discuss major differences between universalizing and ethnic religions.
9. Discuss major differences in the origin, diffusion, and distribution of folk and popular social customs.

Thinking Geographically

1. At least sixteen U.S. states have passed laws mandating English as the language of all government functions. In 1990, Arizona's law making English the official language was ruled an unconstitutional violation of free speech. Should the use of English be encouraged in the United States to foster cultural integration, or should bilingualism be encouraged to foster cultural diversity? Why?
2. Many countries now receive Cable News Network (CNN) broadcasts that originate in the United States, but even English-speaking viewers in other countries have difficulty understanding some American English. A recent business program on CNN created a stir outside the United States when it reported that McDonalds was a major IRA contributor. Viewers in the United Kingdom thought that the American hamburger chain was financing the purchase of weapons by the Irish Republican Army. However, McDonalds, in fact, was contributing to Individual Retirement Accounts for its employees. Can you think of other examples where the use of a word could cause a British-American misunderstanding? How is American English different from British English as a result of contributions by African-Americans and immigrants who speak languages other than English?
3. A widespread view outside of the Middle East is that peace between Israel and its neighbors can be achieved by trading land for peace. In other words, Israel would return the occupied territories in exchange for enforceable peace treaties. What are some of the obstacles that make implementation of this formula difficult for either side?
4. Sharp differences in demographic characteristics, such as natural increase, crude birth, and migration rates, can be seen among Jews, Christians, and Muslims in the Middle East and between Roman Catholics and Protestants in Northern Ireland. How might demographic differences affect future relationships among the groups in these two regions?
5. In what ways might gender affect the impact of social customs on the landscape?
6. What images of social customs do countries depict in campaigns to promote tourism? To what extent do these images reflect local social customs realistically?

Suggestions for Further Reading

Aitchison, J. W., and H. Carter. "The Welsh Language in Cardiff: A Quiet Revolution." *Transactions of the Institute of British Geographers, New Series* 12 (no. 4, 1987): 482–92.

Al Faruqi, Isma'il R., and Lois Lamaya' Al Faruqi. *The Cultural Atlas of Islam*. New York: Macmillan, 1986.

Al Faruqi, Isma'il R., and David E. Sopher. *Historical Atlas of the Religions of the World*. New York: Macmillan, 1974.

Bapat, P. V., ed. *2500 Years of Buddhism*. Delhi: Government of India Ministry of Information and Broadcasting, 1959.

Barrett, David B., ed. *World Christian Encyclopedia*. Oxford: Oxford University Press, 1982.

Baugh, Albert C., and Thomas Cable. *A History of the English Language,* 3d ed. Englewood Cliffs, NJ: Prentice-Hall, 1978.

Bennett, Merril K. *The World's Foods*. New York: Harper and Bros., 1954.

Bhardwaj, Surinder M. *Hindu Places of Pilgrimage in India*. Berkeley: University of California Press, 1973.

Cardona, George, Henry M. Hoeningswald, and Alfred Senn, eds. *Indo-European and Indo-Europeans*. Philadelphia: University of Pennsylvania Press, 1970.

Carlson, Alvar W. "The Contributions of Cultural Geographers to the Study of Popular Culture." *Journal of Popular Culture* 11 (Spring 1978): 830–31.

Carney, George O. "Bluegrass Grows All Around: The Spatial Dimensions of a Country Music Style." *Journal of Geography* 73 (April 1974): 34–55.

_____. "From Down Home to Uptown: The Diffusion of Country-Music Radio Stations in the United States." *Journal of Geography* 76 (March 1977): 104–10.

Chakravarti, A. K. "Regional Preference for Foods: Some Aspects of Food Habit Patterns in India." *Canadian Geographer* 18 (Winter 1974): 395–410.

Chubb, Michael, and Holly R. Chubb. *One Third of Our Time?* New York: Wiley, 1981.

Cooper, Adrian. "New Directions in the Geography of Religion." *Area* 24 (June 1992): 123–29.

Crowley, William K. "Old Order Amish Settlement: Diffusion and Growth." *Annals of the Association of American Geographers* 68 (June 1978): 249–65.

Curry-Roper, Janet M. "Contemporary Christian Eschatologies and their Relation to Environmental Stewardship." *Professional Geographer* 42 (May 1990): 157–69.

Delgado de Carvalho, C. M. "The Geography of Languages." In *Readings in Cultural Geography*, Philip L. Wagner and Marvin W. Mikesell, eds. Chicago: University of Chicago Press, 1962.

Dugdale, J. S. *The Linguistic Map of Europe*. London: Hutchinson University Library, 1969.

Fickeler, Paul. "Fundamental Questions in the Geography of Religions," In *Readings in Cultural Geography*. Edited by Philip L. Wagner and Marvin W. Mikesell. Chicago: University of Chicago Press, 1962.

Gaustad, E. S. *Historical Atlas of Religion in America*. New York: Harper and Row, 1962.

Hardon, John A. *Religions of the World*. 2 vols. Garden City, NY: Image Books, 1968.

Hughes, Arthur and Peter Trudgill. *English Accents and Dialects*. Birkenhead: Edward Arnold, 1979.

Jackson, Richard H., and Roger Henrie. "Perception of Sacred Space." *Journal of Cultural Geography* 3 (Spring/Summer 1983): 94–107.

Jakle, John A. "Roadside Restaurants and Place-Product-Packaging." *Journal of Cultural Geography* 3 (1982): 76–93.

Kaplan, David H. "Population and Politics in a Plural Society: The Changing Geography of Canada's Linguistic Groups." *Annals of the Association of American Geographers* 84 (1994): 46–67.

Karan, Pradyumna P., and Cotton Mather. "Art and Geography: Patterns in the Himalayas." *Annals of the Association of American Geographers* 66 (December 1976): 487–515.

Kay, Jeanne. "Human Dominion over Nature in the Hebrew Bible." *Annals of the Association of American Geographers* 79 (June 1989): 214–32.
Kirk, John M., Stewart Sanderson, J. D. A. Widdowson, eds. *Studies in Linguistic Geography: The Dialects of English in Britain and Ireland.* London: Croom Helm, 1985.
Kniffen, Fred B. "Folk-Housing: Key to Diffusion." *Annals of the Association of American Geographers* 55 (December 1965): 549–77.
Kong, L. "Geography and Religion: Trends and Prospects." *Progress in Human Geography* 14 (1990): 355–71.
Kurath, Hans. *Word Geography of the Eastern United States.* Ann Arbor: University of Michigan Press, 1949.
Levine, Gregory J. "On the Geography of Religion." *Transactions of the Institute of British Geographers, New Series* 11 (no. 4, 1987): 428–40.
Lind, Ivan. "Geography and Place Names." In *Readings in Cultural Geography.* Philip L. Wagner and Marvin W. Mikesell, eds. Chicago: University of Chicago Press, 1962.
Lomax, Alan. *The Folk Songs of North America.* Garden City, NY: Doubleday, 1960.
Lornell, Christopher, and W. Theodore Mealor, Jr. "Traditions and Research Opportunities in Folk Geography." *Professional Geographer* 35 (February 1983): 51–56.
McColl, Robert W. "By Their Dwellings Shall We Know Them: Home and Setting Among China's Inner Asian Ethnic Groups." *Focus* 39 (Winter 1989): 1–6.
Meillet, Antoine, and Marcel Cohen. *Les langues du monde.* Paris: Centre National de la Recherche Scientifique, 1952.
Moseley, Christopher, and R. E. Asher. *Atlas of the World's Languages.* London: Routledge, 1994.
Muller, Siegfried H. *The World's Living Languages.* New York: Frederick Ungar, 1964.
Opie, Iona, and Peter Opie. *Children's Games in Street and Playground.* London: Clarendon Press, 1969.
Renfrew, Colin. *Archaeology and Language.* Cambridge: Cambridge University Press, 1988.
Rooney, John F., Jr. *A Geography of American Sport.* Reading, MA: Addison-Wesley, 1974.
_____, Wilbur Zelinsky, and Dean R. Louder, eds. *This Remarkable Continent: An Atlas of United States and Canadian Society and Culture.* College Station, TX: Texas A and M University, 1982.
Shortridge, James R. "Patterns of Religion in the United States." *Geographical Review* 66 (October 1976): 420–34.
Sopher, David E. *The Geography of Religions.* Englewood Cliffs, NJ: Prentice-Hall, 1967.
_____. "Geography and Religions." *Progress in Human Geography* 5 (1981): 510–24.
Thompson, Jan, and Mel Thompson. *The R. E. Atlas: World Religions in Maps and Notes.* London: Edward W. Arnold, 1986.
Trudgill, Peter. "Linguistic Geography and Geographical Linguistics." *Progress in Geography* 7 (1975): 227–52.
Wagner, Philip L. "Remarks on the Geography of Language." *Geographical Review* 48 (January 1958): 86–97.
Williams, Colin H., ed. *Language in Geographic Context.* Clevedon, U.K.: Multilingual Matters, 1988.
Zelinsky, Wilbur. "An Approach to the Religious Geography of the United States: Patterns of Church Membership in 1952." *Annals of the Association of American Geographers* 51 (June 1961): 139–67.
_____. "North America's Vernacular Regions." *Annals of the Association of American Geographers* 70 (March 1980): 1–16.
We also recommend these journals: *International Folk Music Council Journal, Journal of American Culture, Journal of American Folklore, Journal of American Studies, Journal of Cultural Geography, Journal of Leisure Research, Journal of Popular Culture, Journal of Sport History, Landscape, Leisure Science.*

Let's go West

9 Political Geography

- Differences between states and nations
- Boundaries between states
- Problems matching states and nations
- Cooperation among states

East Berliners arrive back home at Invalidenstrasse crossing point from a shopping trip to West Berlin, November 12, 1989, carrying a stereo set and Let's Go West bags.

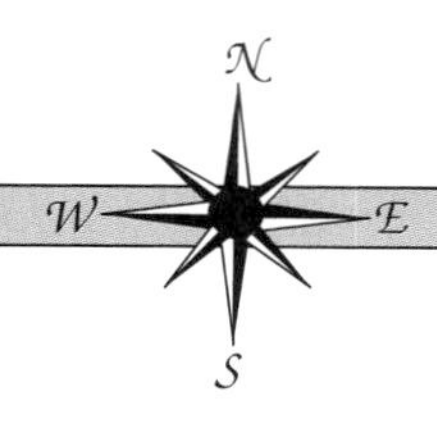

1. Differences between states and nations
The world is divided into about 200 states, varying widely in size. States are created where possible to match the territory occupied by a nation, or nationality.

2. Boundaries between states
States are separated from each other by boundaries, marking their outer limits of territorial control and giving them distinctive shapes. Boundaries can be physical or cultural.

3. Problems matching states and nations
Political unrest is concentrated in locations where boundaries between nations do not coincide with boundaries between states.

4. Cooperation among states
States cooperate with each other by joining international and regional organizations. During the Cold War era, states joined alliances primarily for military reasons, but in recent years economic cooperation has become more important.

INTRODUCTION

The global political landscape has been altered fundamentally in the 1990s by the end of the Cold War, a period dating from about the mid-1950s that might best be described as the polarization of countries into two major camps, one allied with the Soviet Union and the other allied with the United States. Geographic concepts help us understand the changing political organization of Earth's surface. We can also use geographic factors to examine some of the causes of political change and instability and to anticipate potential trouble spots around the world.

Political geographers observe two significant global patterns in the post-Cold War era. First, conflicts in places where cultural groups demand more control over the territory they inhabit have increased in areas that previously were held in place by Communist governments. With the focus no longer on communism, we are more aware of the serious civil wars and political and ethnic conflicts in many developing nations.

As we saw in the previous chapter, the world is inhabited by peoples who have embraced a variety of languages, religions, and other social customs. The territory occupied by a cultural group does not always coincide with the territory governed as an independent country. Conflicts develop among and within countries as a result of the imposition of political boundaries on the cultural and physical landscapes.

The second pattern is that regional and worldwide collections of states have gained greater power compared to individual countries. States have traditionally possessed different amounts of power as a result of their distinctive geographic characteristics. Between the 1940s and 1980s, two countries, the United States and the Soviet Union, were dominant in world political events. With the Soviet Union no longer in existence and the United States less overtly dominant in the political landscape of the 1990s, it is easier to see how power is exercised by collections of states organized primarily for economic cooperation—as well as competition with countries in other regions.

Differences Between States and Nations

We sometimes confuse two important political geography concepts: *state* and *nation*. Before examining issues that arise from the distribution of political units across Earth's surface, we must understand the differences between these two fundamental concepts. In simple terms, a *state* is a political unit, whereas a *nation* is a collection of people having a common culture. The two do not necessarily coincide.

Characteristics of States

A **state** is an area organized into a political unit and ruled by an established government with control over its internal and foreign affairs. It occupies a specific territory on Earth's surface and contains

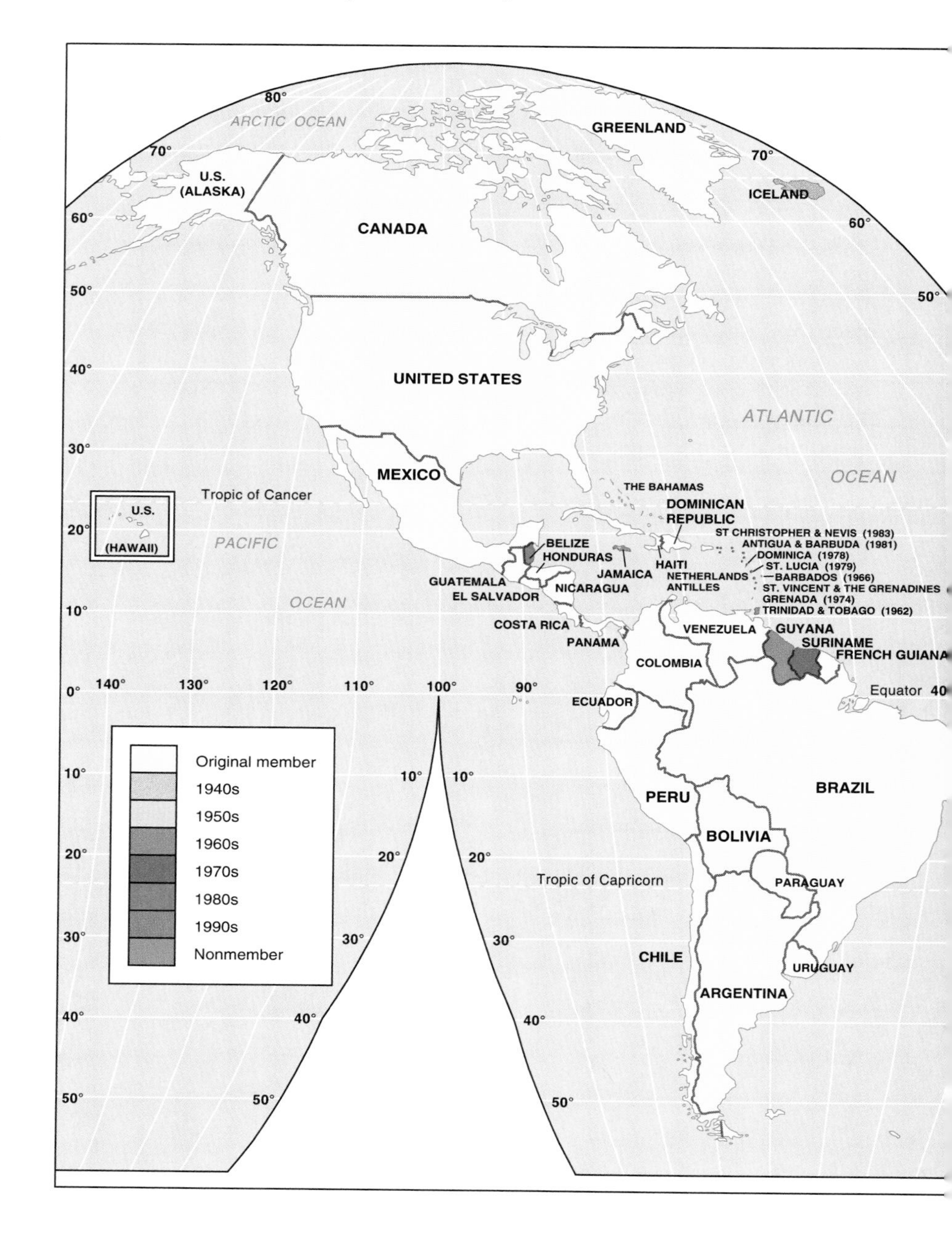

FIGURE 9–1
United Nations members. When it was organized in 1945, the United Nations had only fifty-one members, including forty-nine states, plus Belarus and Ukraine, which were then part of the Soviet Union. The United Nations added eight more states in the late 1940s, twenty-four states during the 1950s, forty-two states during the 1960s, twenty-five states during the 1970s, seven during the 1980s, and twenty-seven during the 1990s. The greatest increase in sovereign states has occurred in Africa. Only four African states were original members of the United Nations—Egypt, Ethiopia, Liberia, and South Africa—and only six more joined during the 1950s. Beginning in 1960, however, a collection of independent states was carved out of most of the remainder of Africa. In 1960 alone, sixteen newly independent African states became members of the United Nations. The process of creating new sovereign states slowed during the late 1980s. Among the states that joined the United Nations during the 1980s, only Zimbabwe had a population in excess of 200,000. The breakup of the Soviet Union and Yugoslavia stimulated the formation of new states during the early 1990s.

a permanent population. A state has **sovereignty,** that is, independence from control of its internal affairs by other states. Because the entire area of a state has the same national government, laws, army, and leaders, it is a good example of a uniform or homogeneous region. The term *country* is a synonym for state.

Earth's land surface is divided into a collection of nearly 200 states (Figure 9–1). All but a handful of states are members of the United Nations. When it was created in 1945, at the end of World War II, the United Nations comprised 49 states, but by 1993 membership had grown to 184 (Table 9–1). The number of countries in the United Nations has increased rapidly on three occasions: 1955, 1960, and 1990–93. Sixteen countries joined in 1955, mostly European countries that had been liberated from Nazi Germany during World War II. Seventeen new members were added in 1960, all but one former British or French African colonies.

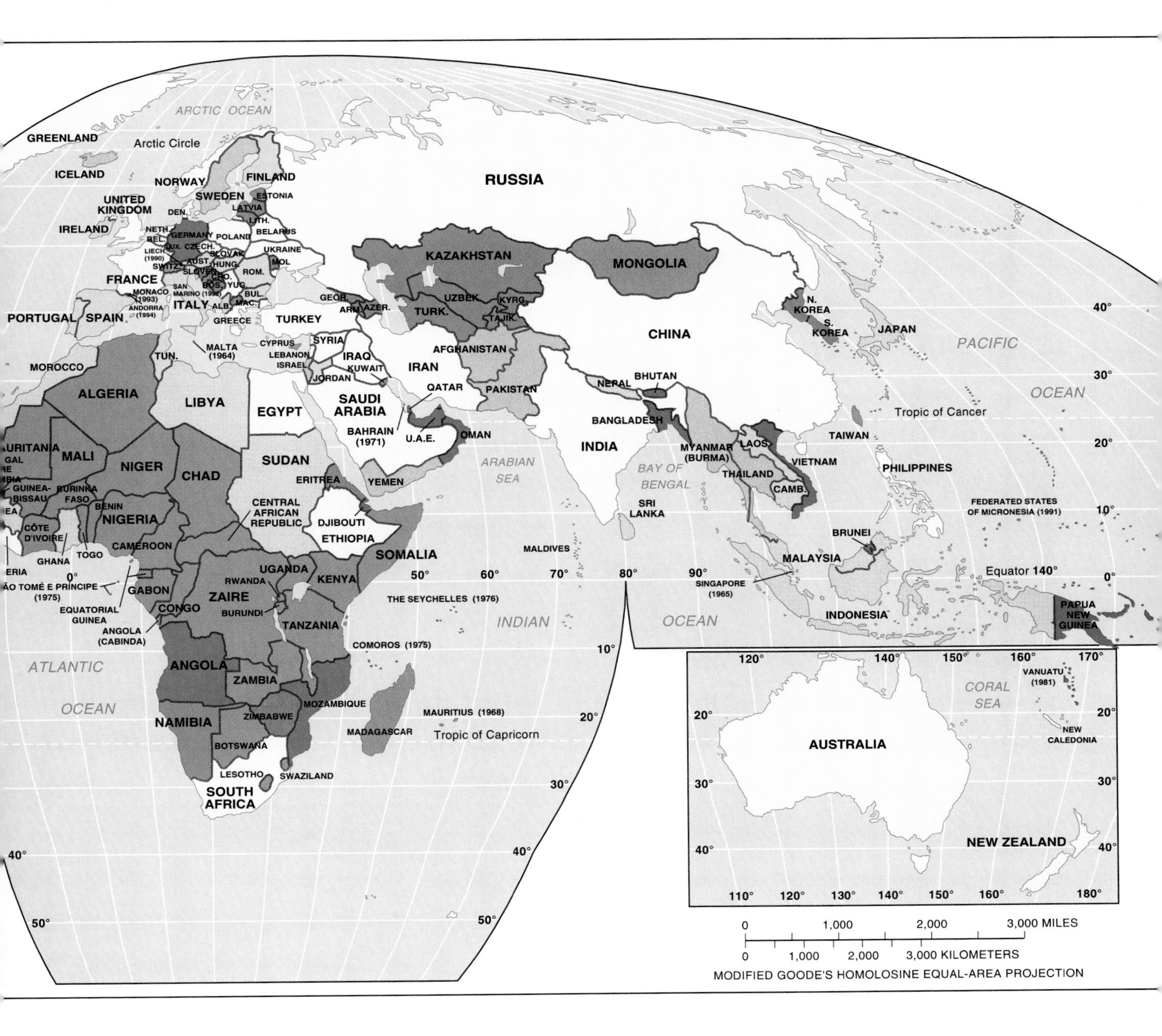

TABLE 9–1

Sovereign States

Members of the United Nations (184)			
Afghanistan	Djibouti	Lebanon	Saint Christopher–Nevis
Albania	Dominica	Lesotho	Saint Lucia
Algeria	Dominican Republic	Liberia	Saint Vincent and the Grenadines
Andorra	Ecuador	Libya	Samoa (Western)
Angola	Egypt	Liechtenstein	San Marino
Antigua and Barbuda	El Salvador	Lithuania	São Tomé e Principe
Argentina	Equatorial Guinea	Luxembourg	Saudi Arabia
Armenia	Eritrea	Macedonia	Senegal
Australia	Estonia	Madagascar	Seychelles
Austria	Ethiopia	Malawi	Sierra Leone
Azerbaijan	Fiji	Malaysia	Singapore
Bahamas	Finland	Maldives	Slovakia
Bahrain	France	Mali	Slovenia
Bangladesh	Gabon	Malta	Solomon Islands
Barbados	Gambia	Marshall Islands	Somalia
Belarus	Germany	Mauritania	South Africa
Belgium	Georgia	Mauritius	Spain
Belize	Ghana	Mexico	Sri Lanka
Benin	Greece	Micronesia	Sudan
Bhutan	Grenada	Moldova	Suriname
Bolivia	Guatemala	Monaco	Swaziland
Bosnia and Herzegovina	Guinea	Mongolia	Sweden
Botswana	Guinea-Bissau	Morocco	Syria
Brazil	Guyana	Mozambique	Tajikistan
Brunei	Haiti	Myanmar (Burma)	Tanzania
Bulgaria	Honduras	Namibia	Thailand
Burkina Faso	Hungary	Nepal	Togo
Burundi	Iceland	Netherlands	Trinidad and Tobago
Cambodia	India	New Zealand	Tunisia
Cameroon	Indonesia	Nicaragua	Turkey
Canada	Iran	Niger	Turkmenistan
Cape Verde	Iraq	Nigeria	Uganda
Central African Republic	Ireland	Norway	Ukraine
Chad	Israel	Oman	United Arab Emirates
Chile	Italy	Pakistan	United Kingdom
China	Jamaica	Panama	United States
Colombia	Japan	Papua New Guinea	Uruguay
Comoros	Jordan	Paraguay	Uzbekistan
Congo	Kazakhstan	Peru	Vanuatu
Costa Rica	Kenya	Philippines	Venezuela
Côte d'Ivoire	Korea, North	Poland	Vietnam
Croatia	Korea, South	Portugal	Yemen
Cuba	Kuwait	Qatar	Yugoslavia*
Cyprus	Kyrgyzstan	Romania	Zaire
Czech Republic	Laos	Russia	Zambia
Denmark	Latvia	Rwanda	Zimbabwe

Not members of the United Nations (7)			
Kiribiti	Switzerland	Tonga	Vatican
Nauru	Taiwan	Tuvalu	

*The U.N. General Assembly voted to expel Yugoslavia from membership September 22, 1992

Twenty-seven countries were added between 1990 and 1993, primarily as a result of the breakup of the Soviet Union and Yugoslavia.

The only large land mass on Earth's surface not part of a sovereign state is Antarctica. Several states claim portions of Antarctica, including Argentina, Australia, Chile, France, New Zealand, Norway, and the United Kingdom (Figure 9–2). Argentina, Chile, and the United Kingdom make conflicting claims to Antarctica. The United States, Russia, and a number of other states do not recognize the claims of any of these countries to Antarctica. Several states have established stations on Antarctica to conduct scientific experiments and monitoring, but as yet no permanent settlements exist there.

Size of states. The land area occupied by the states of the world varies considerably. The largest state is Russia, which encompasses 17.1 million square kilometers (6.6 million square miles), 11 percent of the world's entire land area. The distance between the country's borders with Eastern European countries and the Pacific Ocean is more than 7,000 kilometers (4,300 miles).

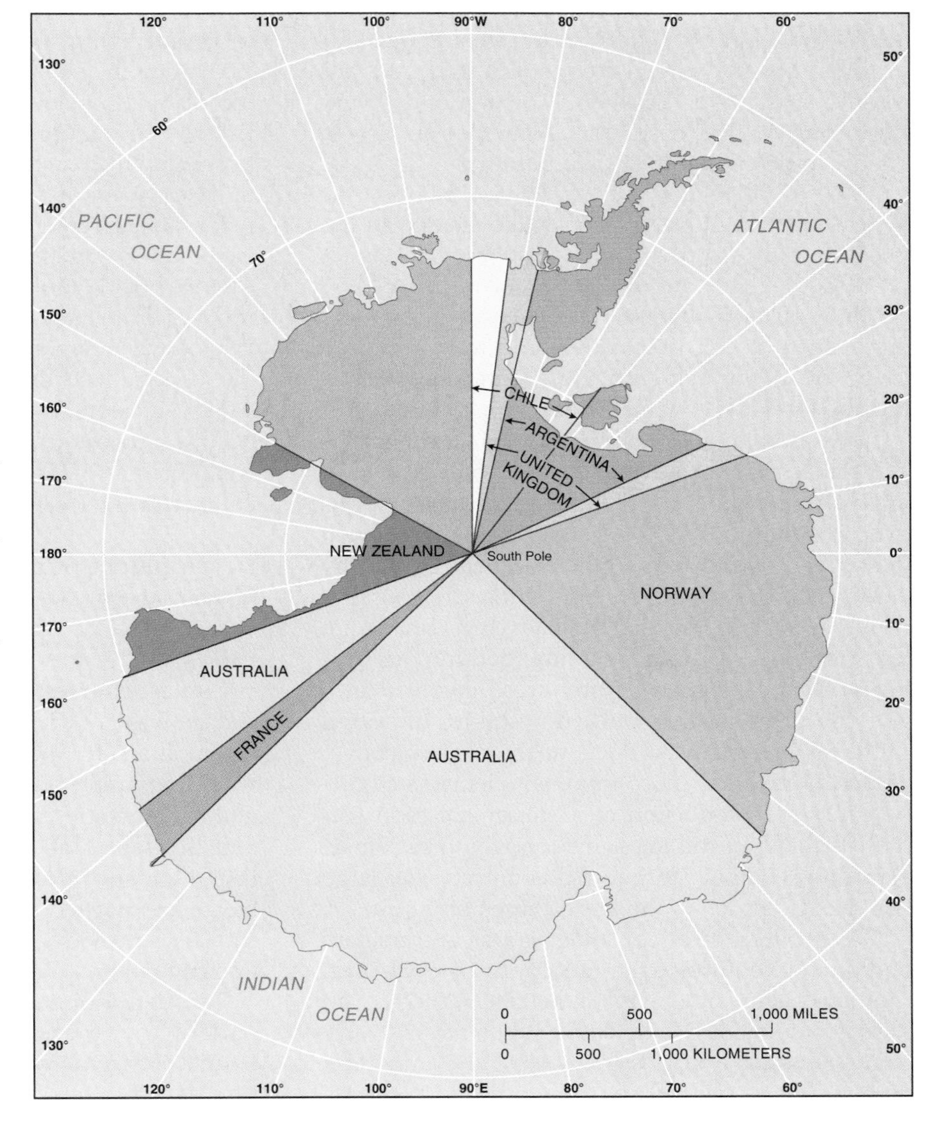

FIGURE 9–2
Natonal claims to Antarctica. Antarctica is the only large land mass in the world not part of a sovereign state. The land mass comprises 14 million square kilometers (5.4 million square miles), 50 percent larger than Canada. Portions of Antarctica are claimed by Argentina, Australia, Chile, France, New Zealand, Norway, and the United Kingdom, and the claims of Argentina, Chile, and the United Kingdom are conflicting. In 1959, these seven countries, along with Belgium, Japan, South Africa, the Soviet Union, and the United States, signed a treaty suspending any territorial claims for thirty years and establishing guidelines for scientific research. In 1991, twenty-four countries agreed to extend the treaty for fifty years. The 1991 treaty also established new pollution control standards and banned mining activities and oil exploration for fifty years.

At the other extreme, the smallest state in the United Nations, Monaco, encompasses only 1.5 square kilometers (0.6 square miles), about the size of downtown Reno, Nevada. Other U.N. member states that are smaller than 1,000 square kilometers include Andorra, Antigua and Barbuda, Bahrain, Barbados, Dominica, Grenada, Liechtenstein, Maldives, Malta, Micronesia, St. Christopher–Nevis, St. Lucia, St. Vincent and the Grenadines, San Marino, São Tomé and Principe, the Seychelles, and Singapore.

Historically, very large size was not an important element in defining power. In the past, some states found that very large size was a liability because they could not effectively guard the entire territory against invasion by neighboring states. People living far from the capital frequently displayed less intense loyalty to the state. Consequently, very large states in the past were compelled to devote energy to defending their own territory.

Since World War II, very large size has become a clear asset in global power, while small size has become more of a hardship. Two principal assets of very large size are availability of resources and stronger military defense. The larger the state, the more likely it can obtain the resources necessary to be relatively self-sufficient. Critical resources include food and raw materials.

A country that can grow enough food for its population has an advantage over one that must import food. The United Kingdom, which imports over two-thirds of its food supply, was threatened during both world wars in the twentieth century by enemy attacks on its supply ships. In contrast, countries such as the United States and Canada are less concerned with safeguarding food imports during wartime. If either country were involved in a war, its population would not starve, although shortages of certain products might cause inconvenience. One of the most important elements of U.S. power is the production of a substantial food surplus. Inability of the Russian economy to maximize agricultural output, despite the country's very large size, has been one of the basic causes of unrest.

The production of consumer goods and military equipment requires a large supply of coal, iron ore, and many other raw materials. Although no state is entirely self-sufficient in raw materials, very large states have access to most of the important ones. Russia is particularly well-endowed with raw materials and ranks among the world's leaders in the mining of a wide variety of them.

As very large states, the United States and the former Soviet Union could quickly deploy armed forces in different regions of the world. To maintain strength in regions that were not contiguous to their own territory, the United States and the Soviet Union established military bases in other countries. From these bases, ground and air support gained proximity to local areas of conflict. Naval fleets patrolled the major bodies of water.

However, the United States and the Soviet Union became military superpowers primarily because of the development of nuclear weapons. Only a very large state could plan for the possibility of surviving a nuclear war. Many observers assert that a nuclear war would inevitably destroy the planet, but during the Cold War leaders in the United States and the Soviet Union planned for the possibility that each side might launch a handful of nuclear weapons (possibly by accident) and then reach an agreement before completely destroying each other. On the other hand, most European, African, and South American states could not recover from a few well-placed bombs. Nuclear war between superpowers has become less probable, especially as the United States and Russia have signed agreements to destroy certain weapons caches; however, the proliferation of nuclear weapons has increased the possibility that another country or a terrorist group could initiate their use.

Development of state concept. The concept of dividing the world into a collection of independent states is recent. Prior to the seventeenth century, Earth's surface was organized in other ways, such as city-states, empires, and tribes, and much of Earth's surface was not organized into any specific territory.

Traditionally, most people lived in groups and were constantly on the move to hunt animals and gather vegetation. These groups were too small and isolated to allow for an independent political institution such as state to govern them. As people began to practice settled agriculture, they formed states to provide protection, store records, and transmit culture.

The development of states can be traced to the part of the ancient Middle East known as the Fertile Crescent, which formed an arc extending between the Persian Gulf and the Mediterranean

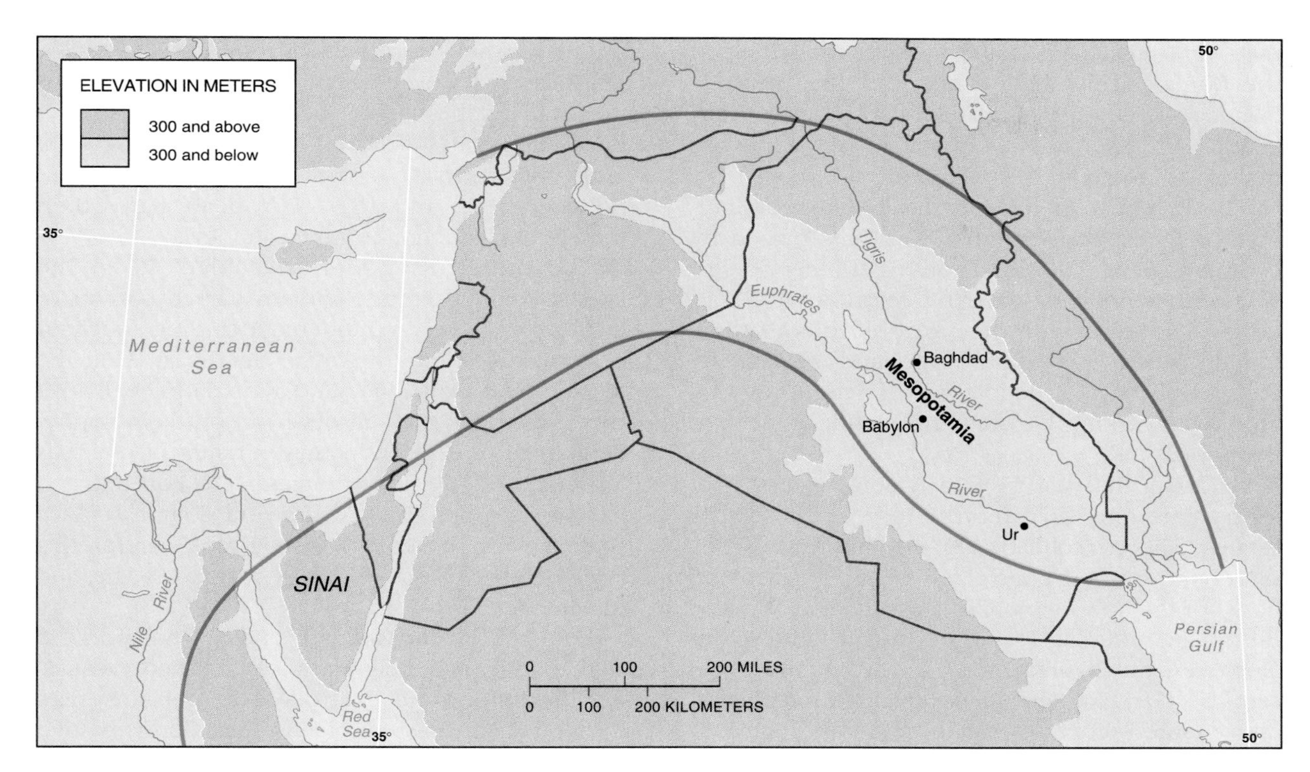

FIGURE 9–3
The Fertile Crescent is a crescent-shaped area of relatively fertile land situated between the Persian Gulf and the Mediterranean Sea. The Nile River Valley of Egypt is sometimes included as an extension of the Fertile Crescent. Several thousand years ago, the territory was organized into a succession of empires. As shown in Chapter 11, many important early developments in agriculture also originated in the region.

Sea. The eastern end of the region, known as Mesopotamia, was centered in the valley defined by the Tigris and Euphrates rivers, in present-day Iraq. The Nile River valley of Egypt is sometimes regarded as an extension of the Fertile Crescent. Situated at the crossroads among Europe, Asia, and Africa, the Fertile Crescent was a center for land and sea communications in ancient times (Figure 9–3).

The first states in Mesopotamia were known as city-states. A **city-state** is a sovereign state that comprises a town and the surrounding countryside. Walls clearly delineated the boundaries of the city, and outside the walls the city controlled agricultural land in order to produce food for the urban residents. The countryside also provided the city with an outer line of defense against attack by other city-states. Periodically, one city or tribe in Mesopotamia would gain military dominance over the others and form an empire. Mesopotamia was organized into a succession of empires by the Sumerians, Assyrians, Babylonians, and Persians.

Meanwhile, the state of Egypt emerged as a separate empire at the western end of the Fertile Crescent. Egypt controlled a long, narrow region along the banks of the Nile River, extending from the Nile Delta at the Mediterranean Sea south several hundred kilometers. Egypt's empire lasted from approximately 3000 B.C. until the fourth century B.C.

Political unity in the ancient world reached its height with the establishment of the Roman Empire, which controlled most of Europe, North Africa, and Southwest Asia, from modern-day Spain to Iran and from Egypt to England. At its maximum extent, the empire comprised thirty-eight provinces, each using the same set of laws that were created in Rome. Massive walls helped the Roman army defend many of the empire's frontiers. The Roman Empire collapsed in the fifth century A.D. after a series of attacks by people living on the frontiers of the empire.

The European portion of the empire was fragmented into a large number of estates owned by competing kings, dukes, barons, and other nobles. A victorious noble would seize control of a defeated rival's estate, but after a noble died, others fought to take possession of the land. Meanwhile, the vast majority of people were forced to live on an estate working and fighting for the benefit of the noble.

By about the year 1100, a handful of powerful kings emerged as rulers of large numbers of estates. The consolidation of neighboring estates under the unified control of a king formed the basis for the development of such modern Western European states as France, Portugal, and Spain. However, much of central Europe, notably present-day Germany and Italy, remained fragmented into a large number of estates and were not consolidated into states until the nineteenth century.

Colonies. A **colony** is a territory that is legally tied to a sovereign state rather than completely independent. A sovereign state in some cases runs only the colony's military and foreign policy, while in other cases it controls the colony's internal affairs as well. At one time, much of the world's political organization was made up of colonies, but today only a handful remain.

European states came to control much of the rest of the world through the process of colonialism. **Colonialism** is the attempt by one country to establish settlements and to impose its political, economic, and cultural principles in another territory.

European states established colonies elsewhere in the world for a variety of reasons. First, European missionaries established colonies to promote Christianity. Second, colonies provided resources that helped the economy of European states. Third, European states considered the number of colonies to be an indicator of relative power. The three motives can be summarized as God, gold, and glory.

The colonial era began in the fifteenth century, when European explorers sailed west for Asia but settled in the Western Hemisphere instead. The European states eventually lost most of the colonies they had established in the Western Hemisphere: the United States declared independence in 1776, and most Latin American states did so as well between 1800 and 1824. European states then turned their attention to Africa and Asia.

The United Kingdom assembled by far the largest colonial empire. Britain gained control of territory on every continent, including much of eastern and southern Africa, Southeast Asia, the Middle East, Australia, and Canada. The British proclaimed that the sun never set on their empire. France had the second-largest amount of overseas territory, although its colonies were concentrated in two locations, West Africa and Southeast Asia. Both the British and the French took control of a large number of islands in the Atlantic, Pacific, and Indian oceans as well.

Portugal, Spain, Germany, Italy, Denmark, the Netherlands, and Belgium all established colonies outside Europe but controlled less territory than the British and French (Figure 9–4). Germany tried to compete with Britain and France by obtaining African colonies that would interfere with communications in the rival European holdings.

The colonial practices of the European states varied. France attempted to assimilate its colonies into French culture and educate an elite group to provide local administrative leadership. After independence, most of these leaders retained close ties with France.

The British created different government structures and policies for various territories of the empire. The decentralized approach helped to protect the diverse cultures, local customs, and educational systems found in the extensive empire. British colonies generally made peaceful transitions to independence, although exceptions can be found in the Middle East and Southern Africa, as well as in Ireland.

Most African and Asian colonies became independent after World War II. Only fifteen African and Asian states were members of the United Nations when it was established in 1945, compared to 102 in 1994. The boundaries of the new states frequently but not always coincide with former colonial provinces.

The most populous remaining colony is Hong Kong, with nearly 6 million inhabitants. Hong

European countries carved up much of Africa into colonies during the late nineteenth century. The United Kingdom assembled the largest collection of colonies in Africa. This 1891 photograph shows the British commanders and governors asserting control over people of West Africa. (Mary Evans Picture Library/Photo Researchers Inc.)

Kong is a British colony situated on several islands and a small portion of the mainland of China. Britain gained control of portions of Hong Kong in a variety of ways. China surrendered Hong Kong Island in 1841, and then in 1860 Britain annexed Stonecutters Island and the Kowloon Peninsula. In 1898, China leased the New Territories, which are mainly agricultural lands, to Britain for a period of ninety-nine years. The lease expires in 1997 (Figure 9–5).

In anticipation of the end of the lease, China and Britain reached an agreement in 1985 whereby the entire Hong Kong colony returns to Chinese sovereignty in 1997. But China agreed that Hong Kong would continue its status as a free port and its separate social, economic, and legal systems for another fifty years. China also guaranteed that the people of Hong Kong could retain their freedoms of speech, religion, and unrestricted travel.

However, the return of Hong Kong to China has provoked concern in Hong Kong that the former colony may not continue to enjoy political liberties and unrestricted international trade. Many Hong Kong residents have migrated to Britain, as well as to Canada and the United States, but those who have been denied visas claim that Britain has an obligation to take in anyone who wishes to emigrate from its soon-to-be relinquished colony before the Chinese regain control.

Characteristics of Nations

In the modern world, the most important principle in the creation of a state is that it *should encompass the same area occupied by a nation.* A **nation**, or **nationality**, is a collection of people occupying a particular portion of Earth's surface who have a strong sense of unity based on a set of shared beliefs and cultural characteristics. The concept of a nation differs from that of a state: *Nation* refers to people, whereas *state* refers to a political structure.

There is no precise method of distinguishing among nationalities. In general, two elements are important in identifying a nationality. First is a common set of cultural characteristics, such as lan-

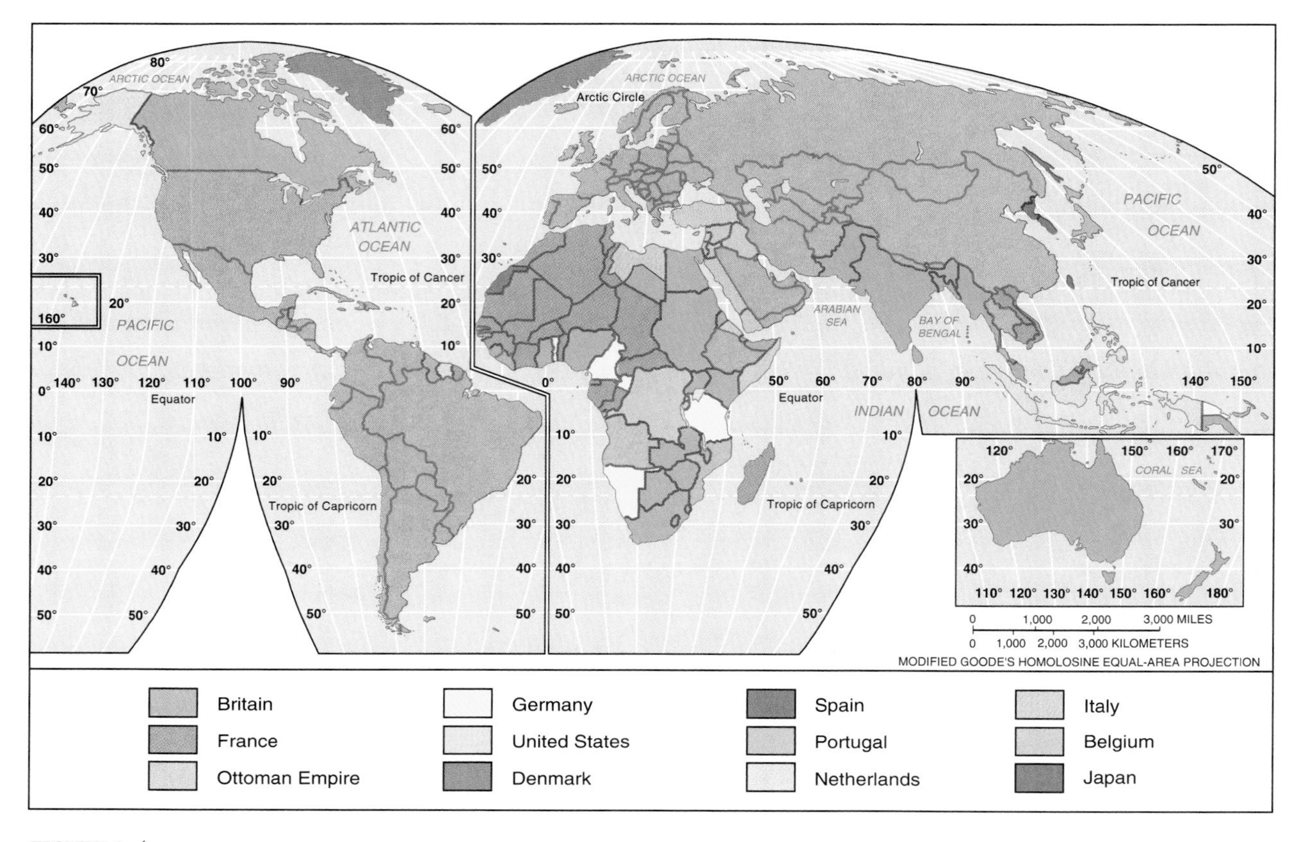

FIGURE 9–4
Colonial possessions. In 1914, at the outbreak of World War I, European states held colonies in much of the world, especially in Africa and Asia. Most of the countries in the Western Hemisphere at one time had been colonized by Europeans but gained their independence in the eighteenth or nineteenth century.

guage and religion. For example, in Europe, a strong sense of national unity exists among speakers of Italian. Shared culture also encompasses appreciation of distinctive forms of creative arts, such as music, theater, and painting.

The second element that distinguishes a nation is a unique set of shared attitudes and emotions. The people of a particular nation share a common ancestry and take pride in their own historical events. Many nationalities exist in Latin America, for example, despite the fact that most people are Roman Catholic and speak Spanish. The people of each Latin American country have their own cultural identity, gained in large measure through a unique history, beginning with the series of events leading to national independence. Being a nation involves being different, and being proud of one's differences.

Desire for self-rule is a particularly important shared attitude for many nations. To preserve and enhance distinctive cultural characteristics, nationalities seek the ability to govern themselves without interference from other groups. The concept that nationalities have the right to govern themselves is known as **self-determination**.

Most of Western Europe had achieved self-determination by the end of the nineteenth century. During the twentieth century, world political leaders have generally supported the right of self-determination for nationalities in other regions and have attempted to organize Earth's surface into a collection of nation-states. A **nation-state** is a state whose territory corresponds to that occupied by a particular nation. Yet, despite the attempt to create nation-states, boundaries of states rarely correspond precisely to the boundaries of a nation.

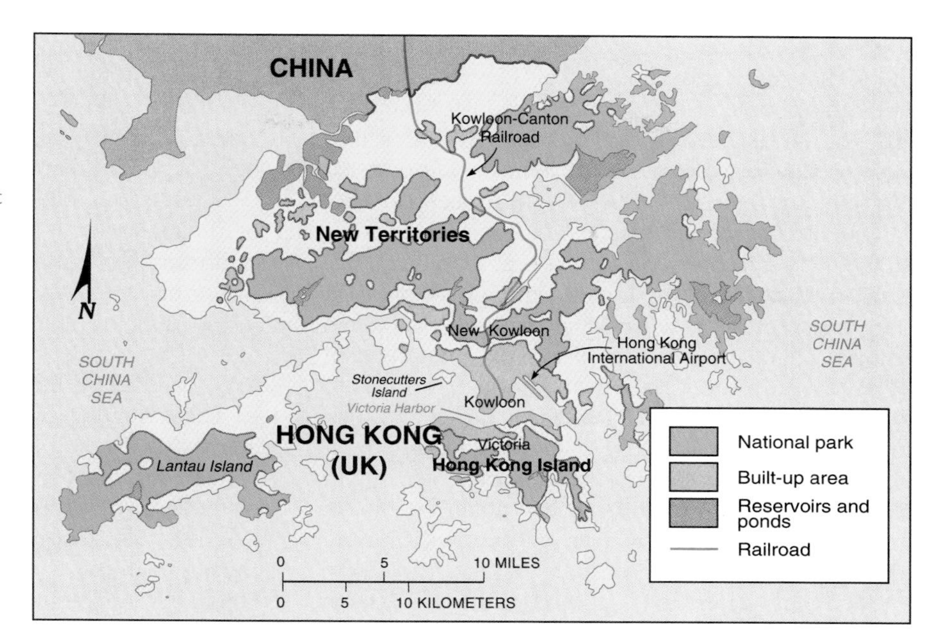

FIGURE 9–5
Hong Kong is the most populous remaining colony. The 1,060-square-kilometer (410-square-mile) territory has been controlled by the British under a ninety-nine-year lease, which expires in 1997. At that time, Hong Kong will be returned to China and will no longer constitute a colony. The view below is from Victoria Peak, on Hong Kong island, looking northeast toward Kowloon. (Michelle Burgess/ The Stock Market)

Slovenia, formerly part of Yugoslavia, is a good example of a nation-state established in the 1990s. More than 90 percent of the residents of Slovenia are Slovenes, and nearly all of the world's 2 million Slovenes live in Slovenia. The relatively close coincidence between the boundaries of the Slovene people and the country of Slovenia has been a major factor in the country's relative peace and stability compared to the other former republics of Yugoslavia.

Centripetal forces. A state, once established, tries to hold the loyalty of its citizens. Most states find that the most effective means to achieve the support of its citizens is to emphasize shared attitudes that unify the people. Attitudes that tend to unify the people and enhance support for the state are known as **centripetal forces**. Attitudes that tend to divide a people and diminish support for the state are known as **centrifugal forces.**

One of the most significant centripetal forces shared by citizens of a state is nationalism. **Nationalism** involves loyalty and devotion to a state that represents a particular nation's distinctive cultural characteristics. People display nationalism by supporting the creation and growth of a state that preserves and enhances the culture and attitudes of their nationality. Nationalism had been effectively suppressed by the Communists, but in recent years nationalism has once again become a major factor in the formation of peoples' cultural identities, especially in Europe.

For many states, mass communications systems are the most effective means of fostering nationalism. Most countries regard an independent source of news as more of a risk than a benefit to the stability of the government. Consequently, only a few states permit the mass communications systems to operate without government interference. Nearly all countries control or at least regulate most forms of communications, including mail, telephone, telegraph, television, radio, and satellite transmissions. The government either owns or controls newspapers in many countries.

States foster nationalism by promoting symbols of the nation-state, such as flags and songs. The symbol of the hammer and sickle on a field of red was long synonymous with the beliefs of communism. After the fall of communism, one of the first acts in a number of Eastern European countries was to redesign flags without a hammer and sickle. Legal holidays were changed from dates associated with Communist victories to those associated with historical events that preceded takeovers. One of the strongest forms of political protest is to burn a state's flag; for example, Americans widely support laws in the United States to make burning the Stars and Stripes illegal.

Nationalism is also instilled through the creation of songs extolling the country's virtues. Nearly every state has a national anthem, which usually combines respect for the state with references to the nation's significant historic events or symbols of unity. Nationalism can have a negative impact.

Latvians celebrate independence from the Soviet Union with a massive demonstration in the streets of the capital, Riga, in 1991. (Ints Kalnins/Woodfin Camp & Associates)

The sense of unity within a nation-state is sometimes achieved through the creation of negative images of other nation-states.

The Soviet Union, Yugoslavia, and Czechoslovakia were dismantled in large measure because minority cultural groups opposed the long-term dominance of local economic, political, and cultural institutions by the country's most numerous nationality. These dominant groups were Czechs in Czechoslovakia, Serbs in Yugoslavia, and Russians in the Soviet Union (Figure 9–6). No longer content with being the majority within a unit of local government in a country, cultural groups sought to constitute the majority in a completely independent nation-state.

More numerous nationalities have had a chance to organize nation-states in Eastern Europe, but the less numerous nationalities still find themselves minorities in other countries. Slovaks, who comprised one-third of the population in the former Czechoslovakia, long complained that the more numerous Czechs treated them unfairly. The Slovaks now constitute a majority in the independent country of Slovakia, but Magyars (Hungarians), who comprise 10 percent of the population of Slovakia, complain that the Slovaks have enacted laws suppressing their ability to use their language in daily activities.

Republics that had once constituted local government units within the Soviet Union, Yugoslavia, and Czechoslovakia have generally made peaceful transitions into independent countries when their boundaries have corresponded reasonably well with the territory occupied by clearly defined cultural groups. Problems have arisen in regions where the former local government units did not match distributions of nationalities.

FIGURE 9–6
The former Soviet Union included 15 republics, named for the country's largest nationalities. Russians comprised about half of the Soviet Union's population, followed by Ukrainians, Uzbeks, and Kazaks. With the breakup of the Soviet Union, the fifteen republics became independent states, but in many cases the new countries contained large percentages of minorities.

Boundaries Between States

A state is separated from its neighbors by a **boundary**, which is an invisible line marking the extent of a state's territory. Boundaries completely surround an individual state, marking the outer limits of its territorial control and giving it a distinctive shape.

Boundaries interest us because the process of selecting their location is frequently difficult and can lead to conflict. The boundary line, which must be shared by more than one state, is the only location where direct contact must take place between two neighboring states. Therefore, the boundary has the potential to become the focal point of conflict between them.

Historically, frontiers rather than boundaries separated two states. A **frontier**—a zone where no state exercises political control—is a geographic area, whereas a boundary is a thin, invisible line. A frontier provides an area of separation, but a boundary brings two neighboring states into direct contact, increasing the potential for violent face-to-face meetings. A frontier area is either uninhabited or sparsely settled by a few isolated pioneers seeking to live outside organized society.

Almost universally, frontiers between states have been replaced by boundaries. Modern communications systems permit countries to monitor and guard boundaries effectively, even in previously inaccessible locations. Once-remote frontier regions have become more attractive for agriculture and mining.

Other than Antarctica, the only region of the world that still has frontiers rather than boundaries is the Arabian peninsula. Frontiers separate Saudi Arabia from Qatar, the United Arab Emirates, Oman, and Yemen. These frontier areas are inhabited by a handful of nomads who cross freely with their herds from one country to another.

Shapes of States

The shape of a state affects the length of its boundaries with other states and therefore the potential for both communications and conflict with neighbors. Countries display a wide variety of shapes, and the shape of a particular state, such as the boot of Italy or the narrow finger of Chile, is part of its unique identity. Beyond its symbolic value, the shape of a state can influence the ease of internal administration and affect social unity.

Countries follow one of five types of shapes—compact, prorupted, elongated, fragmented, and perforated (Figure 9–7). Each displays distinctive characteristics and problems.

Compact. In a **compact state**, the distance from the center to any boundary does not vary significantly. A compact shape is beneficial to most smaller states, because good communications can be more easily established to all regions, especially if the capital is located near the center. Examples of compact states include Bulgaria, Hungary, and Poland.

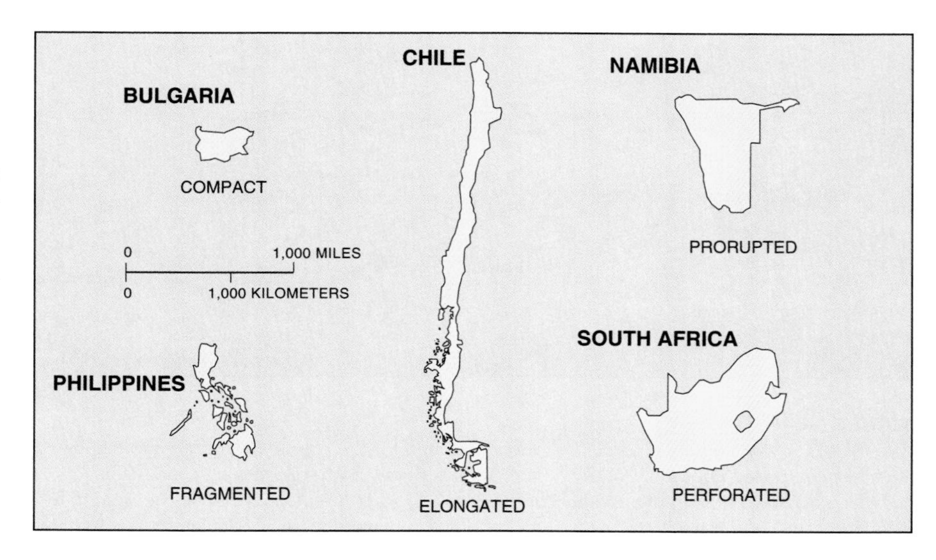

FIGURE 9–7
Shapes of states. Examples are shown of states that are compact (Bulgaria), prorupted (Namibia), elongated (Chile), fragmented (Philippines), and perforated (South Africa). The five states are drawn to the same scale. In general, compactness is an asset, because it fosters good communications and integration among all regions of the country.

Prorupted. An otherwise compact state with a large projecting extension is a **prorupted state**. Proruptions are created for two principal reasons. First, proruptions separate two states that otherwise would share a boundary. When the British ruled the otherwise compact state of Afghanistan, they created a long, narrow proruption to the east, approximately 300 kilometers (200 miles) long and as narrow as 20 kilometers (12 miles) wide. The proruption prevented Russia from sharing a border with Pakistan.

A proruption can also provide a state with access to a resource, such as water. When the Belgians gained control of the Congo (now Zaire), they carved out a 500-kilometer (300-mile) proruption to the west. The proruption, which followed the Zaire (Congo) River, gave the colony access to the Atlantic Ocean. The proruption also divided the Portuguese colony (now the independent state) of Angola into two discontinuous fragments 50 kilometers (30 miles) from each other. The northern region, called Cabinda, constitutes less than 1 percent of Angola's total land area.

In their colony of South West Africa (now Namibia), the Germans carved out a 500-kilometer (300 mile) proruption to the east. The proruption, known as the Caprivi Strip, provided the Germans with access to one of Africa's most important rivers, the Zambezi. The Caprivi Strip also disrupted communications within the British colonies of southern Africa. In recent years, South Africa, which controlled Namibia until its independence in 1990, stationed troops in the Caprivi Strip to fight enemies located in Angola, Zambia, and Botswana.

Elongated. There are a handful of **elongated states**, or states with a long, narrow shape. The clearest example is Chile, which stretches north-south for more than 4,000 kilometers (2,500 miles) but rarely exceeds an east-west distance of 150 kilometers (90 miles). Chile is wedged between the Pacific Coast of South America and the rugged Andes Mountains, which rise more than 6,700 meters (20,000 feet).

Somewhat less extreme examples of elongated states are found elsewhere in the world. Italy extends more than 1,100 kilometers (700 miles) from northwest to southeast but is only approximately 200 kilometers (120 miles) wide in most places. In Africa, Malawi is approximately 850 kilometers (530 miles) north-south but only 100 kilometers (60 miles) east-west.

Elongated states may suffer from poor internal communications. A region located at an extreme end of the elongation may be isolated from the capital, which is usually placed near the center.

Perforated. A state that completely surrounds another one is a **perforated state**. The one good example of a perforated state is South Africa, which completely surrounds the state of Lesotho. Lesotho must depend almost entirely on South Africa for the import and export of goods.

Fragmented. A **fragmented state** includes several discontinuous pieces of territory. Technically, all states that have offshore islands as part of their territory are fragmented. However, fragmentation is particularly significant for some states.

There are two kinds of fragmented states. In one case, the discontinuous areas are separated by water. For example, Indonesia comprises 13,677 islands extending more than 5,000 kilometers (3,000 miles) across the Indian Ocean. Although more than 80 percent of the country's population live on two of the islands, Java and Sumatra, the fragmentation hinders communications and makes integration of people living on remote islands nearly impossible. To foster national integration, the Indonesian government has encouraged migration from the relatively dense islands to some of the sparsely inhabited ones. Japan, the Philippines, and New Zealand are also states that comprise more than one island.

A more difficult type of fragmentation occurs if the two pieces of territory are separated by another state. Picture the difficulty of communicating between Alaska and the lower forty-eight states if Canada were not a friendly neighbor, since all land connections between Alaska and the rest of the United States must pass through Canada. The division of Angola into two pieces by Zaire's proruption creates a fragmented state.

Panama—otherwise an example of an elongated state, 700 kilometers (450 miles) long and 80 kilometers (50 miles) wide—is fragmented by the canal, built and owned by the United States. U.S. ownership of the canal and the surrounding Canal Zone was a source of tension for many years, but the United States and Panama signed a treaty in the late 1970s that transfers the canal to Panama on December 31, 1999. The treaty guarantees the neutrality of the canal and permits the United States to use force if necessary to keep the canal operating.

Types of Boundaries

Boundaries are of two types: physical and cultural. Physical boundaries coincide with significant features of the natural landscape, whereas cultural boundaries follow the distribution of cultural characteristics.

Physical boundaries. Important physical features on Earth's surface can make good boundaries because they are easily seen both on a map and on the ground. Three types of physical elements serve as boundaries between states: mountains, deserts, and water.

Mountains can be effective boundaries if they are difficult to cross. Contact between nationalities living on each side may be limited and completely impossible if passes are closed by winter storms. Mountains are also useful boundaries because they are basically permanent and usually sparsely inhabited.

A boundary drawn in a desert can also effectively divide two states. Like mountains, deserts are hard to cross and sparsely inhabited. Desert boundaries are common in Africa and Asia. In North Africa, the Sahara has generally proved to be a stable boundary separating Algeria, Libya, Egypt, Mauritania, Mali, Niger, Chad, and the Sudan. In the early 1980s the Libyan army moved south across the desert to invade Chad, but retreated in 1987 following French intervention.

Rivers, lakes, oceans, and other bodies of water are the physical features most commonly used as boundaries. Water boundaries are readily visible on a map and are relatively unchanging although erosion and deposition occasionally shift boundaries. Of course, no permanent human settlement exists on water.

Water boundaries offer good protection against attack from another state. An invading state must transport its troops by airplane or ship and secure a landing spot in the country it is attacking. The invaded state can concentrate its defense at the landing point.

Water boundaries also cause problems because states generally claim that the boundary lies not at the coastline but out at sea. Beginning in the late eighteenth century, some states recognized a boundary, known as the territorial limit, that extended 3 nautical miles (3.45 land miles or 5.56 kilometers) from land into the ocean. Some states claimed more extensive territorial limits, and others identified a contiguous zone of influence beyond the territorial limits.

The Law of the Sea, signed by 117 countries in 1983, standardized the territorial limits for most countries at 12 nautical miles (13.81 land miles or 22.22 kilometers). Under the Law of the Sea, states also have exclusive rights to the fish and other marine life within 200 miles (320 kilometers). Countries separated by less than 400 miles of sea must negotiate the location of the boundary between areas of exclusive fishing rights. Disputes can be taken to a Tribunal for the Law of the Sea or to the International Court of Justice.

Water boundaries are critical to states because they facilitate international trade. Relatively bulky goods are normally transported long distances by ship, but to send or receive goods in this way means that a country needs a port where the goods can be transferred between land and sea. A state that does not have a direct outlet to the sea is called a **landlocked state**. To send and receive goods by sea, a landlocked state must depend on another country's port.

Landlocked states are most common in Africa, where fifteen of the continent's fifty-five states have no direct access to the ocean. The prevalence of landlocked states in Africa is a remnant of the colonial era, when Britain and France controlled extensive regions. The European powers built railroads, for the most part in the early twentieth century, to connect the interior of Africa with ports. Railroads helped to move minerals from the interior and to import mining equipment and supplies to the interior. Now that independent states have become divorced from the British and French empires, some of the important colonial railroad lines pass through several independent countries. Newly created landlocked states must cooperate with neighboring states that have ports (Figure 9–8).

Cultural boundaries. The boundaries between some states coincide with differences in cultural characteristics, especially language and religion. Other cultural boundaries are drawn according to geometry: they simply constitute straight lines drawn on a map.

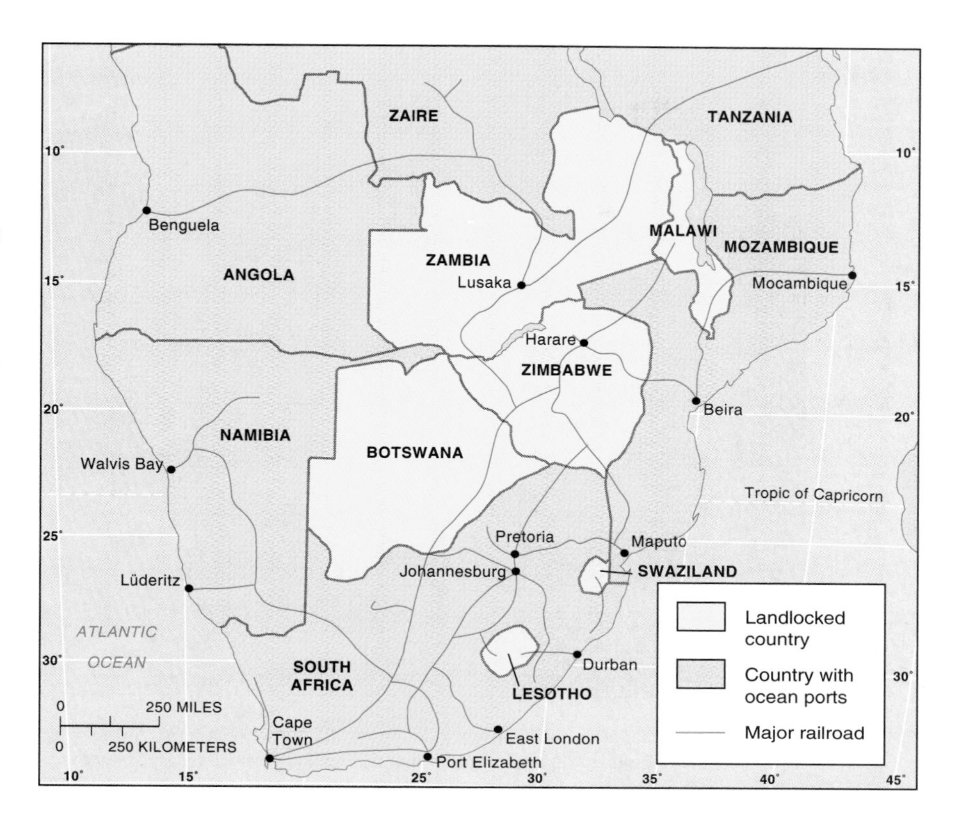

FIGURE 9–8
Landlocked states of southern Africa. To trade with countries in North America, Europe, or Asia, African states must ship goods through ocean ports. Landlocked African states must import and export goods by land-based modes of transportation, primarily rail lines, to reach ocean ports located in neighboring states. Cooperation has been especially difficult until recently in southern Africa because neighboring states opposed South Africa's policy of apartheid.

The 2,100-kilometer (1,300-mile) boundary between the United States and Canada between 95° west longitude and the Strait of Georgia (near the Pacific Ocean) is a straight line along 49° north latitude. The line (more precisely an arc) was established in 1846 by treaty between the United States and Great Britain, which still controlled Canada at the time.

Some people in the United States wanted the boundary to be fixed 600 kilometers (400 miles) farther north at 54°40' north latitude. Before a compromise was reached, some U.S. militants proclaimed "fifty-four forty or fight." The United States and Canada share an additional 1,100-kilometer (700-mile) geometric boundary between Alaska and the Yukon Territory along the north-south line of 141° west longitude.

The 1,000-kilometer (600-mile) boundary between Chad and Libya is a straight line drawn across the desert in 1899 by the French and British to set the northern limit of French colonies in Africa. But subsequent actions by European countries have created confusion concerning the location of the boundary. In 1912, Italy seized control of Libya from the Turks and demanded that the boundary with French-controlled Chad be changed. In 1935, France agreed to move the boundary 100 kilometers (60 miles) to the south, but the Italian government was not satisfied with the settlement and never ratified the treaty. The land that the French would have ceded is known as the Aozou Strip, named for the only settlement in the 100,000 square-kilometer (36,000 square-mile) area (Figure 9–9).

When Libya and Chad became independent countries, the boundary was set at the original northern location. Claiming that the president of Chad had secretly sold it the Aozou Strip, Libya seized the territory in 1973, as well as a tiny bit of northeastern Niger that may contain uranium. In 1987, Chad expelled the Libyan army and regained control of the strip.

Religious differences coincide with boundaries between states in some regions, but in only a few cases has religion been used to select the actual boundary line. A notable example took place in 1947, when the British partitioned India into two states on the basis of religion. The predominantly Muslim portions were allocated to Pakistan, while

The Great Wall of China historically served as one of the world's most visible cultural boundaries. Originally built in the third century B.C. during the Qin (Ch'in) dynasty, the wall was extended the following centuring during the Han dynasty to keep out nomadic horsemen. The wall was partially reconstructed between the fourteenth and sixteenth centuries A.D. during the Ming dynasty. (Simon Holledge/The Stock Market)

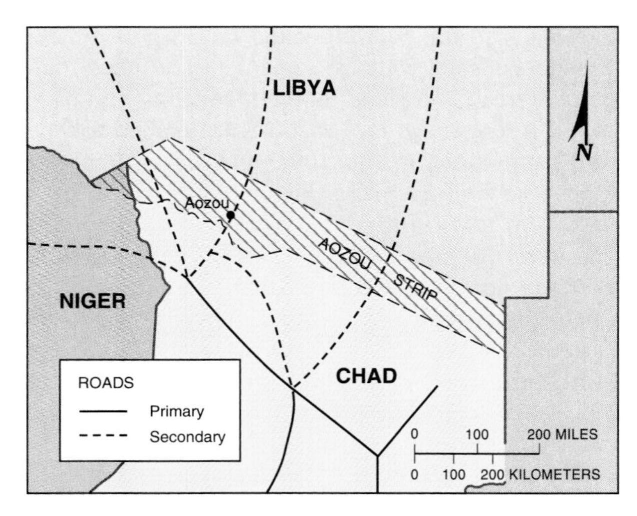

FIGURE 9–9
The boundary between Libya and Chad is a straight line, drawn by European countries early in the twentieth century when the area comprised a series of colonies. Libya, however, claims that the boundary should be located 100 kilometers to the south and that it should have sovereignty over the Aozou Strip.

the predominantly Hindu portions became the independent state of India.

Religion was also used to some extent to draw the boundary between two states on the island of Eire (Ireland), as we discussed in Chapter 8. Most of the island became an independent country, while six counties in the northeast, now known as Northern Ireland, remained part of the United Kingdom. Roman Catholics comprise approximately 95 percent of the population in the twenty-six counties that joined the Republic of Ireland, while Protestants constitute the majority in the six counties of Northern Ireland (see Figure 8-28).

Language is an important cultural characteristic for drawing boundaries, especially in Europe. The French language was a major element in the development of France as a unified state in the seventeenth century. The states of England, Spain, and Portugal coalesced around distinctive languages. In the nineteenth century, Italy and Germany also emerged as states that unified the speakers of particular languages.

The movement to identify states on the basis of language spread throughout Europe in the twentieth century. After World War I, leaders of the victorious countries met at the Versailles Peace Conference to redraw the map of Europe. One of the chief advisors to President Woodrow Wilson, the geographer Isaiah Bowman, played a major role in the decisions. Language was the most important criterion the allied leaders used to create new states in Europe and to adjust the boundaries of existing ones.

Although the boundaries imposed by the Versailles conference on the basis of language were adjusted somewhat after World War II, they proved to be relatively stable and peaceful for most of the twentieth century. However, since the beginning of the 1990s, the map of Europe drawn at Versailles has collapsed. Despite speaking similar languages, Czechs and Slovaks and Croats, Macedonians, Serbs, and Slovenes found that they could no longer live together peacefully in the same state (Figure 9–10).

Problems Matching States and Nations

As people divide Earth's surface into a collection of states, they try to correlate the boundaries of states with the distribution of nations to create a series of nation-states. Many of the problems faced by the states of the world derive from the fact that boundaries of nations and states do not coincide. In general, problems concerning the boundaries of nations and states develop for two reasons. In some cases, the boundaries of a state encompass more than one nationality. In other cases, the population of a nation is split among more than one state.

One State with More Than One Nationality

A state that contains more than one nationality is known as a **multinational state**. Relationships among nationalities vary in different multinational

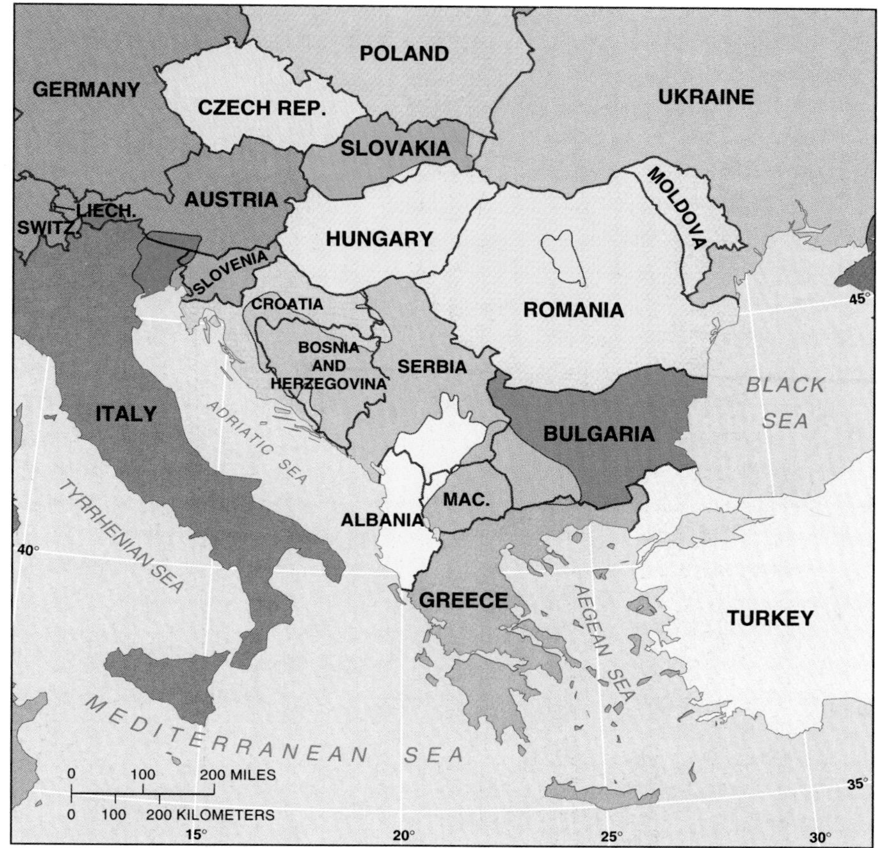

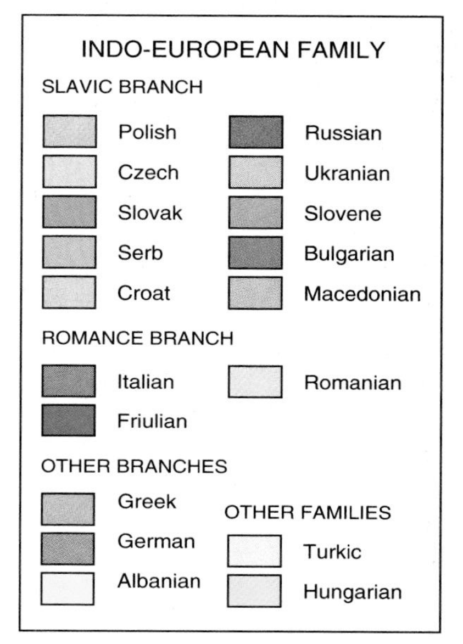

FIGURE 9–10
States and languages in southeastern Europe. The region has traditionally been politically unstable because of a lack of correspondence between the boundaries of states and of nationalities. In the 1990s, the region has been a center of conflict among speakers of different languages and adherents of different religions.

states. In some states, one nationality tries to dominate another, especially if one of the nationalities is much more numerous than the other, while in other states nationalities coexist peacefully. The people in one nation may be assimilated into the cultural characteristics of another nation, but in other cases, the two nationalities remain culturally distinct.

Russia. Russia officially recognizes the existence of thirty-nine nationalities, many of which are eager for independence. Minorities are clustered between the Volga River basin and the Ural Mountains, as well as along southern borders (Figure 9–11).

When it was part of the Soviet Union, the Russian Socialist Federative Soviet Republic was divided into sixteen Autonomous Soviet Social Republics, forty-nine Oblasts (regions), and six Krays (territories). Seven of the Oblasts and one Kray contained a total of ten Autonomous Okrugs (districts), while five of the Krays each contained an Autonomous Oblast.

After the breakup of the Soviet Union, Russia signed a treaty with nineteen of its twenty-one local government units populated principally by minorities to give them increased administrative responsibilities. Two of the twenty-one refused to sign. Chechen-Ingush A.S.S.R. officially declared its independence from Russia, a move being considered by several other local government units. Tatar A.S.S.R. negotiated a separate treaty.

Cyprus. In some multinational states, nationalities coexist only by occupying geographically distinct regions. Cyprus, the third-largest island in the Mediterranean Sea, is a state that contains two nationalities: Greek and Turkish. In 1974, several Greek Cypriot military officers who favored unification of Cyprus with Greece seized control of the government. Shortly after the coup, Turkey invad-

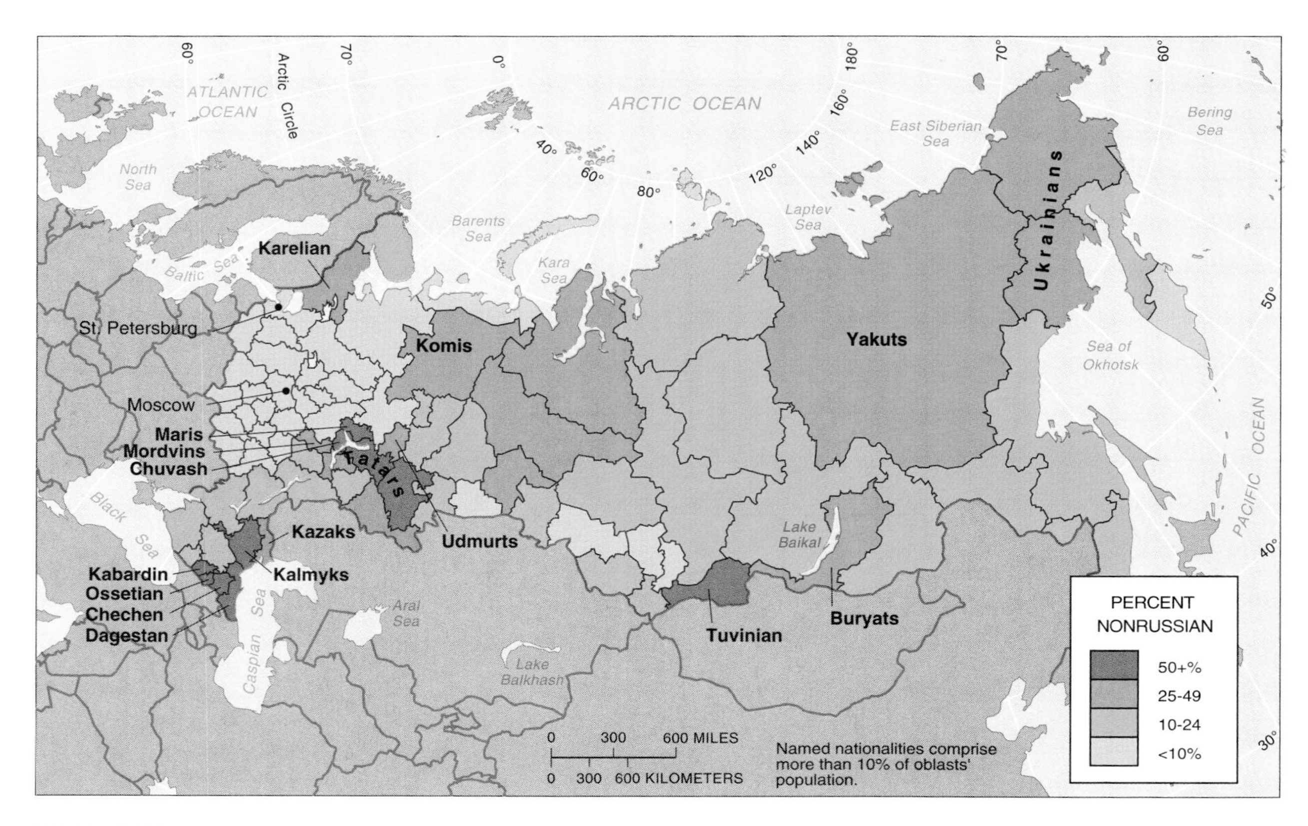

FIGURE 9–11
Russians are clustered in the western portion of Russia, and the percentage declines to the south and east. The largest numbers of non-Russians are found in the center of the country between the Volga River and the Ural Mountains and near the southern boundaries.

ed Cyprus to protect the Turkish Cypriot minority, occupying 37 percent of the island. The Greek coup leaders were removed within a few months, and an elected government was restored, but the Turkish army remained on Cyprus.

Traditionally, the Greek and Turkish Cypriots mingled, but after the coup and invasion, the two nationalities became geographically isolated. The northeastern part of the island is now overwhelmingly Turkish, while the southern part is overwhelmingly Greek. Approximately one-third of the island's Greeks were forced to move from the region controlled by the Turkish army, while nearly one-fourth of the Turks moved from the region now considered the Greek side. The percentage of one nationality living in the region dominated by the other nationality is now very low. The Turkish sector declared itself the independent Turkish Republic of Northern Cyprus in 1983, but only Turkey recognizes it as a separate state (Figure 9–12).

A buffer zone patrolled by U.N. soldiers stretches across the entire island of Cyprus to prevent Greeks and Turks from reaching the other side. The barrier even runs through the center of the capital, Nicosia. Only one official crossing point has been erected, and crossing is difficult except for top diplomats and U.N. personnel. Nevertheless, some cooperation continues between sectors: The Turks supply the Greek side with water and in return receive electricity.

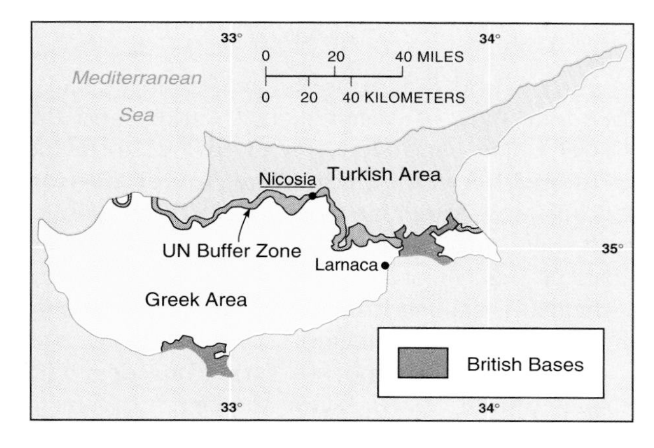

FIGURE 9–12
Since 1974 Cyprus has been divided into Greek and Turkish portions, with little mingling between the two groups. The Turkish sector has declared itself to be the Turkish Republic of Northern Cyprus, but only Turkey recognizes it as an independent country.

South Africa. South Africa is a multinational state where the government for many years divided the population into nationalities according to race. When it was controlled by whites, the South African government created a legal system called **apartheid**, which was the physical separation of races into different geographic areas. A newborn baby was assigned to one of the four races: black, white, Asian, or colored (mixed white and black). Blacks constitute approximately 73 percent of South Africa's population, whites 18 percent, and Asian and colored 3 percent each.

Under apartheid, each of the four races had a different legal status in South Africa. The apartheid laws determined where different races could live, attend school, work, shop, and own land. Separate transportation systems, public bathrooms, and beaches were maintained. Blacks were restricted to certain occupations and were paid far lower wages than were whites for similar work. Blacks could not vote or run for office in national elections.

To assure further geographic isolation of different races, the white-controlled South African government designated ten so-called homelands for blacks. The white minority government expected every black to become a citizen of one of the homelands and to move there (Figure 9–13).

The apartheid laws have been repealed, and South Africa now has a government controlled by the black majority. However, nonwhites still suffer from lower incomes, limited job prospects, and poor housing conditions.

Nations Divided Among More Than One State

Conflicts may develop when one nation finds itself divided among more than one state. Some nationalities consider that their right to self-determination has not been achieved in the current division of the world into nation-states. A nationality split among more than one state—unable to control the government of any state—may seek to carve out a new nation-state from portions of existing ones.

Caucasus. The division of nationalities among more than one state has proved especially troublesome in the rugged, isolated Caucasus mountains once controlled by the Soviet Union. Three former Soviet republics—Armenia, Azerbaijan, and Georgia—

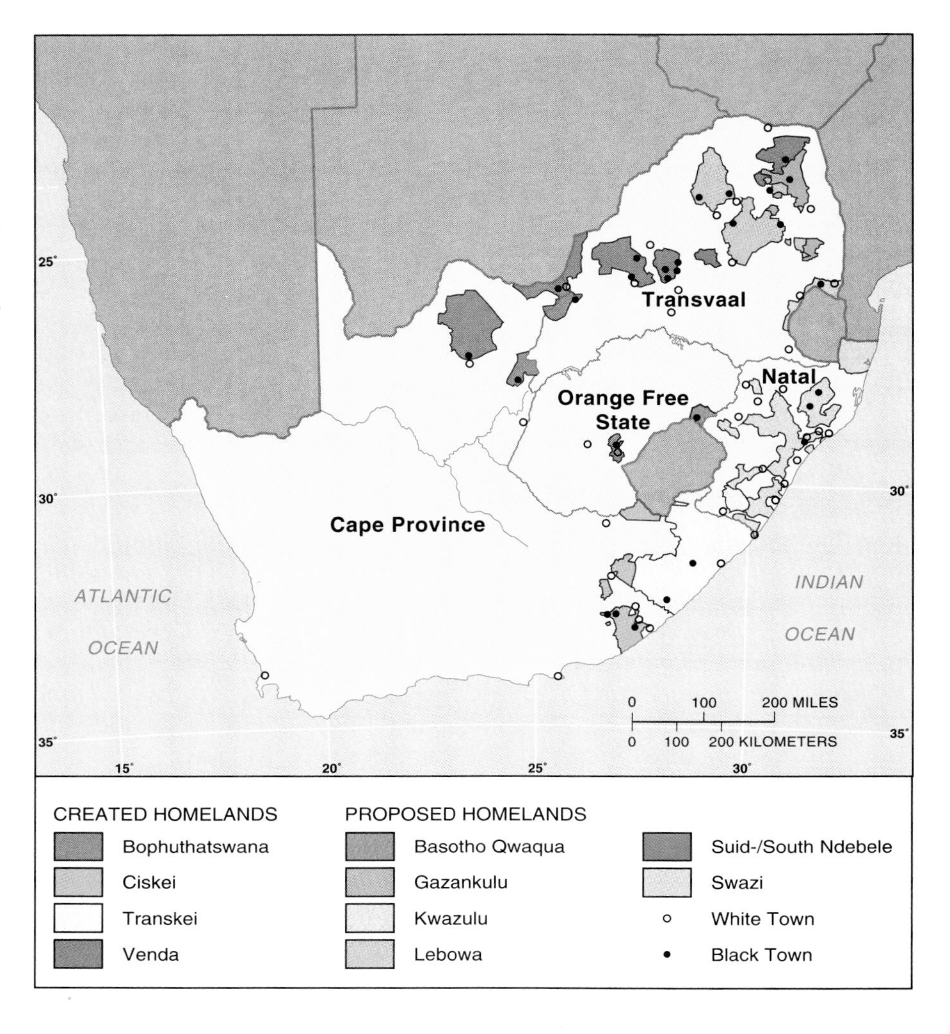

FIGURE 9–13
As part of its apartheid system, the government of South Africa designated ten so-called homelands, expecting that ultimately every black would become a citizen of one of them. South Africa declared four of these homelands to be independent states, but no other country recognized the action. With the dismantling of the apartheid system, the homelands will be integrated back into the rest of South Africa.

have become independent countries, but the boundaries do not match the territories occupied by the Armenian, Azeri, and Georgian nationalities.

Armenians are Eastern Orthodox Christians who speak an Indo-European language. They have lived in the region for thousands of years and controlled an independent kingdom before the founding of either Christianity or Islam. Converted to Christianity in the year 301, they lived for many centuries as an isolated enclave of Christians under the rule of Turkish Muslims. During the late nineteenth and early twentieth centuries, hundreds of thousands of Armenians were killed in a series of massacres organized by the Turks. Others were forced to migrate to Russia, which had gained possession of the eastern portion of Armenia in 1828.

After World War I, the allies created an independent state of Armenia, but Turkey and the Soviet Union concluded a treaty in 1921 to divide Armenia between them. The Soviet portion became the Armenian Soviet Socialist Republic, and then an independent country in 1991. More than 90 percent of the population in Armenia are Armenians, making it the most ethnically homogeneous country in the region.

Azeris trace their roots to Turkish invaders who migrated from central Asia in the eighth and ninth centuries and merged with the existing, mostly Persian, population. They are Shiite Muslims who speak an Altaic language similar to Turkish. A treaty in 1828 allocated the northern portion of the Azeris' territory to Russia and the southern portion to Persia (now Iran). After the Russian Revolution in 1917, the Russian portion briefly operated as an independent state, and then it became the Azerbaijan Soviet Socialist Republic within the Soviet Union in 1923. With the breakup of the Soviet Union in 1991, Azerbaijan became an independent country again.

Approximately 6 million Azeris now live in Azerbaijan, nearly 80 percent of the country's total population. Another 6 million Azeris are clustered in northwestern Iran, where they constitute 10 percent of that country's population. Azeris hold positions of responsibility in Iran's government and economy, but Iran restricts teaching of the Azeri language.

Armenians and Azeris have achieved long-held aspirations of forming nation-states, but neither Armenia nor Azerbaijan matches the territories inhabited by the two groups. Both groups make centuries-old claims to Nagorno-Karabakh, a 5,000-square-kilometer (2,000-square-mile) enclave about the size of Delaware within Azerbaijan that is inhabited primarily by Armenians. Armenians and Azerbaijanis fought a war early in the twentieth century, and fighting continued during the brief period immediately after the 1917 Russian Revolution when Armenia and Azerbaijan were independent countries. After Armenia and Azerbaijan were incorporated into the Soviet Union in the early 1920s, Nagorno-Karabakh was placed in the Socialist Republic of Azerbaijan.

For decades, conflict between the Armenians and Azerbaijanis was suppressed by the Soviet Union, but fighting broke out again in 1988. The Armenians claimed that unrest in Nagorno-Karabakh was justified because their language, religion, and other cultural activities were being suppressed by the predominantly Muslim, Turkish-speaking Azerbaijanis. The two groups have unleashed an unending cycle of escalating violent retaliations as they both seek to control Nagorno-Karabakh.

The 5 million Georgians speak a Caucasian language unrelated to others in the region. Most Georgians adhere to Eastern Orthodox Christianity, although some Georgians, now known as Adhzars or Ajars, were converted to Islam by the Ottomans. Georgia, like Armenia and Azerbaijan, was independent for a few years before becoming part of the Soviet Union in the early 1920s, and then independent again in 1991.

The population of Georgia is more diverse than in Armenia and Azerbaijan. Only 69 percent of the people living in Georgia are ethnic Georgians. The country includes about 9 percent each Armenian and Russian and 5 percent Azeri, 3 percent Ossetian, and 2 percent Abkhazian. This cultural diversity has been a source of unrest in Georgia, especially among the Ossetians and Abkhazians, who have not been able to form nation-states.

Other nationalities in the Caucasus aspire to forming nation-states. One example is the Ossetians, Eastern Orthodox Christians who speak an Iranian language. About 300,000 Ossetians live in Russia, in an area known as North Ossetia, while another 65,000 Ossetians live in South Ossetia, an autonomous region within Georgia. To the east of the Ossetians are the Chechens, Muslims who speak a Turkic language. They, too, seek independence from Russia.

The Kurds are an example of a nationality that has failed to achieve a nation-state in the Caucasus. The Kurds are a non-Arab group of Sunni Muslims who speak a language similar to Pashto and have distinctive literature, dress, and other cultural traditions. The Kurdish population is split among six countries, including 10 million in eastern Turkey, 5 million in western Iran, 4 million in northern Iraq, and smaller numbers in Armenia, Azerbaijan, and northeastern Syria. Kurds comprise a fifth of the population in Iraq, a sixth in Turkey, and nearly a tenth in Iran (Figure 9–14).

When the victorious allies carved up the Ottoman Empire after World War I, they created an independent state of Kurdistan to the south and west of Van Gölü (Lake Van) under the 1920 Treaty of Sevres. Before the treaty was ratified, however, the Turks, under the leadership of Mustafa Kemal (later known as Kemal Ataturk), fought successfully to expand the territory under their control beyond the small area the allies had allocated to them. The Treaty of Lausanne in 1923 established the modern state of Turkey, with boundaries nearly identical to the current ones. Kurdistan became part of Turkey and disappeared as an independent state.

Since then, the Turks have tried repeatedly to suppress Kurdish culture. Use of the Kurdish language was illegal in Turkey until 1991, although laws banning its use in broadcasts and classrooms have remained. Kurdish nationalists for their part have waged a guerrilla war since 1984 against the Turkish army.

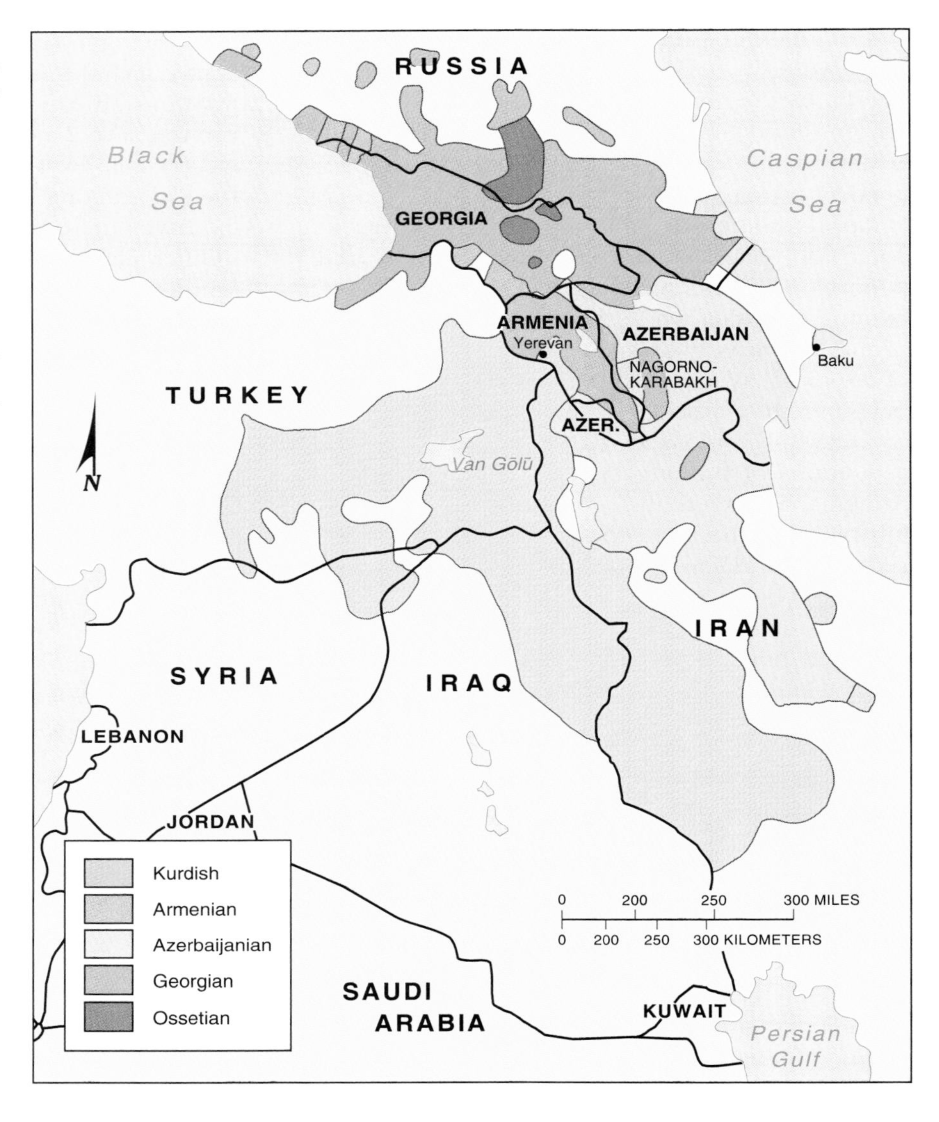

FIGURE 9–14
In the Caucasus, Zagros, and Taurus mountains of southwestern Asia, the Armenians, Azeris, and Georgians are examples of nationalities that have been able to dominate nation-states during the 1990s, following the breakup of the Soviet Union. However, the boundaries of the states of Armenia, Azerbaijan, and Georgia do not match the territories occupied by the Armenian, Azeri, and Georgia people. The Kurds and Ossetians are examples of nationalities that have not been able to organize nation-states. Instead, they have been divided among more than one state, where they constitute minorities.

Kurds in other countries have fared just as poorly as those in Turkey. Iran's Kurds secured an independent republic in 1946 that lasted less than a year. Iraq's Kurds have made several unsuccessful attempts to gain independence, including in the 1930s, 1940s, and 1970s. A few days after Iraq was defeated in the 1991 Gulf War, the country's Kurds launched another unsuccessful rebellion. The United States and its allies decided not to resume their recently concluded fight against Iraq on behalf of the Kurdish rebels, but after the revolt was crushed they did send troops to protect the Kurds from further attacks by the Iraqi army.

Political Refugees

Boundaries may be drawn to create nation-states, but lines cannot be drawn to segregate two cultural groups completely. To facilitate the creation of nation-states, large numbers of a cultural group caught on the "wrong" side of a boundary line may be forced to migrate. People forced to migrate from a particular country for political reasons are known as **refugees**. The United Nations defines political refugees as people who have fled their home country and cannot return for fear of persecution because of their race, religion, nationality, membership in a social group, or political opinion.

Refugees have no home until another country agrees to allow them in. In 1994, the United Nations estimated that there were nearly 23 million refugees in the world, a ten-fold increase in two decades. An additional 26 million people have been forced to relocate from their homes but have not crossed an international boundary (Figure 9–15).

Afghanistan. As a result of the Soviet Union's invasion of Afghanistan in 1979, more than 4 million Afghans fled to refugee camps set up in neighboring countries. Because of a very high natural increase, an average of 2.6 percent since 1979, the population in the refugee camps swelled to more than 6 million by the late 1980s.

The Soviet Union withdrew its troops from Afghanistan in 1989, and the Soviet-installed government collapsed in 1992. Since then, many refugees have returned home, trading the security of the camps for the possibility of reclaiming their farms. However, 3.3 million refugees have not yet returned to Afghanistan because of fighting among rival ethnic groups for control of the country.

The Horn of Africa. Wars in four states in the Horn of Africa—Eritrea, Ethiopia, Somalia, and Sudan—have forced several million people to migrate. Eritrea fought a thirty-year war for independence from Ethiopia. During the civil war, an estimated 665,000 Eritrean refugees fled to Sudan. Beginning in the 1970s, Tigre, the province south of Eritrea, also fought for greater autonomy from Ethiopia.

In 1991 Tigrean-led rebels defeated the Ethiopian army and took control of the national government. Eritrea became an independent state in 1993, and many Eritrean refugees returned home. But a reduction in the number of refugees in Eritrea has been offset by increases elsewhere in Ethiopia, including Tigre in the north, Harar province in the east, and among the Oromo people in the south.

Ethiopia also suffered a war in the eastern region of Ogaden, a desert area also claimed by Somalia. The Ethiopian army uprooted several million native Somalis living in the Ogaden who traditionally brought their animals from the coast to the region during the wet season. According to inter-

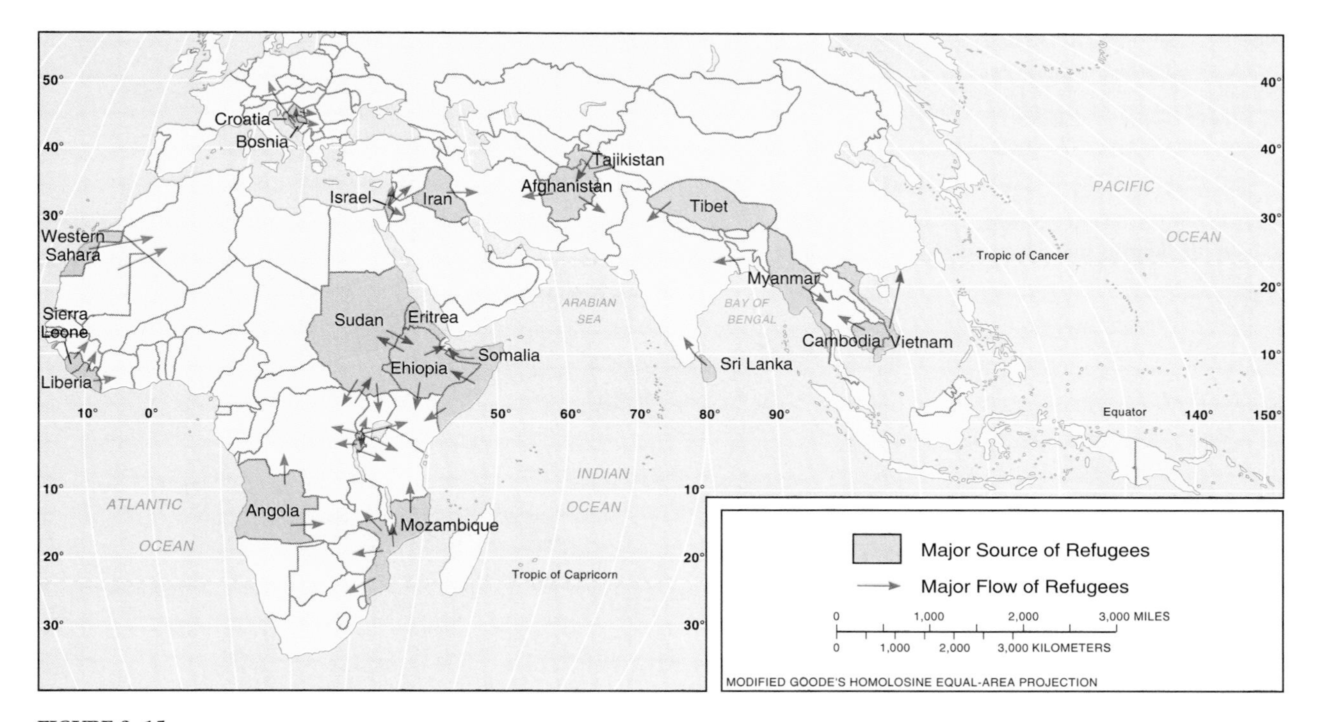

FIGURE 9–15
Political refugees. The largest numbers of refugees were generated in four locations: East Africa, the Middle East, Southeast Europe, and Afghanistan.

national refugee organizations, approximately 365,000 Ethiopians fled to Somalia, although Somali officials argue that more than 800,000 actually arrived (Figure 9–16).

A civil war in Sudan has raged since the late 1960s between largely Christian rebels in the southern provinces and the Muslim-dominated government forces in the north. More than 1 million Sudanese have been forced to migrate from the south to the north and another 350,000 have fled to Ethiopia.

The number of refugees increased in Somalia after the overthrow of the country's long-time dictator Mohammed Siad Barre. Barre brutally suppressed long-standing rivalries among Somalia's six major clans and dozens of sub-clans, but with the collapse of a national government in Somalia in 1991, clans and sub-clans fought each other for control over various regions of the country. As the armies of the individual clans and sub-clans seized food, property, and weapons, members of less-powerful clans and sub-clans were forced to migrate to seek safety and food.

Farther south along the coast of East Africa, Mozambique has been the scene of a civil war between the rebel Mozambique National Resistance force (also known as Renamo) and the Frelimo party, which controls the country's government. The civil war, which began in 1976, has generated 1.4 million refugees. Nearly two-thirds of the refugees have fled to neighboring Malawi, while the remainder have gone to South Africa, Swaziland, Tanzania, Zambia, and Zimbabwe. In addition, several million people have been forced to migrate within Mozambique as a result of the civil war.

In Africa, drought, famine, and other environmental push factors have also caused forced international migration. In countries like Ethiopia, political refugees are not always clearly distinguished from forced migrants seeking food and water.

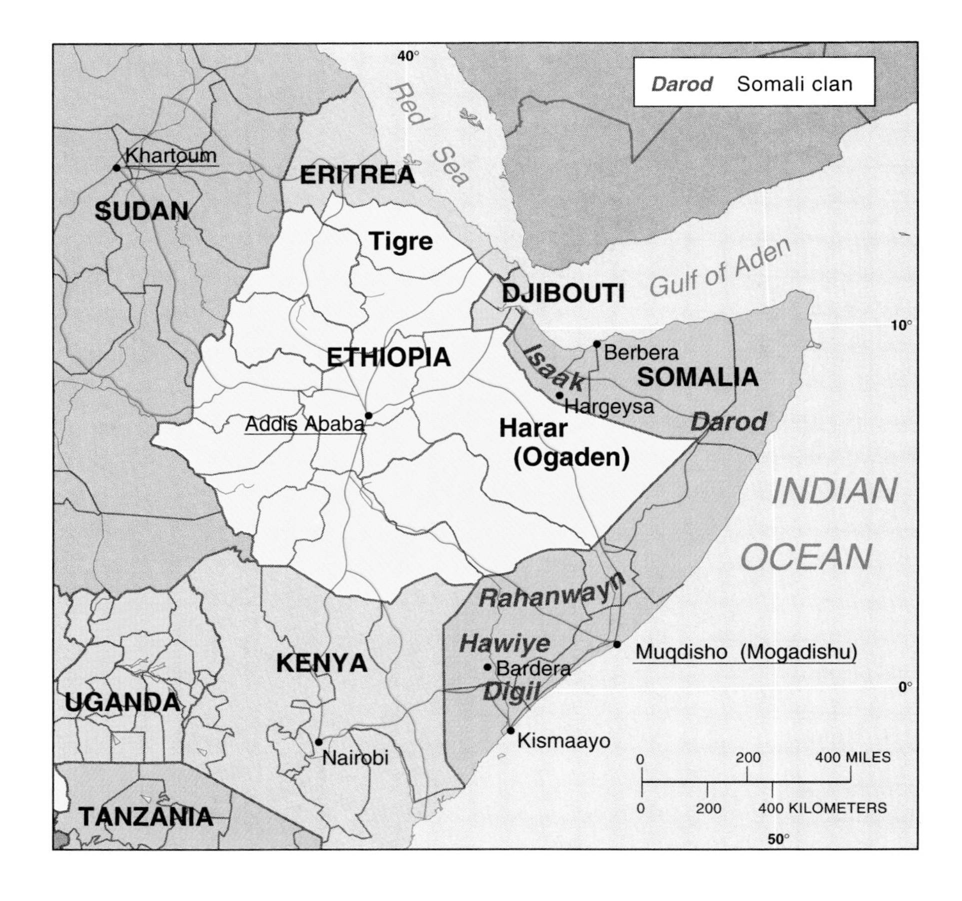

FIGURE 9–16
Refugees in the Horn of Africa. Native Somalis living in Ogaden have been forced to migrate to Somalia, while residents of Eritrea have been forced to migrate to Sudan. Somalis and Sudanese have been forced to move because of civil wars. Famine and drought have also forced many Eritreans, Ethiopians, Somalis, and Sudanese to migrate.

Somalis were forced to migrate during the early 1990s as a result of civil war among rival clans and sub-clans. The United States intervened militarily in 1992 and 1993 when international organizations were unable to deliver food to Somalis living in refugee camps. (Wendy Stone/Gamma-Liaison, Inc.)

Ethnic Cleansing to Create Nation-States

In recent years, the process of forcibly relocating less powerful nationalities in order to transform a multinational region into a region containing only one nationality has been termed **ethnic cleansing**. These political refugees may be placed on buses or trains and herded into detention camps. Others must walk across a border to a neighboring country willing to accept them. Many refugees from ethnic cleansing are killed by the more powerful nationality or die during the forced migration or in the camps.

Although the term was not used at the time, ethnic cleansing was widely practiced after World War II, when Germans, Poles, and other nationalities were forced to migrate. Altogether, approximately 45 million people migrated, primarily because of German and Japanese military expansions during the 1930s and early 1940s, the Allies' counterattack during World War II beginning in 1942, and post-war changes in the boundaries of the Soviet Union, Poland, Germany, and other Eastern European countries (Figure 9–17).

The 1947 partition of South Asia into two states, India and Pakistan, resulted in a massive level of forced migration, because the boundaries between the two did not correspond to the territory inhabited by the two main nationalities. The basis for the separation from India of West Pakistan and East Pakistan (now Bangladesh) was primarily religious. The people living in the two areas of Pakistan were predominantly Muslim, while those in India were predominantly Hindu. Antagonism between the two religious groups was so great that the British, who had held the region as a colony, decided to place the Hindus and Muslims in separate states. Approximately 17 million people caught on the wrong side of a boundary were forced to migrate (Figure 9–18).

Ethnic cleansing in Yugoslavia. In recent years, ethnic cleansing has been most widespread in Southeast Europe and East Africa. In Southeast Europe, about 4 million people have been forced to migrate as a result of ethnic cleansing following the carving up of Yugoslavia into a collection of independent states. The term has been applied primarily to the uprooting of Muslims by Serbs in Bosnia-Herzegovina.

Yugoslavia was created by the victorious allies after World War I as a union of several groups in the Balkans whose people spoke similar South Slavic languages but shared few other cultural characteristics. The most numerous cultural groups brought into Yugoslavia were the Serbs and Croats; others included Slovenes, Macedonians, and Montenegrens. The prefix *Yugo* in the country's name derived from the Slavic word for south.

The Balkans had long been on the frontier of conflict between the Roman Catholic Austria-Hungary Empire to the north and the Muslim Ottoman Empire to the south. The Croats and Slovenes were incorporated into the Austria-Hungary Empire, while the southern portion of the future Yugoslavia was ruled by the Ottomans. The Serbs, having won independence in 1817, considered themselves the leading Slavic defenders against the two large empires to the north and south. Austria-Hungary extended its rule farther south in 1878 to include Bosnia-Herzegovina, even

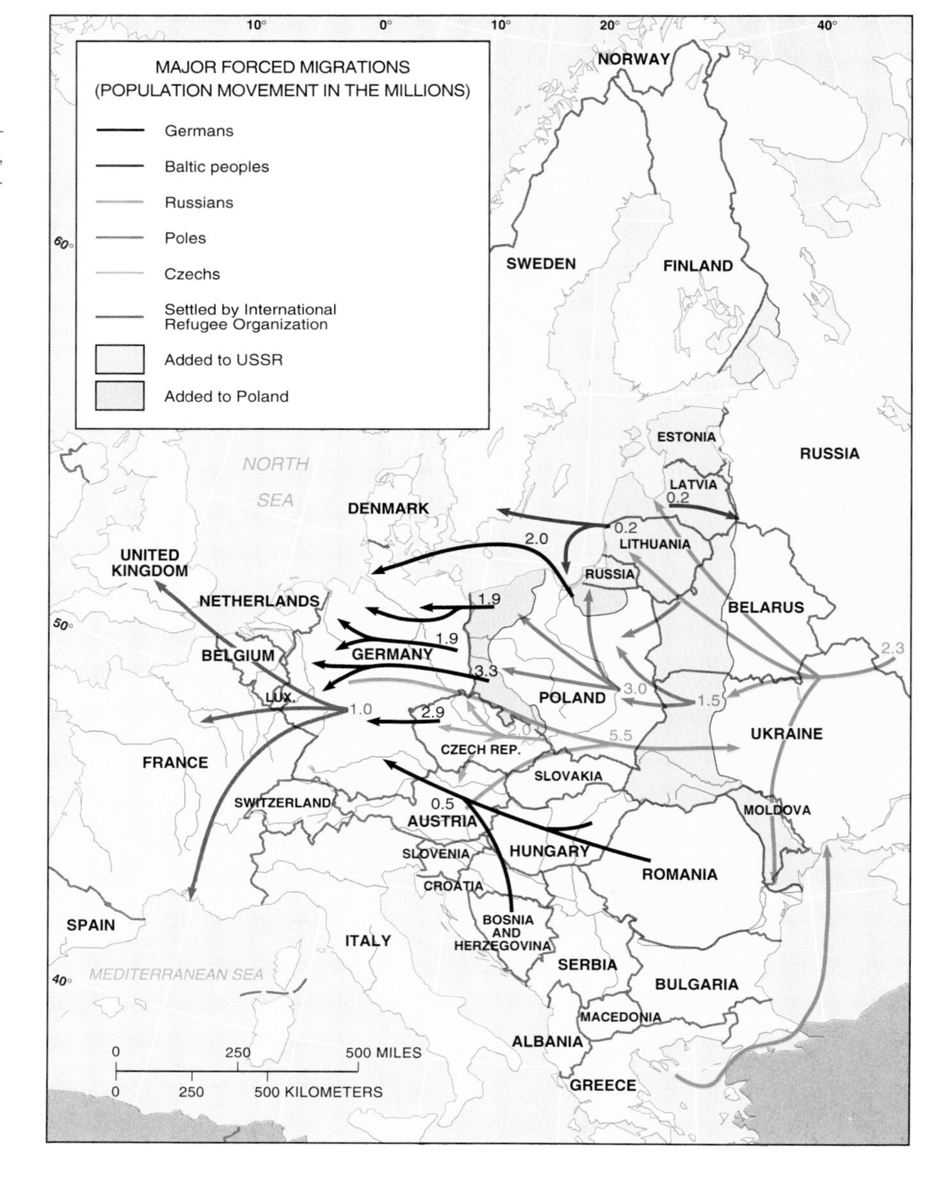

FIGURE 9–17
Forced migration as a result of territorial changes after World War II. The largest number of refugees were Poles forced to move from territory occupied by the Soviet Union, Germans forced to migrate from territory taken over by Poland and the Soviet Union, and Russians forced to return to the Soviet Union from Western Europe.

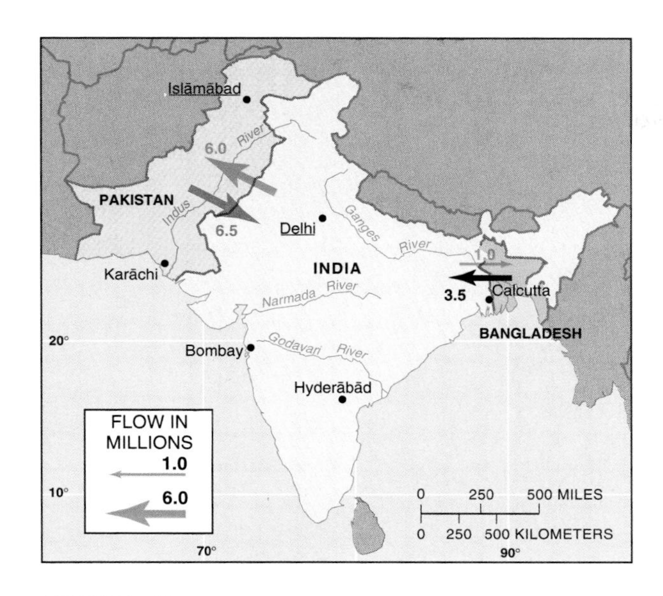

FIGURE 9–18
Forced migration in South Asia. The 1947 partition of British India into two independent states, India and Pakistan, resulted in the forced migration of an estimated 17 million. The creation of Pakistan as two territories nearly 1,600 kilometers (1,000-square miles) apart proved unstable, and in 1971 East Pakistan became the independent country of Bangladesh.

though the majority of the people had been converted to Islam by the Ottomans. In June 1914, the heir to the throne of Austria-Hungary was assassinated in Sarajevo by a Serb who sought independence for Bosnia. The incident ultimately led to World War I (Figure 9–19).

The creation of the country of Yugoslavia after World War I brought the region relative stability that lasted for most of the twentieth century. In particular, under the long period of leadership of Josip Broz Tito, who governed Yugoslavia from 1953 until his death in 1980, old animosities among nationalities were submerged, and younger people in particular began to identify themselves primarily as Yugoslavs rather than as Serbs, Croats, or Montenegrens.

Rivalries among nationalities resurfaced in Yugoslavia during the 1980s after Tito's death, leading to the breakup of the country in the early 1990s. Not only did the boundaries of Yugoslavia's six republics fail to match the territory occupied by nationalities, the country contained other important nationalities that had not received official recognition (Figure 9–20). As long as Yugoslavia comprised one

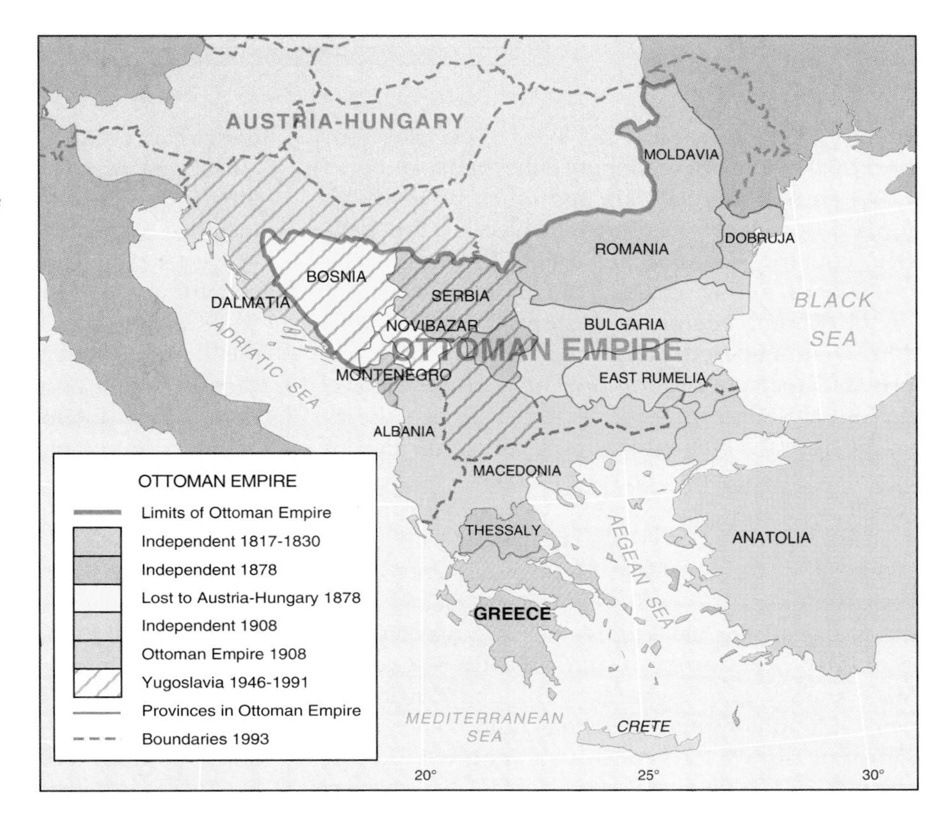

FIGURE 9–19
The Balkans in 1914. At the outbreak of World War I, Austria-Hungary controlled the northern part of the region, including all or part of Croatia, Slovenia, and Romania. The Ottoman Empire controlled some of the south, although during the nineteenth century it had lost control of Albania, Bosnia-Herzegovina, Greece, Romania, and Serbia.

FIGURE 9–20
Until its breakup in 1992, Yugoslavia comprised six republics (plus Kosovo and Vojvodina autonomous regions within the Republic of Serbia). Bosnia-Herzegovina, Croatia, Macedonia, and Slovenia have split off into independent countries.

country, nationalities had not been especially troubled by the name of the republic they inhabited.

When Yugoslavia's six republics were transformed from local government units into independent countries, nationalities fought to redefine the boundaries. Serbia attacked Croatia to gain control of the Krajina region of eastern Croatia, where Serbs outnumbered Croats. Albanians, who comprised 77 percent of the population in the Kosovo region of southern Serbia, fought to free themselves from the cultural and political domination of Serbs, who retained control of Kosovo because it was their historic homeland.

The creation of a viable country proved especially difficult in the case of Bosnia-Herzegovina, because Bosnian was never defined clearly as a nationality in the old Yugoslavia. Yugoslavia's five officially recognized nationalities—Croats, Macedonians, Montenegrens, Serbs, and Slovenes—were able to constitute majorities in the other independent states carved out of Yugoslavia. In contrast, at the time of the breakup of Yugoslavia, the largest group of people in Bosnia-Herzegovina, 40 percent, were classified not by nationality but as Muslims, while the remainder of the republic included 32 percent Serbs and 18 percent Croats.

Rather than live in an independent multinational country with a Muslim plurality, Bosnia-Herzegovina's Serbs and Croats fought to unite the portions of the republic that they inhabited with Serbia and Croatia. Aided by troops and material from neighboring Serb-dominated Yugoslavia, Bosnian Serbs gained control of much of Bosnia-Herzegovina, killing Muslims and forcing them from their homes in the captured territory.

Ethnic cleansing in East Africa. Ethnic cleansing has also been extensive in East Africa, primarily in Rwanda, where 40 percent of the country's 8.5 million inhabitants were killed or forced to migrate in a civil war between ethnic groups beginning in 1994. Violence followed a plane crash in April 1994, in which the presidents of Rwanda and neighboring Burundi were killed.

Muslims living in Bosnia-Herzegovina have been victims of ethnic cleansing by Serbs. Muslims were uprooted from their villages and forced to migrate. The Serbs have herded Muslim men considered capable of fighting back into camps, such as this one at Manjaca, near Banja Luka. (Daniels Marleen/Gamma-Liaison, Inc.)

The plane crash destroyed a peace accord being negotiated at the time between Rwanda's two ethnic groups, the Hutus and Tutsis. Hutus comprised 86 percent of the population, but when Rwanda was a colony of Belgium, Tutsis enjoyed positions of power. After Rwanda gained its independence in 1962, the Hutus gained control of the government and suppressed several uprisings by the Tutsis.

After the plane crash, the Hutu-dominated military killed many Tutsis, as well as moderate Hutu members of the government who favored reconciliation with the Tutsis. Tutsi rebels formed the Rwanda Patriotic Front to fight the Hutu-controlled army. Within a few months, the Tutsi rebels had won the civil war, and millions of Hutus had fled into neighboring Zaire and Tanzania (Figure 9–21).

Violence spread during the summer of 1994 to neighboring Burundi, where 85 percent of the population was Hutu and 15 percent Tutsi. As in Rwanda, Hutus controlled the government before the president died in the April 1994 plane crash. Attempts failed to create a broad-based government with both Hutu and Tutsi representatives.

Cooperation Among States

Most nationalities live peacefully with their neighbors. States encourage peaceful cooperation among nationalities through adoption of governmental systems that give power to their various nationalities and through joining other states to form international economic and military organizations.

Internal Organization of States

The governments of states are organized according to one of two approaches—the unitary system or the federal system. The **unitary state** places most power in the hands of central government officials, whereas the **federal state** allocates strong power to units of local government within the country.

In principle, the unitary government system works best in countries that have relatively few internal cultural differences and a strong sense of national unity. Therefore, states whose boundaries coincide closely with the boundaries of nations are more likely to consider a unitary system of

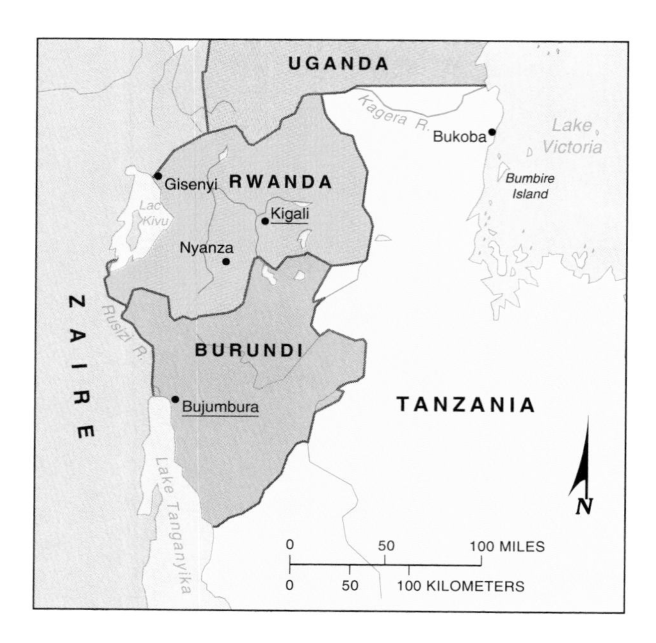

FIGURE 9–21
Burundi and Rwanda, countries with smaller land areas and populations than Maryland, have suffered from ethnic cleansing in recent years, as a result of clashes between the Hutus—who comprise about 85 percent of the population—and the Tutsis.

government. In addition, because the unitary system requires effective communications with all regions of the country, smaller states are more likely to adopt it. If the country is very large or has isolated regions, strong national control is more difficult.

In reality, multinational states often have unitary systems, so that the values of one nationality can be imposed on others. In a number of African and Asian countries, for instance, the mechanisms of a unitary state have enabled one ethnic group to extend dominance over weaker groups. In other cases, a minority group has been able to impose its values on the majority of the population. When Communist parties controlled the governments, Eastern European countries had unitary systems, so that the uniform cultural values could be imposed in otherwise multinational societies.

In a federal state, local governments possess more authority to adopt their own laws. Multinational states usually adopt a federal system of government in order to give power to different nationalities, especially if they live in separate regions of the country. Under a federal system, local government boundaries can be drawn that correspond to the regions inhabited by different nations. The federal system is also more suitable for very large states. The national capitals of very large states may be too remote to provide effective control over isolated regions.

In recent years, there has been a strong global trend in favor of the federal government alternative. Most of the world's largest states were already federal, including Russia, Canada, the United States, Brazil, and India. During the late 1980s and 1990s, unitary systems have been sharply curtailed in a number of countries and scrapped altogether in others.

International Organizations

Few countries choose to stand alone in the modern world. In addition to the United Nations, most countries have joined regional organizations.

United Nations. The most important international organization is the United Nations, created at the end of World War II by the victorious Allies. Switzerland and Taiwan are the most populous territories on Earth's surface not in the United Nations. Switzerland has traditionally avoided membership in most international organizations. Taiwan resigned when the United Nations voted to admit the People's Republic of China in 1971, because the government of Taiwan still considered itself the proper ruler of the Chinese mainland.

The United Nations replaced another organization known as the League of Nations, established after World War I. The League was never an effective peace-keeping organization. The United States did not join, despite the fact that President Woodrow Wilson initiated the idea. The United States Senate refused to ratify the membership treaty because of widespread preference in the country to remain isolated from entanglements and alliances with other countries. By the 1930s, Germany, Italy, Japan, and the Soviet Union all withdrew, and the League could not stop aggression by these states against neighboring countries.

United Nations members can vote to establish a peace-keeping force and request states to contribute soldiers. During the Cold War era, U.N. peace-keeping efforts were often stymied because any one of the five permanent members of the Security

Council—China, France, Russia (formerly as the Soviet Union), the United Kingdom, and the United States—could veto the operation. During the 1990s, the United Nations has played an increasingly important role in separating warring groups in a number of countries, including in Eastern Europe, Africa, Asia, and the Middle East. In the past, one of the superpowers or permanent members usually vetoed U.N. action, preferring to intervene directly.

Regional organizations. The Organization of American States (OAS) includes thirty-two of the thirty-three states in the Western Hemisphere; Canada is not a member. Cuba is a member but was suspended from most OAS activities in 1962. The organization's headquarters, including the permanent council and general assembly, are located in Washington, DC. The OAS promotes social, cultural, political, and economic links among member states.

A similar organization exists in Africa, known as the Organization for African Unity (OAU). Founded in 1963, OAU includes every African state with the exception of South Africa. The organization's major issue has been the elimination of minority white-ruled governments in southern Africa, which was accomplished in the early 1990s.

The Commonwealth of Nations includes the United Kingdom and 48 other states that were once British colonies, including Australia, Bangladesh, Canada, India, Nigeria, and Pakistan. Most of the other members are African states or island countries in the Caribbean or Pacific. Commonwealth members seek economic and cultural cooperation.

The Nonaligned Movement, founded in 1961, has approximately 100 members, including nearly every country in Africa and the Middle East. The organization was established as a haven to countries that did not wish to be forced into an alliance with one of the superpowers. With the end of the Cold War, the initial purpose of the organization has disappeared. Instead, the principal focus has become to represent the interests of countries that are less developed economically. Such a change reflects the growing importance of economic competition among blocs of countries rather than military conflict. The Nonaligned Movement may merge with the Group of 77, another organization of developing countries established in the 1960s.

Military alliances in Europe. The importance of the nation-state has diminished in Western Europe, the world region most closely associated with the development and implementation of the concept during the past 200 years. European nation-states have put aside their centuries-old rivalries to forge the world's most powerful economic union, as we will discuss later.

After World War II, most European states joined one of two military alliances dominated by the two superpowers of the United States and U.S.S.R. The North Atlantic Treaty Organization (NATO) was a military alliance among 16 states, including the U.S. and Canada, plus fourteen European states. Twelve European states—Belgium, Denmark, West Germany, Greece, Iceland, Italy, Luxembourg, the Netherlands, Norway, Portugal, Turkey, and the United Kingdom—participated fully in NATO, while France and Spain were members that did not contribute troops. NATO headquarters, originally in France, were moved to Belgium when France reduced its involvement.

The Warsaw Pact was a military agreement among Eastern European countries to defend each other in case of attack. Seven members joined the Warsaw Pact when it was founded in 1955, including the Soviet Union, Bulgaria, Czechoslovakia, East Germany, Hungary, Poland, and Romania. In 1956, some of Hungary's leaders asked for the help of Warsaw Pact troops to crush an uprising that threatened Communist control of the government. Warsaw Pact troops also invaded Czechoslovakia in 1968 to depose a government committed to creating "communism with a human face."

NATO and the Warsaw Pact were designed to maintain a bipolar balance of power in Europe. For NATO allies, the principal objective was to prevent the Soviet Union from overrunning West Germany and other smaller countries. The Warsaw Pact provided the Soviet Union with a buffer of allied states between it and Germany to discourage a third German invasion of the Soviet Union in the twentieth century.

In a Europe no longer dominated by military confrontation between two blocs, the Warsaw Pact and NATO became obsolete. The number of troops under NATO command was sharply reduced, and the Warsaw Pact was disbanded. The Conference on Security and Cooperation in Europe (CSCE), which had been founded in 1975 by Western European countries, was expanded to more than 50 countries, including the United States, Canada, and former republics of the Soviet Union. The CSCE had played a limited role during the Cold War era, but during the 1990s it became a forum for all countries concerned with preventing or ending conflicts in Europe, especially in the Balkans and Caucasus. Although the CSCE does not directly command armed forces, it can call upon member states to supply troops if necessary.

Economic alliances in Europe. With the decline in the military-oriented alliances, European states increasingly turned to economic cooperation. In 1949, the same seven Eastern European states in the Warsaw Pact formed the Council for Mutual Economic Assistance (COMECON). Cuba, Mongolia, and Vietnam were also members of the alliance, which was designed to promote trade and sharing of natural resources. Like the Warsaw Pact, COMECON disbanded in the early 1990s after the fall of Communism in Eastern Europe.

Western Europe's most important economic organization is the European Union. When it was established in 1958, the European Economic Community, as it was then called, included six countries: Belgium, France, West Germany, Italy, Luxembourg, and the Netherlands. Membership was widened to include Denmark, Ireland, and the United Kingdom in 1973, Greece in 1981, Portugal and Spain in 1986, and Austria, Finland, Norway, and Sweden in 1995 (Figure 9–22). In addition to the European Union, a European Parliament is elected by the people in each of the member states simultaneously.

The main task of the European Union is to promote development within the member states through economic cooperation. At first, the European Union played a limited role, such as providing subsidies to farmers and to depressed regions such as southern Italy. However, the organization has taken on more importance in recent years, as member states seek greater economic and political cooperation. Virtually all barriers to free trade among member states have been removed, as have barriers to the movement of people and capital. In the future, Western Europeans are likely to carry European Union passports and calculate costs in European currency units (ECUs) rather than in pounds, francs, and marks. The effect of these actions has been to turn Western Europe into a wealthier and more populous market than the United States.

Travelers crossing international borders in Western Europe can still observe cultural differences in the organization of the landscape. For examples, highways in the Netherlands are more likely than those in neighboring Belgium to be flanked by well-manicured vegetation and paths reserved for bicycles. From economic and political perspectives, though, boundaries between Western European countries, where hundreds of thousands of soldiers once stood guard, now have little more significance than boundaries between states inside the United States.

While achieving a high degree of economic integration among its members, the European Union faces a challenge of working with other European countries, many of which have applied for membership. The European Union recognizes an obligation to foster economic development throughout Europe, but admitting a large number of relatively poor countries in Southern and Eastern Europe would create an administrative nightmare and dilute the economic benefits that current members enjoy.

At the same time that residents of Western European countries display increased tolerance for the cultural values of neighboring nationalities, opposition has increased to the immigration of people from the south and east, especially those who have darker skins and adhere to Islam. Immigrants from poorer regions of Europe, Africa, and Asia, fill low-paying jobs such as cleaning streets and operating buses that Western Europeans are not willing to perform, as we discussed in Chapter 7. Nonetheless, many Western Europeans fear that large-scale immigration will transform their nation-states into multinational societies.

Underlying this fear of immigration is recognition that natural increase rates are higher in most African and Asian countries than in Western Europe as a result of higher crude birth rates. Many Western Europeans believe that Africans and Asians who immigrate to their countries will continue to maintain relatively high crude birth rates and consequently will constitute even higher percentages of the population in Western Europe in the future.

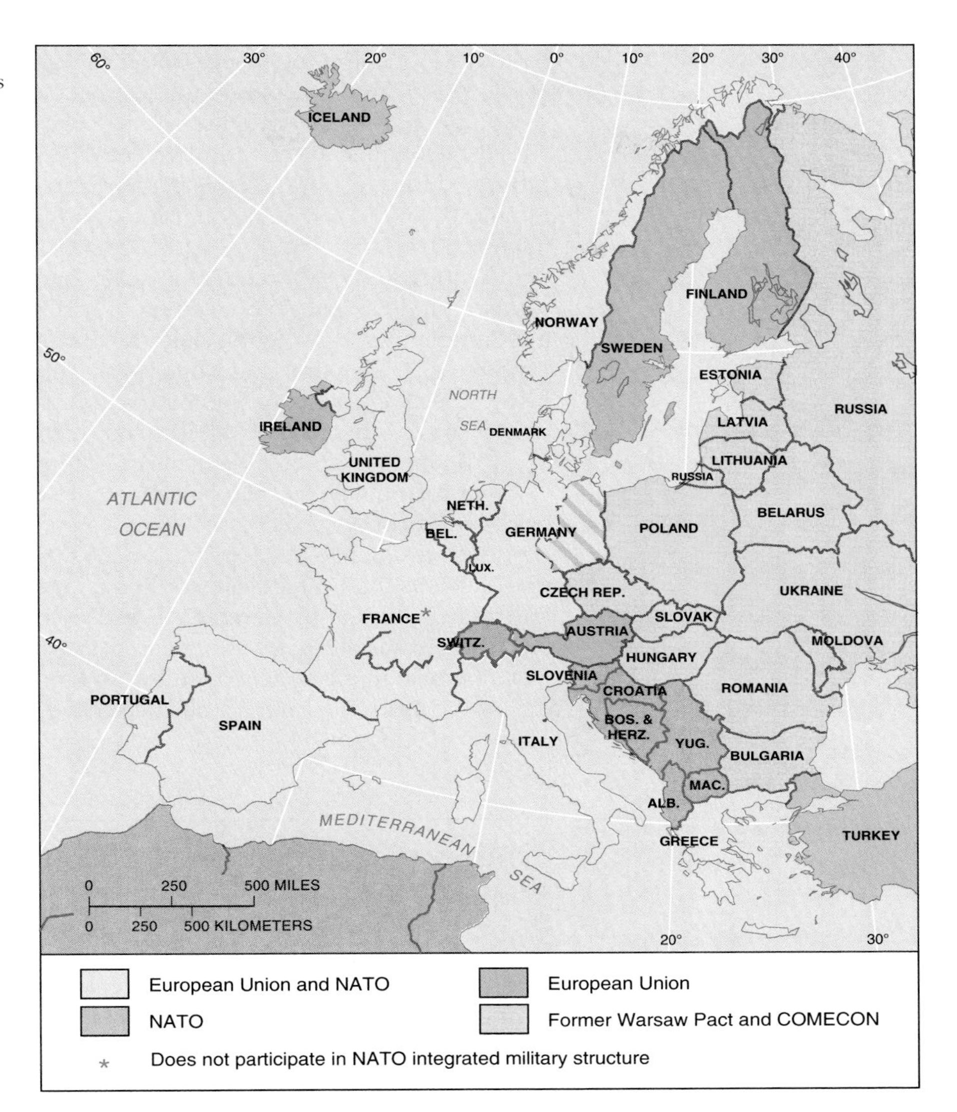

FIGURE 9–22
Twelve Western European countries joined both the European Union (EU) and the North Atlantic Treaty Organization. Austria, Finland, Ireland, and Sweden are members of the European Union but not NATO, while Iceland and Turkey are in NATO but not the EU. Seven Eastern European countries joined an economic alliance, known as COMECON, and a military alliance, known as the Warsaw Pact. Once East Germany, the Soviet Union, and Czechoslovakia ceased to exist and the other Eastern European states adopted noncommunist governments, the Warsaw Pact and COMECON were dismantled. A number of eastern and southern European countries would like to join the EU but must enact economic and political reforms before their applications can be seriously considered.

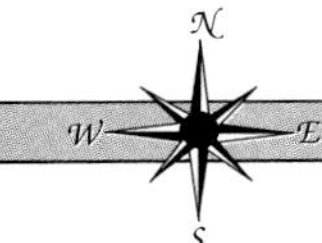

CONCLUSIONS: CRITICAL ISSUES FOR THE FUTURE

The era of a bipolar balance of power formally ended when the Soviet Union was disbanded in 1992. After a half-century dominated by the Cold War between two superpowers, the United States and the Soviet Union, the world has entered a period characterized by an unprecedented increase in the number of new states created to satisfy the desire of nationalities for self-determination.

From the end of World War II in 1945 until the early 1990s, support for or opposition to Communism and economic cooperation were more important political factors than the nation-state principle in Europe. Until they lost power in the late 1980s and early 1990s, Communist leaders in

Eastern Europe and the Soviet Union strongly discouraged nationalities from expressing their cultural uniqueness. Writers and artists were pressured to conform to a style known as "socialist realism," which emphasized socialist economic and political values. Use of the Russian language was promoted throughout the Soviet Union and taught as the second language in other Eastern European countries, and the role of organized religion was minimized.

During the 1990s, the nation-state concept has revived, especially in southern and Eastern Europe, where it had been suppressed for 40 years by the Communists. New nation-states have been carved out of the Soviet Union, Czechoslovakia, and Yugoslavia. Turmoil has resulted because in many cases the boundaries of the new states do not match— and in fact, cannot match—the territories occupied by distinct nationalities.

The most important elements of state power are increasingly economic rather than military. Japan has joined the ranks of major powers on the basis of its economic successes, while Russia has slipped largely due to economic problems. The leading superpower in the 1990s is not a single state, such as the United States or Russia, but an economic union of European states led by Germany. Freed from their history of military conflict, European states now cooperate with each other. At the same time, they compete in a global economy with Japan and North America. With the end of the Cold War, military alliances have become less important than patterns of economic cooperation and competition among states.

Chapter Summary

1. Differences between states and nations
A state is a political unit, with an organized government and sovereignty. Most of Earth's surface is allocated to states, and only a handful of colonies and tracts of unorganized territory remain. A nation or nationality is a group of people with a strong sense of cultural unity. In the modern world, states have been created to match the distribution of nations whenever possible.

2. Boundaries between states
Boundaries between states, where possible, are drawn to coincide either with physical features, such as mountains, deserts, and bodies of water, or with such cultural characteristics as geometry, religion, and language. Boundaries affect the shape of countries and affect the ability of a country to live peacefully with its neighbors.

3. Problems matching states and nations
Problems arise when the boundaries of states do not coincide with the boundaries of nationalities. In some cases, one state contains more than one nation, while in other cases, one nationality is split among two or more states. In some regions, more powerful nationalities have forced other nationalities to migrate, creating large numbers of refugees.

4. Cooperation among states
Nearly all states have joined regional and international organizations in order to settle differences peacefully and to restore order in regions with

conflicts among nationalities. Following World War II, the United States and the Soviet Union formed military alliances in Europe. With the end of the Cold War, nationalities are cooperating with each other in Western Europe, primarily to promote economic growth.

Key Terms

Apartheid, 367
Boundary, 360
Centrifugal forces, 358
Centripetal forces, 358
City-state, 353
Colonialism, 354
Colony, 354
Compact state, 360
Elongated state, 361
Ethnic cleansing, 373
Federal state, 377
Fragmented state, 361
Frontier, 360
Landlocked state, 362
Multinational state, 365
Nationalism, 358
Nationality (or **nation**), 355
Nation-state, 356
Perforated state, 361
Prorupted state, 361
Refugees, 370
Self-determination, 356
Sovereignty, 349
State, 348
Unitary state, 377

Questions for Study and Discussion

1. Outline characteristics of a state.
2. Outline characteristics of a nation. What is the difference between a state and a nation? What is a nation-state?
3. Describe systems of controlling territory in ancient and medieval times.
4. What is colonialism?
5. What is the difference between centrifugal forces and centripetal forces?
6. Outline the two types of boundaries.
7. Describe the five shapes of countries.
8. Discuss problems faced by landlocked states.
9. Why is Russia not a good example of a nation-state?
10. What was apartheid?
11. What are refugees?
12. What is ethnic cleansing?
13. What is the difference between a unitary state and a federal state?
14. What are the functions of the major regional organizations in Europe?

Thinking Geographically

1. In his book *1984*, George Orwell envisaged the division of the world into three large unified states, held together through technological controls. To what extent has Orwell's vision of a global political arrangement been realized?
2. In the winter 1992–1993 issue of *Foreign Policy*, Gerald Helman and Steven Ratner identified countries that they called failed nation-states, including Cambodia, Liberia, Somalia, and Sudan, and others that they predicted would fail. Helman and Ratner argued that the governments of these countries were maintained in power during the Cold War era

through massive military and economic aid from the United States or the Soviet Union. With the end of the Cold War, these failed nation-states have sunk into civil wars, fought among groups of people who have different languages, religions, and other cultural characteristics. What obligations do other countries, especially countries that formerly supported these states, have to restore order in failed nation-states?

3. Given the movement toward increased local government autonomy on the one hand and increased authority for international organizations on the other hand, what is the future of the nation-state? Have political and economic trends in the 1990s strengthened the concept of nation-state, or weakened it?
4. To what extent does national identity derive from economic interests rather than from such cultural characteristics as language and religion?
5. A century ago, the British geographer Halford J. Mackinder identified a heartland in the interior of Eurasia (Europe and Asia) that was isolated by mountain ranges and the Arctic Sea. Surrounding the heartland was a series of fringe areas, which the geographer Nicholas Spykman later called the rimland, oriented towards the oceans. Mackinder argued that whoever controlled the heartland would control Eurasia and hence the entire the world. To what extent has Mackinder's theory been validated during the twentieth century by the creation and then the dismantling of the Soviet Union?
6. The world has been divided into a collection of countries on the basis of the principle that nationalities have the right of self-determination. National identity, however, derives from economic interests as well as from such cultural characteristics as language and religion. To what extent should a country's ability to provide its citizens with food, jobs, economic security, and material wealth, rather than the principle of self-determination, become the basis for dividing the world into independent countries?

Suggestions for Further Reading

Arlinghaus, Sandra L., and John D. Nystuen. "Geometry of Boundary Exchanges." *Geographical Review* 80 (January 1990): 21–31.

Bascom, Johnathan. "The Peasant Economy of Refugee Resettlement in Eastern Sudan." *Annals of the Association of American Geographers* 83 (1993): 320–46.

Bennett, D. Gordon, ed. *Tension Areas of the World: A Problem Oriented World Regional Geography*. Champaign, IL: Park Press, 1982.

Boal, Frederick W., and J. Neville H. Douglas, eds. *Integration and Division: Geographical Perspectives on the Northern Ireland Problem*. London and New York: Academic Press, 1982.

Brown, Curtis M., Walter G. Robillard, and Donald A. Wilson. *Boundary Control and Legal Principles*. New York: John Wiley and Sons, 1986.

Burghart, A. F. "The Bases of Territorial Claims." *Geographical Review* 63 (April 1973): 225–45.

Burnett, Alan D., and Peter J. Taylor, eds. *Political Studies from Spatial Perspectives*. Chichester: John Wiley and Sons, 1981.

Busteed, M. A., ed. *Developments in Political Geography*. London: Academic Press, 1983.

Christopher, A. J. *The British Empire at Its Zenith*. London: Croom Helm, 1988.

Cohen, Saul B. "Global Political Change in the Post-Cold War Era." *Annals of the Association of American Geographers* 81 (December 1991): 551–80.

Cox, Kevin R. *Location and Public Problems: A Political Geography of the Contemporary World*. Chicago: Maaroufa Press, 1979.

Dale, E. H. "Some Geographical Aspects of African Land-Locked States." *Annals of the Association of American Geographers* 58 (September 1968): 485–505.

Dikshit, R. D. "Geography and Federalism." *Annals of the Association of American Geographers* 61 (March 1971): 97–130.

Gottmann, Jean, ed. *Centre and Periphery: Spatial Variation in Politics*. Beverly Hills: Sage, 1980.

Johnston, R. J. *Geography and the State*. New York: St. Martin's Press, 1982.

Kidron, Michael, and Ronald Segal. *The New State of the World Atlas*, 4th ed. New York: Simon and Schuster, 1991.

Kliot, Nurit, and Stanley Waterman, eds. *Pluralism and Political Geography—People, Territory and State*. New York: St. Martin's Press, 1983.

Mathieson, R. S. "Nuclear Power in the Soviet Bloc." *Annals of the Association of American Geographers* 70 (June 1980): 271–79.

Murphy, Alexander B. "Historical Justifications for Territorial Claims." *Annals of the Association of American Geographers* 80 (December 1990): 531–48.

______, "Territorial Policies in Multiethnic States." *Geographical Review* 79 (October 1989): 410–21.

Newland, Kathleen. "Refugees: The New International Politics of Displacement." *Worldwatch Paper* 43. Washington, DC: Worldwatch Institute, March 1981.

O'Loughlin, John, and Herman van der Wusten. "Political Geography of Panregions." *Geographical Review* 80 (January 1990): 1–20.

O'Sullivan, Patrick. *Geopolitics*. New York: St. Martin's Press, 1986.

______and Jesse W. Miller. *The Geography of Warfare*. London: Croom Helm, 1983.

Pacione, Michael, ed. *Progress in Political Geography*. London: Croom Helm, 1985.

Parker, W. H. *Mackinder: Geography as an Aid to Statecraft*. Oxford: The Clarendon Press, 1982.

Pickles, John, and Jeff Woods. "South Africa's Homelands in the Age of Reform: The Case of QwaQwa." *Annals of the Association of American Geographers* 82 (December 1992): 629–52.

Prescott, J. R. V. *Boundaries and Frontiers*. London: Croom Helm, 1978.

______, *The Geography of State Politics*. Chicago: Aldine Publishing, 1968.

______, *Political Geography*. London: Methuen, 1972.

______, *The Political Geography of the Oceans*. Newton Abbot, England: David and Charles, 1975.

Richmond, Anthony H. "Ethnic Nationalism: Social Science Paradigm." *International Social Science Journal* 39 (February 1987): 3–18.

Rogge, John R., ed. *Refguees: A Third World Dilemma*. Totowa, NJ: Rowman and Littlefield, 1987.

Rose, Richard. "National Pride in Cross-National Perspective." *International Social Science Journal* 37 (February 1985): 85–96.

Soffer, Arnon, and Julian V. Minghi. "Israel's Security Landscapes: The Impact of Military Considerations on Land Uses." *Professional Geographer* 38 (February 1986): 28–41.

Taylor, Peter J. *Political Geography of the Twentieth Century: A Global Analysis*. London: Belhaven, 1993.

______, and John W. House, eds. Political Geography: *Recent Advances and Future Directions*. London: Croom Helm, 1984.

Williams, Allan M. *The European Community: The Contradictions of Integration*. Cambridge, MA: Blackwell, 1991.

Zelinsky, Wilbur. *Nation into State*. Chapel Hill: University of North Carolina Press, 1988.

We also recommend these journals: *American Journal of Political Science; American Political Science Review; Foreign Affairs; Foreign Policy; International Affairs; International Journal; International Journal of Middle East Studies; Political Geography; Post-Soviet Geography*.

10 Development

- Indicators of development
- Regional differences in development
- Promoting development
- Obstacles to development

Brickmaking in China.

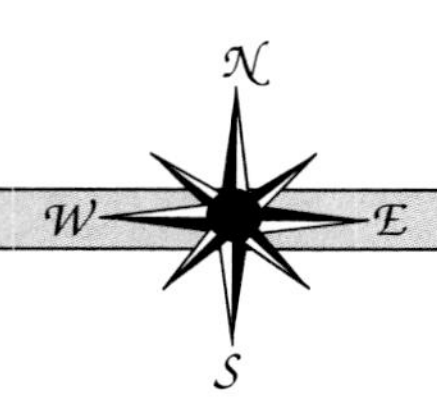

1. Indicators of development
Countries can be classified according to level of development, which is a measure of their economic and social growth. Economic, social, and demographic indicators help to distinguish between relatively developed and developing regions.

2. Regional differences in development
Anglo-America, Western Europe, and Japan form a relatively developed core. Developing regions include Latin America, East Asia, Southeast Asia, South Asia, the Middle East, and Sub-Saharan Africa.

3. Promoting development
Countries generally take one of two approaches to promote their development—international trade or self-sufficiency. International trade has been the alternative increasingly adopted by countries in recent years, because it has been viewed as the more successful approach.

4. Obstacles to development
Adopting the international trade approach has generated two types of problems for many developing countries: problems in paying for development and adverse environmental impacts from some development projects.

INTRODUCTION

The end of the Cold War, the dismantling of the Soviet Union, and the demise of communism in Eastern Europe and central Asia have fundamentally altered the nature of global competition among states. World-wide confrontation between superpowers armed with nuclear weapons has been replaced by economic competition among blocs of countries. Political unrest continues in localities where the boundaries between states and nationalities do not correspond, rather than at a global scale.

The most fundamental economic distinction among world regions is their level of **development,** which is a process of improvement in the material conditions of people through diffusion of knowledge and technology. Economic development is not a process with a beginning and end. Rather, it is a continuous process involving a never-ending series of actions intended to promote constant improvement in the health and prosperity of the people.

Every country falls at some point in a continuum of economic development. However, countries tend to cluster at one of two ends of the continuum. A more developed (MDC) or **relatively developed country** has progressed further along the development continuum. A country at an earlier stage in the process of development frequently is called a less-developed country (LDC), but many analysts prefer to refer to it as **developing**. The term *developing* implies that the country has already made some progress and expects to continue the development process in the future.

To some geographers, the economies of relatively developed core regions have appeared to be exploiting the people and resources of developing peripheral regions. But from the perspective of people in developing regions, integration into a global economy through trade with relatively developed countries may be a worthwhile price to pay to receive material benefits of development, such as steady work, imported goods, and connection to the world through television.

Geographers find that countries that have achieved relatively high levels of development cluster in some regions of the world, while developing countries are located in other regions. The regions geographers identify according to prevailing level of development correspond closely to the nine regions distinguished according to cultural characteristics in Chapter 8.

Relatively developed regions include Anglo-America, Western Europe, and Eastern Europe, plus three other areas—South Pacific, Japan, and South Africa. Developing regions include Latin America, East Asia, Southeast Asia, the Middle East, South Asia, and Sub-Saharan Africa.

In geographic terms, developing countries form a periphery in the global economy, surrounding a core that consists of the relatively developed regions of Anglo-America, Western Europe, and Japan. To create conditions that encourage trade, developing countries adopt economic policies consistent with those in the relatively developed core regions, such as the sale of public utilities to private corporations and measures to control inflation. Industries in developing regions may be producing primarily for sale in relatively developed countries, and developing countries may be more likely to buy products made in relatively developed countries. Developing countries may be forced to reduce spending on social welfare programs and accept greater differentials between their wealthiest and poorest people.

Indicators of Development

The United Nations attempts to measure the level of development each year for every country with more than 1 million inhabitants. The U.N. figure, called the Human Development Index (HDI), combines several indicators of development, including economic, social, and demographic. The highest HDI possible is 1.0. Any measure of development is arbitrary, but the HDI is useful because it recognizes that the process of development includes social and demographic as well as economic changes. According to the 1993 rankings, Japan had the highest HDI, while Guinea had the lowest (Figure 10–1).

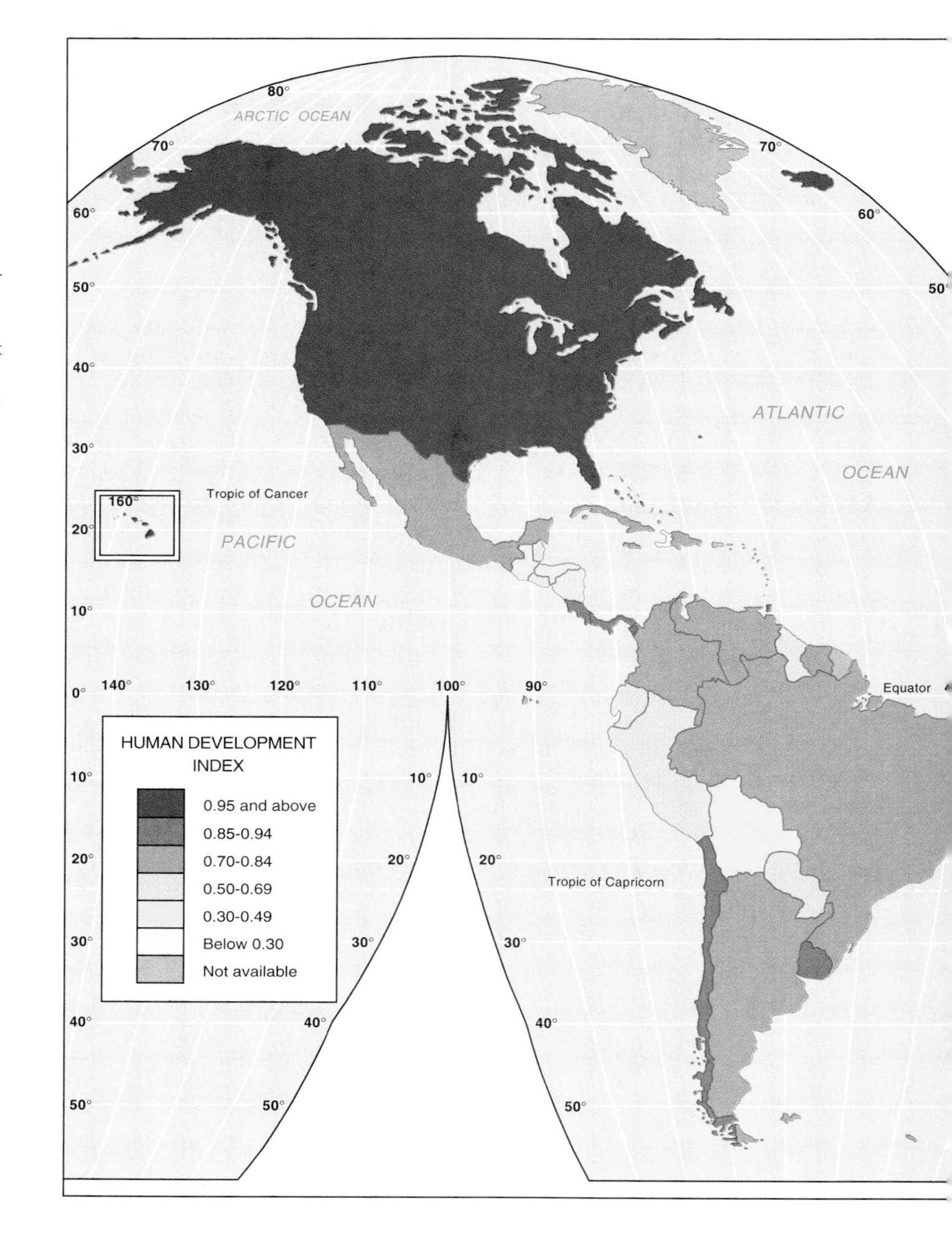

FIGURE 10–1
The Human Development Index, developed by the United Nations, combines several measures of development, including life expectancy at birth, adjusted GDP per capita, and knowledge (one-third mean years of schooling and two-thirds literacy). Each country received an index figure for the various measures, which ranged between minimum and desirable levels. The minimum for each index was set at the lowest actually observed. The desirable levels were 100 percent for literacy and the maximum observed for GDP per capita, life expectancy, and mean years of schooling.

The Human Development Index varies among regions. Two of the nine regions, Anglo-America and Western Europe, score an HDI of at least .95, as do Japan and the South Pacific (Table 10–1). Eastern Europe has a somewhat lower HDI because of its legacy of Communist rule, as well as restructuring after the fall of communism. South Africa, also classified as a relatively developed area, has a lower HDI because of the poorer conditions endured by its nonwhite population.

Among the six regions classified as developing—Latin America, East Asia, Southeast Asia, Middle East, South Asia, and Sub-Saharan Africa—the HDI varies widely. The HDI is approaching .8 in Latin America, closer to the level of Eastern Europe than those of other developing regions, whereas in Sub-Saharan

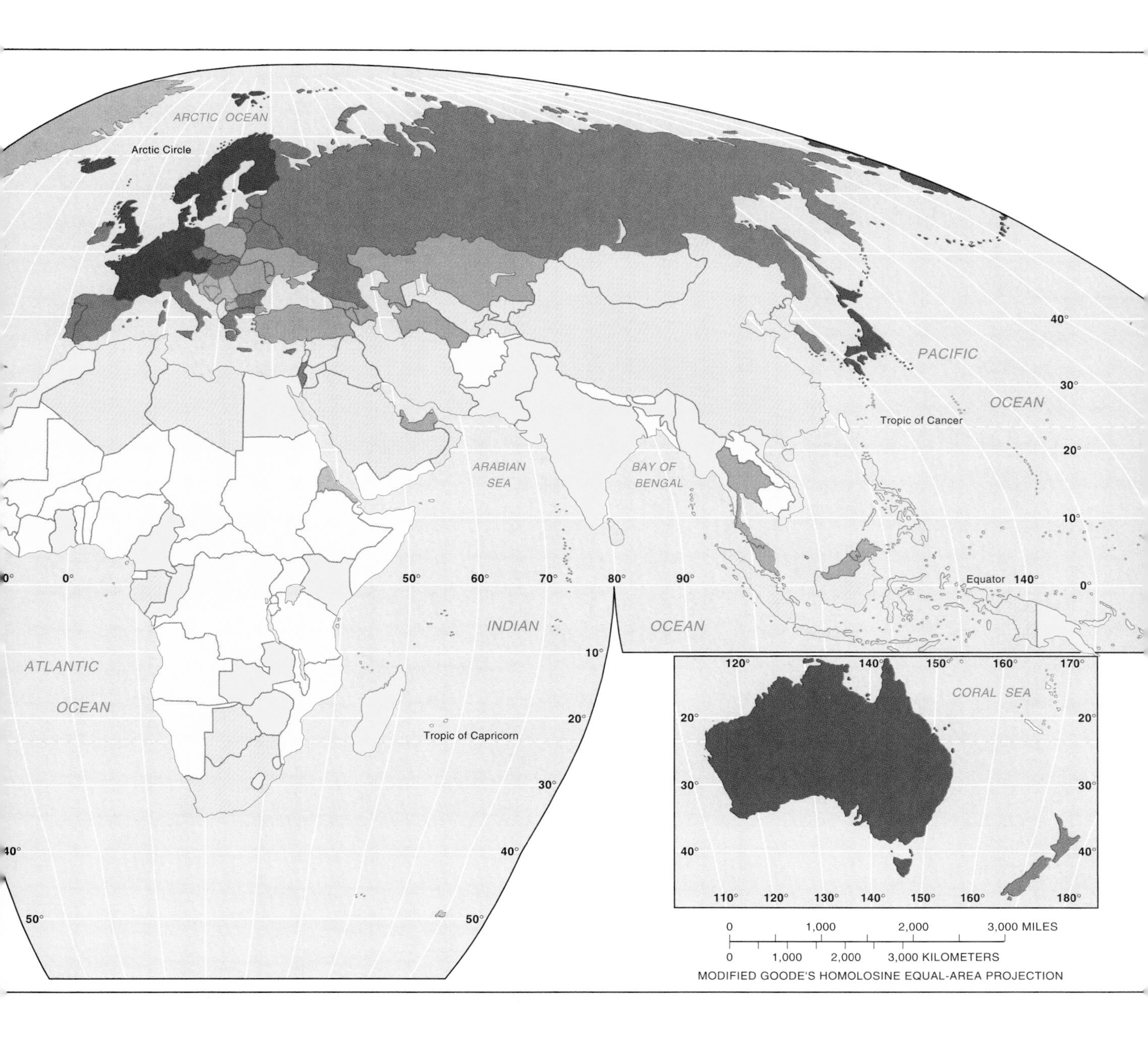

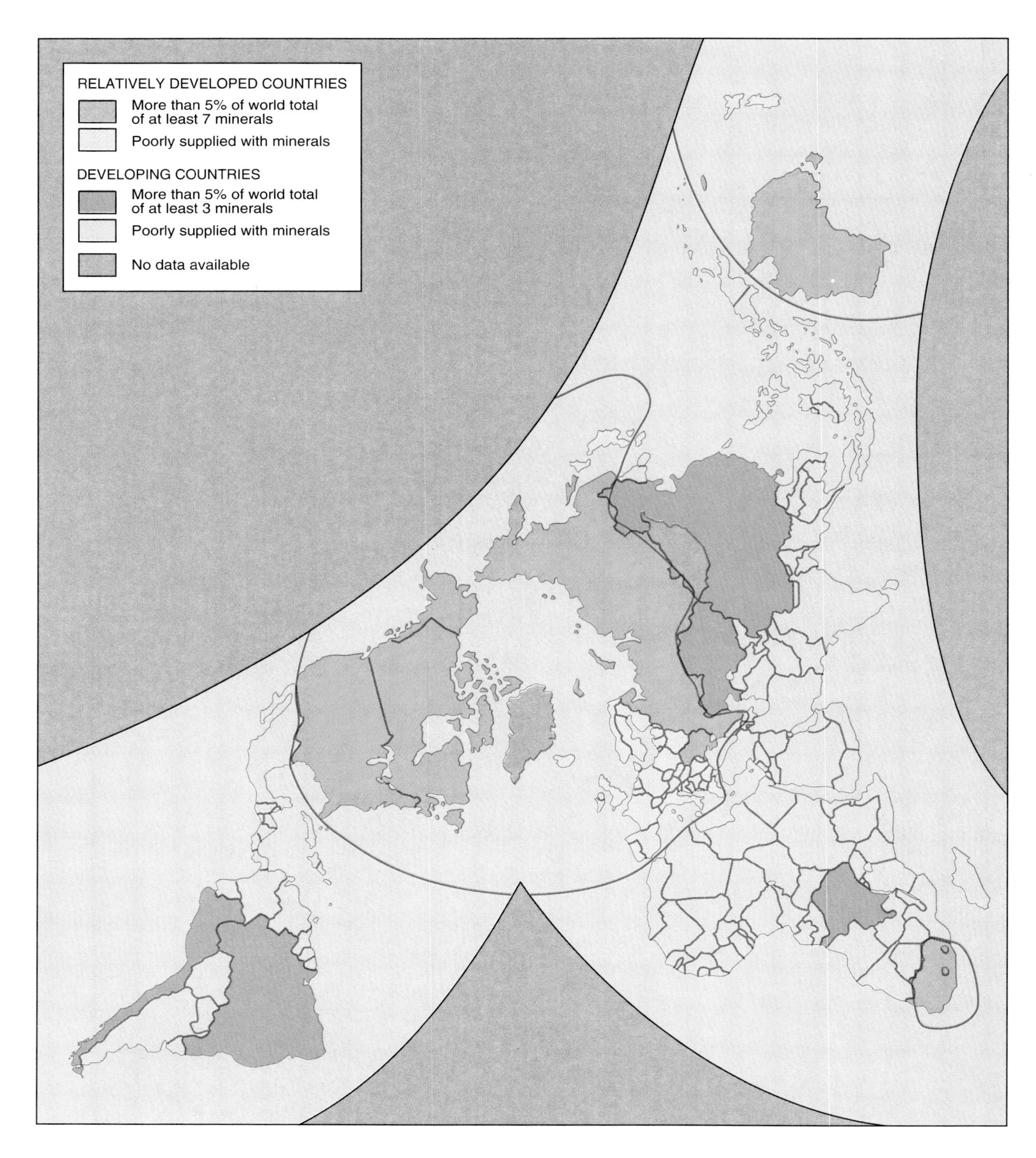

FIGURE 10–2
Core and periphery. Most of the countries that have achieved relatively high levels of development are located north of 30° north latitude. Viewed from a north polar projection, most of the relatively developed countries appear clustered in an inner core, whereas developing countries are generally relegated to peripheral or outer ring locations. Wealthier nations have invested more in mineral exploration than have poorer nations, and this investment in turn has enhanced their level of development.

TABLE 10–1
Human development index by world regions

Developing Regions		Relatively Developed Regions and Areas	
Latin America	.76	Japan	.98
East Asia	.61	Anglo-America	.98
Southeast Asia	.52	South Pacific	.97
Middle East	.51	Western Europe	.95
South Asia	.29	Eastern Europe	.87
Sub-Saharan Africa	.23	South Africa	.67

Africa the level is about .2. The relatively wide variation in HDI among the developing regions reflects future potential as well as different levels of progress in achieving economic development.

The distribution of relatively developed and developing regions reflects a clear global pattern. If we draw a circle at 30° north latitude, we see that the three major relatively developed regions, plus Japan, are all situated to the north, while every developing region lies predominantly, if not entirely, south of the circle. This division of the world between relatively developed and developing regions is known as the north-south split.

Relatively developed and developing regions display different distributions on a north polar projection (Figure 10–2). Most of the relatively developed countries form a core region, while developing countries occupy peripheral locations. Countries located in the periphery have less access to the world centers of consumption, communications, and political power, which are clustered in the core region. This perspective demonstrates the development of an increasingly unified global economy in which the relatively developed regions clustered in the core play dominant roles in forming the economies of the developing regions on the periphery.

Distinguishing between relatively developed and developing regions is made easier by the fact that the spread is increasing between the regions at the relatively high and low ends of the development continuum. Developing regions face considerable difficulties in achieving levels of development comparable to those of relatively developed regions.

Economic Indicators of Development

Several economic and social factors contribute to a society's level of development. Economic factors include income, structure of the economy, worker productivity, access to raw materials, and availability of consumer goods. Geographers measure most of these characteristics in dollars or other numerical indicators.

People in a relatively developed country are generally wealthier, and they have more material goods, than those in a developing country because they are more productive—defined in terms of output of more manufactured goods and services. Higher economic productivity results from use of technology to earn a living by working in a factory, office, or shop, rather than working every day to grow food. Basically the difference lies in whether the economy of a country emphasizes agriculture, as in a developing country, or manufacturing and services, as in a relatively developed country.

The wealth that productive industries generate goes partly to pay for the expensive technology required to operate efficient economic enterprises. Factories, offices, and shops also depend on the creation of costly communications and transportation networks and the availability of large supplies of energy and other resources. Consumers in relatively developed countries use part of the wealth generated by economic activities to buy goods and services. Demand for these products in turn increases the number of needed factories, offices, and shops.

Part of the wealth produced in relatively developed countries goes to make people healthier, safer, and better educated. As a result, infants are more likely to survive, and adults are more likely to live longer. The social benefits available in relatively developed countries in turn provide people with more tools, or at least more chances to be productive.

Per capita GNP. The average individual earns a much higher income in a relatively developed country than in a developing one. The typical worker receives \$5 to \$10 per hour in relatively developed countries, compared to less than \$0.50 per hour in most developing countries. Relatively developed countries generally mandate a minimum wage of at least several dollars per hour.

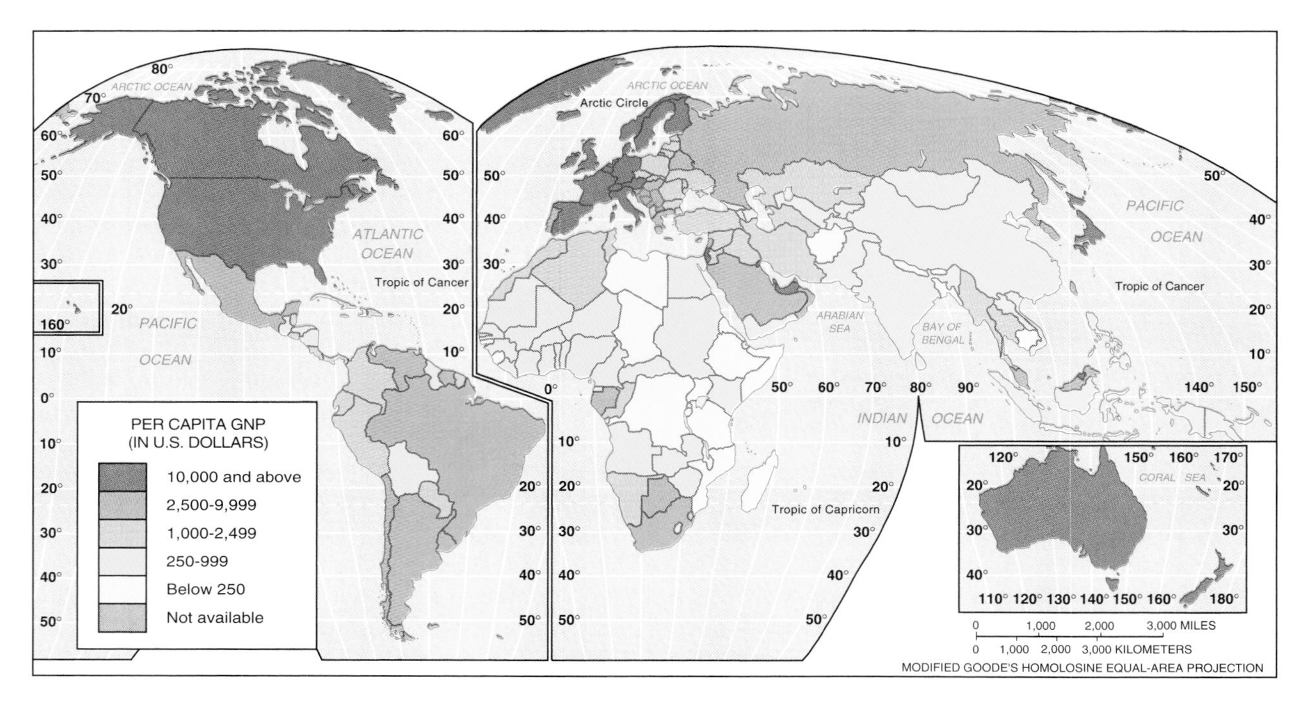

FIGURE 10–3
Annual gross national product per capita. GNP per capita exceeds $15,000 per year in most relatively developed countries, compared to less than $1,000 in most developing countries. The difference in annual GNP per capita between relatively developed and developing countries has grown in the past decade.

Per capita income is a difficult figure to obtain in many countries. Therefore, to get a sense of average incomes in various countries, geographers frequently substitute per capita gross national product, a more readily available indicator. The **gross national product (GNP)** is the value of the total output of goods and services produced in a country in a given time period (normally one year). Dividing the GNP by total population measures the contribution the average individual makes to generating a country's wealth in a year.

Annual per capita GNP exceeds $15,000 in most relatively developed countries, compared to less than $1,000 per year in most developing countries (Figure 10–3). Switzerland has the world's highest per capita GNP, nearly $40,000, while the figure exceeds $20,000 in most other Western European countries, Anglo-America, and Japan. As recently as the late 1980s, several oil-rich states bordering the Persian Gulf had the world's highest per capita GNPs, but the level is lower now because of declining petroleum prices and the lingering effects of conflict in the region. The lowest per capita GNPs are found in Sub-Saharan Africa and South and Southeast Asia. Nearly every country in these regions has a per capita GNP of less than $500 per year.

The gap in per capita GNP between relatively developed countries and developing countries has been increasing. Since the early 1980s, per capita GNP has increased by more than $8,000 in relatively developed countries, compared with less than $300 in developing countries. Per capita GNP has actually declined over the past decade in many African and Latin American countries.

Per capita GNP—or, for that matter, any other single indicator—does not measure perfectly the level of a country's economic development. Not everyone is starving in a developing country with a per capita GNP of a few hundred dollars; and one-sixth of all people live in poverty in the United States, which has a per capita GNP of more than $20,000. Per capita GNP measures average (mean) wealth, not its distribution. If only a few people receive a high percentage of the GNP, then the standard of living for the majority of people may be lower than the average figure implies. On the other hand, the higher the per capita GNP, the greater is the potential for ensuring that all citizens may at some point have a comfortable standard of life.

Economic structure. Average per capita income is higher in relatively developed countries because people typically earn their living by different means than in developing countries. Jobs fall into three categories: primary sector, secondary sector, and tertiary sector. To compare the common types of economic activities of relatively developed and developing countries, we can compute the percentage of people working in each of these three sectors.

Jobs in the **primary sector** are concerned with the direct extraction of materials from Earth's surface, generally through agriculture, although sometimes by mining, fishing, and forestry. The **secondary sector** includes manufacturing jobs that process, transform, and assemble raw materials into useful products. Other secondary-sector industries take manufactured goods and fabricate them into finished consumer goods. The **tertiary sector** involves the provision of goods and services to people in exchange for payment. Jobs in this sector include work in offices, shops, government agencies, law firms, entertainment facilities, and universities. Some analysts identify a **quaternary sector**, which includes the processing of information, especially through computer technology.

The distribution of workers among the primary, secondary, and tertiary sectors varies sharply between relatively developed and developing countries. The percentage of people working in agriculture exceeds 75 percent in many developing countries of Africa and Asia, compared to less than 5 percent in Anglo-America and many Western European countries (Figure 10–4).

The first priority for all people is to secure food they need for survival. A high percentage of agricultural workers indicates that most people in a

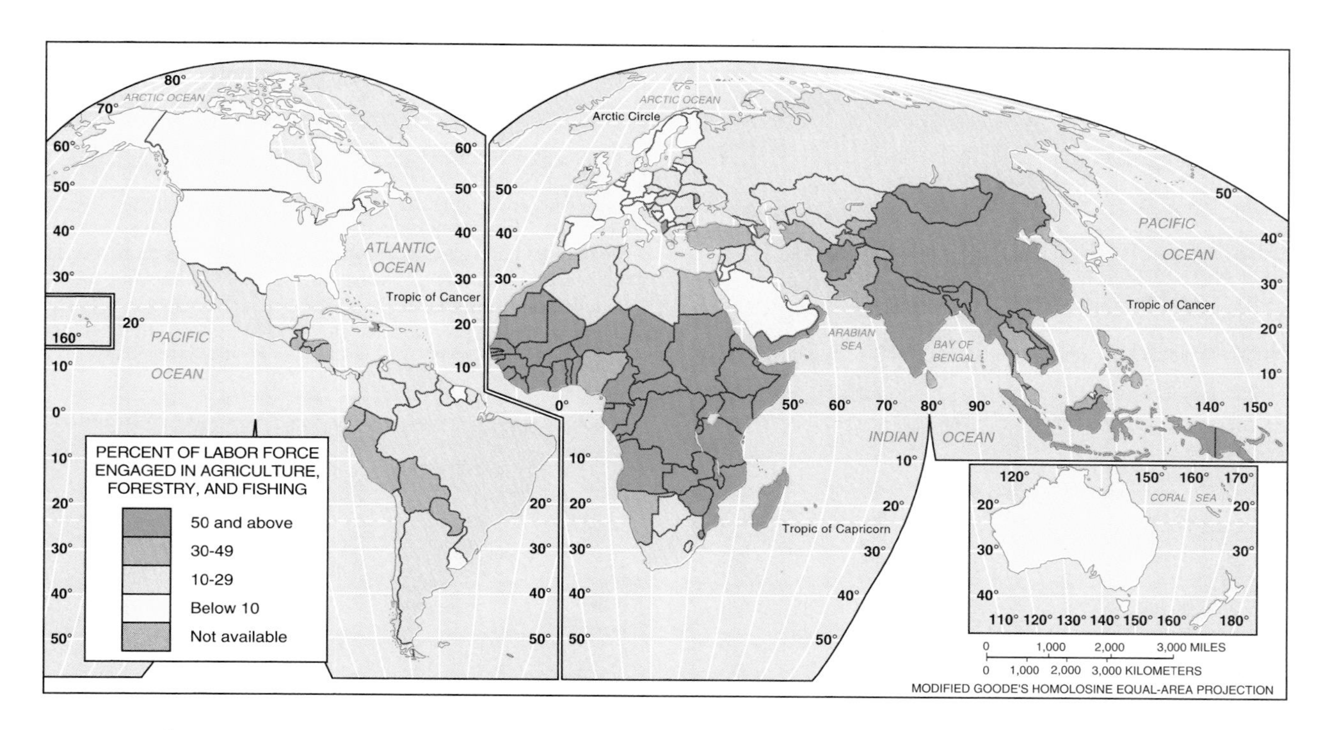

FIGURE 10–4
Percent of labor force who are primary sector workers. A priority for all people is to secure the food they need to survive. In developing countries, most people work in agriculture to produce the food they and their families need. In relatively developed countries, few people are farmers, and most people buy food with money earned by working in factories, offices, or other services.

country must spend their days producing food for their own survival. In contrast, a low percentage of primary sector workers indicates that a relatively small number of farmers can produce enough food for the rest of society. Freed from the task of growing their own food, most people in a relatively developed country can contribute to an increase in the national wealth by working in the secondary and tertiary sectors.

Within relatively developed countries, the number of jobs has decreased in the secondary sector and increased in the tertiary sector. The decline in manufacturing jobs reflects greater efficiency inside the factories as well as increased global competition in many industries. At the same time, employment in the tertiary sector continues to expand as a result of increased consumer demand for many goods and services.

Productivity. In terms of economic development, as mentioned above, **productivity** is defined as the value of a particular product compared to the amount of labor needed to make it. Workers in relatively developed countries are more productive in this sense than those in developing countries.

Workers in relatively developed countries produce more with less effort because they have access to more machines, tools, and equipment to perform much of the work. On the other hand, production in developing countries must rely more on human and animal power. The larger per capita GNP in relatively developed countries in part pays for the manufacture and purchase of machinery, which in turn makes workers more productive and generates more wealth.

Productivity can be measured by the **value added** per worker. The value added in manufacturing is the gross value of the product minus the costs of raw materials and energy. The value added per worker is thirty times higher in relatively developed countries than in developing countries. The average production worker generates a value added of nearly $50,000 in the United States compared to only a few hundred dollars in developing countries.

Raw materials. Development requires access to raw materials that can be fashioned into useful products and provide energy to operate the factories.

Workers in developing countries are less productive than those in relatively developed countries, partly because they must rely on human and animal power to perform much of their work. This Indonesian farmer has placed rice in a tray and is allowing the lighter chaff to be blown away by the wind (see Chapter 11 for description of rice growing). In contrast, workers in relatively developed countries make use of high-tech machinery, such as this computer-programmed steel welder. Photo left: (Harvey Lloyd/The Stock Market) Photo below: (Ted Horowitz/The Stock Market)

Great Britain is considered the first country to be transformed into an economically developed society, beginning in the late eighteenth century. It had abundant supplies of coal and iron ore, which were the most important raw materials for industry at the time. During the nineteenth century, other European countries also took advantage of domestic coal and iron ore supplies to promote industrial development.

European countries in the nineteenth century ran short of many raw materials essential for economic development and began to import them from other regions of the world. To ensure an adequate supply of these materials, European countries established colonies, especially in Africa and Asia. The international flow of raw materials helped sustain development in Europe and retard it in Africa and Asia. Although most of the one-time colonies have become independent states, they still export raw materials to relatively developed countries and import finished goods and services. In the twentieth century, the United States and the former Soviet Union became powerful industrial states, in part because they both possessed a wide variety of raw materials essential for development.

As certain raw materials become more important, a country's level of development can advance. Developing countries that possess energy resources, especially petroleum, have been able to utilize revenues from the sale of these resources to finance development. Prices for other raw materials, such as cotton and copper, have fallen because of excessive global supply and declining industrial demand. Developing countries that have depended on the sale of these resources have been less successful in pursuing development.

In a global economy, availability of raw materials and energy resources measures a developing country's development potential rather than its actual level of development. A country with abundant raw materials has a better chance of achieving greater development. Yet, some countries that lack them—such as Japan, Singapore, and South Korea—have developed through world trade.

Consumer goods. Part of the wealth generated in relatively developed countries goes to purchasing goods and services in addition to the minimal human needs of food, clothing, and shelter. The purchase of these so-called nonessential goods and services promotes expansion of manufacturing, which in turn generates additional wealth in society.

The quantity and type of goods and services produced and consumed in a society is a good measure of the level of development. Among the thousands of products consumers buy, three are particularly good indicators of a society's development: motor vehicles, telephones, and televisions.

Television is significant for two reasons. First, watching television is the most popular leisure activity in relatively developed countries through-

Watching television is an increasingly popular activity in developing countries, although many people must share a television set there. These Chinese viewers are watching the first live television broadcast from Mount Everest, the world's highest mountain peak, on the border of China and Nepal. A team of Chinese, Japanese, and Nepalese climbers carried a television camera to the top of Mount Everest in May 1988. (Reuters/Bettmann)

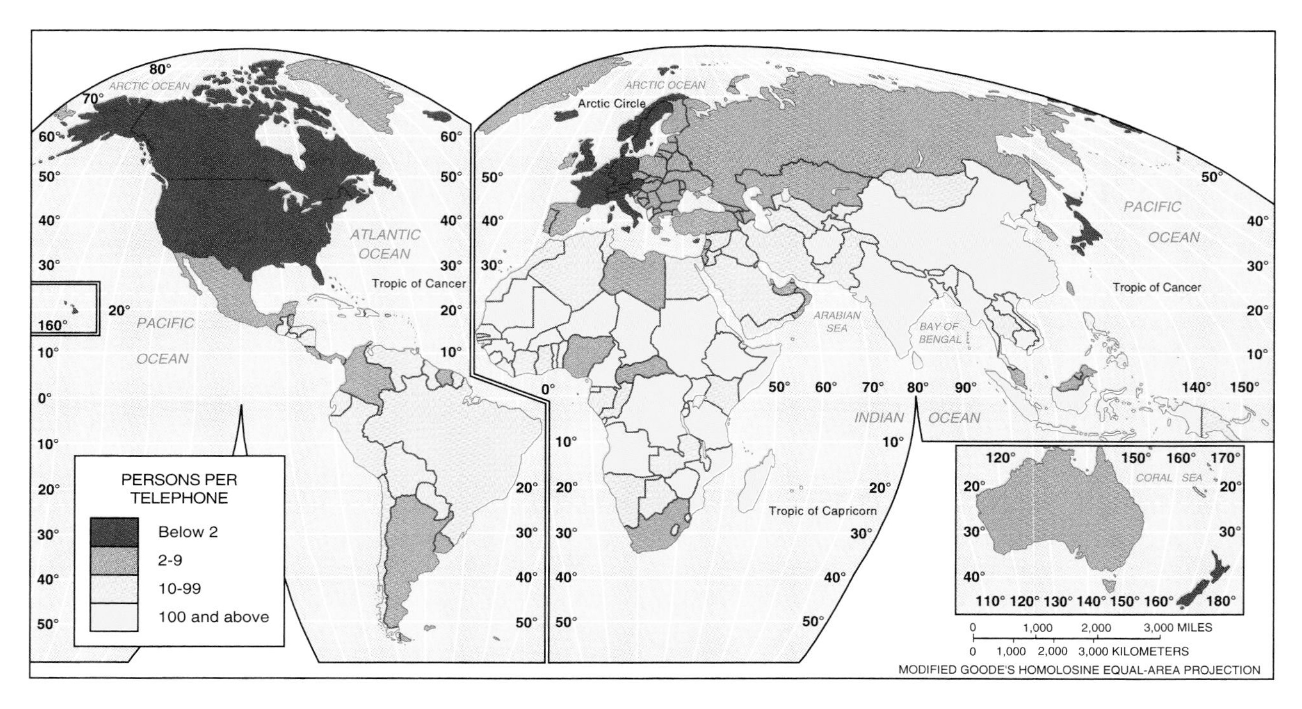

FIGURE 10–5
Persons per telephone. Relatively developed countries have one telephone for every one or two persons, whereas developing countries have one telephone for several hundred people.

out the world. Second, recall from Chapter 8 that television is the most important mechanism by which knowledge of popular culture rapidly is diffused (see Figure 8-25).

In relatively developed countries, the ratio of people to motor vehicles and telephones is approaching 1:1 (Figure 10–5). In other words, there are nearly one motor vehicle and one telephone for each person in relatively developed countries. The number of persons per telephone or motor vehicle exceeds 100 in most developing countries, indicating that people are much less likely to have access to these products in developing countries than in relatively developed ones.

The motor vehicle, telephone, and television are accessible to virtually all residents in relatively developed countries and are vital to the economy. All three of these items provide communication links to other locations and cultures that foster industrial growth. Motor vehicles provide individuals with transportation to reach better jobs and services, and businesses can distribute their products to a larger area. Telephones make possible or enhance communications with suppliers and customers of goods and services. Televisions provide exposure to news and entertainment in different locations around the world.

In contrast, these products do not play a central role in the daily lives of many people in developing countries. Motor vehicles are not essential to people who live in a small village and work all day growing food in nearby fields. Telephones are not essential for those who live in the same village as their friends and relatives. Televisions are not essential to persons who have little leisure time.

Most people in developing countries are familiar with these consumer goods even if they cannot afford them. These objects may be desired as symbols of development rather than as essential elements in the functioning of daily life. Because possession of consumer goods is not universal in developing countries, a gap may emerge between the "haves" and the "have-nots." The minority who have these goods may include the political leaders, landowners, and other elites, while the majority who are denied access to these goods may stand against those who are better off economically. Unrest and dissension can result.

In many developing countries, the "haves" are concentrated in urban areas, while the "have-nots" live in the countryside. Technological innovations tend to diffuse from urban to rural areas. Access to consumer goods is more important in urban areas because these goods are more central to the daily activities of urban dwellers—they produce them and consume them in homes, factories, offices, and shops. Motor vehicles, telephones, and televisions also contribute to social and cultural elements of development, through accessibility and communication. As a result of greater exposure to cultural diversity, people in relatively developed countries display different social characteristics from people in developing countries.

Social Indicators of Development

Relatively developed countries use their greater wealth in part to provide schools, hospitals, and welfare services. As a result, people are better educated, healthier, and better protected from hardships in relatively developed countries than in developing countries. A well-educated, healthy, and secure population in turn is an economically productive population.

Education and literacy. In general, the higher the level of development, the greater are both the quantity and the quality of education. A measure of the quantity of education is the number of years that the average person attends school. The assumption is that no matter how poor the school, the more years pupils attend, the more likely they are to learn something. The average pupil attends school for about ten years in relatively developed countries, compared to only a couple of years in developing countries.

Women are less likely than men to attend school in developing countries. In relatively developed countries, 99 women attend secondary school for every 100 men, compared to only 60 women for 100 men in developing countries. Globally, the overall average is 73 women for 100 men. In other words, females constitute 40 percent of the secondary school students in developing

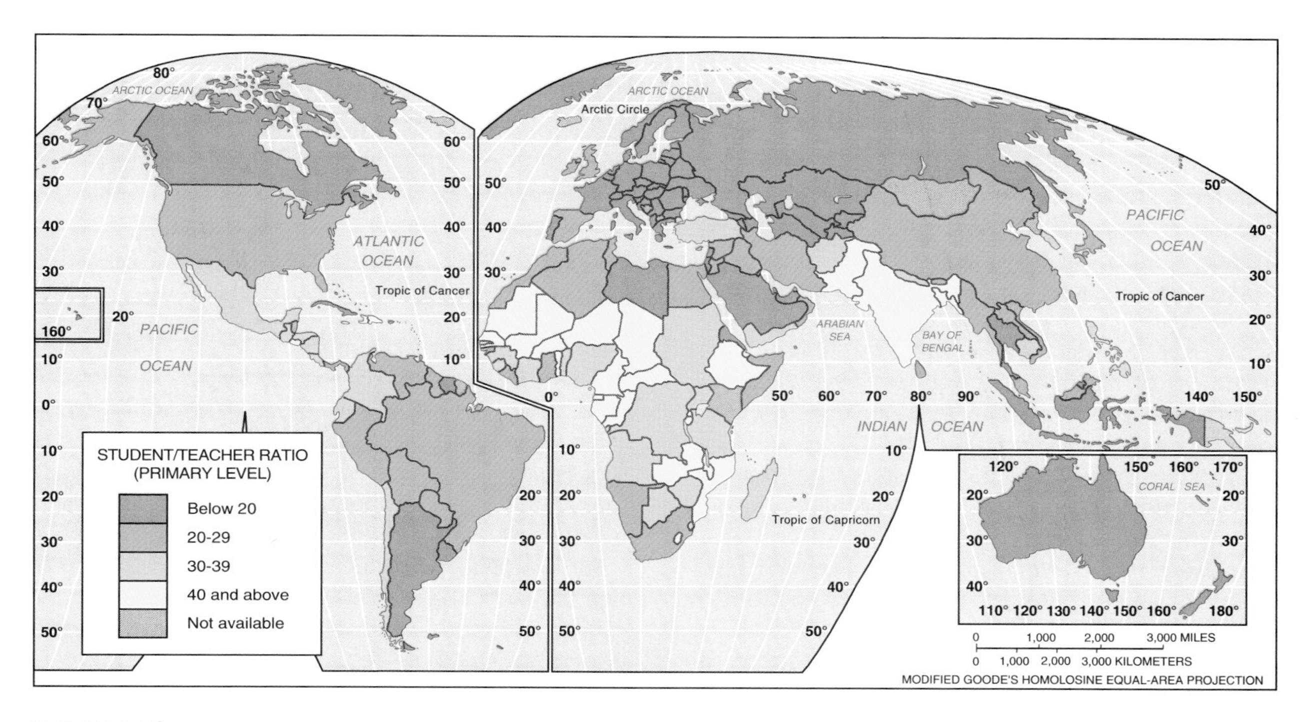

FIGURE 10–6
Students per teacher, primary school. Primary school teachers must deal with much higher average class sizes in developing countries than in relatively developed ones.

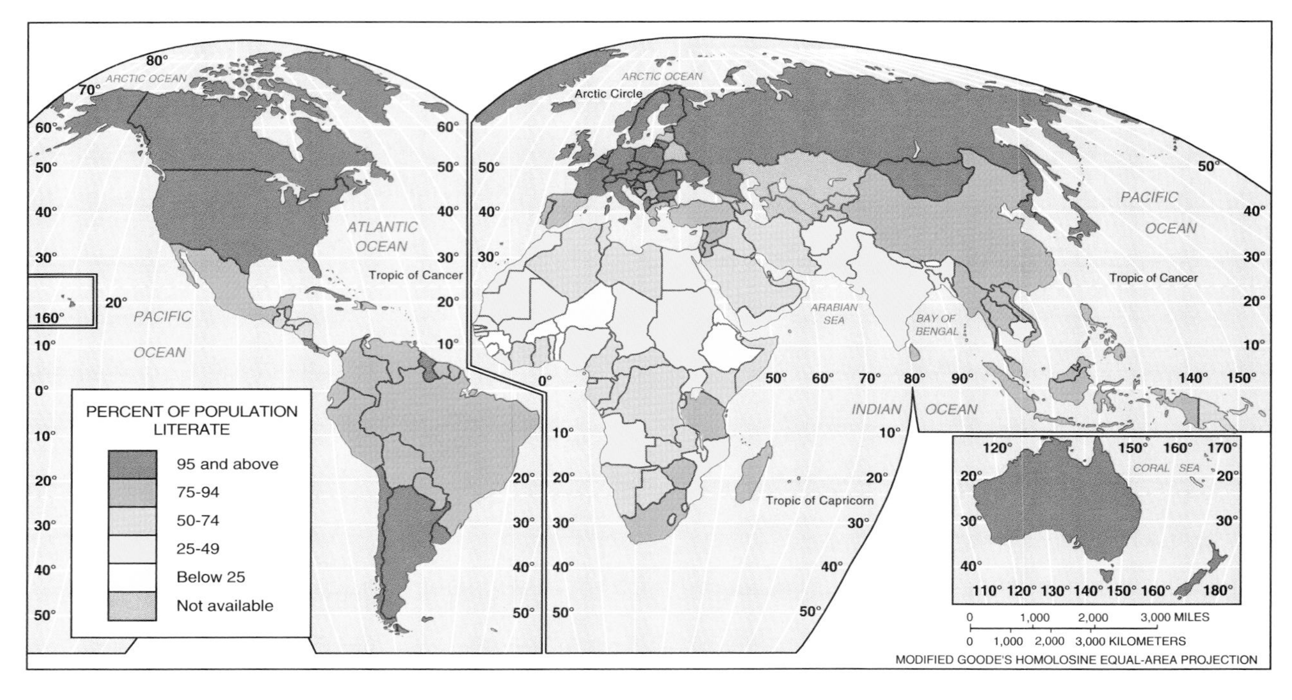

FIGURE 10–7
Literacy rate. In most of the relatively developed countries, at least 95 percent of adults are able to read and write. The percentage ranges from about 50 percent to only 10 percent in most developing countries, especially in Africa and Asia.

countries while comprising roughly half of the total population in that age group. But the ratio in developing countries has improved over the past quarter-century; in 1970, only 45 women attended secondary school for 100 men.

The quality of education is measured in two ways. One indicator is the ratio of teachers to pupils. The fewer pupils a teacher has, the more likely it is that each pupil will receive instruction. The ratio of teachers to pupils is twice as high in developing countries than in relatively developed ones (Figure 10–6).

Another measure of the quality of education is the **literacy rate**, which is the percentage of a country's people who can read and write. The literacy rate exceeds 95 percent in relatively developed countries compared to less than one-third in many developing countries (Figure 10–7).

The gap between relatively developed and developing countries is greater if literacy rates are compared for women rather than both sexes. In a number of countries in the Middle East and South Asia, literacy rates fall between 25 and 75 percent for both sexes combined but are less than 25 percent for women (Figure 10–8). Elsewhere in Asia and Latin America, gender differences are lower, but in nearly every developing country the literacy rate is higher for men than for women. In contrast, literacy rates for men and women are virtually the same in relatively developed countries. A low literacy rate among women is a serious obstacle to development for a country.

Because a higher percentage of people can read and write, publishers in relatively developed countries print more books, newspapers, and magazines per person. Relatively developed countries dominate the worldwide distribution of scientific and other nonfiction publications. Students in developing countries must learn technical information by reading books in English, German, Russian, or French.

For many people in developing countries, education is the ticket to better jobs and higher status in society. Improving education is a major goal of many developing countries, but funds often are scarce. Education may receive a higher percentage of the GNP in developing countries than in relatively

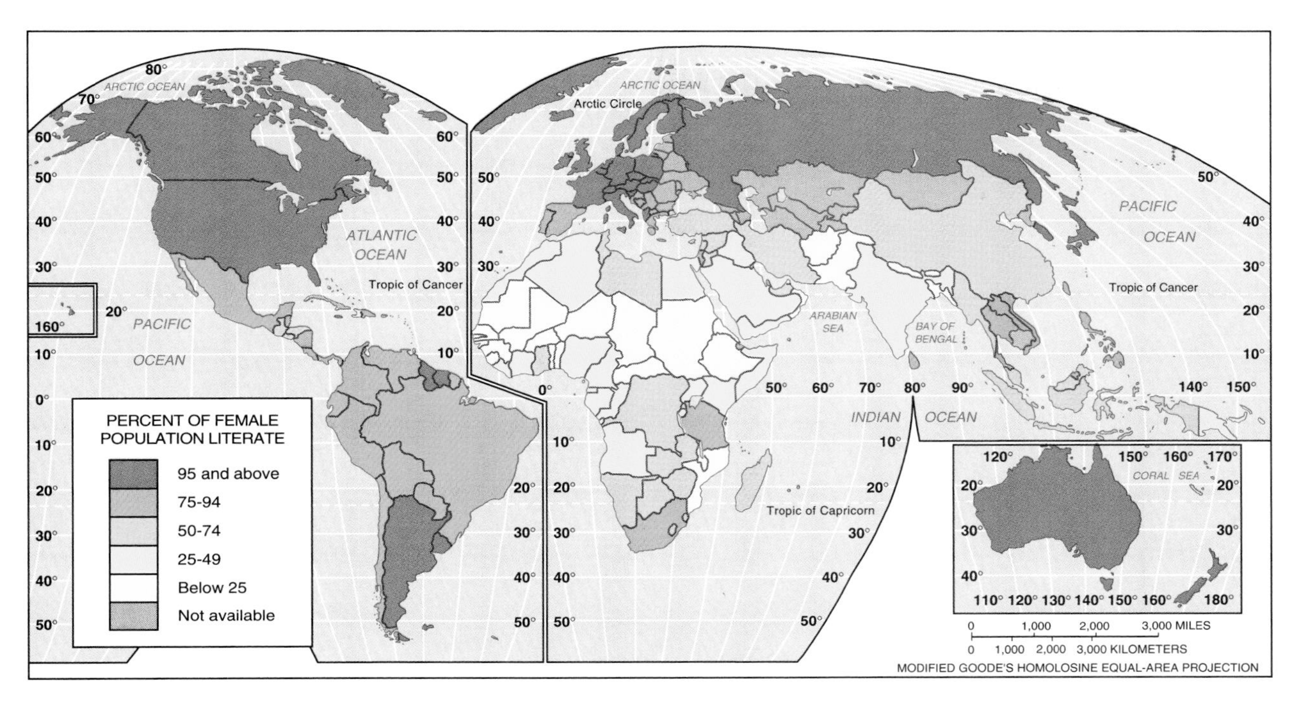

FIGURE 10–8
Female literacy rate. In some developing countries, the percentage of women who can read and write is much lower than that for men. Compare with total literacy (Figure 10–7). The gender gap is relatively high in South Asia and the Middle East.

developed countries, but because relatively developed countries have much higher GNPs, they far outspend developing countries on a per pupil basis.

Health and welfare. People are healthier in relatively developed countries than in developing ones. The health of a population is influenced by diet. Most people in developing countries of Africa and Asia do not receive the U.N. recommended daily minimum allowance of calories and proteins (Figure 10–9).

When people get sick, communities in relatively developed countries possess the resources to take care of them. Relatively developed countries have more favorable ratios of people to hospitals, doctors, and nurses (Figure 10–10). In many relatively developed countries, health care is a public service that is available at little or no cost. The United States still considers health care as an activity primarily met by private profit-making enterprises.

Relatively developed countries use part of their wealth to protect people who, for various reasons, are unable to work. In relatively developed countries, some public financial assistance is offered to people if they are sick, elderly, poor, disabled, orphans, war veterans, widows, unemployed, or single parents. Countries in northwestern Europe typically provide the highest level of public assistance payments.

Relatively developed countries are hard-pressed to maintain the current levels of public assistance. Since World War II, rapid economic growth in the United States, for example, permitted it to finance these programs with little hardship, and other relatively developed countries have had similar experiences since the industrial revolution. In recent years, economic growth has slowed, whereas the percentage of people needing public assistance has increased. Governments have faced a choice between reducing benefits and increasing taxes to pay for them.

Demographic Indicators of Development

Relatively developed countries display many demographic differences in comparison to developing countries. Chapter 7 discussed a number of

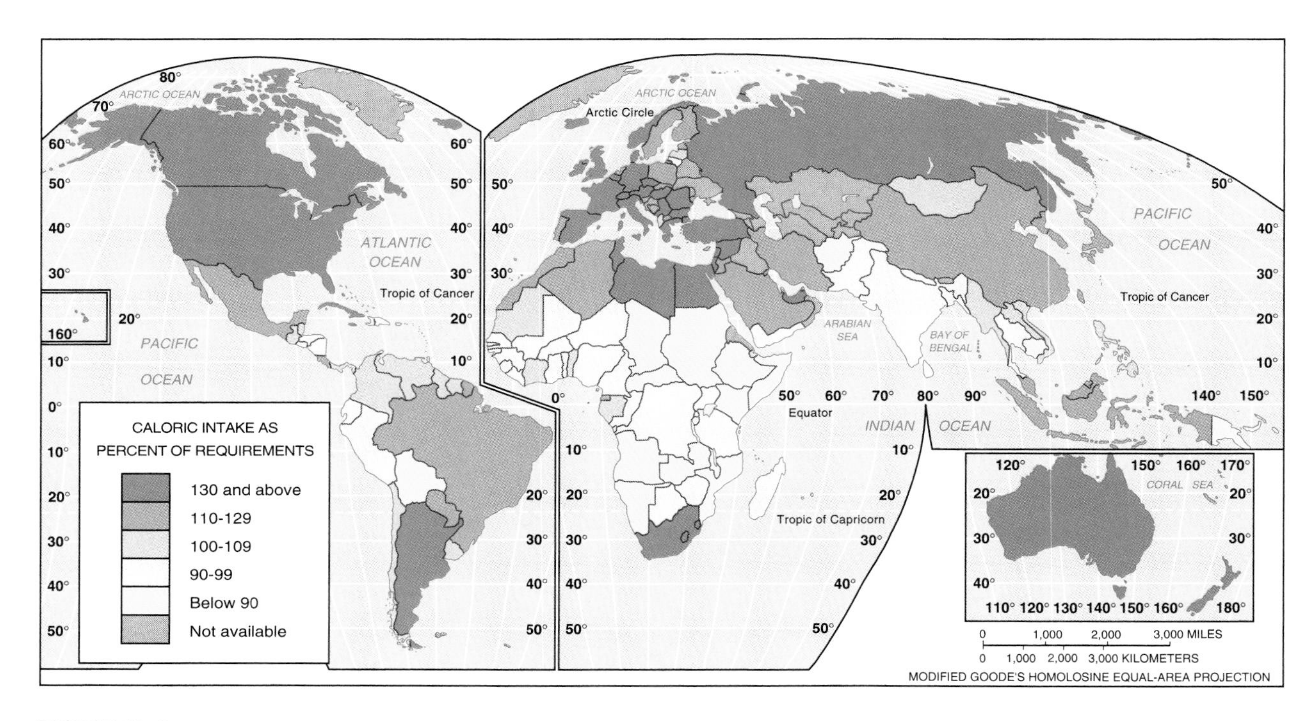

FIGURE 10–9
Daily available calories per capita as a percentage of requirements. Daily available calories per capita (food supply) is the domestic agricultural production plus imports, less exports and nonfood uses. In order to maintain a moderate level of physical activity, an average individual requires at least 2,360 calories per day, according to the United Nations Food and Agricultural Organization. The figure must be adjusted according to age, sex, and region of the world. In relatively developed countries, the average citizen consumes about one-third more calories than the minimum needed. The typical resident of a developing country receives almost precisely the minimum number of calories needed to maintain moderate physical activity. At first glance, the map concerning caloric intake in developing countries may not reveal a serious problem. However, remember that the figures represent means, and a substantial proportion of the population therefore gets less than the necessary daily minimum.

these demographic characteristics. The four major demographic characteristics that distinguish relatively developed and developing countries are infant mortality, natural increase, crude birth rates, and age structure.

Infant mortality rate. The higher levels of health and welfare in relatively developed countries permit more babies to survive infancy. The number of children who die before reaching one year is fewer than 10 per 1,000 per year in many relatively developed countries, compared to more than 100 in many developing countries (see Figure 7-6).

The infant mortality rate is greater in developing countries for a number of reasons. Children may die because of malnutrition or lack of medicine needed to survive illness such as dehydration from diarrhea. Babies also may die from poor medical practices arising from lack of education. For example, in India, the use of a dirty knife to cut the umbilical cord is a major cause of fatal tetanus.

Natural increase rate. The natural increase rate averages more than 2 percent per year in developing countries and less than 1 percent in relatively developed ones. A higher natural increase rate strains the ability of a country to support hospitals, schools, jobs, and other services that can help people be healthier and more productive. Many developing countries must allocate increasing percentages of their GNPs just to care for the rapidly expanding population rather than to improve the condition of the current population (see Figure 7-4).

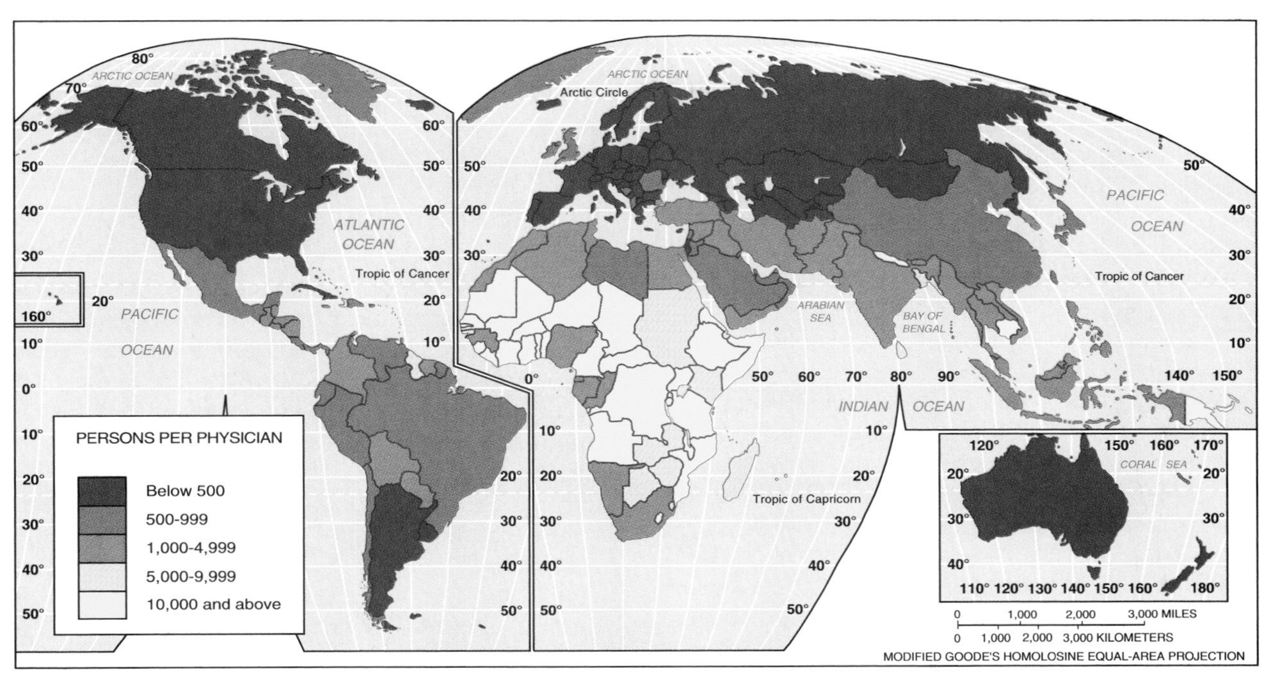

FIGURE 10–10
Persons per physician. People in relatively developed countries have more access to health care, such as hospital beds, nurses, and doctors. In relatively developed countries, for example, there is 1 doctor for about 500 people, but in developing countries each doctor is shared by thousands of people.

Crude birth rate. Developing countries have higher natural increase rates because they have higher crude birth rates. The annual crude birth rate exceeds 40 per 1,000 population in many developing countries, compared to less than 15 in relatively developed countries. People in relatively developed countries choose to have fewer children for a variety of economic and cultural reasons, and they have access to a variety of birth control devices to achieve their goal (see Figure 7-5).

The crude death rate is not an indicator of a society's level of economic development. Relatively developed and developing countries both have crude death rates around 10 per 1,000 per year. There are two reasons for the lack of difference. First, diffusion of medical technology from relatively developed countries has eliminated or sharply reduced the incidence of several diseases in developing countries. Second, relatively developed countries have higher percentages of older people, who have high mortality rates, as well as lower percentages of children, who have low mortality rates once they survive infancy.

The mortality rate is significantly higher in developing countries for women in childbirth. For every 100,000 babies born, fewer than 10 mothers die giving birth in most relatively developed countries, compared with several hundred in developing countries. Again, this phenomenon may be related to poor medical and health care practices that result from poverty or lack of education. Doctors are not required to assist in childbirth, but prenatal check-ups and anticipation of serious problems can make a great difference in maternity care.

Age structure. The higher crude birth rates in developing countries result in differing age structures compared to relatively developed countries. Developing countries have a higher percentage of children under age 15, who are too young to work and must be supported by employed adults and government programs. Relatively developed countries have a higher percentage of older people who have retired from jobs and also need some public assistance. The overall percentage of young and old dependents is lower in relatively developed countries than in developing ones (see Figure 7-10).

Correlation of Development Indicators

A country's level of development is a relative concept, because every country is at some position on a continuous scale. However, many countries fall into one of two extreme positions on the scale. Geographers justify dividing countries into relatively developed and developing groups because a wide variety of economic, social, and demographic characteristics tend to coincide.

The correlation of economic, social, and demographic indicators of development is clearly demonstrated by comparing the United States and India. By every measure of level of development, the United States ranks among the world's most developed countries, while India falls among the world's lowest ranked countries by most development indicators (Figure 10–11).

Per capita GNP in India is less than 2 percent of the level in the United States, and the gap between the two countries has increased during the past two decades. The lower GNP per capita in India reflects the fact that three-fourths of the people are farmers, compared to less than 5 percent in the United States. The preponderance of non-primary-sector workers in the United States produces a variety of goods and services that increase the country's wealth.

The United States has an abundant supply of many raw materials needed for industrial production and can purchase remaining resources, such as copper and petroleum, from other countries. India has many raw materials essential for manufacturing but needs a number of other critical resources and lacks the funds to import them.

Consumer goods are scarcer in India. The numbers of people per motor vehicle, telephone, and television are several hundred times larger in India than in the United States. Because of its much higher GNP, the United States can afford to pay much more for education, as well as for public assistance programs, such as social security. Underlying the social and economic differences are demographic contrasts: the natural increase rate is much higher in India than in the United States, because of a higher crude birth rate. In India, women especially suffer from relatively high rates of illiteracy and mortality.

The same economic, social, and demographic indicators that distinguish the United States and India can be used at a global scale. Geographers can divide the world into regions based on a correlation of the various development measures.

Regional Differences in Development

Geographers and other analysts often group countries according to level of development into three "worlds" based on a concurrence of the factors described so far. The *First World* includes the relatively developed, highly industrialized regions of Anglo-America and Western Europe, plus Japan and the South Pacific. The *Second World* includes Eastern Europe and the former Soviet Union, where economies until recently were centrally planned by Communist national governments. The *Third World* includes the developing regions of Latin America, Africa, and Asia.

From a more purely financial perspective, the World Bank subdivides the Third World into four groups: low-income, lower-middle-income, upper-middle-income countries, and capital-surplus exporters of petroleum. Other observers designate the lowest income developing regions as the Fourth World and the oil-rich Middle East states as the Fifth World.

Dividing Earth into several "worlds" according to level of development may help to highlight differences in economic, social, and demographic characteristics. However, identifying distinct worlds can be misleading, because a central element in understanding why a region is at a particular level of development is its linkages to other regions. The relatively low development level of one region may be in large measure the result of interaction with a relatively developed region as part of a global economy. Also, change is constant, and the concept of development on a continuum is important.

Core Regions

A more useful division according to level of development is between the relatively developed regions located at the core of the global economy and the developing regions located in the periphery. Economic power, as well as wealth, is concentrated in the relatively developed core regions.

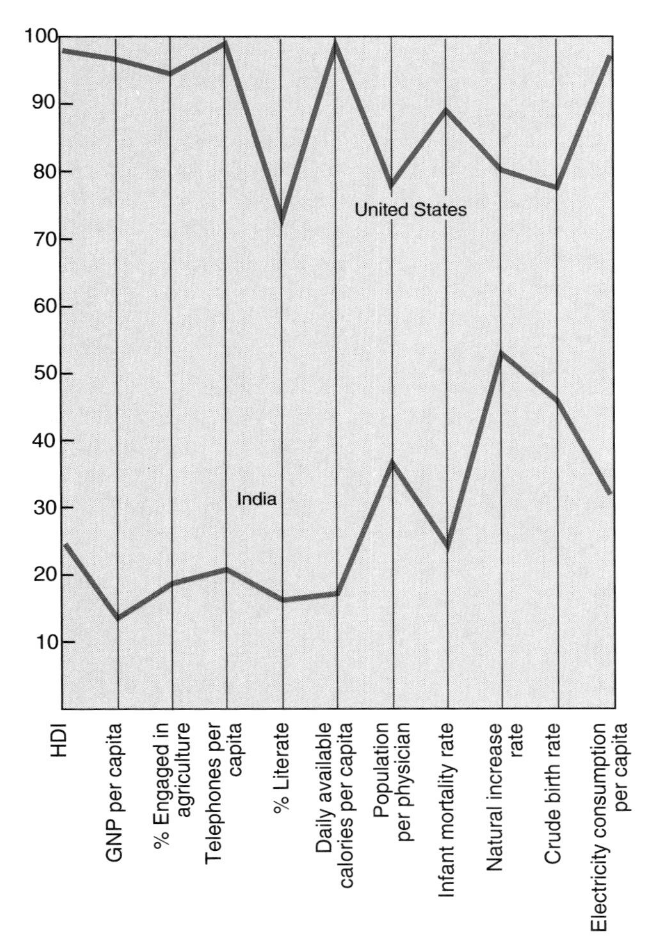

FIGURE 10–11
The graph shows the percentage of all countries that rank below the United States or India according to the particular measure. As a relatively developed country, the United States ranks among the highest percentages in a wide variety of development indicators.

Anglo-America, Western Europe, and Japan form the three major core areas in the global economy (Figure 10–2).

Anglo-America. The United States and Canada, the two largest countries in Anglo-America, rank among the world's most developed countries, according to the various economic, social, and demographic characteristics presented earlier. This region has the highest per capita GNP and is well-endowed with most raw materials and minerals needed for industry.

Although fewer than 5 percent of the workers are engaged in agriculture, Anglo-America is the world's leading food exporter and the only one with a significant amount of idle agricultural land. On the other hand, the region has the world's highest percentage of tertiary sector employees. Anglo-America is the leading provider of word processing services, media, computer analysis, and information monitoring. The region also specializes in entertainment, mass media, sports, recreation equipment, and other industries that promote use of leisure time.

New jobs have been created at a faster rate in the United States than in the other core regions since the early 1990s, primarily through expansion of service industries, such as health-care, sports and entertainment, hotels, transportation, and restaurants. Average wages have not increased in the United States during the period, and workers in several Western European countries as well as Japan are now compensated at higher levels. Productivity has increased rapidly in the United States as well, especially in manufacturing and agriculture, partly through large-scale investment in new technology and partly through reduction in the work force.

Prospects for future economic growth in the United States have been clouded by the country's large budget deficit, a result of Americans' reluctance to raise taxes to cover spending for desired social welfare and military programs. To cover the budget deficit, the United States borrows large sums of money, and as a result owes far more money to foreigners than any other country.

Western Europe. The level of development is especially high in this region's core area, which includes the western part of Germany, northeastern France, northern Italy, Switzerland, southern Scandinavia, Belgium, the Netherlands, and Luxembourg. Western Europe's peripheral areas—Ireland, southern Italy, Portugal, Spain, and Greece—rank somewhat lower on the development scale.

To maintain a relatively high level of development, Western Europe must import food, energy, and minerals. In past centuries, Western Europeans explored and mapped the rest of the world and established colonies on every continent. These colonies supplied many of the resources that Europeans needed to foster development. Colonization also promoted the diffusion of Western European languages, religions, and other social customs throughout the world.

Now that most colonies are independent, Western Europeans must buy raw materials from other countries. To pay for their imports, Western Europeans provide high-value goods and services, such as insurance, banking, and luxury motor vehicles, including the Mercedes-Benz and Rolls-Royce.

Competition among Western European states for control of territory has led to many wars, most recently World War II. Since the end of World War II, however, most Western European states have joined multinational organizations that promote economic and military cooperation. The elimination of economic barriers within the European Union makes Western Europe the world's largest and richest market, as we discussed in Chapter 9.

Japan. The level of development in Japan, the fifth most populous state in Asia, contrasts sharply with other large Asian countries. Japan, with a per capita GNP of nearly $30,000, ranks among the world's most productive states, whereas the GNP per capita is generally less than $1,000 per year in the rest of East, South, and Southeast Asia. Japan has ranked first or second each year on the Human Development Index.

Japan's development is especially remarkable because it has an extremely unfavorable ratio of population to resources. The country has some of the world's most intensively farmed land and one of the highest physiological densities (see Table 7-1). The Japanese consume relatively little meat and grain but still must import these products. Japan also lacks many of the key raw materials for basic industry. For example, although Japan is the world's leading steel producer, it must import virtually all of the coal and iron ore needed for steel production.

Japan first become an industrial power by taking advantage of the country's major asset, its labor force. Japan had an abundant supply of people willing to work hard for low to modest wages. The Japanese government encouraged manufacturers to sell their products in other countries at lower prices than domestic competitors. Having gained a foothold in the global economy by selling low-cost products, Japan then began to specialize in high-quality, high-value products, such as electronics, motor vehicles, and cameras. Wages in Japan are now comparable to rates in Western Europe and North America, but Japanese workers have developed a reputation for high-quality production.

Japan achieved economic dominance in part by concentrating its resources in rigorous educational systems and training programs to create a skilled labor force. Japanese companies spend 8 percent of their revenues on research and development, twice as high as the level spent by U.S. firms, and the government provides further assistance to develop new products and manufacturing processes.

Eastern Europe and Former Soviet Union

Winston Churchill declared in a 1946 speech that an "Iron Curtain" had descended across Europe, from the Baltic Sea on the north to the Adriatic Sea on the south. East of 15° east longitude, most of the European states came under Communist control in the late 1940s, while to the west most were democratic. Between the late 1940s and late 1980s, Communist-dominated Eastern Europe comprised eight relatively small countries—Albania, Bulgaria, Czechoslovakia, the German Democratic Republic (East Germany), Hungary, Poland, Romania, and Yugoslavia—and the Soviet Union, which occupied 15 percent of the world's land area.

Centralized development planning. Early communist theorists, such as Karl Marx and Friedrich Engels, believed that communism would triumph in relatively developed countries where exploited factory workers would lead the revolution. Their social and economic programs were based on conditions in advanced industrial societies. But when Communist parties gained control of Russia in 1917 and other Eastern European countries after World War II, few of these states were advanced industrial powers. Instead, the Communists had to promote socialism in poor, agricultural societies.

As centrally planned societies, Eastern European countries typically had economies directed by government officials rather than private entrepreneurs. The government allocated resources and labor and, in effect, decided what should be bought and sold. Communists believed that socialism would eliminate the greedy acquisition of wealth by a handful of people. In principle, socialism would promote an egalitarian society and improved conditions among working class people.

In the Soviet Union, for example, a national planning commission known as Gosplan developed a series of five-year plans to guide economic

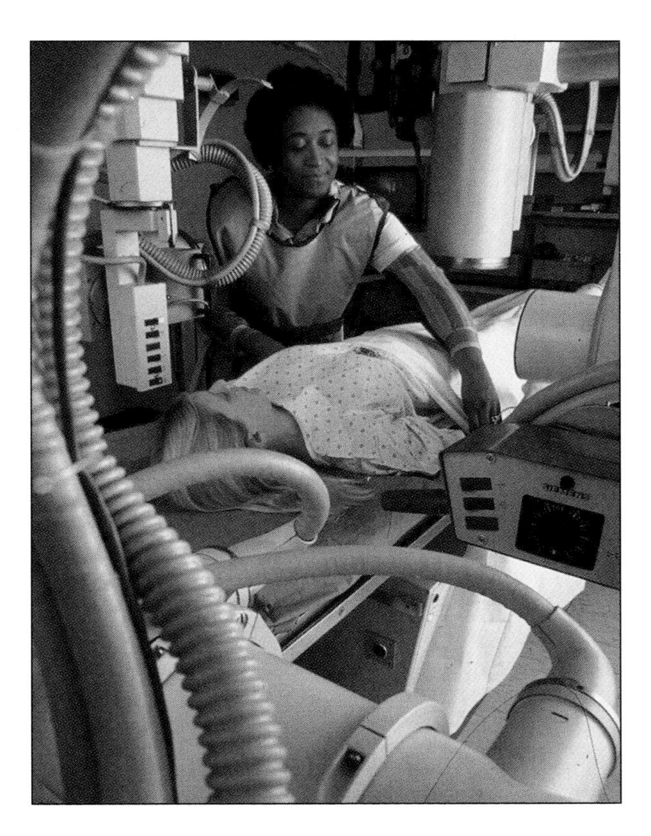

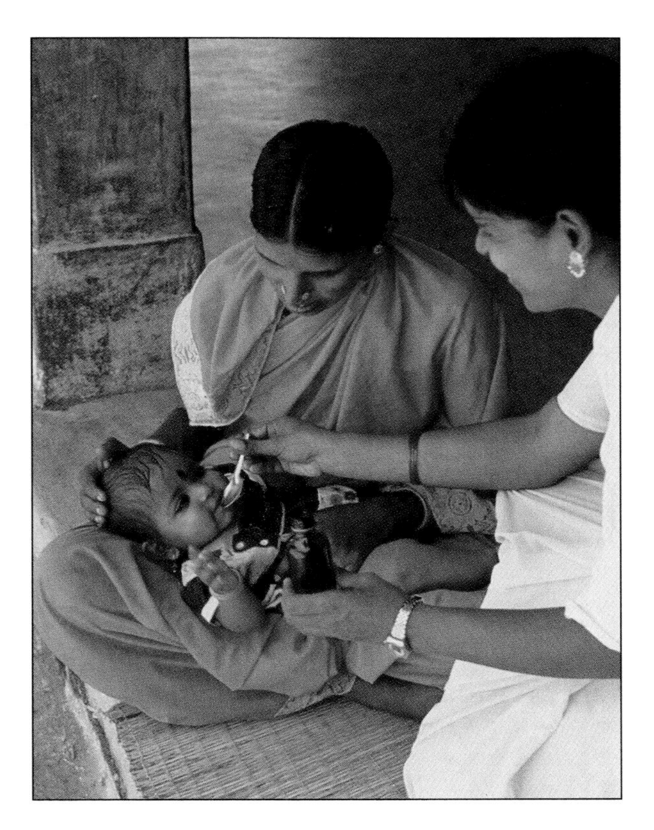

Relatively developed countries possess better equipped hospitals and more extensive medical technology to diagnose and treat people's illnesses, such as this cat-scan, than is the case in developing countries, such as this clinic in India. (Gabe Palmer/The Stock Market)

(S. Nagendra/Photo Researchers, Inc.)

development. The plans prescribed the production goals for the whole country by economic sector and region. They specified the type and quantity of minerals, manufactured goods, and agricultural commodities to be produced and the factories, railways, roads, canals, and houses to be built in each part of the country.

The five-year plans featured three main development policies. First, Soviet planners placed emphasis on so-called heavy industries, such as iron and steel, machine tools, petrochemicals, mining equipment, locomotives, and armaments. To allow industrial growth, the country also promoted development of mining, electric power, and transportation facilities.

Second, plans promoted dispersal of production facilities from the European to the Asian portion of the Soviet Union. Soviet decision-makers considered the concentration of industry in the west to be a liability, because the country had been invaded from the west by Napoleon in the nineteenth century and Hitler in the twentieth century. Planners also wished to promote more equal levels of economic development throughout the country and believed that dispersal of industries would accomplish this goal.

Third, Soviet planners preferred to locate manufacturing facilities as close as possible to sources of raw materials rather than near markets. This policy reflected in part the needs of the particular industries emphasized in the Soviet plans and in part the lack of effective consumer demand. By emphasizing heavy industry located near sources of raw materials, Soviet planners gave lower priority to the production of consumer goods, such as telephones, washing machines, shoes, and dishes.

By centrally planning their economies, Eastern European states significantly improved their level of development, according to some measures, especially during the 1950s and 1960s. Annual per capita GNPs increased from a few hundred to several thousand dollars, and some social and demographic indicators rose to levels comparable to Western European countries.

The post-Communist era. During the late 1980s and early 1990s, the Communist parties lost power throughout Eastern Europe, and the national governments now exercise less control over the economies. Aside from the desire for freedom, the principal reason that Eastern Europeans rejected communism was that central planning proved to be inefficient at running national economies. Scarce funds were used to meet annual production targets rather than to invest in long-term improvements in productivity, such as installing more modern equipment and redeploying workers to other tasks. Despite an abundant supply of productive farmland, Eastern Europe had to import food from the West because of inefficient agricultural practices.

Orders sent from national government offices hundreds of kilometers away were often not implemented in the factories. Some targets were impossible to achieve, but others were simply ignored: why work hard when your job is guaranteed and your supervisor can't fire you? Factories polluted the air and water, and citizens were unable to pressure their governments into investing in pollution control devices.

For many Eastern Europeans, the most fundamental problem was that, by concentrating on production in basic industries, the Communists neglected consumer-oriented products, such as automobiles, refrigerators, and clothing. An oversupply of some goods and an undersupply of others created a paradox: much unsold and yet much wanted. Severe shortages of housing forced entire families to live in dwellings the size of a college dormitory room. Although restricted from visiting western countries, many Eastern Europeans could see on television the much higher level of comfort on the other side of the Iron Curtain.

During the 1990s, Eastern European countries dismantled the economic structure inherited from the Communists. Government-owned shops have been sold to private individuals, and factories have been turned over to private corporations, in many cases foreign ones. Citizens have been able to buy shares in some of the privatized companies at low prices, and stock exchanges have been established to buy and sell the shares.

The level of development varies widely among Eastern European countries. The Czech Republic, Hungary, and Slovenia have converted relatively rapidly to market economies, taking advantage of their proximity to the relatively developed core region of Western Europe. Because workers in these countries are comparably skilled yet much lower-paid compared to their counterparts in Western Europe, some manufactured goods can be successfully exported to wealthier countries in the West. As painful memories of the Communist era fade, these countries will display social and economic characteristics similar to such Western European countries as Greece, Ireland, and Portugal. On the other hand, other former Communist countries in Eastern Europe, such as Albania, Belarus, Bulgaria, and Romania have few factories that can compete in global markets.

Converting to market economies has proved painful in some Eastern European countries. The closure of inefficient businesses has increased unemployment, and prices for many goods have skyrocketed with the elimination of government subsidies. Disagreement on the pace of economic reform rather than cultural animosity was the most important factor contributing to the breakup of Czechoslovakia. Czechs were willing to bear a short-term decline in their standard of living, because they believed that rapid conversion to a market economy would bear long-term benefits. Slovaks wanted to slow down the pace of change: they feared high levels of unemployment in the large, inefficient factories that the Communists had clustered there to promote economic development during the 1950s.

Similarly, the Soviet Union and Yugoslavia split in part because republics such as Russia and Slovenia preferred more rapid economic change than Ukraine and Serbia. However, the end of communism in these countries created a vacuum into which poured long-suppressed clashes among nationalities.

Peripheral Regions

Regions outside the core include Latin America, East Asia, Southeast Asia, South Asia, the Middle East, and Sub-Saharan Africa. Levels of development vary widely among these regions, as well as within them.

Latin America. According to the Human Development Index, Latin America has a higher level of development than any other developing region (Table 10–1). Per capita GNP is relatively high along the South Atlantic coast from Curitiba, Brazil, to Buenos Aires, Argentina. This area enjoys high agricultural productivity and ranks among the world's leaders in production and export of wheat and corn (maize). Venezuela's per capita GNP is higher than the region's average, in part because it is the only South American country with extensive petroleum reserves. The region's lowest per capita GNPs are concentrated in Central America, several Caribbean islands, and the interior of South America.

Latin Americans are more likely to live in urban areas than people in other developing regions. Mexico City, São Paulo, and Buenos Aires rank among the ten largest, according to the United Nations. The region's population is highly concentrated along the Atlantic coast, while population density remains low in most of the region, especially the tropical interior of South America. Large areas of the interior rainforest are being destroyed, either to sell the timber or to clear the land for settled agriculture.

Development in Latin America is hindered by inequitable income distribution. In many countries, a handful of wealthy families control much of the land and rent parcels to individual farmers. Many tenant farmers grow coffee, tea, and fruits for export to relatively developed countries rather than food for domestic consumption. Latin American governments encourage redistribution of land to peasants but do not wish to alienate the large property owners, who generate much of the national GNP.

During the 1970s, Latin America achieved the highest growth rate of GNP of any region outside the petroleum-rich Middle East, but development has slowed in the region since the 1980s. To finance development, the region's countries have borrowed large sums of money from international organizations and banks in relatively developed countries, but have been unable to repay some of them.

East Asia. China, the largest country in East Asia, ranks among the world's poorest in per capita GNP. Despite its very low per capita GNP, China's total GNP ranks third in the world behind those of the United States and Japan, a reflection of the country's large population as well as its potential for further development. South Korea, Taiwan, and Hong Kong are three entities in East Asia that have achieved relatively high levels of development compared to other countries in the region.

Traditionally, most farmers in China were forced to pay high rents and turn over a percentage of the crops to a property owner. Farmers in a typical year produced enough food to survive but frequently suffered from famines, epidemics, floods, and other disasters. Exploitation of the country's resources by Europe and Japan further retarded China's development.

Following its victory in a civil war, the Communist party declared the creation of the People's Republic of China on October 1, 1949. Since then, dramatic changes have been made in the country's economy. To ensure the production and distribution of enough food, China's government took control of much of the agricultural land. In some villages, officials assigned specific tasks to each farmer, distributed food to each family according to individual needs, and sold any remaining food to urban residents. In other cases, farmers rented land from the local government, received orders to grow specific amounts of particular crops, and sold for their own profit any crops above the minimum production targets.

In recent years, strict governmental control of agriculture has been loosened, and individuals again are able to own land and control production. Farmers have an incentive to work hard because the sale of surplus crops is the main source of revenue to buy desired household goods. However, agricultural land must be worked intensively in order to produce enough food for China's large population, and farmers in the country's less-fertile areas may not be able to produce a large surplus.

China has large reserves of fossil fuels, especially coal, and it has been using these resources—along with its low-cost labor force—to promote rapid industrial development. Interaction with other countries has substantially increased in recent years, as a result of both more imports and more exports.

The Chinese people are subject to more government control over daily life than in other countries, and they have difficulty obtaining some goods. Nonetheless, most Chinese recognize that they are better off now than before the revolution, because they have less fear of famines than in the past. Because China has a much lower natural increase rate than other developing regions, more of the country's growing GNP can contribute to increasing the standard of living of the existing population rather than meeting the needs of a rapidly expanding population.

Southeast Asia. The Southeast Asia region comprises five countries situated entirely on the Asian mainland—Cambodia, Laos, Myanmar (Burma), Thailand, and Vietnam—and six others scattered across thousands of islands in the Indian and Pacific oceans. Most of these islands are part of Indonesia or the Philippines.

The region's tropical climate limits intensive cultivation of most grains. The heat is nearly continuous, the rainfall abundant, and the vegetation dense. Soils are generally poor, because the heat and humidity rapidly destroy nutrients when land is cleared for cultivation; volcanic soils in the Philippines and Indonesia are quite fertile, though. Development is also limited in Southeast Asia by several mountain ranges, active volcanoes, and frequent typhoons.

As a result of the inhospitable environment, population was traditionally low in Southeast Asia. The injection of Western medicine and technology has resulted in one of the most rapid rates of population increase in the world, approximately 2.5 percent per year since the 1940s.

Southeast Asia's most populous country, Indonesia, includes 13,667 islands, but nearly two-thirds of the population live on the island of Java, which has one of the world's highest arithmetic densities. People concentrated on Java in part because the island's soil, derived from volcanic ash, is more fertile than elsewhere in the region and in part because the Dutch established their colonial headquarters on the island.

Rice, the region's most important food, is exported in large quantities from some countries, such as Thailand and Vietnam, but must be imported to other countries in the region, such as Malaysia and the Philippines. Because of distinctive vegetation and climate, farmers in Southeast Asia concentrate on harvesting products that are used in manufacturing. The region produces a large percentage of the world's supply of palm oil and copra (coconut oil), natural rubber, kapok (fibers from the ceiba tree used for insulation and filling), and abaca (fibers from banana leafstalks used in fabrics and ropes).

Southeast Asia contains a large percentage of the world's tin as well as some petroleum reserves. However, economic development lags because of an unfavorable ratio of population to most resources essential for manufacturing.

The region has suffered from a half-century of nearly continuous warfare. Japan, the Netherlands, France, and the United Kingdom were all forced to withdraw from colonies they had established in the region. In addition, France and the United States both fought unsuccessfully to prevent Communists from controlling Vietnam. Wars have also devastated neighboring Laos and Cambodia.

South Asia. South Asia is the region with the second-highest population and the second-lowest per capita GNP. The population density is very high throughout the region, and the natural increase rate is among the world's highest.

India, the region's largest country, is the world's leading producer of jute (used to make burlap and twine), peanuts, sugar cane, and tea and contains reserves of minerals such as uranium, bauxite, coal, manganese, iron ore, and chromite. However, the overall ratio of population to resources is unfavorable because of the magnitude of the region's population.

India is one of the world's leading rice and wheat producers, and the region was one of the principal beneficiaries of the Green Revolution, a series of inventions beginning in the 1960s that dramatically increased agricultural productivity. As a result of the Green Revolution, "miracle" rice and wheat seeds were widely diffused through South Asia (see Chapter 11).

Agricultural productivity in South Asia also depends on climate. The region receives nearly all of its precipitation during the monsoon season between May and August. Agricultural output declines sharply when the monsoon rains fail to arrive. In a typical year, farmers in South Asia produce a surplus of grain that is stored for distribution during dry years. But several consecutive years without monsoon rains produces widespread hardship in South Asia.

The Middle East. Much of the Middle East consists of deserts that can sustain only sparse concentrations of animal life, and most products must be imported. However, the Middle East possesses one major economic asset: a large percentage of the world's petroleum reserves.

Because of petroleum exports, the Middle East is the only one of the nine major world regions that enjoys a trade surplus. The value of imports exceeds exports in every other major region, to a considerable extent because other countries must purchase large quantities of petroleum from Middle Eastern states.

Government officials in many of the Middle Eastern states, such as Saudi Arabia and the United Arab Emirates, have used the billions of dollars generated from petroleum sales to finance development. The Middle East is the only region in which development is not hindered by lack of capital for new construction. To the contrary, many governments in the region have access to more money than they can use to finance development.

On the other hand, not every country in the region has abundant petroleum reserves, because most of the petroleum reserves are concentrated in states that border the Persian Gulf. Development possibilities are limited in Egypt, Jordan, Syria, and other Middle East countries that lack significant petroleum reserves.

The large gap in per capita income between the petroleum-rich countries and those that lack resources is a major source of tension in the Middle East. People in the poorer Middle Eastern states held little sympathy for Kuwait after Iraq's invasion in August 1990. Kuwait was charged with not sharing its petroleum-generated wealth and failing to provide good living conditions for guest workers from the poorer Arab countries.

The challenge for many Middle East states is to promote development without abandoning the traditional cultural values of Islam, which is followed by more than 95 percent of the region's population. Many countries in the Middle East sharply restrict the role of women in business and prevent the diffusion of financial practices considered incompatible with Islamic religious principles. The low level of literacy among women is the main reason that the United Nations considers the level of development among the petroleum-rich states of the Middle East to be lower than the per capita GNP would indicate.

The region also suffers from serious internal cultural disputes, as discussed in Chapter 8. Iraq's long war with Iran and attempted annexation of Kuwait split the Arab world. Countries dominated by Shiite Muslims, especially Iran, have promoted revolutions elsewhere in the region in order to sweep away elements of development and social customs they perceive as influenced by Europe or Anglo-America.

Israel, the region's only state controlled by Jews, has successfully repelled several attacks by neighboring states and, since 1967, has occupied territory captured from its adversaries. Palestinians living in the occupied territories prefer to establish an independent state rather than live under Israeli rule, and other neighboring countries are reluctant to normalize relations with Israel until the status of the Palestinians is settled. Peace will be difficult to achieve as long as substantial numbers of Palestinians refuse to recognize the existence of Israel, and Israelis oppose relinquishing control of all occupied territories. Money that could be used to promote development is diverted to military funding and rebuilding war-damaged structures.

Consumers in relatively developed countries are more likely to obtain their food in supermarkets. Farmers in developing countries who have produced a surplus may sell products in a market, such as this one at Xishuangbahna, in Yunnan Province, China.
Photo left: (Roy Morsch/The Stock Market) Photo right: (Michele Burgess/The Stock Market)

Sub-Saharan Africa. Per capita GNP in Sub-Saharan Africa is comparable to the level in South, East, and Southeast Asia, and population density is lower than in any other developing region. Sub-Saharan Africa contains many resources important for economic development, including bauxite (aluminum ore) in Guinea, cobalt and copper in Zaire and Zambia, iron ore in Liberia, manganese in Gabon, petroleum in Nigeria, and uranium in Niger.

Despite these assets, Sub-Saharan Africa is the region with the least-favorable prospects for increasing the level of development. Some of the region's economic problems are a legacy of the colonial era. Mining companies and other businesses were established to supply European industries with needed raw materials rather than to promote overall development in Sub-Saharan Africa. In recent years, African countries have suffered because world prices for their resources have fallen.

Poor leadership has also plagued Sub-Saharan Africa. After independence, leaders of many countries in the region pursued personal economic gain and local wars rather than policies to promote development of the national economy. Frequent wars within and between countries in Sub-Saharan Africa have retarded economic development.

The fundamental problem in many countries of Sub-Saharan Africa is a dramatic imbalance between the number of inhabitants and the capacity of the land to feed the population. Nearly all of the region consists of either tropical or dry climate. Dry regions can support some human life, but not large population concentrations. Tropical humid climates support large concentrations of people in Southeast Asia, but much of the soil in tropical Sub-Saharan Africa is infertile. Yet, because Sub-Saharan Africa has by far the world's highest rate of natural increase, the region's land is more and more overworked, and agricultural output per person has declined.

Promoting Development

Developing countries in every region share the same priority: to increase their level of development. This means increasing the per capita GNP and using the additional funds to improve the social and economic conditions of the people. Developing countries have chosen one of two policies to promote development. One approach emphasizes international trade, while the other advocates self-sufficiency. Each has important advantages and serious problems. Examples can be cited of countries that have successfully and unsuccessfully used each alternative.

International Trade Approach to Development

The first model of development calls for a country to identify its distinctive or unique economic assets. What animal, vegetable, or mineral resource does the country have in abundance that other countries are willing to buy? What product can the country manufacture and distribute at a higher quality and a lower cost than other countries?

According to the international trade approach, a country can promote development by concentrating scarce resources into the expansion of its distinctive local industries. The sale of these products in the world market brings funds into the country that can be used to finance other development projects.

Rostow's development model. A leading advocate of this approach was W. W. Rostow, who proposed a five-stage model of economic development in the 1950s that a number of countries have adopted. According to Rostow, development should proceed in the following steps:

1. *The traditional society.* Rostow has employed this term to define a country that has not yet started a process of development. A traditional society contains a very high percentage of people engaged in agriculture and a high percentage of national wealth allocated to what Rostow called "nonproductive" activities, such as the military and organized religion.
2. *The preconditions for take-off.* According to Rostow, the process of economic development begins when an elite group of people initiates innovative economic activities. Under the influence of these well-educated leaders, the country starts to invest in new technology and infrastructure, such as water supplies and transportation systems. These projects ultimately will stimulate an increase in productivity.
3. *The take-off.* Rapid growth is generated in a limited number of economic activities, such as textiles or food products. These few take-off industries achieve technical advances and become productive, while other sectors of the economy remain dominated by traditional practices.

4. *The drive to maturity.* Modern technology, previously confined to a few take-off industries, diffuses to a wide variety of industries, which then experience rapid growth comparable to the take-off industries. Workers become more skilled and specialized.
5. *The age of mass consumption.* The economy shifts from production of heavy industry, such as steel and energy, to consumer goods, such as motor vehicles and refrigerators.

According to Rostow's model, each country is in one of the five stages of the development process. Relatively developed countries are in stage 4 or 5, while developing countries are in one of the three earlier stages. The model also asserts that today's relatively developed countries have already passed through the early stages. The United States, for example, which was in Stage 1 prior to independence, Stage 2 during the first half of the nineteenth century, Stage 3 during the middle of the nineteenth century, and Stage 4 during the late nineteenth century, before entering Stage 5 during the early twentieth century.

A country that concentrates on international trade to promote development benefits from exposure to consumers in other countries. To remain competitive, the take-off industries must constantly evaluate changes in international consumer preferences, marketing strategies, production engineering, and design technologies. This concern for international competitiveness in the exporting take-off industries will filter through less-advanced economic sectors.

Rostow's optimistic projection for development was based on two factors. First, the relatively developed countries of Western Europe and Anglo-America had been joined by others, notably Japan. If Japan could become more economically developed by following this model, why not other countries?

Second, many developing countries contain an abundant supply of raw materials sought by manufacturers and producers in relatively developed countries. In the past, European colonial powers extracted many of these raw materials without compensation. In a global economy, the sale of these raw materials could generate funds for developing countries to promote economic development.

Rostow's model has been heavily criticized in recent years by some geographers and other analysts for making inaccurate generalizations about the development process based on the historical experiences of Anglo-America and Western Europe. Critics charge that the model does not accurately account for current conditions in peripheral regions that influence prospects for development. Nonetheless, the international trade model is important because it has been adopted as the basis for promoting development by large lending institutions, as well as by government officials in an increasing number of developing countries.

Example: Persian Gulf states. One group of countries oriented toward international trade is located along the Arabian Peninsula near the Persian Gulf. Saudi Arabia is the most prominent member of this group; others include Bahrain, Oman, and the United Arab Emirates.

Until the 1970s, this region was the one of the world's least developed, but escalation of petroleum prices transformed these countries overnight into some of the wealthiest per capita. These countries use their petroleum revenues to finance large-scale projects, such as housing, highways, airports, universities, and telecommunications networks. Recently built steel, aluminum, and petrochemical factories compete on world markets with the help of government subsidies.

The landscape has been further changed by the diffusion of consumer goods. Large motor vehicles, color televisions, audio equipment, and motorcycles are readily available and affordable. Supermarkets are stocked with food imported from Europe and Anglo-America.

While the region's economy has changed dramatically in a short period of time, people's social customs have changed more slowly. Daily life is dominated by Islamic religious principles, some of which conflict with Western business practices. Women are excluded from holding most jobs and visiting public places. In some places they are expected to wear traditional black clothes, including a veil, although women may dress as they please in their own homes. All business halts several times a day when Muslims are called to prayers. Shops suspend business and permit people to unwrap their prayer rugs and prostrate themselves on the floor. Yet a cultural underground is present: many people buy videocassette recorders so that they can watch publicly banned movies and television programs at home, and of course people do have private social gatherings.

Four Asian dragons. Another group of countries that has utilized the international trade alternative includes South Korea, Singapore, and Taiwan, plus the British colony of Hong Kong. These four areas have been given several nicknames, including the "four dragons," the "four little tigers," and "the gang of four."

Singapore (a British colony until 1965) and Hong Kong have virtually no natural resources; both comprise large cities surrounded by very small amounts of rural land. South Korea and Taiwan have traditionally taken their lead from Japan, which occupied both of them until after World War II; Japan's success with the international trade approach strongly influenced their adoption of the strategy.

Lacking natural resources, the four dragons have promoted development by concentrating on the production of a handful of manufactured goods, especially clothing and electronics. Low labor costs enable these countries to sell these products inexpensively in relatively developed countries.

Problems with international trade approach. Two problems have hindered countries outside the Persian Gulf and the four Asian dragons from increasing their level of development through the international trade approach:

1. *Resource distribution.* Resources are not distributed uniformly among developing countries. Several Middle East countries have successfully moved to more advanced stages of development because the price of petroleum products skyrocketed during the 1970s. Other countries find that the prices of their commodities have not increased—and in some cases actually decreased—in recent years. Developing countries that depend on the sale of one product have suffered because the price of their leading commodity has not increased as rapidly as the cost of the products they need to buy. For example, Zambia's economy, which depends on the sale of copper, has suffered in recent years because of declining world prices for that commodity.
2. *Market stagnation.* Countries such as the four dragons that depend on selling low-cost manufactured goods find that the world market for many products is expanding less rapidly than in the past. Relatively developed countries have limited growth in population, consumer purchasing power, and market size. To increase sales, developing countries may need to capture sales from established competitors rather than share in an expanding market.

Self-sufficiency Approach to Development

The second approach to promoting development is the self-sufficiency alternative. According to the self-sufficiency approach, a country should spread investment throughout all sectors of the economy rather than concentrate on one or two take-off industries. This approach promotes balanced growth, because people and enterprises throughout the country receive a fair share of resources.

Countries that adopt the self-sufficiency approach encourage businesses to make goods for domestic consumption rather than for export. Economic growth may be modest, but in the long run the country benefits because it is not dependent on changing policies in other countries and fluctuations in the price of commodities.

States promote self-sufficiency by setting barriers that limit the import of goods from other places. These barriers may include setting high taxes on imported goods to make them more expensive than domestically produced goods, fixing quotas to limit the quantity of imported goods, and requiring licenses to restrict the number of legal importers.

India. For many years, the two most populous developing countries, China and India, were strong advocates of the self-sufficiency approach. Businesses were discouraged from exporting goods, and barriers prevented the import of many goods.

India made heavy use of all three types of barriers to importing—high taxes, quotas, and licenses. To import goods into India, most foreign companies had to secure a license. The process was long and cumbersome, because several dozen government agencies had to approve the request for a license. Once a company received a license, the government severely restricted the quantity of the product it could sell in India. The government also imposed heavy taxes on imported goods, which could double or triple the price to consumers.

At the same time, Indian businesses were discouraged from producing goods for export to relatively developed or other developing countries. Instead, priority went to making goods for domestic consumption. If private companies could not make a profit from selling goods only inside India, the government provided subsidies or took over direct operation of the company. As a result, India has produced more steel and motor vehicles per capita than most other developing countries produce, but the products have sold at twice the world market price.

In recent years, India's government has moved away from complete adherence to the self-sufficiency approach. The government has lowered the tax on some imported goods and eliminated the license requirement for some importers. Indian companies are encouraged to become more competitive with foreign firms. India's GNP has increased by more than 4 percent per year over the past quarter century, but population has increased by more than 2 percent per year. Therefore, more than half of India's economic growth goes to accommodating the additional people.

Problems with self-sufficiency approach. India's experience illustrates some of the problems with the self-sufficiency approach. First, the approach may encourage inefficient industries. In a global economy, firms increasingly find that the domestic market is too small to make a profit. Unable to make a profit through increased overseas sales, companies may need government subsidies to remain in operation. Companies protected from international competition may not experience pressure to keep abreast with rapid technological changes.

The second problem with the self-sufficiency alternative is the need for a large bureaucracy to administer the controls. A complex administrative system encourages abuse and corruption. Potential entrepreneurs find that struggling to produce goods or services may be less rewarding financially than advising others on how to manipulate the controls. Other potential entrepreneurs may earn more money by illegally importing goods and selling them at inflated prices on the black market.

Comparing International Trade and Self-sufficiency Approaches

Developing countries have been influenced by studies conducted by the World Bank to determine whether international trade or self-sufficiency has been more successful at promoting development. Forty-one developing countries were classified into four groups: strongly oriented toward international trade, moderately oriented toward international trade, strongly oriented toward self-sufficiency, and moderately oriented toward self-sufficiency. The World Bank then compared the growth in the per capita GNP achieved by the countries in the four groups between 1963 and 1973 and between 1974 and 1985.

Between 1963 and 1973, the per capita GNP generally increased in all four groups, but the countries strongly oriented toward international trade registered the largest increases (more than 7 percent per year), followed by those moderately oriented toward international trade (nearly 4 percent per year); countries strongly oriented toward self-sufficiency had the lowest increases (less than 2 percent per year).

The rate of change in the GNP per capita was lower among all four groups of countries between 1974 and 1985, a period when higher petroleum prices triggered a worldwide economic slowdown, but the relative performance of the four groups remained the same as during the earlier period.

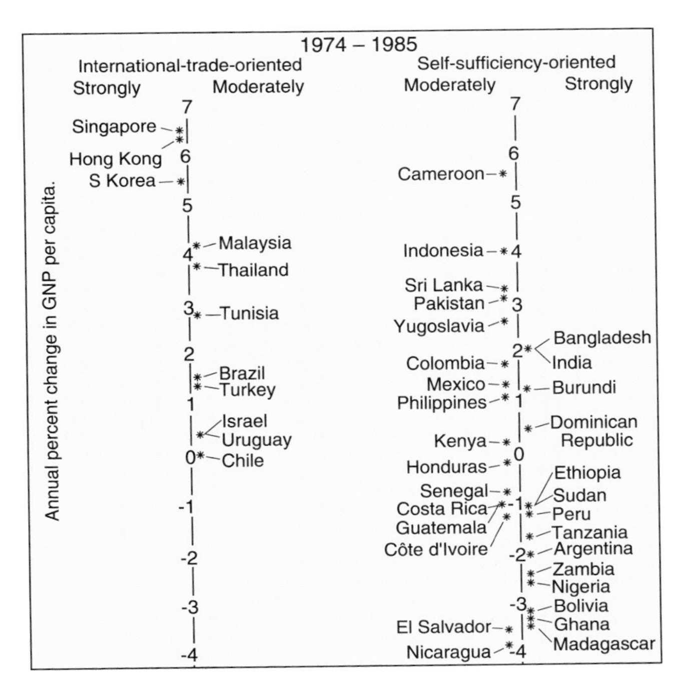

FIGURE 10–12
Countries that have adopted the international trade approach to development have generally had higher levels of economic growth than those that have adopted the self-sufficiency approach.

The per capita GNP increased more than 6 percent per year in countries strongly oriented toward international trade, and it declined by more than 1 percent per year in countries strongly oriented toward self-sufficiency (Figure 10–12).

More recently, the World Bank compared the growth in GNP per capita for twenty-nine countries in Sub-Saharan Africa between 1987 and 1991. The World Bank found that the six countries that had most vigorously adopted economic reforms consistent with the international trade approach—Ghana, Tanzania, Gambia, Burkina Faso, Nigeria, and Zimbabwe—had the region's highest growth rates in GNP per capita during the period.

Evidence such as the World Bank studies has convinced government officials in Eastern Europe, Latin America, Asia, and other regions that development can be best promoted through adoption of the international trade approach. By trading with other countries, developing countries can be more fully integrated into the world-economy and gain some of the benefits produced by that system.

Obstacles to Development

Adopting the international trade approach has produced economic benefits for many developing countries, but it has also generated social and environmental problems. Social problems stem from increased dependency of developing countries on the financial resources of relatively developed countries.

Financing Development

To pay for development, developing countries generally must obtain funds from relatively developed countries. Developing countries receive some of the money in the form of direct grants but must borrow much of it from financial institutions in relatively developed countries, including commercial banks and international lending organizations, such as the World Bank and International Monetary Fund.

Developing countries use much of the money to build new infrastructure projects, such as hydro-

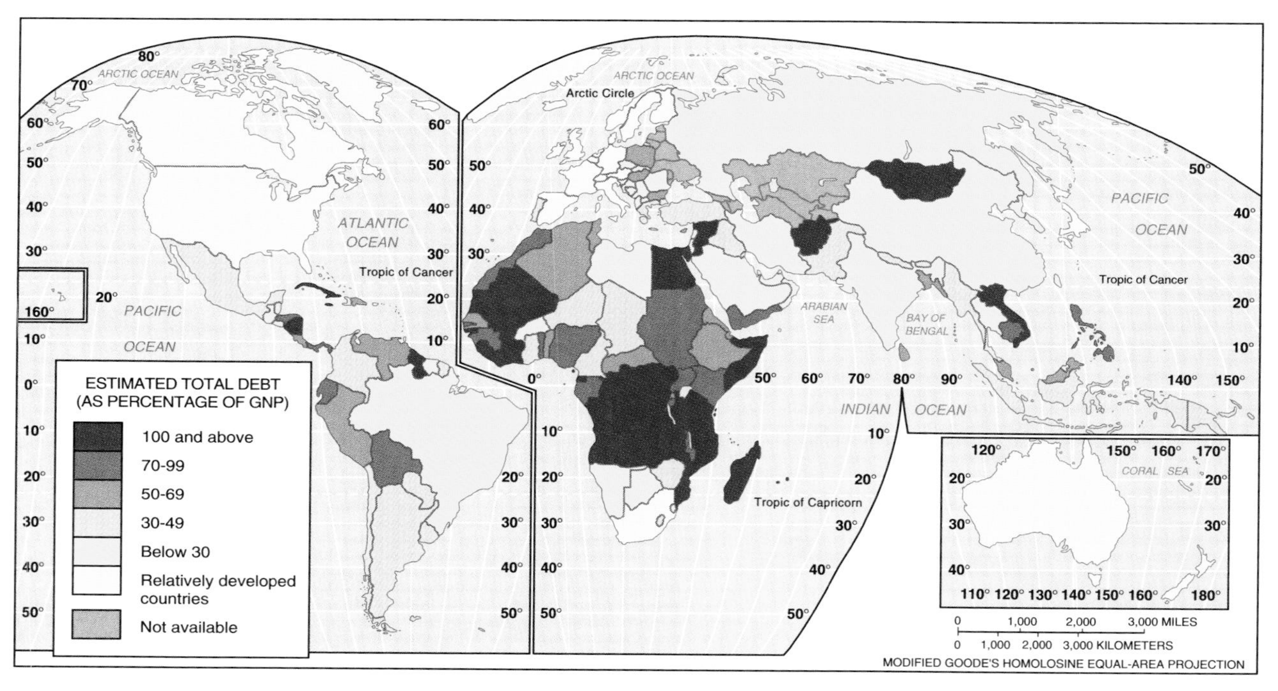

FIGURE 10–13
Debt as percentage of GNP. To finance development, many developing countries have accumulated large foreign debts in relation to their annual gross national products. As a result, a large percentage of the national budget must be used for repayment of the loans, unless financial institutions in relatively developed countries permit delays in repayment.

electric dams, electric transmission lines, flood protection systems, water supplies, roads, and hotels. New infrastructure can improve people's living conditions and promote economic growth.

New projects do not always succeed. Tanzania, for example, built a new railroad line in the 1970s to transport copper from neighboring landlocked Zambia to the port of Dar es Salaam. More than a decade later, Tanzanians were still not trained to operate the system, little copper had been hauled out, and the trains ran only with foreign engineers. In Mali, a French-sponsored project to pump water from the Niger River through solar energy functioned for only one month. Even when it worked, the project, which cost over $1 million, produced no more water than could two diesel pumps that together cost $6,000.

In principle, the new economic activities attracted to the area provide additional revenue needed to repay the loans. However, in recent years, many developing countries have been unable to repay the interest on their loans, let alone the principal. Brazil, Mexico, Argentina, and several other Latin American countries have accumulated the largest debts, although several African countries have very high ratios of debt to GNP (Figure 10–13).

When countries are unable to repay their debts, financial institutions in relatively developed countries refuse to make further loans, and development of needed infrastructure stops. The inability of many developing countries to repay loans also hurts the relatively developed countries, whose financial institutions suffer losses.

The relatively developed and developing countries run the risk of increasing confrontation. Although economically dominant, the relatively developed countries represent a minority of Earth's population and have been pressed by developing countries to share the world's wealth more evenly. Developing countries point out that prices have declined for many of their resources but increased for most of the goods manufactured in relatively developed countries. As a result, developing countries argue that they receive less for exporting their raw materials but must pay more to import manufactured goods. Developing countries also demand an increased role in the decisions to issue loans made by international agencies. A 1974 U.N. declaration called for creation of a "new international economic order," based on greater equality and economic interdependence between relatively developed and developing countries. In the two decades since the U.N. declaration, differences between relatively developed and developing countries in many indicators of development, such as GNP per capita, have widened, not narrowed.

For their part, relatively developed countries are increasingly concerned about their own economic health and are more careful in providing grants or loans to developing states. In exchange for canceling or refinancing the debts, the international lending agencies, which are dominated by the relatively developed countries, require the governments of developing countries to impose economic austerity programs. Developing countries are required to raise taxes, reduce government spending, and increase charges for using public services. These programs may be unpopular with the voters and encourage political unrest.

Transnational corporations. The global economy is increasingly influenced by large **transnational corporations**, sometimes called multinational corporations, or multinationals. A transnational corporation operates factories in countries other than the one in which its headquarters are located. Initially, transnational corporations were nearly all American-owned, but in recent years Japanese, German, French, and British companies have been active as well (Figure 10–14).

Some transnational corporations locate factories in other countries to expand their markets. Manufacturing the product where it is to be sold overcomes the restrictions that some developing countries place on imports. Furthermore, given the lack of economic growth in many relatively developed countries, a corporation may find that the only way it can increase sales is to move into a developing country. Transnational corporations also open factories in developing countries in order to reduce production costs, especially through hiring workers who are paid much lower wages than in relatively developed countries.

Electronic communications systems allow transnational corporations to centralize decision-making in a relatively developed country while decentralizing some day-to-day operations to developing countries. Governments have facilitated locational flexibility by reducing controls over the transfer of money from one country to another.

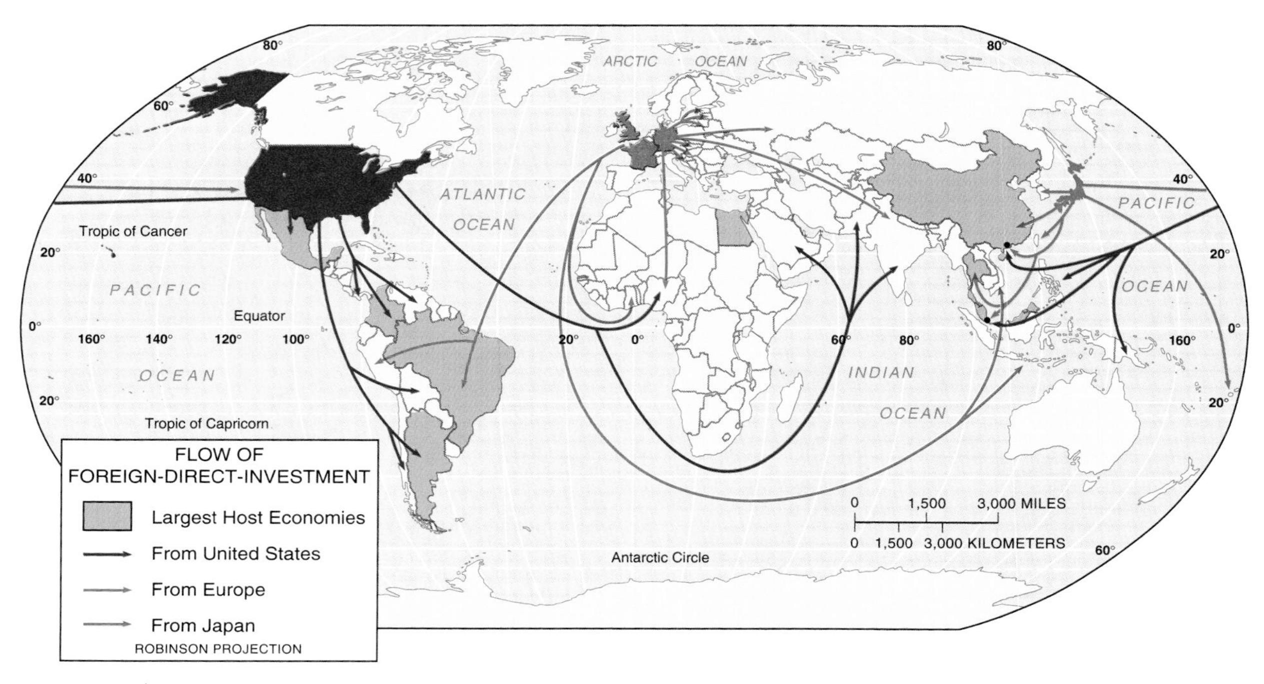

FIGURE 10–14
Flow of foreign investment, transnational corporations. Most transnational corporations originated in one of five countries: the United States, Japan, the United Kingdom, Germany, and France. These companies have invested most of their resources in the other relatively developed countries. In addition, U.S. transnationals are more likely than European or Japanese ones to invest in Latin America, whereas European transnationals are more likely to invest in Eastern Europe, and Japanese transnationals are more likely to invest in Asia.

Transnational corporations can reduce risk by having the ability to increase or decrease production in various places, depending on local conditions. Maintaining a factory in a developing country may account for a small percentage of a transnational corporation's total investment, but increasing or decreasing production in this factory could have a major impact on the economy of a developing country.

Uneven Development

Not everyone in a developing country has access to the benefits of an increasing GNP. The international trade approach calls for concentrating investment to promote the country's key assets. As a result, regions of the country where the assets are concentrated may receive most of the investment funds, and people involved in the particular sector of the economy may receive most of the benefits of development.

Not only do people in other economic sectors and in other regions of the country fail to benefit from the international trade approach; they may actually suffer reduced standards of living, as national governments eliminate subsidies on food and health care.

Adverse environmental impacts. Development can improve the standard of living and reduce poverty and hunger, but these benefits often come at a high environmental cost, both to the country and to the entire world. The challenge for today's developing countries is to promote development and environmental protection at the same time.

Traditional approaches to development have resulted in construction of new infrastructure without taking into account environmental, economic, and social consequences. Once it completes a project, the World Bank or other international lender typically departs from the scene and turns over its management to the local community. The local community, though, may not benefit from the new machinery if it lacks the expertise to operate it, the money to pay for its maintenance and repair, and the additional support equipment to efficiently utilize it.

For example, to promote more productive agriculture, international development agencies have paid for dams, reservoirs, and canals that distribute much-needed water to farms. However, some of these projects have caused diffusion of water-borne diseases such as schistosomiasis, while others remain underutilized because local farmers cannot afford to buy fertilizers, insecticides, and water pumps.

A United Nations commission on Environment and Development headed by the Prime Minister of Norway, Gro Harlem Brundtland, has called for a new form of development known as sustainable development. **Sustainable development** is the level of development that can be maintained in a country without depleting resources to the extent that future generations will be unable to achieve a comparable level of development. The commission's 1987 report, *Our Common Future*, defined sustainable development as "development that meets the needs of the present without compromising the ability of future generations to meet their own needs."

The concept of sustainable development is based on the current practice of sustained yield management of renewable resources, such as forests and fisheries. In some places, the amount of timber cut down in a forest or the number of fish removed from a body of water is controlled at a level that does not reduce future supplies.

The U.N. commission urged development agencies to consider the full range of environmental and social costs of a project before building it. The costs and benefits of new infrastructure should be examined by ecologists and public health officials as well as by engineers and economists. A new system should be designed so that it can be operated and maintained by local residents, not just foreign experts.

In recent years, the World Bank and other international development agencies have embraced the concept of sustainable development. Planning for development involves consideration of many more issues today than was the case in the past. However, one important recommendation of the Brundtland report has not been implemented: increased international cooperation to reduce the gap between relatively developed and developing countries.

One prominent difference between a relatively developed country and a developing country is access to motor vehicles. In the United States, most people commute to work by car, often driving on freeways, such as I-93 in Boston. Because few people can afford a car, walking and biking are more common in developing regions, such as Kochi (Cochin), India, a city of nearly 1 million inhabitants.
Photo left: (David Sailors/The Stock Market)
Photo right: (Blain Harrington III/The Stock Market)

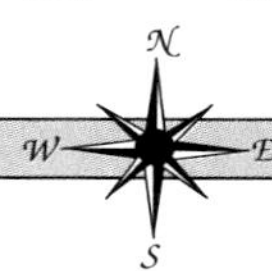

CONCLUSIONS: CRITICAL ISSUES FOR THE FUTURE

In geographic terms, development in a country depends on access. People need access to financial resources; access to information, knowledge, and skills; access to transport, machinery, and other technology; and access to health and security. These conditions are not available to two-thirds of the world's inhabitants who live in conditions of underdevelopment. Underdevelopment is the presence of economic and social barriers that limit access to the means of promoting development.

To achieve the benefits of development, developing countries must overcome the barriers of underdevelopment. In a global economy, where development is achieved through adopting the international trade alternative, securing access to resources has proven difficult for many developing countries.

Countries in some regions of the world, especially in Asia, have made considerable progress in recent years at promoting development, especially in terms of the main social and economic indicators of development. A major challenge for these developing countries is to view the atmosphere, biosphere, hydrosphere, and lithosphere as fragile and valuable resources in need of careful handling.

Other regions—notably Eastern Europe and Sub-Saharan Africa—have seen the standard of living decline in recent years. Even if these regions receive a massive investment of international development funds, they will be fortunate to see development built back to the level of the 1970s.

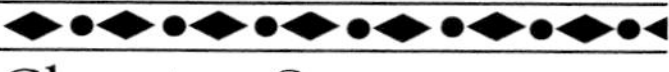

Chapter Summary

1. Indicators of development

Development is the process by which the material conditions of a country's people are improved. A relatively developed country has a higher level of per capita GNP, achieved through a transformation in the structure of the economy from a predominantly agricultural to an industrial and service-providing society. Relatively developed countries use their wealth in part to provide better health, education, and welfare services. Developing countries must use much of their additional wealth to meet the needs of a rapidly growing population.

2. Regional differences in development

Anglo-America, Western Europe, and Japan form a core of relatively developed areas. Six developing regions include Latin America, East Asia, Southeast Asia, South Asia, the Middle East, and Sub-Saharan Africa. These developing regions face different prospects for promoting development.

3. Promoting development

Developing countries choose between international trade and the self-sufficiency paths toward development. According to the international trade approach, a country promotes development by concentrating scarce resources on distinctive local assets. According to the self-sufficiency approach, a country spreads investment throughout all sectors of the economy and encourages domestic production rather than imports and exports.

4. Obstacles to development

The international trade approach to development has been adopted by an increasing number of developing countries, because it has been viewed as more successful than the self-sufficiency approach at increasing GNP per capita. However, the international trade approach can increase a developing country's dependency on financial institutions and markets in relatively developed countries, and the approach can foster uneven development with the country. Unless sustainable development is practiced, development may come at the expense of depletion of future resources.

Key Terms

Developing country, 389
Development, 389
Gross national product (GNP), 394
Literacy rate, 400
Primary sector, 395
Productivity, 396
Quaternary sector, 395
Relatively developed country, 389
Secondary sector, 395
Sustainable development, 419
Tertiary sector, 395
Transnational corporation, 417
Value added, 396

Questions for Study and Discussion

1. What is the Human Development Index?
2. Identify economic, social, and demographic indicators of development.
3. Why is the percentage of people working in the primary sector an important indicator of a country's level of development?
4. What is meant by *productivity?* Explain why someone in a developing country can work very hard but be much less productive than someone in a relatively developed country?
5. Why is the crude birth rate a good indicator of a country's level of development, whereas crude death rate is not?
6. Compare the major economic assets of the three core regions—North America, Western Europe, and Japan.
7. What has been the impact of the fall of communism on the level of development in Eastern European countries?
8. Compare the level of development between Latin America and Sub-Saharan Africa. Are differences in level of development greater between developing and relatively developed regions or greater among the various developing regions?
9. Describe the major differences between the international trade and self-sufficiency approaches to development. Why have countries switched to the international trade approach in recent years?
10. How do countries finance development?
11. What is meant by *sustainable development?*

Thinking Geographically

1. Review the major economic, social, and demographic characteristics that contribute to a country's level of development. Which indicators can vary significantly by gender within countries and between countries at various levels of development?
2. Some geographers have been attracted to the concepts of Immanuel Wallerstein, who argued that the modern world consists of a single entity, the capitalist world-economy, that is divided into three regions: the core, semi-periphery, and periphery. How have the boundaries among these three regions changed?
3. China has relied on the self-sufficiency approach to promote development, whereas Hong Kong has been a prominent practitioner of the international trade approach. What problems might China and Hong Kong face in reconciling these two approaches, once Hong Kong becomes part of China in 1997?
4. Some developing countries claim that the requirements placed on them by lending organizations such as the World Bank impede rather than promote development. Should developing countries be given a greater role in deciding how much the international organizations should spend, and how such funds should be spent?
5. What obstacles do Eastern European countries face as they dismantle forty years of socialism and convert to market economies?

Suggestions for Further Reading

Ballance, R., J. Ansari, and H. Singer. *The International Economy and Industrial Development: Trade and Investment in the Third World*. Totowa, NJ: Allanheld, Osmun, 1982.

Barker, Randolph, and Robert W. Herdt. *The Rice Economy of Asia*. Washington: Resources for the Future, 1985.

Bater, James H. *The Soviet Scene: A Geographical Perspective*. New York: Routledge, 1989.

Bebbington, Anthony J., Hernan Carrasco, Lourdes Peralbo, Galo Ramon, Jorge Trujillo, and Victor Torres. "Fragile Lands, Fragile Organizations: Indian Organizations and the Politics of Sustainability in Ecuador." *Transactions of the Institute of British Geographers, New Series* 18 (1993): 179–96.

Berry, Brian J. L., Edgar C. Conkling, and D. Michael Ray. *Economic Geography*. Englewood Cliffs, NJ: Prentice-Hall, 1987.

Blakemore, Harold, and Clifford T. Smith, eds. *Latin America: Geographical Perspectives*, 2d ed. London: Methuen, 1983.

Bobek, Hans. "The Main Stages in Socioeconomic Evolution from a Geographic Point of View." In *Readings in Cultural Geography*, edited by Philip L. Wagner and Marvin W. Mikesell. Chicago: The University of Chicago Press, 1962.

Chang, Sen-dou. "Modernization and China's Urban Development." *Annals of the Association of American Geographers* 71 (December 1981): 572–79.

Chisholm, Michael. *Modern World Development: A Geographical Perspective*. Totowa, NJ: Barnes and Noble Books, 1982.

_____. "The Wealth of Nations." *Transactions of the Institute of British Geographers, New Series* 5 (no. 2, 1980): 255–76.

Cole, John P. *The Development Gap: A Spatial Analysis of World Poverty and Inequality*. New York: Wiley, 1980.

Crow, Ben, and Alan Thomas. *Third World Atlas*. Philadelphia: Open University Press, 1985.

Demko, George, ed. *Regional Development: Problems and Policies in Eastern and Western Europe*. New York: St. Martin's Press, 1984.

DeSouza, Anthony R., and Phillip Porter. *The Underdevelopment and Modernization of the Third World*. Washington, DC: Association of American Geographers, 1974.

DeSouza, Anthony R., and Frederick P. Stutz. *The World Economy: Resources, Location, Trade, and Development*. 2d ed. New York: Macmillan, 1994.

Dickenson, J. P., C. G. Clarke, W. T. S. Gould, R. M. Prothero, D. J. Siddle, C. T. Smith, E. M. Thomas-Hope, and A. G. Hodgkiss. *A Geography of the Third World*. New York: Methuen, 1983.

Dott, Ashok K., ed. *Southeast Asia: Realm of Contrasts*. 3d ed. Boulder, CO: Westview Press, 1985.

Flavin, Christopher. "Electricity for a Developing World: New Directions." *Worldwatch Paper* 70. Washington, DC: Worldwatch Institute, June 1986.

Forbes, D. K. *The Geography of Underdevelopment: A Critical Survey*. Baltimore: The Johns Hopkins University Press, 1984.

Fryer, Donald D. "The Political Geography of International Lending by Private Banks." *Transactions of the Institute of British Geographers, New Series* 12 (no. 4, 1987): 413–32.

Ginsburg, Norton S. *Atlas of Economic Development*. Chicago: University of Chicago Press, 1961.

_____, ed. *Essays on Geography and Economic Development*. Chicago: University of Chicago Press, 1960.

Grossman, Larry. "The Cultural Ecology of Economic Development." *Annals of the Association of American Geographers* 71 (June 1981): 220–236.

James, Preston E. *Latin America*. 4th ed. New York: Odyssey House, 1969.

Jones, D. B., ed. *Oxford Economic Atlas of the World*. 4th ed. London and New York: Oxford University Press, 1972.

Jumper, Sidney R., Thomas L. Bell, and Bruce A. Ralston. *Economic Growth and Disparities: A World View*. Englewood Cliffs, NJ: Prentice-Hall, 1980.

Mabogunje, Akinlawon L. *The Development Process: A Spatial Perspective*. London: Hutchinson University Library, 1981.

Momsen, Janet Henshall, and Janet Townsend. *Geography of Gender in the Third World*. Albany: State University of New York Press, 1987.

Myrdal, Gunnar. *Rich Lands and Poor*. New York: Harper and Bros., 1957.

Rostow, Walter W. *The Stages of Economic Growth*. Cambridge: Cambridge University Press, 1960.

Seager, Joni, and Ann Olson. *Women in the World: An International Atlas*. New York: Simon & Schuster, 1986.

Smith, David M. *Where the Grass is Greener: Living in an Unequal World*. London: Croom Helm, 1979.

Szentes, Tamas. *The Political Economy of Underdevelopment*. 4th ed. Budapest, Hungary: Akademiai Kiado, 1983.

Taylor, Peter J. "World-Systems Analysis and Regional Geography." *Professional Geographer* 40 (August 1988): 259–65.

Wallerstein, Immanuel. *The Capitalist World-Economy*. Cambridge: Cambridge University Press, 1979.

_____. *The Politics of the World-Economy*. Cambridge: Cambridge University Press, 1984.

Wheeler, James O., and Peter O. Muller. *Economic Geography*. New York: Wiley, 1981.

Wilbanks, Thomas J. *Location and Well-Being: An Introduction To Economic Geography*. San Francisco: Harper & Row, 1980.

World Commission on Environment and Development. *Our Common Future*. London: Oxford University Press, 1987.

We also recommend these journals: *Economic Development and Cultural Change, Economic Geography, International Development Review, International Economic Review, International Journal of Political Economy, Journal of Developing Areas, Netherlands Journal of Economic and Social Geography, Regional Studies*.

11 Agriculture

- Distribution of agriculture
- Subsistence agriculture in developing regions
- Commercial agriculture in relatively developed regions
- Increasing the world supply of food

Rice on terraced land in the Phillipines.

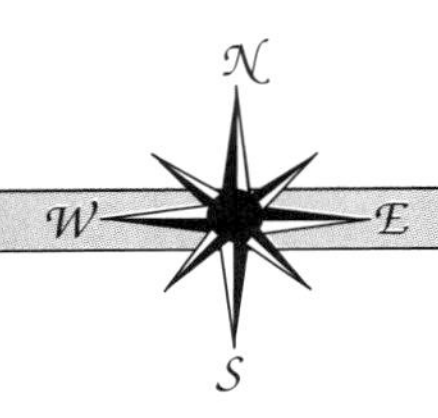

1. Distribution of agriculture
Most people in developing countries practice subsistence agriculture, whereas commercial agriculture predominates in relatively developed countries. Agriculture probably originated independently in multiple locations in both the Eastern and Western Hemispheres.

2. Subsistence agriculture in developing regions
Principal forms of subsistence agriculture are shifting cultivation, pastoral nomadism, and intensive subsistence.

3. Commercial agriculture in relatively developed regions
Compared to subsistence farms, commercial farms are relatively large, make more extensive use of machinery, and are well-integrated with other segments of the food production industry.

4. Increasing the world supply of food
Subsistence farmers in developing countries lack the land and supplies needed to expand crop production to feed a rapidly expanding population. Commercial farmers in relatively developed countries produce food surpluses, which help keep food prices low.

INTRODUCTION

Geographers study the distribution of agriculture across Earth's surface and how that distribution relates to cultural and environmental factors. Elements of the physical environment, such as climate, soil, and topography, set broad limits on agricultural practices, but farmers can observe and modify the environment in a variety of ways.

How farmers work with the physical environment varies according to customary beliefs, preferences, technological capabilities, and other cultural factors. The farmers of a society possess specific knowledge about environmental conditions and particular technological capabilities for modifying the land. Within the limits of technological capability, farmers choose particular agricultural practices based on perceptions of the relative value of various alternatives.

Farmers' values are partly economic and partly cultural. Farmers generally try to undertake the most profitable type of agriculture, although economic calculations can be altered by government programs that subsidize some forms of agricultural production and discourage others. Farmers also select agricultural practices based on cultural perceptions, because a society may hold some foods in high esteem while it avoids others (see Chapter 8).

This chapter examines the predominant types of agriculture practiced around the world. While geographers observe a wide variety of agricultural practices, the most important distinction is what happens to the product of the farm. As discussed in Chapter 10, geographers divide the world into the economically developing regions, where the output is frequently consumed on or near the farm where it was produced, and the relatively developed regions, where the farmer sells the crops and livestock off the farm.

Distribution of Agriculture

As with other cultural and environmental phenomena, geographers document the origin, diffusion, and current distribution of agriculture across Earth's surface. The distribution forms the basis for analyzing processes by which patterns of agricultural practices develop and issues arising from the distribution patterns.

Agricultural Regions

Farmers practice forms of agriculture distinctive to their area of the world. Several attempts have been made to classify the world's major types of agriculture into meaningful groups, but, significantly, few of these classifications include maps that distribute these groups into regions.

Many contemporary geographers accept Derwent Whittlesey's 1936 classification with some

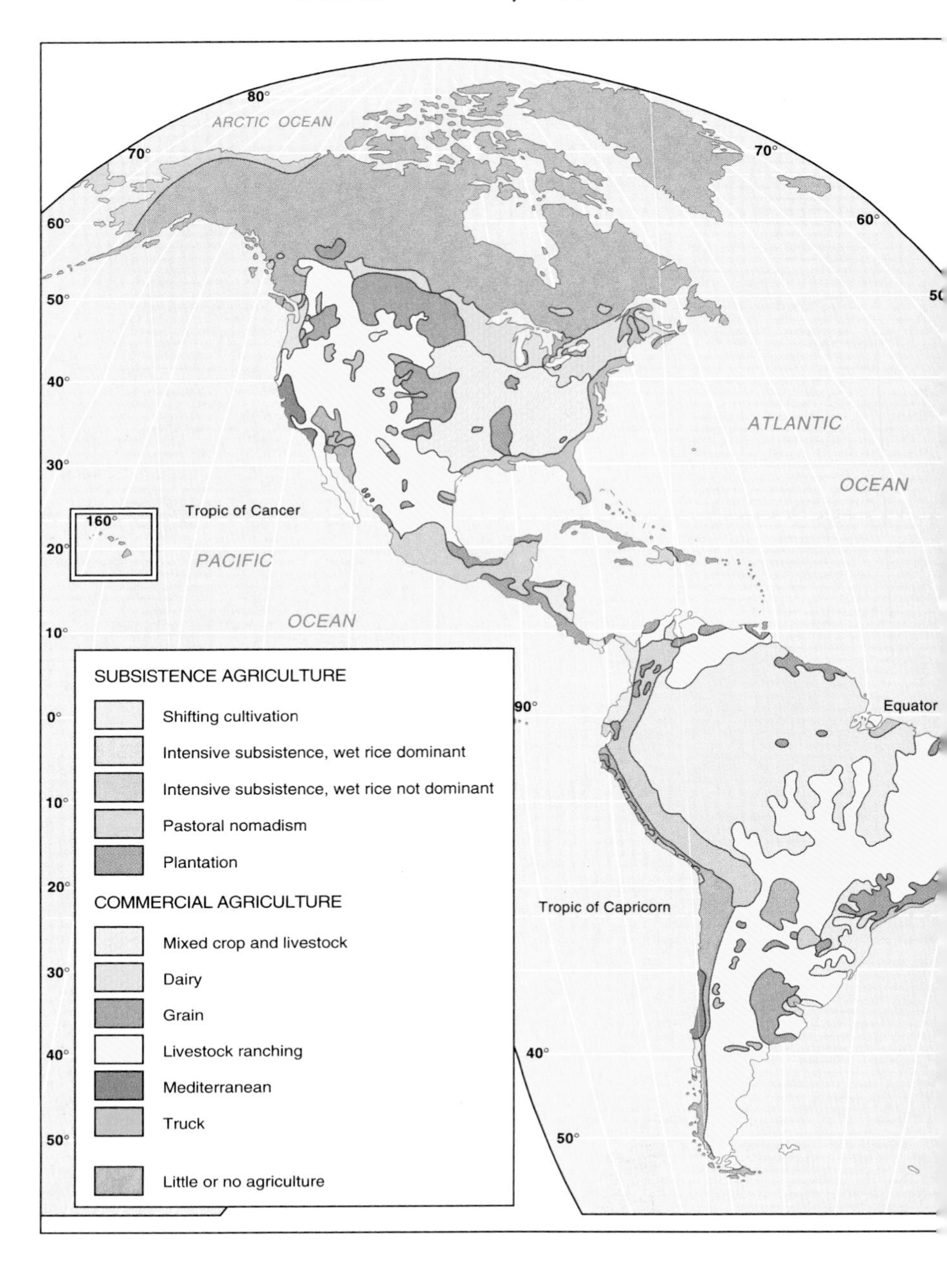

FIGURE 11–1
The major agricultural practices of the world can be divided into subsistence and commercial regions. Subsistence regions include shifting cultivation, intensive subsistence with wet rice dominant, intensive subsistence with wet rice unimportant, and pastoral nomadism. Commercial regions include mixed crop and livestock, dairying, grain, ranching, Mediterranean, commercial gardening and fruit farming, and plantation. (Reproduced by permission from the *Annals of the Association of American Geographers,* Vol. 26, 1936, p. 241, Figure 1., D. Whittlesay)

modification. Whittlesey identified eleven main agricultural regions, plus an area where agriculture was nonexistent at that time (Figure 11–1). These eleven types of agriculture include five that are important in developing countries and six that are important in relatively developed countries.

The major agriculture regions in developing countries include shifting cultivation, pastoral nomadism, intensive subsistence (wet rice dominant), intensive subsistence (crops other than rice dominant), and plantation. Shifting cultivation is found primarily in the tropical regions of South America, Africa, and Southeast Asia, while pastoral nomadism is characteristic of the dry lands of North Africa and Asia. The two forms of intensive subsistence agriculture are found primarily in the large population concentrations of East and South Asia; rice is the preferred crop, with other crops selected where growing rice is difficult. Plantations are found in the tropical and subtropical regions of Latin America, Africa, and Asia.

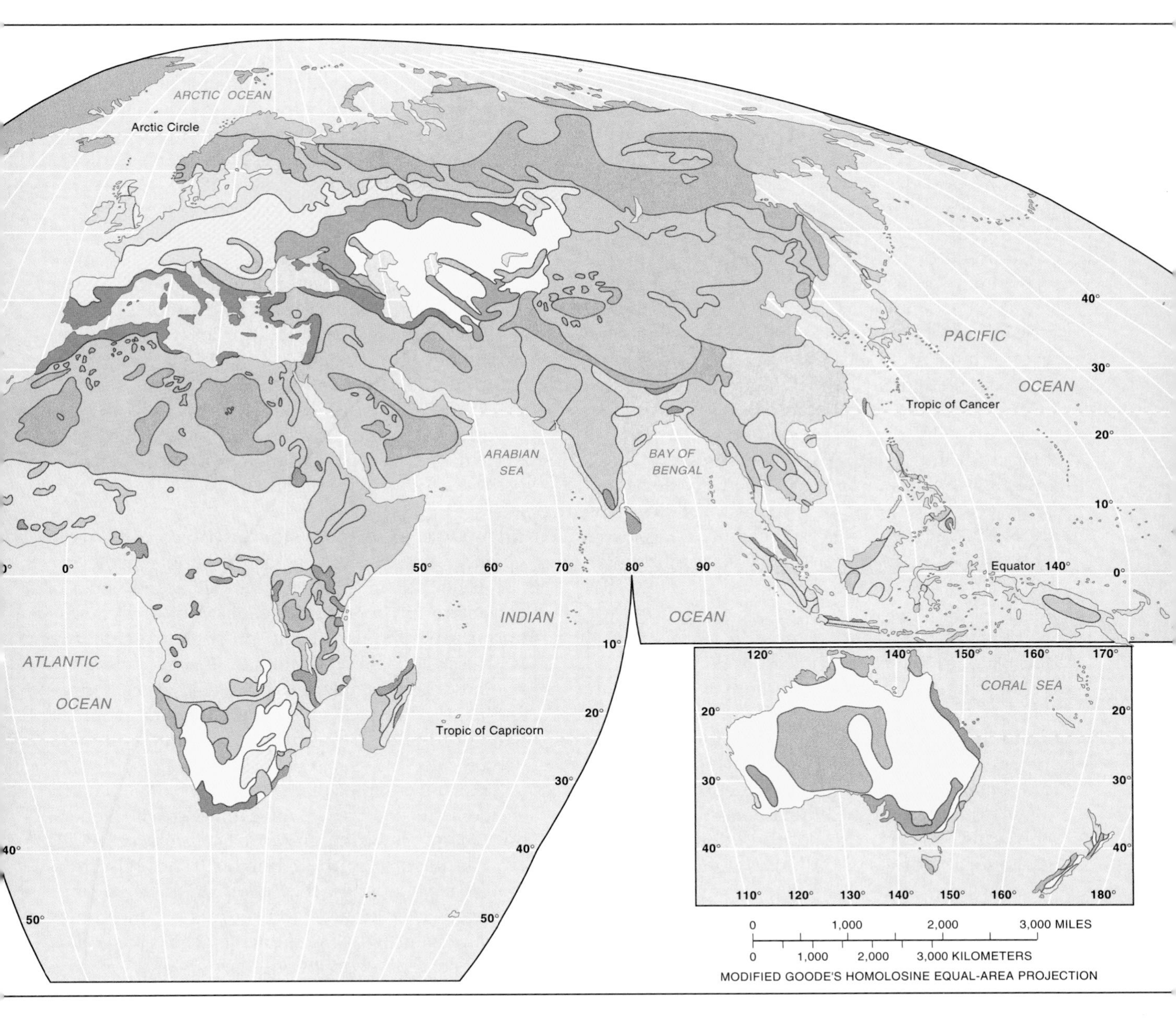

Major agricultural regions in relatively developed countries include mixed crop and livestock (primarily in the U.S. Midwest and central Europe); dairying (primarily near population clusters in northeastern United States, southeastern Canada, and northwestern Europe); grain (primarily north central United States and eastern Europe); and ranching (primarily the dry lands of western United States, southeastern South America, central Asia, southern Africa, and Australia). Mediterranean agriculture is characteristic of lands surrounding the Mediterranean Sea, as well as in the western United States and Chile, while truck farming (also known as commercial gardening and fruit farming) is found primarily in the southeastern United States and southeastern Australia.

The largest percentage of people in developing countries practice **subsistence agriculture,** which primarily provides food for direct consumption by the farmer and the farmer's family and neighbors. In developing regions where subsistence agriculture predominates, most people work in agricultural rather than industrial or service jobs. Farmers in developing countries may sell some of their output to the government or to private firms for distribution in cities or foreign markets. However, the surplus product is not the farmer's primary purpose and may not be available some years because of poor harvests.

In relatively developed countries, farmers sell their crops and livestock off the farm. Few people engage in full-time farming in relatively developed countries, but these farmers are able to grow enough food to feed the vast majority of residents, who work in industrial or service sector jobs.

Origin and Diffusion of Agriculture

Because agriculture began before recorded history, we cannot document its origins with certainty. Scholars try to reconstruct a logical sequence of events based on fragments of information about ancient agricultural practices and historical environmental conditions. Improvements in cultivating crops and domesticating animals evolved over thousands of years. This section offers one explanation for the origin and diffusion of agriculture.

Hunters and gatherers. Before the invention of agriculture, all humans probably obtained the food they needed for survival through hunting for animals, gathering wild plants, or fishing. Hunters and gatherers lived in small groups, usually fewer than fifty, because a larger number would quickly exhaust the available resources. They survived by collecting food often, perhaps every day. The daily search for food could take only a short amount of time or much of the day, depending on conditions in the particular location. The men went out to hunt wild animals or to fish, and the women collected fruits, berries, nuts, and roots.

The group traveled frequently, establishing new home bases or camps. The direction and frequency of migration depended on the movement of wild animals and the growth of seasonal plants at various locations. Groups communicated with each other concerning hunting rights, intermarriages, and other specific subjects. For the most part, groups tried to steer clear of each other's territory.

Today, only about 250,000 people, or less than 0.005 percent of the world's population, still survive by hunting and gathering. These people live in a number of isolated locations, including Arctic regions and parts of the interiors of Africa, Australia, and South America. Examples include the Bushmen of Namibia and Botswana and the Aborigines of Australia.

Contemporary hunting and gathering societies are isolated groups living on the periphery of world settlement. But they provide insight into human customs prevailing in prehistoric times, before the invention of agriculture.

Adoption of settled agriculture. Why did nomadic groups convert from hunting, gathering, and fishing to agriculture? Determining the origin of agriculture first requires a definition of agriculture—not an easy term to define precisely. **Agriculture** involves the deliberate human effort to modify a portion of Earth's surface through cultivation of crops and rearing livestock for sustenance or for economic gain. Agriculture thus originated when humans domesticated plants and animals for their use.

The cultivation of plants may have originated by accident. In the process of gathering wild vegetation, members of the group may have accidentally cut plants or dropped berries. These hunters would have noticed that after a period of time, damaged or destroyed food sources produced new plants. Eventually, members of the group may have decided to deliberately cut plants or drop berries on the ground to see if they produced new plants. Subsequent generations may have poured water over the site, permitted animal droppings to fertilize the soil, and performed other experiments. Over hundreds or thousands of years, plant cultivation evolved from a series of accidents and deliberate experiments, according to this theory.

Prehistoric people may have originally domesticated animals for noneconomic reasons, such as sacrifices and other religious ceremonies. Other animals probably were domesticated as household pets, surviving on the group's food scraps.

The earliest form of plant cultivation, according to the prominent cultural geographer Carl Sauer, consisted of vegetative planting. Sauer defined **vegetative planting** as the reproduction of plants by direct cloning from existing plants, such as cutting stems and dividing roots. Plants found growing wild were deliberately divided and transplanted.

Seed agriculture, practiced by most contemporary farmers, came later, according to Sauer. **Seed agriculture** involved the reproduction of plants through annual introduction of seeds, which result from sexual fertilization.

Location of first agriculture. Agriculture probably did not originate in one location, but in multiple, independent *hearths,* or points of origin. From these hearths, agricultural practices diffused to other portions of Earth's surface.

Vegetative planting probably originated in Southeast Asia, according to Sauer (Figure 11–2). The diversity of climate and topography in the region probably encouraged the growth of a wide variety of plants suitable for dividing and transplanting. Also, because the people in this region obtained food primarily by fishing, rather than hunting and gathering, they may have been more sedentary and therefore able to devote more attention to growing plants. The first plants domesticated in Southeast Asia through vegetative planting probably included roots such as the taro and yam and tree crops such as the banana and palm. The dog, pig, and chicken were probably first domesticated in Southeast Asia.

Early hearths of vegetative planting may have also emerged independently in West Africa and northwestern South America. Vegetative planting may have been based on the oil-palm tree and yam

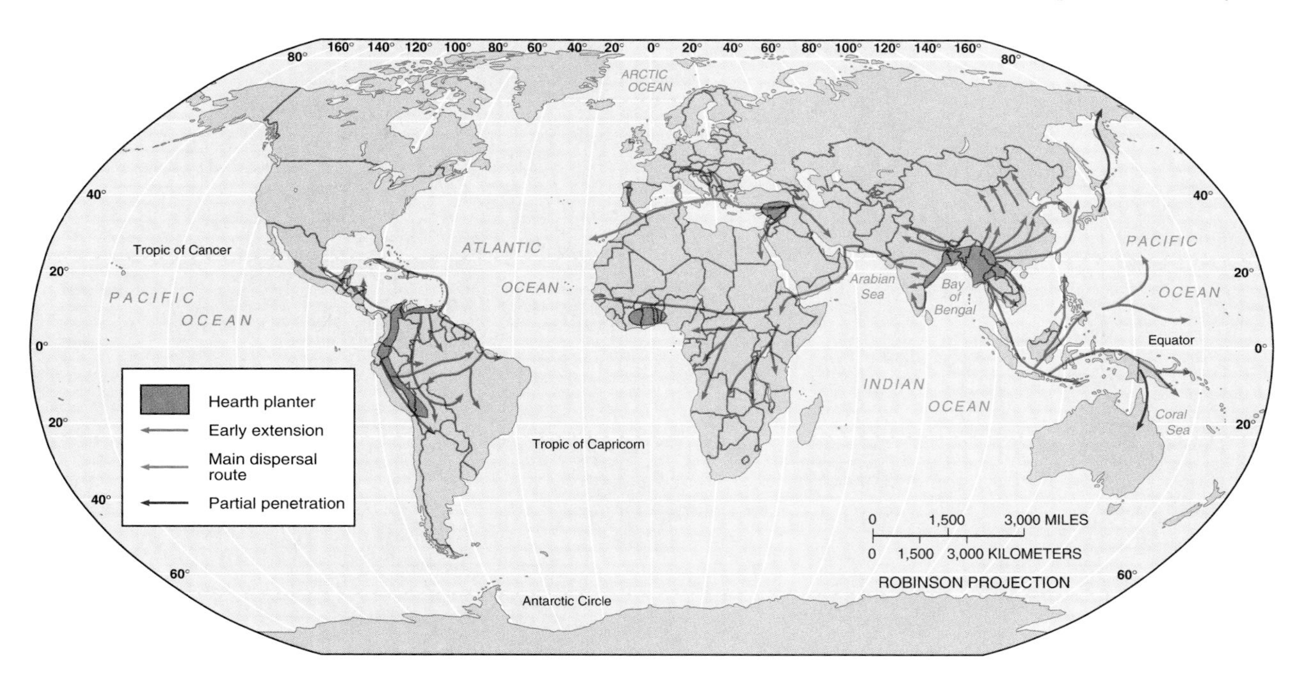

FIGURE 11–2
Origin and diffusion of vegetative planting. Vegetative planting, which is the reproduction of plants by direct cloning from existing plants, originated primarily in Southeast Asia, according to Carl Sauer. Two other early centers of vegetative planting were in West Africa and northwestern South America. From these hearths, the practice diffused to other regions. (Adapted from Carl O. Sauer, *Agricultural Origins and Dispersals,* with the permission of the American Geographical Society.)

in West Africa and the manioc, sweet potato, and arrowroot in South America. Vegetative planting diffused from the Southeast Asian hearth north and east to China and Japan and west through India to Southwest Asia, tropical Africa, and the Mediterranean lands. The practice spread from northwestern South America to Central America and eastern portions of South America.

Seed agriculture also originated in more than one hearth. Sauer identified three hearths in the Eastern Hemisphere: western India, northern China, and Ethiopia. Seed agriculture diffused quickly from western India to Southwest Asia, where important early advances were made. Wheat and barley, two grains that became particularly important thousands of years later in European and American civilizations, were domesticated in this region (Figure 11–3).

Inhabitants of Southwest Asia also apparently were the first to integrate seed agriculture with the domestication of herd animals such as cattle, sheep, and goats. These animals were used to plow the land before planting seeds and, in turn, were fed part of the harvested crop. Other animal products, such as milk, meat, and skins, were first exploited at a later date, according to Sauer. This integration of plants and animals is a fundamental element of modern agriculture.

Early diffusion of agriculture. Seed agriculture diffused from Southwest Asia to the west across Europe and through North Africa. Greece, Crete, and Cyprus display the earliest evidence of seed agriculture in Europe. From these countries, agriculture may have diffused northwestward through the Danube basin, eventually to the Baltic and North seas, and northeastward to Ukraine. Most of the plants and animals domesticated in Southwest Asia spread into Europe, although barley and cattle became more important further north, perhaps because of cooler and moister climatic conditions.

Seed agriculture also diffused eastward from Southwest Asia to northwestern India and the Indus River plain. Again, a variety of domesticated plants and animals were brought from Southwest Asia, although other plants, such as cotton and rice, arrived in India from different hearths.

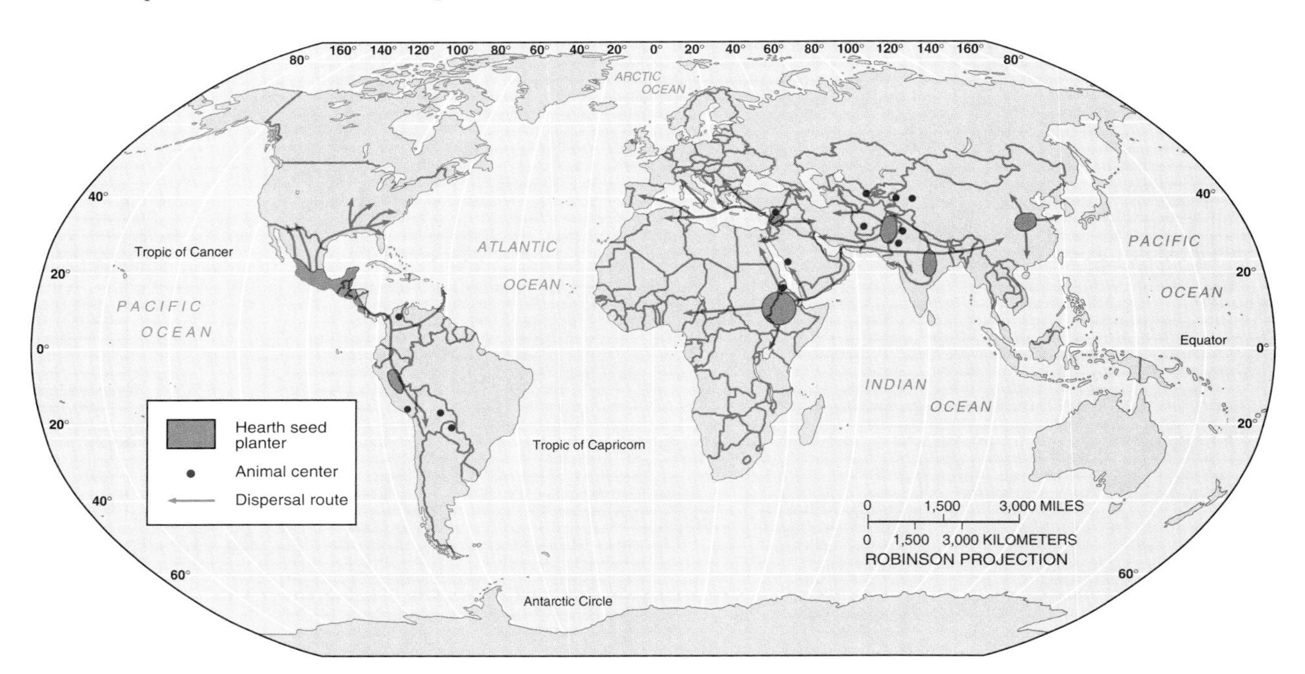

FIGURE 11–3
Origin and diffusion of seed agriculture and livestock herding. Seed agriculture may have originated in several hearths, including western India, northern China, and Ethiopia. Southern Mexico and northwestern South America may have been other early hearths. Early advances were made in Southwest Asia. (Adapted from Carl O. Sauer, *Agricultural Origins and Dispersals,* with the permission of the American Geographical Society.)

From the northern China hearth, millet diffused to South and Southeast Asia. Rice, which ultimately became the most important crop in much of Asia, has an unknown hearth, although some geographers consider Southeast Asia to be the most likely location. Sauer identified a third independent hearth in Ethiopia, where millet and sorghum were domesticated at an early time. However, Sauer argued that the agricultural advances made in Ethiopia did not diffuse widely to other locations.

Two independent seed agriculture hearths originated in the Western Hemisphere: southern Mexico and northern Peru. The hearth in southern Mexico, which extended into Guatemala and Honduras, was the point of origin for squash and maize (popularly known as corn in the United States). Squash, beans, and cotton may have been domesticated in northern Peru. From these two hearths, agricultural practices diffused to other parts of the Western Hemisphere, although agriculture was not widely practiced until European colonists began to arrive some 500 years ago. The only domesticated animals were the llama, alpaca, and turkey; herd animals were unknown until European explorers brought them in the sixteenth century.

That agriculture had multiple origins suggests that from the earliest times people have produced food in distinctive ways in different regions. This diversity derives from a unique legacy of wild plants, climatic conditions, and cultural preferences in each region. Improved communications in recent centuries have encouraged the diffusion of some plants to a variety of locations in the world. Many plants and animals thrive across a wide portion of Earth's surface, not just in the place of original domestication. Only since A.D. 1500, for example, have wheat, oats, and barley been introduced to the Western Hemisphere and maize to the Eastern Hemisphere.

Subsistence Agriculture in Developing Regions

Subsistence farmers in developing countries employ a wide variety of agricultural practices designed primarily to provide food for direct consumption. This section examines four types of subsistence agriculture: *shifting cultivation, intensive subsistence* of two types—wet rice dominant and wet rice not dominant, and *pastoral nomadism.* Plantation agriculture, a form of commercial agriculture in developing regions, is considered along with commercial agriculture in relatively developed countries.

Shifting Cultivation

Shifting cultivation is the main form of agriculture found in much of the world's tropical, or *A,* climate regions, which have relatively high temperatures and abundant rainfall (see Chapter 3 for a review). Shifting cultivation predominates primarily in three tropical regions: the Amazon area of South America, Central and West Africa, and Southeast Asia, including Indochina, Indonesia, and New Guinea.

We use the term *cultivation* rather than *agriculture* to describe this means of obtaining food, because agriculture implies greater use of tools and animals and more sophisticated attempts to modify the landscape. Shifting cultivation bears little relationship to the agriculture found in the relatively developed regions of Western Europe and North America or even in other developing countries such as China.

Characteristics of shifting cultivation. Compared to other types of agriculture, **shifting cultivation** has two particularly important characteristics. First, farmers usually clear the land for planting in part by slashing the vegetation and burning the debris. As a result, shifting cultivation is sometimes known as *slash-and-burn agriculture.* Second, farmers grow crops on a cleared field for only a few years and then leave it fallow for many years.

People who practice shifting cultivation generally live in a small village and grow food in the surrounding land, which the village controls. Well-recognized boundaries usually separate a village from its neighbors.

Each year, the villagers designate for planting an area surrounding the settlement. Before planting, they must remove the dense vegetation that typically covers tropical land. The villagers cut down most of the trees with axes, sparing only those that are economically useful. One strategy is to chop down a handful of large trees situated at key junctures; as they fall, the larger trees will bring down smaller ones that may have been weakened by the notching the people performed earlier.

These shifting cultivation farmers in Peru are preparing fields for planting by slashing and burning the vegetation. The dense vegetation is chopped down, and the debris is burned in order to provide the soil with needed nutrients. (Asa C. Thoresen/Photo Researchers, Inc.)

The undergrowth is cleared away with a machete or other long knife. On a windless day, the debris is then burned under carefully controlled conditions. The rains wash the fresh ashes into the soil, providing needed nutrients. The cleared area is known by a variety of names in different regions of the world, including **swidden,** ladang, milpa, chena, and kaingin.

Before planting, fields are prepared by hand, perhaps with the help of a simple implement such as a hoe; plows and animals are rarely used. The only fertilizer generally available is potash (potassium) from burning the debris when the site is cleared. Relatively little weeding is done the first year that a cleared patch of land is farmed, although weeds may need to be cleared away with a hoe in subsequent years.

The cleared land can be used to grow crops for only a short time, usually three years or less. In many regions, the most productive harvest comes in the second year after burning. Thereafter, the nutrients in the soil are rapidly depleted, and the land is no longer sufficiently fertile to grow crops. Rapid growth of weeds also contributes to the abandonment of a swidden after a few years.

When the swidden is no longer fertile, the villagers identify a new site and begin the process of clearing the field. They leave the old site uncropped for many years, allowing it to be overrun again by natural vegetation. The field is not actually abandoned: the villagers will return to the site some day, perhaps as few as six years or as many as twenty years later, to begin the process of clearing the land again. In the meantime, they may still care for fruit-bearing trees on the site.

If a cleared area outside a village is too small to provide food for the entire village population, then some of the people may establish a new village and practice shifting cultivation there. In some cases, farmers may have to move temporarily to another settlement if the field they are clearing that year is relatively distant from the village.

The precise crops grown by each village vary by local custom and taste. The predominant crops include upland rice in Southeast Asia, maize (corn) and manioc (cassava) in South America, and millet and sorghum in Africa. Yams, sugar cane, plantain, and vegetables may also be grown in some regions. These crops may have been originally associated with one region of shifting cultivation but have diffused to other areas in recent years.

The Kayapo people of Brazil's Amazon tropical rainforest arrange crops in concentric rings. At first, they plant sweet potatoes and yams in the inner area and corn and rice, manioc, and more yams in successive rings. The outermost ring contains papaya, banana, pineapple, mango, cotton, and beans. Plants that require more nutrients are located in the outer ring, where the leafy crowns of the large trees hit the ground when they are cut to clear the field. In subsequent years, the inner area of potatoes and yams expands to replace corn and rice.

Because most families grow for their own needs, one swidden may contain a large variety of intermingled crops, which are harvested individually at the appropriate time. The field in shifting cultivation appears much more chaotic than a farm in a relatively developed region, where one crop, such as corn or wheat, may grow over an extensive area. In some cases, families may specialize in a few crops and trade with villagers who have a surplus of others.

Traditionally, land is owned by the village as a whole rather than separately by each resident. The chief or ruling council allocates a patch of land to each family and allows it to retain the output. Individuals may also have the right to own or protect specific trees surrounding the village. This land-tenure system has been replaced by private ownership in some communities, especially in Latin America.

Brazil's Amazon rainforest is being cleared for cattle ranching and other forms of agriculture. The timber is sold to builders. Mining activities often provided stimulus for deforestation. (Antonio Ribeiro/Gamma-Liaison, Inc.)

Future of shifting cultivation. The percentage of land devoted to shifting cultivation is declining in the tropics, and its future role in world agriculture is not clear. Shifting cultivation is being replaced by logging, cattle ranching, and cultivation of cash crops.

All of the alternatives to shifting cultivation require cutting down vast expanses of forest. In recent years, tropical rainforests have disappeared at the rate of 10 to 20 million hectares (25 to 50 million acres) per year. The amount of Earth's surface allocated to tropical rainforests has already been reduced to less than half of its original area, and unless drastic measures are taken, the area will be reduced by another 20 percent within a decade (see Chapter 5).

Governments in developing countries have supported the destruction of rainforests, because they view activities such as selling timber to builders or raising cattle for fast-food restaurants as more effective strategies for promoting economic development than shifting cultivation. Until recent years, the World Bank has provided loans to finance development proposals that require clearing forests. Furthermore, shifting cultivation is regarded as a relatively inefficient approach to growing food in a hungry world. The problem with shifting cultivation compared to other forms of agriculture is that it can support only a low level of population in an area without causing environmental damage.

To its critics, shifting cultivation is at best a preliminary step in the process of economic development for a society. Pioneers use shifting cultivation to clear forests in the tropics and to open land for development in places where permanent agriculture never existed. People unable to find agricultural land elsewhere can migrate to the tropical forests and initially practice shifting cultivation. It then should be replaced by other forms of agriculture that produce greater yields per land area.

Defenders of shifting cultivation consider it the most environmentally sound form of agriculture for the tropics. Practices associated with other forms of agriculture, such as introduction of fertilizers and permanent clearing of fields, may damage the soil and upset the ecological balance in the tropics. Destruction of the rainforests may also contribute to global warming, as discussed in chapters 3 and 5.

Elimination of shifting cultivation could upset traditional culture as well. The activities involved in shifting cultivation may be intertwined with other social, religious, political, and other folk customs. A drastic change in the agricultural economy could disrupt other activities of daily life.

In recognition of the importance of tropical rainforests to the global environment, developing countries have been pressured into restricting further destruction of them. In one innovative strategy, Bolivia agreed to set aside 1.5 million hectares (3.7 million acres) in a reserve in exchange for cancellation of $650,000,000 of its debt to relatively developed countries. Deforestation of Brazil's Amazon rainforest has declined from 2 million hectares (5.2 million acres) per year during the 1980s—including a peak of 2.9 million hectares (7.4 million acres) in 1985—to 1.1 million hectares (2.8 million acres) per year during the 1990s.

Pastoral Nomadism

Pastoral nomadism is a form of subsistence agriculture based on the herding of domesticated animals. It is adapted to dry climates where intensive subsistence agriculture is difficult or impossible. Pastoral nomads live primarily in the large belt of arid and semiarid land that includes North Africa, the Middle East, and parts of Central Asia. The Bedouins of Saudi Arabia and North Africa and the Maasai of East Africa are examples of nomadic groups. Only approximately 15 million people are pastoral nomads, but they occupy approximately 20 percent of Earth's land area.

In contrast with other subsistence farmers, pastoral nomads depend primarily on animals rather than crops for their survival. They take milk from the animals for food and skins and hairs for clothing and tents. Like other subsistence farmers, though, pastoral nomads consume mostly grain rather than meat. Their animals are commonly not slaughtered for meat, because nomads consider the size of their herd to be not only an important measure of power and prestige but also the main element of security in periods of adverse environmental conditions. When food supplies are limited, though, they will sell some of their animals for food, or consume meat.

Some pastoral nomads obtain grain from sedentary subsistence farmers in exchange for animal products. More often, part of a nomadic group, perhaps the women and children, may plant crops at a fixed location while the rest of the group wanders with the herd. Nomads may hire workers to practice sedentary agriculture in return for grain and protection. Other nomads may sow grain in recently flooded areas and return later in the year to harvest the crop. Yet another strategy is to remain in one place and cultivate the land when rainfall is abundant; then, during periods that are too dry to grow crops, the group can increase the size of the herd and migrate in search of food and water.

Nomads select the type and number of animals for the herd according to local cultural and physical characteristics. The choice depends on the relative prestige of animals and the ability of species to adapt to a particular climate and vegetation. The camel is most frequently desired in North Africa and the Middle East, and sheep and goats are used as well. In Central Asia, the horse is particularly important.

The camel is well suited to arid climates, because it can go long periods without water, carry heavy baggage, and move rapidly. However, the camel is particularly bothered by flies and sleeping sickness and has a relatively long gestation period, twelve months, from conception to birth. Goats need more water than camels but are tough and agile and can survive on virtually any vegetation, no matter how poor. Sheep are relatively slow-moving and are more affected by climatic changes. They require more water and are more selective as to which plants they will consume. The minimum number of animals necessary to support each family adequately varies according to the particular group and animal. The typical nomadic family

In the dry lands of developing regions, pastoral nomads, such as these in Algeria, herd camels and other animals adapted to the dry conditions. The size of a herd is a traditional measure of wealth among pastoral nomads. (Tom McHugh/Photo Researchers, Inc.)

needs between twenty-five and sixty goats or sheep or between ten and twenty-five camels.

Agricultural experts once regarded pastoral nomadism as a stage in the evolution of agriculture, from the hunters and gatherers who migrated across Earth's surface in search of food to sedentary farmers who cultivated grain in one place. Because they had domesticated animals but not plants, pastoral nomads were considered more advanced than hunters and gatherers but less advanced than settled farmers.

Pastoral nomadism now generally is recognized to be an offshoot of sedentary agriculture, not a primitive precursor of it. It is a practical way of surviving on land that receives too little rain for cultivation of crops. The domestication of animals, the basis for pastoral nomadism, probably was achieved originally by sedentary farmers, not by nomadic hunters. Pastoral nomads therefore had to be familiar with sedentary farming, and in many cases practice it.

Pastoral nomads do not wander randomly across the landscape but have a strong sense of territoriality. Every group controls a piece of territory and will invade another group's territory only in an emergency or if war is declared. The goal of each group is to control a territory large enough to contain the vegetation and water needed for survival. The actual amount of land a group controls depends on its wealth and power.

A group's precise migration patterns evolve from intimate knowledge of the area's physical and cultural characteristics. Groups frequently divide into herding units of five or six families and choose routes depending on experience concerning the most likely water sources during the various seasons of the year. The selection of routes will vary in unusually wet or dry years and will be influenced by the condition of the group's animals and the area's political stability.

Some pastoral nomads practice **transhumance,** which is seasonal migration of livestock between mountains and lowland pastures. Sheep or other animals may graze in alpine meadows in the summer and be herded back down into valleys for the winter.

Pastoral nomadism is a declining form of agriculture, a victim in part of modern technology. Before recent transportation and communications inventions, pastoral nomads played an important diffusion role as carriers of goods and information across the sparsely inhabited dry lands. Nomads used to be the most powerful inhabitants of the dry lands, but now, with modern weapons and other methods of coercion, national governments can control the nomadic population more effectively.

Governments force groups to give up pastoral nomadism because they want the land for other uses. Land that can be irrigated is converted from nomadic to sedentary agriculture. In some instances, the mining and petroleum industries now operate in dry lands formerly occupied by pastoral nomads. Government efforts to resettle nomads have been particularly vigorous in China, Kazakhstan, and several Middle Eastern countries, including Egypt, Israel, Saudi Arabia, and Syria. As nomads are reluctant to cooperate, these countries have experienced difficulties in trying to force settlement in collectives and cooperatives.

Some nomads are encouraged to try sedentary agriculture or to work for mining or petroleum companies. Others are still allowed to move somewhat but only on ranches with fixed boundaries. In the future, pastoral nomadism will be increasingly confined to areas that cannot be irrigated or that lack valuable raw materials.

Intensive Subsistence Agriculture

The greatest number of farmers living in the large population concentrations of East, South, and Southeast Asia practice **intensive subsistence agriculture.** The term *intensive* implies that farmers in this region must expend a relatively large amount of effort to produce the maximum feasible yield from a given parcel of land.

The typical farm in the intensive subsistence agriculture regions of Asia is much smaller than elsewhere in the world. Many Asian farmers own several fragmented plots, frequently a result of dividing individual holdings among children over a several-hundred-year period.

Because the agricultural density—the ratio of farmers to arable land—is so high in parts of East and South Asia, families must produce enough food for their survival from a very small area of land. They do this through careful agricultural practices, refined over thousands of years in response to local environmental and cultural patterns. Most of the work is done by hand or with animals rather than with machines, in part because of the abundance of labor but largely because of shortage of funds to buy equipment.

In order to maximize food production, intensive subsistence farmers waste virtually no land. Corners of fields and irregularly shaped pieces of land are planted rather than left idle, while paths and roads are kept as narrow as possible to minimize the loss of arable land. Livestock are rarely permitted to graze on land that could be used to plant crops, and little grain is grown to feed the animals.

Wet rice dominant. The intensive agriculture region of Asia can be divided between areas in which wet rice dominates and areas where other crops predominate. Wet rice occupies a relatively small percentage of Asia's agricultural land but is the region's most important source of food. Intensive wet rice farming is the dominant type of agriculture in Southeast China, East India, and much of Southeast Asia (Figure 11–4).

Successful production of large yields of rice is an elaborate, time-consuming process, done mostly by hand. Each family member, including the children, contributes to the effort.

The process of growing rice involves a number of steps. First, a farmer prepares the field for planting using a plow drawn by water buffalo or oxen. As stated earlier, the use of a plow and animal power is one characteristic that distinguishes subsistence agriculture from shifting cultivation.

The next step in the rice-growing process is to flood the plowed land with water collected from either rainfall, river overflow, or irrigation. Too much or too little water can damage the crop—a particular problem for farmers in South Asia who depend on monsoon rains, which do not always arrive at the same time each summer. Before the rice can be planted, dikes and canals must be repaired to assure that the quantity of water in the field is appropriate. The flooded field is called a **sawah** in the Austronesian language widely spoken in Indonesia, including Java. Europeans and North Americans frequently, but incorrectly, call it a **paddy,** the Malay word for wet rice.

Rice is introduced to a field in one of two ways. One is to broadcast dry seeds by scattering them through the field, a method used to some extent in South Asia. The more customary way to introduce

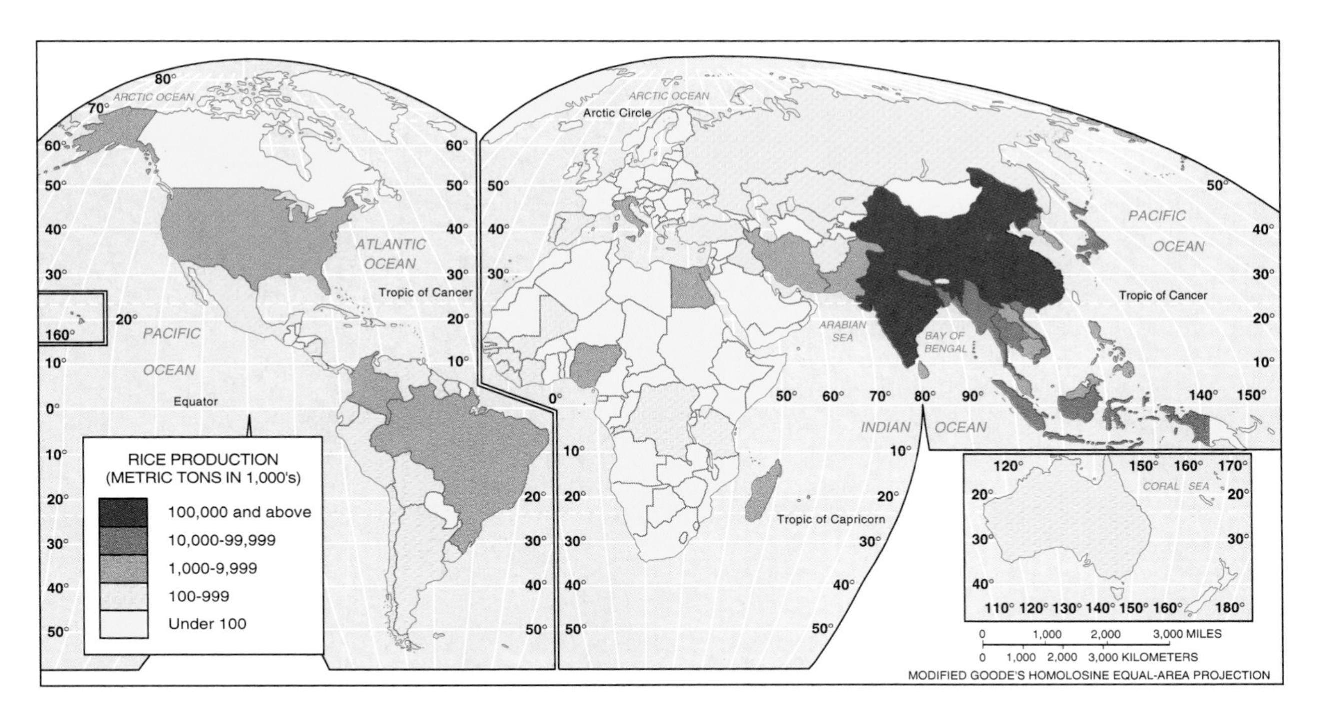

FIGURE 11–4
Rice production. Rice is the most important crop in the large population concentrations of East, South, and Southeast Asia. Asian farmers grow more than 90 percent of the world's rice, and two countries—China and India—account for more than half of world production.

Growing rice, such as in Yunnan Province, China, is a labor-intensive operation, done mostly by hand. Rice seedlings grown in a nursery are transplanted to the field. (Ken Straiton/The Stock Market)

the rice is by transplanting seedlings, which are first grown in a nursery. Typically one-tenth of a sawah is devoted to the cultivation of seedlings. After about a month the seedlings are transferred to the rest of the field. Rice plants grow submerged in water for approximately three-fourths of the growing period.

Rice plants are harvested by hand, usually with knives. To separate the husks, known as chaff, from the seeds, the heads are threshed by beating them on the ground or treading on them barefoot. The threshed rice is placed in a tray, and the lighter chaff is winnowed, that is, allowed to be blown away by the wind. If the rice is to be consumed directly by the farmer, the hull, or outer covering, is removed by mortar and pestle. Rice that is sold commercially is frequently whitened and polished, a process that removes some nutrients but leaves rice more pleasing in appearance and taste to many consumers.

Wet rice should be grown on flat land, because the plants are submerged in water much of the time. Therefore, most wet-rice cultivation is located in river valleys and deltas. The pressure of population growth in parts of East Asia has resulted in expansion of the land area under rice cultivation. One method of developing additional land suitable for growing rice is to terrace the hillsides of river valleys.

Land is used even more intensively in parts of Asia by obtaining two harvests per year from one field, a process known as **double cropping.** Double cropping is common in places with relatively warm winters, such as South China and Taiwan, but is relatively rare in India, where most areas have dry winters. Normally, double cropping involves alternating between wet rice, grown in the summer, when precipitation is higher, and wheat, barley, or another dry crop, grown in the drier winter season. Crops other than rice may be grown in the wet rice region in the summer on nonirrigated land.

Wet rice is the dominant crop in much of East, South, and Southeast Asia. Because wet rice should be grown on flat lands, hillsides may be terraced in places such as Bali to increase the area of rice production. (Pedro Coll/The Stock Market)

Wet rice not dominant. Climate prevents farmers from growing wet rice in portions of Asia, especially where summer precipitation levels are too low and winters are too harsh. Agriculture in much of interior India and northeast China is devoted to crops other than wet rice.

This region shares most of the characteristics of intensive subsistence agriculture with the wet-rice region. Land is intensively used and worked primarily by human power with the assistance of some hand implements and animals. Wheat is the most important crop, followed by barley. A wide variety of other grains and legumes is grown, including millet, oats, corn, kaoliang, sorghum, and soybeans, as well as some cash crops such as cotton, flax, hemp, and tobacco.

In milder parts of the region where wet rice does not dominate, more than one harvest can be obtained some years through skilled use of crop rotation. In colder climates, wheat or another crop is planted in the spring and harvested in the fall, but no crops can be sown through the winter.

Since the 1950s, private individuals have not owned most of the agricultural land in China. Instead, the Communist government of China organized agricultural-producer communes, which typically consisted of several villages containing several hundred people. By combining several small fields into a single large unit, the government hoped to promote agricultural efficiency, because scarce equipment and animals could be shared, and larger improvement projects, such as flood control, water storage, and terracing, could be completed. In reality, productivity did not increase as much as the government had expected, because people did not work as efficiently when they worked for the commune rather than for themselves.

China has begun to dismantle the agricultural communes. The communes still technically hold the legal title to the ownership of the agricultural land, but villagers sign contracts entitling them to farm equal portions of the land as private individuals. Chinese farmers will be allowed to sell the right to use the land to others and to pass on the right to their children. Reorganization has been difficult because irrigation systems, equipment, and other infrastructure were developed to serve large communal farms rather than small individually managed ones, which cannot afford as independent units to operate and maintain the machinery.

Commercial Agriculture in Relatively Developed Regions

The type of agriculture practiced on a commercial farm depends on site factors, such as climate, soil, and slope of the land. However, a commercial farmer also considers situation factors, especially the location of the farm in relation to the markets where its products are sold.

Characteristics of Commercial Agriculture

The most important difference between **commercial agriculture** and subsistence agriculture is that farmers grow crops and raise animals primarily for sale off the farm rather than for their own consumption. Agricultural products are not sold directly to consumers but to food processing companies.

Large processors, such as General Mills and Ralston Purina, typically sign contracts with commercial farmers to buy their chickens, cattle, and grain. Farmers may have contracts to sell sugar beets to sugar refineries, potatoes to distilleries, and oranges to manufacturers of concentrated juices.

Small percentage of farmers. Several other characteristics distinguish commercial agriculture from subsistence agriculture. First, the percentage of people who work as farmers is less than 5 percent in relatively developed countries such as the United States, Canada, and the United Kingdom, compared to 60 percent in developing countries. Yet the small percentage of farmers can produce enough food not only for themselves and the rest of the country but also a surplus to help feed people elsewhere in the world.

Commercial farmers raise animals primarily for sale to large food processing companies and restaurant chains. Chickens are being processed at Holly Farms in North Wilksboro, North Carolina. (Charles Gupton/The Stock Market)

The number of farmers has declined in relatively developed societies during the twentieth century. Both push and pull migration factors have been responsible: people have moved away from farms because they were no longer needed and, at the same time, have been lured to higher-paying jobs in urban areas. The United States had approximately 2 million farms in 1990, compared to 5.6 million in 1950 and 4 million in 1960.

Although there are fewer farms and farmers, the amount of land devoted to agriculture has remained fairly constant in Western Europe and Anglo-America since 1900. The annual loss of farmland in the United States is only 0.1 percent—primarily because of the growth of urban land uses—but this loss has been offset by the creation of new agricultural land through irrigation and reclamation. A more serious problem in the United States is the loss of the most productive farmland, known as *prime agricultural land,* because of the sprawl of suburban areas into the surrounding countryside (see Chapter 13).

Heavy use of machinery and chemicals. A second distinctive characteristic of commercial agriculture is the high degree of reliance on technological and scientific improvements. A small number of farmers are able to feed a large number of people in relatively developed societies because commercial farmers rely on machinery rather than people or animals to supply power (Figure 11–5).

Traditionally, the farmer or local craftsworker made the equipment from wood, but beginning in the late eighteenth century factories produced farm machinery. The first all-iron plow was produced in the 1770s and was followed in the nineteenth and twentieth centuries by a series of inventions designed to make farming less dependent on human or animal power. Tractors, combines, corn pickers, planters, and other factory-made farm machines have replaced or supplemented manual labor.

Transportation improvements also aid commercial farmers. The coming of railroads in the nineteenth century and trucks in the twentieth century have enabled farmers to transport crops and live-

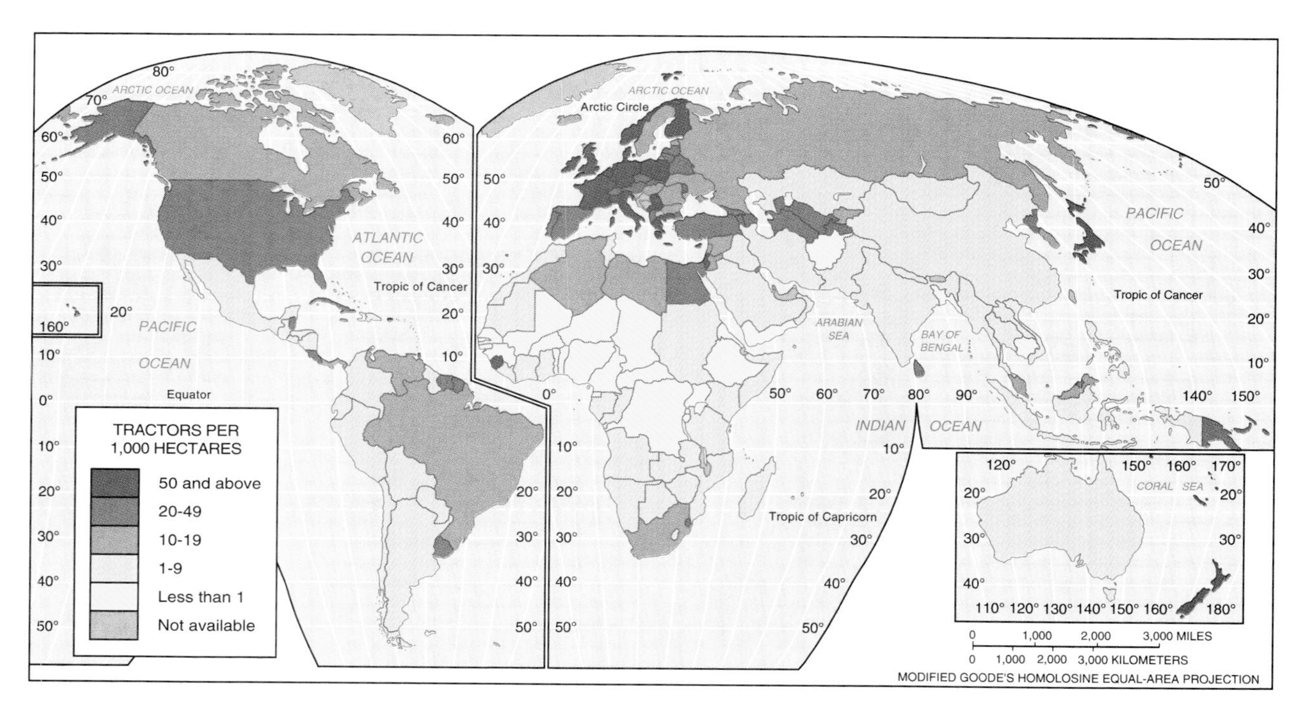

FIGURE 11–5
Tractors per 1,000 hectares. Farmers in relatively developed countries possess more machinery, such as tractors, than in developing ones. The machinery makes it possible for commercial farmers to farm extensive areas, a necessary practice to pay for the expensive machinery.

stock more rapidly. Cattle arrive at the destination heavier and in better condition when transported by truck or train than when driven on foot. Crops reach markets without spoiling.

Commercial farmers also make extensive use of scientific advances to increase productivity. Experiments conducted in the laboratories of universities, industries, and research organizations generate new fertilizers, herbicides, hybrid plants, pesticides, and animal breeds. The result is higher crop yields and healthier animals. Access to other scientific information has enabled farmers to make more intelligent decisions concerning proper agricultural practices. However, as a result of large-scale application of chemicals to the soil, commercial agriculture is the largest single source of water pollution in the United States.

Large farm size. A third distinctive characteristic of commercial agriculture is the relatively large size of the average farm, especially in the United States and Canada. The average U.S. farm is approximately 187 hectares (468 acres). Despite the large size, most commercial farms in relatively developed countries are family owned and operated. Commercial farmers frequently expand their holdings by renting nearby fields.

Large size is partly a consequence of mechanization. Combines, pickers, and other machinery perform most efficiently at very large scales, and their considerable expense would not be justifiable on a small farm. As a result of the large size and the high level of mechanization, commercial agriculture is an expensive business. Farmers must spend hundreds of thousands of dollars to buy or rent land and machinery before beginning operations. This money is frequently borrowed from a bank and repaid after the output is sold.

Integration with other businesses. A fourth distinctive characteristic of commercial farming is its close ties to other businesses. The system of commercial farming practiced in the United States and other relatively developed countries has been called **agribusiness,** because the family farm is not an isolated activity but is integrated into a large food production industry. Less than 5 percent of the U.S. labor force are farmers, but more than 20 percent are engaged in other activities related to agribusiness. These activities include processing, packaging, storing, distributing, and retailing food. Agribusiness encompasses such diverse enterprises as tractor manufacturers, fertilizer producers, and seed distributors, as well as those businesses that receive the produced food for distribution, such as restaurants, home-delivered pizzas, and other fast-food sources. While most farms are owned by individual families, many other aspects of agribusiness are controlled by large corporations.

Types of Commercial Agriculture

Commercial agriculture in relatively developed countries can be divided into six main types: mixed crops and livestock; dairying; livestock ranching; grain farming; gardening and fruit culture; and Mediterranean agriculture. Each of these agricultural types is predominant in distinctive regions within relatively developed countries, depending on a combination of situation, site, and cultural factors (see Focus Box 11–1).

Mixed crop and livestock. The most common form of commercial agriculture is mixed crop and livestock. Mixed commercial farming is found in much of Europe from Ireland to Russia, North America west of the Appalachians and east of the Great Plains, South Africa, northeastern Argentina, southeastern Australia, and New Zealand.

The most distinctive characteristic of mixed crop and livestock farming is the integration of crops and livestock. Most of the crops grown on a mixed commercial farm are fed to animals rather than consumed directly by humans. In turn, the livestock can supply manure to improve soil fertility to grow more crops. A typical mixed commercial farm devotes nearly all of the land area to growing crops but derives more than three-fourths of its income from the sale of animal products, such as meat and eggs. In the United States, pigs are often bred directly on the farms, while cattle may be brought in to be fattened on corn.

Mixed crop and livestock farming permits farmers to distribute the work load more evenly through the year. Fields require less attention in the winter than in the spring, when crops are planted, and in the fall, when they are harvested. Livestock, on the other hand, require attention throughout the year. A mix of crops and livestock also reduces seasonal variations in income; most income from

FOCUS BOX 11-1

Von Thünen's Model

A model emphasizing the importance of transport costs in agriculture was first proposed in 1826 by Johann Heinrich von Thünen, a farmer in northern Germany, in a book titled *The Isolated State*. According to von Thünen's model, which was later modified by geographers, a commercial farmer initially considers alternative crops to cultivate and animals to raise on the basis of the location of the market.

In making a choice, a commercial farmer compares two costs: the price of the land and the expense of transporting the products to the market. First, a commercial farmer identifies the crops that can be sold for more than the rent. Assume that a farmer rents land at a cost of $100 per hectare. In this example, a farmer would consider planting wheat if the output from one hectare of land can be sold at the market for more than $100, minus the nonrent expenses. The farmer would consider another crop, such as corn, if the yield from one hectare of land can be sold for more than $100.

A farmer may not always plant the crop that sells for the highest price per hectare. The choice also depends on the distance of the land from the central market city. Distance to the market is a critical factor because each crop has a unique transportation cost.

Von Thünen based his general model of the spatial arrangement of different crops on his experiences as owner of a large estate in northern Germany during the early nineteenth century (Figure 1). He found that specific crops were grown in different rings around the cities in the area. Market-oriented gardens and milk producers were located in the first ring out from the cities. These products are expensive to deliver and must reach the market quickly because they are perishable.

The next ring out from the cities contained wood lots, where timber was cut for construction and fuel. The next rings were used for various crops and for pasture, although the precise choice was rotated from one year to the next. The outermost ring was devoted exclusively to animal grazing.

Von Thünen did not consider site or cultural factors in his model. The model assumed that all land in a study area had similar physical characteristics and was of uniform quality, although he recognized that the model would vary according to topography and other distinctive physical conditions. For example, a river might modify the shape of the rings because transportation costs change when products are shipped by water routes rather than over roads. The model also failed to understand that tradition, taste, and other social customs influence the attractiveness of plants and animals for a commercial farmer.

crops comes during the harvest season, but livestock products can be sold throughout the year.

A mixed crop and livestock farmer typically practices crop rotation. The farm is divided into a number of fields, and the crop planted in a particular field differs from one year to the next. Crop rotation enables a farmer to maintain the fertility of a field, because various crops deplete the soil of certain nutrients and restore others. The practice of crop rotation contrasts with the system used in shifting cultivation, in which nutrients depleted from a field are restored by leaving the field uncropped for many years. Because, in any given year, crops cannot be planted in most of an area's fields, the over-

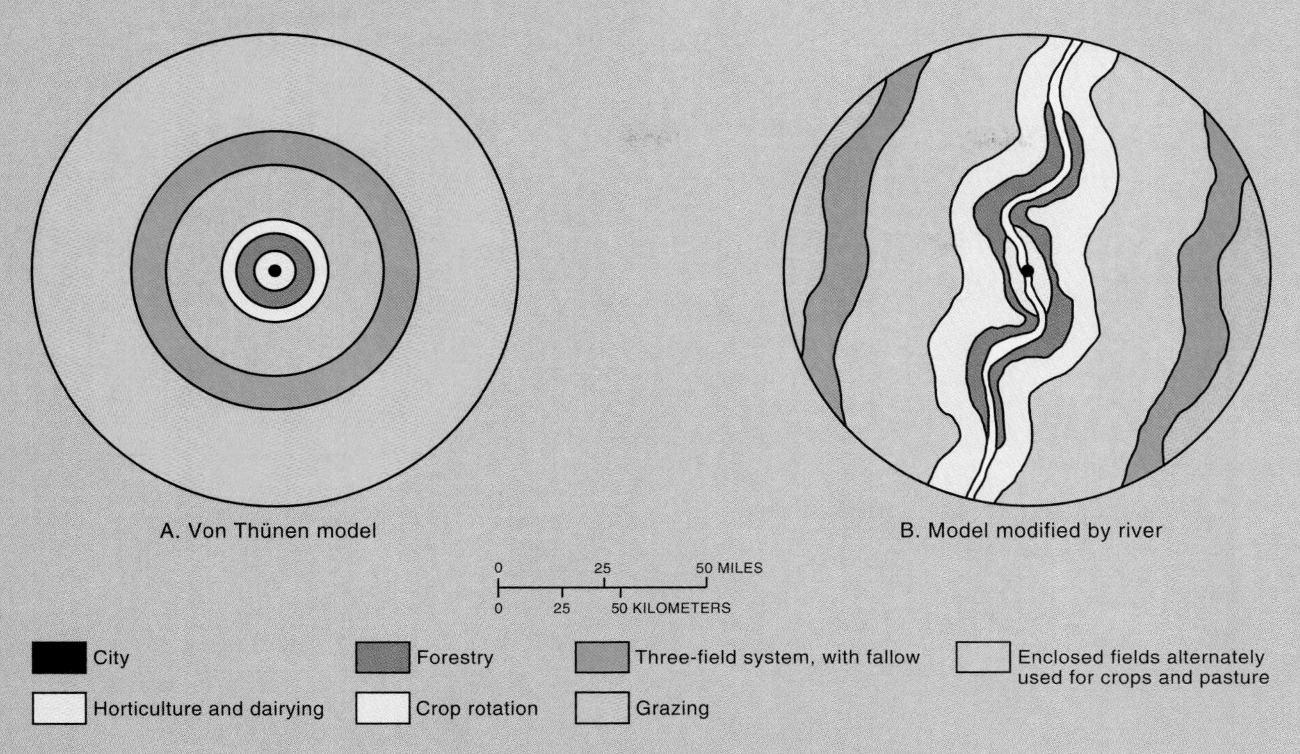

According to the von Thünen model, in the absence of topographic factors, different types of farming would appear at different distances from a city, depending on the cost of transportation and the value of the product. Von Thünen recognized that his model would be modified by physical features, such as a river. Because a river changes the pattern of relative accessibility of different parcels of land to the market center, agricultural uses that seek highly accessible places could locate closer to the river.

Although von Thünen developed the model for a small region with a single market center, it also applies with considerable modification at a national or global scale. Farmers in relatively remote locations who wish to sell their output in the major markets of Western Europe and North America, for example, are less likely to grow highly perishable and bulky products.

Within the United States, however, site factors are not constant. For example, vegetables can be grown most of the year in the southern parts of Florida and Texas but not in New Jersey. Decreasing transport costs have made it possible for growers in the southern United States to ship vegetables to distant northern markets. Because improved transportation has reduced the importance of distance in the United States, site factors such as climate now are relatively more important.

all production level is much lower in shifting cultivation than in mixed commercial farming.

In the United States, the most important mixed crop and livestock farming region extends from Ohio to the Dakotas with its center in Iowa. The region is frequently called the Corn Belt, because approximately half of the crop land is planted in corn (maize). Mixed commercial farmers select corn most frequently because of higher yields per area than other crops. Some of the corn is consumed by people either directly or as oil, margarine, and other food products, but most is fed to pigs and cattle (Figure 11–6).

Soybeans have become the second most impor-

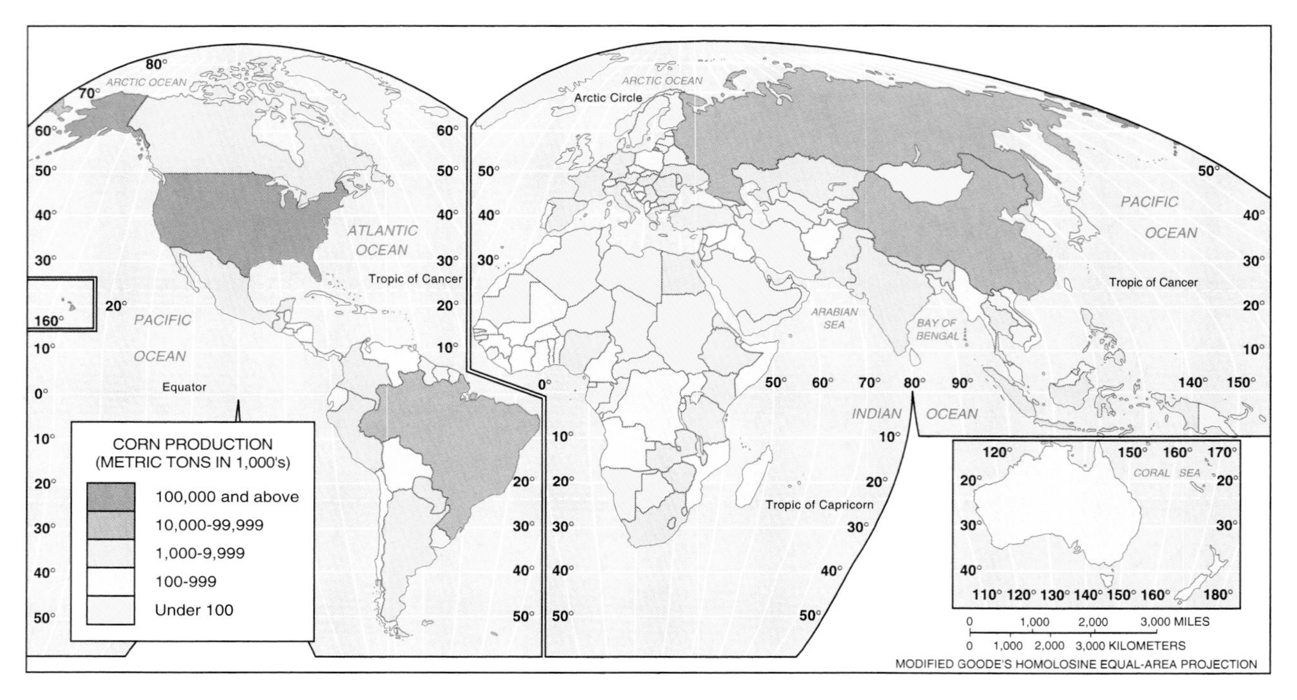

FIGURE 11–6
Corn (maize) production. The United States accounts for about 40 percent of the world's production of corn, which is known as maize outside North America. China is the second leading producer.

tant crop in the U.S. mixed commercial farming region. Like corn, soybeans are used sometimes to make products consumed directly by people but mostly to make animal feed. Tofu, made from soybean milk, is a major food source, especially for people in China and Japan. Soybean oil is widely used in U.S. foods, but as a hidden ingredient.

Dairy farming. Dairy farming is the most important type of commercial agriculture practiced on farms outside the large urban areas of the northeast United States, southeast Canada, and northwest Europe. It accounts for approximately 20 percent of the total value of agricultural output throughout Western Europe and North America. Russia, Australia, and New Zealand also have extensive areas devoted to dairy farming. Nearly 90 percent of the world's supply of milk is produced and consumed in these relatively developed regions (Figure 11–7).

Traditionally, fresh milk rarely was consumed except directly on the farm or in nearby villages. With the rapid growth of cities in relatively developed countries during the nineteenth century, demand for the sale of milk to urban residents increased. Rising incomes permitted urban residents to buy milk products that were once considered luxuries. Average weekly consumption of milk per person in England, for example, rose from 0.8 liters (0.2 U.S. gallons) in the 1870s to 2.8 liters (0.7 U.S. gallons) by the 1950s.

Dairying is an especially important type of commercial agriculture outside large cities because of transportation factors. Dairy farms need to be located closer to the market than other products because milk is highly perishable and must reach the consumer quickly to prevent spoiling. The region surrounding a city from which milk can be supplied without spoiling is known as the **milkshed.**

Improvements in transportation have permitted dairying to be undertaken farther away from the market. Until the 1840s, when railroads were first used for transporting dairy products, milksheds rarely extended beyond a radius of more than 50 kilometers (30 miles). In the twentieth century, milk can be driven quickly from the farm to the railroad station, and refrigerated rail cars and trucks enable farmers to ship milk more than 500

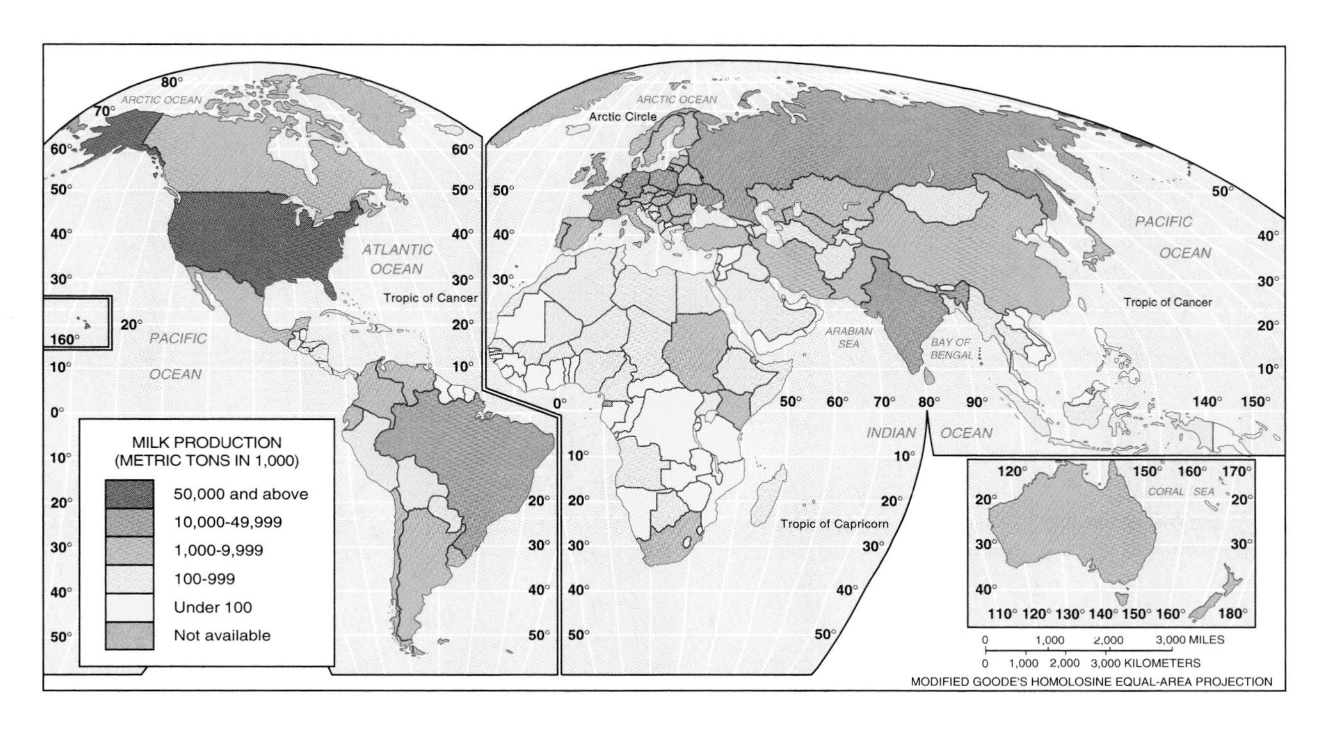

FIGURE 11–7
Milk production. The distribution of milk production closely matches the division of the world into relatively developed and developing regions. Consumers in relatively developed countries have the income to pay for milk products, and farmers in these countries can afford the high cost of establishing dairy farms. Two very populous countries—Brazil and India—rank among the world leaders in total milk production but not in production per capita.

kilometers (300 miles). As a result, nearly every farm in the U.S. northeast and northwest Europe is within the milkshed of at least one urban area.

Some dairy farms specialize in products other than milk. Originally, butter and cheese were made directly on the farm, primarily from the excess milk produced in the summer, before modern agricultural methods evened the flow of milk through the year. In the twentieth century, dairy farmers have generally chosen to specialize either in milk production or other products such as butter and cheese.

In general, the farther the farm is from large urban concentrations, the lower is the percentage of the output devoted to fresh milk. Farms located farther from consumers are more likely to sell their output to processors who make butter, cheese, or dried, evaporated, and condensed milk. The reason is that butter and cheese keep longer than milk and therefore can be safely shipped from remote farms. In the United States, for example, virtually all milk from East Coast dairy farms is sold in liquid form to consumers living in New York, Philadelphia, Boston, and the other large urban areas, whereas virtually all of the milk in Wisconsin is processed.

Countries likewise tend to specialize in certain products. New Zealand, the world's largest producer of dairy products, devotes only 8 percent to liquid milk, compared to 68 percent in the United Kingdom. New Zealand farmers do not sell much liquid milk because the country is too far from North America and northwest Europe, the two largest relatively wealthy population concentrations.

Dairy farmers, like other commercial farmers, usually do not sell their products directly to consumers. Instead, they generally sell milk to wholesalers, who distribute it in turn to retailers. Retailers then sell milk to consumers in shops or at home. Farmers also sell milk to butter and cheese manufacturers.

Like other commercial farmers, dairy farmers face economic problems because of declining revenues and rising costs. Distinctive features of dairy farming have exacerbated the economic difficulties. First, dairy farming is labor-intensive, because the cows must be milked twice a day, every day. Although the actual milking can be done by machines, dairy farming nonetheless requires constant attention throughout the year.

Dairy farmers also face the expense of feeding the cows in the winter, when they may be unable to graze on grass. In northwestern Europe and New England, farmers generally purchase hay or grain for winter feed. In the western part of the U.S. dairy region, crops are more likely to be grown in the summer and stored for winter feed on the same farm.

Grain farming. Some form of grain is the major crop on most farms. Commercial grain agriculture is distinguished from mixed crop and livestock farming, because crops on a grain farm are grown primarily for consumption by humans rather than by livestock. Subsistence farmers in developing countries also grow crops for human consumption, but the farmers directly consume the output. Commercial grain farms, on the other hand, sell the output to manufacturers of food products.

Large-scale commercial grain production is found in only a few countries, including the United States, Russia, France, and Canada. Commercial grain farms are generally located in regions that are too dry for mixed crop and livestock agriculture (Figure 11–8).

The most important crop grown is hard wheat, used to make bread flour. Wheat generally can be sold for a higher price than other grains, such as rye, oats, and barley, and has more uses as human food. It can be stored relatively easily without spoiling and can be transported a long distance. Because wheat has a relatively high value per unit weight, it can be shipped profitably from remote farms to markets.

Large-scale grain production is concentrated in three areas of North America. The first is the winter wheat belt that extends through Kansas, Colorado, and Oklahoma. In the **winter wheat** area, the crop is planted in the autumn and develops a strong root system before growth stops for the winter. The wheat survives the winter, especially if it is under a snow blanket, and is ripe by the beginning of summer.

The second important grain-producing region in North America is the spring wheat belt of the Dakotas, Montana, and southern Saskatchewan, Canada. Because winters are usually too severe for winter wheat in this region, **spring wheat** is planted in the spring and harvested in the late summer. Approximately two-thirds of the wheat grown in the United States comes either from the winter or the spring wheat belt. A third important grain growing region is the Palouse region of Washington State.

Large-scale grain production, like other commercial farming ventures in relatively developed

Commercial agriculture depends heavily on expensive machinery to manage large farms efficiently. These combine machines are reaping, threshing, and cleaning wheat in Sherman County, Kansas. (Cotton Coulson/Woodfin Camp & Associates)

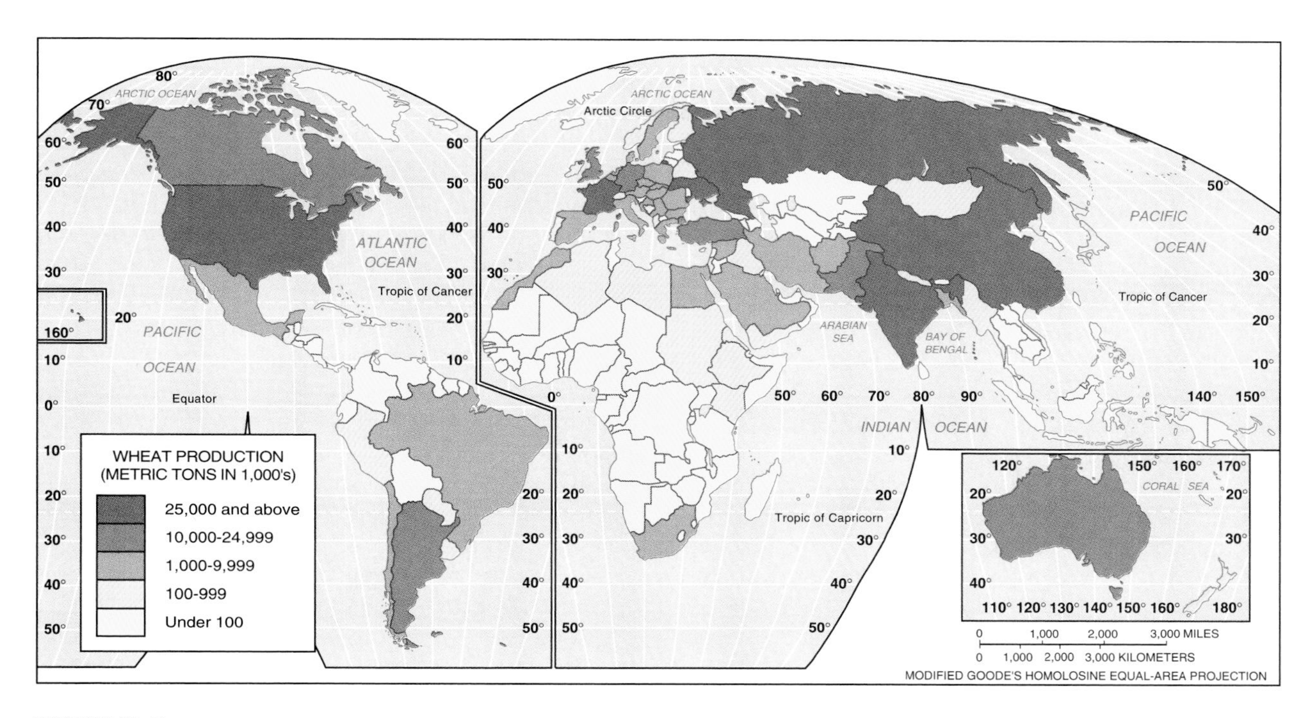

FIGURE 11–8
Wheat production. China is the world's leading producer of wheat. The United States, Russia, and India are other major producers. Wheat grown in the Aisan countries is used principally to feed the local population, whereas a large percentage of the wheat grown in North America is sold to other countries.

countries, is heavily mechanized, conducted on large farms, and oriented to consumer preferences. The McCormick reaper, invented in the 1830s, first permitted large-scale wheat production. Today, the combine machine performs in one operation the three tasks of reaping, threshing, and cleaning.

Unlike work on a mixed crop and livestock farm, the effort required to grow wheat is not uniform throughout the year. Some individuals or firms may therefore have two sets of fields, one in the spring-wheat belt and one in the winter-wheat belt. Because the planting and harvesting in the two regions occur at different times of the year, the work load can be distributed throughout the year. In addition, the same machinery can be used in the two regions, thus spreading out the cost of the expensive equipment. Combine companies start working in Oklahoma in early summer and move north.

Wheat's significance extends beyond the amount of land or number of people involved in growing it. Unlike other agricultural products, wheat is grown to a considerable extent for international trade and is the world's leading export crop. As the United States and Canada account for about half of the world's wheat exports, the North American prairies are accurately labeled the world's "breadbasket." The ability to provide food for many people elsewhere in the world is a major source of economic and political strength for the United States and Canada.

Ranching. **Ranching,** the commercial grazing of livestock over an extensive area, is a form of agriculture adapted to semiarid or arid land. It is practiced in relatively developed countries where the vegetation is too sparse and the soil too poor to support crops.

Cattle were first brought to the Americas by Columbus on his second voyage, because they were sufficiently hardy to survive the ocean crossing. Living in the wild, the cattle multiplied and thrived on the abundant supply of grazing land on the frontiers of North and South America. Immigrants from Spain and Portugal, the only

European countries with a tradition of cattle ranching, began ranching in the Americas. They taught the practice to settlers from Northern Europe and the eastern United States who moved to Texas and other frontier territories in the nineteenth century.

Cattle ranching in Texas, as glamorized in popular culture, actually dominated commercial agriculture for a short period of time, from 1867 to 1885. This period of cattle ranching began because demand for beef in the eastern United States increased during the 1860s, and transportation improvements, especially construction of railroads, made delivery of cattle to eastern markets feasible.

Cattle ranching declined in importance during the 1880s, after it came into conflict with sedentary agriculture. The U.S. government, which owned most of the land used for open grazing, began to turn it over to farmers to grow crops, leaving ranchers with no legal claim to it. For a few years, the ranchers tried to drive out the farmers by cutting fences and then illegally erecting their own fences on public land. The farmers' most potent weapon proved to be barbed wire, first commercially produced in 1873. The farmers eventually won the battle, and ranchers were compelled to buy or lease land to accommodate their cattle. Large cattle ranches were established, primarily on land that was too dry to support crops, although 60 percent of the cattle grazing done today is on land leased from the U.S. government.

With the spread of irrigation techniques and hardier crops, land in the United States has been converted from ranching to crop growing. Ranching generates lower income per area of land, although it has lower operating costs. Cattle are still raised on ranches but are frequently sent for fattening to farms or to local feed lots along major railroad and highway routes rather than directly to meat processors. The average size of a ranch is large, because the capacity of the land to support cattle is low in much of the semiarid West. Large ranches may be owned by meat-processing companies rather than individuals.

Commercial ranching is also found in other relatively developed regions of the world (Figure 11–9). Ranching is rare in Europe, except in Spain and Portugal.

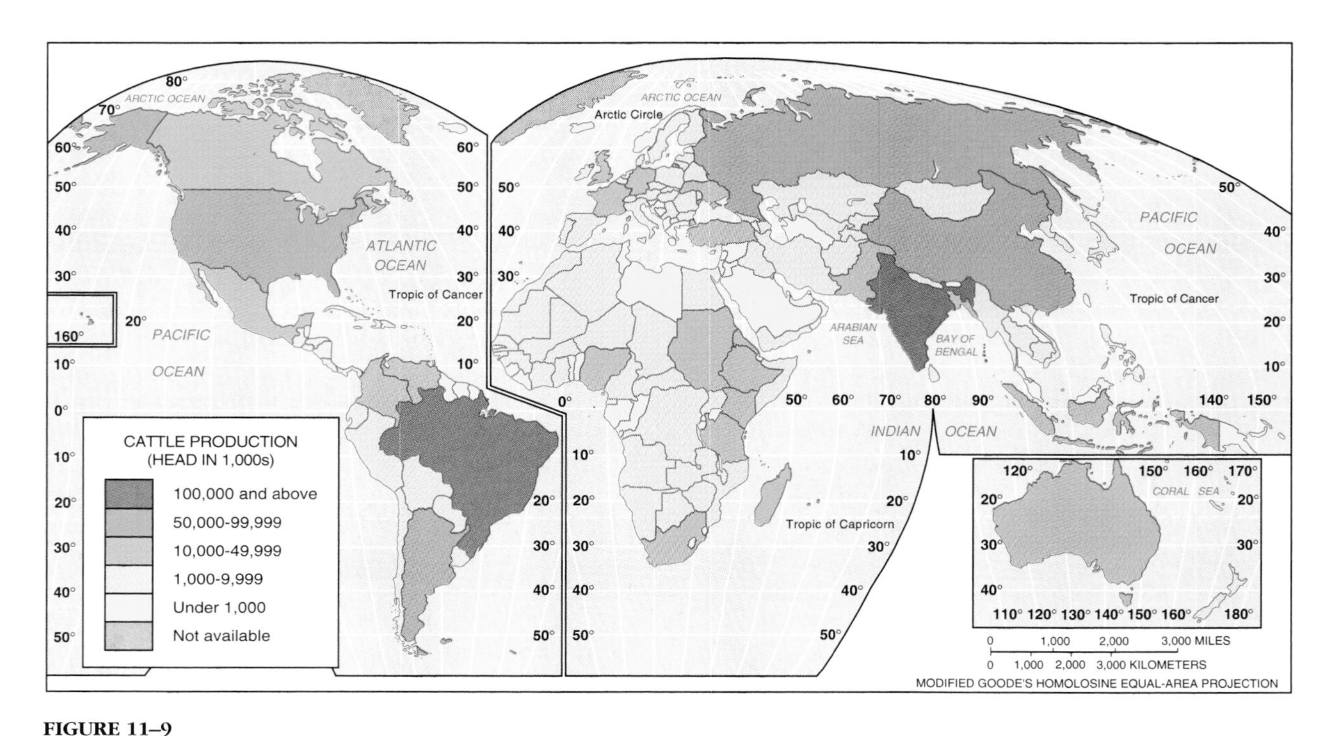

FIGURE 11–9
Cattle production. Cattle outnumber people in Argentina, Australia, and New Zealand, where commercial ranching is an important type of agriculture. Commercial ranching is not widely practiced in the relatively developed countries of Western Europe, a region that lacks extensive areas of dry lands.

In South America, a large portion of the pampas of Argentina, southern Brazil, and Uruguay are devoted to grazing cattle and sheep. The cattle industry grew rapidly in Argentina in part because the land devoted to ranching was relatively accessible to the ocean, and meat could be transported to overseas markets.

The relatively humid climate on the pampas also stimulated the growth of ranching in South America, because more cattle can graze on a given area of land than in the U.S. West. Land was divided into large holdings in the nineteenth century, in contrast to the U.S. practice of permitting common grazing on public land. Ranching has declined in Argentina, as in the United States, because growing crops is more profitable except on very dry lands.

The interior of Australia was opened for grazing in the nineteenth century, although sheep are more common than cattle. Ranches in the Middle East, New Zealand, and South Africa are also more likely to have sheep. Like the U.S. West, Australia's dry lands went through several land-use changes. Until the 1860s, shepherding was practiced on the open range. Then large ranches with fixed boundaries were established, stock was improved, and water facilities were expanded. Eventually, ranching was confined to drier lands, and wheat, which yielded greater profits per hectare than ranching, was planted where precipitation levels permitted.

Thus, ranching has followed similar stages around the world. First came the herding of animals over open ranges, in a semi-nomadic style. Then, ranching was transformed into fixed farming by dividing the open land into privately held fenced-in tracts. Many of the farms converted to growing crops, and ranching was confined to the drier lands. To survive, the remaining ranches experimented with new methods of breeding and sources of water and feed. Ranching became part of the meat processing industry rather than an economic activity carried out on isolated farms. In this way, commercial ranching differs from pastoral nomadism.

Mediterranean agriculture. Mediterranean agriculture exists primarily in the lands which border the Mediterranean Sea in southern Europe, northern Africa, and western Asia. Farmers in California, Central Chile, the southwestern part of South Africa, and Southwest Australia practice Mediterranean agriculture as well. Every Mediterranean area borders a sea and—with the exception of some of the lands surrounding the Mediterranean Sea—is all located on the west coast of a continent. Recall from Chapter 3 the distinctive Mediterranean climate: warm, dry summers and cool, rainy winters. Prevailing winds from the sea help to provide moisture and to moderate the temperatures during the winter. Summers are hot and dry, but sea breezes provide some relief. Most of this region is very hilly, and mountains frequently plunge directly down to the sea, leaving very narrow strips of flat land along the coast.

Farmers derive a smaller percentage of income from animal products in the Mediterranean region than in the mixed crop and livestock region. Livestock production is hindered during the summer by the lack of water and good grazing land. Some farmers living along the Mediterranean Sea traditionally used transhumance to raise animals, although the practice is now less common. In the practice of transhumance, animals, primarily sheep and goats, are kept on the coastal plains in the winter and transferred to the hills in the summer.

Most crops in Mediterranean lands are grown for human consumption rather than for animal feed. Tree crops and horticulture form the commercial base of the Mediterranean farming areas. Most of the world's olives, grapes, and other fruits and vegetables are grown in the Mediterranean agriculture area. A combination of local physical and cultural characteristics determines which crops will be grown in which Mediterranean farming areas. The hilly landscape encourages farmers to plant a variety of crops within one Mediterranean farming area.

In the lands bordering the Mediterranean Sea, the two most important cash crops are olives and grapes. Two-thirds of the world's wine is produced in countries that border the Mediterranean Sea, especially Italy, France, and Spain, while Mediterranean agricultural regions elsewhere in the world produce most of the remaining one-third. The lands near the Mediterranean Sea are also responsible for a large percentage of the world's supply of olives, an important source of cooking oil.

Although olives and grapes are the most important sources of income to commercial farms bordering the Mediterranean Sea, approximately half of the land is devoted to growing cereals, especially wheat for pasta and bread. As in the U.S. winter-wheat belt, the seeds are sown in the fall and the crops harvested in early summer. After cultivation, cash crops are planted on approximately 20 percent of the land, while the remainder is left fallow for a year or two to conserve moisture in the soil.

Cereals, or grains, occupy a much lower percentage of the cultivated land in California than in other Mediterranean climates. Instead, 30 percent of California farmland is devoted to fruits and vegetables. California supplies most of the citrus fruits, tree nuts, and deciduous fruits consumed in the United States. Fruits and vegetables are found in other Mediterranean climates, but not to the extent found in California.

The rapid growth of urban areas in California, especially the Los Angeles area, has resulted in the conversion of high-quality agricultural land to housing developments. Thus far, the loss of farmland has been offset by the expansion of agriculture into arid lands. However, farming in dry lands requires massive irrigation projects to provide an adequate water supply. In the future, agriculture may face stiffer competition as efforts continue to divert the U.S. Southwest's increasingly scarce water supply for other uses.

Commercial gardening and fruit farming. Commercial gardening and fruit farming is the predominant type of agriculture in the U.S. Southeast. The region has a long growing season and is accessible to the large markets of New York, Philadelphia, Washington, and the other eastern U.S. urban areas. The type of agriculture practiced in this region is frequently called truck farming, because *truck* was a Middle English word meaning bartering or the exchange of commodities.

Truck farms grow many of the fruits and vegetables that consumers demand in relatively developed societies, such as apples, broccoli, cherries, lettuce, onions, and tomatoes. Some of these fruits and vegetables are sold fresh to consumers, but most are sold to large processors for canning or freezing.

Truck farms are highly efficient, large-scale operations that take full advantage of available machines at every stage of the growing process. Truck farmers are willing to experiment with new varieties, seeds, fertilizers, and other inputs in order to maximize efficiency. Labor costs are kept down by hiring migrant farm workers, some of whom are undocumented immigrants from Mexico, to work at very low wages. Farms tend to specialize in a few crops, and a handful of farms may dominate national output of some fruits and vegetables.

A form of truck farming called specialty farming has spread to New England. Farmers are profitably growing crops that have limited but increasing demand among affluent consumers, such as asparagus, peppers, mushrooms, strawberries, and nursery plants. Specialty farming represents a profitable alternative for New England farmers, at a time when dairy farming is declining because of relatively high operating costs and low milk prices.

Plantation Farming

The plantation is a form of commercial agriculture found in the tropics and subtropics, especially in Latin America, Africa, and Asia. Although generally situated in developing countries, plantations are often owned or operated by Europeans or North Americans and grow crops for sale primarily in relatively developed countries.

A **plantation** is a large farm that specializes in the production of one or two crops for sale. Among the most important crops found on plantations are cotton, sugar cane, coffee, rubber, and tobacco. Cocoa, jute, bananas, tea, coconuts, and palm oil are also produced in large quantities. Latin American plantations are more likely to grow coffee, sugar cane, and bananas, while Asian plantations may provide rubber and palm oil.

Because plantations are usually situated in sparsely settled locations, they must import workers and provide them with food, housing, and social services. Plantation managers try to spread the work as evenly as possible throughout the year to make full use of the large labor force. Where the climate permits, more than one crop is planted and harvested during the year. Rubber tree plantations can try to spread the task of tapping the trees through the year. Crops such as tobacco, cotton, and sugar cane, which can be planted only once a year, are less likely to be grown on large plantations today than they were in the past. Crops are normally processed at the plantation before shipping. Processed goods are less bulky and therefore cheaper to ship long distances to the North American and European markets.

Sugar cane is being harvested on a plantation in Colombia. (Vivaine Moos/Stock Montage, Inc.)

Until the Civil War, plantations were important in the U.S. South, where the principal crop was cotton, followed by tobacco and sugar cane. Demand for cotton increased dramatically after the establishment of textile factories in England at the start of the industrial revolution in the late eighteenth century. Cotton production was stimulated by the invention of the cotton gin by Eli Whitney in 1793 and the development of new varieties which were hardier and easier to pick. Africans brought to America as slaves performed most of the labor, until the defeat of the South in the Civil War and the abolition of slavery. Thereafter, plantations declined in the United States; they were subdivided and either sold to individual farmers or worked by tenant farmers.

Increasing the World Supply of Food

Ironically, both subsistence and commercial farmers face a similar problem: farming is not producing a sufficiently high income for the desired standard of living. However, the underlying cause of low incomes differs significantly between developing and relatively developed countries.

Problems for Subsistence Farmers

The fundamental problem in developing countries is the need to assure an adequate supply of food for the people. Traditional subsistence farming can continue to produce enough food for people living in rural villages to survive, assuming there is no drought, flood, or other natural disaster. But developing countries must provide enough food for a rapidly increasing population as well as for the growing number of urban residents who cannot grow their own food.

The green revolution. During the 1960s, population began to grow more rapidly than the expansion of agricultural land, especially in developing countries. At the time, many experts issued grim forecasts of massive global famine within a decade. However, these dire predictions have not come true, because new agricultural practices have permitted farmers throughout the world to achieve much higher yields from the same amount of land.

The invention and rapid diffusion of more productive agricultural techniques during the 1970s and 1980s is known as the **green revolution.** The green revolution involves two main practices: the introduction of new higher-yield seeds and the greater use of fertilizers. Because of the green revolution, agricultural productivity has increased at a global scale faster than population growth.

During the 1950s, scientists began an intensive series of experiments to develop a higher-yield form of wheat. A decade later, the new so-called miracle wheat seed was ready. Shorter and stiffer than traditional breeds, the new wheat was less sensitive to variations in the length of days, responded better to fertilizers, and matured more rapidly. The Rockefeller and Ford foundations sponsored many of the studies, and the program's director, Dr. Norman Borlaug, won the Nobel Peace Prize in 1970.

The International Rice Research Institute, established in the Philippines by the Rockefeller and Ford foundations, concentrated on developing a miracle rice seed. During the 1960s, scientists produced a hybrid between Indonesian rice and Taiwan dwarf rice that was hardier and increased yields. More recently, scientists have developed new high-yield maize (corn).

The new miracle seeds were diffused rapidly around the world. India's wheat production, for example, more than doubled in five years. After importing 10 million tons of wheat per year in the mid-1960s, India by 1971 had a surplus of several million tons. Other Asian and Latin American countries recorded similar productivity increases.

To take full advantage of the new miracle seeds, farmers need to use more fertilizer and machinery. Farmers have known for thousands of years that application of manure, bones, and ashes somehow increases or at least maintains the fertility of the land. Not until the nineteenth century did scientists identify nitrogen, phosphorus, and potassium (potash) as the critical elements that caused the rise in fertility. Today, these three elements form the basis for fertilizers, or products that farmers apply on their fields to enrich the soil by restoring lost nutrients.

Nitrogen, the most important fertilizer, constitutes 78 percent of the atmosphere (Chapter 2). But gaseous nitrogen is not immediately available to plants, so it has to be converted to other forms. Europeans most commonly produce a fertilizer known as *urea,* which contains 46 percent nitrogen. In North America, nitrogen is generally converted to ammonia gas, which is 82 percent nitrogen but, as a gas, more awkward than urea to transport and store.

Both urea and ammonia gas combine nitrogen and hydrogen. The problem is that the cheapest methods for producing both types of nitrogen-based fertilizers obtain hydrogen from natural gas or petroleum products. As fossil fuel prices increase, so do the prices for nitrogen-based fertilizers, which then become too expensive for many farmers in developing countries.

In contrast with nitrogen, phosphorous and potash reserves are not distributed uniformly across Earth's surface. Two-thirds of the world's proven phosphate rock reserves are clustered in Morocco and the United States. Proven potash reserves are concentrated in Canada, Germany, Russia, and Ukraine.

Farmers need tractors, irrigation pumps, and other machinery to make most effective use of the new miracle seeds. In developing countries, farmers cannot afford to buy this equipment, nor, in view of rising energy costs, can they pay for the fuel to operate the equipment. To maintain the benefits of the green revolution, governments in developing countries must allocate scarce funds for subsidizing the cost of seeds, fertilizers, and machinery.

The green revolution has not stopped after the initial generation of miracle seeds. Scientists have continued to produce higher-yield hybrids that are adapted to environmental conditions in specific regions. Because of the green revolution, Dutch scientists calculate that the maximum annual crop yields currently have reached 6,000 kilograms of grain per hectare (5,000 pounds per acre) in parts of Asia and Latin America. Although still far lower than the maximum possible yields of 15,000 kilograms per hectare (13,000 pounds per acre) in Asia and 20,000 kilograms per hectare (18,000 pounds per acre) in Latin America, this output is improving. The green revolution was largely responsible for preventing a food crisis in these regions during the 1970s and 1980s, but will these scientific breakthroughs continue in the twenty-first century?

New food sources. More effective use of Earth's resources for food can be promoted in at least three ways: cultivate the oceans, develop higher-protein cereals, and increase the palatability of rarely consumed foods.

At first glance, increased reliance on food from the oceans is an attractive alternative. Although they cover about seven-tenths of Earth's surface and lie near most population concentrations, oceans historically have provided a relatively small percentage of the world food supply. Approximately two-thirds of the fish caught from the ocean is consumed directly, while the remainder is converted to fish meal and fed to poultry and hogs.

Hopes were raised during the 1950s and 1960s that increased fish consumption could meet the needs of a rapidly growing global population. The world's annual fish catch increased from 22 million tons in 1954 to more than 100 million tons in 1991. However, the population of some fish species has declined because they have been caught at a more rapid rate than they can reproduce. Overfishing has been particularly acute in the North Atlantic and Pacific oceans. The U.S. National Marine Fisheries Service estimates that 65 of 153 species of fish that it monitors off the Atlantic and Pacific coasts are overfished.

To protect fishing areas, several countries have claimed control of the oceans beyond the long-accepted limit of 3 nautical miles. Peru and Ecuador have declared a 200-mile off-shore limit in the Pacific, while Iceland has established a 50-mile jurisdiction in the North Atlantic. These countries seize foreign fishing boats that ignore the extended boundary claims.

Peru has been especially sensitive to the overfishing problem after the country's catch of anchovies, its most important fish, declined by more than 75 percent between 1970 and 1973. To prevent further overfishing, the government nationalized its fish meal production industry, but the Peruvian experience demonstrates that the ocean is not a limitless source of fish.

Increased use of fish for food may depend on the development of fish farming, rather than fishing in the ocean. Fish farming accounts for approximately 6 percent of the world's total fish catch, but 40 percent of China's fish is raised on farms. India and Russia rank behind China as other leading countries for fish farming. In the United States, only catfish and trout are produced in any quantity on farms.

A second possible new source of food is the development of higher-protein crops. People in relatively developed countries obtain proteins by consuming meat, but people in developing countries generally rely on wheat, corn, and rice, which lack certain proteins. Scientists are experimenting with hybrids of the world's major cereals that have higher protein content.

People can also obtain needed nutrition by consuming foods that are fortified during processing with vitamins, minerals, and protein-carrying amino acids. This approach achieves better nutrition without changing food consumption habits. However, fortification has limited application in developing countries, where most people grow their own food rather than buy processed food.

One way to make more effective use of existing global resources is to encourage consumption of foods that are avoided for social reasons. To fulfill

Fish farming is a potential source of additional food. This fish hatchery near Tampa, Florida, raises mostly catfish. (Robert Frerck/The Stock Market)

basic nutritional needs, people consume types of food adapted to their community's climate, soil, and other physical characteristics. People also select foods on the basis of religious values, taboos, and other social customs that are unrelated to nutritional or environmental factors.

Loss of productive farmland. Historically, world food production has increased primarily from expansion of the land area used for agriculture. When the world's population began to increase more rapidly in the late eighteenth and nineteenth centuries, pioneers could migrate to uninhabited territory and convert the land to agricultural use. Sparsely inhabited land suitable for agriculture was available for migrants in areas such as western North America, central Russia, and Argentina's pampas. People believed that good agricultural land would always be available for pioneers willing to migrate.

Today, few scientists believe that further expansion of agricultural land can feed the growing world population. Since approximately 1950, the world's population has increased more rapidly than the expansion of agricultural land.

At first glance, the traditional alternative should still be available, because only approximately 11 percent of the world's land is currently cultivated. Growth is possible in North America, where some arable land is not cultivated for economic reasons, and the tropics of Africa and South America offer some hope for developing new agricultural land. However, prospects for expanding the percentage of cultivated land are not good in much of Europe, Asia, and Africa.

In some regions, the amount of available agricultural land is actually declining rather than increasing. Farmland is frequently abandoned because of an excess or shortage of water. Human actions are causing land, especially in semiarid regions, to become more deserts-like. This is the process of *desertification,* or semiarid land degradation, discussed in Chapter 4.

Arid lands that can support a handful of pastoral nomads are overutilized because of rapid population growth. Excessive crop planting, animal grazing, and tree cutting exhaust the soil's nutrients and preclude any agricultural activity. The United Nations estimates that desertification removes 27 million hectares (104,000 square miles) of land from agricultural production each year, an area roughly equivalent to the size of the state of Colorado.

Excessive water threatens other agricultural areas, especially drier lands that receive water from human-built irrigation systems. If the irrigated land has inadequate drainage, the underground water level rises to the point where roots become waterlogged. The United Nations estimates that 10 percent of all irrigated land is waterlogged, mostly in Asia and South America. If the water is salty, the plants may also be damaged by excessive salinity. The ancient civilization of Mesopotamia may have collapsed in part because of waterlogging and excessive salinity in the agricultural lands near the Tigris and Euphrates Rivers.

Urbanization may also contribute to reducing the amount of agricultural land. As urban areas grow in population and land area, farms on the periphery are replaced by homes, roads, shops, and other urban land uses. In North America, nonworking farms outside urban areas are left idle until the speculators who own them can sell them at a profit to builders and developers, who actually convert the land to urban uses.

Paying for agricultural improvements. Subsistence farmers lack the land and supplies needed to expand crop production; they need higher-yield seeds, fertilizer, and tools. How can they obtain these aids? Farmers in developing countries could pay for some of the needed improvements by growing more crops and selling the surplus to people living in cities. However, because of low food prices, farmers are discouraged from producing a surplus. And if they did produce a surplus, farmers would have difficulty in transporting the food to urban markets because of poor roads and lack of distribution and storage facilities.

For many African and Asian countries, agricultural supplies must be obtained primarily by importing from other countries. These countries do not have enough money to buy agricultural equipment and replacement parts from relatively developed countries. How can developing countries generate the funds they need to buy agricultural equipment? They must produce something they can sell in relatively developed countries.

The United States remains the largest exporter of grain. This vessel being loaded in New Orleans is bound for Rotterdam, Europe's largest port. (John Olson/ The Stock Market)

Developing countries try to sell manufactured goods, but most raise funds through the sale of crops in relatively developed countries. Consumers in relatively developed countries are willing to pay high prices for fruits and vegetables that would otherwise be out of season or for crops such as coffee and tea that cannot be grown there because of the climate.

The sale of export crops brings a developing country foreign currency, a portion of which can be used to buy agricultural supplies. But governments in developing countries face a dilemma: the more land that is devoted to growing export crops, the less is available to grow crops for domestic consumption. Rather than helping to increase productivity, the funds generated through the sale of export crops may be needed to feed the people who switched from subsistence farming to growing export crops.

Problems for Commercial Farmers

Commercial farmers suffer from low incomes because they produce too much rather than too little food. A surplus of food has been produced because of widespread adoption of efficient agricultural practices and lack of growth in market due to low population growth and lack of funds abroad to import mass quantities of U.S. food. The diffusion of new seeds, fertilizers, and pesticides, as well as mechanical equipment, has enabled farmers to obtain greatly increased yields per area of land. Thus, commercial farmers have dramatically increased the capacity of the land to produce food.

Commercial farmers are burdened by very high capital costs. One machine can cost as much as $100,000, and farmland in the U.S. Midwest can cost more than $2,500 per hectare ($1,000 per acre). The total capital investment necessary to acquire and operate a modern 200-hectare farm can be well over $500,000. Farmers typically finance these operations by borrowing money.

The level of debt can be critical to a farmer's survival. During the 1970s, farming was relatively profitable in the United States, and the price of farmland increased. Many farmers bought additional land at the time, using their existing holdings as collateral to borrow money. But during the 1980s farm profits declined, because production costs increased more rapidly than crop prices. Because land values declined, many farmers did not have enough collateral to cover their debts and were forced out of business.

While the supply of food has increased in the relatively developed countries, demand has remained constant, because the market for most products is already saturated. In relatively developed countries, consumption of a particular commodity may not change significantly if the price changes. Americans, for example, do not switch from wheat to corn products if the price of corn falls relative to the price of wheat. Demand is also stagnant for most agricultural products in relatively developed countries because of low population growth.

The U.S. government has adopted two types of policies to attack the problem of excess productive capacity. First, farmers are encouraged to avoid producing crops that are in excess supply. Because soil erosion is a constant threat, the government encourages planting fallow crops, such as clover, to restore nutrients to the soil. These crops can be used for hay or producing seeds for sale.

Second, the government buys surplus food production. The main method by which the policy is carried out is a system of price supports, or farm subsidies. The government guarantees to buy certain commodities at fixed prices, which are calculated to give the farmer the same price for the commodity today as in the past, when compared to other consumer goods and services. The government encourages consumption of excess grain by selling or donating it to foreign governments.

The three most important export grains are wheat, maize (corn), and rice. Before World War II, Western Europe was the only region in the world that imported a large quantity of grain. Prior to independence, colonies supplied food to the European population concentrations. Asia became a net grain importer in the 1950s, Africa and Eastern Europe in the 1960s, and Latin America in the 1970s. Population increases in these regions largely accounted for the need to import grain. By 1980, North America was the only major exporting region in the world.

In response to the increasing global demand for food imports, the United States passed the Agricultural, Trade, and Assistance Act of 1954, frequently referred to as P.L. 480 (Public Law 480). Title I of the act provided for the sale of grain at low interest rates, and Title II gave grants to needy groups of people.

The largest beneficiary of U.S. food aid has been India. In 1966 and 1967, when the monsoon rains failed, 60 million Indians were fed entirely by U.S. grain. At the height of the rescue, 600 ships

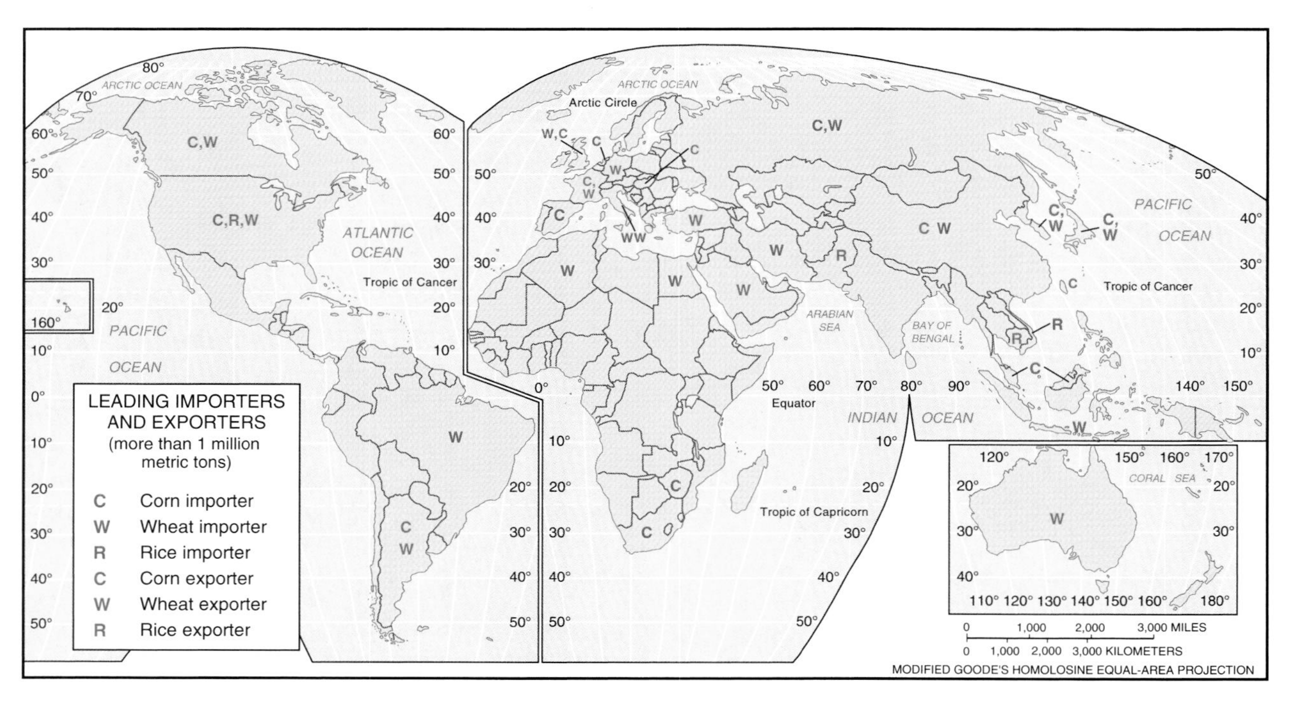

FIGURE 11–10
Most countries in the world must import more food than they export. The United States has by far the largest excess of food exports compared to imports. Argentina, Australia, Canada, and France are the other leading food exporters.

Animals, such as blue wildebeest and impala, cluster at a water hole in Africa. The dry lands of Africa can support limited populations, but the number of people and animals has exceeded the region's supply of food and water. (Leonard Lee Ruf II/Photo Reseachers, Inc.)

filled with grain sailed to India, the largest maritime maneuver since the Allied invasion of Normandy on D-Day, June 6, 1944. The United States allocated 20 percent of its wheat crop those years to feed India's population.

The United States remains the largest exporter of grain and accounts for nearly two-thirds of all corn and soybean exports, one-third of wheat, and one-fifth of rice. However, since 1980, the United States has decreased its grain exports while other countries have increased theirs. Thailand replaced the United States as the leading rice exporter, and other Asian countries, such as Pakistan, Vietnam, and India, account for most of the remaining rice exports. Australia and France have joined the United States and Canada as major wheat exporters (Figure 11–10).

Russia is by far the leading grain importer and ranks at or near the top in wheat, corn, and rice imports. Russia and Japan together account for one-half of the world's corn imports, while Russia and China together account for one-fourth of the wheat imports. Asian countries account for nearly all of the rice imports.

The United States spent $134 billion during the 1980s on farm subsidies, including $59 billion for feed grains, such as corn and soybeans, $22 billion for wheat, and $16 billion for dairy products. Annual spending has varied considerably, from more than $25 billion in 1986 and $22 billion in 1987 to less than $5 billion in 1981 and $10 billion in 1989. Subsidy payments are lower in years when market prices rise and production is down, typically as a result of poor weather conditions in the United States or political problems in other countries.

U.S. policies point out a fundamental irony in worldwide agricultural patterns. In a relatively developed country such as the United States, some government programs attempt to reduce production, while developing countries struggle to increase food production at the same rate as the growth in population.

Africa's Food Supply Problems

The ratio of population to food supply has become more favorable in many regions, especially in Asia, where population growth has slowed, while food supply has grown. The challenge in these regions is to continue recent progress by further expanding food resources. In other regions, especially Sub-Saharan Africa, the problem is getting worse. Food supply must be quickly expanded to meet the needs of Africa's rapidly expanding population.

Some countries that previously depended on imported grain have become self-sufficient in recent years. Higher productivity generated by the green revolution is primarily responsible for reducing dependency on imports, especially in

Asia. India no longer ranks as a major wheat importer, and China no longer imports rice. As long as population growth continues to decline and agricultural productivity continues to increase, the large population concentrations of Asia should be able to maintain a delicate balance between population and resources.

In contrast, Sub-Saharan Africa is losing the race to keep food production ahead of population growth. The United Nations Food and Agricultural Organization estimates that 70 percent of the African people do not have enough to eat, and famine is widespread in half of the African countries. By all estimates, the problem will get worse in the coming years.

Per capita production of most food crops is lower today in Africa than in the 1960s. At the same time, population is increasing more rapidly than any other world region. As a result, production of food per person has declined in the last three decades in all but a handful of the region's countries, in several cases by more than 20 percent. At current population growth rates, by the end of the century, agriculture in Sub-Saharan Africa will be able to feed little more than half of the region's population.

The problem is particularly severe in the Horn of Africa, including Somalia, Ethiopia, and Sudan. Also facing severe food shortages are countries in the Sahel region, a 400- to 550-kilometer-wide (250- to 350-mile-wide) belt in West Africa which marks the southern border of the Sahara Desert. The most severely affected countries in the Sahel include Gambia, Senegal, Mali, Mauritania, Burkina Faso, Niger, and Chad.

Traditionally, the region supported a limited amount of agriculture. Pastoral nomads moved their herds frequently, permitting vegetation to regenerate. Farmers grew groundnuts for export and used the receipts to import rice. With rapid population growth, the size of herds increased beyond the capacity of the land to support animal life. Animals destroyed the limited vegetation through overgrazing and clustered at the scarce water sources. Many died of hunger.

Farmers exhausted the nutrients in the soil by overplanting and reducing the amount of time the land remained fallow. Soil erosion increased after most of the remaining trees were cut down to provide urban residents with wood and charcoal for cooking and heating. Productivity declined further following several unusually dry years in the 1970s, 1980s, and 1990s.

Government policies have aggravated the food shortage crisis. To make food affordable for urban residents, governments keep agricultural prices low. Constrained by price controls, farmers are unable to sell their commodities at a profit and therefore have little incentive to increase productivity.

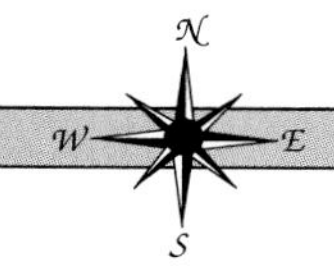

CONCLUSIONS: CRITICAL ISSUES FOR THE FUTURE

The world needs to produce much more food to keep up with continued growth in the world's population, as well as to feed those who are currently hungry. However, long-term expansion of the world's food resources is hampered by poor agricultural management practices. Soil erosion is a significant problem around the world, including roughly one-fourth of the cropland in the United States. Agricultural productivity is suffering from declining inherent fertility of soil. Widespread use of fertilizer masks the problem of declining soil fertility.

Conservation strategies generally require farmers to limit cultivation of the most erodible lands and to incorporate soil-conserving cover crops in their rotations. Farmers are reluctant to till and plant less land, because of financial pressures. In the short run, adopting conservation practices can reduce income needed to repay debts, to further expand production, or to enjoy a more comfortable lifestyle. However, a farmer who ignores conservation practices in pursuit of maximizing immediate profits may incur higher costs in the long run.

In response to the need for conservation, some farmers have adopted a new form of agriculture known as sustainable agriculture. **Sustainable agriculture** involves farming methods that preserve the long-term productivity of the land and minimize pollution of the soil, ground water, and streams that drain the land. In most cases, sustainable agriculture means rotating soil-restoring crops with cash crops and reducing inputs of fertilizer and pesticides.

Farmers practicing sustainable agriculture typically generate lower revenues than conventional farmers, but they also have lower costs. On balance, sustainable agriculture may enjoy greater net benefits when costs are subtracted from revenues, especially if long-term environmental costs are included. It remains the world's best hope for increased long-term productivity in developing regions, consistent with sound environmental management.

Chapter Summary

1. Distribution of agriculture

Agricultural regions can be divided between subsistence and commercial farming. The predominant form of agriculture is subsistence in developing countries and commercial agriculture in relatively developed countries. Agriculture was not invented simply but was the product of thousands of years of experiments and accidents. Vegetative planting apparently originated primarily in Southeast Asia and diffused to the north and west. Significant advances in settled agriculture, including domestication of wheat and barley and integration of herd animals with crop grazing, originated in Southwest Asia. Agricultural practices probably originated independently in the Western Hemisphere.

2. Subsistence agriculture in developing regions
Most people in the world, especially those in developing countries, are subsistence farmers, growing crops primarily to feed themselves. Several important kinds of subsistence agriculture can be identified, including shifting cultivation, pastoral nomadism, and intensive subsistence farming. Regions where subsistence agriculture is practiced are characterized by a large percentage of the labor force engaged in agriculture, with few mechanical aids.

3. Commercial agriculture in relatively developed regions
Few people in relatively developed regions are farmers, but, through the use of expensive machinery and scientific techniques, these few farmers are able to produce an abundant supply of food. Farmers in relatively developed countries who grow crops and raise livestock for sale off the farm to processors are only one part of a large food production industry. The most common type of farm found in relatively developed societies is mixed crop and livestock. Where mixed crop and livestock farming is not suitable, commercial farmers practice a variety of other types of agriculture, including dairying, commercial grain, and ranching.

4. Increasing the world supply of food
The green revolution has stimulated increased productivity in developing countries, but subsistence farmers lack funds to benefit fully from the green revolution. Commercial farmers produce more food than the market demands, especially in the United States. Because of increased productivity, most regions of the world are generating enough food to meet the needs of their population. The exception is Sub-Saharan Africa, where the gap between food supply and population size is growing wider.

Key Terms

Agribusiness, 443
Agriculture, 440
Commercial agriculture, 441
Double cropping, 440
Green revolution, 454
Intensive subsistence agriculture, 438
Milkshed, 446
Paddy, 438
Pastoral nomadism, 436
Plantation, 452
Ranching, 449
Sawah, 438
Seed agriculture, 431
Shifting cultivation, 433
Spring wheat, 448
Subsistence agriculture, 430
Sustainable agriculture, 461
Swidden, 434
Transhumance, 437
Vegetative planting, 431
Winter wheat, 448

Questions for Study and Discussion

1. What is the difference between agriculture and hunting and gathering?
2. What is the difference between seed agriculture and vegetative planting? How did each diffuse?
3. What are the principal forms of subsistence agriculture? In what climate regions do they predominate?
4. Why is the amount of land under shifting cultivation declining?
5. What are the differences in the distinctive forms of agriculture in the dry lands of developing regions and the dry lands of relatively developed regions?
6. Why is wet rice not dominant throughout the intensive subsistence region?
7. What are the principal characteristics of commercial agriculture?
8. Outline the main agricultural regions within the United States.
9. What is meant by the *green revolution?*
10. Outline some of the obstacles to increasing agricultural production in developing countries.
11. Why does the United States pay its farmers not to grow crops?

Thinking Geographically

1. Assume that the United States constitutes one agricultural market, centered around New York City, the largest metropolitan area. To what extent can the major agricultural regions of the United States be viewed as irregularly shaped rings around the market center, as von Thünen applied to southern Germany?
2. New Zealand once sold nearly all of its dairy products to the British, but since the United Kingdom joined the European Community in 1973 New Zealand has been forced to find other markets. Are there other examples of countries that have restructured their agricultural production in the face of increased global interdependence and regional cooperation?
3. Review the concept of overpopulation (the number of people in an area exceeds the capacity of the environment to support life at a decent standard of living). What agricultural regions have relatively limited capacities to support intensive food production? Which of these regions face rapid population growth?
4. Compare world distributions of corn, wheat, and rice production. To what extent do differences derive from environmental conditions and to what extent from food preferences and other social customs?
5. How might the loss of farmland on the edge of rapidly growing cities alter the choice of crops that other farmers make in a commercial agricultural society?
6. Malthus argued two hundred years ago that overpopulation was inevitable, because population increased geometrically, while food supply increased arithmetically. Was Malthus correct about changes in food supply? Why, or why not?

Suggestions for Further Reading

Babbington, Anthony, and Judith Carney. "Geography in the International Agricultural Research Centers: Theoretical and Practical Concerns." *Annals of the Association of American Geographers* 80 (March 1990): 34–48.

Bascom, Johnathan B. "Border Pastoralism in Eastern Sudan." *Geographical Review* 80 (October 1990): 416–30.

Brown, Lester R., and Edward C. Wolf. "Reversing Africa's Decline." *Worldwatch Paper* 65. Washington, DC: Worldwatch Institute, 1985.

Brown, Lester R., and Pamela Shaw. "Six Steps to a Sustainable Society." *Worldwatch Paper* 48. Washington, DC: Worldwatch Institute, March 1982.

Chakravarti, A. K. "Green Revolution in India." *Annals of the Association of American Geographers* 63 (September 1973): 319–30.

Cochran, Willard W., and Mary E. Ryan. *American Farm Policy 1948–73*. Minneapolis: University of Minnesota Press, 1976.

Cromley, Robert G. "The Von Thünen Model and Environmental Uncertainty." *Annals of the Association of American Geographers* 72 (September 1982): 404–10.

Cusack, David F., ed. *Agroclimate Information for Development: Reviving the Green Revolution*. Boulder, CO: Westview Press, 1983.

Dahlberg, Kenneth A., ed. *New Directions for Agriculture and Agricultural Research: Neglected Dimensions and Emerging Alternatives*. Totowa, NJ: Rowman and Allanheld, 1986.

Dove, Michael R. *Swidden Agriculture in Indonesia: The Subsistence Strategies of the Kalimantan Kantu*. Amsterdam: Mouton, 1985.

Duckham, A. N., and G. B. Masefield. *Farming Systems of the World*. New York: Praeger, 1970.

Durand, Loyal, Jr. "The Major Milksheds of the Northeastern Quarter of the United States." *Economic Geography* 40 (January 1964): 9–33.

Ebeling, Walter. *The Fruited Plain: The Story of American Agriculture*. Berkeley, CA: University of California Press, 1979.

Falkenmark, Malin, and Carl Widstrand. "Population and Water Resources: A Delicate Blance." *Population Bulletin* 47 (3). Washington, DC: Population Reference Bureau, November 1992.

Furuseth, Owen J., and John T. Pierce. *Agricultural Land in an Urban Society*. Washington, DC: Association of American Geographers, 1982.

Grigg, David B. *An Introduction to Agricultural Geography*. London: Hutchinson Education, 1984.

_____. *The Agricultural Systems of the World: An Evolutionary Approach*. London: Cambridge University Press, 1974.

_____. *The World Food Problem 1950–1980*. New York: Basil Blackwell, Ltd., 1985.

Hart, John Fraser. "Change in the Corn Belt." *Geographical Review* 76 (January 1986): 51–73.

Heathcote, R. L. *The Arid Lands: Their Use and Abuse*. London: Longman, 1983.

Hewes, Leslie, and Christian I. Jung. "Early Fencing on the Middle Western Prairie." *Annals of the Association of American Geographers* 71 (June 1981): 177–201.

Ilbery, Brian W. *Agricultural Geography: A Social and Economic Analysis*. New York, 1985.

Knight, C. G., and P. Wilcox. *Triumph or Triage? The World Food Problem in Geographical Perspective*. Washington, DC: Association of American Geographers, 1975.

Lewthwaite, G. R. "Wisconsin and the Waikato: A Comparative Study of Dairy Farming in the United States and New Zealand." *Annals of the Association of American Geographers* 54 (March 1974): 59–87.

Pannell, Clifton. "Recent Chinese Agriculture." *Geographical Review* 75 (April 1985): 170–85.

Peters, William J., and Leon F. Neuenschwander. *Slash and Burn: Farming in the Third World Forest.* Moscow, ID: University of Idaho Press, 1988.

Pierce, John T. *The Food Resource.* New York: Longman Scientific and Technical, 1990.

Sauer, Carl O. *Agricultural Origins and Dispersals.* 2d ed. Cambridge, MA: M.I.T. Press, 1969.

Smith, Everett G., Jr. "America's Richest Farms and Ranches." *Annals of the Association of American Geographers* 70 (December 1980): 528–41.

Symons, Leslie. *Agricultural Geography.* rev. ed. London: G. Bell, 1979.

Tarrant, John R. *Agricultural Geography.* New York: Wiley, 1974.

Turner, B. L., II, and Stephen B. Brush, eds. *Comparative Farming Systems.* New York: Guilford, 1987.

von Thünen, Johann Heinrich. *von Thünen's Isolated State: An English Edition of "Der Isolierte Staat".* Translated by Carla M. Wartenberg. Elmsford, NY: Pergamon Press, 1966.

Whittlesey, Derwent. "Major Agricultural Regions of Earth." *Annals of the Association of American Geographers* 26 (September 1936): 199–240.

12 Industry

- Distribution of manufacturing
- Situation factors in industrial location
- Site factors in industrial location
- Industrial restructuring

Port Newark, New Jersey.

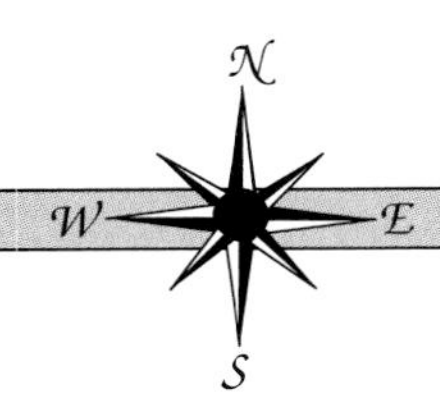

1. Distribution of manufacturing
Manufacturing is clustered in four regions—Western Europe, Northeastern United States/Southeastern Canada, Eastern Europe, and Japan. Subareas have emerged within these regions.

2. Situation factors in industrial location
Two types of costs—site and situation—influence industrial location decisions. Situation factors include the costs of shipping inputs to the factory and shipping finished products to consumers. Firms seek locations that minimize aggregate transportation costs.

3. Site factors in industrial location
Site factors include the costs of land, labor, and capital. Modern industries require large tracts of land. Labor costs are important for some industries. Access to capital, such as through government subsidies, induces other locational choices.

4. Industrial restructuring
From a global perspective, industrial decline is caused by a combination of stagnant demand for many products and excessive capacity to produce these products.

INTRODUCTION

In 1993, Mercedes-Benz revealed that it was designing an entirely new sport utility vehicle, and it would open a new factory somewhere in the United States to build it. The announcement touched off a fierce competition among states and localities to become the home for the Mercedes-Benz plant.

Swamped with material from competing communities, Mercedes-Benz took several months to select a factory site. The choice was Vance, Alabama, then a village of a few hundred inhabitants, 25 kilometers (15 miles) east of Tuscaloosa and 55 kilometers (35 miles) southwest of downtown Birmingham.

Mercedes-Benz's process of selecting a location for its factory raises several issues that geographers address. First, what factors did Mercedes-Benz consider in evaluating alternative locations? Geographers recognize that transportation costs are critical in locating some factories. Materials must be brought into the factory, and the finished products must be shipped to consumers. Other factories are attracted to locations where suitable land, workers, and financing are available.

Second, why did communities throughout the United States compete to get the Mercedes-Benz factory? Government officials throughout the world recognize the critical role such an industry plays in the economic health of a community. In the global competition to attract new industries—or, in many places, to retain existing ones—a community possesses distinctive locational characteristics. Geographers identify a community's assets that enable it to compete successfully for industries, as well as its locational handicaps that must be overcome to retain older companies.

Third, why did the Mercedes-Benz decision makers feel compelled to build a new plant at all? The company could have built the vehicles at a factory it owned in another country or in a U.S. factory recently closed by other motor vehicle firms. To succeed in an intensely competitive global market for products such as motor vehicles, corporations are under increasing pressure find the best locations for their factories.

Distribution of Manufacturing

The modern concept of industry—manufacturing goods in a factory—began in Great Britain in the late eighteenth century. The process of change, known as the industrial revolution, transformed both the way in which goods are produced for a society and the way people obtain food, clothing, and shelter. Effects of the industrial revolution have penetrated virtually all economic, social, and political elements of society.

Approximately three-fourths of the world's industrial production is concentrated in four regions—eastern North America, Western Europe, Eastern Europe, and Japan (Figure 12–1). The distribution of industry differs from that of agriculture. Agriculture occupies one-fourth of Earth's land

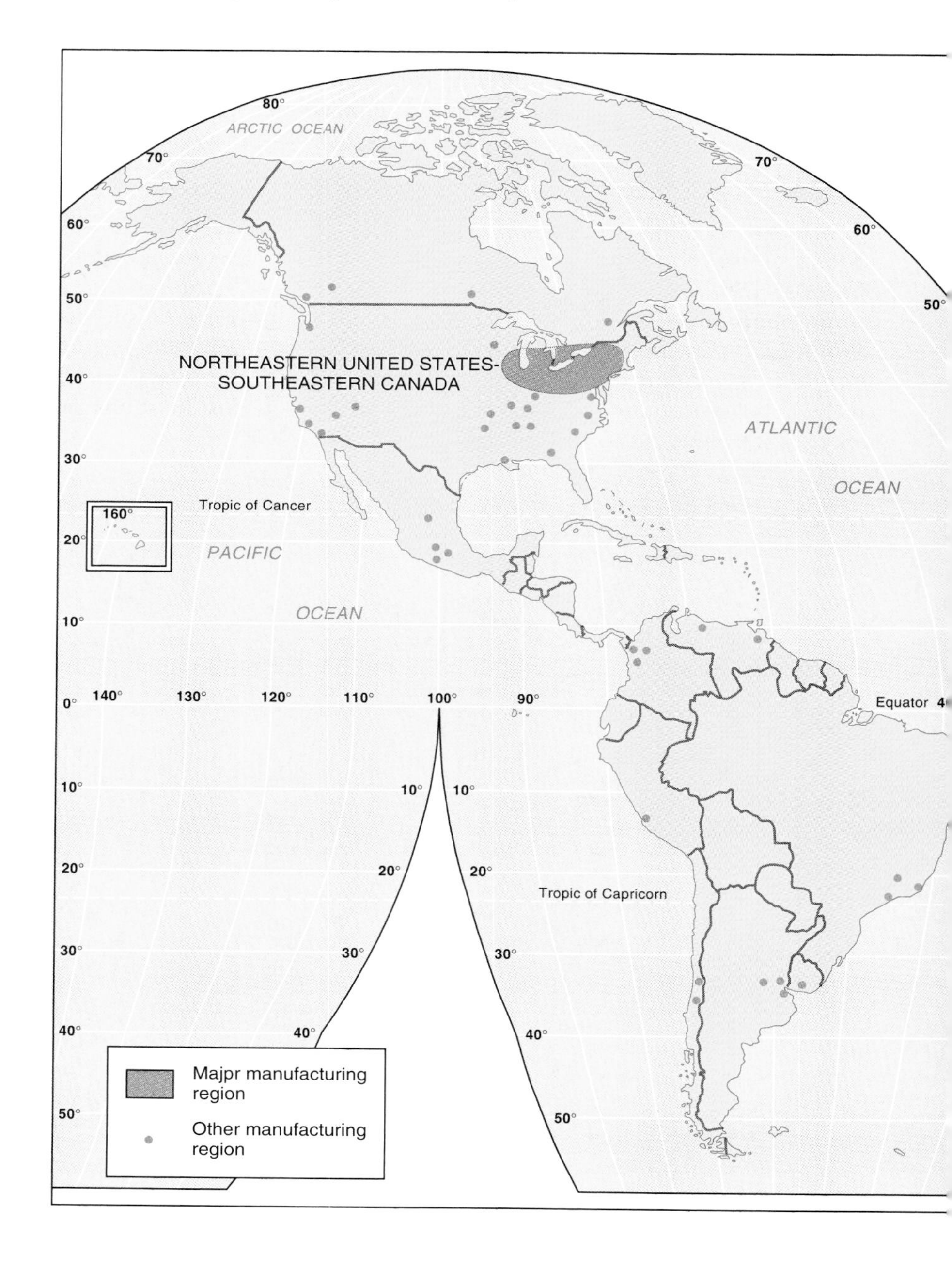

FIGURE 12–1
Manufacturing is clustered in four main regions: Northeastern United States–Southeastern Canada, Western Europe, Eastern Europe, and Japan.

area and covers extensive areas throughout the inhabited areas of the world. In contrast, less than 1 percent of Earth's land is devoted to industry.

The Industrial Revolution

The first country to be transformed by the industrial revolution, Great Britain had a system of production that far outpaced that of the rest of the world. Great Britain was the world's dominant industrial power from the late eighteenth century until the middle of the nineteenth century. The country was responsible for producing more than half of the world's cotton fabric and iron and for mining two-thirds of the coal.

From Britain, the industrial revolution diffused to other countries in two main directions during the nineteenth century: to the east through Europe and to the west across the Atlantic Ocean to North

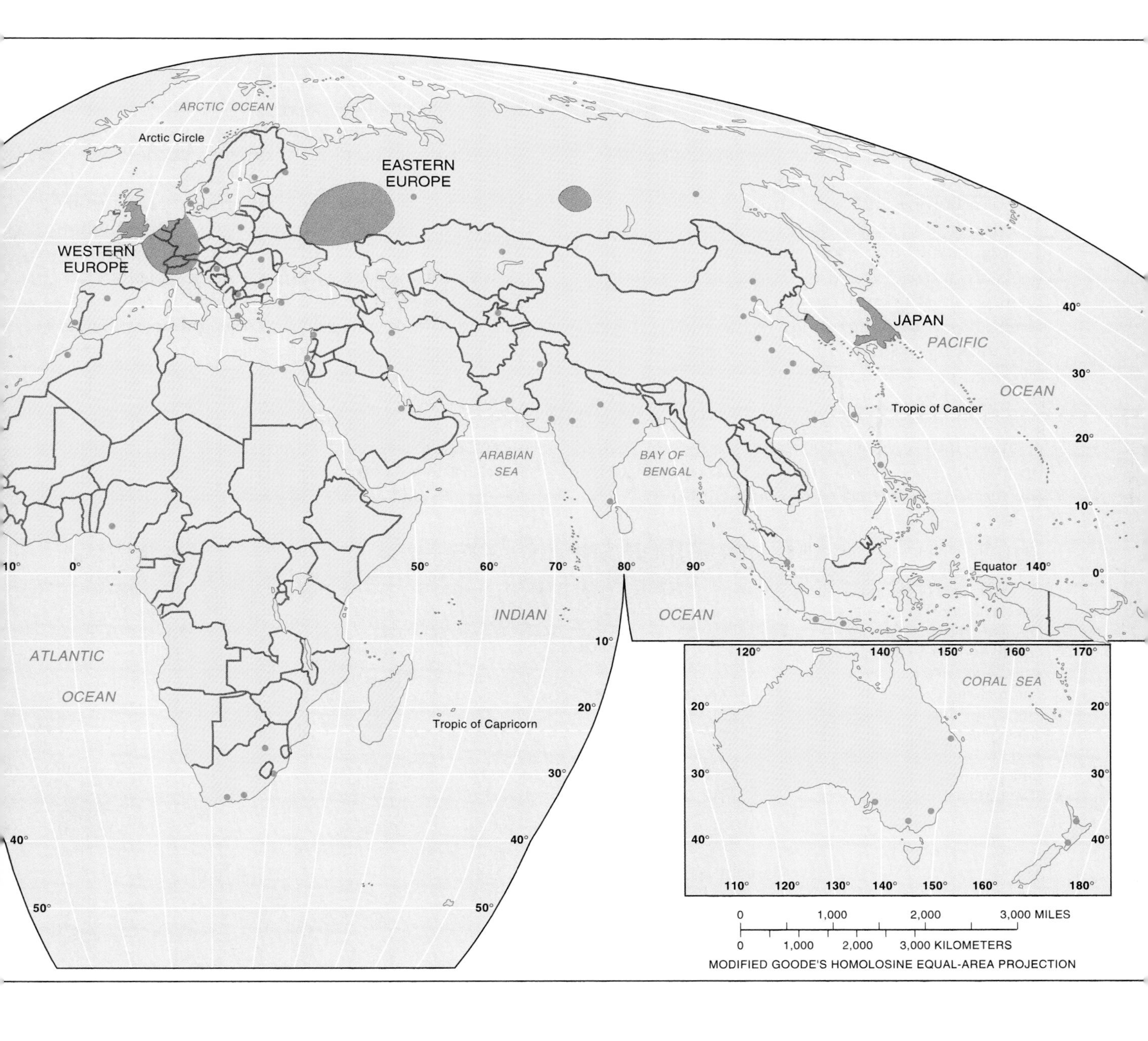

America. From these places, industrial development subsequently reached other parts of the world.

The industrial revolution involved a series of inventions that transformed the way goods were manufactured. These improvements in industrial technology created an unprecedented expansion in human productivity, resulting in substantially higher standards of living.

The term *industrial revolution* is somewhat misleading. The industrial revolution resulted in new inventions in social, economic, and political arenas, not just in industry. The changes involved a gradual diffusion of new ideas and techniques rather than an overnight revolution. Nonetheless, the term industrial revolution is commonly used to define the events of a certain time period—the late eighteenth and early nineteenth centuries in Western Europe and North America.

Prior to the industrial revolution, industry was geographically dispersed. People made household tools and agricultural equipment in their own homes or obtained them in the local village. Home-based manufacturing was known as the cottage industry system. One important cottage industry was textile manufacturing. People known as *putters-out* were hired by merchants to drop off wool at homes, where women and children sorted, cleaned, and spun it. The putters-out then picked up the finished work and paid according to the number of pieces that were completed.

The steam engine. The industrial revolution was characterized by the invention of hundreds of mechanical devices, but the one invention most important for the development of factories was the steam engine. In 1765, James Watt, a mathematical instrument maker in Glasgow, Scotland, was asked to repair a broken model of a Newcomen engine. The Newcomen engine worked by alternate injection and condensation of steam. To prevent the steam from condensing before the piston had completed its upward stroke, the cylinder was heated. The cylinder then had to be cooled to condense the steam for the return stroke. The Newcomen engine was not very efficient because most of its generated energy was used to constantly warm and then cool the cylinder.

Watt's steam engine solved the problem. He introduced a separate condenser kept permanently cool while the cylinder could be permanently hot. The cylinder was drilled with precision to provide a tight fit for the piston, using a device patented by James Wilkinson in 1774 for boring the barrel of a cannon. The steam engine could pump water far more efficiently than watermills or the Newcomen engine, let alone human or animal power.

The iron and textile industries were the first to increase production through extensive use of the steam engine and other new inventions. From the needs of these two pioneer industries, new industrial techniques diffused during the nineteenth century.

The textile industry was one of the first to cluster in large factories in the early years of the industrial revolution. The steam engine permitted the simultaneous operation of many machines, such as these looms for the weaving of cotton cloth in a factory in Lancashire, England. The colored engraving, from 1834, makes working conditions appear more attractive than they actually were. (The Granger Collection)

Distribution of Manufacturing Within Western Europe

Europeans were responsible for many early inventions of the industrial revolution in the late eighteenth century. The Belgians led the way in new coal mining techniques, the French had the first coal-fired blast furnace for making iron, and the Germans had the first industrial cotton mill. However, the industrial revolution did not make a significant impact elsewhere in Europe until the late nineteenth century.

Political instability delayed the diffusion of the industrial revolution in Europe. The French Revolution and Napoleonic Wars disrupted Europe during the late eighteenth and early nineteenth centuries, and Germany did not become a unified country until the 1870s. Other revolutions and wars plagued Europe throughout the nineteenth century.

Political problems in Europe retarded the development of modern transportation systems, especially the railway. Cooperation among small neighboring states was essential to build an efficient rail network and raise the money needed to construct and operate the system. Because such cooperation could not be attained, construction of railways in some parts of Europe did not occur until fifty years after their first appearance in Britain (Figure 12–2).

The industrial revolution reached Italy, the Netherlands, Russia, and Sweden in the late nineteenth century, although industrial development in these countries did not match the level in Belgium, France, and Germany until the twentieth century. Other Southern and Eastern European countries joined the industrial revolution during the twentieth century.

The principal industrial region of Western Europe is situated in the northwest and encompasses a portion of several countries (Figure 12–3). Several distinct districts have emerged, primarily because European countries competed with each other to develop their own industrial areas.

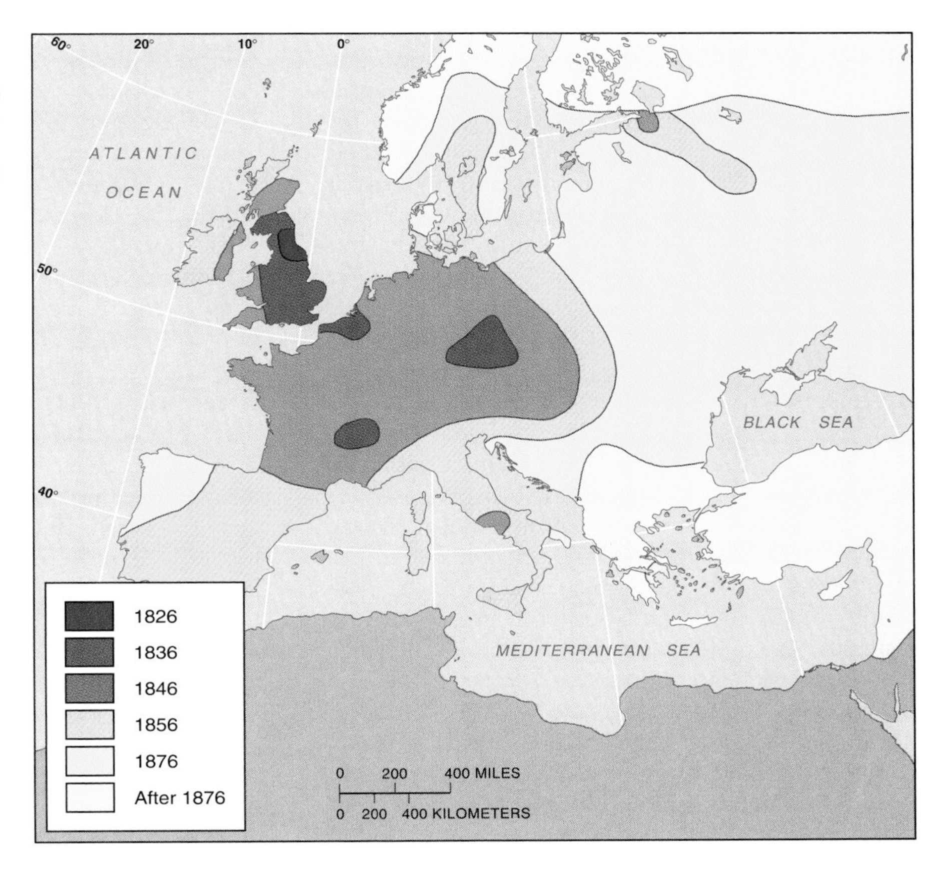

FIGURE 12–2
More than fifty years passed between the construction of the first railways in Britain and the first ones in eastern Europe. The diffusion of the railway from Great Britain to the European continent reflects the diffusion of the industrial revolution. (From Peter A. Gould, *Spatial Diffusion*. Washington D.C.: Association of American Geographers, 1969. Reprinted by permission.)

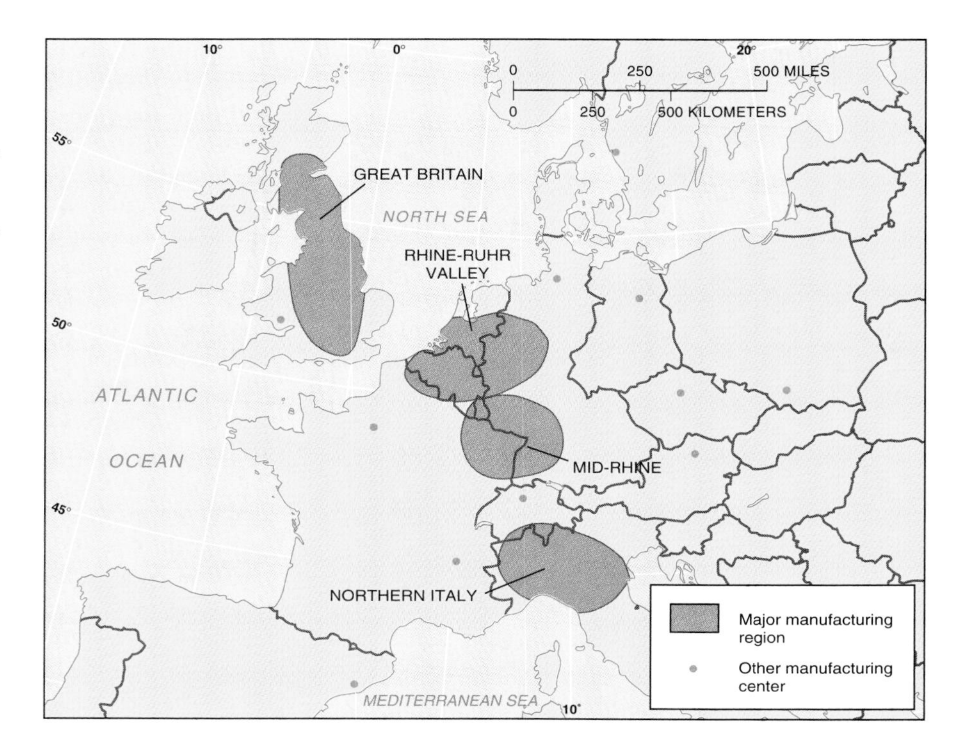

FIGURE 12–3
Manufacturing in Western Europe. A large percentage of manufacturing in Western Europe extends in a north-south belt, from Great Britain on the north to Italy on the south. At the core of the European manufacturing region lie the Rhine-Ruhr Valley and Mid-Rhine regions in Germany, France, and the Benelux countries (Belgium, the Netherlands, and Luxembourg).

Rhine-Ruhr Valley. Europe's most important industrial area is the Rhine-Ruhr Valley. The region lies for the most part in northwestern Germany but also extends into nearby areas of Belgium, France, and the Netherlands. Within the region, industry is dispersed rather than concentrated in one or two cities. Iron and steel manufacturing has concentrated in the Rhine-Ruhr Valley because of proximity to large coalfields. Access to iron and steel production stimulated the location of other industries, such as locomotives, machinery, and armaments, which made heavy use of the metals.

Mid-Rhine. The second most important industrial region in Europe, the Mid-Rhine, includes southwestern Germany, northeastern France, and the small country of Luxembourg. In contrast to the Rhine-Ruhr region, the German portion of the Mid-Rhine region lacks an abundant supply of raw materials. The Mid-Rhine region is the most centrally located industrial area within the European Union.

The three largest cities in the German portion of the region are Frankfurt, Stuttgart, and Mannheim. Frankfurt became West Germany's most important financial and commercial center and the hub of its road, rail, and air networks. As a result, Frankfurt attracted industries that produce goods for distribution to consumers throughout the country. Frankfurt is well-situated to play a comparable role in the European Union. Stuttgart specializes in industries that produce high-value goods and require skilled labor. The Mercedes-Benz and Audi automobiles rank among the city's best-known products. Mannheim, an inland port along the Rhine, has a large chemical industry that manufactures products such as synthetic fibers, dyes, and pharmaceuticals.

Great Britain. During the nineteenth century, the north of Great Britain dominated world production of iron and steel, textiles, and coal-mining. In the twentieth century, the region has lost its preeminent global position. British industries are more likely to locate in southern England, near the country's largest concentrations of population and wealth. International competition has hurt the British industrial region even more, because the world suffers from an oversupply of steel and textiles, the industries traditionally associated with the area.

British industries have faced an especially difficult challenge in regaining global competitiveness. As the first country to enter the industrial revolution, Britain is saddled with relatively outmoded and deteriorating factories and supporting services. The British sometimes refer ironically to their "misfortune" of winning World War II. The losers, Germany and Japan, have become industrial powers in part because they received American financial assistance to build modern factories in place of the ones devastated during the war.

Northern Italy. A fourth European industrial region of some importance is found in the Po Basin of northern Italy. The region contains about one-fifth of Italy's land area but approximately half of the country's population and two-thirds of its industries.

Modern industrial development in the Po Basin began with the establishment of textile manufacturing during the nineteenth century. The Po Basin has attracted textile manufacturers and other industries because the region possesses two principal assets compared to Europe's other industrial regions. First, the region has a larger supply of people willing to work for lower wages than workers in northern Europe. Second, the nearby Alps provide the region with inexpensive sources of hydroelectricity. Industries that have concentrated in the region include processors of raw materials and assemblers of mechanical parts.

Distribution of Manufacturing Within North America

Industry arrived a bit later in the United States than in Western European countries such as France and Belgium, but it grew at a much more rapid rate. At the time of independence, the United States was a predominantly agricultural society, dependent on the import of manufactured goods from Great Britain. Manufacturing was more expensive in the United States than in Great Britain, because of a scarcity of labor and capital and the high cost of shipping to European markets.

The first U.S. textile mill was built in Pawtucket, Rhode Island in 1791 by Samuel Slater, a former worker at Arkwright's factory in England. The textile industry grew rapidly after 1808, when the U.S. government imposed an embargo on European trade in order to avoid entanglement in the Napoleonic Wars. The textile industry grew from 8,000 spindles in 1808 to 31,000 in 1809 and 80,000 in 1811.

By 1860, the United States had become a major industrial nation, second only to Great Britain. However, with the exception of textiles, the leading industries at the time in the United States did not make widespread use of the new industrial processes. Instead, many engaged in processing North America's abundant food and lumber resources. Industries such as iron and steel did not apply new manufacturing techniques on a large scale in the United States until the final third of the nineteenth century.

Manufacturing in North America is concentrated in the northeastern portion of the United States and in southeastern Canada. The region comprises only 5 percent of the land area of these countries but contains one-third of the population and nearly two-thirds of the manufacturing output.

This manufacturing belt has achieved its dominance through a combination of historical and environmental factors. As the first area of settlement, the East Coast of the United States was tied to European markets and industries during the first half of the nineteenth century. The early date of settlement gave the eastern cities an advantage in creating the infrastructure needed to become the country's dominant industrial center.

The northeast U.S. and southeast Canadian region became the manufacturing center of North America in part because of availability of essential raw materials, such as iron and coal. Good transportation facilitated the movement of raw materials to the factories and manufactured goods to markets. The Great Lakes and the Mississippi, Ohio, St. Lawrence, and other rivers were supplemented in the nineteenth century by canals, railways, and highways, which helped to connect the frontier with the manufacturing centers.

New England. Within the North American manufacturing belt several subareas have emerged (Figure 12–4). The oldest industrial area in North America is southern New England. The area developed as an industrial center in the early nineteenth century, beginning with cotton textiles. Cotton was imported from southern states, and finished products were shipped to Europe. European immigrants provided an abundant supply of inexpensive labor throughout the nineteenth century. Today, New England is known for relatively skilled but expensive labor.

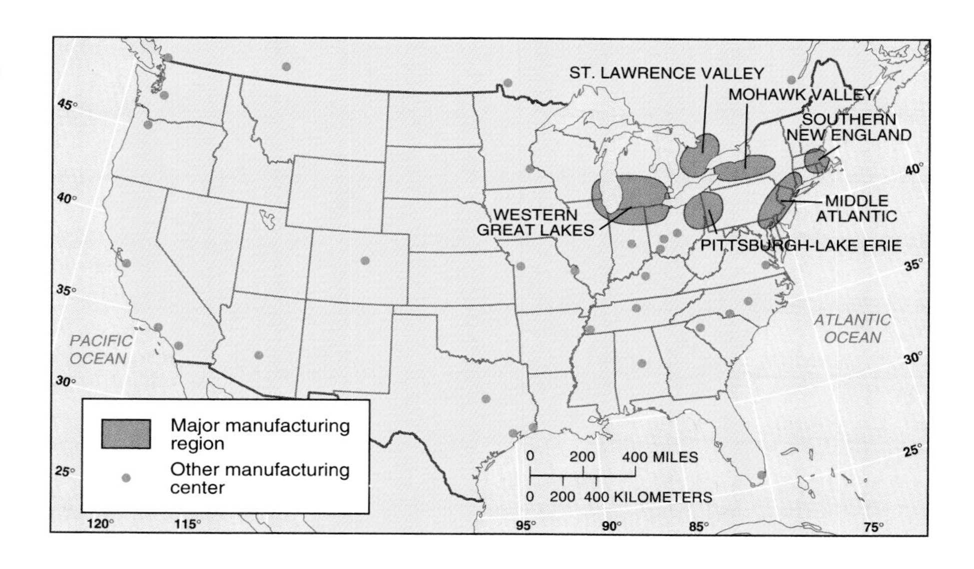

FIGURE 12–4
Manufacturing in North America is highly clustered in several regions within the northeastern United States and southeastern Canada, although important manufacturing centers exist elsewhere in the two countries.

Middle Atlantic. The Middle Atlantic region, which includes New York City and several other important urban areas, is the United States' largest market and has long attracted industries that need proximity to a large number of consumers. Many industries that depend on foreign markets or sources of raw materials have selected a location near one of the region's main ports, including New York City (the nation's largest port), Baltimore, Philadelphia, and Wilmington, Delaware. Other firms seek locations near the financial, communications, and entertainment industries, which are highly concentrated in New York.

Mohawk Valley. A linear industrial belt developed in upper New York State along the Hudson River and Erie Canal, the only water route at the time between New York City and the Great Lakes. Buffalo, near the confluence of the Erie Canal and Lake Erie, is the region's most important industrial center, especially for steel and food processing. Inexpensive, abundant electricity, generated at nearby Niagara Falls, has attracted aluminum, paper, and electrochemical industries to the region.

Pittsburgh–Lake Erie. The area between Pittsburgh and Cleveland is the nation's oldest steel-producing area. Steel manufacturing originally concentrated in the region because of proximity to Appalachian coal and iron ore. When northern Minnesota became the main source of iron ore, the Pittsburgh–Lake Erie region could bring in the ore through the Great Lakes.

Western Great Lakes. The western Great Lakes region extends from Detroit and Toledo, Ohio, on the east to Chicago and Milwaukee, Wisconsin, on the west. Chicago, the third largest U.S. urban area, is the dominant market center between the Atlantic and Pacific coasts and the hub of the nation's transportation network. Because road, rail, air, and water routes converge on Chicago, the city has become a transfer point between transportation systems or between routes within the same type of transportation system. Automobile manufacturers and other industries that have a national market locate in the western Great Lakes region to take advantage of the convergence of transportation routes. The region's industries are also the main suppliers of machine tools, transportation equipment, clothing, furniture, agricultural machinery, and food products to people living in the interior of the country.

St. Lawrence–Ontario Peninsula. Canada's most important industrial region is the St. Lawrence Valley–Ontario Peninsula area, which stretches across southern Canada along the U.S. border. The region has several assets: centrality to the Canadian market, proximity to the Great Lakes, and access to inexpensive electric power at Niagara Falls. Most of Canada's steel production is concentrated in Hamilton, while most automobiles are assembled in the Toronto area. Inexpensive electricity has attracted aluminum manufacturing, paper making, flour mills, textile manufacturing, and sugar refining.

Other regions. Industry has grown in the United States outside the main manufacturing belt. Steel, textiles, tobacco products, and furniture industries have been dispersed through smaller communities in the South. The Gulf Coast is becoming an important industrial area because of access to oil and natural gas. Oil refining, petrochemical manufacturing, food processing, and aerospace product manufacturing are located along the Gulf coast.

Los Angeles is the largest industrial area on the West Coast for aircraft, electronics, oil refining, and sportswear. Other important West coast industrial concentrations include the aerospace industry in Seattle, food processing in the San Francisco Bay area, and naval services in San Diego.

Distribution of Manufacturing Within Eastern Europe

Eastern Europe has five major industrial regions, including three located entirely or principally in Russia and one in Ukraine. Two of the regions, Central (around Moscow) and Eastern Ukraine, became manufacturing centers in the nineteenth century, while the other two, Volga and Urals, were established by the Communists during the twentieth century. Russia contains another important industrial region, Kuznetsk, in the far eastern or Asian portion of the country (Figure 12–5).

Central industrial district. Russia's oldest industrial region is centered around Moscow, the country's capital and largest city. Although not well-endowed with natural resources, the central industrial district produces one-fourth of Russian industrial output, primarily because it is situated near the country's largest market. Products manufactured in the region tend to be of high value relative to their bulk and require a large pool of skilled labor. Thirty percent of Moscow's industrial work force is employed in making linen, cotton, wool, and silk fabrics. Moscow factories also specialize in chemicals and light industrial goods.

Eastern Ukraine industrial district. The Donetsk coalfield, in the far eastern portion of Ukraine, contains one of the world's leading reserves of coal. Eastern Ukraine also possesses large deposits of iron ore, manganese, and natural gas. As a result of these assets, the region is Eastern Europe's largest producer of iron and steel. Major plants are located at Krivoy Rog, near iron ore fields, and Donetsk, near coalfields.

Volga industrial district. Situated along the Volga and Kama rivers, the district grew rapidly during World War II, when many plants in the Central and Eastern Ukraine districts were occupied by the invading German army. The Volga district contains Russia's largest petroleum and natural gas fields. Within the district, the motor vehicle industry is concentrated in Togliatti, oil refining in Kuybyshev, chemicals in Saratov, metallurgy in Volgograd, and leather and fur in Kazan.

Urals industrial district. The Ural mountain range contains more than 1,000 types of minerals, the most varied collection found in any mining region in the world. Valuable deposits include iron, copper, potassium, manganese, bauxite, salt, and tungsten. Proximity to these raw materials encouraged the Soviet government to locate in the region such industrial activities as iron and steel manufacturing, chemicals, and machinery and metal fabricating. Although well endowed with metals, industrial development in the district is hindered by a lack of nearby energy sources. Coal must be shipped nearly 1,500 kilometers (900 miles) from Kuznetsk, and oil and natural gas are piped in from the Volga-Ural, Bukhara, and central Siberian fields. Russia controls nearly all of the Urals minerals, although the southern portion of the region extends into Kazakhstan.

Kuznetsk industrial district. Kuznetsk is Russia's most important manufacturing district east of the Ural Mountains. The region contains the country's largest reserves of coal and an abundant supply of iron ore. Soviet planners took advantage of these natural assets to invest considerable capital in constructing iron, steel, and other factories in the region.

Outside the former Soviet Union, Eastern Europe's leading manufacturing area is in Silesia, in southern Poland and northern Czech Republic. The area is an important center of steel production, based on proximity to coalfields, although iron ore must be imported.

Manufacturing in Japan

The most important industrial area outside Europe and North America, Japan may appear to have few geographic assets. Because it lacks many natural resources, the country must import nearly all of its energy and raw materials. For example, Japan possesses only 0.2 percent of the world's iron ore, yet it is one of the world's leading steel producers. The country is far from the major concentrations of wealthy consumers in North America and Western Europe, yet it has become the world's leading exporter of consumer goods.

Faced with isolation from world markets and a shortage of nearly all essential resources, Japan has taken advantage of its one abundant resource: a large and dedicated labor force. Its industries devastated during World War II, Japan became an industrial power in the 1950s and 1960s, initially by producing goods that could be sold in large quantity at cut-rate prices to consumers in other countries. Prices were kept low despite the high cost of shipping goods to overseas markets, because workers received much lower wages in Japan than in North America or Western Europe.

Japanese planners, aware that other countries were building industries based on even lower-cost labor, began to train workers for highly skilled jobs. At the same time, because wages remained lower than in other relatively developed countries, Japan could build high-quality products at a lower cost than those in North America or Western Europe. As a result, during the 1970s and 1980s Japan gained a reputation for high-quality electronics, precision instruments, and other products that required well-trained workers. The country became the world's leading manufacturer of products such as automobiles, ships, cameras, radios, and televisions.

As in other countries, industry is not distributed uniformly within Japan. Manufacturing is concentrated in the central region between Tokyo and Nagasaki, especially the two large urban areas of Tokyo-Yokohama and Osaka-Kobe-Kyoto.

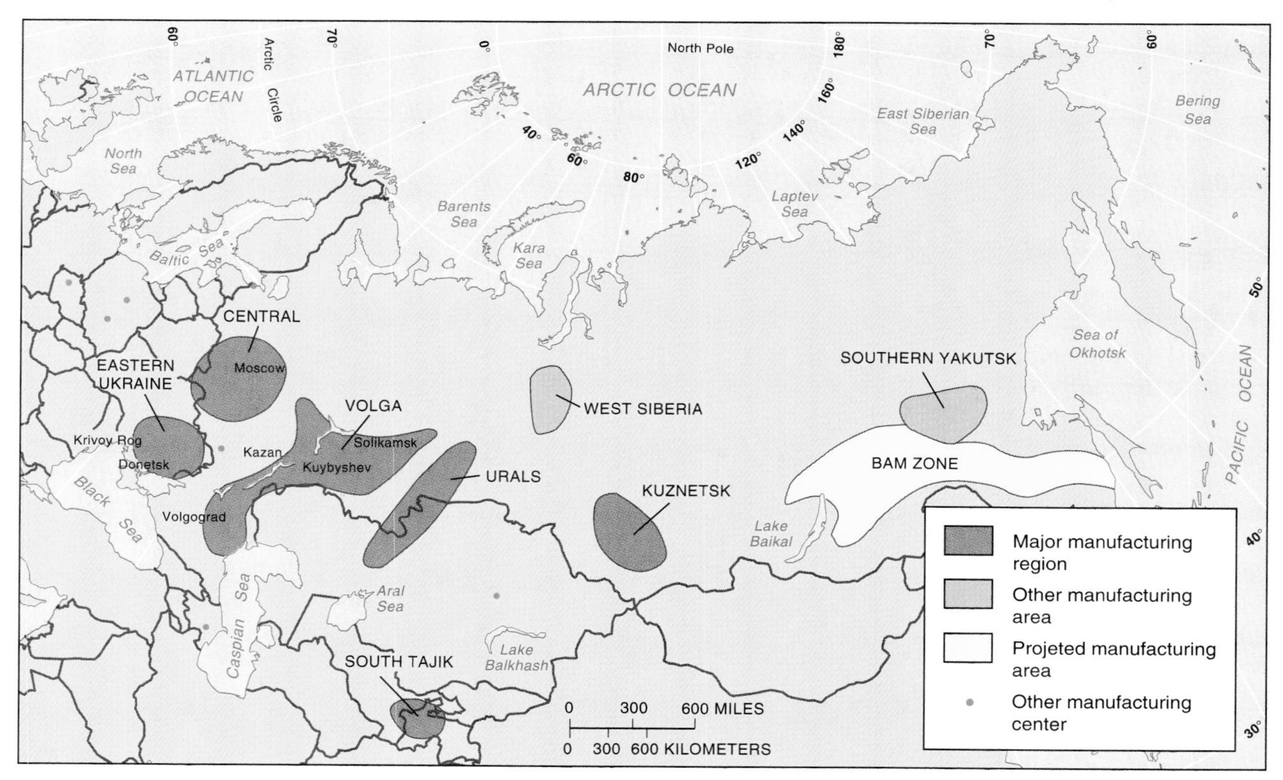

FIGURE 12–5
Manufacturing in Eastern Europe is clustered in the western or European portion of Russia, as well as Ukraine. The Soviet government encouraged growth of manufacturing regions in the center of the country east of the Ural Mountains.

Situation Factors in Industrial Location

Industry seeks to maximize profits by minimizing production costs. Geographers try to explain why one location may prove more profitable for the location of a factory or other site of industry than other places.

A company ordinarily faces two kinds of geographical costs: situation and site. **Situation factors** relate to transportation of materials into and from a factory. A location that minimizes the costs of transporting inputs to the factory and finished goods to the consumers is ideal. **Site factors** relate to the costs of factors of production inside the plant resulting from distinctive characteristics of a particular location. Land, labor, and capital are the three traditional production factors, and these factors may vary among locations. While a variety of situation and site costs explain the location of factories, the particular combination of critical factors varies among firms.

All manufacturers have buyers and sellers—companies and individuals who buy the product and other companies and individuals who supply the firm with inputs needed to manufacture it. One of the objectives of every company is to minimize the aggregate cost of transporting inputs to its factory and of transporting finished products from its plant to consumers. The farther something is transported, the higher is the cost so a manufacturer prefers to locate its factory as close as possible to its buyers and sellers.

A company that obtains all of its needed inputs from one source and sells all of its products to one customer can easily compute the optimal location for its factory. If the cost of transporting the product to the customer is greater than the cost of bringing in the necessary inputs, the optimal plant location is as close as possible to the customer. Conversely, if the inputs are relatively expensive to transport, the preferred location for the factory would be near the source of inputs. We can find examples of industries for which one or the other of the two situation costs is more critical in locational decisions.

Location Near Inputs

Every industry uses inputs. For some firms, these inputs include minerals and other raw materials found in the physical environment that are useful in the production process. For others, the most important inputs may be parts or materials that other companies manufactured.

Bulk-reducing industries. Minimizing the problem of access to inputs is especially critical for **bulk-reducing industries,** which produce output that is less bulky than their inputs. If the weight and volume of one particular input is sufficiently great, the firm may locate near the source of that input to minimize transportation costs.

Wood furniture-making is an example of an industry that may locate near its principal source of raw materials. In the United States, Grand Rapids became a center for wood-furniture manufacturing during the nineteenth century, largely because of its proximity to Michigan's extensive forests. With depletion of timber resources in Michigan, furniture-makers have turned to wood from Georgia and North Carolina, and those two states have become furniture-manufacturing centers in recent decades. (See discussion in Chapter 6 of the impacts of deforestation.)

The North American copper industry provides a good example of a bulk-reducing industry that locates near its sources of inputs to minimize transportation costs. The first step in the production of copper products is the mining of copper ore. Much of the copper mined in North America is low-grade, with less than 1 percent of the ore being copper. The rest constitutes waste, known as *gangue.*

The next step in the production process is to concentrate the copper. The process, which removes 98 percent of the waste from the ore, takes place in concentration mills. These mills are located near the copper mines because the process of concentration transforms a relatively bulky raw material, copper ore, into a product of higher value per weight.

The concentrated copper then becomes the main input for smelters, which separate the copper found in the concentrated ore from other metals. Smelting plants further reduce the weight of copper by about 60 percent. As a bulk-reducing industry, smelters are also located close to their main source of inputs—the concentration mills—in order to minimize transportation costs.

The purified copper produced by smelting plants, known as blister copper, is further treated at refineries. Because there is no further weight loss in the copper, the refineries do not consider that proximity to the mines, mills, and smelters is a critical factor in the location decision.

A map of North America demonstrates the locational needs for the various steps in copper processing. The most important locations for copper mining—Arizona, Utah, and Ontario—are also centers for concentration mills and smelting plants. However, the largest refining centers, including Baltimore, Morristown in Pennsylvania, and Perth Amboy in New Jersey, are on the U.S. East Coast, 1,000 kilometers (600 miles) from the nearest copper mine (Figure 12–6).

Iron and steel production. Steel-making is another bulk-reducing industry that traditionally has been located primarily to minimize the cost of transporting inputs. The U.S. steel industry also demonstrates how the location can change if the source and cost of raw materials change.

The main inputs in the production of steel are iron ore and coal. Both raw materials are extremely bulky, contain a high percentage of impurities, and must be used in large quantities. These characteristics have traditionally necessitated that steel factories be located where the cost of transporting inputs is minimized.

The U.S. steel industry concentrated in the mid-nineteenth century in southwestern Pennsylvania, where iron ore and coal were both mined. Later in the nineteenth century, steel mills were more likely to be built in Cleveland, Youngstown, and Toledo, Ohio, Detroit, and other communities near Lake Erie.

The shift to locations near Lake Erie was largely influenced by the discovery of iron ore in the Mesabi range of Minnesota, which soon became the source of virtually all iron ore used in the U.S. steel industry. Iron was transported to the new steel mills by way of Lake Superior, Lake Huron, and Lake Erie, while coal was shipped from Appalachia by train.

In the late nineteenth century, new steel mills began to be located farther west—Gary in Indiana, Chicago, and other communities near the southern end of Lake Michigan. The main raw materials continued to be iron ore and coal, but changes in the steel production process meant that relatively more iron ore was needed compared to coal. As a result, new steel mills were pulled closer to the Mesabi Range in order to minimize the aggregate costs of transporting the two types of raw materials. Coal was available from nearby southern Illinois as well as from Appalachia (Figure 12–7).

Most of the large steel mills built in the United States during the first half of the twentieth century were located in communities near the East and West Coasts, such as Trenton in New Jersey, Baltimore, and Los Angeles. These coastal locations were partly

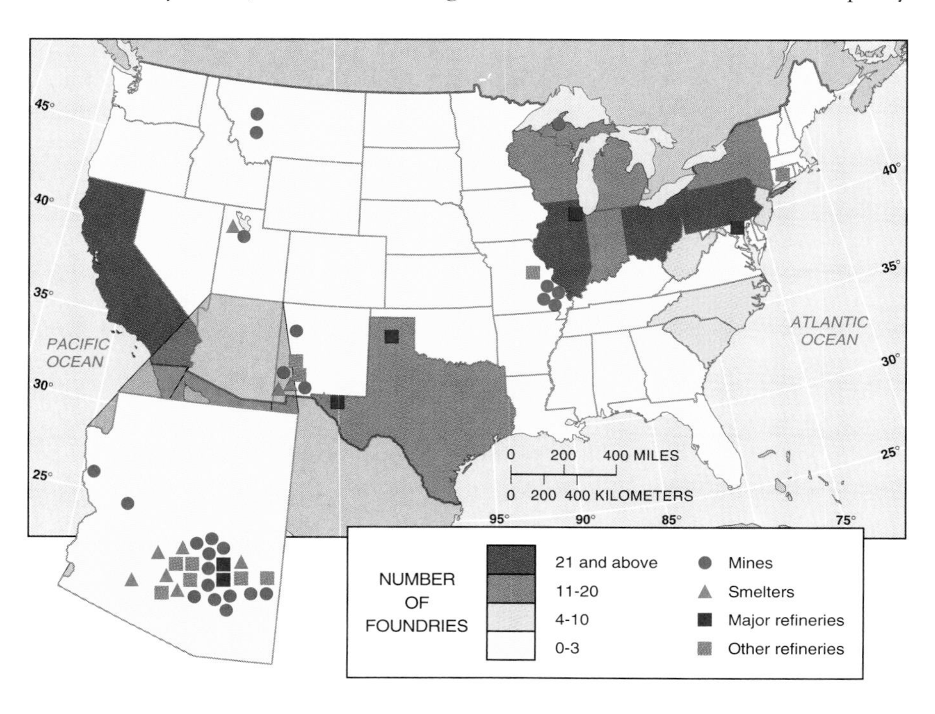

FIGURE 12–6
Copper production, an example of a bulk-reducing industry. Most copper concentration, milling, and refining plants in the United States are clustered in southwestern states, especially Arizona, to take advantage of proximity to copper mines. A few copper refining plants in coastal locations use imported material.

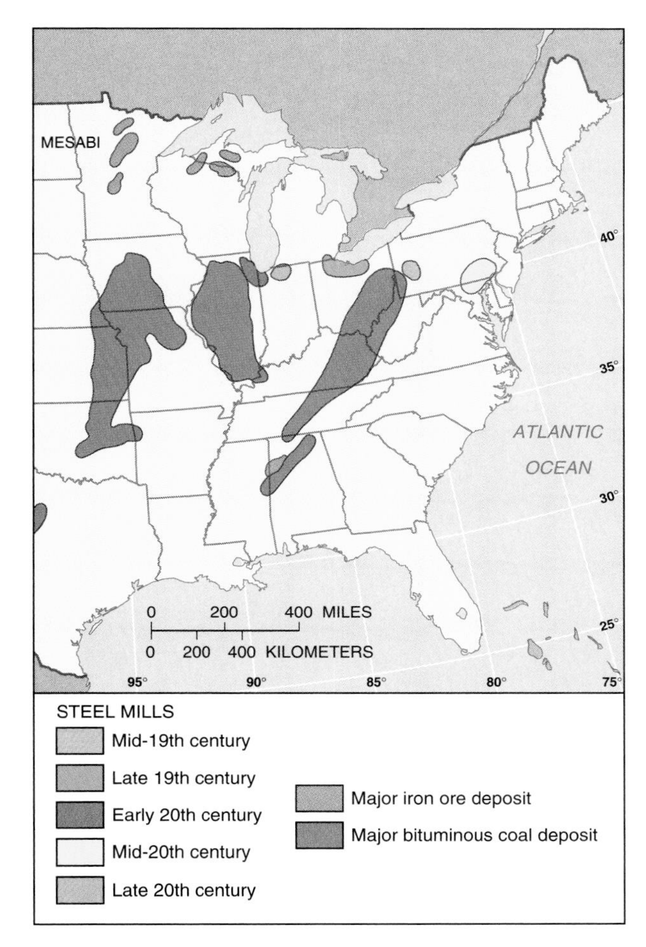

FIGURE 12–7
U.S. steel production centers. Integrated steel mills are highly clustered near the southern Great Lakes, especially Lake Erie and Lake Michigan. Historically, the most critical factor in the selection of locations for steel mills was minimizing the cost of transporting raw materials, especially iron ore and coal, to the factories. In recent years, many integrated steel mills have closed, and most of the survivors are located in the Midwest in order to maximize access to customers.

a reflection of further changes in the cost of transporting raw materials. Iron ore was increasingly obtained from other countries, especially Canada and Venezuela, and communities near the Atlantic and Pacific oceans were more accessible to foreign sources than those near the Great Lakes. Furthermore, scrap metal—widely available in the large metropolitan areas of the East and West Coasts—has become an important input in the steel-production process.

Recently, more steel plants have been closed than opened in the United States. Among the survivors, the share of national production concentrated in the southern Lake Michigan area has significantly increased, while East Coast plants have also captured an increasing percentage of national steel production. The success of steel mills in these regions derives primarily from their accessibility to markets rather than to inputs. In contrast with past locational decisions, successful steel mills are increasingly located near major markets. The coastal plants can provide steel to customers in the large East Coast population centers, while the southern Lake Michigan plants are centrally located to distribute their products throughout country.

The growth of *minimills*, which are smaller, specialized mills, also demonstrates the increasing importance in steel production of access to markets rather than to inputs. Traditionally, most steel was produced at large, integrated mills, which processed iron ore, converted coal into coke, manufactured iron in a blast furnace, converted the iron into steel, and pressed the steel into sheets or other shapes. Minimills, generally limited to one step in the process of steel production, have captured one-fourth of the U.S. steel market. Less expensive than integrated mills to build and operate, minimills can locate near their markets because their main input, scrap metal, is widely available.

Location Near Markets

For many firms, the optimal location is near the market, the place where the good is sold. The cost of transporting goods to consumers is a critical locational factor for three types of industries: bulk-gaining, communications-oriented, and single-market.

Bulk-gaining industries. A product that has a higher volume or a heavier weight after production than before is a **bulk-gaining industry**. One example of a bulk-gaining industry is soft drink bottling. One of the two main inputs in soft drinks, syrup, is easy to transport; the other, water, is available in every community. The output, bottled or canned beverages, has a greater volume and weighs more than the input and is consumed in large quantities by people throughout the world.

Given these characteristics of inputs and output, soft drink bottlers minimize their aggregate trans-

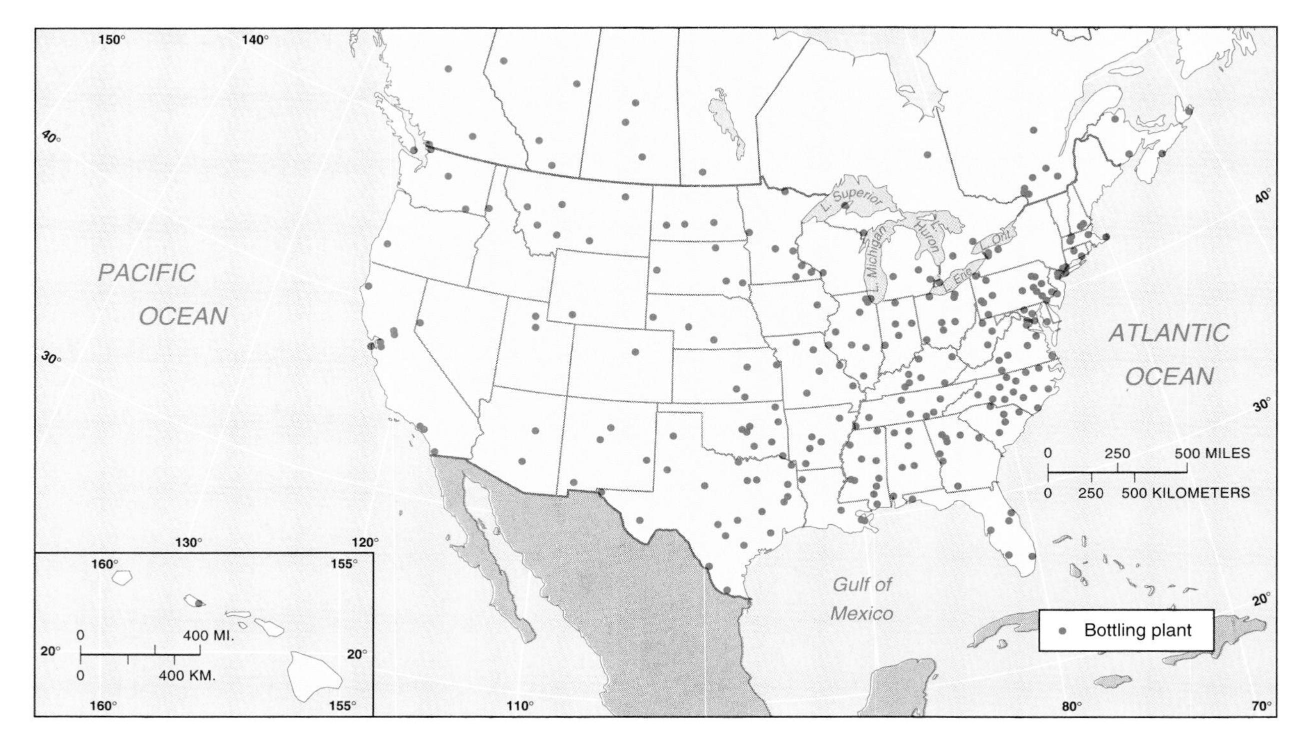

FIGURE 12–8
U.S. and Canadian Coca-Cola bottling plants, an example of bulk-gaining industries. A soft drink bottling plant needs to be located near consumers. Consequently, Coca-Cola is bottled at more than 200 locations in the United States and Canada, situated near all major population concentrations.

portation costs by locating near markets rather than near inputs. Locating near a large population concentration minimizes the cost of shipping the relatively bulky output to customers. Consequently, the major soft drink companies, such as Coca-Cola and Pepsico, manufacture syrups according to secret recipes and ship them to bottlers located in thousands of communities around the world (Figure 12–8).

More commonly, bulk-gaining industries manufacture products that gain volume but not weight. One of the most prominent examples is the fabricated metals industry. A fabricated metals factory brings together a number of previously manufactured parts as the main inputs and assembles them into a more complex product. Many commonly used products are so fabricated, including television sets, refrigerators, and automobiles. If the fabricated product occupies a much larger volume than the individual parts, then the cost of shipping the final product to the consumers is likely to be a critical factor. A fabricated industry seeks a location that minimizes the cost of shipping the relatively bulky product to the market.

Communications-oriented industries. To deliver their products to consumers as rapidly as possible, **communications-oriented industries** must be located near their markets. One type of product that must be sold quickly is fresh food. Bakeries and dairies must locate near their customers in order to assure rapid delivery, because no one wants stale bread or sour milk.

The daily newspaper is an example of a product other than food that is highly perishable and therefore must locate near markets to minimize transportation costs. People demand delivery of their newspaper as soon after its printing as possible. Difficulty with timely delivery is one of the main factors in the demise of afternoon newspapers, which have disappeared from most cities. Morning newspapers are printed between 9 P.M. and 6 A.M. and delivered during the night, when traffic is light. Afternoon newspapers, published between 9 A.M. and 5 P.M., must be delivered in heavy daytime traffic, which slows delivery and thereby raises production costs.

A daily newspaper is an example of a communications-oriented industry that locates near its customers for rapid delivery. Delivery is difficult in many cities, but this truck driver was able to stop illegally next to a fire hydrant. (Jan Halaska/Photo Researchers , Inc.)

In European countries, national newspapers are printed in the largest city during the evening and delivered by train throughout the country overnight. In the past, the United States was too large to make a national newspaper feasible. With satellite technology, however, *The New York Times, The Wall Street Journal,* and *USA Today* have moved in the direction of national delivery. These newspapers are composed in New York or Washington and the page images sent electronically to other locations, such as Atlanta and Chicago, where the papers are printed. The papers are then delivered by air and surface transport to consumers around the country.

Many tertiary-sector activities are communications-oriented industries, although they do not engage in manufacturing. Financial services, such as stock investors, insurance companies, and lending institutions, must receive and send out information rapidly. Despite the diffusion of computers and other electronic means of communication, face-to-face interaction between service providers and customers is still essential for many service-oriented industries (see Chapter 13).

Specialized manufacturers. Specialized component manufacturers also cluster near the market. Single-market manufacturers make products that are designed to be sold primarily in one location. For example, several times a year, buyers from individual clothing stores and department store chains come to New York from all over the United States to select the high-style clothing they will sell in the coming season. As a result of decisions by these store buyers, manufacturers of fashion clothing receive orders for the production of a large quantity of certain garments in a short time. Consequently, high-style clothing manufacturers concentrate in or near New York.

The New York-based high-style clothing manufacturers in turn demand rapid delivery of large quantities of specialized components, such as clasps, clips, pins, and zippers. The specialized component manufacturers therefore also concentrate in New York.

Automobile production. The location of automobile production within the United States and Canada reflects the importance of situation factors. The industry also demonstrates that market changes can alter the optimal plant locations.

The automotive industry comprises two types of factories. Several thousand components plants manufacture one or more of the parts that go into vehicles. These parts are then combined into finished vehicles at about seventy assembly plants across the United States and Canada.

For a fabricated product, such as an assembled automobile, the critical factor in locating the factory is to minimize transportation of the finished products to customers throughout North America. Manufacturers of automotive components also try to minimize the cost of transporting their products to the customers; however, most are specialized manufacturers that sell

Clothing manufacturers cluster in New York City's Garment District, near Seventh Avenue and 34th Street. The area is a center for production of fashionable knitwear. Manufacturers of specialized products, such as zippers and buttons, also cluster in the Garment District to be near their customers, the clothing makers. (Roy Morsch/The Stock Market)

to only one or two customers—automobile producers such as General Motors and Ford.

Historically, General Motors and Ford divided North America into regions and located an assembly plant in or near a large metropolitan area within each region. For example, during the 1950s, GM operated 11 Chevrolet assembly plants, located in or near New York City; Baltimore, Maryland; Atlanta, Georgia; Flint, Michigan; Cincinnati, Ohio; Janesville, Wisconsin (80 kilometers northwest of Chicago); St. Louis and Kansas City, Missouri; Los Angeles and Oakland, California; and Oshawa, Ontario (50 kilometers east of Toronto). These 11 plants assembled the identical Chevrolet automobile model for distribution within a designated region. Ford had a similar geographic arrangement for production of its low-priced model. Luxury cars, such as Cadillac and Lincoln, were assembled at only one plant, in both cases located in Detroit.

Since 1980, the long-time distribution of automobile assembly plants has changed. Factories located near major East- and West-Coast population centers have closed, while new ones have been built in the interior of the country, especially near interstate highways 65 and 75 (Figure 12–9).

Interior locations have resulted from an increase in the variety of automobile models produced in North America. In the past, all of the models produced under one nameplate, such as Chevrolet, were substantially the same and differed mainly in minor details, such as body trim and seat covers. Beginning in the 1960s, the models of a particular nameplate began to vary in size, ranging from subcompacts less than 150 inches long to full-sized vehicles more than 210 inches long.

Instead of producing the same model for regional distribution at several assembly plants, automobile companies now operate assembly plants that specialize in production of one of their many models for distribution throughout the United States and Canada. In geographic terms, if a company has a product that is made at only one plant, and the critical locational factor is to minimize the cost of distributing the product to consumers throughout the United States and Canada, then the optimal location for the factory is in the interior of the United States.

The diversity of products sold in North America further increased by the expansion of sales of foreign vehicles, especially Japanese. U.S. and Canadian customers were first attracted to Japanese vehicles during the 1970s, because they were less expensive to purchase and operate in a time of high global gasoline prices (see Chapter 6). Japanese companies more than offset the added transportation costs of shipping vehicles to North America by paying substantially lower prices for parts and labor in Japan.

The gap in wage rates between North American and Japanese auto workers has disappeared, and many North Americans are concerned about the

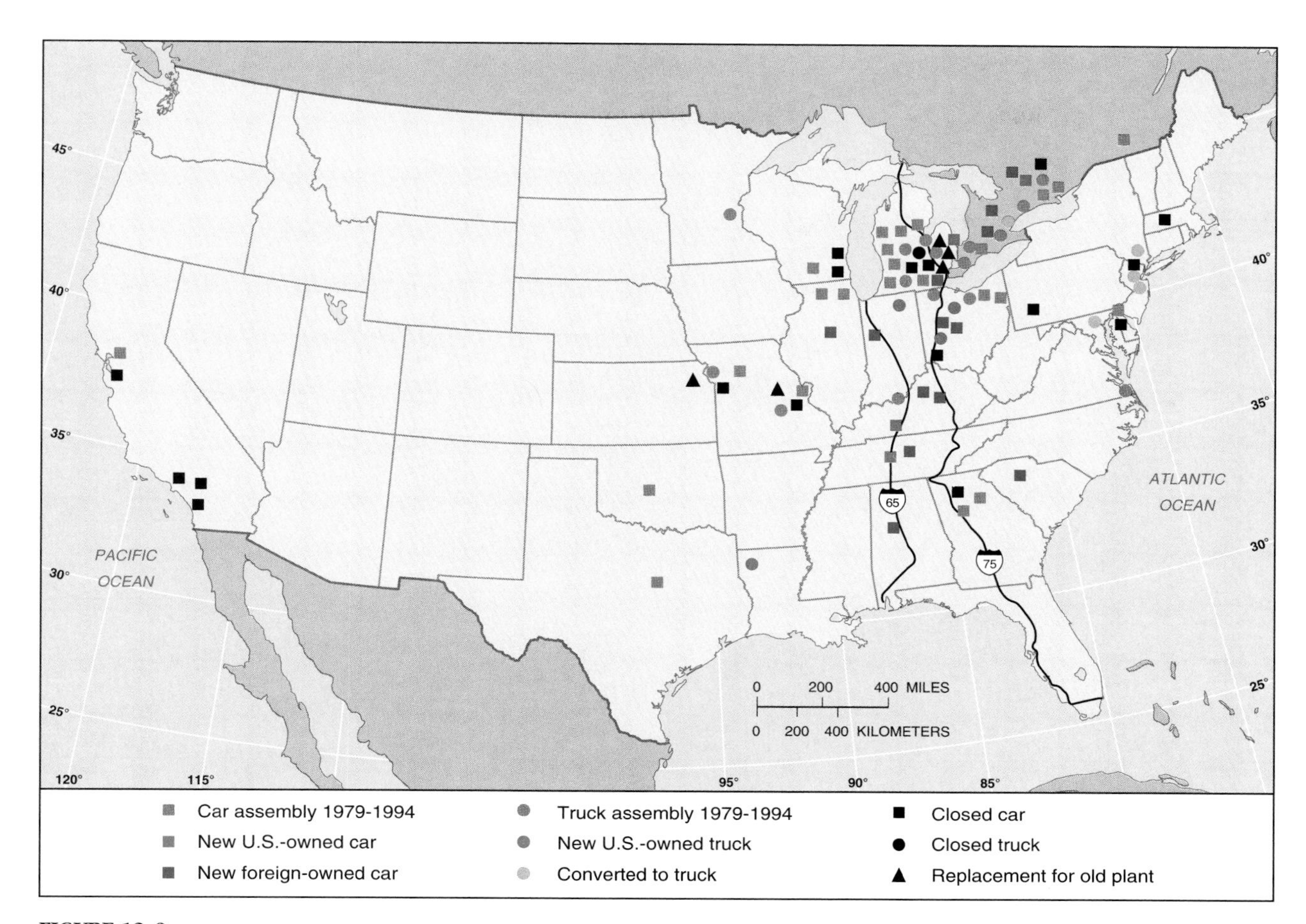

FIGURE 12–9
U.S. and Canadian car and truck assembly plants. Producers of fabricated products, such as cars and trucks, select locations that minimize the cost of transporting the product to consumers. During the 1980s and 1990s, motor vehicle producers selected locations for new assembly plants in the interior of the United States rather than near East and West Coast population concentrations, as was the case in the past. Most coastal plants were closed during the period.

size of the trade deficit with Japan. To protect their North American markets, Japanese auto makers opened assembly plants in the United States and Canada during the 1980s. Most cars sold in North America with Japanese nameplates are actually assembled in the United States or Canada rather than in Japan, although many of the parts are still made in Japan. The Japanese-built assembly plants have all been located in the interior of the country. During the 1990s, the German car makers BMW and Mercedes-Benz have built assembly plants in the United States as well.

While assembly plants were located around the country, most components plants clustered in Michigan and adjacent states. These parts manufacturers typically sent their products to the automobile companies' warehouses and distribution centers in Michigan. Parts producers also clustered near the southern Great Lakes, because the region was the center for production of the industry's most important input, steel. Today, many parts are produced in factories near the assembly plant where they will be attached to the automobiles.

Proximity to the assembly plant is increasingly important for parts producers because of the diffusion of "just-in-time" delivery. Under the "just-in-time" inventory system, parts are delivered to the assembly plant within minutes of actually being used, rather than weeks or months in advance. This reduces the amount of inventory stock firms

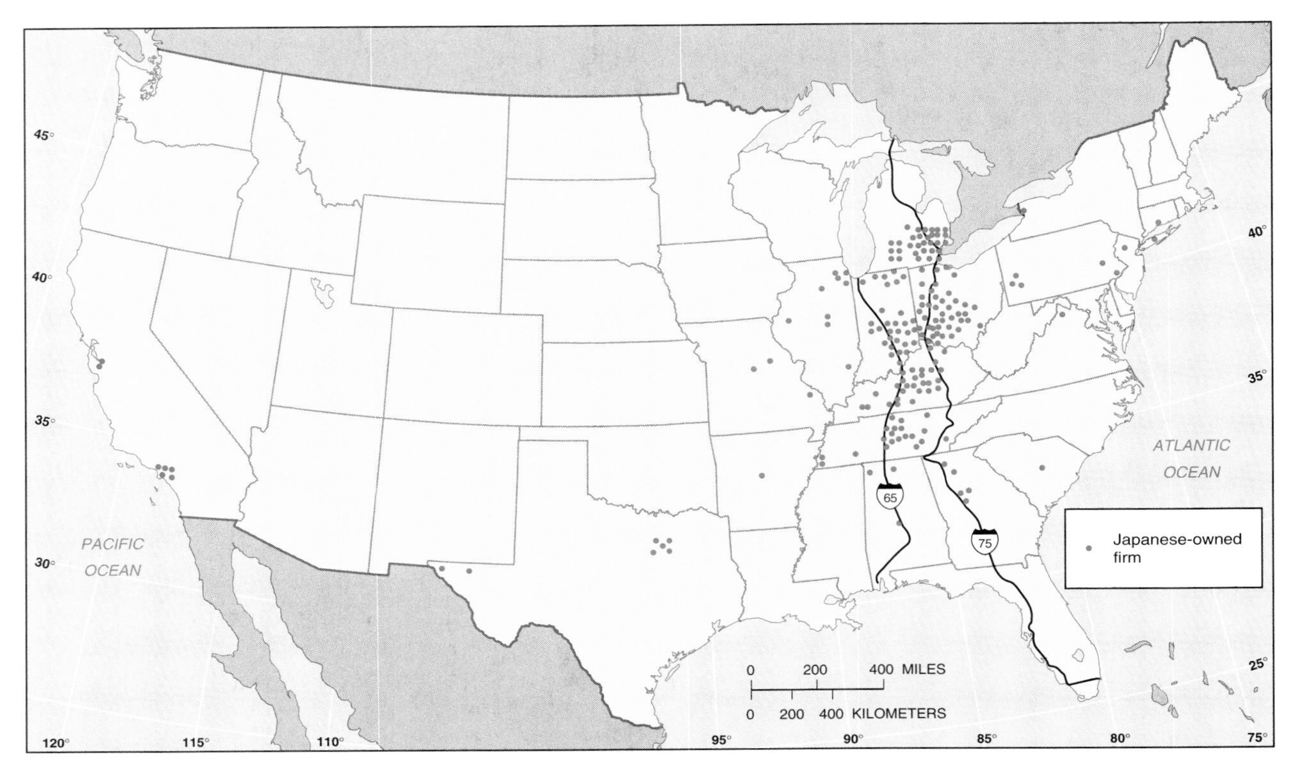

FIGURE 12–10
Japanese-owned manufacturers of automotive parts have clustered in the interior of the United States, especially along interstates 65 and 75. These locations facilitate rapid delivery of parts to final assembly plants, which are also clustered in the two corridors (see Figure 12–9).

must hold, and frees finances for other uses. The clustering of parts manufacturers around their customers—the new Japanese-operated assembly plants in the United States—clearly illustrates the influence of this system (Figure 12–10).

Alternative modes of transport. Transportation inventions played a critical role in the diffusion of the industrial revolution. New transportation systems enabled factories to attract large numbers of workers, bring in bulky raw materials, and ship finished goods to consumers.

Inputs and products are transported in one of four ways: ship (or barge), rail, truck, and airplane. Firms seek the lowest-cost mode of transport, but the cost of each of the four alternatives varies, depending on the distance that the goods are being sent.

The farther goods are transported, the lower are the costs per kilometer or mile. Longer-distance transportation is cheaper per kilometer or mile in part because firms must pay workers to load goods onto a vehicle and unload them at their destination, regardless of the distance the vehicle travels. The cost per kilometer or mile decreases at different rates for various modes, because the loading and unloading expenses differ for each mode.

The lowest-cost form of land transport is generally by truck for relatively short distances and by train for longer distances, because trucks can be loaded and unloaded more quickly and cheaply than trains. Trucks can deliver materials and pick up products at the large percentage of customers who are not situated next to a rail line. Therefore, trucks are most often used for short-distance delivery, and trains are most often used for longer distances. If a water route is available, transporting by ship is attractive for very long distances, because the cost per kilometer or mile is even lower than for a train.

The airplane is normally the most expensive alternative for all distances, but an increasing number of firms transport by air to ensure speedy delivery of small bulk, high-value packages. Air

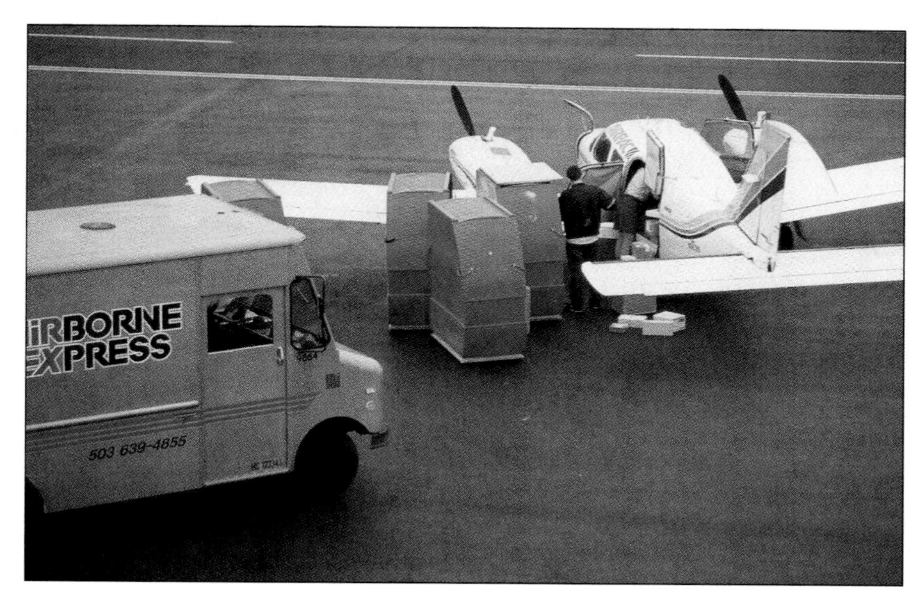

An air freight company collects a package by truck, takes it to the nearest airport, flies it to a hub airport in the interior of the United States, transfers it during the middle of the night to another airplane parked at the hub airport, and delivers it by truck to the recipient the next day. (Jeff Zaruba/The Stock Market)

transport companies promise overnight delivery for most packages. They pick up packages in the afternoon and transport them by truck to the nearest airport. Late at night, an airplane filled with packages is flown to an airport located in the interior of the country, such as Memphis, Tennessee, or Dayton, Ohio. The packages are then transferred to other planes, flown to another airport, transferred to trucks, and delivered the next morning to the desired destination. This is how companies like Federal Express and Airborne Express operate.

Break-of-bulk points. Regardless of the form of transportation, costs rise each time the inputs or products are transferred from one mode of transport to another. For example, workers must unload the goods from a truck and then reload them onto an airplane. The company may need to build or rent a warehouse to store the goods after unloading from one mode and before loading to another mode. Some companies may calculate that the cost of one mode is lower for some inputs and products, while another mode may be cheaper for other goods.

Many companies that use more than one mode of transport locate at a break-of-bulk point. A **break-of-bulk point** is a location where transfer among transportation modes is possible. Important break-of-bulk points include seaports and airports.

With increased cooperation among operators of different modes of transport, the attraction of a break-of-bulk point has diminished somewhat. Many ships are designed to carry cargo in special box-like containers, which can be transferred to land-based forms of transport much more quickly than in the past.

Situation factors remain important for many firms, but the relative importance of different situation factors has changed. Locations near markets or break-of-bulk points have become more important than locations near raw materials for firms in relatively developed countries. Consumers concentrated in large urban areas have greater wealth with which to buy products. Communications improvements have increased demand for rapid access to products.

Situation factors do not explain the growing importance of Japanese and other Asian manufacturers. Japan lacks key raw materials needed by industries and is relatively far from the most important North American and European markets. Japan and other Asian countries have become increasingly important industrial centers because of site factors. Manufacturing has grown in Japan and other Asian countries primarily because site factors have become increasingly important in industrial location decisions.

Site Factors in Industrial Location

The cost of conducting business varies among locations depending on the price of a firm's factors of production. As discussed, the three types of production factors are land, labor, and capital.

Containers waiting to be loaded onto ships in the port of Hong Kong. Shipping in containers reduces the time needed to load, unload, and transfer between water-based and land-based transportation. The chapter opener shows containers lined up in the port of Newark, New Jersey. (James Marshall/The Stock Market)

Land as a Factor in Industrial Location

Modern factories are more likely to be located in suburban or rural locations rather than near the center of a city. Contemporary factories tend to require large tracts of land, because they usually operate more efficiently when laid out in one-story buildings. The land needed to build one-story factories is more likely to be available in suburban or rural locations.

Firms are also encouraged to select suburban or rural locations because land is much less costly there than near the center of the city. A hectare or an acre of land in the United States may cost only a few thousand dollars in a rural area, or tens of thousands dollars in a suburban location, but hundreds of thousands of dollars near the center of a city. Mercedes-Benz located its U.S. factory in rural Alabama in part to take advantage of relatively low land costs.

Access to source of energy. Industries may be attracted to specific parcels of land that are accessible to energy sources. Prior to the industrial revolution, many economic activities were located near rivers and close to forests, because water and wood were the two most important sources of energy. When coal became the predominant form of industrial energy in the late eighteenth century, location near coalfields became more important. Because coalfields were less ubiquitous than streams or forests, industry began to be concentrated in a relatively few locations.

In the twentieth century, electricity has become an important source of energy for industry. Electricity is generated in a variety of ways, using coal, oil, natural gas, hydroelectric plants, and nuclear fuel. In the United States, it is normally purchased from a utility company, a privately owned monopoly regulated by the state government.

Like home consumers, industries are charged a certain rate per kilowatt hour of electricity consumed, although large industrial users may pay a lower rate than home consumers. Each utility company sets its own rate schedule, subject to approval by the state regulatory commission. Industries with a particularly high demand for energy may select a location with lower electrical rates.

The aluminum industry, for example, requires a large amount of electricity to remove pure aluminum from aluminum oxide. The first aluminum plant in the United States was located near Niagara

Falls to take advantage of the large amount of cheap hydroelectric power generated there. Aluminum plants have been built near other sources of inexpensive hydroelectric power, including the Tennessee Valley and the Pacific Northwest.

Industry may also be attracted to a particular location because of the amenities available at the site. Not every location has the same climate, topography, recreational opportunities, cultural facilities, and living costs. Some executives select locations in the U.S. South and West because they are attracted to the relatively mild climates and opportunities for year-round outdoor recreation activities. Others prefer locations that are accessible to cultural facilities or major-league sports franchises.

Labor Costs as a Factor in Industrial Location

The cost of labor varies considerably, not only from one country to another but within regions of one country. Types of industries for which labor costs comprise a high percentage of expenses are known as **labor-intensive** industries. Some labor-intensive industries require less skilled, inexpensive labor to maximize profits, while others may need skilled labor.

Textile and clothing industries. Textile and clothing production are prominent examples of labor-intensive industries that generally require less skilled, low-cost workers. Textile production involves three principal steps, first the spinning of fibers to prepare strands of yarn, second the weaving or knitting of yarn to manufacture fabric, and third the finishing of fabric by bleaching or dyeing. The finished fabric is then used to manufacture clothing or other products, such as carpets and towels.

Fibers can be spun from natural or synthetic elements. To some extent, fiber production is concentrated in countries where cotton, the principal source of natural fibers, is grown. China, India, Pakistan, the United States, and Uzbekistan grow more than half of the world's cotton fiber production (Figure 12–11). The other major source of natural fiber, wool, is not produced in proximity to sheep farms.

Synthetic fibers, which are produced from petroleum and other chemical processes, account for an increasing share of textile production. Production was once dominated by a handful of relatively developed countries where the chemical industry was concentrated, but developing countries now account for about half of global production of synthetic fibers. Fiber production has expanded especially rapidly in China and Indonesia.

Developing countries are responsible for three-fourths of the woven cotton fabric, compared to two-thirds of the spun yarn. Cotton textile weaving is more likely than spinning of cotton fibers to locate in developing countries, because labor accounts for a higher percentage of total production costs (Figure 12–12). Despite their remoteness from European and North American markets, Asian countries have become especially important producers of woven fabrics, because relatively low labor costs offset the added expenses of shipping inputs and products long distances.

Most of the world's production of cotton clothing, such as shirts, pants, and underwear, is still found in the relatively developed countries of Europe and North America (Figure 12–13). During the 1980s, production of shirts declined by more than one-third in the United States and Europe while remaining about the same in the developing countries. As a result, the percentage of shirts produced in developing countries increased during the 1980s from about 45 percent to 55 percent.

U.S. textile and clothing industries. Within the United States, textile weavers and clothing manufacturers have changed locations in order to be near sources of low-cost employees. During most of the nineteenth century, U.S. textile and clothing firms were concentrated in the Northeast. The region's major attraction was a large supply of European immigrants willing to work long hours sewing in so-called *sweatshops* for low pay. In the late nineteenth and early twentieth centuries, textile and clothing workers began to demand better working conditions and higher wages and formed unions to represent their interests. Their claim was bolstered by tragic events, such as the 1911 Triangle Shirtwaist Company fire in New York, when 146 workers, mostly women, died because the owners locked the doors to the eighth-floor workroom, where the fire originated, to prevent them from taking breaks and stealing company property.

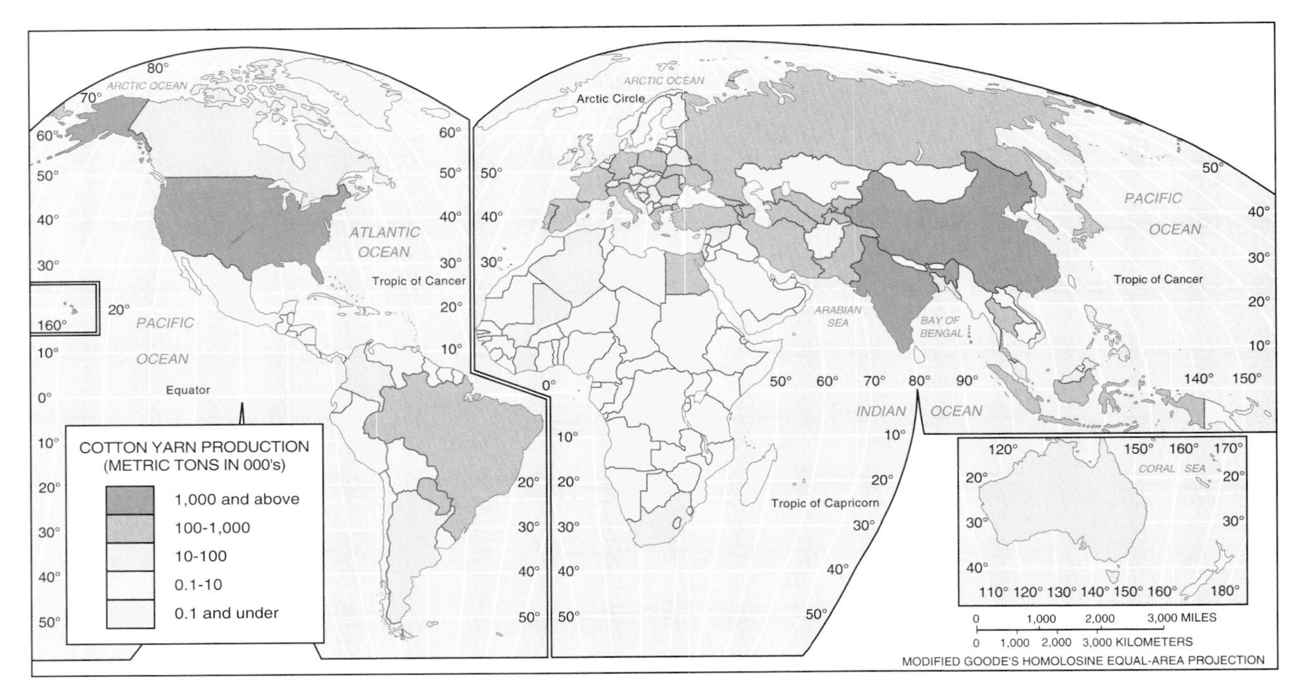

FIGURE 12–11
Production of cotton yarn is clustered in a handful of countries where cotton is grown, including China (shown in the photograph), as well as the United States, Russia, China, India, and Pakistan.

(Doug Handel/The Stock Market)

Employers argued that they could not afford to pay high wages and still make a profit. Because so many workers were needed in the industry, the wages of each individual worker had to be kept low. Faced with union demands for higher wages in the Northeast, cotton textile and clothing manufacturers moved to the Southeast, where people were willing to work longer hours for lower wages. Although they earned less than their northeastern counterparts, workers were eager to work in the southeastern textile and clothing factories because wages were higher than in other types of work in the region. With better working conditions and higher wages than previously found in the region, workers were less likely to vote to join a union, thus keeping costs to industry low.

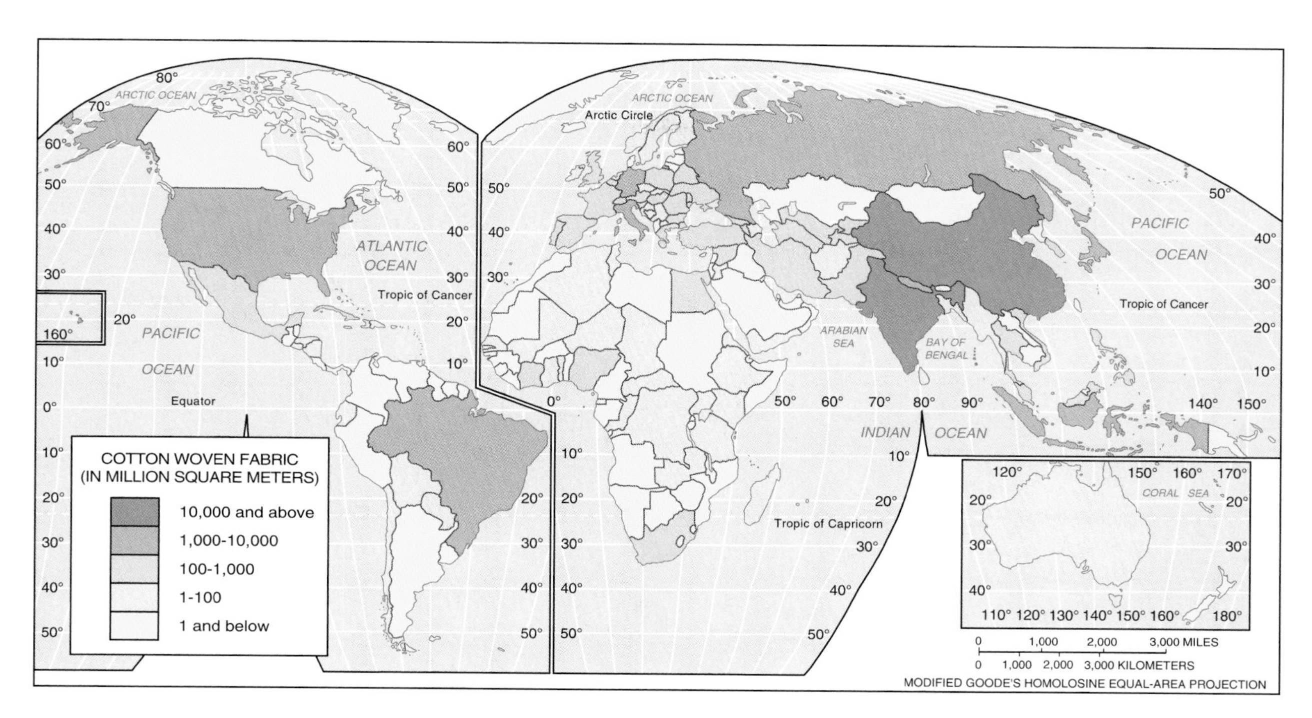

FIGURE 12–12
Production of woven cotton fabric is likely to be found in developing countries, because the process is more labor-intensive than the other major processes in textile and clothing manufacturing.

The Southeast attracts clothing and textile manufacturers, as well as firms in other sectors, who seek a location where few workers have joined labor unions. Southeastern states are known as **right-to-work states,** because they have passed laws preventing a union and company from negotiating a contract that requires workers to join a union as a condition of employment.

Cotton textile and clothing manufacturing in the United States is now located in the Appalachian Mountains and Piedmont of the Southeast, especially the western parts of North and South Carolina and the northern parts of Georgia and Alabama (Figure 12–14). Firms are dispersed among a large number of communities rather than concentrated in a few cities. They are in the same general region to take advantage of lower labor costs and consequently do not need to be located in the same city.

The clothing industry has not completely abandoned the Northeast. The wool industry has remained there because its labor demands are different from those of the cotton textile industry.

Early in the twentieth century, most U.S. clothing manufacturers were located in New England, such as this shoe factory, seen in a photograph taken around 1910. Recent immigrants from Europe supplied much of the low-cost labor. Most clothing manufacturing has moved to the southeast, including North Carolina, shown in the photograph accompanying Figure 12–13. (The Granger Collection)

The manufacture of wool clothing, such as knit outerwear, requires more skill to cut and assemble the material, and many skilled textile workers are available in the Northeast.

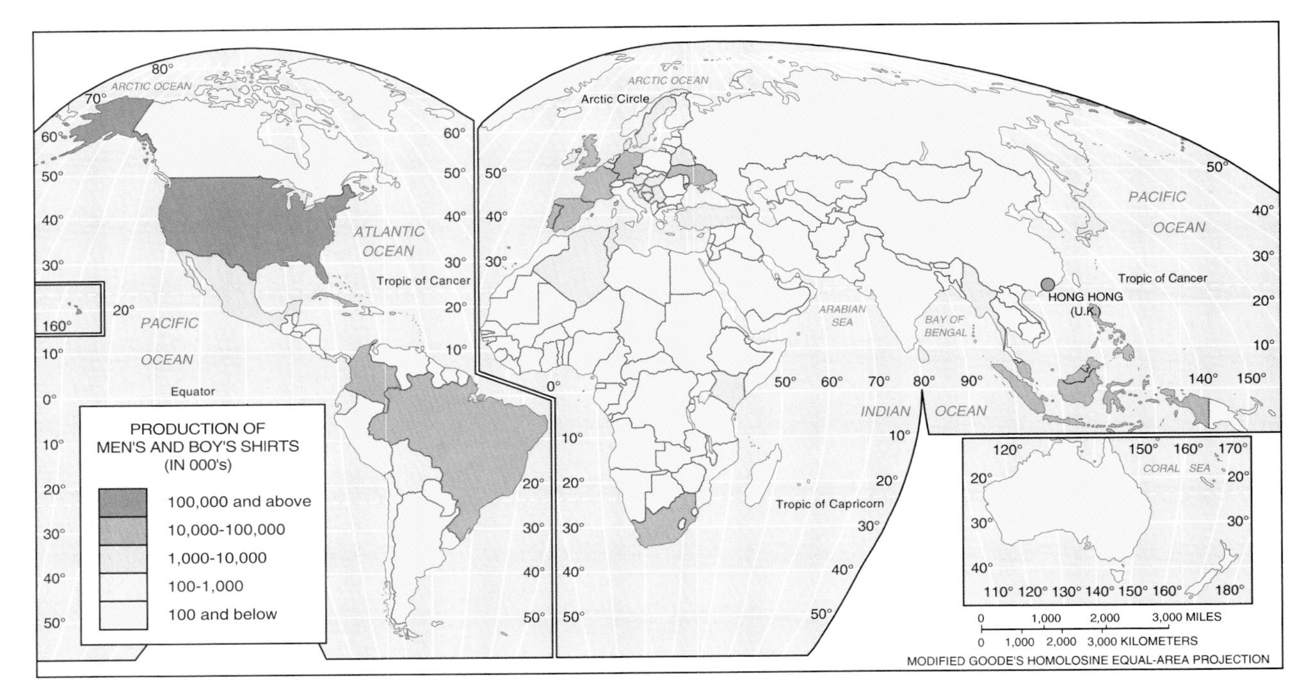

FIGURE 12–13
Manufacturers of men's and boys' shirts. Most men's and boys' shirts are produced in relatively developed countries, although some production has moved to developing countries in recent years. Clothing producers seek a balance between the need for low-wage workers and the need for proximity to customers. The photo shows a clothing factory in North Carolina, which has become the U.S. center for producing low-cost clothes (see Figure 12–14).

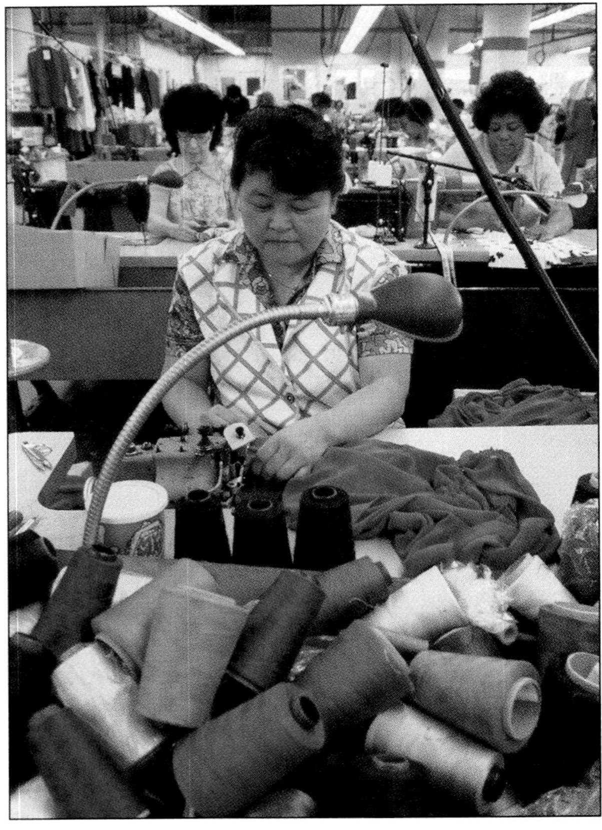
(Joseph Nettis/Photo Researchers, Inc.)

Mexican maquiladoras. A number of U.S. manufacturers have relocated production to Mexico to take advantage of low-cost labor. Under U.S. and Mexican laws, companies receive tax breaks if they ship materials from the United States, assemble components at a participating plant in Mexico, and import the finished product back to the United States. These plants are known as **maquiladoras,** derived from the Spanish verb "maquilar," which means to take a measure of payment for grinding or processing corn; the term originally applied to a tax when Mexico was a colony of Spain.

Several thousand U.S. companies have established maquiladoras in Mexico. U.S. firms are eager to locate production in Mexico because hourly wages are much lower than in the United States, typically less than $1 per hour. The implementation of the North American Free Trade Agreement (NAFTA) in 1994 encouraged further investment by U.S. companies in Mexican factories.

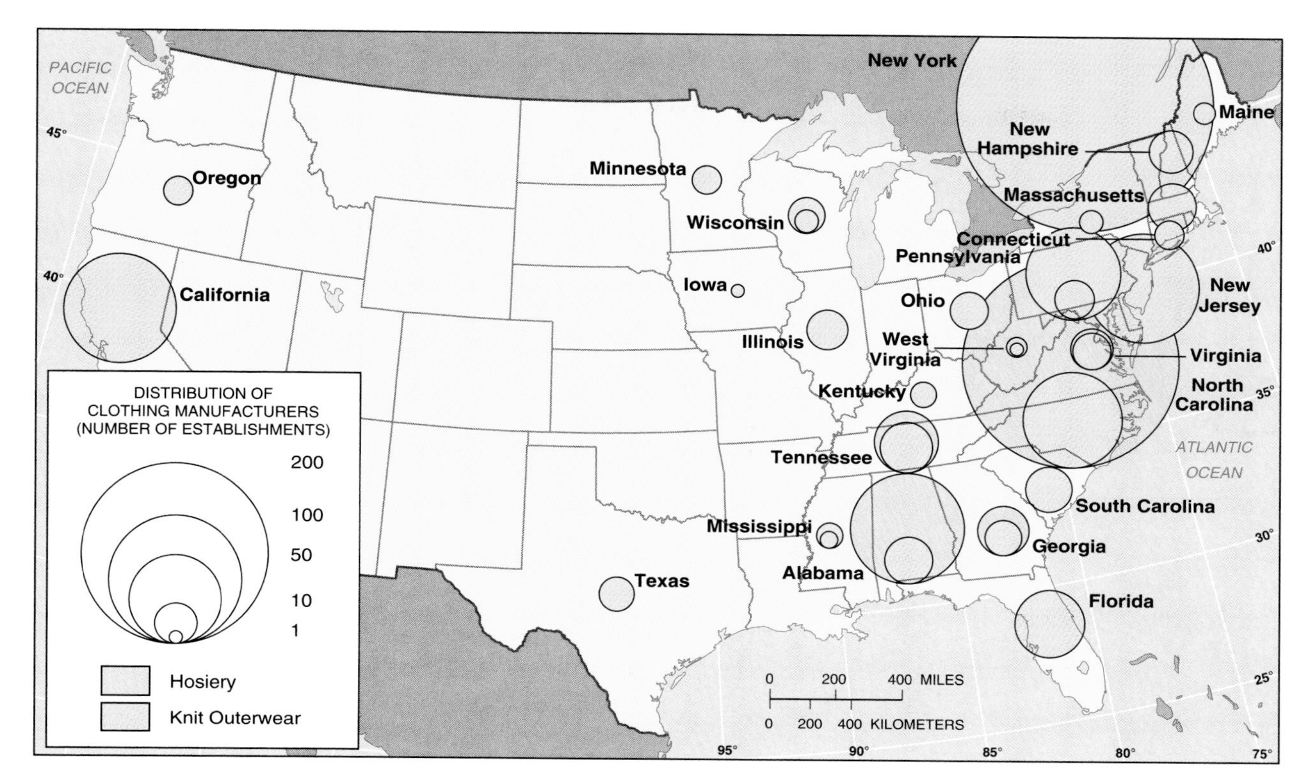

FIGURE 12–14
Manufacturers of hosiery and knit outerwear by state. As a labor-intensive industry, manufacturers of hosiery select locations where they can obtain relatively low-cost workers. In the United States, the lowest-cost labor is concentrated in the Southeast. Products that require more skilled workers, such as knit outerwear, are still produced primarily in or near New York City, the country's largest market and center of skilled clothing workers.

Demand for skilled labor. An increasing number of firms require workers to perform highly skilled tasks, such as working with complex equipment or performing precise cutting and drilling operations. Companies may be more successful by paying higher wages for skilled labor than by producing an inferior product made with lower-paid, less-skilled workers.

One industry that demands highly skilled workers is electronics. Computer manufacturers have concentrated in the highest wage regions in the United States, especially New York, Massachusetts, and California. These regions have a large concentration of skilled workers, because of proximity to major university centers (Figure 12–15).

Many industries are attracted to locations with relatively skilled labor in order to introduce new work rules. Traditionally in large factories, each worker was assigned one specific task to perform repeatedly. Some geographers call this **Fordist production,** because the Ford Motor Company was one of the first to organize its production this way early in the twentieth century. In recent years, companies have adopted more flexible rules, such as the allocation of workers to teams that must perform a variety of tasks. Relatively skilled workers are needed to master the wider variety of assignments given them under flexible, or so-called "post-Fordist" work rules.

Availability of Capital as a Factor in Industrial Location

Manufacturers typically need to borrow funds in order to establish new factories or to expand existing ones. The U.S. motor vehicle industry concentrated in Michigan in the early part of the twentieth century in part because financial institutions there were more willing than those in the East to loan money to the industry's pioneers.

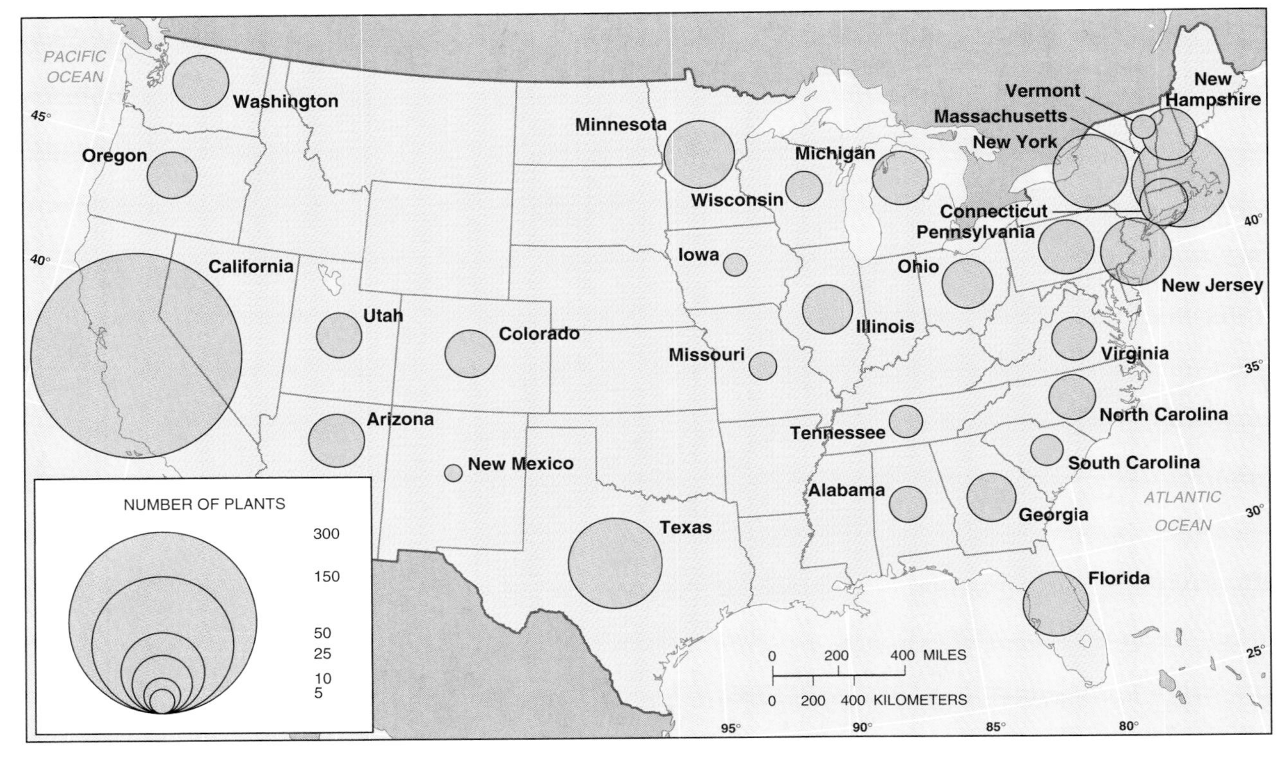

FIGURE 12–15
Manufacturers of electronic computing equipment by state. Manufacturers of computing equipment need access to highly skilled workers to perform precise tasks and are willing to pay relatively high wages to attract the workers. The largest clusters of skilled workers are in the Northeast and West Coast.

The ability to borrow money has become a critical factor in the distribution of industries in developing countries. Because financial institutions in many developing countries are short of funds, new industries seek loans from banks in relatively developed countries. Enterprises may not receive loans if they are located in a country that is perceived to have an unstable political system, a high debt level, an unfavorable exchange rate, or ill-advised economic policies.

Local and national governments increasingly attempt to influence the location of industry by providing a variety of financial incentives, including grants, low-cost loans, and tax breaks. Communities compete with each other to offer new factories the most attractive financial package. Generally, the cost of the financial package is less than the additional revenues the new firm will generate in taxes and employment.

Obstacles to Optimal Location

The location chosen by a firm cannot always be explained by situation and site factors. Many industries have become "footloose," or able to locate in a wide variety of places without a significant change in the costs of transportation, land, labor, and capital.

Because specific factory sites are selected by real people, the processes by which these individuals make decisions can explain the location of industries. Individuals with high levels of knowledge and power may be able to identify sites close to the point that maximizes profits, whereas less skilled entrepreneurs might select inferior locations. Individuals may choose locations on the basis of a corporate goal other than to maximize profit—for example, to promote growth or to assure survival of the firm. Personal preferences of the owner are especially important in influencing the location of a smaller firm. The location may be dictated by where the owner was born, went to school, or participates in leisure activities.

As the search for the optimal location may be time-consuming and costly, the selected plant site may be the first satisfactory alternative encountered rather than the best possible one. The firm may select its location on the basis of inertia and history. Once a firm is located in a particular community, expansion in the same place is likely to be cheaper than moving operations to a new one. Large corporations may operate plants in inferior locations that were inherited through mergers and acquisitions.

Industrial Restructuring

Government officials throughout the world consider that stimulating the economy of their community or country is one of their most fundamental priorities. They take steps to attract new industries and services and to retain existing ones. Geographers point out that the problems of economic growth faced by one community or country are related to economic restructuring elsewhere.

Industrial Problems from a Global Perspective

From a global perspective, the fundamental industrial problem is a gap between the demand for industrial products and the capacity of the world's factories to produce the goods. Global industrial capacity has increased more rapidly than demand for many products.

Stagnant demand. Except during periods of major wars or economic depressions, demand for a wide variety of industrial goods generally increased from the beginning of the industrial revolution in the late eighteenth century until the mid-1970s. Industrial growth in relatively developed countries was fueled by long-term increases in population and wealth. More people with more wealth demanded more industrial goods.

Since the mid-1970s, demand for many industrial goods has not continued to expand at the same rate in the relatively developed countries. The energy crisis played a critical role in triggering industrial restructuring, as discussed in Chapter 6, but even without the energy crisis, changing patterns of supply and demand were causing problems for manufacturers.

Most relatively developed countries have little, if any, population growth. Because salaries have not risen as fast as prices during the past twenty years, individuals typically have not increased their level of spending, when adjusted for inflation. Demand has also been flat for many consumer goods in relatively developed countries because of market saturation. Nearly every household that desires them has already purchased so-called consumer durable goods, such as a color television, a refrigerator, and an automobile. Most contemporary purchasers of these products replace older models rather than buy for the first time.

Industrial output also is stagnant because of increased demand for high-quality goods. In the market for durable goods, consumers in the relatively developed societies increasingly select specific models on the basis of quality and reliability rather than low price, and they replace them less frequently.

During the 1980s, Japanese companies expanded their share to more than one-fourth of the North American automobile market by selling products that were comparably priced with American models but widely acknowledged to be built better. In recent years, the gap in quality between American and Japanese products has narrowed, if not disappeared altogether, although some American consumers still perceive that Japanese models are superior. Japanese automakers have retained their share of the North American market during the 1990s by selling larger, more luxurious cars to customers who had positive experiences driving relatively small and inexpensive models in the past. The children of these purchasers are now part of the market share, too.

Changing technology has resulted in declining demand for some industrial products. For example, global demand for steel today is lower than it was in the mid-1970s. Today's typical automobile uses one-fourth less steel than those manufactured twenty years ago. Automobile manufacturers now build smaller and lighter vehicles and have replaced steel with plastic and ceramic products in the body, chassis, passenger compartment, and trim.

Increased capacity. While demand for products such as steel has stagnated since the 1970s, global capacity to produce them has increased. Higher industrial capacity is primarily a result of two trends: the global diffusion of the industrial revolution and the desire by individual countries to maintain production despite a global overcapacity.

Historically, manufacturing was concentrated in a few locations. From the beginning of the industrial revolution until recently, demand for products manufactured in the relatively developed countries increased in part through sales to countries that lacked competing industries. Industrial growth through increased international sales was feasible when most of the world was organized into colonies and territories controlled by the relatively developed countries.

For much of the nineteenth century, output in some industrial sectors was higher in the United Kingdom than in the rest of the world combined. From the late nineteenth century until recent years, the United States, Russia (or the Soviet Union), Germany, and several other countries in Europe and North America joined the British in dominating global industrial production. Then, Asian countries, such as Japan and South Korea, became major industrial producers. Now, few colonies remain in the world, and nearly every independent country wants to establish its own industrial base.

The steel industry illustrates the changing distribution of the global economy. In the mid-1970s, the relatively developed countries of North America, Western Europe, and Japan accounted for nearly two-thirds of the world's steel production, compared to approximately one-fourth in Eastern Europe and the Soviet Union, and less than 10 percent in developing countries.

Global steel production in the early 1990s was virtually the same as it was in the mid-1970s, but production levels changed significantly in various regions. During the period, production declined by nearly one-fourth in the relatively developed countries and more than doubled in the developing countries. In two decades, the share of the world's steel production concentrated in the relatively developed countries (other than Eastern Europe) has declined from nearly two-thirds to less than one-half, while production in the developing countries has increased from less than 10 percent to more than 20 percent of the world's output. Developing countries such as Brazil, South Korea, Taiwan, India, and the People's Republic of China have substantially increased steel production, while the relatively developed countries, including even Japan, have reduced production (Figure 12–16).

As a result of the global diffusion of steel mills, capacity exceeds demand by a wide margin. Many companies have been unable to sell enough steel to make a profit and have gone out of business. Because the governments of many relatively developed countries have been reluctant to let their steel mills close, the problem of excessive capacity and unprofitable operations persists.

Steel mills in many countries receive substantial government financial support in order to remain open. Many European governments heavily subsidize the continued operation of their steel mills because, if the mills closed, governments would have to pay unemployment compensation to the laid-off workers and deal with the social problems caused by increased unemployment. Maintaining a steel industry also ensures countries of a domestic steel source in times of crisis.

Industrial Problems in Relatively Developed Countries

Countries at all levels of development face a similar challenge: to make their industries competitive in an increasingly integrated global economy. While they may share the same overall goal, though, countries face distinctive geographical issues in ensuring that their industries compete effectively. Industries in relatively developed regions must protect their markets from new competitors, whereas developing countries of Africa, Asia, and Latin America must identify new markets and sources of revenue to generate industrial growth.

Competition among blocs. Industrial competition in the relatively developed world increasingly takes place among blocs of countries. Countries within North America cooperate with each other, as do those within Western Europe and East Asia, but each bloc competes against the other two regions to promote industrial growth.

NAFTA has eliminated most trade barriers among the United States, Canada, and Mexico, while similar efforts have been made among the members of the European Union. Cooperation is less extensive in East Asia, where Japanese indus-

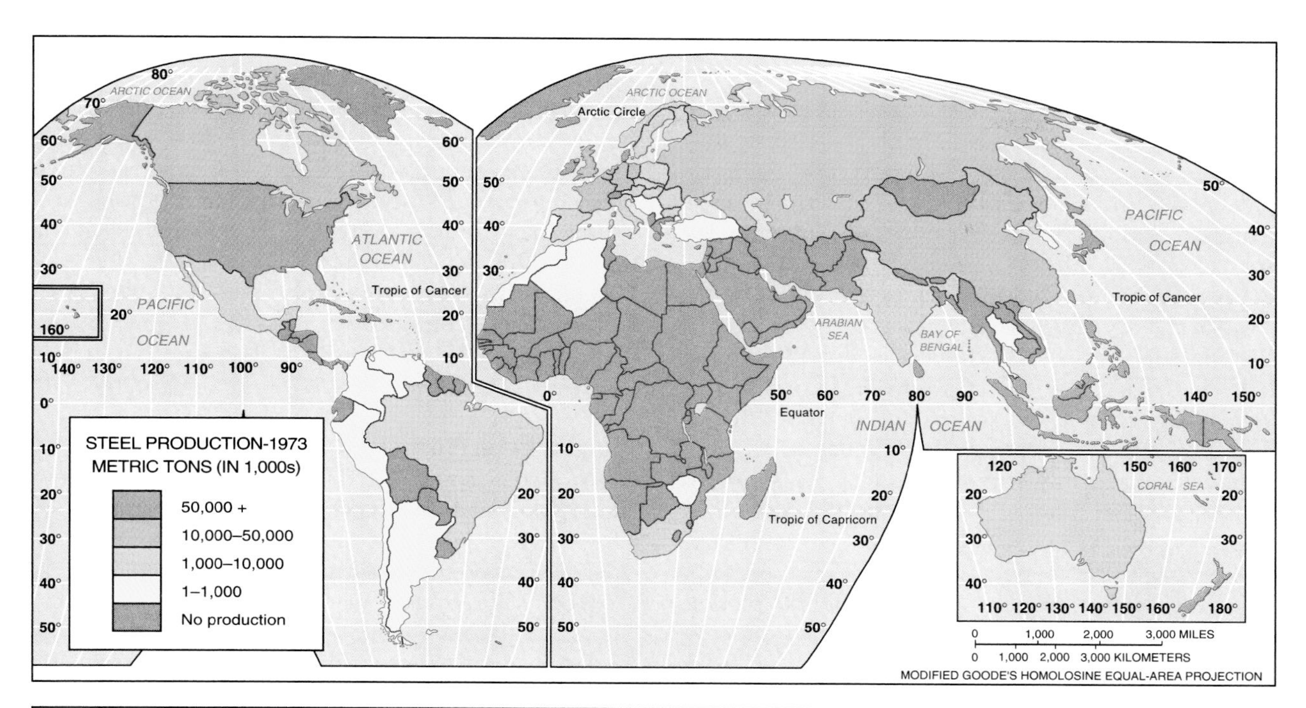

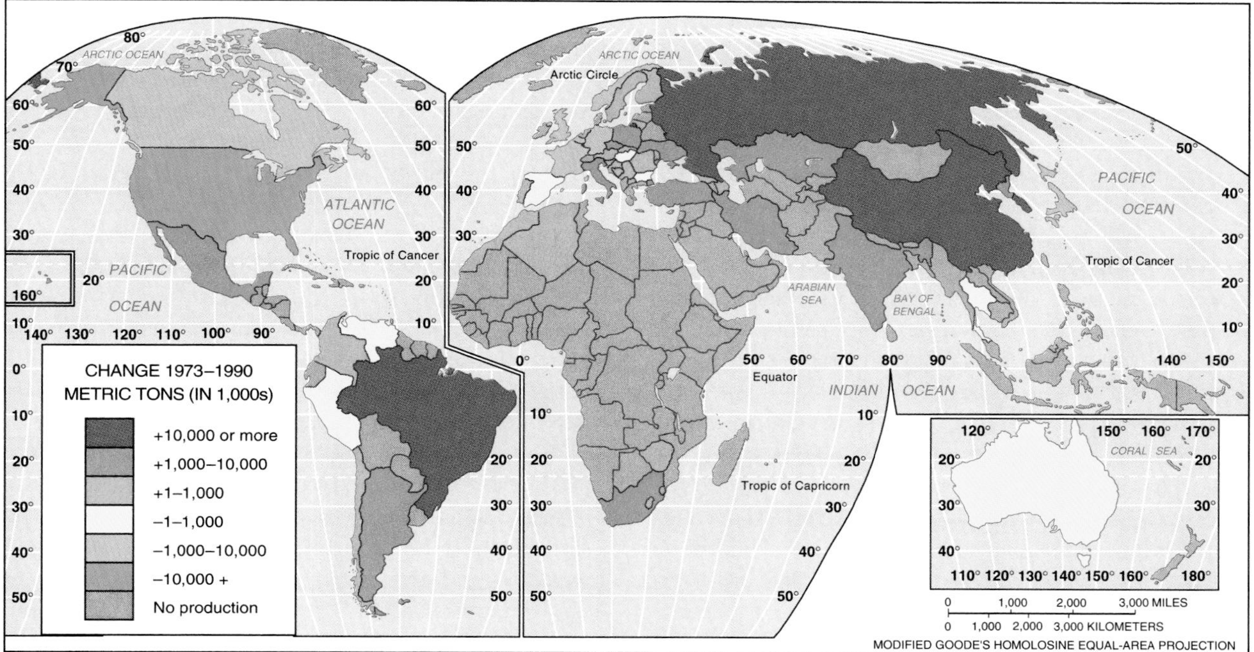

FIGURE 12–16

World steel production 1973 and change between 1973 and 1990.

(Top) In 1973, the United States was the largest steel producer, followed by the Soviet Union and Japan. These three countries, plus other relatively developed countries in Europe and North America, accounted for 90 percent of global production. (Bottom) Steel production has diffused from Western Europe and North America to other regions. Global production has remained virtually constant in recent years, but the distribution has changed. Production has declined in the relatively developed countries and increased in developing ones, especially China and Brazil.

tries tend to take the lead in exporting industrial goods to other countries.

The free movement of most products across borders has led to closer integration of industries within North America and Western Europe. For example, traditionally most automobiles sold in Canada were manufactured in Canada, but now most automobiles sold in Canada are assembled in the United States. Every Ford Taurus sold in Canada actually is assembled in the United States, while every Ford Crown Victoria sold in the United States is actually assembled in Canada. On balance, Canada exports twice as many automobiles to its southern neighbor than it imports. Canadians, though, argue that while they may assemble a disproportionately large share of North American automobiles, the United States has virtually all of the high-skilled engineering, design, and executive jobs in the auto industry.

At the same time they have promoted internal cooperation, the three trading blocs have erected barriers to restrict the ability of industries from other regions to compete effectively. European Union members slap a tax on goods that were produced in other countries. Japan has lengthy permit procedures that effectively hinder foreign companies from selling there. The Japanese government maintains quotas on the number of automobiles its companies can export to the United States in order to counter charges of unfair competition.

Faced with a decline in domestic steel production of about one-third during the late 1970s, the U.S. government negotiated a series of voluntary export restraint agreements with other major steel-producing countries. These quotas limited the sales of foreign-made steel to about 20 percent of the U.S. market.

During the period that the quotas were in effect, from 1982 until 1992, U.S. steel companies spent $24 billion to modernize their plants and buy more efficient equipment. As a result of restructuring, production levels have stabilized in the U.S. steel industry, but the number of steelworkers has fallen by two-thirds. Because of declining employment, the number of hours of labor needed to produce a ton of steel, a widely used measure of industrial efficiency, is lower in the United States than in Japan or Europe.

Communities where the steel mills were built have suffered severely from the decline in employment. The USX steel mill at Gary, Indiana—the country's largest—employed nearly 30,000 workers during the 1970s, compared to less than 8,000 in the early 1990s. Youngstown, Ohio, had more than 26,000 steel industry jobs in the mid-1970s; it has lost 80 per cent of them since then. Some of the unemployed steel workers have taken lower-paying jobs in other businesses, some have migrated to other parts of the country in search of jobs, and some have retired or remained unemployed. The steel industry's problems have affected the economic well-being and morale of communities such as Gary and Youngstown. Declines in other manufacturing sectors in relatively developed countries have had similar impacts in other communities.

Regional disparities. The distribution of industry has been a source of difficulty for many Western European countries. One region of a country may have lower levels of income and amenities as a result of having less industry than other regions.

For example, the United Kingdom has a disparity between the relatively poor north and west and the relatively prosperous south and east. Unemployment is 50 percent higher in the north and west, while average incomes are 25 percent higher in the south and east. The northern regions of the United Kingdom were the first in the world to enter the industrial revolution in the eighteenth century. Today, many of the region's industries are no longer competitive in the global economy. On the other hand, industries in the south and east are healthier, because of proximity to consumers in the country's largest urban area, London, as well as in Western Europe.

Other European countries have similar regional differences arising from industrial location patterns. French industry and wealth are concentrated in the Paris region, whereas the south and west suffer. Per capita income is three times higher in the north of Italy than the south, while Sweden's south is much more developed than its north. In each case, industry is concentrated in the regions most accessible to the large concentrations of population and wealth found in western and central Europe.

Germany has had a particularly difficult problem with regional disparities. The eastern portion of the country has required massive financial assistance to modernize its industries, a legacy of the 40-year period when the region comprised the separate, Communist-run German Democratic Republic.

A number of relatively developed countries have adopted policies to lure industries to poorer regions, and to discourage growth in the richer

The number of steel workers has declined by two-thirds in the United States since the 1970s. Since this photograph was taken in 1978, this steel mill in Youngstown, Ohio, has closed, and several thousand steelworkers have lost their jobs. (Roy Morsch/ The Stock Market)

regions. In the United Kingdom, for example, the south and east are much wealthier than the north and west. To aid development in the less prosperous regions, the government has designated several Development Areas and Intermediate Areas. Industries that locate in one of these two types of areas may be entitled to receive loans, grants, tax reductions, and other government aid. On the other hand, industries may be required to obtain government permission to locate in an unassisted area. Other European countries also have regional policies that employ financial incentives and regulations to encourage industrial location in peripheral regions and to discourage it in the congested cores.

The European Union provides assistance to regions that suffer from lack of industrial investment, such as Greece, Ireland, Portugal, southern Italy, and most of Spain. Funds are also available for declining industrial areas, including the northern areas of Denmark, England, France, Italy, and Spain. Regions that are eligible for support are required to submit five-year development plans explaining how the funds would be used.

The problem of regional disparity is somewhat different in the United States. The South, historically the poorest U.S. region, has had the most rapid growth since the 1930s, stimulated partly by government policies and partly by changing site factors. The selection of Alabama by Mercedes-Benz is an example of that trend. The Northeast, traditionally the wealthiest and most industrialized region, claims that development in the South has been at the expense of the old industrialized communities of New England and the Great Lakes states.

Regional development policies in relatively developed countries were reasonably successful as long as the national economy as a whole was expanding, because the lagging regions could share in some of the national growth. In the 1990s, an era of limited economic growth for relatively developed countries, governments increasingly question the wisdom of policies that strongly encourage industrial location in poorer regions. Excessive controls on industrial location could harm the overall national economy. Relatively developed countries have not completely abandoned policies that aid poorer regions, but the level of financial commitment has been severely reduced.

Industrial Problems in Developing Countries

The poorer countries of Africa, Asia, and Latin America seek to reduce the disparity in wealth between themselves and European and North American countries. Because the extent to which economic growth can be built on agriculture is limited, the leaders of virtually every developing state encourage the construction of new industries. Industrial development could not only raise the value of exports, which these countries need to generate the funds to buy other products, but also could supply local people with goods that are currently imported. If Western countries built their wealth on industrial modernization, why can't other countries do it?

In some respects, developing countries face industrial problems that are similar to the past experiences of today's relatively developed countries. In a fashion similar to countries that industrialized in the past, developing countries in the contemporary world must overcome two obstacles.

First, today's newly industrializing countries are relatively distant from the main markets—the wealthy consumers in the relatively developed countries. In the early nineteenth century, factories in the United States and central Europe were far from England, then the world's most important concentration of wealthy consumers. In the twentieth century, the major concentrations of wealthy consumers are in North America and Western Europe, distant from the developing countries of Africa, Asia, and Latin America. To minimize the obstacle of geographic isolation, countries that wish to develop their industries must invest scarce resources in constructing and subsidizing transportation facilities.

Second, as in the past, today's developing countries lack support services which are critical for industrial development. These support services include adequate transportation and communications systems and domestic sources of equipment, tools, and machines that are needed to build and operate new factories. Developing countries also lack universities capable of training the large number of factory managers, accountants, and other experts needed for industrial development. Support services are obtained either by importing advisers and materials from other countries or by borrowing money to develop domestic sources.

Countries currently industrializing also face a new obstacle to development. Traditionally, newly established factories in Europe, then North America, and more recently Asia have depended to some degree on selling products in countries that lacked competing industries. Because there are few untapped foreign markets left to exploit, new industries must either sell primarily to consumers inside the country or take away customers from existing businesses in other countries. Frequently, the domestic market is too small and poor to support large-scale industrial development.

In view of the obstacles in launching new industries for which access to markets is a critical locational factor, what kind of factories can developing countries attract? According to principles of economic geography, two other critical locational factors remain: access to raw materials and site costs.

In fact, new factories in Africa generally have been those for which either access to raw materials or a site factor is critical. Some African countries take advantage of proximity to raw materials. Mineral resources, such as bauxite in Guinea, uranium in Niger, iron ore in Mauritania and Liberia, and copper in Zambia, are processed for industrial uses elsewhere in the world. African countries also process food and agricultural products, such as palm and peanut oil, flour, and beer. Fertilizer is produced from phosphate in Mozambique, Senegal, Uganda, and Zimbabwe. However, manufacturing has expanded slowly in Africa, in part because world prices for processed raw materials have increased more slowly than for other products.

Also attracted to locations in less developed countries are industries for which the most critical site factor is access to abundant, low-cost labor. For example, the textile and clothing industries still consider low-cost labor to be their most critical site selection factor. Consequently, textile and clothing manufacturers that migrated within the United States from New England to the Southeast earlier in the twentieth century because of lower-cost labor have migrated again in recent years to Asia, Latin America, and Africa. Workers in developing countries are willing to work in textile and clothing mills for a fraction of the wages paid in any region of the United States, including the Southeast. The Bata Company, for example, has shoe factories in Sudan, Zambia, Nigeria, Côte d' Ivoire, and Cameroon.

Transnational corporations have been especially aggressive at taking advantage of low-cost labor in developing countries. To remain competitive in the global economy, transnational corporations carefully review all steps in their production and identify those processes that can be performed by low-paid, low-skilled workers in developing countries. Given the substantial difference in the level of wages paid to workers in relatively developed and developing countries, transnational corporations find it advantageous to transfer some work to developing countries despite higher transportation costs. At the same time, operations that require highly skilled workers are still performed in factories in relatively developed countries. The transfer of some types of jobs to developing countries is known as the **new international division of labor.**

Structuring for the Future

Three recent changes in the structure of manufacturing have geographic consequences. First, factories have become more productive through the introduction of new machinery and processes. The factory may continue to operate at the same location but require fewer workers to produce the same output. Faced with meager prospects of getting another job in the same community, workers laid off at these factories migrate to other regions.

Second, companies are locating production in communities where workers are willing to adopt more flexible work rules. Firms are especially attracted to smaller towns where low levels of union membership and high visibility reduce vulnerability to work stoppages, even if wages are kept low and lay-offs become necessary.

Third, large corporations, by spreading production among many countries or among many communities within one country, have increased their bargaining power with local governments and labor forces. Production can be allocated to locations where the local government is especially helpful and generous in subsidizing the costs of expansion, and the local residents are especially eager to work in the plant.

Regional competition. Competition to attract new industries and to retain existing ones extends across international borders. The governments of Canada, Mexico, and the United States have eliminated tariffs and other barriers to free trade among the three countries. As competition increases among regional blocs of countries, U.S. and Canadian business and government leaders have seen substantial benefits to including Mexico in a free trade zone. With the addition of Mexico, the North American Free Trade area rivals the European Union as the world's most populous and wealthy market.

Creating an integrated North American economy is a formidable task, given the substantially lower standard of living in Mexico than in the United States and Canada. U.S. and Canadian labor union leaders are concerned that with the removal of bar-

U.S. industries feel a tension between increasing global cooperation and competition. (Top) A Japanese corporate executive is welcomed in the United States because his company plans to invest in a factory in the United States. (John Feingersh/The Stock Market). (Bottom) Japanese cars are not welcomed at a U.S. auto parts manufacturer because their production is seen as taking jobs away from American workers. (Craig Hammell/The Stock Market).

riers, more manufacturers will relocate production to Mexico to take advantage of lower wage rates. Such labor-intensive industries as electrical and textile manufacturing may be especially attracted to a region where prevailing wage rates are lower.

Environmentalists fear that under a free trade agreement, firms will move production to Mexico, where laws governing air and water quality standards are less stringent than in the United States and Canada. Mexico has adopted regulations to reduce air pollution in Mexico City; catalytic converters were required on Mexican automobiles beginning in 1991.

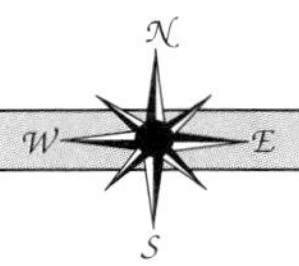

CONCLUSIONS: CRITICAL ISSUES FOR THE FUTURE

According to industrial location theory, firms select locations for a variety of situation and site factors. Wage rates and environmental controls constitute two of the important site factors, but such factors as access to markets and to skilled workers are also critical. Geography's global perspective in analyzing industrial location reinforces the fact that the problems of an unemployed steel worker in Gary or Youngstown are related to worldwide characteristics of the steel industry. The future health of industry in the United States depends on a national commitment to a combination of competition and cooperation in a global economy.

To recapture competitiveness with other nations' industries, North American business leaders must learn more about other nations' culture, politics, and economy. The success enjoyed by Japanese and Korean businesses in North America derives to a considerable extent from the fact that executives in those countries know more about U.S. society than Americans know about Asia. Asian officials are likely to speak English and are familiar with the tastes and preferences of American consumers, whereas few American officials speak Japanese or Korean and have relatively little knowledge of the buying habits of Asians.

At the same time, global industrial development depends on increased cooperation among different nations. As a result of lower transportation costs, more people in the world have access to more goods at lower prices than in the past. Given this trend, consumers in industrialized countries are increasingly challenged to choose between the purchase of the highest-quality, lowest-cost goods regardless of where they were made and the support for local industries against foreign competitors at any price.

Chapter Summary

1. Distribution of manufacturing

In contrast to agriculture, which covers a large percentage of Earth's land area, industry is highly concentrated. Approximately three-fourths of the world's industrial output is concentrated in four regions—the North American manufacturing belt, Western Europe, Eastern Europe, and Japan.

2. Situation factors in industrial location

In situating a factory, firms try to identify the location where production costs are minimized. Critical industrial location costs are situation factors for some firms and site factors for others. Situation factors involve the cost of transporting inputs into the factory and products from the factory to consumers. Bulk-reducing industries select locations that minimize the cost of obtaining inputs. Bulk-gaining industries, such as fabricators, and communications-oriented industries tend to locate near their customers.

3. Site factors in industrial location

Site factors involve the cost of doing business at a particular location because of the characteristics of different production factors, especially land, labor, and capital.

4. Industrial restructuring

The world faces a problem with industry because global capacity to produce many goods exceeds the demand. Relatively developed countries in North America and Western Europe have a distinctive problem which results from an uneven internal distribution of industry and wealth. Developing countries, located further from markets, must attract industries for which access to inputs or low-cost labor are critical.

Key Terms

Break-of-bulk point, 487
Bulk-gaining industry, 481
Bulk-reducing industry, 479
Communications-oriented industry, 482
Fordist production, 493
Labor-intensive industry, 489
Maquiladora, 492
New international division of labor, 501
Right-to-work state, 491
Site factors, 479
Situation factors, 479

Questions for Study and Discussion

1. How did the industrial revolution change the distribution of manufacturing?
2. What are the main industrial regions within Western Europe? What are the principal locational assets of each?
3. What are the main industrial regions within North America? What are the principal locational assets of each?
4. What are the main industrial regions within Russia? What are the principal locational assets of each?
5. What are the two main situation factors?
6. How has the location of steel production changed within the United States? What situation factors help to explain the changes?
7. What types of industries locate near their markets?
8. What is meant by a *break-of-bulk point?* What is its importance in industrial location decisions?
9. What are the three main site factors?
10. What are the three main steps in the manufacture of textiles and clothing? How does the spatial distribution of each differ?
11. What supply and demand factors contribute to global industrial problems?
12. What problems do developing countries face in trying to achieve industrial growth?

Thinking Geographically

1. To induce Toyota to build its U.S. production facilities in Kentucky, the state spent $49 million to buy the 600-hectare (1,500-acre) site, $40 million to construct roads and sewers, and $68 million to train the new workers. Kentucky also agreed to spend up to $168 million to pay the interest on loans should Toyota decide to borrow money to finance the project. Did Kentucky overpay for Toyota? Explain.
2. Foreign cars account for one-fourth of the sales in the midwestern United States, compared to half in California and other West Coast states. What factors might account for this regional difference?
3. Draw a large triangle on a map of Russia, with one point near Moscow, one point in the Ural Mountains, and one point in Central Asia. What are the principal economic assets of the three regions at each side of the triangle? How do the distributions of markets, resources, and surplus labor vary within Russia?
4. What are the principal manufacturers in your community? How have they been affected by increasing global competition?
5. What have been the benefits and costs to Canada, the United States, and Mexico of the North American Free Trade Agreement?

Suggestions for Further Reading

Amin, Ash, and John Goddard, eds. *Technological Change, Industrial Restructuring, and Regional Development*. London: Allen and Unwin, 1986.

Ashton, Thomas S. *The Industrial Revolution*. New York: Oxford University Press, 1964.

Behrman, Jack N. *Industrial Policies: International Restructuring and Transnationals*. Lexington, MA: Lexington Books, 1984.

Bell, Michael E., and Paul S. Lande, eds. *Regional Dimensions of Industrial Policy*. Lexington, MA: Lexington Books, 1982.

Birdsall, Stephen S., and John W. Florin. *Regional Landscape of the United States and Canada*. 2d ed. New York: Wiley, 1981.

Blackbourn, Anthony, and Robert G. Putnam. *The Industrial Geography of Canada*. New York: St. Martin's Press, 1984.

Bluestone, Barry, and Bennett Harrison. *The Deindustrialization of America: Plant Closings, Community Abandonment, and the Dismantling of Basic Industry*. New York: Basic Books, 1982.

Brotchie, John F., Peter Hall, and Peter W. Newton, eds. *The Spatial Impact of Technological Change*. London: Croom Helm, 1987.

Casetti, Emilio, and John Paul Jones III. "Spatial Aspects of the Productivity Slowdown: An Analysis of U.S. Manufacturing Data." *Annals of the Association of American Geographers* 77 (March 1987): 76–88.

Caves, Richard E. *Multinational Enterprise and Economic Analysis*. Cambridge, England: Cambridge University Press, 1982.

Dicken, Peter. *Global Shift: Industrial Change in a Turbulent World*. 2d ed. London: Harper & Row, 1991.

Earney, F. C. F. "The Geopolitics of Minerals." *Focus* 31 (May–June 1981): 1–16.

Gillespie, A. E., ed. *Technological Change and Regional Development*. London: Pion Ltd., 1983.

Glasmeier, Amy, Jeffery W. Thompson, and Amy J. Kays. "The Geography of Trade Policy: Trade Regimes and Location Decisions in the Textile and Apparel Complex." *Transactions of the Institute of British Geographers, New Series* 18 (1993): 19–35.

Gould, Peter. *Spatial Diffusion*. Washington, DC: Association of American Geographers, 1969.

Habakkuk, H. J., and M. M. Postan, eds. *The Cambridge Economic History of Europe*. Vol. 6. Cambridge, England: Cambridge University Press, 1965.

Hamilton, F. E. Ian, ed. *Contemporary Industrialization*. London and New York: Longman, 1978.

Hamilton, F. E. Ian, and G. J. R. Linge, eds. *Spatial Analysis, Industry and the Industrial Environment: Progress in Research and Applications. Volume I: Industrial Systems* (1979); *Volume II: International Industrial Systems* (1981); *Volume 3: Regional Economies and Industrial Systems* (1983). Chichester, England: Wiley.

Harris, C. D. "The Market as a Factor in the Localization of Industry in the United States." *Annals of the Association of American Geographers* 44 (December 1954): 315–48.

Hoare, Anthony G. "What Do They Make, Where, and Does It Matter Any More? Regional Industrial Structures in Britain Since the Great War." *Geography* 7 (October 1986): 289–304.

_____. *The Location of Industry in Britain*. New York: Cambridge University Press, 1983.

Hoffman, George W., ed. *Eastern Europe: Essays in Geographical Problems*. London: Methuen, 1971.

Hogan, William T. *Minimills and Integrated Mills: A Comparison of Steelmaking in the United States*. Lexington, MA: Lexington Books, 1987.

_____. *Global Steel in the 1990s: Growth or Decline*. Lexington, MA: Lexington Books, 1991.

Langton, John. "The Industrial Revolution and the Regional Geography of England." *Transactions of the Institute of British Geographers, New Series* 9 (no. 2, 1984): 145–67.

Langton, John, and R. J. Morris, eds. *Atlas of Industrializing Britain, 1780–1914*. London: Methuen, 1986.

Law, Christopher M., ed. *Restructuring the Global Automobile Industry*. London: Routledge, 1991.

Massey, Doreen, and Richard Meegan, eds. *Politics and Method: Contrasting Studies in Industrial Geography*. New York: Methuen, 1986.

Oakey, Raymond P. *High Technology Small Firms: Innovation and Regional Development in Britain and the United States*. New York: St. Martin's Press, 1984.

Oxford University Cartographic Department. *Oxford Economic Atlas: The United States and Canada*. London: Oxford University Press, 1975.

Pattie, Charles J., and R. J. Johnston. "One Nation or Two? The Changing Geography of Unemployment in Great Britain, 1983–1988." *Professional Geographer* 42 (August 1990): 288–298.

Peet, Richard, ed. *International Capitalism and Industrial Restructuring*. Boston: Allen & Unwin, 1987.

Rubenstein, James M. *The Changing US Auto Industry*. London: Routledge, 1992.

Schmenner, Roger W. *Making Business Location Decisions*. Englewood Cliffs, NJ: Prentice-Hall, 1982.

Scott, Allen J., and Michael Storper, eds. *Production, Work, Territory*. Boston: Allen and Unwin, 1986.

Smith, David M. *Industrial Location: An Economic Geographical Analysis*. 2d ed. New York: Wiley, 1981.

_____. "A Theoretical Framework for Geographical Studies of Industrial Location." *Economic Geography* 42 (April 1966): 95–113.

South, Robert B. "Transnational 'Maquiladora' Location." *Annals of the Association of American Geographers* 80 (December 1990): 529–70.

Storper, Michael, and Richard Walker. *The Capitalist Imperative: Territory, Technology, and Industrial Growth*. New York: Basil Blackwell, 1989.

Toyne, Brian, Jeffrey S. Arpan, David A. Ricks, Terence A. Shimp, and Andy Barnett. *The Global Textile Industry*. London: Allen and Unwin, 1984.

Warren, Kenneth. "World Steel: Change and Crisis." *Geography* 70 (March 1985): 106–17.

Webber, Michael J. *Industrial Location*. Beverly Hills: Sage Publications, 1984.

ZumBrunnen, Craig, and Jeffrey Osleeb. *The Soviet Iron and Steel Industry*. Totowa, NJ: Rowman and Allanheld, 1986.

We also recommend these journals: *Journal of Industrial Economics, Journal of International Economics, Journal of Marketing, Journal of Transport Economics and Policy, Journal of Transport History, Journal of Urban Economics.*

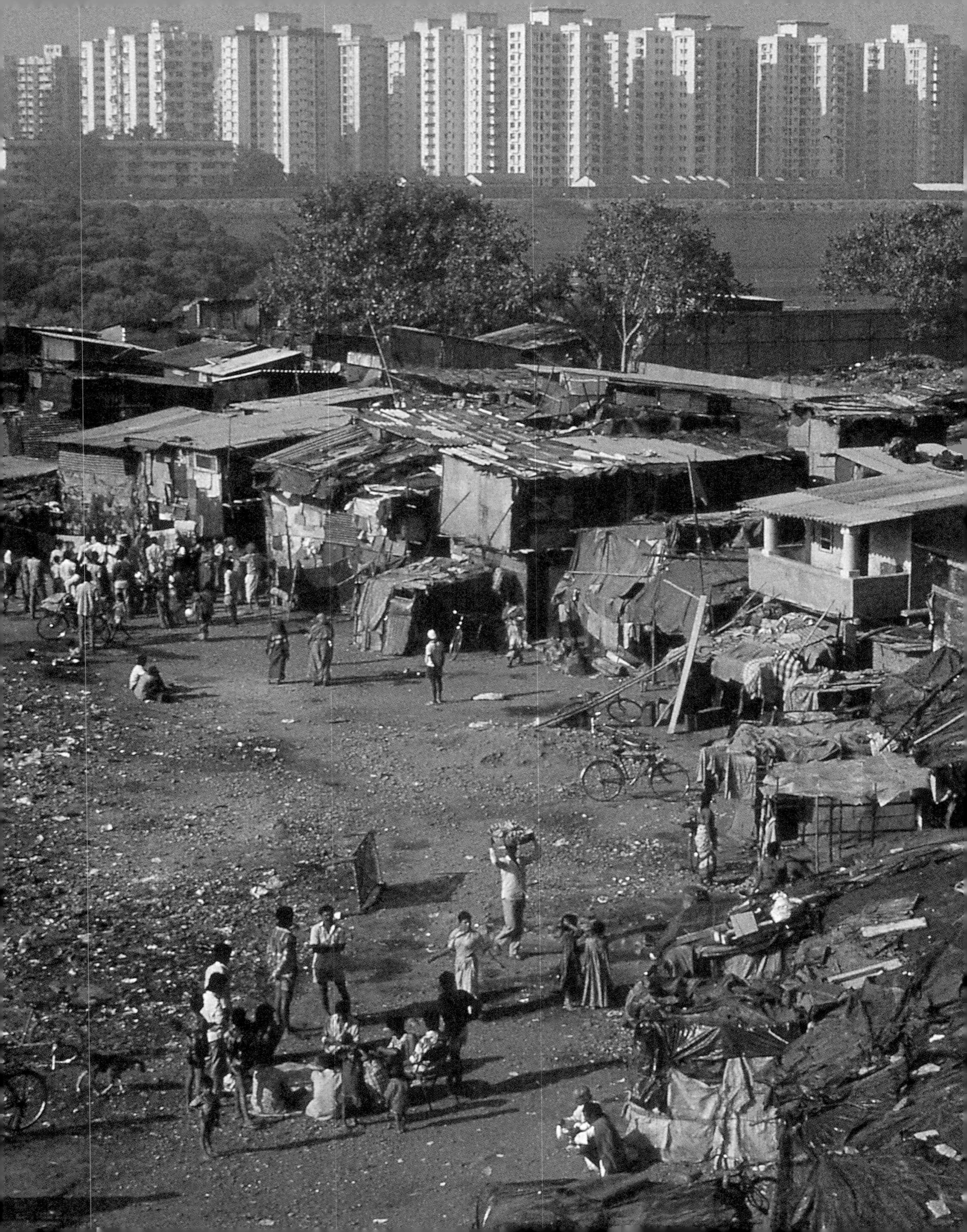

13 Settlements

- Origin of settlements
- Urbanization
- The central city
- Cities and suburbs

Rich and poor areas in Bombay, India.

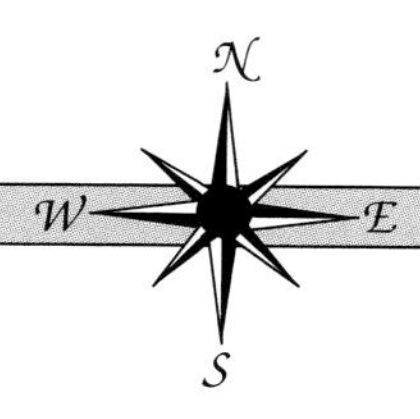

1. Origin of settlements
Settlements originated for a combination of cultural and economic reasons. Rural settlements, where agriculture is the principal economic activity, can be arranged in a clustered or dispersed pattern.

2. Urbanization
Urbanization is the increase in both population and the percentage of a society's population that lives in urban settlements. Urbanization has been especially important in the past 200 years. Today, urban growth is most rapid in developing countries.

3. The central city
Urban areas contain central business districts surrounded by extensive residential areas. Economic conditions have improved in many central business districts, but surrounding them are segregated neighborhoods containing large numbers of poor nonwhite residents who have limited prospects for economic improvement.

4. Cities and suburbs
Three models—the concentric zone, sector, and multiple nuclei—help to explain the distribution of different types of people and activities within cities.

INTRODUCTION

Most people reside in some form of **settlement,** a fixed collection of buildings and inhabitants. Taken together, urban and rural settlements occupy a very small percentage of Earth's surface, yet they exert a great influence on the world's economy and culture. Settlements are places to find jobs, goods, and services. Beyond that, settlements are storage nodes for the world's cultural and economic wealth, and points of origin from which economic and cultural innovations are diffused. This chapter examines where settlements occur across Earth's surface and what factors explain their distribution.

Urban settlements occupy a tiny percentage of the world's land area, well under one percent. Yet nearly half of the world's people live in an urban area. Geographers describe where different activities are located in a city and where different types of people live. They also try to explain why these activities and people are located where they are within the city.

In developing countries, most people live in rural settlements having few goods and services. This poses a hardship, for they may have to travel great distances to obtain them. In relatively developed countries, virtually everyone is functionally tied to an urban settlement. Not all residents of relatively developed countries actually live, work, and shop in an urban settlement, but virtually all can drive to a major metropolitan area if they wish.

Access to goods and services is denied to some people in relatively developed countries, not by distance, but by economic inequality and racial segregation. Poor and minority residents may be unable to use nearby stores and services because they lack the financial resources or social standing.

Origin of Settlements

The world contains about 200 settlements that have more than 2 million inhabitants (Figure 13–1). It is difficult to rank the world's cities in order of population, because the definition of what constitutes a city varies from one country to another, and up-to-date figures are rarely available, even in relatively developed countries. A few years ago, the United Nations applied a consistent definition to estimate the population of cities around the world. They concluded that the Tokyo-Yokohama region in Japan is the world's largest urban area, followed by São Paulo, Brazil.

Reasons for Establishing Settlements

Contemporary settlements exist primarily to serve economic functions, but we believe that the earli-

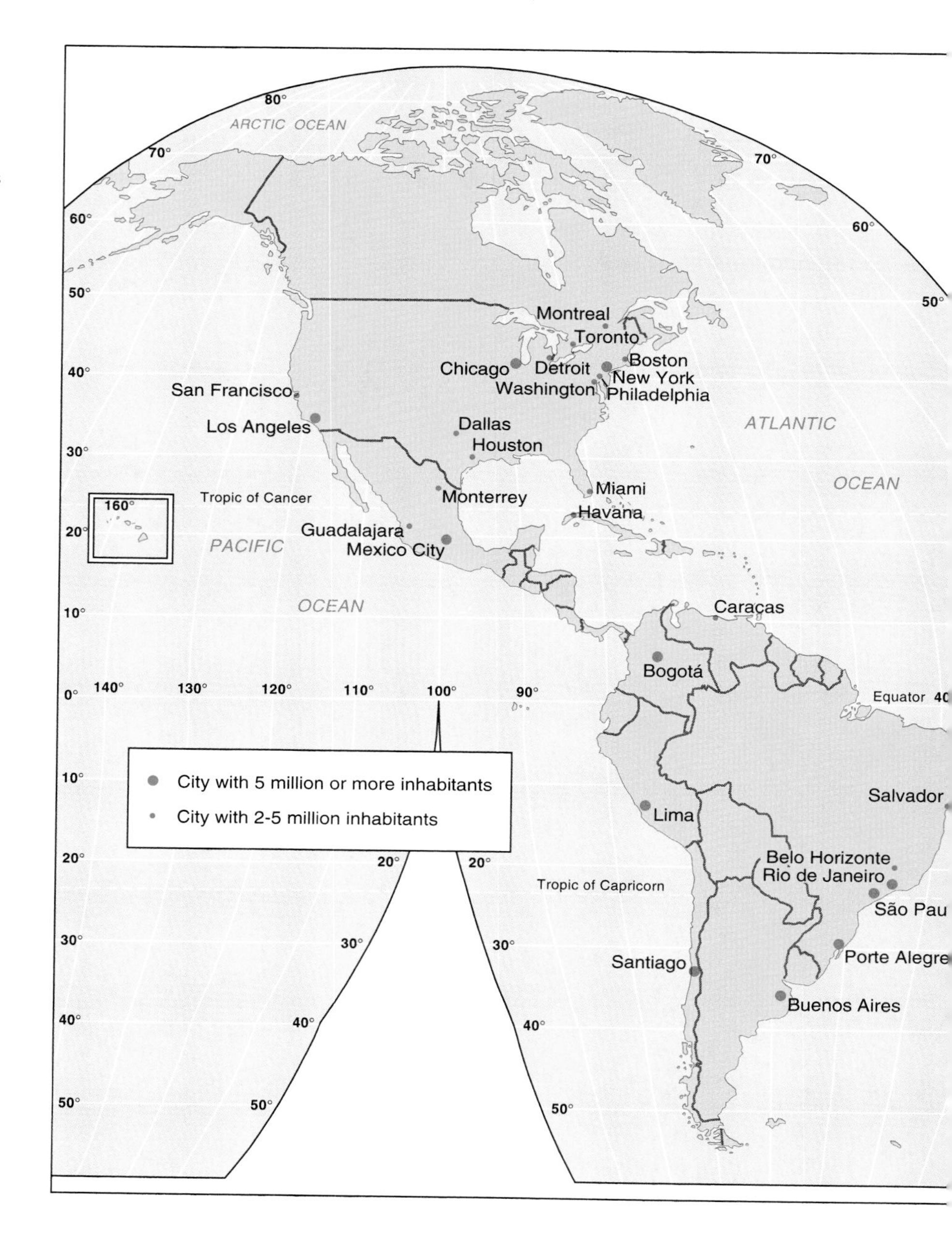

FIGURE 13–1
Cities having a population of 2 million or more. Although the percentage of people living in urban areas is greater in relatively developed countries, most of Earth's largest urban areas are now located in developing countries. The rapid growth of cities in developing countries reflects both an increase in the overall national population and migration from rural areas.

est settlements were established for other reasons. Precisely what those reasons were is shrouded in mystery, as they occurred before recorded history. To understand why early peoples might have created permanent settlements, picture their condition at the time. People were nomadic, migrating in small groups across the landscape in search of food and water. They gathered wild berries and roots, or killed wild animals for food. Why would they need to settle in one place?

Cultural reasons for establishing settlements. Based on archeological research, settlements probably originated for cultural rather than economic reasons. The first permanent settlements may have been places to bury the dead. Perhaps nomadic groups had rituals honoring the dead, maybe including memorial services on the anniversary of a death. Having established a permanent resting place for the dead, the group might install priests at the site to perform ceremonies.

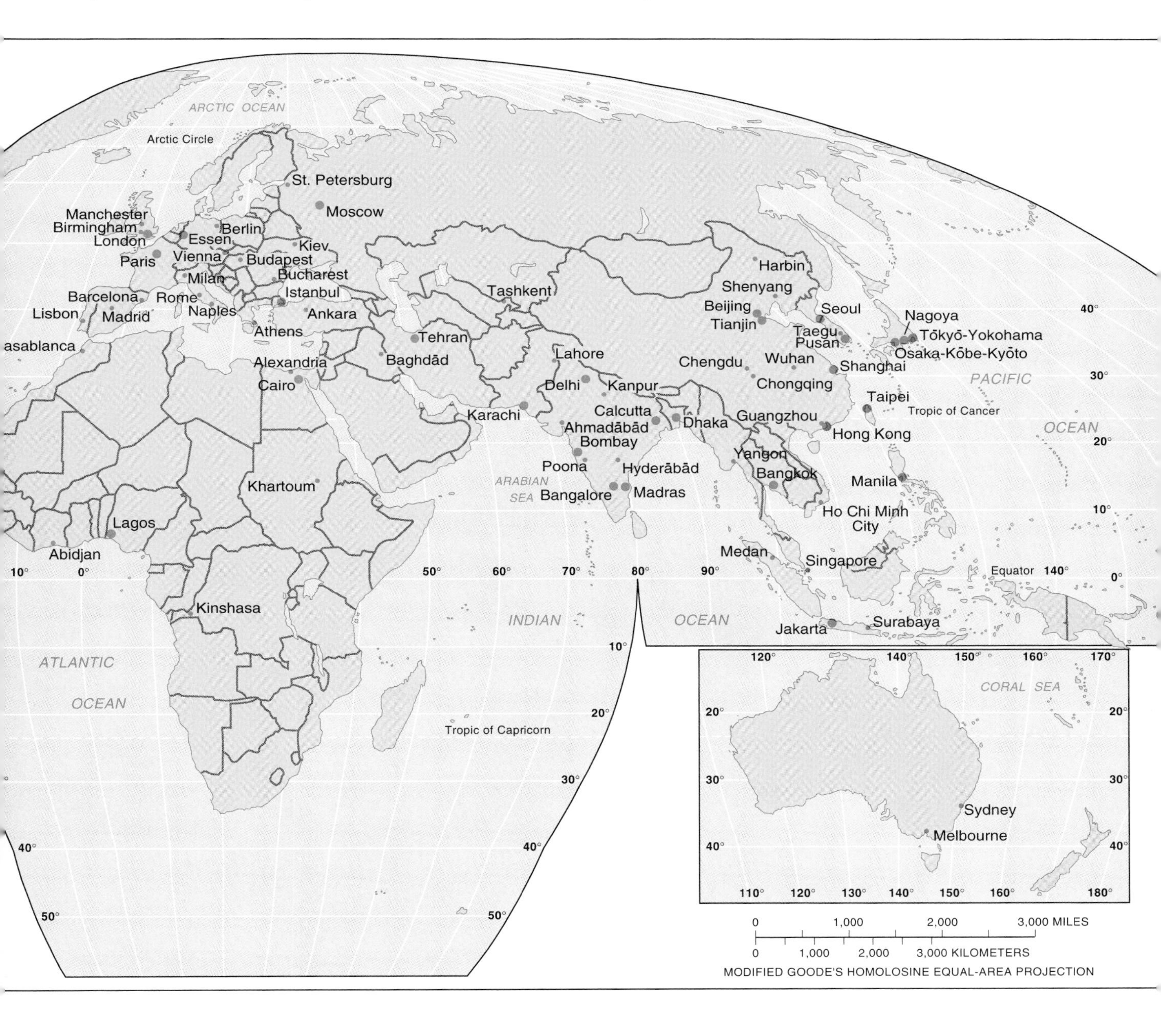

All this would have encouraged the building of structures—places for ceremonies and dwellings. By the time recorded history began around 5,000 years ago, many settlements featured a temple. In fact, until the invention of skyscrapers in the late nineteenth century, religious buildings were often the tallest structures in the community. Settlements continue to be important as religious centers.

Settlements also may have served as places to house women and children, permitting the men, unburdened, to travel farther and faster in their search for food. Women made household objects, such as pots, tools, and clothing, and children were educated in settlements. Making pots and educating children may have originated for practical reasons, but over thousands of years these activities provided the basis for creation and transmission of a group's values and heritage. Today, settlements contain society's schools, libraries, museums, and archives—the permanent repositories for passing on knowledge from one generation to the next.

In developing countries today, rural settlements often contain a high percentage of women and children. Men are more likely to migrate to large cities or to other countries to seek employment, leaving behind their wives and children in the villages.

The location of early settlements may have been chosen to protect the group's land claims and food sources against competitors. The group's political leaders lived in the settlement, and elaborate structures were built to house them. Collecting the group's priests, leaders, teachers, women, and children into settlements made them vulnerable to attacks from other groups. For protection, some members of the group became soldiers.

For defense, the group might surround the settlement with a wall. Defenders were stationed at small openings or atop the wall, giving them a strategic advantage over the attackers. Thus, settlements became citadels—centers of military power. Walls proved an extremely effective defense for thousands of years, until warfare was revolutionized by the introduction of gunpowder in Europe in the fourteenth century. Even though cannonballs could destroy walls, they continued to be built around cities. Paris, for example, surrounded the city with new fortifications as recently as the 1840s and did not completely remove them until 1932 (Figure 13–2).

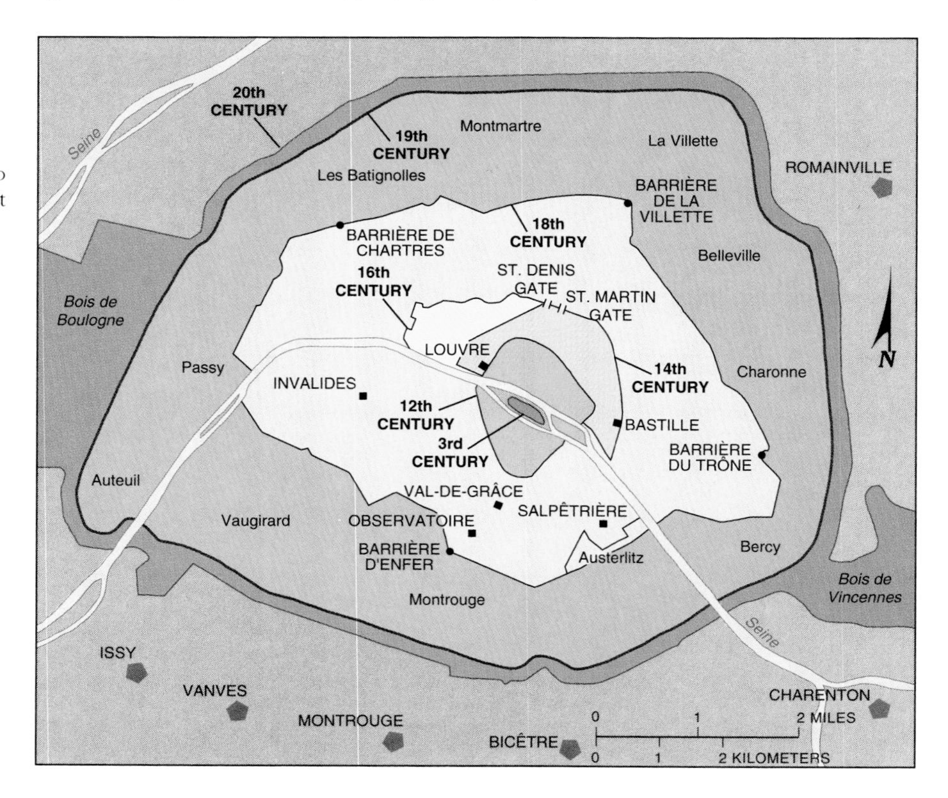

FIGURE 13–2
The growth of Paris from the third century to present. Paris was surrounded by a wall, originally for protection. Periodically, a new wall ("barrière") would be constructed to encompass new neighborhoods that had grown on the periphery. Highways and parks have been built on the sites of the nineteenth century walls. The old gates of St. Denis and St. Martin still stand, although the walls have been demolished.

Medieval European cities were often surrounded by walls for protection. The walls have been demolished in most places, but they still stand around the old center of Carcassonne, in southwestern France. A small portion of this modern industrial city of 40,000 inhabitants can be seen at far right. (Jonathan Blair/Woodfin Camp & Associates)

Although settlements no longer have walls, their military and political functions continue to be important. The largest structure in our nation's capital—the Pentagon—houses the U.S. Department of Defense. Similarly, Russian military leaders work in the Kremlin, the medieval walled area of central Moscow.

Economic reasons for establishing settlements. Everyone in the settlements needed food, which was supplied by the group through hunting or gathering. At some point, someone probably wondered: why not bring in extra food in case of hard times, such as drought or conflict? This is likely how the economic role of settlements began—as a *warehousing center* to store extra food.

Through centuries of experiments and accidents, people realized that some of the wild vegetation they had gathered could generate food if deliberately placed in the ground and nursed to maturity—in other words, agriculture. The settlement might then become an *agricultural center.* Eventually, people were able to produce most of their food through deliberate agricultural practices. They no longer had to survive through hunting and gathering.

In addition to food, people needed tools, clothing, shelter, containers, fuel, and other material goods. Settlements therefore became *manufacturing centers.* Men gathered the materials needed to make a variety of objects, including stones for tools and weapons, grass for containers and matting, animal hair for clothing, and wood for shelter and heat. Women used these materials to make household objects.

Not every group had access to the same resources because of the varied distribution of vegetation, animals, fuelwood, and mineral resources across the landscape. The settlement therefore was likely to become a *trading center.* People brought objects and materials they had collected or produced into the settlement and exchanged them for items brought by others. The settlement served as neutral ground where several groups could safely come together to trade goods. To facilitate trade, officials in the settlement regulated the terms of transactions, set fair prices, kept records, and created a currency system.

Central place theory. According to **central place theory,** a settlement is a market center for the exchange of goods and services by people who are attracted from the surrounding area. The area from which they are attracted is the **market area** or *hinterland*. A market area is a good example of a *nodal region*—one with a core where the characteristic is most intense. For analysis, geographers represent market areas with hexagons, for reasons explained in Figure 13–3.

Central places compete with one another, and this creates a regular pattern of settlements. This is most evident in simpler regions like the Great Plains, which are neither heavily industrialized nor interrupted by major physical features like rivers or mountain ranges.

Range is the maximum distance people are willing to travel to obtain a good or service. **Threshold** is the minimum number of people needed in the market area to support the good or service. A small settlement provides goods and services that characteristically have small ranges and thresholds, like groceries and gasoline. A large settlement provides goods and services that have larger ranges and thresholds, like BMW dealerships. Small settlements are more numerous and closer together, while larger settlements are fewer and farther between.

In many relatively developed societies, geographers observe that the ranking of settlements from largest population to smallest tends to produce a regular pattern. This is the **rank-size rule**, in which the *n*th largest settlement in the country is $1/n$ the population of the largest settlement. In other words, the second-largest city is ½ the size of the largest, the fourth-largest city is ¼ the size of the largest, and so on.

Developing countries rarely follow the rank-size

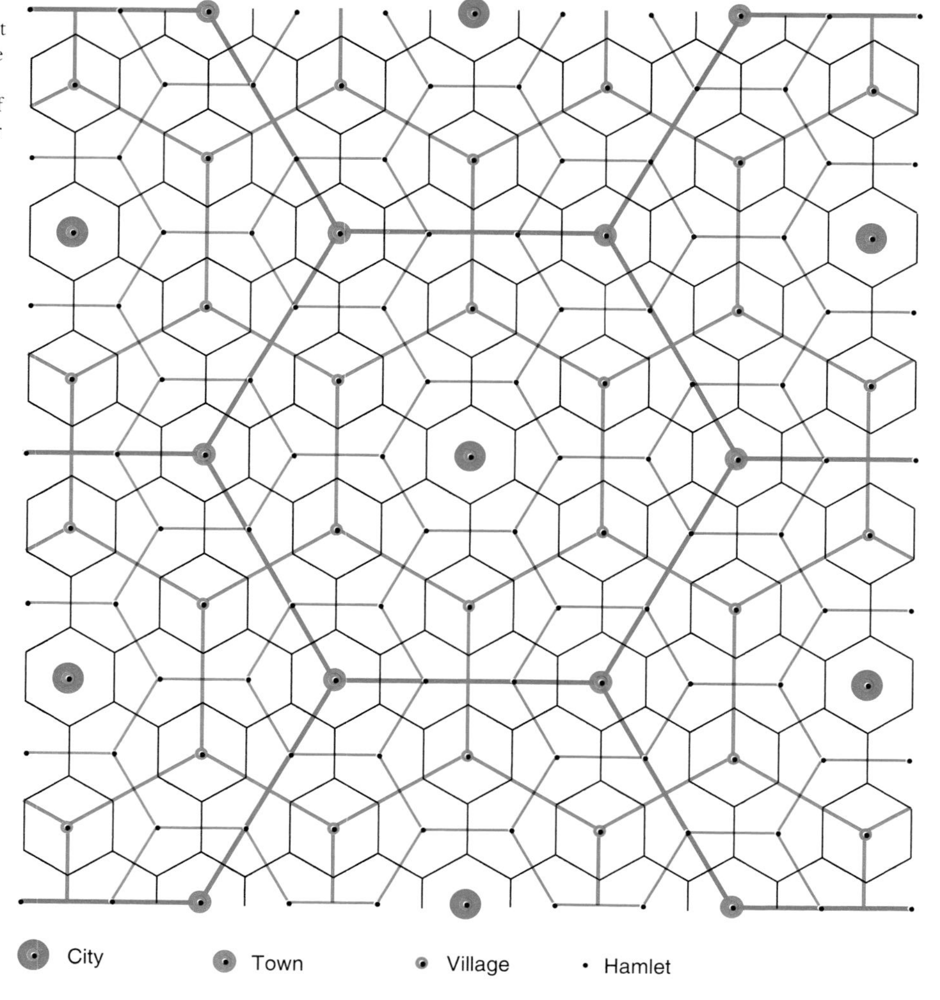

FIGURE 13–3
Geographers use hexagons to depict the market area of a good or service because they offer a compromise between the geometric properties of circles (equidistance from the center to the edge) and squares (nesting together without gaps). (From Walter Christaller, *The Central Places in Southern Germany*, Prentice Hall, 1966. Used by permission)

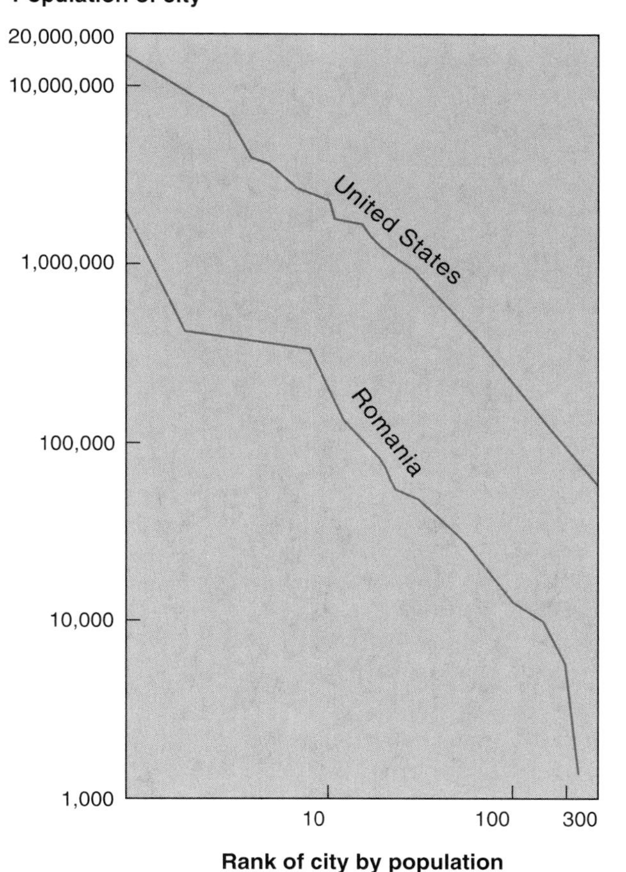

FIGURE 13–4
U.S. settlements generally follow the rank-size distribution, as reflected by their nearly straight line on this logarithmic display. In contrast, Romania has a shortage of settlements in two size groups: between 350,000 and 2,000,000 population, and fewer than 10,000 inhabitants.

rule. Instead, their largest settlement follows the **primate city** rule, in which the largest settlement has more than twice as many people as the next-largest. The absence of a rank-size distribution of settlements indicates that a society is not sufficiently wealthy to provide goods and services throughout the country (Figure 13–4).

Economic base. Each community is a general market center. But it also conducts individual economic activities, including basic and nonbasic industries. **Basic industries** sell most of their products or services *outside* the settlement. A community's basic industries taken together form its **economic base.** The significance of basic industries and the economic base they form is (1) products exported from the settlement bring money into the local economy, and (2) a new basic industry will attract workers, families, and local services to the community, helping it to grow (Figure 13–5). A settlement's *nonbasic industries* provide services, mostly within the community, and do not bring in outside money.

Rural Settlements

Settlements have evolved primarily to provide two types of economic goods and services: some have become centers for agriculture, while others have become centers for manufacturing, warehousing, and retailing. Agriculture is the predominant economic activity in communities that we now call **rural settlements,** while manufacturing, warehousing, and retailing are the main economic activities in **urban settlements.** We will now take a look at each type.

Clustered rural settlements. Worldwide, most settlements are rural rather than urban, because most people survive by farming rather than by manufacturing or trading. A number of families may live in a clustered rural settlement and work in the surrounding fields. Such a **clustered rural settlement** typically includes barns, tool sheds, and other farm structures, plus homes, religious structures, schools, and supporting services. In common language, such a settlement is called a hamlet or village.

Each person living in a clustered rural settlement is allocated land in the surrounding fields. The fields must be accessible to the farmers and are thus generally limited to a radius of 1 or 2 kilometers (0.5 mile to 1 mile) from the buildings. The land is allocated in different ways. In some places, individual farmers own or rent the land; in other places, the land is owned collectively by the settlement or by a lord, and farmers do not control the choice of crops or use of the output.

Parcels of land surrounding the settlement may be allocated to specific agricultural activities, either because of land characteristics or because of decisions by the inhabitants. Consequently, farmers typically own, or have responsibility for, a collection of scattered parcels in several fields. This pattern of controlling several fragmented parcels of land has encouraged living in a clustered rural settlement to minimize travel time to the various fields.

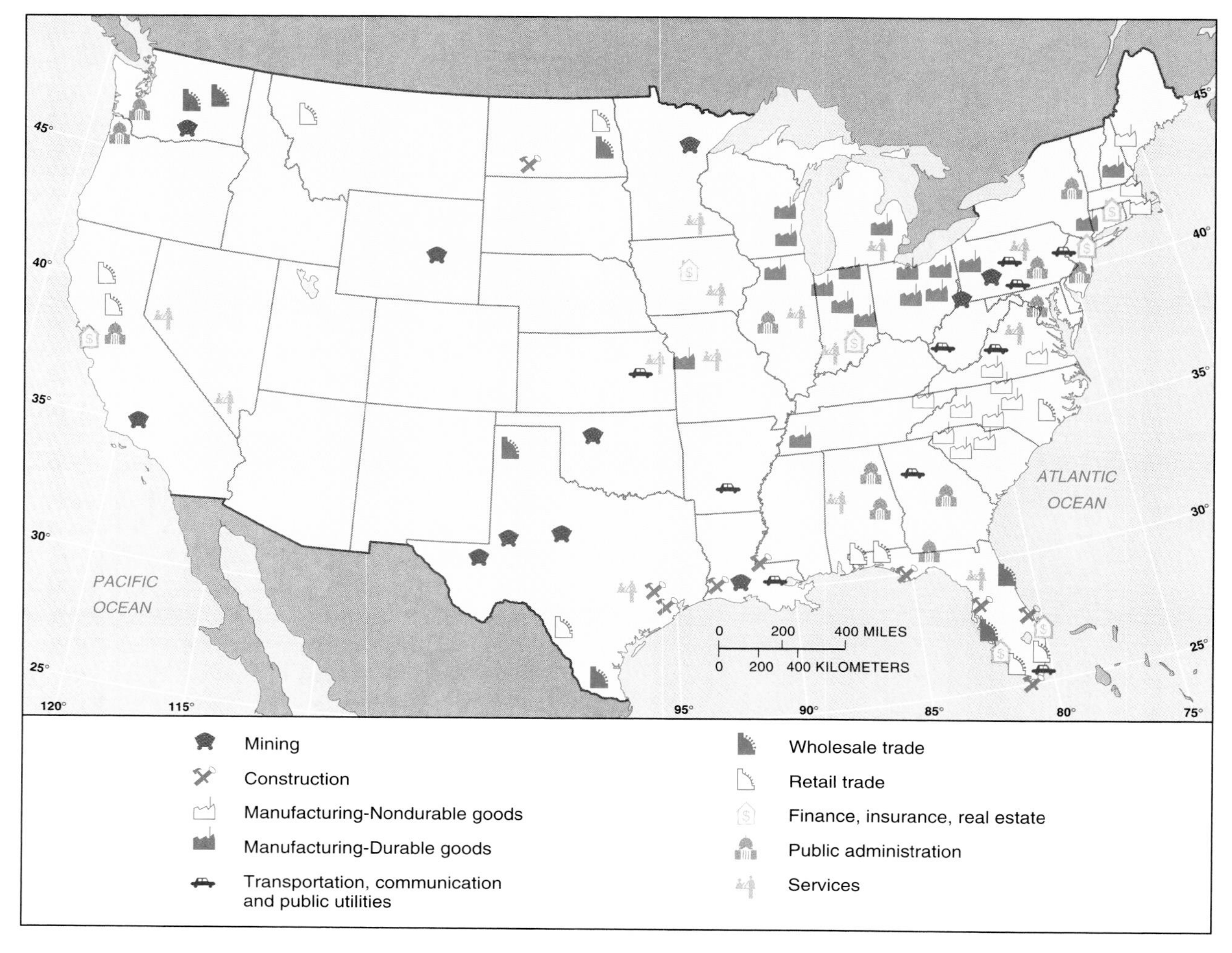

FIGURE 13–5
Basic industries of U.S. cities. Symbols on this map represent cities that have a significantly higher percentage of their labor force engaged in the type of economic activity shown. Other cities also engage in such activities, but are not shown because they specialize in multiple activities, or are near the national average for all sectors. (Mathematically, a city was included if the percentage of its labor force in one sector was more than two standard deviations above the mean for all U.S. cities.)

Traditionally, when the population of a settlement grew too large for the capacity of the surrounding fields, new settlements were established nearby. This was possible because not all land was under cultivation. The establishment of satellite settlements is often reflected in their place names (Figure 13–6).

Homes, public buildings, and fields in a clustered rural settlement are arranged according to local cultural and physical characteristics. Clustered rural settlements are often arranged in one of two types of patterns: circular and linear:

- *Circular rural settlements* have a central open space surrounded by structures. The central space is used as a common grazing land, protected enclosure, or site for public buildings, such as the church. Fields are arranged around the outer perimeter of the settlement. Examples of circular settlement patterns include the Kraal villages of Africa and the Rundling settlements of southern Germany (Figure 13–7).
- *Linear rural settlements* feature buildings clustered along a road, river, or dike to facili-

Most rural settlements in Africa, like this one in Côte d'Ivoire, are clustered. Houses and farm structures are built close to each other, and the fields and grazing land surround the settlement. (Marc & Evelyne Bernheim/Woodfin Camp & Associates)

tate communications. The fields extend behind the buildings in long, narrow strips. Linear rural settlements can be found in the portions of North America settled by the French. The French long-lot or "seigneurial" settlement pattern was commonly used along the St. Lawrence River in Québec and along the lower Mississippi River.

In the French long-lot system, houses were erected along a river, which was the principal water source and means of communication. Narrow lots from 5 to 100 kilometers deep (3 to 60 miles) were established perpendicular to the river, so that each original settler had river access. This created a linear settlement along the river. Eventually, these long, narrow lots were subdivided. French law required that each son inherit an equal portion of an estate, so the heirs established separate farms in each division. Roads were constructed parallel to the river for access to inland farms. In this way, a new linear settlement emerged along each road, parallel to the original riverfront settlement.

Dispersed rural settlements. In the past 200 years, dispersed rural settlements have become more common, especially in Anglo-America and the United Kingdom, because in relatively developed societies they are generally considered more efficient than clustered settlements. **Dispersed rural settlement** patterns are characterized by farmers living on isolated farms rather than together in villages.

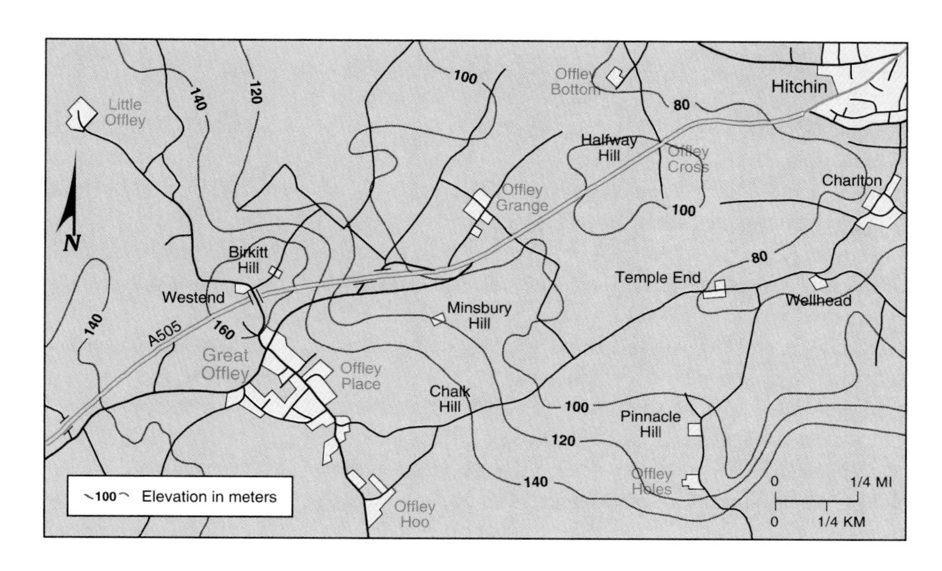

FIGURE 13–6
The rural landscape reflects the historical pattern of growth through establishment of satellite settlements. On the map, note the numerous places with "Offley" in their name: Great Offley (the largest and the original settlement), Little Offley, Offley Grange (barn), Offley Cross, Offley Bottom, Offley Place, Offleyhoo (house), and Offley Holes. These are satellite rural settlements in the parish of Offley, in Hertfordshire, England.

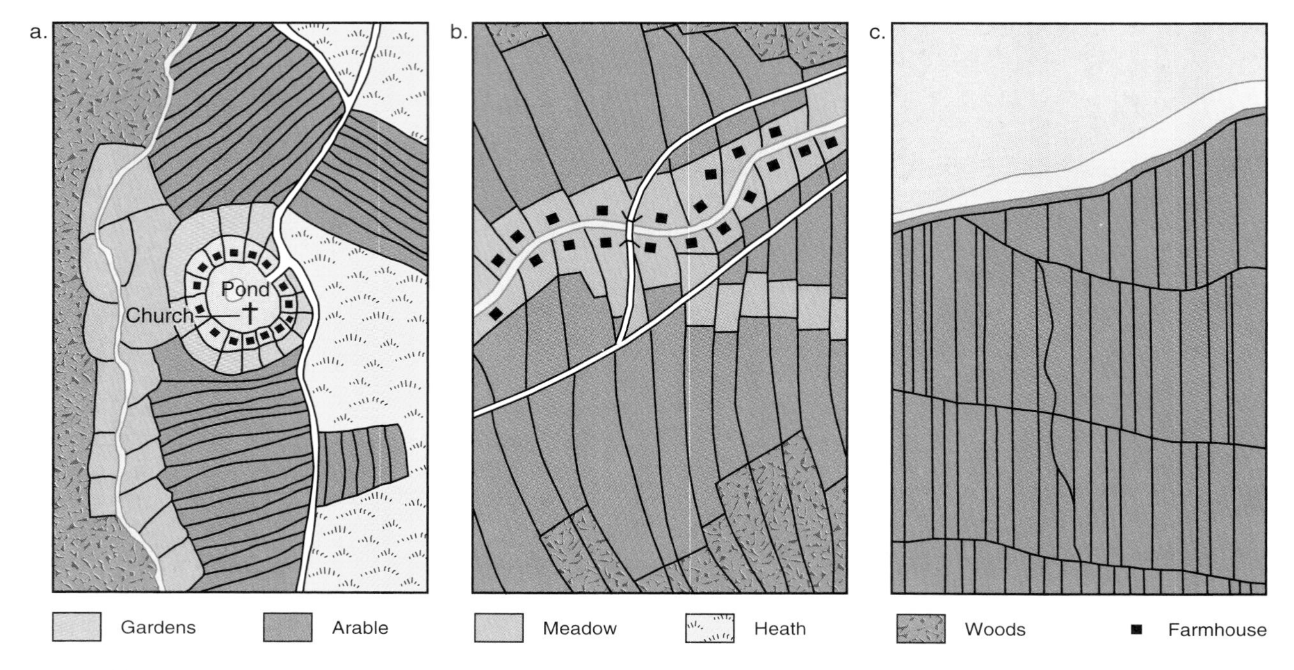

FIGURE 13–7
Two distinctive rural settlement patterns. At left is the circular arrangement common in Germany. At center is a linear arrangement called "long-lot," used in France, which gives everyone access to the river. When French settlers came to North America, the long-lot system came with them, as shown at right.

A number of European countries converted their rural landscapes from clustered to dispersed patterns to improve agricultural production. A prominent example was the **enclosure movement** in Great Britain between 1750 and 1850. The British government transformed the rural landscape by consolidating individually owned strips of land surrounding a village into a large farm owned by a single individual. When necessary, the government forced people to surrender their former holdings.

The benefit of enclosure was greater agricultural efficiency, because a farmer did not have to waste time scurrying between discontinuous fields. With the introduction of farm machinery, farms operated more efficiently at a larger scale. Because the enclosure movement coincided with the industrial revolution, villagers who were displaced from farming moved to urban settlements and became workers in factories and services.

The enclosure movement brought greater agricultural efficiency, but it destroyed the self-contained world of village life. Village population declined drastically as displaced farmers moved to urban settlements. Some villages became the centers of the new, larger farms, but villages that were not centrally located to a new farm's extensive land holdings were abandoned and replaced with entirely new farmsteads at more strategic locations. As a result, the isolated, dispersed farmstead, unknown in medieval England, is now a common feature of that country's rural landscape.

The first European colonists in what would become the United States settled the East Coast in three regions: New England, the Southeast, and the Middle Atlantic. The colonists in each area came from different places in Europe, and for different reasons. Each brought their distinctive religion, language, political view, and individual farming experience. These backgrounds resulted in a variety of colonial rural settlement patterns, which we will now examine.

New England rural settlements. The rural landscape in New England reflects its settlement by groups who left England in the seventeenth century. They left primarily to gain religious freedom or for other cultural reasons. Typically, a group was granted land of 4 to 10 square miles (10 to 25 square kilometers) by the English government and then traveled to America to settle the land. In each group, most came from the same English village and belonged to the same church. Once in the colony, members of the group stayed close to each other to reinforce their common cultural and religious values, and for protection.

Such groups simulated the arrangement they knew in England: a clustered settlement, which they built near the center of the land grant. The village center usually had an open area called the *common*. Settlers grouped their homes and public buildings, such as the church and school, around the common. In addition to their houses, each settler had a home lot of 1 to 5 acres (0.4 to 2 hectares), which contained a barn, garden, and enclosures for feeding livestock.

This clustered pattern also was encouraged by the central role of the church in daily activities. The settlement's leader often was an official of the Puritan church. Land was not sold, but rather awarded to an individual after the town's residents felt confident that the recipient would work hard. Outsiders could obtain land in the settlement only through permission of the town's residents. Colonists also favored clustered settlements for defense against Indian attacks.

The diverse topography, soil, and drainage of the New England landscape caused a complex land division among the colonists. To acquire the variety of land types needed for different crops, each villager owned several discontinuous parcels outside the village. Beyond the fields, the town held pasture and woodland for the common use of all residents.

The clustered village system was appropriate for the original small, stable groups. But by the eighteenth century, a more dispersed distribution began to replace the clustered settlements in New England, for two reasons: population increase and economic development. As population increased, through the excess of births over deaths and through net in-migration, the villages had no spare land to offer newcomers. The solution was to establish a new village nearby. As in the older settlements, the new village contained a central common surrounded by houses and public buildings, home lots, and outer fields. However, the shortage of land eventually forced new arrivals to strike out alone and claim farmland on the frontier.

At the same time, demand for more efficient agricultural practices led to a redistribution of farmland. People bought, sold, and exchanged land to create large, continuous holdings instead of several isolated pieces. The old system of discontinuous fields had several disadvantages: farmers lost time moving between fields; villagers had to build more roads to connect the small lots; and farmers had been restricted in what they could plant.

Descendants of the original settlers grew less interested in the religious and cultural values that had unified the original immigrants. They permitted people to buy land regardless of their religious affiliation. The cultural bonds that had created clustered rural settlements had weakened.

The New England landscape today contains remnants of the old clustered rural settlement pat-

New England rural settlements, such as Chelsea, Vermont, a village of 1,000 inhabitants, were originally organized around a common green and church. With population growth and modernization of agriculture, the village and surrounding land were turned over to individual private ownership. (John M. Roberts/Stock Montage, Inc.)

tern. Many New England towns still have a central common surrounded by the church, the school, and various houses. However, the contemporary New England town is little more than a picturesque shell of a clustered rural settlement, because today's residents work in factories, shops, and offices, rather than on farms.

Southeastern rural settlements. The southeastern colonies were first settled in the seventeenth century with small, dispersed farms. Then a different style emerged, called a plantation, a large farm that used many workers to produce tobacco and cotton for sale in Europe and the northern colonies. Plantations grew more profitable in the eighteenth century when the tobacco and cotton markets expanded and a large supply of labor was identified: indentured whites, who were legally bound to work for the plantation for a period of time, or black slaves forcibly transported from Africa and sold to the plantation owner.

The plantation's wealthy owner lived in a large mansion, frequently fronting on a body of water. Surrounding the mansion were service buildings, including a laundry, kitchen, dairy, and bakery. Other buildings on the estate included a flour mill, carpenter shop, stables, coach house, and living quarters for the slaves.

Middle Atlantic rural settlements. The Middle Atlantic colonies were settled by a more heterogeneous group of people, including immigrants from Germany, Holland, Ireland, Scotland, and Sweden, as well as from England. Further, most Middle Atlantic colonists came as individuals rather than as members of a cohesive religious or cultural group. They bought tracts of land from speculators or directly from individuals who had been given large land grants by the British government, such as William Penn (Pennsylvania), Lord Baltimore (Maryland), and Sir George Carteret (the Carolinas).

Inhabitants of the Middle Atlantic colonies were the main font for pioneers to the American West. They crossed the Appalachian Mountains and established dispersed farms on the frontier. Land was plentiful and cheap, and people bought as much as they could manage. As new agricultural practices favored larger farms, the settlement pattern in the American midwest became more dispersed.

Urban Settlements

Prior to modern times, virtually all settlements were rural rather than urban, because the economy was based on the agriculture of the surrounding fields. Shops and services met the needs of farmers living in the village. However, some urban settlements have existed for thousands of years, primarily as trading, administrative, or military centers.

Ancient and medieval urban settlements. Among the oldest well-documented urban settlements is Ur in Mesopotamia (present-day Iraq). Ur, which means fire, was the settlement that Abraham inhabited prior to his journey to Canaan in approximately 1900 B.C. Archaeologists have unearthed ruins in Ur that date from approximately 3000 B.C. (Figure 13–8).

Settlements also date from the beginning of documented history elsewhere in Mesopotamia and in Egypt, the Indus Valley, and China. Settlements may have developed independently in each of the four areas, or they may have diffused from Mesopotamia. In either case, from these four centers the concept of urban settlements diffused to the rest of the world.

Urban settlements in Europe. Settlements were first established in the Eastern Mediterranean region around 2500 B.C. The oldest settlements in this region include Knossos on the island of Crete, Troy in Asia Minor, and Mycenae in Greece. These settlements were trading centers for the inhabitants of the thousands of islands dotting the Aegean Sea and the eastern Mediterranean. Urban settlements diffused from the eastern Mediterranean toward the west during the eighth and seventh centuries B.C., when hundreds of new towns were founded to fill a gap in trading routes and to open new markets for goods.

Athens was probably the first city to attain a population of 100,000 (Figure 13–9). Rome grew to at least 250,000 inhabitants as the center of the Roman Empire's communications and trading networks. The fall of the Roman Empire in the fifth century A.D. brought a decline in urban settlements. With the empire fragmented into control by hundreds of rulers, trade decreased, and the need for urban settlements diminished.

Urban life revived in Europe beginning in the eleventh century. Feudal lords established new urban settlements and gave the residents charters of rights to establish the settlements as independent

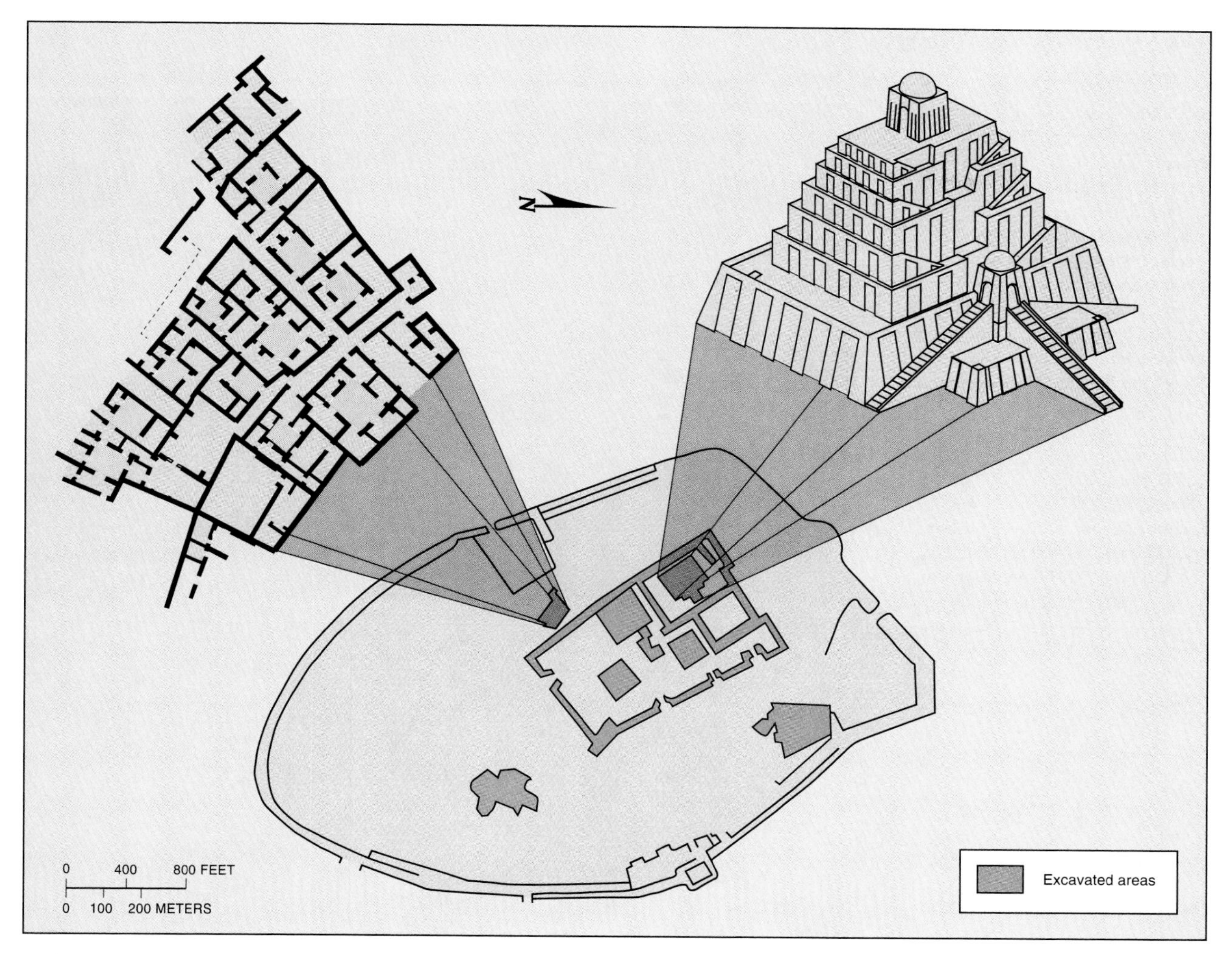

FIGURE 13–8
The remains of Ur, in present-day Iraq, provide evidence of early urban civilization. The most prominent building was the stepped temple at right, called a *ziggurat*. Surrounding the ziggurat was a dense network of small residences built around courtyards and opening onto narrow passageways. The excavation site was damaged during the 1991 war in the Persian Gulf.

cities. In exchange for the charter of rights, urban residents agreed to fight for the lord. Surplus from the countryside was brought into the city for sale or exchange, and markets were expanded through trade with other free cities (Figure 13–10).

From the collapse of the Roman Empire until the diffusion of the Industrial Revolution across Europe during the nineteenth century, most of the world's largest cities were located in Asia rather than Europe. Around A.D. 900, the five most populous cities are thought to have included Baghdad (in present-day Iraq), Constantinople (now called Istanbul, in Turkey), Kyoto (in Japan), and Changan and Hangchow (in China). Beijing (China) competed with Constantinople as the world's most populous city for several hundred years, until London claimed the distinction during the early 1800s. Agra (India), Cairo (Egypt), Canton (China), Isfahan (Iran), and Osaka (Japan) also ranked among the world's most populous cities prior to the Industrial Revolution.

Urbanization

Although a majority of Earth's people still reside in rural settlements, the percentage living in urban settlements is rapidly increasing and, at current growth rates, will constitute a majority within a few years (Table 13–1). This movement from rural to urban settlements reflects a shift in the way peo-

FIGURE 13–9
Dominating the skyline of modern Athens is the original hilltop site of the city, called the Acropolis. Ancient Greeks selected this high place to erect shrines to their gods because it was defensible. The most prominent structure on the Acropolis is the Parthenon, built in the fifth century B.C. to honor the goddess Athena. The structure to the left of the Parthenon, dating from the same time, is the Propylaea, the only opening in the wall surrounding the Acropolis. At the bottom of the hill is the Theater of Herodes Atticus, named for a wealthy Roman who built it as a memorial to his wife in A.D. 161. (David Pollack/The Stock Market)

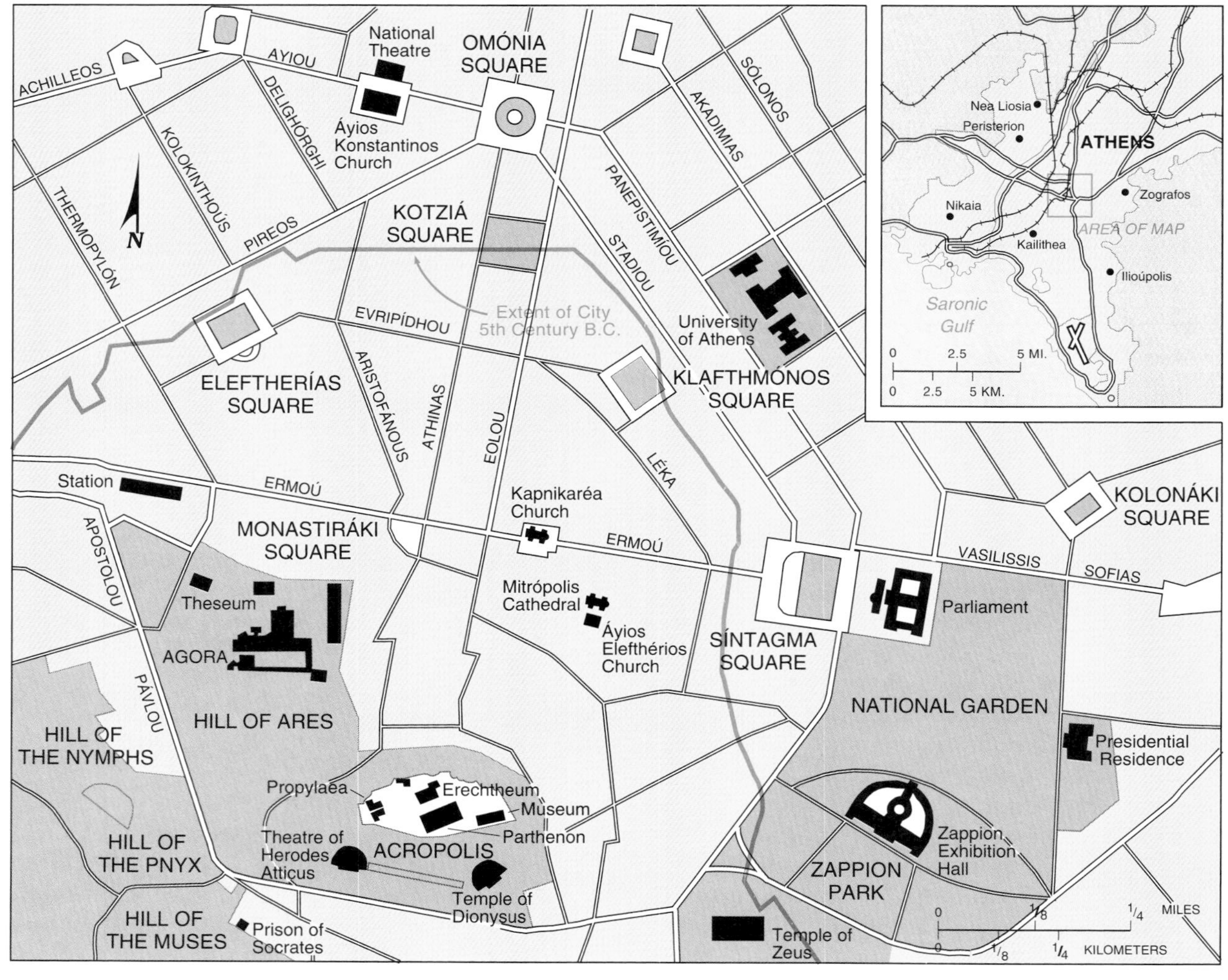

FIGURE 13–10
Modern Brugge (Bruges in French) is a town of more than 100,000 in the western part of Belgium, near the North Sea coast. Beginning in the twelfth century, Brugge was the most important port in northwestern Europe and a major center for manufacturing wool. However, three events forced the city's decline during the fifteenth century: foreign competitors captured much of the wool industry; the Belgian city of Antwerpen developed a better port; and the river Zwin silted, stranding the town 13 kilometers (8 miles) inland from the North Sea. Typical of medieval towns, the center of Brugge is dominated by squares surrounded by public buildings, churches, and markets.

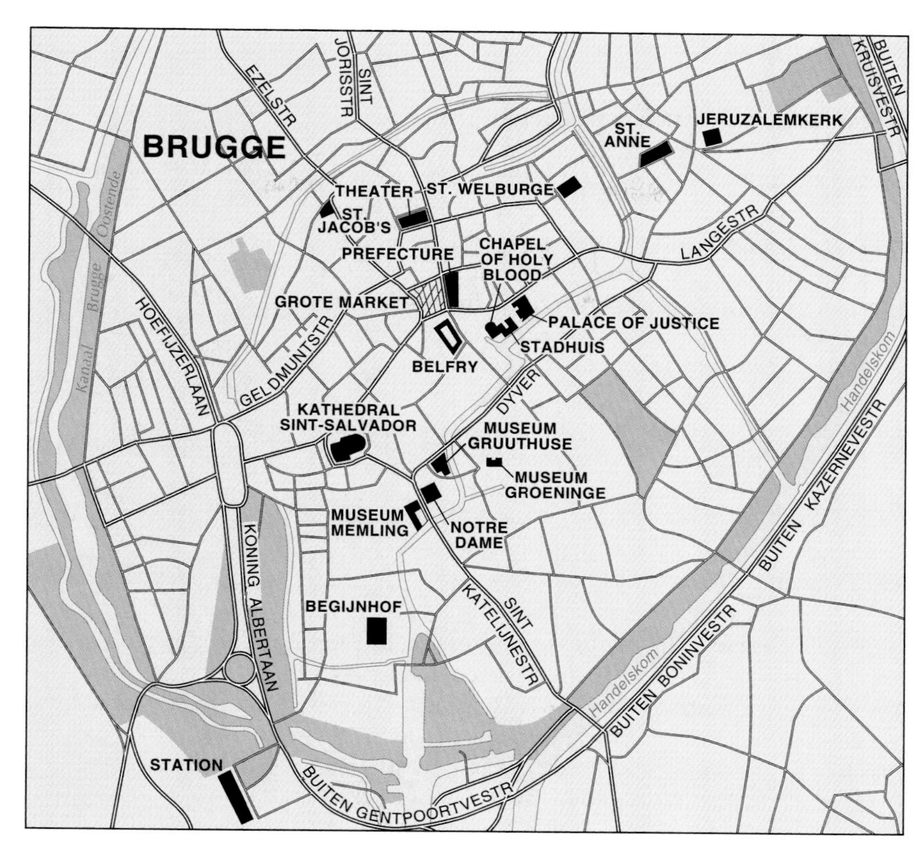

TABLE 13–1

Year	1800	1850	1900	1950	1994
Percentage living in cities	3	6	14	30	43

ple earn a living, from agriculture to manufacturing and services.

Process of Urbanization

Until modern times, urban settlements attracted only a small percentage of a society's total population, and rarely exceeded a few thousand inhabitants. In the past two centuries, relatively developed countries have been transformed from predominantly rural to predominantly urban societies.

The process of **urbanization** involves two changes in the distribution of a country's population: an increase in the *quantity* of people in urban settlements and an increase in the *proportion* of people there. The increase in *quantity* occurs simply because, as a country's population grows, some of the additional people inevitably will live in

(Benjamin Rondel/The Stock Market)

urban settlements. But the increase in the *proportion* of urban dwellers means a corresponding decrease in the proportion of rural residents, indicating a change in the society's economic structure. A higher proportion of urban residents means that most of the jobs are in factories, offices, and services, rather than on farms. Note the increasing proportion of city-dwellers over the past 200 years.

Traditionally, urbanization has been central to a society's development. Urban settlements grew in part because the state's overall population increased, and in part because people moved from the countryside to work in "city jobs." To some extent, this is still true, because the proportion of people in urban settlements is greater in relatively developed societies—nearly three-fourths, compared to only one-third in developing countries (Table 13–2).

But, in Table 13–3, note that seven of the ten most populous cities now are located in *developing* countries, rather than in industrialized societies. People are migrating from rural to urban areas because of very poor economic conditions in rural settlements, rather than because of realistic prospects for jobs in the cities. The rapid growth of urban areas in developing countries also partly reflects the overall world population increase. Thus, the rapid growth of cities in developing countries does not indicate improved economic development, as was the case historically in North America and Western Europe.

Annexation. Until recently in the United States, as cities grew, their legal boundaries were simply changed to include the new territory, a process called **annexation.** Normally, land can be annexed into a city if a majority of residents in the affected area vote to do so. In the nineteenth century, peripheral residents generally desired annexation because the city offered services: municipal water, sewage disposal, trash pickup, paved streets, public transportation, street lighting, police and fire protection, and libraries. Thus, while U.S. cities were growing rapidly in the nineteenth century, few definition problems arose, because the legal boundaries frequently changed to accommodate the newly developed areas.

Now, however, cities are less likely to annex peripheral land because the residents prefer to organize their own services rather than pay city taxes for them. As a result, U.S. cities in the twentieth century are surrounded by a collection of suburban jurisdictions, whose residents prefer to remain legally independent of the large city. Originally, some of these peripheral jurisdictions were small, isolated towns that decided to keep their independent local government as they became surrounded by urban growth. Others are newly created communities whose residents wish to live close to the large city, but not to be legally a part of it.

The number of local governments exceeds 1,400 in the New York City area, 1,100 in the Chicago area, and 20,000 throughout the United States. Approximately 40 percent of these 20,000 local governments are general local governments, and the remainder are special districts for schools, sanitation, transportation, water, and fire protection.

TABLE 13–2

Percentage of population that is urban, 1994.

Developing Regions		Relatively Developed Regions and Areas	
Latin America	71	Western Europe	82
Middle East	48	Japan	77
Southeast Asia	30	North America	75
South Asia	27	South Pacific	70
East Asia	26	Eastern Europe	67
Sub-Saharan Africa	24	South Africa	57

Source: Population Reference Bureau, *1994 World Population Data Sheet.*

TABLE 13–3

World's most populous urban areas.

1992 Rank	Urban Area	Country	Population (millions)	1950 Rank
1	Tokyo-Yokohama	Japan	25.8	3
2	São Paulo	Brazil	19.2	21
3	New York	U.S.A.	16.2	1
4	Mexico City	Mexico	15.3	13
5	Shanghai	China	14.1	6
6	Bombay	India	13.3	12
7	Los Angeles	U.S.A.	11.9	11
8	Buenos Aires	Argentina	11.8	7
9	Seoul	South Korea	11.6	45
10	Beijing	China	11.4	24

Source: 1950 rank from Kingsley Davis, *World Urbanization 1950–1970: Volume I.* 1992 rank from United Nations, *World Urban Agglomerations.*

U.S. urban definitions. The combination of rapid growth and political fragmentation has required new urban definitions. One now used in the United States is the **urbanized area,** which includes the largest city in the area—called the **central city**—plus its contiguous built-up suburbs where population density exceeds 1,000 persons per square mile (400 persons per square kilometer). Approximately 60 percent of the U.S. population live in urban areas, divided about equally between central cities and surrounding jurisdictions.

"Urbanized areas" have limited use in the United States because few statistics are available about them. Information about urbanized areas is difficult to obtain because data in most countries are published at the level of local government units, and urbanized areas do not correspond to these.

The urbanized area also has limited applicability because it does not accurately reflect the influence that an urban settlement has in contemporary society. A city's area of influence extends beyond its legal boundaries and adjacent built-up jurisdictions. Commuters may travel far to work and shop in the city or built-up suburbs. People in a wide area watch the city's television stations, read the city's newspapers, and support the city's sports teams and arts. Therefore, we need a better definition of an urban settlement to account for its more extensive zone of influence.

The U.S. Bureau of the Census has created a method of measuring the functional area of a city. It is the **metropolitan statistical area (MSA),** which includes a central city of at least 50,000 population, the city's surrounding county, and adjacent counties that are functionally tied to the central city according to indicators of community, density, and growth. Studies of metropolitan areas in the United States are usually based on information about MSAs, because many statistics are published for counties, and they are the basic building blocks of MSAs (see the St. Louis example in Figure 13–11).

MSAs have two geographical problems, however. First, they include a considerable amount of land that is not urban by most other definitions. For example, the Great Smokies National Park is located partly in the MSA of Knoxville, Tennessee, and part of Sequoia National Park is in the MSA of Visalia-Tulare-Porterville in California. Typically,

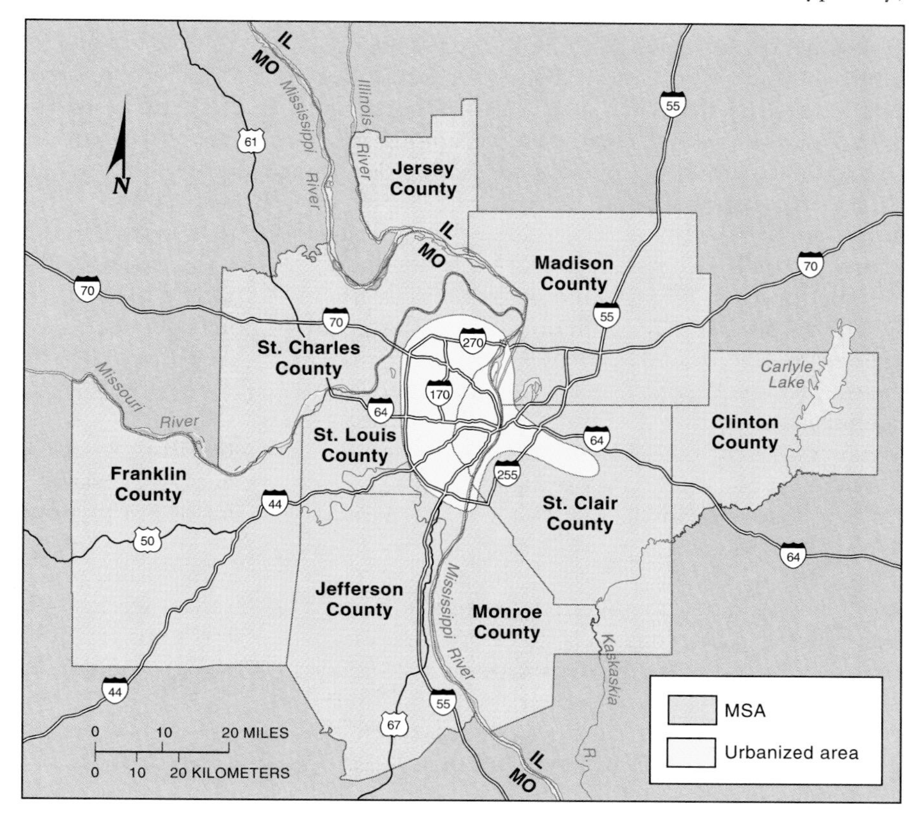

FIGURE 13–11
Surrounding the city of St. Louis is an urbanized area that spreads westward into St. Louis County and eastward across the Mississippi River into Illinois. The St. Louis Metropolitan Statistical Area (MSA) includes four Missouri counties and five in Illinois, as well as the city of St. Louis. The situation of St. Louis makes it a diversified trade center, for it is at the confluence of the Missouri and Mississippi Rivers and several federal highways.

the urbanized area occupies only 10 percent of the MSA's land area, but includes over 75 percent of its population.

The second geographical problem is overlapping of adjacent MSAs. A county located between two central cities may send a large number of commuters to jobs in each. If two adjacent metropolitan statistical areas have overlapping commuting patterns, they may be combined into a **consolidated metropolitan statistical area (CMSA).** CMSAs include New York–Northern New Jersey–Long Island, Los Angeles–Anaheim–Riverside, and Chicago–Gary–Lake County.

Daily urban systems. Virtually all residents of relatively developed countries are functionally tied to a metropolitan area. In the northeastern United States, large metropolitan areas are situated so close together that they form one continuous urban complex extending from north of Boston to south of Washington. In 1961, the geographer Jean Gottmann named this region *Megalopolis,* a Greek word meaning great city; others have called it the *Boswash* corridor.

Within *Megalopolis,* the downtown areas of individual cities, such as Baltimore, New York, and Philadelphia, retain their distinctive identities. Within each metropolitan area, sharp physical and social differences are clearly visible among inner city, suburban, and peripheral neighborhoods. The cities are visibly separated from each other by open space used as parks, military bases, and dairy or truck farms. But the boundaries among the metropolitan areas within *Megalopolis* overlap. Washingtonians attend major league baseball games in Baltimore, Baltimoreans attend major league football games in Washington, and both attend major league hockey and basketball games in an arena situated between the two cities.

Other continuous urban complexes exist in the United States, including the southern Great Lakes between Chicago and Milwaukee on the west and Pittsburgh on the east, and in southern California from Los Angeles to Tijuana. Among important examples in other relatively developed countries are the German Ruhr (including the cities of Dortmund, Düsseldorf, and Essen), Randstad in the Netherlands (including the cities of Amsterdam, the Hague, and Rotterdam), and Japan's Tokaido (including the cities of Tokyo and Yokohama).

Not all residents of relatively developed countries choose to work and shop in a metropolitan area, read a big-city newspaper, and watch the television programs emanating from the nearest metropolitan area. But at least the opportunity is available to virtually all residents in relatively developed societies. Recognizing that virtually every American has proximity to a major metropolitan area, the U.S. Department of Commerce divided the 48 contiguous states into 171 functional regions centered on commuting centers. These regions are called "daily urban systems" (Figure 13–12).

The Central City

Within urban settlements, two distinctive areas can be identified. The central city includes a central commercial core and inner residential neighborhoods. Surrounding the central city are suburbs containing residences, manufacturing, and commercial districts. Geographers identify patterns within these two areas and analyze the reasons for them.

The Central Business District

An urban settlement has a central city, and within this is a core area, known in the United States as the **central business district (CBD),** where retail and office activities are concentrated. The CBD is the best-known and most visually distinctive area of most cities. It is usually one of the oldest districts in a city and is often located on or near the original site of the settlement. The CBD is compact—less than 1 percent of the urban land area—but contains a large percentage of the urbanized area's shops, offices, and public institutions (Figure 13–13).

Retailing in the CBD. Retailers are attracted to the CBD because of its accessibility. The center is the easiest part of the city to reach from the rest of the region and is the focal point of the region's transportation network. Three types of retail activity generally concentrate in the CBD, because it is important for them to be accessible to everyone in the region:

1. Shops that require a large threshold (a large minimum number of customers). These have traditionally desired accessibility for a large

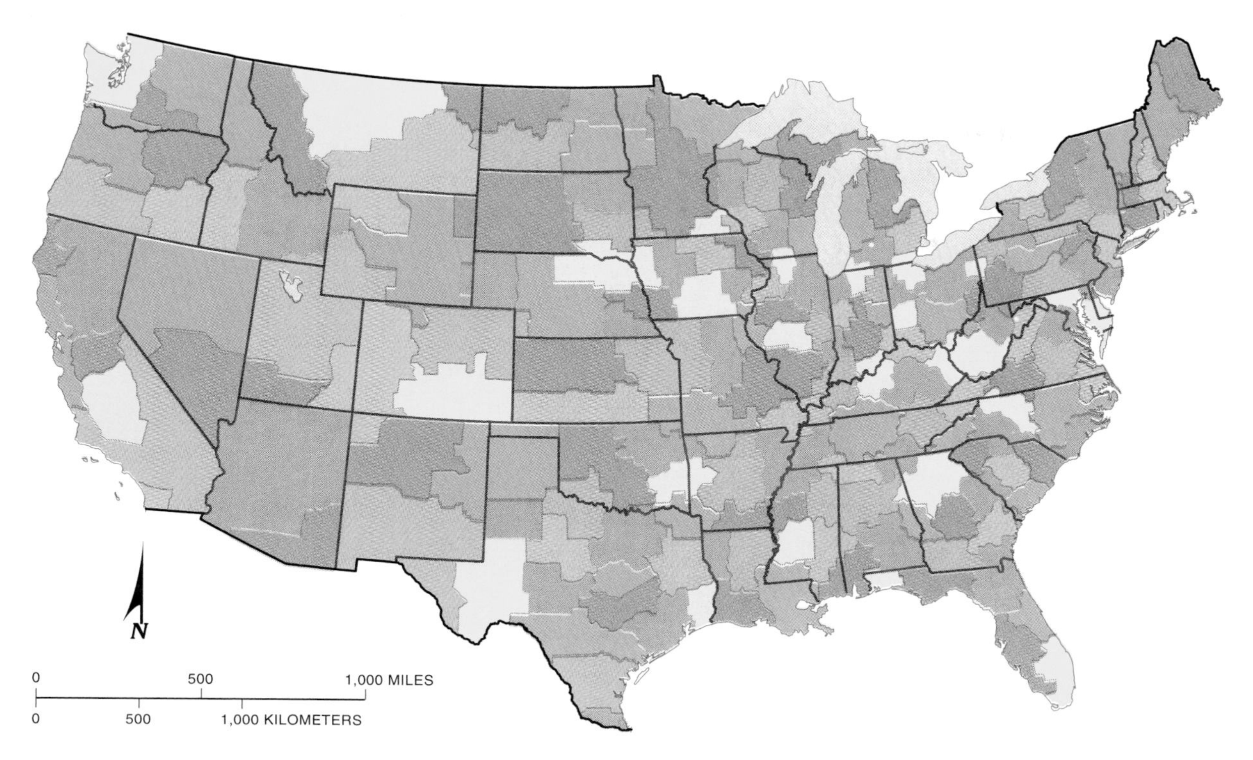

FIGURE 13–12
The U.S. Department of Commerce divided the 48 contiguous states into "daily urban systems." Regions are delineated by functional ties, especially commuting, to the nearest metropolitan area. Dividing the country into "daily urban systems" demonstrates that everyone in the United States has access to shops, jobs, and cultural activities in at least one metropolitan area.

number of customers. Examples include large furniture or department stores.

2. Shops that have a high range (a great maximum distance that customers are willing to travel to obtain the service). These include highly specialized stores such as expensive jewelers or clothing shops.
3. Shops that provide goods and services needed by people who work in the CBD and shop during lunch or working hours. These include office supplies, computers, photocopying, restaurants, shoe repair, film processing, and dry cleaning.

CBDs in cities outside of North America are more likely to contain supermarkets, bakeries, butchers, and other food stores. Many of their customers have difficulty buying groceries closer to home because these stores may have limited hours in the evenings or on weekends. The twenty-four-hour supermarket is rare outside of North America, because of shopkeepers preferences, government regulations, and long-time shopping habits.

Many shops with high ranges and thresholds have moved to suburban shopping malls for better space and accessibility. In contrast to these retailers, shops that appeal to office workers are expanding in CBDs, in part because the number of downtown office workers has increased and partly because downtown offices require more services.

Offices in the CBD. Offices also cluster in the central business district because of accessibility. Officials in the professions of advertising, banking, finance, journalism, and law depend on proximity to their professional colleagues. Lawyers, for example, locate downtown to be near government offices and courts. Services such as temporary secretarial agencies and quickprint/copy services locate downtown to be near lawyers, forming a chain of interdependency that continues to draw offices to the CBD.

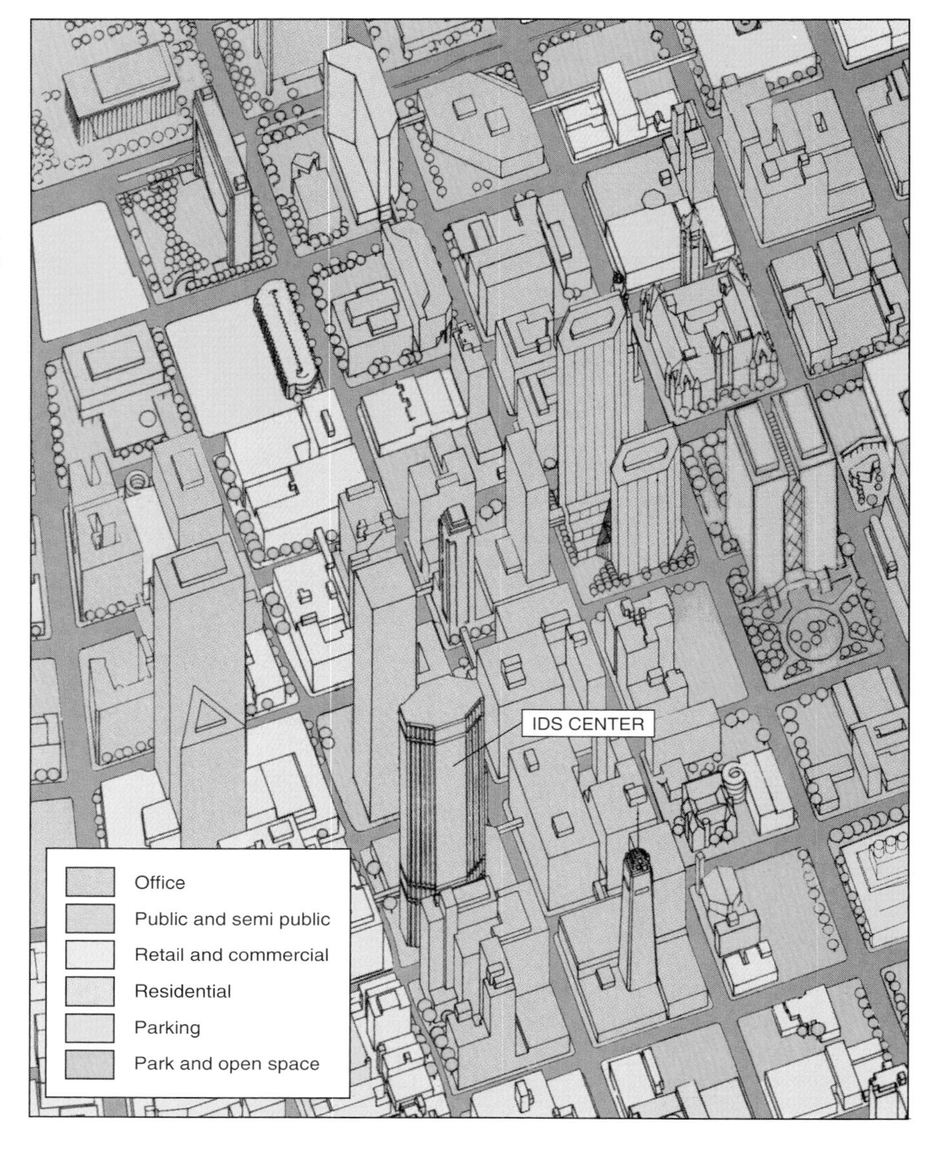

FIGURE 13–13
The central business district of Minneapolis, Minnesota, is dominated by retail and office buildings. Also clustered in the downtown area are a number of public and semipublic buildings, such as City Hall, government office buildings, and the central post office. The city's largest stadium, the Hubert H. Humphrey Metrodome, is located a few blocks to the east. The Target Center Arena, in the foreground of the photograph, is a few blocks to the west of the map. The light-colored tower to the right is the IDS Center.
(Adapted from University of Minnesota, Cartography Laboratory, Department of Geography, 1986)

Despite the diffusion of telecommunications, many professionals still exchange information primarily through face-to-face contact. Financial analysts exchange information about attractive stocks or impending corporate takeover attempts. Lawyers meet with each other to settle their clients' disputes out-of-court. Offices are centrally located to facilitate rapid communication of fast-breaking news through spatial proximity. Face-to-face contact also helps to establish a relationship of trust based on shared professional values.

A central location is also helpful for offices that employ workers who live in a variety of neighborhoods. The top executives may live in one neighborhood, the junior executives in another, the secretaries in another, and the custodians in another. Only a central location is readily accessible to all groups. Firms that need highly specialized employees are more likely to find new talent in the central area, perhaps currently working for another company located downtown.

Consequences of high accessibility. The accessibility of the center produces extremely strong competition for the limited sites available. As a result, the value of land in the CBD is very high. A hectare may cost several thousand dollars in a rural area and tens of thousands in a suburb. Even if a hectare of land were available in a large CBD such as New York or London, it would cost several hundred million dollars. In the CBD of Tokyo, land transactions have exceeded $1 billion per acre; if this page were a land parcel in Tokyo, it would sell for over $12,000!

Two distinctive characteristics of the central city follow from these very high land values. First, land is used more intensively in the center than elsewhere in the city. Second, some activities are excluded from the center because of the high cost of space.

The intensive demand for space has given the central city a three-dimensional character, adding the vertical dimension, both up and down. Compared to other parts of the city, the CBD makes extensive use of space below and above ground level. The typical "underground city" includes multistory parking garages, loading docks for deliveries to offices and shops, and infrastructure, such as water and sewer lines and wiring for electricity and communications. Subways run under the streets of larger central cities, and cities in colder climates have underground pedestrian passages and shops.

The demand for space in the central city has also made high-rise structures economically feasible. Downtown skyscrapers give a city one of its most important visual images and unifying symbols. The nature of an activity influences which floor a business occupies in a typical high-rise. Some shop owners may demand street-level locations to entice the most customers and are willing to pay higher rent for this prime space. Professional offices are less dependent on walk-in trade and can occupy the higher levels at lower rents. Hotel rooms and apartments may be included in the upper floors of a skyscraper to take advantage of less noise and the panoramic views.

An interesting exception to the vertical use of space is Washington, DC. It is the one large U.S. CBD without skyscrapers, because no building is permitted to be higher than the U.S. Capitol dome. This limits downtown Washington buildings to thirteen stories. As a result, the typical office building in Washington D.C. uses more land than in other cities, and the city's CBD has spread out horizontally, instead of growing vertically.

Land uses excluded from the CBD. High rents and the shortage of land discourage two principal activities from concentrating in the central area: manufacturing and residential. The typical modern industry requires a large area to permit spreading operations over one-story buildings. Such land is more available and cheaper on the periphery of the urban area.

Few people live in the CBD, because offices and shops can afford to pay higher rents for the

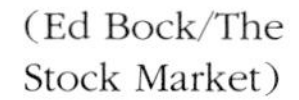
(Ed Bock/The Stock Market)

scarce space. Monthly rents in downtown high-rise office buildings are several dollars per square foot, versus less than $1 per square foot for a typical apartment. Also, a combination of push and pull factors are operating here. People have migrated away from central cities to escape the dirt, crime, congestion, and poverty of the central city. They have been lured to suburbs, which offer larger homes, private yards, and modern schools.

European CBDs. The central area is less dominated by commerce in Europe than in the United States. In addition to retail and office functions, many European cities have a legacy of low-rise structures and narrow streets, some built in medieval times, which today are protected from the intrusion of contemporary development. The most prominent structures may be churches and former royal palaces, situated on the most important public squares, at road junctions, or on hilltops. Many parks in the center of European cities were first laid out as private gardens for aristocratic families and later were opened to the public.

European cities have tried to preserve their historic core by limiting the number of cars and high-rise buildings. During the early 1970s, several high-rise offices were built in Paris, including Europe's tallest office building, the 210-meter (688-foot) Tour Montparnasse. The public outcry over the damage to the city's historic skyline was so great that officials reenacted lower maximum height restrictions. Officials in Rome periodically try to ban all private automobiles from the center of the city, because they cause pollution and congestion, and damage ancient monuments by contributing to acid precipitation.

Although constructing large new buildings is difficult, many shops and offices still wish to be in the center of European cities. The alternative to new construction is renovation of older buildings. However, renovation is more expensive and does not always produce enough space to meet the demand. As a result, rents are much higher in the center of European cities than in U.S. cities of comparable size.

Revitalization of CBDs. Downtowns that were generally considered dead and beyond help as recently as the 1970s are now filled with local residents and tourists, even during evenings and weekends, when offices are closed. U.S. CBDs experienced an unprecedented boom in new office construction during the 1980s. Entirely new downtown shopping malls have successfully attracted suburban shoppers as well as out-of-town tourists because they offer unique recreation and entertainment, not merely shops. Once-rotting downtown waterfronts have become major tourist attractions in North American cities like Boston, Toronto, Baltimore, and San Francisco, and in European cities like Barcelona and London.

The building boom in U.S. CBDs has created some problems, however. Skyscrapers block light, and high winds can be channeled through the deep

Quincy Market, along with the adjacent Faneuil Hall, in downtown Boston, was the city's market center in the eighteenth century. The renovated market buildings attract customers from throughout the Boston metropolitan area, as well as tourists. (Geri Engberg/The Stock Market)

canyons between buildings. Recently built skyscrapers in some CBDs are only partially occupied because of overbuilding by developers. In other CBDs, tenants have filled the new skyscrapers, leaving a high percentage of vacancies in the older ones.

The Zone in Transition

The atmosphere of animation and prosperity found in many CBDs does not extend to surrounding residential areas of the central city, known as the **zone in transition.** Except for a handful of renovated neighborhoods, the zone in transition in U.S. central cities is inhabited by numerous people who are trapped in an unending cycle of economic and social problems and who are unable to share in the revival of the CBDs. These inner-city residents are often called a permanent **underclass.**

Characteristics of the underclass. The underclass suffers from high rates of unemployment, alcoholism, drug addiction, illiteracy, juvenile delinquency, and crime. In the zone of transition, schools are deteriorated and affordable housing is increasingly difficult to find. Neighborhoods lack adequate police and fire protection, clinics, and shops.

The prospects are bleak for the underclass because they are increasingly unable to compete for jobs. Inner-city residents lack technical skills because fewer than half complete high school. The gap between the skills typically demanded by employers and the training of inner-city residents is growing. In the past, people with limited education could become factory workers or filing clerks, but today these jobs require knowledge of computing and electronics. Inner-city residents don't even have access to low-skilled jobs as fast-food servers or custodians, because these are increasingly located in the suburbs.

Despite the importance of education in obtaining jobs, many people in the underclass live in an atmosphere that does not emphasize the value of good study habits, such as regular school attendance and completion of homework assignments. The household may have only one parent, who often is forced to choose between working to generate income and staying at home to provide child care.

Trapped in such a hopeless environment, many inner-city residents turn to drugs. While drug use is a problem in many suburbs and rural areas as well, rates of use in recent years have increased most rapidly in the inner-cities. Some users resort to crime to obtain money needed for drugs. Gangs form in inner-city neighborhoods to control the lucrative distribution of drugs. Violence erupts when two gangs fight over the boundaries between their drug distribution areas.

An increasing number of the underclass are homeless. Several million Americans sleep in doorways, on warm street grates, and in bus and subway stations. Los Angeles County alone has an estimated 35,000 homeless individuals, attracted by the area's relatively mild climate. Homelessness is

Hundreds of thousands of homeless people sleep on sidewalks and in parks in U.S. cities. An especially large number are attracted to Sunbelt cities with warm climates, such as Phoenix, Arizona. (Thomas Ives/The Stock Market)

more extensive in some developing countries. In Calcutta, India, an estimated 300,000 people sleep, bathe, and eat on sidewalks and traffic islands.

Roughly one-third of the homeless in the United States are individuals who are unable to cope in society after being released from hospitals or other institutions. However, the other two-thirds are homeless because they are unable to afford housing and have no regular source of income. Homelessness may have its roots in family problems or loss of employment.

Racial segregation. Most inhabitants of inner-city residential areas in U.S. cities are low-income nonwhites. This pattern is largely a product of race relations. Many U.S. neighborhoods remain racially segregated, with African-Americans or Hispanics concentrated in one or two large continuous areas of the inner city, and whites living in the suburbs.

The zone in transition in many American cities includes extensive areas of vacant and abandoned housing, such as this area of Brooklyn, New York. The population of Brooklyn has declined by about 15 percent since 1950, as people move to neighborhoods and suburban communities farther from the zone of transition. (Geri Engberg/The Stock Market)

Within U.S. metropolitan areas, two-thirds of whites live in suburbs and one-third in central cities, while two-thirds of African-Americans live in central cities and one-third in suburbs. African-Americans comprise one-fourth of the U.S. central-city population, but 7 percent of the suburbs. Two-thirds of Latinos live in central cities.

As the number of low-income and minority families increases, the territory they occupy expands. Neighborhoods can change from predominantly white middle-class occupants to low-income nonwhites within a few years. Many of the large houses built by wealthy families in the nineteenth century are subdivided by absentee landlords into smaller dwellings for occupancy by low-income families. This subdivision of houses and occupancy by successive waves of lower-income people is called **filtering.** The ultimate result of filtering is abandonment of the dwelling, when the cost to maintain the building exceeds the rent that can be charged. Thousands of vacant and abandoned houses stand in the inner areas of American cities.

This deterioration in urban neighborhoods is aggravated by blockbusting and redlining. Real estate agents who practice **blockbusting** start by convincing white property owners to sell their houses at low prices, preying on their fear that nonwhite families will soon move into the neighborhood. The agent then sells or rents the houses to nonwhite families at a considerable profit. Some banks engage in **redlining,** so-named because they draw lines on a map and refuse to loan money for property within those boundaries. Families who wish to fix up houses in the inner city therefore have difficulty borrowing money. Although both blockbusting and redlining are illegal, enforcing laws against them is frequently difficult.

Public housing. In the zone in transition of European and North American cities, many old, substandard houses have been demolished and replaced with public housing. In the United States, **public housing** is government-owned and reserved for low-income households, which must pay 30 percent of their income for rent. Public housing accounts for only 2 percent of all dwellings in the United States, but it may total

Two high-rise public housing projects in the Clapton Park Estate, in the East End of London, England, are demolished simultaneously. The high-rise buildings (known in England as council tower blocks) will be replaced by town houses with gardens, a more suitable environment for families with children. (Gill Allen/AP/Wide World Photos)

more than 10 percent within the zone in transition of many cities. More than one-third of all housing is publicly owned in the United Kingdom, and the percentage is even higher in northern cities, such as Liverpool, Manchester, and Glasgow.

A number of the high-rise public housing projects built in the United States and Europe during the 1950s and early 1960s are now unsatisfactory for families with children. The elevators are frequently broken, juveniles terrorize others in hallways, and drug use and crime rates are high. Some observers claim that high-rise buildings caused the problem, because too many low-income families are concentrated into a high-density environment. The public housing authorities in several U.S. and European cities have demolished high-rise public housing projects in recent years because of poor conditions. Cities that have done so include Dallas, Newark, St. Louis, Liverpool, and Glasgow.

In recent years, the U.S. government has stopped providing funds to construct new public housing. Some federal support is available to renovate older buildings and to help low-income households pay their rent, but the overall level of funding is much lower today than in the late 1970s. As a result, the supply of public housing and other government-subsidized housing diminished by approximately 1 million dwelling units between the early 1980s and the early 1990s. During the same period, however, the number of households needing low-rent dwellings increased by more than 2 million.

Some neighborhoods in the zone of transition have been renovated by middle-class families who wish to live near the central business district in cities such as Boston. (Bill Buchman/Photo Researchers, Inc.)

Renovated Housing. In the zone of transition, some older neighborhoods never saw decline. They contained some of the city's social elite, who maintained an enclave of expensive property near the center of the city. In other cases, inner-city neighborhoods have only recently achieved high status as a result of renovation by the city and private investors. Some middle-class families are moving back into the zone in transition, a process called **gentrification.**

Gentrification may occur because middle-class families are attracted to older houses, which may be larger, better built, more visually appealing, and less expensive than suburban homes. Renovated homes in the zone in transition may also attract middle-class individuals who work downtown or use cultural and recreational opportunities clustered in the central area.

Suburbs

Although inner-city neighborhoods attract an increasing number of middle-class families, this movement is quite small compared to the process of suburbanization. The contemporary city is surrounded by increasingly extensive residential suburbs, made possible by transportation improvements. The suburbs are inhabited by families who wish to live at a lower density than is possible near the central area. To serve these suburban residents, commercial activities have expanded as well.

Suburban Land Uses

People are attracted to suburbs by the prospect of occupying more space than is available or afford-

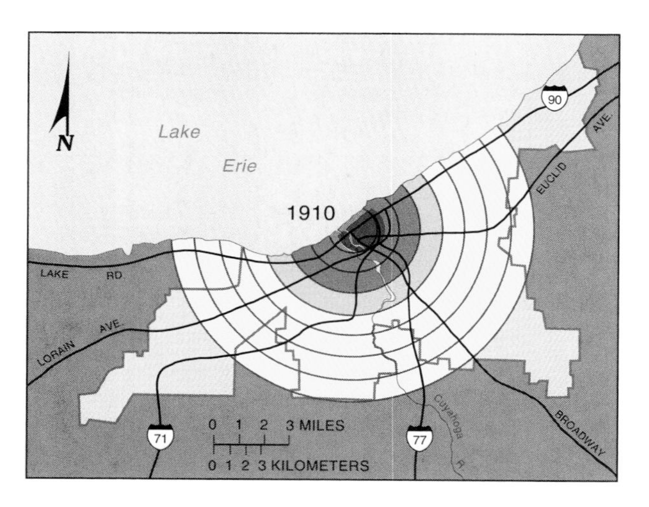

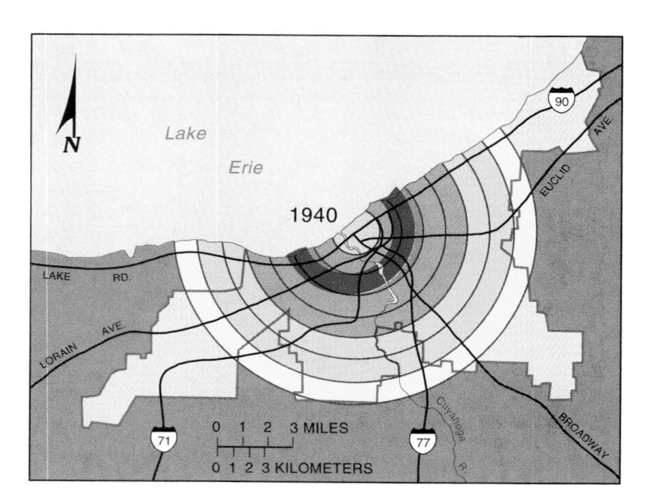

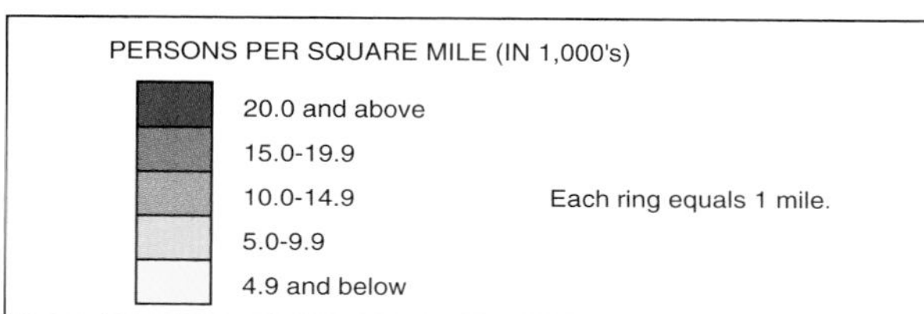

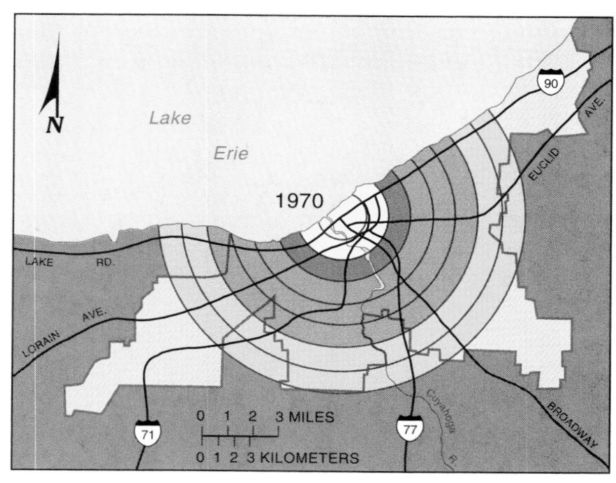

FIGURE 13–14
The density gradient has changed in Cleveland during the twentieth century. In 1910, population was highly clustered in and near the central business district. By 1970, population was distributed over a much larger area, the variation in densities were found in rings near the CBD. The maps show the present boundaries of the city of Cleveland. (Data from Avery M. Guest, "Population Suburbanization in American Metropolitan Areas, 1940–1970," *Geographical Analysis*, July 1975. Used by permission of the publisher.)

FOCUS BOX 13-1

Edge Cities: A New Cooperation with the Center City

Some U.S. suburbs are now called **edge cities.** These are large nodes of office and retail activity on the urban periphery, spread out in ribbons along boulevards and interstate highways at the edge of a metropolitan area. Typically, they are built on land that was farmed only a couple of decades ago. Edge cities originated as suburban communities for central city workers. To provide services to these new residents, shopping malls were built. Edge cities are a new stage in urban development, because they are nodes where offices, factories, and warehouses, not just residences and shops, are clustered.

An edge city has become both a residential area that people travel from to work and an employment center that attracts people to work. So many new jobs have been attracted to edge cities that a major obstacle to further expansion is a shortage of qualified workers. Ironically, many unemployed inner-city residents lack transportation to reach these jobs in edge cities, and the skills to hold such jobs.

For years, many suburbanites have held negative feelings toward their central cities. These are reflected in their unwillingness to venture into the center city, citing crime, traffic, or that the city has nothing to offer that cannot be obtained in the suburb. This lack of concern for the condition of central cities was captured in a *New York Daily News* headline in 1976 when then-President Gerald Ford refused to offer aid to financially strapped New York City: "Ford to New York: Drop Dead!"

Edge cities are a more positive expression of the competition that exists between cities and their suburbs. Confrontation is being replaced with cooperation. Politicians increasingly recognize that the economic health of their communities is based on a strong regional economy and in shared resources, such as airports, stadiums, and recreational areas.

able in the center city. The density at which people live generally declines with increasing distance from the center city, a phenomenon called the **density gradient.** This can be seen in the upper left portion of Figure 13–14.

However, two factors have altered the density gradient in recent years. First, the number of people living in the center itself has decreased. The density gradient thus has a gap in the center, where few people live. Second is the trend toward less density difference within urban areas. These effects can be seen in the lower right portion of Figure 13–14. The number of people living on a hectare of land has decreased in the central residential areas through population decline and abandonment of old housing. At the same time, density has increased on the periphery through construction of apartment and row house projects and diffusion of suburbs across a larger area.

As long as demand for separate single-family homes remains high, land on the fringe of urbanized areas must be converted from open space to residential land use. The current system for developing land on urban fringes is inefficient, especially in the United States. Land is not transformed systematically from farms to housing developments. Instead, developers buy farms for future construction of houses by individual builders. Developers frequently reject land adjacent to built-up areas in favor of detached isolated sites, which may be cheaper to buy and more physically attractive. The rural-urban fringe in U.S. cities therefore looks like Swiss cheese, with pockets of development and gaps of open space.

Such sprawling low-density suburbs creates inefficiencies:

- They require roads, water, and sewer lines, which are expensive and taxpayer-funded.
- They nibble away farmland, reducing availability of local produce.
- They waste energy, because automobile trips become necessary to do everything.

In Europe, the land supply along urban peripheries is more strictly regulated. This avoids sprawl, but has driven up housing prices. Restrictions on new building sites in the United states could further decrease the ability of low-income families to find affordable housing.

Growth in the number of people living in suburbs has stimulated nonresidential construction as well. Suburban shops have been built to serve the needs of nearby residents. Industry has moved to the suburbs to find space for expansion. Many offices have relocated in suburbs to take advantage of lower land costs (Figure 13–15).

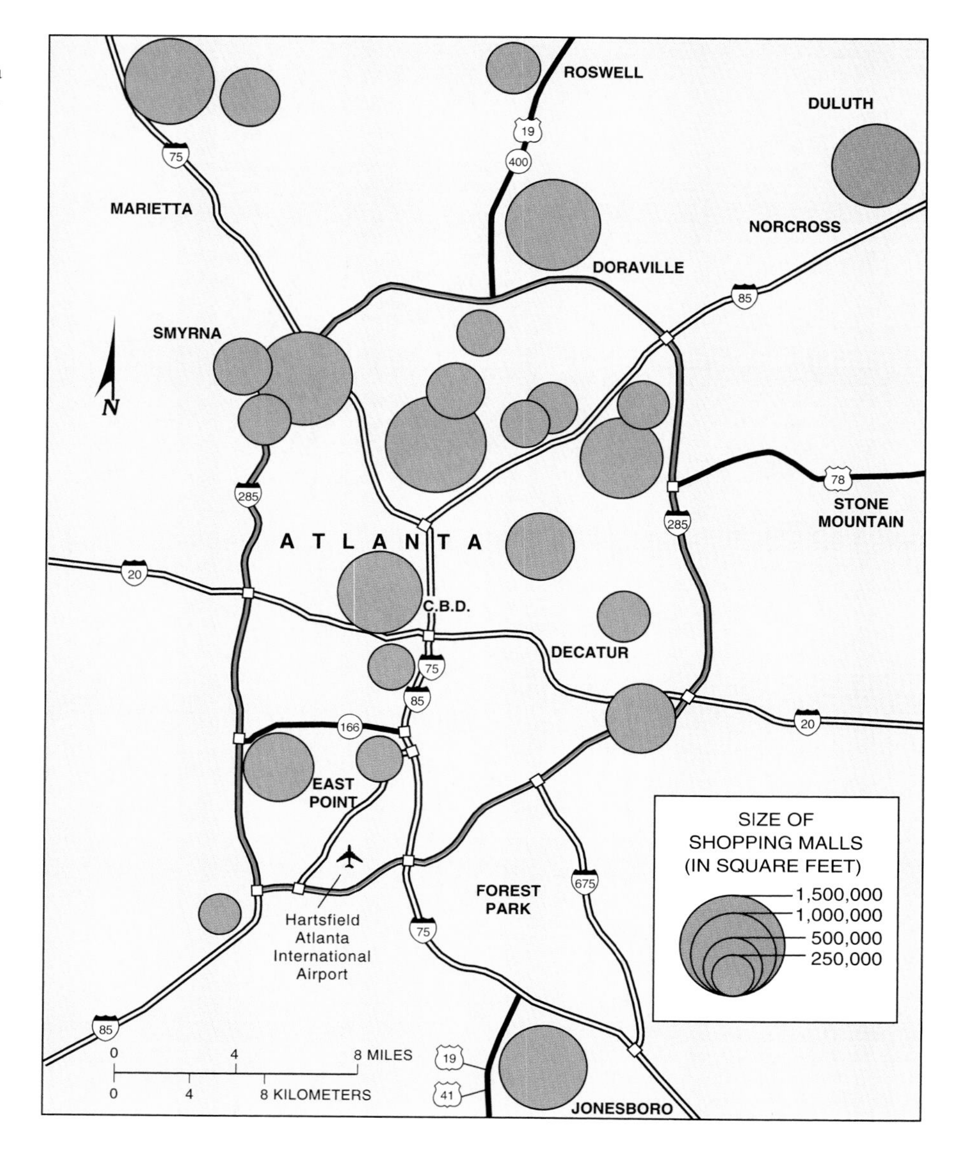

FIGURE 13–15
Most shopping malls in the Atlanta metropolitan area, as elsewhere in North America, are located in the suburbs rather than in the inner city. The optimal location for a large shopping mall is near an interchange on the interstate highway that encircles many American cities, such as I-285 around Atlanta.

In very large cities, such as Tokyo, subways and other forms of public transportation move large numbers of people into the CBD in the morning and home again in the evening. Attendants are hired in Tokyo to jam as many passengers as possible into the trains, and to make certain that the doors can close. (Ken Straiton/The Stock Market)

Urban Transportation

Because few people live in the center, urban areas are characterized by high levels of commuting into the CBD in the morning and out in the evening. The intense concentration of people in the center during working hours strains the transportation system, because a large number of people must reach a small area of land at the same time in the morning and disperse at the same time in the afternoon. The four consecutive 15-minute periods that have the heaviest traffic are called the **rush hour.**

In larger cities, public transportation may carry a substantial number of people during these peak travel hours. Such transportation is better suited than cars to the movement of large numbers of people, because each traveler takes up far less space, and is moved at the same speed. However, commuters prefer the independence of traveling by car. One-third of the high-priced central land is devoted to streets and parking lots, and multistory and underground garages also are constructed.

Historically, the growth of suburbs was constrained by transportation problems. People lived in crowded cities because they had to be within walking distance of shops and places of employment. The invention of the railroad in the nineteenth century enabled people to live in suburbs and work in the central city. Cities then built street railways (called trolleys, streetcars, or trams) and underground railways (subways) to accommodate commuters. In the nineteenth century, rail and trolley lines restricted suburban development to narrow ribbons within walking distance of the stations.

The suburban explosion in the twentieth century has relied on automobiles rather than railroads, especially in the United States. Automobiles have permitted large-scale development of suburbs at greater distances from the center, in the gaps between rail lines. Automobile drivers have much greater flexibility in the choice of residence than was ever before possible.

Public transportation is particularly suited to bringing a large number of people into a small area in a short period of time. It is cheaper, less polluting, and more energy-efficient than the automobile. Also, using an automobile has costs beyond those of ownership and operation: delays imposed on others, increased need for highway maintenance, construction of new highways, and pollution. Most people overlook these indirect costs because the car offers privacy and flexibility of schedule. The use of public transportation is generally confined in the United States to rush-hour commuting by workers to and from the CBD. Some cities have constructed new subways and light rail lines to encourage suburban commuters to leave their cars at home.

Ironically, people who are too poor to own an automobile may still be unable to reach their employment by public transportation. Low-income people often live in inner-city neighborhoods, but the job opportunities—especially those requiring minimal training and skill—are located in suburban areas not well-served by public transportation. Inner-city neighborhoods have high unemployment rates at the same time that suburban firms have difficulty attracting workers. In some cities, governments and employers subsidize vans to carry low-income inner-city residents to suburban jobs.

Three Models of Urban Structure

Geographers use three models to help explain the distribution of different social groups within urban areas—the concentric zone, sector, and multiple nuclei models. They were developed in Chicago, a city located on a flat prairie, with few physical features, other than Lake Michigan, to interrupt the region's growth. The three models were later applied to cities elsewhere in the United States and in other countries.

- The **concentric zone model,** created in 1923 by sociologist E. W. Burgess, shows that a city grows outward from a central area in concentric rings, like the growth rings of a tree. The size and width of the rings vary from one city to another, but the same basic

FIGURE 13–16
According to the concentric zone model, a city grows in a series of rings surrounding the central business district.

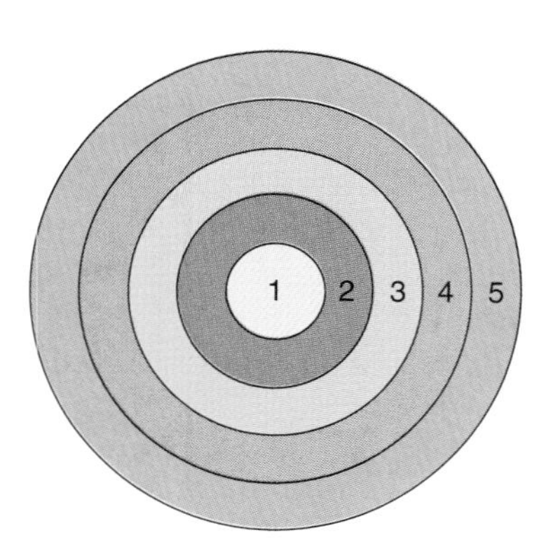

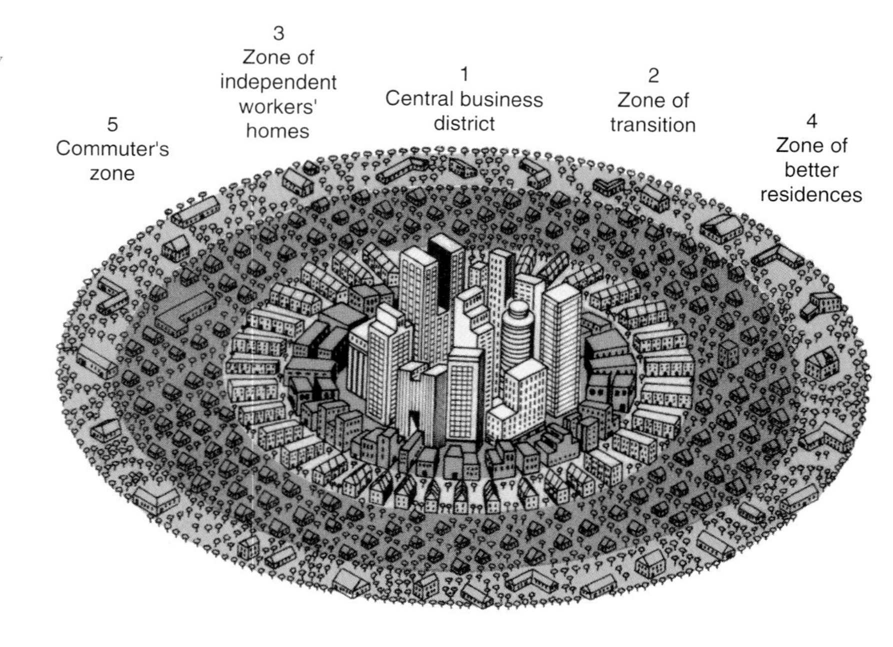

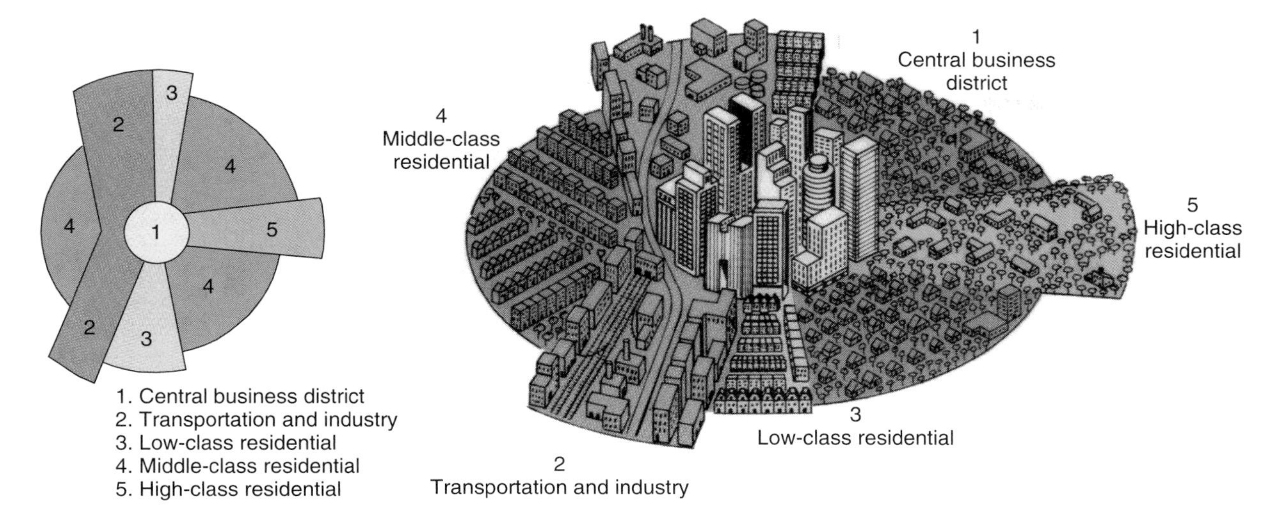

FIGURE 13–17
According to the sector model, a city grows in a series of wedges or corridors, which extend outward from the central business district.

rings appear in all cities, in the same order (Figure 13–16).

- The **sector model,** developed in 1939 by land economist Homer Hoyt, shows that a city develops in sectors, not rings. Certain areas of a city are more attractive for certain activities, because of an environmental factor or even by chance. As a city grows, activities expand outward in a sector, or wedge, from the center. Once a district with high-class housing is established, the most expensive new housing is built on the outer edge of that district, extending the sector further from the center. The best housing is therefore found in a corridor extending from downtown to the outer edge of the city. Industrial and retailing activities develop similarly in other sectors, usually along good transportation lines (Figure 13–17).
- The **multiple nuclei model,** developed in 1945 by geographers C. D. Harris and E. L. Ullman, shows that a city is a complex structure. It includes more than one center, or node, around which activities revolve. Examples of such nodes include a port, neighborhood business center, university, airport, and park. Some activities are attracted to particular nodes while others avoid them (Figure 13–18).

Applications of the models. All three models help us understand where people with different social characteristics live within an urban area, and why. However, none of the three models by itself completely explains why different social groups live in distinctive parts of the city. Critics point out that the models are too simple and fail to consider the variety of reasons that lead people to select particular residential locations. Because the three models are all based on conditions that existed in U.S. cities between the two world wars, critics also question their relevance to contemporary urban patterns in the United States or in other countries.

If the models are combined, however, they do help geographers explain where different types of people live in a city. People's choice of living location depends on their personal characteristics. This does not mean that everyone with the same characteristics will live in the same neighborhood, but the models say that most people prefer residing near others who have similar characteristics.

- Figure 13–19, using the concentric zone model, suggests that, if we compare two households with the same income and ethnic background, but one contains a married couple and the other does not, we are much more likely to find the married household liv-

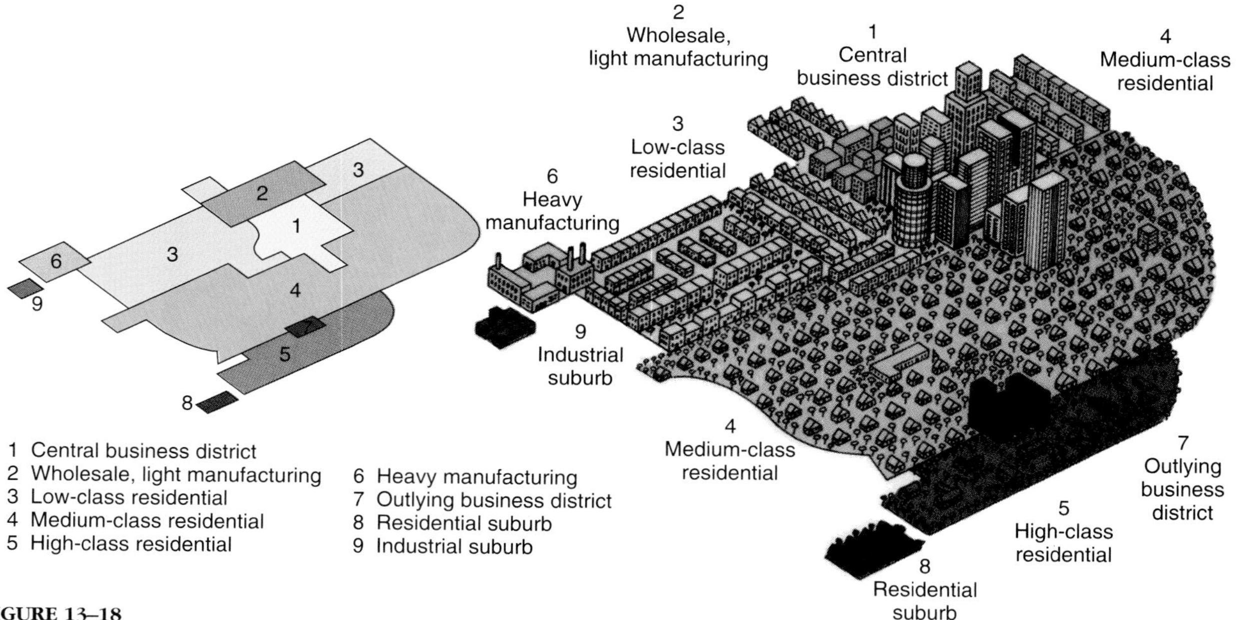

FIGURE 13–18
According to the multiple nuclei theory, a city consists of a collection of individual nodes or centers, around which different types of people and activities cluster.

ing in an outer ring and the unmarried one in an inner ring.

- Figure 13–20, using the sector model, suggests that two families of the same age and same number of children will live in different sectors based on income, with the higher-income family living in a different sector form the poorer one.
- Figure 13–21, using the multiple nuclei theory, suggests that people of the same ethnic or racial background want to live near each other.

Putting the three models together, we can identify, for example, the neighborhood in which a childless, wealthy, Asian-American family is most likely to live.

Use of the models outside North America. The three models may describe the spatial distribution of social classes in the United States, but what about cities elsewhere in the world? The spatial distribution of social characteristics elsewhere is different. These differences do not invalidate the models, but point out that social groups in other countries may have different reasons for selecting particular neighborhoods.

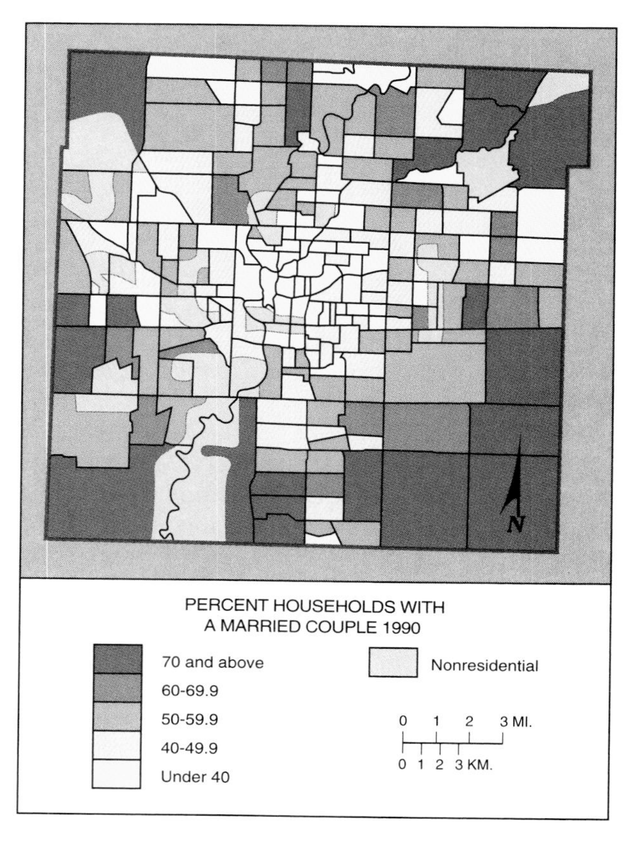

FIGURE 13–19
The distribution of married couples follows the concentric zone model in Indianapolis. The percentage of households with a married couple is lower near the central business district and higher in the outer rings of the city.

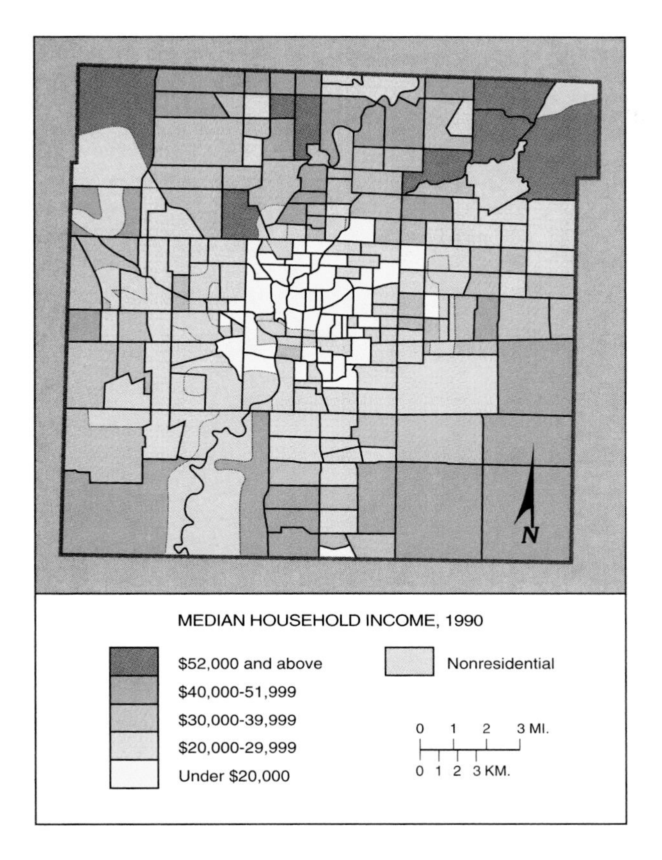

FIGURE 13–20
The distribution of high-income households follows the sector model in Indianapolis. The median household income is the highest in a sector to the north, which extends beyond the city limits into the adjacent county.

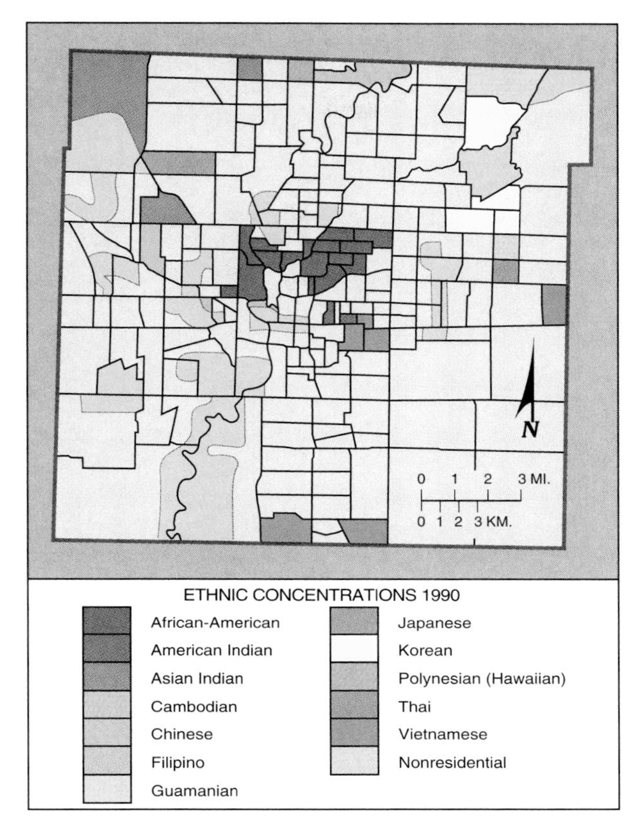

FIGURE 13–21
The distribution of minorities follows the multiple nuclei model in Indianapolis. The African-American concentration consists of census tracts that are 90 percent or more African-American. The other groups are clustered in tracts that contain at least 5 percent of the total population of that ethnic group in Indianapolis.

In European cities, wealthier people often live in townhouses and apartments near the center, while poorer people are relegated to the outskirts (Figure 13–22). A central location offers the region's best shops, restaurants, cafes, and cultural facilities. Wealthy people are also attracted by the opportunity to occupy elegant residences in carefully restored, beautiful old buildings.

Poorer people are housed in high-rise apartment buildings in the suburbs and face the prospect of long commutes by public transportation to reach jobs and other urban amenities. High-density apartment buildings for poorer people are less expensive to build in Europe than in the United States and help to preserve the countryside from development.

The rich also live in the center of cities in developing countries, for reasons similar to those in Europe (Figure 13–23). A central location is also attractive because basic utilities are more available and reliable (Figure 13–24). As in Europe, the poor live in the suburbs, but developing countries are unable to provide enough housing for the burgeoning poor.

Due to the housing shortage, many poor immigrants to urban areas in developing countries live in **squatter settlements.** These are known by a variety of local names, including *barrios, barriadas,* and *favelas* in Latin America, *bidonvilles* in North Africa, *bustees* in India, *gecekondu* in Turkey, *kampongs* in Malaysia, and *barung-barong* in the Philippines. Typically, a squatter settlement is initiated by a group that moves together, literally overnight, onto land outside the city, owned either by a private individual or more often by the gov-

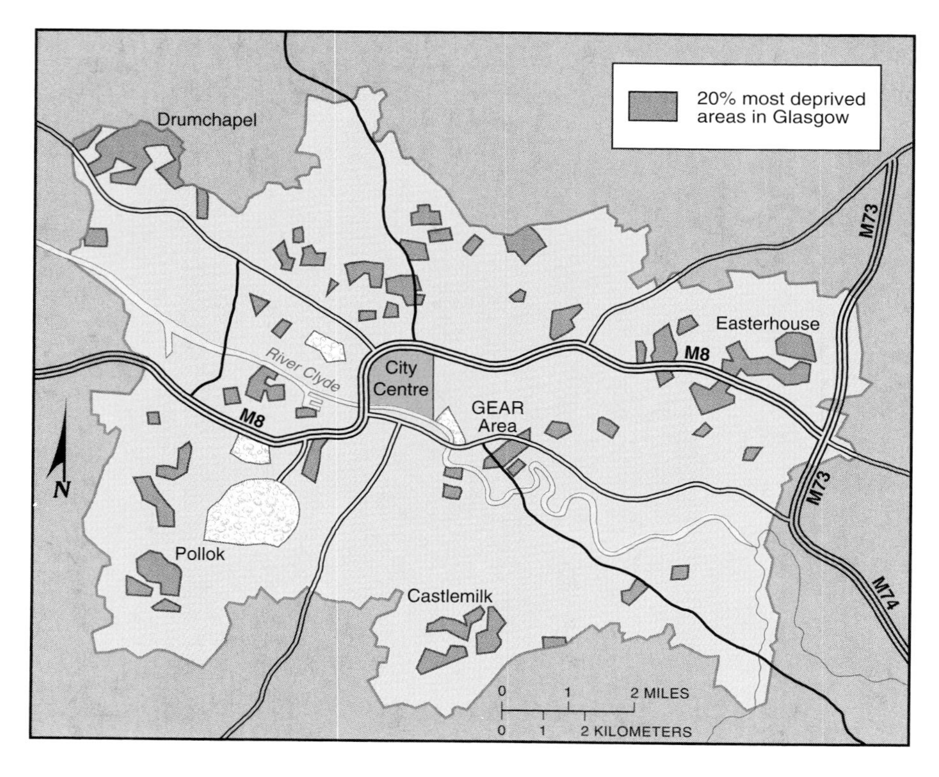

FIGURE 13–22
Neighborhoods classified as deprived in Glasgow, Scotland, are primarily in outer areas, as is the case in U.S. cities. Areas of social deprivation contain high concentrations of unemployed people who receive public assistance. Most areas of high social deprivation consist of massive housing projects built after World War II.

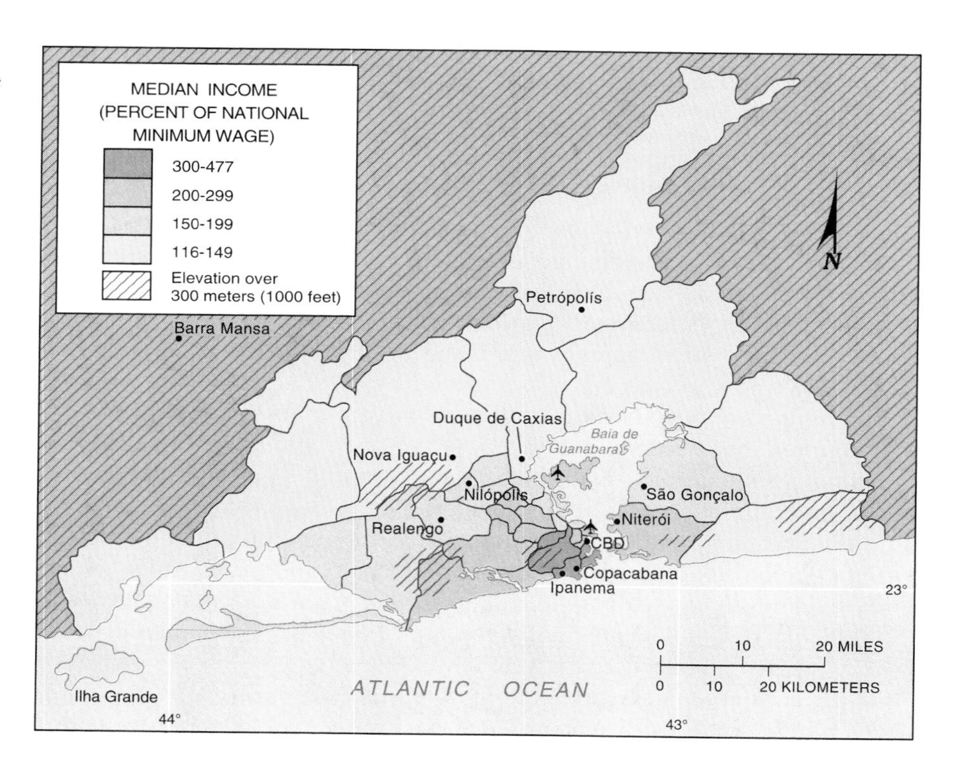

FIGURE 13–23
The highest-income areas in Rio de Janeiro, Brazil, are near the central business district. Low-income people are more likely to live in peripheral areas.

FIGURE 13–24
High-income people are attracted to central areas in developing cities like Rio de Janeiro because services, such as municipal sewers, are more widely available than in peripheral areas.

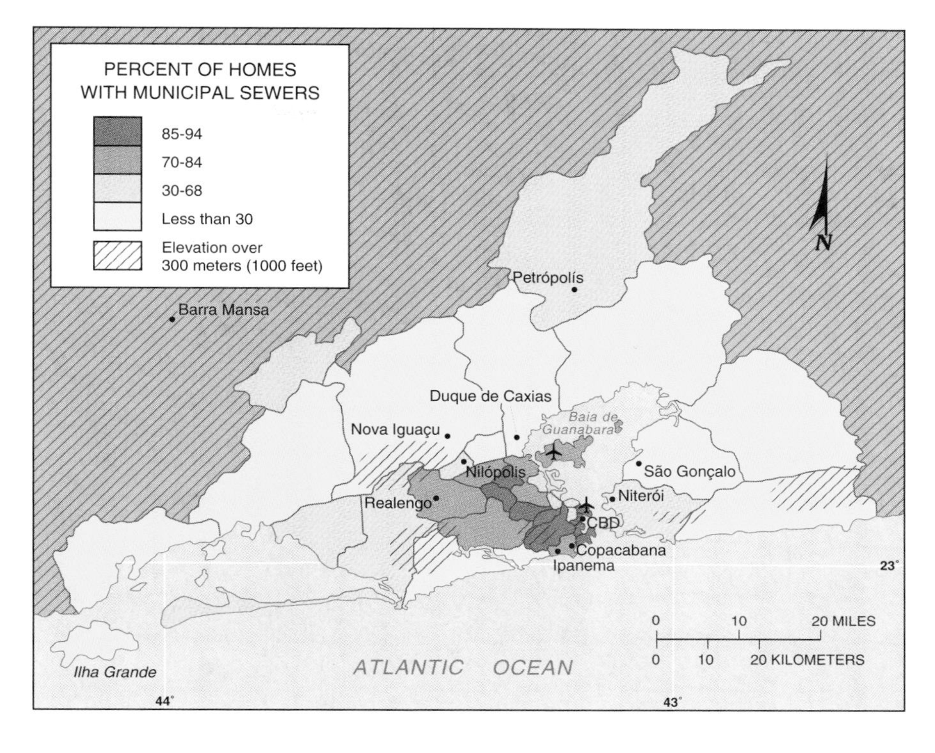

ernment. They bring all their possessions, which usually are few enough to be easily carried. The leaders of such an "invasion" allocate small parcels of the seized land to each participating family.

At first, squatters do little more than camp out on the land or sleep in the street; in severe weather, they take shelter in markets and warehouses. Families then erect primitive shelters with scavenged cardboard, wood boxes, sack cloth, and crushed beverage cans. As they find new bits of material, they add them to their shacks. After a few years, they may build a tin roof and partition the space into rooms, and the structure acquires a more permanent appearance.

Squatter settlements have few services because neither the city nor the residents can afford them. Latrines are designated by the settlement's leaders, and water is carried from a central well or dispensed from a truck. The settlements lack schools,

A large percentage of people in the rapidly growing cities of developing countries live in squatter settlements, such as this *favela* in Rio de Janeiro, Brazil, in the foreground. Rio's wealthier people live in high-rise apartment buildings nestled between the hills and the Atlantic Ocean, while the *favelas* are built on the hillsides, where services are difficult to provide. (Stephanie Maze/Woodfin Camp & Associates)

The suburbs of a European city are more likely to consist of apartment buildings, whereas the center city contains low-rise buildings, the opposite pattern of that found in U.S. cities. Massive apartment complexes were built by the Communists in the suburbs of Eastern European cities, such as Budapest, Hungary. (Bill Weems/Woodfin Camp & Associates)

paved roads, electricity, telephones, or sewers. In the absence of bus service or available private cars, a resident may have to walk two hours to reach a place of employment.

Governments in developing countries face a difficult choice regarding squatter settlements. If the government sends in the police or army to raze the settlement, it risks sparking a violent confrontation with the displaced people. On the other hand, if the government decides that improving and legalizing squatter settlements is cheaper than building the necessary new apartment buildings, it may encourage other poor rural people to migrate to the city as squatters.

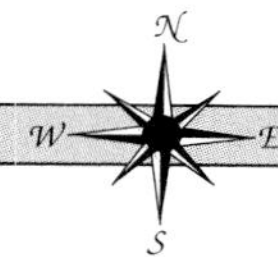

CONCLUSIONS: CRITICAL ISSUES FOR THE FUTURE OF SETTLEMENTS

Urbanization is a process with a beginning, middle, and end. The United Kingdom and the United States are urbanized countries in which the proportion of urban dwellers has remained fairly constant in recent decades.

The traditional advantage of a settlement was its centrality, which provided access to goods and services. Now that virtually all people in relatively developed countries have access to modern transportation and communication, the importance of centrality may have lessened. Diffusion of communication technology, such as cellular phones, video, fax, modem, Internet, and overnight courier services, will further improve communication.

The diffusion of rapid communication systems in relatively developed countries—and presumably in developing countries in the future—poses a challenge to geographic analysis of the current system of settlements. Possessing instant communications tools, people are less restricted to living in a particular settlement because of their employment. More people now *telecommute,* working in their homes and communicating electronically with their employers.

As a greater proportion of the world's population becomes urbanized, urban settlements increasingly reflect the trends and patterns in society as a whole. As this chapter has shown, contradictory trends are at work within cities. Why does one inner-city neighborhood become a slum and another a high-class district? Why does one city attract new shoppers and visitors while another languishes? Geographers help to explain why these patterns arise, and what can be done about them.

Chapter Summary

1. Origin of settlements

Settlements were originally established for noneconomic reasons, including religious, cultural, military, and political. Rural settlements evolved as places for farmers to live, while urban settlements provided goods and services to factory, office, and service workers. Most rural residents originally clustered in small communities, but since the industrial revolution most farmers in relatively developed countries operate dispersed, isolated farmsteads.

2. Urbanization

Since 1800, the percentage of the world's population living in urban settlements has increased from 3 percent to more than 40 percent. Until recently, urban growth was associated with development, but in the past two decades, the world's largest cities have increasingly been located in developing countries. Although urban and rural areas present different visual images, distinguishing between the two is increasingly difficult in relatively developed countries.

3. The central city

The North American central business district is dominated by retailing, offices, and other commercial activities. Surrounding the CBD is an inner residential area containing older low-quality housing and public housing, as well as pockets of high-status renovated housing. Residents of these inner-city neighborhoods disproportionately are poor minorities.

4. Cities and suburbs

The suburban life-style—as exemplified by the separate single-family house with surrounding yard—attracts most people. Transportation improvements, most notably the railroad in the nineteenth century and the automobile in the twentieth century, have facilitated the sprawl of urban areas. To explain where various types of people live in urban areas, three models have been developed: the concentric zone, sector, and multiple nuclei. None fully explains the structure of the city, but the three combined present a useful framework for understanding the distribution of social and economic groups within an urban area.

Key Terms

Annexation, 524
Basic industries, 515
Blockbusting, 532
Central business district (CBD), 526
Central city, 525
Central place theory, 513
Clustered rural settlement, 515
Concentric zone model, 538
Consolidated metropolitan statistical area (CMSA), 526
Density gradient, 535
Dispersed rural settlement, 517
Economic base, 515
Edge city, 535
Enclosure movement, 518
Filtering, 532
Gentrification, 534
Market area, 514
Metropolitan statistical area (MSA), 525
Multiple nuclei model, 539
Primate city, 515
Public housing, 532
Range, 514
Rank-size rule, 514
Redlining, 532
Rural settlement, 515
Rush (or peak) hour, 537
Sector model, 539
Settlement, 509
Squatter settlement, 541
Threshold, 514
Underclass, 531
Urban settlement, 515
Urbanization, 523
Urbanized area, 525
Zone in transition, 531

Questions for Study and Discussion

1. Discuss cultural and economic reasons for the origin of settlements.
2. What are the differences between rural and urban settlements?
3. Outline the principles of central place theory.
4. What is the difference in the distribution of cities of different sizes according to the rank-size rule and the primate city rule?
5. What is meant by the *economic base* of a community?
6. What is the difference between a clustered and a dispersed rural settlement?
7. What is meant by *urbanization?*
8. What is the difference between the urbanized area and the metropolitan statistical area?
9. What is meant by the *daily urban system?*
10. What activities are clustered in the central business district? Why are they there?
11. What types of residential areas are found in the zone of transition?
12. Describe the three models of internal structure of cities. What types of social characteristics follow each of the three models?
13. Describe differences in the internal structure of cities in Europe and in developing countries, compared to the United States.

Thinking Geographically

1. Compare Toronto and Detroit. How do the central business districts and the inner residential areas compare in the two cities? What might account for these differences?
2. Draw a sketch of your community or neighborhood. In accordance with Kevin Lynch's *The Image of the City,* place five types of information on the map: districts (homogeneous areas), edges (boundaries that separate districts), paths (lines of communication), nodes (central points of interaction), and landmarks (prominent objects on the landscape). How clear an image does your community have for you?
3. To Jane Jacobs, in *The Death and Life of Great American Cities,* an attractive urban environment is one that is animated with an intermingling of a variety of people and activities, such as found in many New York City neighborhoods. What are the attractions and drawbacks to living in such environments?
4. Determine the economic base of your community. Consult the *U.S. Census of Manufacturing* or *County Business Patterns*. To make a rough approximation of your community's basic industries, compute the decimal fraction of the nation's population that lives in your community. It will be a small number, such as 0.0005. Then, find the total number of U.S. firms (or employees) in each industrial sector that is present in your community. Multiply these national figures by your local population fraction. Subtract the result from your community's actual number of firms (or employees) for that type of industry. If the difference is positive, you have identified one of your community's basic industries.
5. Two factors can explain the performance of your community's basic industries. One is that the sector is expanding or contracting nationally. The second is that the industry is performing much better or worse in the community than in the nation as a whole. Which of the two factors better explains the performance of your community's basic industries?
6. Officials of rapidly growing cities in developing countries discourage building houses that do not meet international standards for sanitation and construction methods. Also discouraged are privately owned transportation services, because the vehicles generally lack decent tires, brakes, and other safety features. Yet, the residents prefer substandard housing to no housing, and they prefer unsafe transportation to no transportation. What would be the advantages and problems for a city if health and safety standards for housing, transportation, and other services were relaxed?

Suggestions for Further Reading

Archer, Clark J., and Ellen R. White. "A Service Classification of American Metropolitan Areas." *Urban Geography* 6 (1985): 122–51.

Baldasarre, Mark. *Trouble in Paradise: The Suburban Transformation in America*. New York: Columbia University Press, 1986.

Benevolo, Leonardo. *The History of the City*. 2d ed. Cambridge, MA: MIT Press, 1991.

Berry, Brian J. L. *The Geography of Market Centers and Retail Distribution*. Englewood Cliffs, NJ: Prentice Hall, 1988.

______. *The Human Consequences of Urbanization*. New York: St. Martin's Press, 1973.

______. "Transnational Urbanward Migration, 1830–1980." *Annals of the Association of American Geographers* 83 (1993): 389–405.

______, and John D. Kasarda. *Contemporary Urban Ecology.* New York: Macmillan, 1977.
Bourne, Larry S., ed. *Internal Structure of the City.* 2d ed. New York: Oxford University Press, 1982.
______, and J. W. Simmons. *System of Cities.* New York: Oxford University Press, 1978.
Bourne, L. S., R. Sinclair, and K. Dziewonski, eds. *Urbanization and Settlement Systems, International Perspectives.* New York: Oxford University Press, 1984.
Bratt, Rachel G. *Rebuilding a Low-Income Housing Policy.* Philadelphia: Temple University Press, 1990.
Brunn, Stanley D., and Jack L. Williams, eds. *Cities of the World: World Regional Urban Development.* New York: Harper and Row, 1983.
Cervero, Robert. *America's Suburban Centers: The Land Use Transportation Link.* Boston: Unwin and Hyman, 1989.
Chisholm, Michael. *Rural Settlement and Land Use.* 3d ed. London: Hutchinson, 1979.
Christaller, Walter. *The Central Places of Southern Germany.* Englewood Cliffs, NJ: Prentice-Hall, 1966.
Clawson, Marion, and Peter Hall. *Planning and Urban Growth.* Baltimore: The Johns Hopkins University Press, 1973.
Daniels, P. W. *Service Industries: A Geographical Appraisal.* London: Methuen, 1986.
Davis, Kingsley. *Cities: Their Origin, Growth, and Human Impact.* San Francisco: W. H. Freeman, 1973.
Demangeon, Albert. "The Origins and Causes of Settlement Types." In *Readings in Cultural Geography,* edited by P. L. Wagner and M. W. Mikesell. Chicago: University of Chicago Press, 1962.
Detwyler, Thomas, and Melvin Marcus, eds. *Urbanization and Environment.* Belmont, CA: Duxbury Press, 1972.
Dickinson, Robert E. "Rural Settlements in the German Lands." *Annals of the Association of American Geographers* 39 (December 1949): 239–63.
Frieden, Bernard J., and Lynne B. Sagalyn. *Downtown Inc.: How America Rebuilds Cities.* Cambridge: MIT Press, 1989.
Furuseth, Owen J., and John T. Pierce. *Agricultural Land in an Urban Society.* Washington, DC: Association of American Geographers, 1982.
Garreau, Joel. *Edge City: Life on the New Frontier.* New York: Doubleday, 1991.
Goss, Jon. "The 'Magic of the Mall': An Analysis of Form, Function, and Meaning in the Contemporary Retail Built Economy." *Annals of the Association of American Geographers* 83 (1993): 18–47.
Gottmann, Jean. *Megalopolis.* New York: Twentieth Century Fund, 1961.
Guest, Avery M. "Population Suburbanization in American Metropolitan Areas, 1940–1970." *Geographical Analysis* 7 (July 1975): 267–83.
Harris, Chauncey D. "A Functional Classification of Cities in the United States." *Geographical Review* 33 (January 1943): 86–99.
______, and Edward L. Ullman. "The Nature of Cities." *Annals of the American Academy of Political and Social Science* 143 (1945): 7–17.
Harvey, David. *The Urban Experience.* Baltimore: The Johns Hopkins University Press, 1989.
Hauser, Philip M., and Leo F. Schnore, eds. *The Study of Urbanization.* New York: Wiley, 1965.
Hoyt, Homer. *The Structure and Growth of Residential Neighborhoods.* Washington, DC: Federal Housing Administration, 1939.
Jackson, Kenneth T. *Crabgrass Frontier: The Suburbanization of the United States.* New York: Oxford University Press, 1985.
Jacobs, Jane. *The Death and Life of Great American Cities.* New York: Vintage Books, 1961.
______. *The Economy of Cities.* New York: Vintage Books, 1970.
Johnston, R. J. *City and Society: An Outline for Urban Geography.* London: Hutchinson Education, 1984.

King, Leslie J. *Central Place Theory*. Beverly Hills: Sage Publications, 1984.

Kirn, Thomas J. "Growth and Change in the Service Sector of the U.S.: A Spatial Perspective." *Annals of the Association of American Geographers* 77 (September 1987): 353–72.

Knox, Paul L., ed. *The Restless Urban Landscape*. Englewood Cliffs, NJ: Prentice Hall, 1993.

______. *Urbanization: An Introduction to Urban Geography*. Englewood Cliffs, NJ: Prentice Hall, 1994.

Ley, David. *A Social Geography of the City*. New York: Harper and Row, 1983.

Longley, Paul A., Michael Batty, and John Shepherd. "The Size, Shape, and Dimensions of Urban Settlements." *Transactions of the Institute of British Geographers, New Series* 16 (1991): 75–94.

Losch, August. *The Economics of Location*. New Haven: Yale University Press, 1954.

Lowder, Stella. *Inside Third World Cities*. London: Routledge, 1988.

Lynch, Kevin. *The Image of the City*. Cambridge: MIT Press, 1960.

Marshall, J. N. "Services in a Postindustrial Economy." *Environment and Planning A* 17 (1985): 1,155–67.

Mayer, Harold M., and Charles R. Hayes. *Land Uses in American Cities*. Champaign: Park Press, 1983.

Mitchelson, Ronald L., and James O. Wheeler. "The Flow of Information in a Global Economy: The Role of the American Urban System in 1990." *Annals of the Association of American Geographers* 84 (1994): 87–107.

Morrill, Richard. "The Structure of Shopping in a Metropolis." *Urban Geography* 8 (1987): 97–128.

Mumford, Lewis. *The City in History*. New York: Harcourt, Brace, and World, 1961.

Park, Robert E., Ernest W. Burgess, and Roderick D. McKenzie, eds. *The City*. Chicago: University of Chicago Press, 1925.

Rossi, Peter. *Down and Out in America: The Origins of Homelessness*. Chicago: University of Chicago Press, 1989.

Scofield, Edna. "The Origin of Settlement Patterns in Rural New England." *Geographical Review* 28 (October 1938): 652–63.

Scott, Peter. *Geography and Retailing*. London: Hutchinson University Press, 1970.

Short, J. R., L. M. Benton, W. B. Luce, and J. Walton. "Reconstructing the Image of an Industrial City." *Annals of the Association of American Geographers* 83 (1993): 207–24.

Trewartha, Glenn T. "Types of Rural Settlements in North America." *Geographical Review* 36 (October 1946): 568–96.

United States National Advisory Commission on Civil Disorders, Otto Kerner, chairman. *Report*. New York: Dutton, 1968.

Wheeler, James O., and Ronald L. Mitchelson. "Information Flows among Major Metropolitan Areas in the United States." *Annals of the Association of American Geographers* 79 (December 1989): 523–43.

White, Paul. *The West European City, A Social Geography*. London and New York: Longman, 1984.

Whyte, William H. *City: Rediscovering the Center*. New York: Doubleday, 1988.

Wirth, Louis. *On Cities and Social Life*. Chicago: University of Chicago Press, 1964.

We also recommend these journals: *Environment and Planning; Journal of Historical Geography; Journal of Housing; Journal of Regional Science; Journal of Rural Studies; Journal of the American Planning Association; Journal of Urban Economics; Land Economics; Planning; Urban Geography; Urban Land; Urban Studies*.

Appendix
Map Scale and Projections

Phillip C. Muehrcke

Unaided, our human senses provide a limited view of our surroundings. To overcome these limitations, humankind has developed powerful vehicles of thought and communication, such as language, mathematics, and graphics. Each of these tools is based on elaborate rules, each has an information bias, and each may distort its message, often in subtle ways. Consequently, to use these aids effectively, we must understand their rules, biases, and distortions. The same is true for the special form of graphics we call maps: we must master the logic behind the mapping process before we can use maps effectively. A fundamental issue in cartography, the science and art of making maps, is the vast difference between the size and geometry of what is being mapped—the real world, we will call it—and that of the map itself. Scale and projection are the basic cartographic concepts that help us understand this difference and its effects.

Map Scale

Our senses are dwarfed by the immensity of our planet; we can sense directly only our local surroundings. Thus, we cannot possibly look at our whole state or country at one time, even though we may be able to see the entire street where we live. Cartography helps us expand what we can see at one time by letting us view the scene from some distant vantage point. The greater the imaginary distance between that position and what we are looking at, the larger the area the map can cover but the smaller the features on the map will appear. This reduction is defined by the *map scale*, the ratio of the distance on the map to the distance on the earth. Map users need to know about map scale for two reasons: (1) so they can convert measurements on a map into meaningful real-world measures and (2) so they can know how abstract the cartographic representation is.

Real-world measures. A map can provide a useful substitute for the real world for many analytical purposes. With the scale of a map, for instance, we can compute the actual size of mapped features (length, area, or volume). These calculations are helped by three expressions of map scale: a word statement, a graphic scale, and a representative fraction.

A *word statement* of a map scale compares X units on the map to Y units on the earth, often abbreviated *"X units to Y units."* For example, the expression "one inch to ten miles" means that one inch on the map represents ten miles on the earth (Figure A–1). Because the map is always smaller

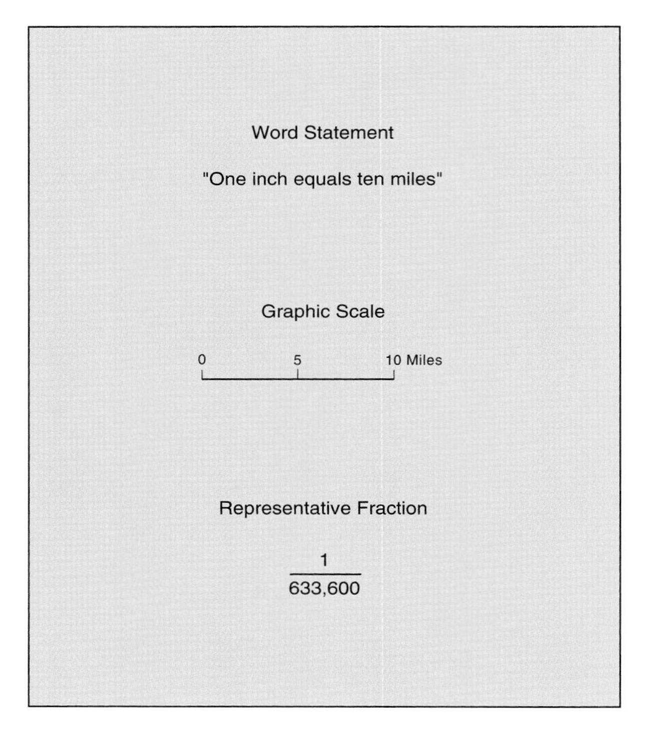

FIGURE A–1
Common expressions of map scale.

than the area that has been mapped, the ground unit is always the larger number. Both units are expressed in meaningful terms, such as inches and miles or centimeters and kilometers. Word statements are not intended for precise calculations but give the map user a rough idea of size and distance.

A *graphic scale*, such as a bar graph, is concrete and therefore overcomes the need to visualize inches and miles that is associated with use of a word statement of scale (see Figure A–1). A graphic scale permits direct visual comparison of feature sizes and the distances between features. No ruler is required; any measuring aid will do. It needs only to be compared with the scaled bar: if the length of one toothpick is equal to two miles on the ground and the map distance equals the length of four toothpicks, then the ground distance is four times two or eight miles. Graphic scales are especially convenient in this age of copying machines, when we are more likely to be working with a copy than with the original map. If a map is reduced or enlarged as it is copied, the graphic scale will change in proportion to the change in the size of the map and thus will remain accurate.

The third form of map scale is the *representative fraction* (RF). An RF defines the ratio between the distance on the earth in fractional terms, such as $\frac{1}{633,600}$ (also written 1/633,600 or 1:633,600). The numerator of the fraction always refers to the distance on the map, and the denominator always refers to the distance on the earth. No units of measurement are given, but both numbers must be expressed in the same units. Because map distances are extremely small relative to the size of the earth, it makes sense to use small units, such as inches or centimeters. Thus, the RF 1:633,600 might be read as "1 inch on the map to 633,600 inches on the earth."

Herein lies a problem with the RF. Meaningful map-distance units imply a denominator so large that is impossible to visualize. Thus, in practice, reading the map scale involves an additional step of converting the denominator to a meaningful ground measure, such as miles or kilometers. The unwieldy 633,600 becomes the more manageable 10 miles when divided by the numbers of inches in a mile (63,360).

On the plus side, the RF is good for calculations. In particular, the ground distance between points can be easily determined from a map with an RF. One simply multiplies the distance between the points on the map by the denominator of the RF. Thus, a distance of five inches on a map with an RF of 1/126,720 would signify a ground distance of 5 x 126,720, which equals 633,600. Because all units are inches and there are 63,360 inches in a mile, the ground distance is 633,600 ÷ 63,360, or 10 miles. Computation of area is equally straightforward with an RF. Computer manipulation and analysis of maps is based on the RF form of map scale.

Guides to generalization. Scales also help map users visualize the nature of the symbolic relation between the map and the real world. It is convenient here to think of maps as falling into three broad scale categories (Figure A–2). (Do not be confused by the use of the words *large* and *small* in this context; just remember that the larger the denominator, the smaller the scale ratio and the larger the area that is shown on the map.) Scale ratios greater than 1:100,000, such as the 1:24,000 scale of the U.S. Geological Survey topographic quadrangles, are large-scale maps. Although these

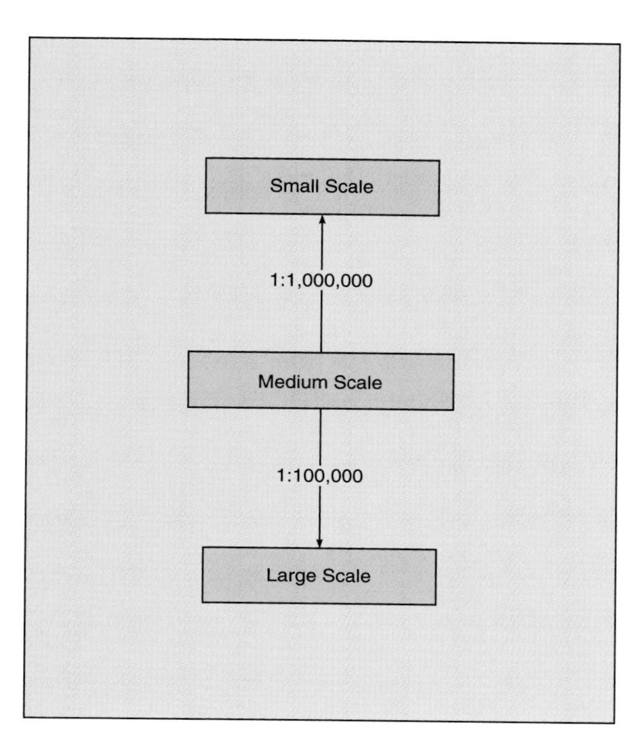

FIGURE A–2
The scale gradient can be divided into three broad categories.

maps can cover only a local area, they can be drawn to rather rigid standards of accuracy. Thus, they are useful for a wide range of applications that require detailed and accurate maps, including zoning, navigation, and construction.

At the other extreme are maps with scale ratios of less than 1:1,000,000, such as maps of the world that are found in atlases. These are small-scale maps. Because they cover large areas, the symbols on them must be highly abstract. They are therefore best suited to general reference or planning, when detail is not important. Medium- or intermediate-scale maps have scales between 1:100,000 and 1:1,000,000. They are good for regional reference and planning purposes.

Another important aspect of map scale is to give us some notion of geometric accuracy; the greater the expanse of the real world shown on a map, the less accurate is the geometry of that map. Figure A–3 shows why. If a curve is represented by straight line segments, short segments *(X)* are more similar to the curve than are long segments *(Y)*. Similarly, if a plane is placed in contact with a sphere, the difference between the two surfaces is slight were they touch *(A)* but grows rapidly with increasing distance *(B)* from the point of contact. In view of the large diameter and slight local curvature of the earth, distances will be well represented on large-scale maps (those with small denominators) but will be increasingly poorly represented at smaller scales. This close relationship between map scale and map geometry brings us to the topic of map projections.

Map Projections

The spherical surface of the earth is shown on flat maps by means of map projections. The process of "flattening" the earth is essentially a problem in geometry that has captured the attention of the best mathematical minds for centuries. Yet no one has ever found a perfect solution: there is no known way to avoid spatial distortion of one kind or another. Many map projections have been devised, but only a few have become standard. Because a single flat map cannot preserve all aspects of the earth's surface geometry, a mapmaker must be

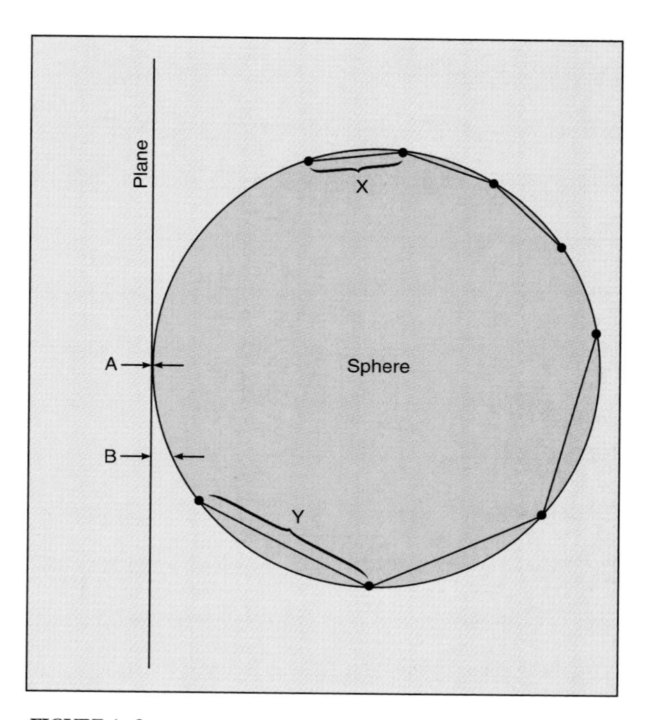

FIGURE A–3
Relationships between surfaces on the round earth and a flat map.

careful to match the projection with the task at hand. To map something that involves distance, for example, a projection should be used in which distance is not distorted. In addition, a map user should be able to recognize which aspects of a map's geometry are accurate and which are distortions caused by a particular projection process. Fortunately, this is not too difficult.

It is helpful to think of the creation of a projection as a two-step process (Figure A–4). First, the immense earth is reduced to a small globe with a scale equal to that of the desired flat map. All spatial properties on the globe are true to those on the earth. Second, the globe is flattened. Since this cannot be done without distortion, it is accomplished in such a way that the resulting map exhibits certain desirable spatial properties.

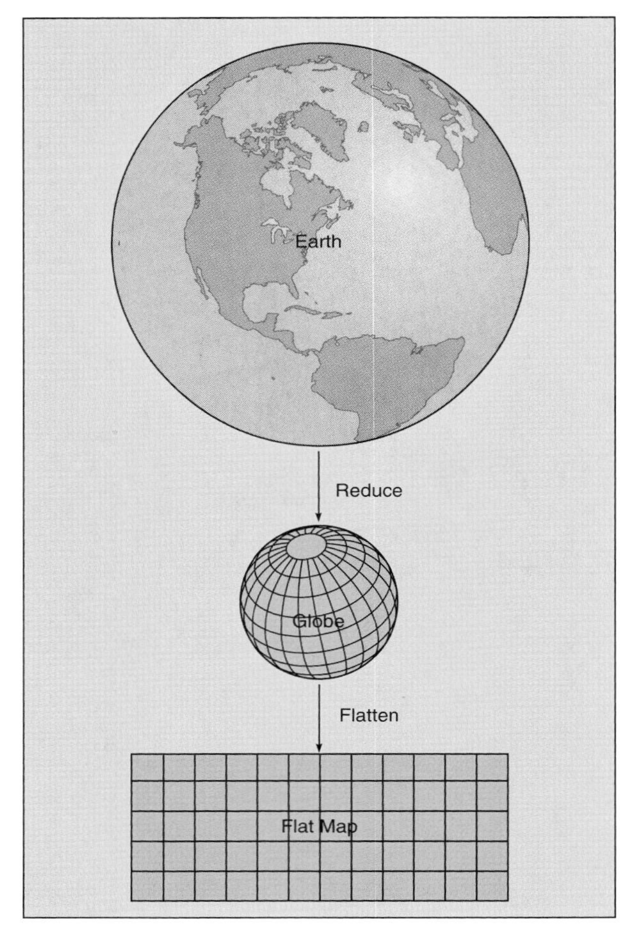

FIGURE A–4
The two-step process of crating a projection.

Perspective models. Early map projections were sometimes created with the aid of perspective methods, but this has changed. In the modern electronic age, projections are normally developed by strictly mathematical means and are plotted out or displayed on computer-driven graphics devices. The concept of perspective is still useful, however, in visualizing what map projections do. Thus, projection methods are often illustrated by using strategically located light sources to cast shadows on a projection surface from a latitude/longitude net inscribed on a transparent globe.

The success of the perspective approach depends on finding a projection surface that is flat or that can be flattened without distortion. The cone, cylinder, and plane possess these attributes and serve as models for three general classes of map projections: *conic, cylindrical,* and *planar* (or azimuthal). Figure A–5 shows these three classes, as well as a fourth, a false cylindrical class with an oval shape. Although the *oval* class is not of perspective origin, it appears to combine properties of the cylindrical and planar classes (Figure A–6).

The relationship between the projection surface and the model at the point or line of contact is critical because distortion of spatial properties on the projection is symmetrical about, and increases with distance from, that point or line. This condition is illustrated for the cylindrical and planar classes of projections in Figure A–7. If the point or line of contact is changed to some other position on the globe, the distortion pattern will be recentered on the new position but will retain the same symmetrical form. Thus, centering a projection on the area of interest on the earth's surface can minimize the

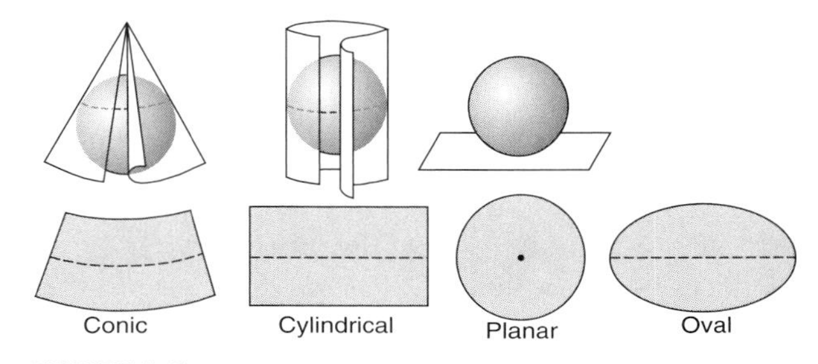

FIGURE A–5
General classes of map projection. (Courtesy of ACSM)

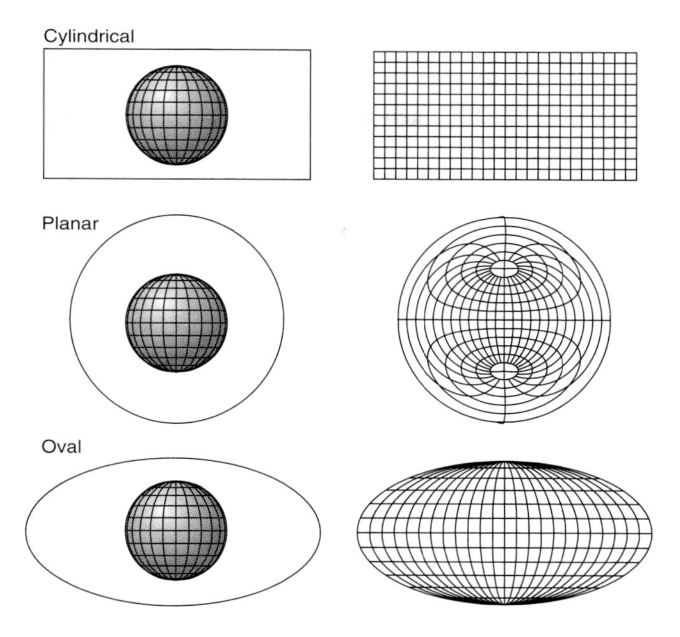

FIGURE A–6
The visual properties of cylindrical and planar projections combined in oval projections. (Courtesy of ACSM)

effects of projection distortion. And recognizing the general projection shape, associating it with a perspective model, and recalling the characteristic distortion pattern will provide the information necessary to compensate for projection distortion.

Preserved properties. For a map projection to truthfully depict the geometry of the earth's surface, it would have to preserve the spatial attributes of *distance, area, shape*, and *proximity*. This task can be readily accomplished on a globe, but it is not possible on a flat map. To preserve area, for example, a mapmaker must stretch or shear shapes; thus, area and shape cannot be preserved on the same map. To depict both direction and distance from a point, area must be distorted. Because the earth's surface is continuous in all directions from every point, discontinuities that violate proximity relationships must occur on all map projections. The trick is to place these discontinuities where they will have the least impact on the spatial relationships in which the map user is interested.

We must be careful when we use spatial terms because the properties they refer to can be confusing. The geometry of the familiar plane is very different from that of a sphere; yet when we refer to a flat map, we are in fact making reference to the spherical earth that was mapped. A shape-preserving projection, for example, is truthful to local shapes—such as the right-angle crossing of latitude and longitude lines—but does not preserve shapes at continental or global levels. A distance-preserv-

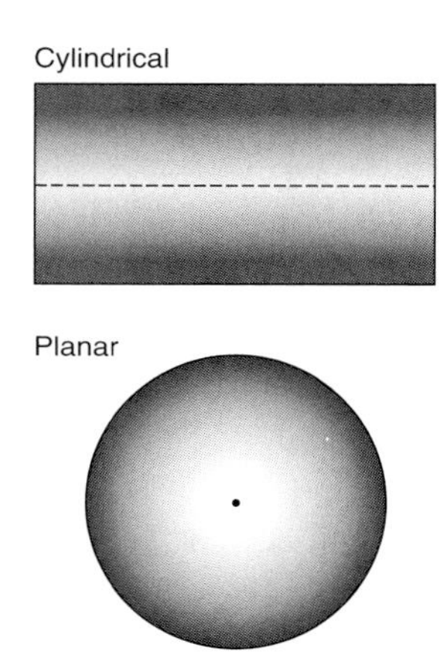

FIGURE A–7
Characteristic patterns of distortion for the cylindrical and planar projection classes. Here, darker shading implies greater distortion. (Courtesy of ACSM)

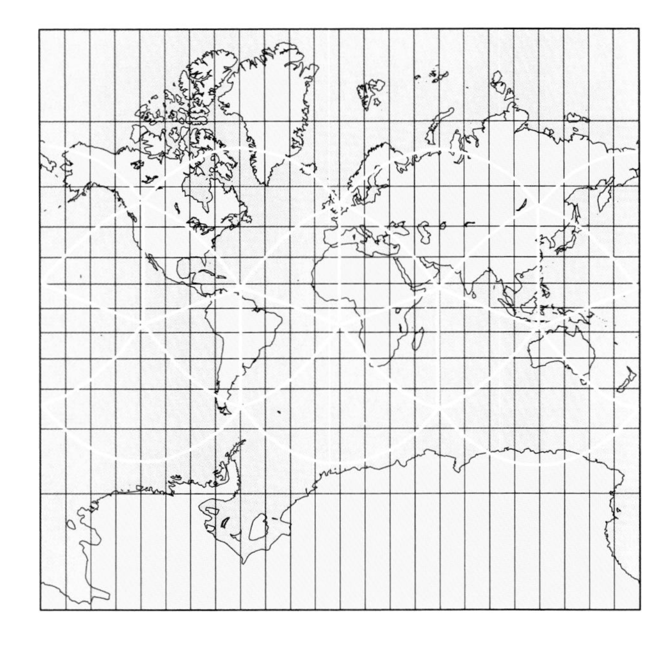

FIGURE A–8
The useful Mercator projection, showing extreme area distortion in the higher latitudes. (Courtesy of ACSM)

ing projection can preserve this property from one point on the map in all directions or from a number of points in several directions, but distance cannot be preserved in the general sense that area can be preserved. Direction can also be generally preserved from a single point or in several directions from a number of points but not from all points simultaneously. Thus, a shape, distance, or direction-preserving projection is truthful to these properties only in part.

Partial truths are not only the consequence of transforming a sphere into a flat surface. Some projections exploit this transformation by expressing traits that are of considerable value for specific applications. One of these is the famous shape-preserving *Mercator projection* (Figure A–8). This cylindrical projection was derived mathematically in the 1500s so that compass bearings (called *rhumb* lines) between any two points of the earth would plot as straight lines on the map. This trait let navigators plan, plot, and follow courses between origin and destination, but it was achieved at the expense of extreme areal distortion toward the margins of the projection (see Antarctica in Figure A–8). Although the Mercator projection is admirably suited for its intended purpose, its widespread but inappropriate use for nonnavigational purposes has drawn a great deal of criticism.

The *gnomonic projection* is also useful for navigation. It is a planar projection with the valuable characteristic of showing the shortest (or great circle) route between any two points on the earth as straight lines. Long-distance navigators first plot the great circle course between origin and destination on a gnomonic projection (Figure A–9, top). Next they transfer the straight line to a Mercator projec-

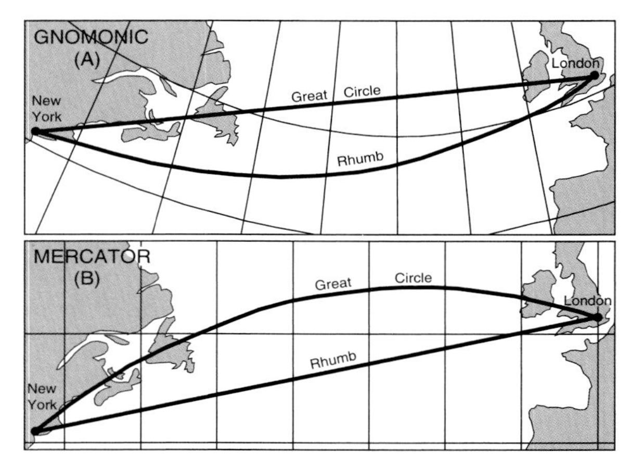

FIGURE A–9
A gnomonic projection (A) and a Mercator projection (B), both of value to long-distance navigators.

tion, where it normally appears as a curve (Figure A–9, bottom). Finally, using straight-line segments, they construct an approximation of this course on the Mercator projection. Navigating the shortest course between origin and destination then involves following the straight segments of the course and making directional corrections between segments. Like the Mercator projection, the specialized gnomonic projection distorts other spatial properties so severely that it should not be used for any purpose other than navigation or communications.

Projections used in textbooks. Although a map projection cannot be free of distortion, it can represent one or several spatial properties of the earth's surface accurately if other properties are sacrificed. The two projections used for world maps throughout this text illustrate this point well. *Goode's homolostine projection*, shown in Figure A–10, belongs to the oval category and shows area accurately, although it gives the impression that the earth's surface has been torn, peeled, and flattened. The interruptions in Figure A–10 have been placed in the major oceans, giving continuity to the land masses. Ocean areas could be featured instead by placing the interruptions in the continents. Obviously, this type of interrupted projection severely distorts proximity relationships. Consequently, in different locations the properties

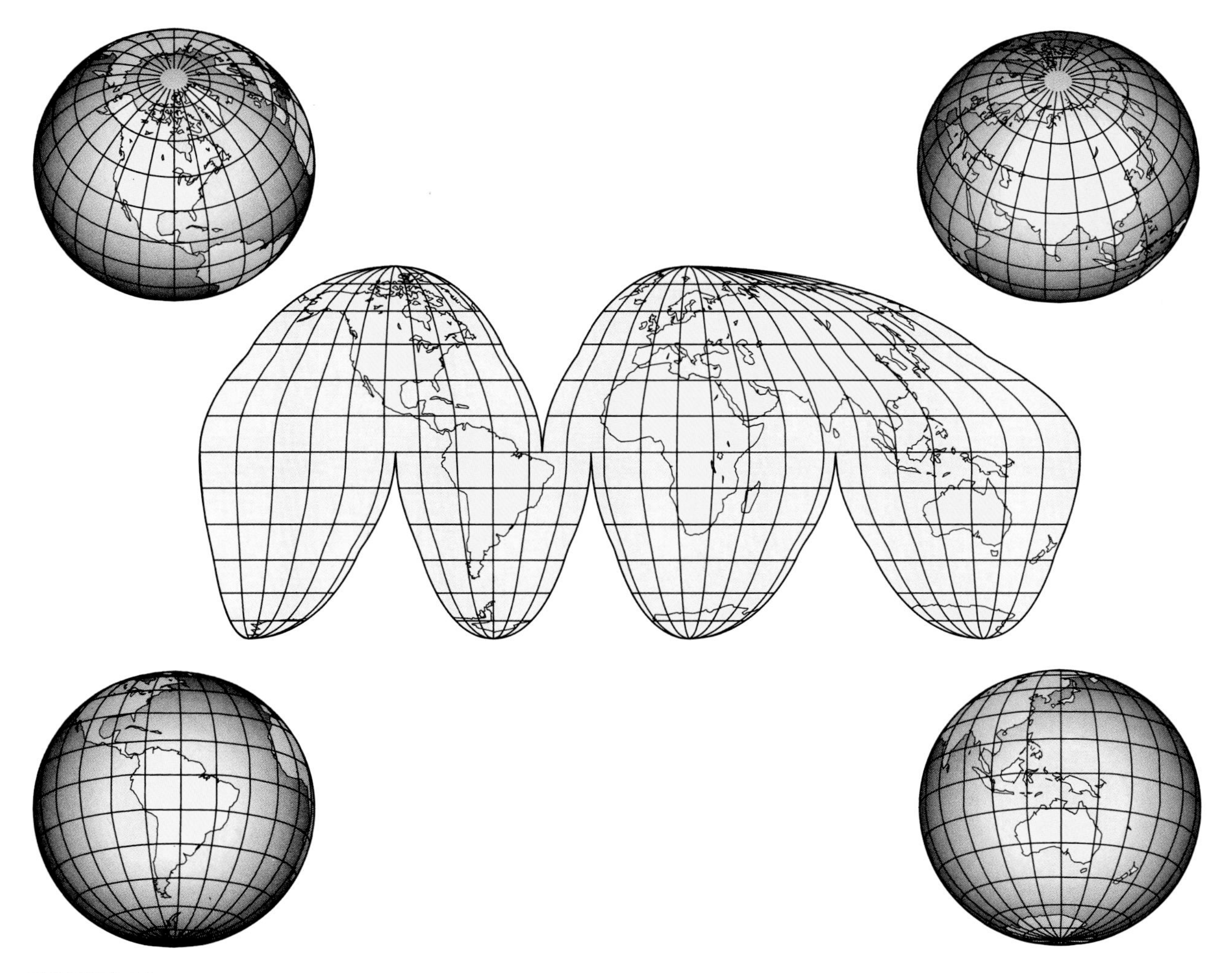

FIGURE A–10
An interrupted Goode's homolosine, an equal-area projection. (Courtesy of ACSM)

of distance, direction, and shape are also distorted to varying degrees. The distortion pattern mimics that of cylindrical projections, with the equatorial zone the most faithfully represented (Figure A–11).

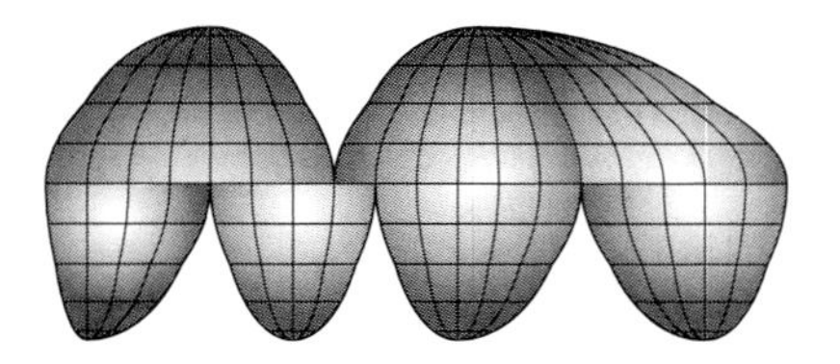

FIGURE A–11
The distance pattern of the interrupted Goode's homolosine projection, which mimics that of cylindrical projections. (Courtesy of ACSM)

An alternative to special-property projections such as the equal-area Goode's homolosine is the compromise projection. In this case no special property is achieved at the expense of others, and distortion is rather evenly distributed among the various properties, instead of being focused on one or several properties. The *Robinson projection*, which is also used in this text, falls into this category (Figure A–12). This oval projection has a global feel, somewhat like that of Goode's homolosine. But the Robinson projection shows the North Pole and the South Pole as lines that are slightly more than half the length of the equator, thus exaggerating distances and areas near the poles. Areas look larger than they really are in the high latitudes (near the poles) and smaller than they really are in

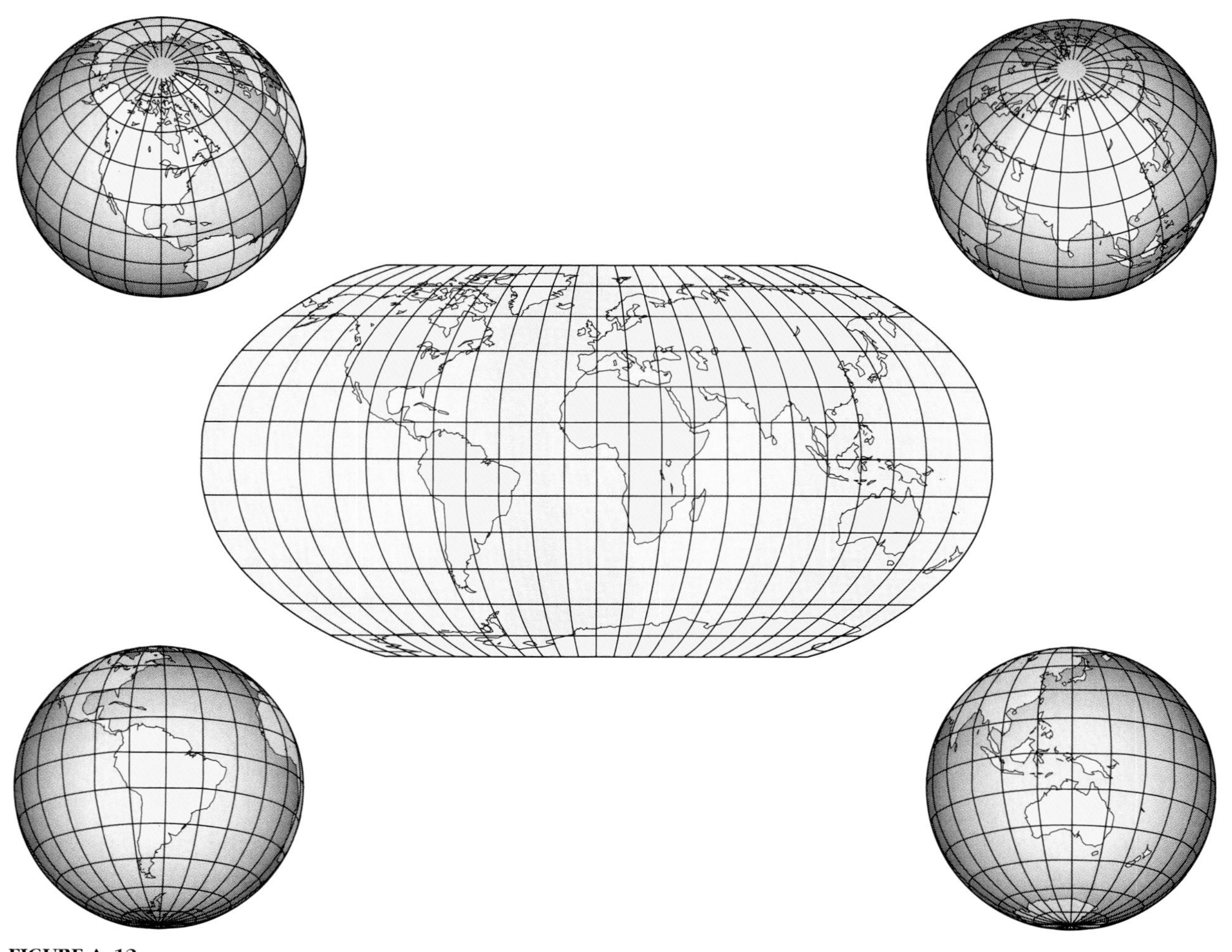

FIGURE A–12
The compromise Robinson projection, which avoids the interruption of Goode's homolosine but preserves no special properties. (Courtesy of ACSM)

the low latitudes (near the equator). In addition, not all latitude and longitude lines intersect at right angles, as they do on the earth, so we know that the Robinson projection does not preserve direction or shape either. However, it has fewer interruptions than Goode's homolosine does, so it preserves proximity better. Overall, the Robinson projection does a good job of representing spatial relationships, especially in the low to middle latitudes and along the central meridian.

Scale and Projections in Modern Geography

Computers have drastically changed the way in which maps are made and used. In the preelectronic age, maps were so laborious, time-consuming, and expensive to make that relatively few were created. Frustrated, geographers and other scientists often found themselves trying to use maps for purposes not intended by the map designers. But today, anyone with access to computer mapping facilities can create projections in a flash. Thus, projections will be increasingly tailored to specific needs, and more and more scientists will do their own mapping rather than have someone else guess what they want in a map.

Computer mapping creates opportunities that go far beyond the construction of projections, of course. Once maps and related geographical data are entered into computers, many types of analyses can be carried out involving map scales and projections. Distances, areas, and volumes can be computed; searches can be conducted; information from different maps can be combined; optimal routes can be selected; facilities can be allocated to the most suitable sites; and so forth. The term used to describe these processes is *geographical information system*, or GIS (Figure A–13). Within a GIS, projections provide the mechanism for linking data from different sources, and scale provides the basis for size calculations of all sorts. Mastery of both projection and scale becomes the user's responsibility because the map user is also the map maker. Now more than ever, effective geography depends on knowledge of the close association between scale and projection.

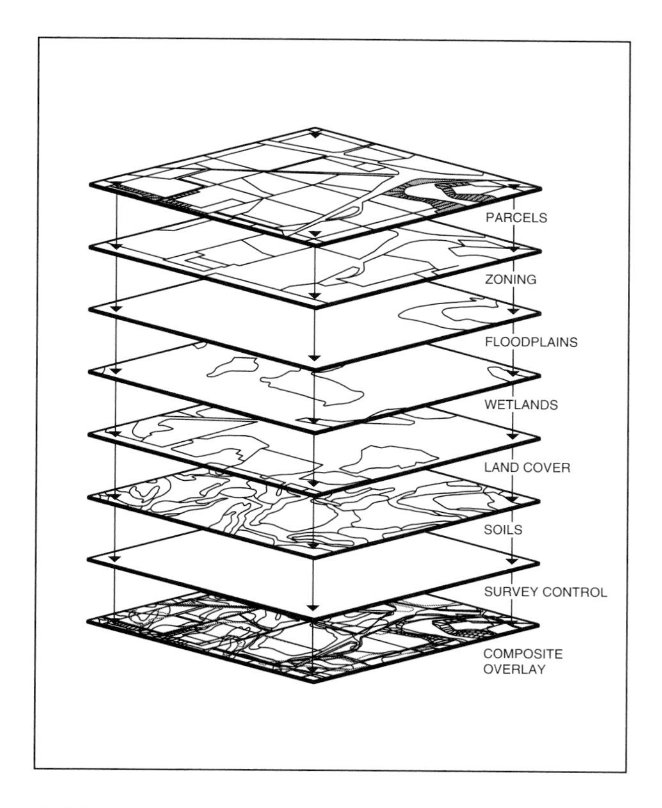

FIGURE A–13
Within a GIS, environmental data attached to a common terrestrial reference system, such as latitude/longitude, can be stacked in layers for spatial comparison and analysis.

Glossary

Abiotic system A physical system composed of nonliving or inorganic matter.

Actual evapotranspiration (ACTET) The amount of water evaporated and/or transpired in a given environment.

Adiabatic cooling Cooling of air as a result of expansion of rising air; *adiabatic* means "without heat being involved."

Advection Horizontal movements of air, heat or substances by wind or ocean currents.

Agribusiness Commercial agriculture characterized by integration of different steps in the food processing industry, usually through ownership by large corporations.

Agricultural density The ratio of the number of farmers to the total amount of land suitable for agriculture.

Agricultural revolution The time when human beings first domesticated plants and animals and no longer relied entirely on hunting and gathering.

Agriculture The deliberate effort to modify a portion of Earth's surface through the cultivation of crops and the raising of livestock for sustenance or economic gain.

Air mass A large region of air with distinctive characteristics of temperature, pressure, and humidity.

Albedo Reflection from Earth's surface of incoming shortwave energy.

Alluvial fan A fan-shaped deposit of sand and gravel formed where a stream emerges from a narrow canyon onto a wider valley floor.

Alpine glacier A glacier occupying a valley in a mountainous area. The movement of an alpine glacier is primarily governed by the underlying topography.

Angle of incidence The angle at which solar radiation strikes a particular place at a point in time.

Animism Belief that objects, such as plants and stones, or natural events, like thunderstorms and earthquakes, have a discrete spirit and conscious life.

Annexation The process of legally adding land area to a city in the United States.

Apartheid Laws in South Africa that physically separated different races into different geographic areas.

Arithmetic density The total number of people divided by the total land area.

Atmosphere A thin layer of gases surrounding Earth to an altitude of less than 480 kilometers (300 miles).

Autumnal equinox September 22 or 23, one of two dates when at noon the perpendicular rays of the sun strike the equator (the sun is directly overhead along the equator).

Barometer An instrument for measuring atmospheric pressure.

Basic industries Industries which sell their products primarily to consumers outside the settlement.

Beach A deposit of wave-carried sediment along a shoreline, on which waves break.

Biochemical oxygen demand The amount of dissolved oxygen consumed in the decomposition of a waste discharged into water.

Biodiversity The amount of variety of living things in a given environment.

Biogeochemical cycles Recycling processes that supply essential substances such as carbon, nitrogen, and other nutrients to the biosphere.

Biomagnification The tendency for substances that accumulate in body tissues to increase in concentration as they are passed to higher trophic levels in a food chain.

Biomass The dry mass of living or formerly living matter in a given environment.

Biome A large region of Earth's surface characterized by particular plant and animal types.

Biosphere All living organisms on Earth and the portions of the three abiotic systems where life can exist.

Biotic system A physical system which encompasses living organisms and the portions of the three abiotic systems where they exist.

Blockbusting A process in which real estate agents convince white property owners to sell their houses at low prices because of fear that nonwhite families will soon move into the neighborhood.

Boreal forest An evergreen needle-leaf forest characteristic of cold continental climates.

Boral forest climate A humid, continental climate with short, warm summers and long, cold winters.

Boundary An invisible line that marks the extent of a state's territory.

Brain drain Large-scale emigration by talented people.

Branch A large and fundamental division within a religion.

Break-of-bulk point A location where transfer is possible from one mode of transportation to another.

Broadleaf deciduous forest A forest with broad-leaved trees that lose their leaves in the winter; characteristic of humid mid-latitude environments.

Bulk-gaining industry An industry in which the final product weighs more or comprises a greater volume than the inputs.

Bulk-reducing industry An industry in which the final product weighs less or comprises a lower volume than the inputs.

Carbon cycle The movement of carbon among the atmosphere, hydrosphere, biosphere, and lithosphere as a result of processes such as photosynthesis and respiration, sedimentation, weathering, and fossil fuel combustion.

Carbon monoxide A compound containing one carbon atom and one oxygen atom, formed by incomplete combustion of hydrocarbons.

Cartel A group of producers of a commodity who cooperate to control the market for that commodity.

Cartography The science of making maps.

Caste The class or distinct hereditary order into which a Hindu is assigned according to religious law.

Central business district (CBD) The area of the city where retail and office activities are clustered.

Central city The largest city in a metropolitan area.

Central place theory A theory that explains the distribution of settlements, based on the fact that settlements serve as market centers for people living in the surrounding area; larger settlements are fewer and further apart than smaller settlements and serve a larger number of people who are willing to travel further for goods and services.

Centrifugal forces Attitudes that tend to divide a people and diminish support for the state.

Centripetal forces Attitudes that tend to unify a people and enhance support for the state.

Chain migration Process by which people are given preference for migrating to another country because a relative was previously admitted.

Chaparral A distinctive shrub woodland dominated by waxy-leaved trees and shrubs.

Chemical weathering The breakdown of rocks or minerals through chemical reactions at Earth's surface.

Chlorofluorocarbon (CFC) A large human-made molecule consisting of chlorine, fluorine, and carbon atoms; used in a wide variety of commerical and industrial applications including coolants in air conditioners and refrigerators. Also a major destroyer of ozone in the ozone layer.

City-state A sovereign state that comprises a town and surrounding countryside.

Climate The totality of weather conditions over a period of several decades or more.

Climax A plant community type that is the ultimate stage of the succession process. Once the climax stage is reached, further succession does not occur without a disturbance.

Closed system A system that is entirely self-contained by being shut off from the surrounding environment, with the exception of energy flows.

Clustered rural settlement A rural settlement in which the houses and farm buildings of each family are situated close to each other, and fields surround the settlement.

Cold front The boundary formed when a cold air mass advances against a warmer one.

Colonialism The attempt by one country to establish settlements and to impose its political, economic, and cultural principles in another territory.

Colony A territory that is legally tied to a sovereign state rather than completely independent.

Commercial agriculture Agriculture undertaken primarily to generate products for sale off the farm.

Communications-oriented industry An industry for which rapid delivery of the product to consumers is a critical factor.

Community succession A process of ecosystem change in which organisms modify their immediate environments in ways that allow other species to establish themselves and dominate.

Compact state A state in which the distance from the center to any boundary does not vary significantly.

Composite cone volcano A volcano formed by a mixture of lava eruptions and more explosive ash eruptions.

Concentration The spread of something over a given study area.

Concentric zone model A model of the internal structure of cities in which social groups are spatially arranged in concentric rings.

Condensation Water changing from a gas state (vapor) to a liquid or solid state.

Conduction The transfer of heat through materials by molecular contact.

Consolidated metropolitan statistical area (CMSA) In the United States, two or more metropolitan statistical areas tied together by overlapping commuting patterns.

Consumer Animal that eats producers (herbivore), or other animals (carnivore), or both (omnivore).

Contagious diffusion The widespread diffusion of a feature or trend throughout a population or physical system.

Continental drift Slow movement of continents riding on Earth's crustal plates; also known as plate tectonics.

Continental glacier A thick glacier hundreds to thousands of kilometers across, large enough to be only partly guided by underlying topography.

Convection Circulation in a fluid caused by temperature—induced density differences, such as the rising of warm air in the atmosphere.

Convergent plate boundary A boundary between tectonic plates in which the two plates move toward one another, destroying or thickening the crust.

Coriolis effect The tendency of an object moving across Earth's surface to be deflected from its apparent path as a result of Earth's rotation.

Crude birth rate (CBR) The total number of live births in a year for every 1,000 people alive in the society.

Crude death rate (CDR) The total number of deaths in a year for every 1,000 people alive in the society.

Culture The body of customary beliefs, social forms, and material traits of a group of people.

Cumulus clouds Tall, puffy clouds with billowy tops, formed by vertical motion in the atmosphere.

Custom The frequent repetition of an act, to the extent that it becomes characteristic of the group of people performing the act.

Cyclone Large low-pressure area in which winds converge in a counterclockwise swirl (in the Northern Hemisphere; clockwise in the Southern Hemisphere).

Decomposers Small organisms that digest and recycle organic debris; they include bacteria, fungi, insects, and worms.

Demographic transition The process of change in a society's population from a condition of high crude birth and death rates and low rate of natural increase to a condition of low crude birth and death rates, low rate of natural increase, and a higher total population.

Demography The scientific study of population characteristics.

Denomination Division within a branch of a religion.

Density The frequency with which something exists within a given unit of area.

Density gradient The change in density in an urban area from the center toward the periphery.

Dependency ratio The percentage of people under the age of fifteen and over age sixty-four, compared to the number of people active in the labor force.

Desert A vegetation type with sparsely distributed plants that are specially adapted for moisture-gathering and moisture-retention. Or, a climate with low precipitation and high potential evaporation typically associated with such vegetation.

Desert pavement The stony surface of a desert soil formed by selective removal of fine particles by surface erosion.

Desertification The process of a region's soil and vegetation cover becoming more desertlike as a result of overuse of the land, usually by overgrazing or cultivation.

Developing country Sometimes called a less developed country (LDC), a country that is at a relatively early stage in the process of economic development.

Development Process of improvement in the material conditions of people through diffusion of knowledge and technology.

Dialect A form of a language spoken in a local area.

Diaspora Forced dispersion of Jews from Palestine to other countries.

Diffusion The process of spread of a feature or trend across the landscape.

Discharge The quantity of water flowing past a point on a stream per unit time.

Dispersed rural settlement A rural settlement pattern characterized by isolated farms rather than clustered villages.

Dissolved oxygen Oxygen in dissolved form in water, critical to aquatic and marine life.

Distance The measurement of amount of separation between two places.

Distance decay The diminishing in importance and eventual disappearance of a phenomenon with increasing distance from its origin.

Divergent plate boundary A boundary between tectonic plates in which the two plates move away from each other, and new crust is created between them.

Double cropping Harvesting twice a year from the same field.

Doubling time The number of years needed to double a population, assuming a constant rate of natural increase.

Drainage basin The geographical area that contributes runoff to a particular stream, defined with respect to a specific location along that stream.

Drainage density The total length of streams in a drainage basin divided by the drainage area.

Dune An accumulation of wind-blown sand, shaped by the wind.

Earthquake A sudden release of energy within Earth producing a shaking of the crust.

Ecology The scientific study of ecosystems.

Economic base A community's collection of basic industries.

Ecosystem (or ecological system) A group of organisms and the nonliving physical and chemical environment with which they interact.

Ecumene The portion of Earth's surface occupied by permanent human settlement.

Edge city A large node of office and retail activities on the edge of an urban area.

El Niño A circulation change in the eastern tropical Pacific Ocean, from westward flow to eastward flow, that occurs every few years, influencing weather over a large area.

Elongated state A state with a long, narrow shape.

Emigration Migration *from* a location.

Enclosure movement The process of consolidating small land holdings into a smaller number of larger farms in England during the eighteenth century.

Endogenic processes Forces within Earth that affect its surface, such as plate tectonics, volcanic eruptions, and earthquakes.

Environmental determinism A nineteenth- and early twentieth-century approach to the human-environment perspective that argued that the general laws sought by geographers could be found in the physical sciences. Geography was therefore the study of how the physical environment caused human activities.

Epicenter The location on Earth's surface immediately above the focus of an earthquake.

Equilibrium A condition in which a system maintains its general character and structure over a given time frame.

Ethnic cleansing The process of forcibly relocating less powerful nationalities in order to transform a multinational region into a region containing only one nationality.

Ethnic religion A religion with a relatively concentrated spatial distribution whose principles are likely to be based on the physical characteristics of the particular location in which its adherents are concentrated.

Evaporation Water changing from a liquid to a gas state (vapor).

Evapotranspiration The sum of evaporation and transpiration.

Exogenic processes Forces originating in the atmosphere which, aided by gravity, shape Earth's surface. Erosion by running water, glaciers, wind, and waves are examples.

Expansion diffusion The spread of a feature or trend among people or physical phenomena from one area to another in a snowballing process.

Federal state An internal organization of a state that allocates most powers to units of local government.

Feedback The return of a system's output as an input, producing a circular flow of information.

Filtering A process of change in the use of a house, from single-family owner occupancy to abandonment.

Focus (of an earthquake) The location in Earth where motion originates in an earthquake.

Folk custom A custom traditionally practiced by a small, homogeneous, rural group living in relative isolation from other groups; also known as a vernacular custom.

Food chain The sequential consumption of food in an ecosystem, beginning with producers (green plants), followed by consumers (herbivores and carnivores), and ending with decomposers.

Fordist Form of mass production in which each worker is assigned one specific task to perform repeatedly.

Formal region (or uniform or homogeneous region) An area in which the selected trend or feature is present throughout.

Fossil fuel A substance found in rocks, containing chemical energy stored in plants or animals in Earth's past.

Fragmented state A state that includes several discontinuous pieces of territory.

Front A boundary between warm air and cold air.

Frontal uplift Forcing of air upward along a front between cold and warm air masses.

Frontier A zone separating two states in which neither state exercises political control.

Functional region (or nodal region) An area in which an activity has a focal point. The characteristic dominates at a central node, diffuses toward the outer part of the region, diminishes in importance, and eventually disappears.

Gaia hypothesis A holistic view that likens Earth to a living organism that has the ability to regulate critical functions such as climate through interactions among the atmosphere, hydrosphere, biosphere, and lithosphere.

Gentrification A process of converting an urban neighborhood from a predominantly low-income renter-occupied area to a predominantly middle-class owner-occupied area.

Geomorphology The study of the shape of Earth's surface and the processes that modify it.

Ghetto During the Middle Ages, a neighborhood in a city set up by law to be inhabited only by Jews; now used to denote a section of a city in which members of any minority group live because of social, legal, or economic pressure.

GIS (Geographic Information System) A computer system that stores, organizes, analyzes, and displays geographic data.

Glacial budget An accounting of the total amount of water added to and lost from a glacier in a given time period.

Glacier A large mass of flowing, perennial ice.

Grade A condition in which a stream's ability to transport sediment is just balanced by the amount of sediment delivered to it.

Green revolution Rapid diffusion of new agricultural technology, especially new high-yield seeds and fertilizers.

Greenhouse effect Atmospheric warming that results from the passage of incoming shortwave energy and the capture of outgoing longwave energy.

Greenhouse gases Trace substances in the atmosphere that contribute to the greenhouse effect; carbon dioxide, water vapor, ozone, methane, and chlorofluorocarbons are important examples.

Greenwich Mean Time The time in the time zone which encompasses the prime meridian or 0° longitude.

Gross national product (GNP) The value of the total output of goods and services produced in a country in a given time period (normally one year).

Groundwater Water beneath Earth's surface at a depth where rocks and/or soils are saturated with water.

Guest workers Workers who migrate to the relatively developed countries of northern and western Europe, usually from southern and eastern Europe or from northern Africa, in search of relatively highly paid jobs.

Gyre A circular ocean current beneath a subtropical high-pressure cell.

Habit A repetitive act performed by a particular individual.

Hearth A center of innovation; the region from which innovative ideas originate.

Hierarchical diffusion The spread of a feature or trend from one key person or node of innovation to another through bypassing of other persons or areas.

Highland climate A climate with generally low temperatures due to high elevation; mountainous regions are mapped as highland climates on a world climate map because local climatic variations are too detailed to be shown at that scale.

Human-made resource Something that is created by humans, and is useful to them.

Humid continental climate A climate characterized by cold winters and warm summers, with moderate levels of precipitation.

Humid subtropical climate A climate with cool winters, hot summers, and moderately high levels of precipitation.

Humid tropical climate A climate with high temperatures and high rainfall amounts all the year.

Hurricane An intense tropical cyclone that develops over warm ocean areas in the tropics and subtropics, primarily during the warm season.

Hydrocarbon A compound of hydrogen and carbon, containing stored chemical energy.

Hydroelectric power Electricity generated by passing water through a turbine, usually at a dam.

Hydrologic cycle The movement of water from the atmosphere to Earth's surface, across that surface, and back to the atmosphere.

Hydrosphere The water realm of Earth's surface, including the oceans, surface waters on land, ground water in soil and rock, glaciers, and water vapor in the atmosphere.

Ice-cap climate A climate with very cold temperatures all the year, with summer temperatures rarely above freezing.

Igneous rocks Rocks formed by crystallization of a magma.

Immigration Migration *to* a new location.

Imperialism The imposition by one state of its culture and political organization on another inhabited territory.

Industrial revolution A series of improvements in industrial technology that transformed the process of manufacturing goods starting in the late 18th century.

Infant mortality rate The total number of deaths in a year among infants under one year old for every 1,000 births in a society.

Infiltration capacity The maximum amount of water that can soak into a soil per unit time.

Insolation The amount of solar energy intercepted by a particular area of Earth.

Intensive subsistence agriculture A form of subsistence agriculture in which farmers must expend a relatively large amount of effort to produce the maximum feasible yield from a parcel of land.

Internal migration Permanent movement within a particular country.

International Date Line An arc that for the most part follows 180° longitude, although it deviates in several places to avoid dividing land areas. When you cross the International Date Line heading east (toward America), the clock moves back 24 hours, or one entire day. When you go west (toward Asia), the calendar moves ahead one day.

International migration Permanent movement from one country to another.

Intertropical convergence zone (ITCZ) A low-pressure zone near the equator where surface winds converge.

Isostatic adjustment A vertical movement of Earth's crust caused by loading or unloading of the buoyant crust.

Labor-intensive industry An industry for which labor costs comprise a high percentage of total expenses.

Landlocked state A state which does not have a direct outlet to the sea.

Language A system of communication through the use of speech, a collection of sounds understood by a group of people to have the same meaning.

Language branch A group of languages that share a common origin, but that have evolved into individual languages. The differences are not as extensive or as old as with language families, which may include several branches.

Language family A collection of individual languages related to each other by virtue of having a common ancestor before recorded history.

Language group Several individual languages within a language branch that share a common origin in the relatively recent past and display relatively few differences in grammar and vocabulary.

Latent heat The energy necessary to change water from one of three states—solid, liquid, or gaseous, to another. It is only not detectable by a thermometer *latent* means "hidden."

Latent heat exchange Release or absorption of latent heat caused by melting, freezing, evaporation or condensation.

Latent heat of vaporization The amount of heat required to make water change state from liquid to vapor.

Latitude The numbering system used to indicate the location of parallels drawn on a globe and measuring distance north and south of the equator (0°).

Lava Magma that reaches Earth's surface and erupts.

Life expectancy The average number of years an individual can be expected to live, given current social, economic, and medical conditions. Life expectancy at birth is the average number of years a newborn infant can expect to live.

Lingua franca A language mutually understood and commonly used in trade by people who have different native languages.

Literacy rate The percentage of a country's people who can read and write.

Lithosphere Earth's crust and a portion of upper mantle directly below the crust extending down to 70 kilometers (45 miles).

Location The position of anything on Earth's surface.

Loess An accumulation of wind-blown silt.

Longitude The numbering system used to indicate the location of meridians drawn on a globe and measuring distance east and west of the prime meridian (0°).

Longshore current A current in the surf zone along a shoreline, parallel to the shore.

Longshore transport Sediment transport by a longshore current.

Longwave energy Energy radiated by Earth in wavelengths of about 5.0 to 30.0 microns.

Magma Molten rock beneath Earth's surface.

Mantle The portion of Earth above the core and below the crust.

Map A two-dimensional, or flat, representation of Earth's surface, or a portion of it.

Maquiladora Factory built by U.S. company in Mexico near the U.S. border to take advantage of much lower labor costs in Mexico

Marine west-coast climate A climate with moderately cool winters, moderately warm summers, and moderate to high rainfall all year.

Market area Also called *hinterland,* the area surrounding a central place, from which people are attracted to use the place's goods and services.

Mass movement Downslope movement of rock and soil at Earth's surface, driven mainly by the force of gravity acting on those materials.

Meandering The tendency of flowing water to follow a sinuous course with alternating right- and left-hand bends.

Mechanical weathering The breakdown of rocks into smaller particles caused by application of physical or mechanical forces.

Medical revolution Diffusion of medical technology invented in Europe and North America to the poorer countries of Latin America, Asia, and Africa. Improved medical and health care practices have eliminated many of the traditional causes of death in poorer countries and enabled more people to live longer and healthier lives.

Mediterranean climate A climate with warm, dry summers and cool, moist winters.

Meltwater channel A river channel carved by water from a melting glacier.

Meridian An arc drawn on a globe between the North and South poles.

Metamorphic rocks Rocks formed by modification of other rock types, usually by intense heat and/or pressure.

Metropolitan statistical area (MSA) In the United States, a central city of at least 50,000 population, the city's surrounding county, and adjacent counties that meet one of several tests indicating a functional connection to the central city.

Micron One millionth of a meter.

Mid-latitude low pressure zones Regions of low pressure and air converging from the subtropical and polar high-pressure zones.

Migration Permanent move to a new location.

Milankovich cycles Regular variations in solar radiation received by Earth, caused by variations in the geometry of Earth's rotation on its axis and revolution around the sun.

Milkshed The area surrounding a city from which milk is supplied.

Missionary An individual who helps to diffuse a universalizing religion.

Monsoon circulation Seasonal reversal of pressure and wind over a large continent, characterized by wind blowing toward the continental interior in summer and toward the ocean in winter. Asia has a strong monsoon circulation pattern.

Moraine An accumulation of rock and sediment deposited by a glacier, usually in or near the melting area.

Multinational state A state that contains more than one nationality.

Multiple nuclei model A model of the internal structure of cities in which social groups are arranged around a collection of nodes of activities.

Nationalism The attitude of the people in a nation in support of the existence and growth of a particular state.

Nationality (or nation) A group of people who occupy a particular area and have a strong sense of unity based on a set of shared beliefs and attitudes.

Nation-state A state whose territory corresponds to that occupied by a particular nation.

Natural increase The percentage growth of a population in a year, computed as the crude birth rate minus the crude death rate.

Natural resource Something that is created through natural processes and used by people.

Negative feedback A feedback which slows or discourages response in a system.

Net migration The difference between the level of immigration and the level of emigration.

Net primary productivity The amount of biomass produced by the vegetation in a given environment, expressed as a dry weight.

Network A system of connected lines forming an interrelated chain.

New international division of labor Transfer of some types of jobs, especially those requiring low-paid less skilled workers, from relatively developed to developing countries.

Nitrogen oxide A compound of nitrogen and oxygen formed through high-temperature combustion. Nitrogen oxide is an important contributor to photochemical smog and acid deposition.

Nonrenewable resource A resource that exists in a finite quantity and is either not replenished or is replenished much more slowly than it is used.

Open system A system that exchanges matter with other systems.

Orographic uplift Forcing of air up and over mountains, often producing precipitation.

Outwash plain An accumulation of sand and gravel carried by meltwater streams from a glacier, usually deposited immediately beyond the terminal moraine from the glacier.

Overland flow Water flowing across the soil surface on a hillslope, usually resulting from precipitation falling faster than the ground can absorb it.

Overpopulation The number of people in an area exceeds the capacity of the environment to support life at a given level of technology and decent standard of living.

Ozone A gas composed of molecules with three oxygen atoms; a highly corrosive gas at ground level, but in the upper atmosphere essential to protecting life on Earth by absorbing ultraviolet (UV) radiation.

Paddy Malay word for wet rice, commonly but incorrectly used to describe a sawah.

Parallel A circle drawn around the globe parallel to the equator and at right angles to the meridians.

Parent material Mineral matter such as rocks or transported sediments from which soil is formed.

Particulate A small solid particle in air, such as a dust particle or fragments of ash in smoke.

Pastoral nomadism A form of subsistence agriculture based on herding domesticated animals.

Pattern The geometric, or other regular arrangement of something in a study area.

Perforated state A state that completely surrounds another one.

Permafrost Soil or rock with a temperature below 0°C (32°F) all the year.

Photochemical smog An air pollution condition in which sunlight causes nitrogen oxides and hydrocarbons to react in the atmosphere, forming other pollutants.

Photosynthesis A chemical reaction that occurs in green plants in which carbon dioxide and water are converted to carbohydrates (a plant's food) and oxygen.

Photovoltaic cell A device for direct conversion of light to electricity.

Physiological density The number of people per unit of area of arable land, which is land suitable for agriculture.

Pilgrimage A journey to a place considered sacred for religious purposes.

Plantation A large farm, located in a tropical or subtropical climate of a developing country, that specializes in the production of one or two crops for sale, usually to consumers in a relatively developed country.

Plate tectonics theory The movement of large, continent-sized slabs of Earth's crust relative to one another.

Polar front A boundary between cold polar air and warm subtropical air circling the globe in the mid-latitudes.

Polar high-pressure zones Regions of high pressure and descending air near the North and South poles.

Pollution Discharge by humans of a substance to the environment in greater concentrations than would occur naturally.

Popular custom A custom found in a large, heterogeneous society that shares certain habits despite differences in other personal characteristics; also known as an international custom.

Population pyramid A bar graph representing the distribution of population by age and sex.

Positive feedback A feedback which amplifies or encourages response in the system.

Possibilism The theory that the physical environment may set limits on human actions, but people have the ability to adjust to the physical environment and choose a course of action from many alternatives.

Potential evapotranspiration (POTET) The amount of water that would be evaporated and/or transpired in a given environment if it were available.

Potential resource Something that is not useful to humans at present, but that might become useful in the foreseeable future.

Prairie A vegetation type characterized by dense grass up to 2 meters high, found in mid-latitude semiarid climates.

Precipitation Water falling from the atmosphere in liquid or solid form.

Pressure gradient The difference in atmospheric pressure per unit distance between two locations.

Primary sector The portion of the economy concerned with the direct extraction of materials from Earth's surface, generally through agriculture, although sometimes by mining, fishing, and forestry.

Primate city The largest settlement in a country, if it has more than twice as many people as the second-ranking settlement.

Prime meridian The meridian, designated as 0° longitude, which passes through the Royal Observatory at Greenwich, England.

Producer A plant; through photosynthesis it produces food for itself and for animals (consumers) that eat the plant.

Productivity The value of a particular product compared to the amount of labor needed to make it.

Projection The system used to transfer locations from Earth's surface to a flat map.

Prorupted state An otherwise compact state with a large projecting extension.

Public housing Housing owned by the government; in the United States, it is rented to low-income residents, and the rents are set at 30 percent of the families' incomes.

Pull factors Factors that induce people to move to a new location.

Push factors Factors that induce people to leave old residences.

Quaternary Period The period of geologic time encompassing approximately the last 3 million years.

Quaternary sector The portion of the economy concerned with processing of information, especially through computer technology.

Quota In reference to migration, a law that places maximum limits on the number of people who can immigrate to a country.

Radiation Energy in the form of electromagnetic waves that radiate in all directions.

Ranching A form of commercial agriculture in which livestock graze over an extensive area.

Range The maximum distance people are willing to travel to obtain a good or use a service.

Rank-size rule A pattern of settlements in a country, such that the *n*th largest settlement is $1/n$ the population of the largest settlement.

Redlining A process by which banks draw lines on a map and refuse to lend money to purchase or improve property within the boundaries.

Refugees People who are forced to migrate from a country for political reasons.

Region An area defined by one or more distinctive trends or features.

Relative humidity The *actual* water content of the air expressed as a percentage of how much water the air *could* hold at a given temperature.

Relatively developed country A country that has progressed relatively far along a continuum of economic development.

Relocation diffusion The spread of a feature or trend through bodily movement of people or physical phenomena from one place to another.

Remote sensing The acquisition of data about Earth's surface from an orbiting satellite or other long-distance method.

Renewable resource A resource that is continually replenished so that it can be used indefinitely if it is managed properly.

Respiration A chemical reaction that occurs in plants and animals in which carbohydrates and oxygen are combined, releasing water, carbon dioxide, and heat.

Right-to-work state A U.S. state that has passed a law preventing a union and company from negotiating a contract that requires workers to join a union as a condition of employment.

Rills Small channels formed by soil surface erosion.

Runoff Flow of water from the land, either on the soil surface or in streams.

Rural settlement A settlement in which the principal occupation of the residents is agriculture.

Rush hour The four consecutive 15-minute periods in the morning and evening with the heaviest volumes of traffic.

Sanitary landfill A facility for disposal of solid wastes through daily burial.

Saturation vapor pressure The maximum water vapor that air can hold.

Savanna A vegetation type characterized by grasses and scattered trees, characteristic of seasonally dry tropical climates.

Sawah A flooded field for growing rice.

Scale The relation between the length of a feature on a map and the length of the actual feature on Earth's surface

Sea level The general elevation of the sea surface, averaging out variations caused by waves, storms, and tides.

Seafloor spreading Creation of a new oceanic crust where two tectonic plates move apart from each other.

Seasonally humid tropical climate A climate with warm temperatures all the year, a season with high rainfall and a pronounced dry season.

Secondary sector The portion of the economy concerned with manufacturing useful products through processing, transforming, and assembling raw materials.

Sect A relatively small denominational group that has broken away from an established church.

Sector model A model of the internal structure of cities in which social groups are arranged around sectors or wedges radiating out from the central business district.

Sediment transport The movement of rock particles by surface erosion.

Sedimentary rocks Rocks formed through accumulation of many small rock fragments at Earth's surface.

Seed agriculture Reproduction of plants through annual introduction of seeds, which result from sexual fertilization.

Seismic waves Vibrations or shock waves originating at the focus of an earthquake and transmitted through Earth.

Seismograph A device for recording movements of Earth's crust, such as earthquakes.

Self-determination The concept that nationalities have the right to govern themselves.

Semiarid climate A climate with precipitation slightly less than potential evapotranspiration for most of the year.

Sensible heat Heat detectable by sense of touch or by a thermometer

Settlement A fixed collection of buildings and inhabitants.

Sex ratio The number of males per 100 females in the population.

Shield The ancient core of a continent.

Shield volcano A volcano with relatively gentle slopes formed by eruption of relatively fluid lavas.

Shifting cultivation A form of subsistence agriculture in which people shift activity from one field to another; each field is used for crops for a relatively few years and left fallow for a relatively long period.

Shortwave energy Radiant energy emitted by the sun in wavelengths about 0.2 to 5.0 microns.

Sial Crust formed of relatively less dense minerals, dominated by silicon and aluminum (for Silicon-Aluminum).

Sima Crust formed of relatively dense minerals, dominated by silicon and magnesium (for Silicon-Magnesium).

Site factors Location factors related to the costs of factors of production inside the plant, such as land, labor, and capital.

Situation factors Location factors related to the transportation of materials into and from a factory.

Soil creep Slow downslope movement of soil caused by many individual, near-random particle movements such as those caused by burrowing animals or freeze and thaw.

Soil fertility The ability of a soil to support plant growth through storing and supplying water, air, and nutrients.

Soil horizon A layer in the soil with distinctive characteristics derived from soil-forming processes.

Soil order A major category in the U.S. soil classification system.

Sovereignty The ability of a state to govern its territory free from control of its internal affairs by other states.

Spatial association The distribution of one phenomenon across the landscape which is systematically related to the distribution of another.

Spatial distribution The regular arrangement of a phenomenon across Earth's surface.

Spatial interaction The movement of physical processes, human activities, and ideas within and among regions.

Specific heat The amount of energy required to raise a unit mass of substance's temperature by a given amount.

Spring wheat Wheat planted in the spring and harvested in the late summer.

Squatter settlement An area within a city in a developing country in which people illegally establish residences on land they do not own or rent and erect homemade structures.

Standard language The form of a language used for official government business, education, and mass communications.

State An area organized into a political unit and ruled by an established government with control over its internal and foreign affairs.

Stationary front A stalled, unmoving boundary between air masses.

Steady-state equilibrium A system whose rates of inputs and outputs are equal, and the amounts of energy and matter are stable or are fluctuating around a stable average.

Steppe A vegetation type characterized by relatively short, sparse grasses, found in mid-latitude semiarid climates.

Stimulus diffusion The spread of an underlying principle, even though a specific characteristic is rejected.

Storm surge An area of elevated sea level in the center of a hurricane that may be several meters high, and which does most of the damage when a hurricane comes ashore.

Stratus clouds Flat layers of clouds often formed along a warm front.

Subarctic climate A high-latitude climate characterized by brief, cool summers and long cold winters.

Subsistence agriculture Agriculture designed primarily to provide food for direct consumption by the farmer and the farmer's family.

Substitutability The ability of one resource to take the place of another in various human uses.

Subtropical high-pressure zones Regions of high pressure and descending air at about 25° north and south latitudes.

Sulfur dioxide A compound of sulfur and oxygen emitted through combustion of sulfur-containing fossil fuels. Sulfur dioxide is hazardous in high concentrations, and contributes to acid deposition.

Summer solstice For places in the Northern Hemisphere, June 20 or 21, the date when at noon the sun is directly overhead at 23.5° north latitude; for places in the Southern Hemisphere, December 21 or 22, the date when at noon the sun is directly overhead at 23.5° south latitude.

Sunspot A cool region on the surface of the sun.

Surface erosion Downslope movement of rock and soil at Earth's surface, driven mainly by air, water, or ice moving across the surface.

Sustainable agriculture Agriculture that uses farming methods intended to preserve the long-term productivity of the land and to minimize pollution of the soil, groundwater, and streams that drain the land.

Sustainable development The level of development that can be maintained in a country without depleting resources to the extent that future generations will be unable to achieve a comparable level of development.

Sustained yield A management strategy for renewable natural resources that indefinitely allows a continuous harvest of goods.

Swidden A patch of land cleared for planting through slashing and burning.

System. An interdependent group of items interacting in a regular way to form a unified whole.

Taboo A restriction on behavior imposed by social custom.

Tectonic plate A large, continent-sized piece of Earth's crust that moves in relation to other pieces.

Temperature inversion A condition in the atmosphere in which warmer air lies above cooler air, limiting vertical circulation and pollutant dispersal.

Terminal moraine An accumulation of rock and sediment at the toe of a glacier.

Tertiary sector The portion of the economy concerned with the provision of goods and services to people in exchange for payment.

Threshold The minimum number of people needed to support a good or service.

Toponym The name given to a portion of Earth's surface.

Total fertility rate The average number of children a woman will have throughout her childbearing years; a future prediction based on today's assumptions .

Toxic substance A chemical that is dangerous to humans and other organisms at small concentrations.

Transform plate boundary A boundary between tectonic plates in which the two plates pass one another in a direction parallel to the plate boundary.

Transhumance The seasonal migration of livestock between mountains and lowland pastures.

Transnational corporation (multinational corporation) A corporation which operates factories in countries other than the one in which its headquarters are located.

Transpiration The use of water by plants, normally drawing it from the soil through roots, evaporating it in the leaves and releasing it to the atmosphere.

Trophic level A position in the food chain relative to other organisms, such as producer, herbivore, or carnivore.

Tropic of Cancer The parallel of 23.5° north latitude.

Tropic of Capricorn The parallel of 23.5° south latitude.

Tropical rainforest Broadleaf-evergreen vegetation characteristic of humid tropical environments.

Tsunami An extremely long wave created by an underwater earthquake; it may travel hundreds of kilometers per hour.

Tundra A low, slow-growing vegetation type found in high-latitude and high-altitude conditions in which snow covers the ground most of the year.

Tundra climate A climate characterized by long, very cold winters and short, cool summers. Summer temperatures are only a few degrees above freezing.

Underclass A group in society prevented from participating in the material benefits of a relatively developed society because of a variety of social and economic characteristics.

Unitary state An internal organization of a state that places most power in the hands of central government officials.

Universalizing religion A religion that attempts to appeal to all people, not just those living in a particular location.

Urban heat island Warmer temperatures in an urban area relative to the surrounding rural area.

Urban settlement A settlement in which the principal economic activities are manufacturing, warehousing, trading, and provision of services.

Urbanization Urbanization is the increase in both population and the percentage of a society's population that lives in urban settlements.

Urbanized area A central city and its contiguous built-up suburbs.

Value added The gross value of the product minus the costs of raw materials and energy.

Vegetative planting Reproduction of plants by direct cloning from existing plants.

Vernacular region (or perceptual region) An area that people believe to exist, as part of their cultural identity.

Vernal (spring) equinox March 20 or 21, one of two dates when at noon the perpendicular rays of the sun strike the equator (the sun is directly overhead along the equator).

Volcano A vent in Earth's surface where lava emerges.

Warm front A boundary formed when a warm air mass advances against a cooler one.

Water budget An accounting of the amounts of precipitation, evapotranspiration, soil moisture storage, and runoff at a given place.

Wavelength In the context of waves on a water surface, the horizontal distance from one wave crest to the next.

Winter solstice For places in the Southern Hemisphere, June 20 or 21, the date when at noon the sun is directly overhead at places along the parallel of 23.5° north latitude; for places in the Northern Hemisphere, December 21 or 22, the date when at noon the sun is directly overhead at places along the parallel of 23.5° south latitude.

Winter wheat Wheat planted in the fall and harvested in the early summer.

Zero population growth (ZPG) The total fertility rate declines to the point where the natural increase rate equals zero.

Zone in transition An area surrounding the central business district that contains primarily low-rent housing and some older industries and warehouses.

Index